Lecture Notes in Computer Science 10385

Commenced Publication in 1973
Founding and Former Series Editors:
Gerhard Goos, Juris Hartmanis, and Jan van Leeuwen

Editorial Board

More information about this series at http://www.springer.com/series/7407

Ying Tan · Hideyuki Takagi
Yuhui Shi (Eds.)

Advances
in Swarm Intelligence

8th International Conference, ICSI 2017
Fukuoka, Japan, July 27 – August 1, 2017
Proceedings, Part I

 Springer

Editors
Ying Tan
Peking University
Beijing
China

Hideyuki Takagi
Kyushu University
Fukuoka
Japan

Yuhui Shi
Southern University of Science
 and Technology
Shenzhen
China

ISSN 0302-9743   ISSN 1611-3349   (electronic)
Lecture Notes in Computer Science
ISBN 978-3-319-61823-4   ISBN 978-3-319-61824-1   (eBook)
DOI 10.1007/978-3-319-61824-1

Library of Congress Control Number: 2017944216

LNCS Sublibrary: SL1 – Theoretical Computer Science and General Issues

Printed on acid-free paper

This Springer imprint is published by Springer Nature
The registered company is Springer International Publishing AG
The registered company address is: Gewerbestrasse 11, 6330 Cham, Switzerland

Preface

This book and its companion volumes, LNCS vols. 10385 and 10386, constitute the proceedings of the 8th International Conference on Swarm Intelligence (ICSI 2017) held from July 27 to August 1, 2017, in Fukuoka, Japan.

The theme of ICSI 2017 was "Serving Life with Intelligence and Data Science." ICSI 2017 provided an excellent opportunity and/or an academic forum for academics and practitioners to present and discuss the latest scientific results and methods, innovative ideas, and advantages in theories, technologies, and applications in swarm intelligence. The technical program covered all aspects of swarm intelligence and related areas.

ICSI 2017 was the eighth international gathering in the world for researchers working on all aspects of swarm intelligence, following successful events in Bali (ICSI 2016), Beijing (ICSI-CCI 2015), Hefei (ICSI 2014), Harbin (ICSI 2013), Shenzhen (ICSI 2012), Chongqing (ICSI 2011), and Beijing (ICSI 2010), which provided a high-level academic forum for participants to disseminate their new research findings and discuss emerging areas of research. It also created a stimulating environment for participants to interact and exchange information on future challenges and opportunities in the field of swarm intelligence research. ICSI 2017 was held in conjunction with the Second International Conference on Data Mining and Big Data (DMBD 2017) at Fukuoka, Japan, to share common mutual ideas, promote transverse fusion, and stimulate innovation.

ICSI 2017 was held in the center of the Fukuoka City. Fukuoka is the fifth largest city in Japan with 1.55 million inhabitants and is the seventh most liveable city in the world according to the 2016 Quality of Life Survey by *Monocle*. Fukuoka is in the northern end of the Kyushu Island and is the economic and cultural center of Kyushu Island. Because of its closeness to the Asian mainland, Fukuoka has been an important harbor city for many centuries. Today's Fukuoka is the product of the fusion of two cities in the year 1889, when the port city of Hakata and the former castle town of Fukuoka were united into one city called Fukuoka. The participants of ICSI 2017 enjoyed traditional Japanese dances, the local cuisine, beautiful landscapes, and the hospitality of the Japanese people in modern Fukuoka, whose sites are part of UNESCO's World Heritage.

ICSI 2017 received 267 submissions from about 512 authors in 35 countries and regions (Algeria, Australia, Austria, Brazil, Brunei Darussalam, Canada, China, Colombia, Germany, Hong Kong SAR China, India, Indonesia, Iran, Italy, Japan, Lebanon, Malaysia, Mexico, New Zealand, Nigeria, Norway, Romania, Russia, Serbia, Singapore, South Africa, South Korea, Spain, Sweden, Chinese Taiwan, Thailand, Tunisia, Turkey, UK, USA) across six continents (Asia, Europe, North America, South America, Africa, and Oceania). Each submission was reviewed by at least two, and on average 2.9 reviewers. Based on rigorous reviews by the Program Committee members and reviewers, 133 high-quality papers were selected for publication in this proceedings volume, with an acceptance rate of 49.81%. The papers are organized in 24 cohesive sections covering all major topics of swarm intelligence and computational intelligence research and development.

On behalf of the Organizing Committee of ICSI 2017, we would like to express our sincere thanks to the Research Center for Applied Perceptual Science of Kyushu University and the Computational Intelligence Laboratory of Peking University for their sponsorship, to the IEEE Computational Intelligence Society for its technical sponsorship, to a Japan Chapter of IEEE Systems, Man and Cybernetics Society for its technical cosponsorship, as well as to our supporters the International Neural Network Society, World Federation on Soft Computing, IEEE Beijing Section, and Beijing Xinghui Hi-Tech Co. and Springer. We would also like to thank the members of the Advisory Committee for their guidance, the members of the international Program Committee and additional reviewers for reviewing the papers, and the members of the Publications Committee for checking the accepted papers in a short period of time. We are particularly grateful to Springer for publishing the proceedings in the prestigious series *Lecture Notes in Computer Science*. Moreover, we wish to express our heartfelt appreciation to the plenary speakers, session chairs, and student helpers. In addition, there are still many more colleagues, associates, friends, and supporters who helped us in immeasurable ways; we express our sincere gratitude to them all. Last but not the least, we would like to thank all the speakers, authors, and participants for their great contributions that made ICSI 2017 successful and all the hard work worthwhile.

May 2017   Ying Tan
                                                                Hideyuki Takagi
                                                                    Yuhui Shi

Organization

General Co-chairs

Ying Tan   Peking University, China
Hideyuki Takagi   Kyushu University, Japan

Program Committee Chair

Yuhui Shi   Southern University of Science and Technology, China

Advisory Committee Co-chairs

Russell C. Eberhart   IUPUI, USA
Gary G. Yen   Oklahoma State University, USA
Hisao Ishibuchi   Osaka Prefecture University, Japan

Technical Committee Co-chairs

Kay Chen Tan   National University of Singapore, Singapore
Xiaodong Li   RMIT University, Australia
Nikola Kasabov   Auckland University of Technology, New Zealand
Ponnuthurai N. Suganthan   Nanyang Technological University, Singapore

Plenary Session Co-chairs

Mengjie Zhang   Victoria University of Wellington, New Zealand
Andreas Engelbrecht   University of Pretoria, South Africa

Invited Session Co-chairs

Yan Pei   University of Aizu, Japan
Chaomin Luo   University of Detroit Mercy, USA

Special Sessions Co-chairs

Shangce Gao   University of Toyama, Japan
Ben Niu   Shenzhen University, China
Qirong Tang   Tongji University, China

Tutorial Co-chairs

Milan Tuba   John Naisbitt University, Serbia
Andreas Janecek   University of Vienna, Austria

Publications Co-chairs

Swagatam Das   Indian Statistical Institute, India
Xinshe Yang   Middlesex University, UK

Publicity Co-chairs

Yew-Soon Ong   Nanyang Technological University, Singapore
Carlos Coello   CINVESTAV-IPN, Mexico
Yaochu Jin   University of Surrey, UK
Shi Cheng   Shanxi Normal University, China
Bin Xue   Victoria University of Wellington, New Zealand

Finance and Registration Chairs

Chao Deng   Peking University, China
Suicheng Gu   Google Corporation, USA

Local Arrangements Chair

Ryohei Funaki   Kyushu University, Japan

Conference Secretariat

Jie Lee   Peking University, China

International Program Committee

Mohd Helmy Abd Wahab   Universiti Tun Hussein Onn Malaysia, Malaysia
Harshavardhan Achrekar   University of Massachusetts Lowell, USA
Miltos Alamaniotis   Purdue University, USA
Peter Andras   Keele University, UK
Esther Andrés   INTA, Spain
Helio Barbosa   Laboratório Nacional de Computação Científica, Brazil
Carmelo J.A. Bastos Filho University of Pernambuco, Brazil
David Camacho   Universidad Autonoma de Madrid, Spain
Bin Cao   Tsinghua University, China
Jinde Cao   Southeast University, China
Kit Yan Chan   DEBII, Australia
Shi Cheng   Shaanxi Normal University, China
Manuel Chica   European Centre for Soft Computing, Spain

Carlos Coello Coello	CINVESTAV-IPN, Mexico
Jose Alfredo Ferreira Costa	Federal University, Brazil
Prithviraj Dasgupta	University of Nebraska, USA
Mingcong Deng	Tokyo University of Agriculture and Technology, Japan
Ke Ding	Tencent Corporation, China
Yongsheng Dong	Xi'an Institute of Optics and Precision Mechanics of CAS, China
Mark Embrechts	RPI, USA
Andries Engelbrecht	University of Pretoria, South Africa
Jianwu Fang	Xi'an Institute of Optics and Precision Mechanics of CAS, China
Shangce Gao	University of Toyama, Japan
Beatriz Aurora Garro Licon	IIMAS-UNAM, Mexico
Maoguo Gong	Xidian University, China
Yinan Guo	China University of Mining and Technology, China
J. Michael Herrmann	University of Edinburgh, UK
Lu Hongtao	Shanghai Jiao Tong University, China
Andreas Janecek	University of Vienna, Austria
Yunyi Jia	Clemson University, USA
Changan Jiang	Ritsumeikan University, Japan
Mingyan Jiang	Shandong University, China
Chen Junfeng	Hohai University, China
Liangjun Ke	Xian Jiaotong University, China
Thanatchai Kulworawanichpong	Suranaree University of Technology, Thailand
Rajesh Kumar	MNIT, India
Hung La	University of Nevada, USA
Germano Lambert-Torres	PS Solutions, Brazil
Xiujuan Lei	Shanxi Normal University, China
Bin Li	University of Science and Technology of China
Xiaodong Li	RMIT University, Australia
Xuelong Li	Chinese Academy of Sciences, China
Andrei Lihu	Politehnica University of Timisoara, Romania
Fernando B. De Lima Neto	University of Pernambuco, Brazil
Ju Liu	Shandong University, China
Qun Liu	Chongqing University of Posts and Communications, China
Wenlian Lu	Fudan University, China
Wenjian Luo	University of Science and Technology of China
Mohamed Arezki Mellal	M'Hamed Bougara University, Algeria
Sanaz Mostaghim	Institute IWS, Germany
Ben Niu	Arizona State University, USA
Quan-Ke Pan	Huazhong University of Science and Technology, China
Shahram Payandeh	Simon Fraser University, Canada

Yan Pei   The University of Aizu, Japan
Somnuk Phon-Amnuaisuk   Universiti Teknologi, Brunei
Radu-Emil Precup   Politehnica University of Timisoara, Romania
Kai Qin   RMIT University, Australia
Quande Qin   Shenzhen University, China
Boyang Qu   Zhongyuan University of Technology, China
Robert Reynolds   Wayne State University, USA
Luneque Silva Junior   Federal University of Rio de Janeiro, Brazil
Pramod Kumar Singh   ABV-IIITM Gwalior, India
Ponnuthurai Suganthan   Nanyang Technological University, Singapore
Hideyuki Takagi   Kyushu University, Japan
Ying Tan   Peking University, China
Qian Tang   Xidian University, China
Qirong Tang   Tongji University, China
Milan Tuba   John Naisbitt University, Serbia
Mario Ventresca   Purdue University, USA
Gai-Ge Wang   Jiangsu Normal University, China
Guoyin Wang   Chongqing University of Posts and
                               Telecommunications, China
Lei Wang   Tongji University, China
Ka-Chun Wong   City University of Hong Kong, SAR China
Shunren Xia   Zhejiang University, China
Bo Xing   University of Johannesburg, South Africa
Benlian Xu   Changshu Institute of Technology, China
Yingjie Yang   De Montfort University, UK
Wei-Chang Yeh   National Tsing Hua University, Taiwan
Kiwon Yeom   NASA Ames Research Center, USA
Peng-Yeng Yin   National Chi Nan University, Taiwan
Zhuhong You   Shenzhen University, China
Zhigang Zeng   Huazhong University of Science and Technology,
                               China
Zhi-Hui Zhan   Sun Yat-sen University, China
Jie Zhang   Newcastle University, UK
Jun Zhang   Waseda University, Japan
Junqi Zhang   Tongji University, China
Lifeng Zhang   Renmin University of China
Mengjie Zhang   Victoria University of Wellington, New Zealand
Qieshi Zhang   Shaanxi Normal University, China
Qiangfu Zhao   The University of Aizu, Japan
Yujun Zheng   Zhejiang University of Technology, China
Cui Zhihua   Complex System and Computational Intelligence
                               Laboratory, China
Guokang Zhu   Shanghai University of Electric Power, China
Xingquan Zuo   Beijing University of Posts and Telecommunications,
                               China

Additional Reviewers

Abdul Rahman, Shuzlina
Ambar, Radzi
Chen, Zonggan
Farias, Felipe
Feng, Jinwang
Figueiredo, Elliackin
Haji Mohd, Mohd Norzali
Huang, Yu-An
Jiao, Yuechao
Joyce, Thomas
Joyce, Thomes
Junyi, Chen
Li, Liping
Li, Xiangtao
Liu, Xiaofang
Liu, Zhenbao
Lu, Quanmao
Lyu, Yueming
Mohamad Mohsin, Mohamad Farhan
Oliveira, Marcos

Qiujie, Wu
Saharan, Sabariah
Shang, Ke
Siqueira, Hugo
Vázquez Espinoza De Los Monteros,
    Roberto Antonio
Wang, Lei
Wang, Lingyu
Wang, Tianyu
Wang, Yanbin
Wang, Zi-Jia
Yahya, Zainor Ridzuan
Yan, Shankai
Yang, Le
Yu, Jun
Zhang, Hao
Zhang, Jianhua
Zhang, Jiao
Zhang, Qinchuan
Zheng, Lintao

Contents – Part I

Particle Swarm Optimization

Applications of Particle Swarm Optimization

Ant Colony Optimization

Artificial Bee Colony Algorithms

Genetic Algorithms

Brain Storm Optimization Algorithm

Cuckoo Search

Firefly Algorithm

Contents – Part II

Community Detection

Multi-agent Systems and Swarm Robotics

Hybrid Optimization Algorithms and Applications

Fuzzy and Swarm Approach

Clustering and Forecast

Classification and Detection

Planning and Routing Problems

Dialog System Applications

Robotic Control

Other Applications

Theories and Models of Swarm Intelligence

Comparative Analysis of Swarm-Based Metaheuristic Algorithms on Benchmark Functions

Kashif Hussain[1], Mohd Najib Mohd Salleh[1(✉)], Shi Cheng[2], and Yuhui Shi[3]

[1] Universiti Tun Hussein Onn Malaysia, Batu Pahat, Malaysia
najib@uthm.edu.my
[2] School of Computer Science, Shaanxi Normal University, Xian, China
[3] Department of Computer Science and Engineering,
Southern University of Science and Technology, Shenzhen, China

Abstract. Swarm-based metaheuristic algorithms inspired from swarm systems in nature have produced remarkable results while solving complex optimization problems. This is due to their capability of decentralized control of search agents able to explore search environment more effectively. The large number of metaheuristics sometimes puzzle beginners and practitioners where to start with. This experimental study covers 10 swarm-based metaheuristic algorithms introduced in last decade to be investigated on their performances on 12 test functions of high dimensions with diverse features of modality, scalability, and valley landscape. Based on simulations, it can be concluded that firefly algorithm outperformed rest of the algorithms while tested unimodal functions. On multimodal functions, animal migration algorithm produced outstanding results as compared to rest of the methods. In future, further investigation can be conducted on relating benchmark functions to real-world optimization problem so that metaheuristic algorithms can be grouped according to suitability of problem characteristics.

Keywords: Swarm-intelligence · Swarm-based algorithms · Metaheuristic · Benchmark functions

1   Introduction

It has been an established fact and has been proved in wide spectrum of experimental research that metaheuristics are able to solve difficult optimization problems effectively. Problems in science, engineering, transportation, and business have been solved with optimum results. Moreover, loosely speaking, the ingenuity of inventors of these methods has not spared any source of inspiration from nature to be metaphorized into an efficient tool. Among these tools are swarm-based metaheuristic algorithms employing collective intelligence for producing optimum results. This collective intelligence of a group of agents in the form of swarm is deemed as a mechanism of generating optimal solutions fast.

© Springer International Publishing AG 2017
Y. Tan et al. (Eds.): ICSI 2017, Part I, LNCS 10385, pp. 3–11, 2017.
DOI: 10.1007/978-3-319-61824-1_1

Each swarm individual (bird, fish, or bee etc.) interacts with others for sharing information in order to make collective decision in subsequent moves with the help of some operators depending on the information (fitness value) with swarm individuals which is featuring a solution to the problem in hand.

Despite of differentiating strategies, one common property of all the swarm-based algorithms is moving towards better solution with collaborative effort. A high level description of swarm-based metaheuristics is given in Algorithm 1.

Algorithm 1. Swarm-based Metaheuristic Algorithm

```
Initialize algorithm parameters and swarm individuals;
while stop condition satisfies do
    Evaluate fitness of all the swarm individuals;
    Find swam individual with best fitness value in the swarm,
    (global best individual);
    Update position of swarm individuals based on collective
    intelligence;
end while
Return global best swarm individual;
```

Following the steps defined in Algorithm 1 and approaches taken from swarm behaviors in nature, metaheuristics are designed on ant [3], wolf [6], elephant [9], crow [2], etc. Because of extensive availability of the algorithms, it creates problem for practitioners to choose the best for their problem. Hence, this experimental study was conducted with a motivation to provide a starting platform for readers to have a glance on the latest swarm-based metaheuristics and their performances on various test functions. Since a careful review of literature shows that researchers used many benchmark functions to analyze the performances of their algorithms, therefore, this study provides an analysis of metaheuristic performances on a set of well-known numerical test functions.

The next section gives an overview of the selected metaeheuristics to merely highlight the main features; while, the reader is encouraged to refer to the related literature for further knowledge. Section 3 conducts experiments followed by Sect. 4 of results analysis. This study is duly concluded in Sect. 5.

2   Swarm-Based Metaheuristic Algorithms

This study selected some of the swarm-based metaheuristics introduced a decade ago. These algorithms are: Firefly Algorithm (FA) [10], Bat Algorithm (BA) [11], Brain Storm Optimization (BSO) [7], Biogeography-Based Optimization (BBO) [8], Animal Migration Optimization (AMO) [4], Elephant Herding Optimization (EHO) [9], Gray Wolf Optimizer (GWO) [6], Chicken Swarm Optimization (CSO) [5], Crow Search Algorithm (CSA) [2]. This section briefs about these algorithm, while the greater detail of the algorithms can be found in the literature cited herewith.

In FA, the swarm of fireflies socially interacts using flashing light for attraction. The FA begins by initializing fireflies with random light intensity. During iterations, FA updates light intensity based on light absorption concept. Fireflies attract other fireflies with light intensity, this attraction decreases with the increase of distance from other fireflies. The firefly with highest intensity of light is considered as the best solution. FA applies the light absorption mechanism to decrease flashing light of fireflies so that an element of exploration of search space can be implemented in the algorithm.

BA is based on the echolocation system of bats. The bats in the swarm communicate with each other via sound waves which allow them to determine the location or distance from other bats and food. These bats fly with certain velocity at certain position, frequency with varying pulse rate, and loudness. BA starts with random initialization of position, velocity, frequency, and loudness for each bat. As the BA proceeds with its iterations, it varies the frequency and loudness in order to implement dynamic behavior of bats.

BSO imitates how the humans sit together and do brainstorming for generating ideas collectively, usually with the assistance of a facilitator. The facilitator starts with grouping the participants (swarm individuals) of the brainstorming session into some clusters; for generating maximum possible scenarios (ideas) for the solution of the problem in hand. These individuals generate ideas while the facilitator weighs and chooses the best idea (having high probability to be chosen) from each cluster. The selected ideas and the remaining ideas (clues) are used to generate better ideas. For introducing diversity in thought process, BSO chooses any cluster(s) with random probability to generate completely new ideas. This process carries until the best solution is found.

In BBO, a swarm individual is a habitat with certain habitat suitability index (HSI) referring to its fitness of a solution. The greater the HSI, the greater is the number of species staying in the habitat, and resultantly, the better is fitness of a solution candidate. BBO starts with the initialization of habitats followed by the HSI evaluation (fitness value). After obtaining HSI for each habitat, it is mapped with the number of species in each habitat, immigration rate and emigration rate. Immigration and emigration rates are now used to modify habitats which later on are mutated based probability of species count in each habitat. This process continues until stop condition is satisfied.

AMO mimics the process of migration of animals from one place to another for survival. During this process, these animals have social interaction and as a result some animals leave the group while others join this seasonal movement led by a leader animal (with best fitness value). AMO starts with initialization of population of animal swarm. After being initiated, the animals have to first migrate from one place to another in the forms of small subgroups called neighborhoods. One of the neighborhood is selected randomly to update the position of an animal accordingly. Whereas, the next process is population update where some of the animals have to leave the herd and new animals join in.

A swarm-based metaheuristic EHO designed on the social animal elephant. The female elephants and calves live in herds consisting of several clans led of a matriarch (usually an old elephant). The male elephants when grown up leave the group. In EHO, the position of each elephant depicts problem solution. The elephants change their positions in accordance with their respected clan leader (with best fitness value) in each iteration. To implement diversification in search process, the worst elephants are replaced with new ones.

Another metaheuristic following the herding behavior of animals is KHA. This algorithm is designed by following the herding behavior of small crustaceans in ocean, called Krill. After attack from predators, the main objective for krill is to get closer to food and create dense herd. Keeping in view this strategy of krill, KHA generates new position of a swarm individual which is influenced by movement from other krill individuals, foraging, and scattering randomly. In order to improve performance, KHA introduces diversification using genetic operator: crossover and mutation.

The GWO employs the analogy of gray wolf while hunting prey. This intelligent prey hunting works in four tier leadership as alpha, beta, delta, and omega wolves with three step strategy: prey searching, prey surrounding, and attacking the prey. Among four groups, alpha, beta, and delta wolves are the top performers who guide other omega wolves to best positions for attacking. Each wolf updates its position based on interaction with the three top wolves.

The swarm intelligence of a group of chicken families, each having one rooster (the family head), several hens, and plenty of chicks, has been metaphorized in CSO. The fittest rooster in the group leads its family to a much better food source as compared to other weaker roosters. Whereas, the stronger hens have better chances of stealing food from others share as compared to weaker hens. The chickens update their position based on competition between chicken groups of earning more and more food. During the course of iterations, CSO ranks all the chickens in the swarm and assigns top N chickens as roosters, M worst ones as chicks, and rest are determined as hens.

Crows live in flocks and follow each other in search of food and then stock it in a hidden place so that it can be consumed when needed. This metaphor of intelligent behavior of crows is simulated in CSA. Here, the flying position of a crow represents a solution. Each crow also maintains memory of the hidden place for food storage. Because crows follow other crows randomly to discover better food source, they generate new position and evaluate the feasibility before moving that direction. After this, the crows update their memory. This process carries in iterations.

3 Experiments

This experimental study employed several of the test functions for testing the performances of swarm-based metaheuristic algorithms. Different evaluation metrics are used such as mean of best values and standard deviation of best solutions found over 30 runs, and also ranks.

Since different test functions imitate different levels of difficulties in optimization problems, these functions help reveal true robustness of an algorithm.

The test functions used in [12] are selected for this paper. Table 1 lists these functions along with their features. Functions from f_1 to f_7 are unimodal with single global optimum, and from f_8 to f_{12} are multimodal functions with multiple local minima that exponentially increase with dimensions. The prior group of functions is suitable for testing exploitation component of metaheuristic algorithms. On the other hand, later ones are helpful in analyzing the explorative component and how the algorithm is able to get rid of local minima [12].

Table 1. Benchmark test function. U-Unimodal, M-Multimodal, S-Separable, N-Nonseparable, C-Continuous, D-Discrete with Dimension = 30 and Global minimum = 0

ID	Function	Class	Range	ID	Function	Class	Range
f_1	Sphere	UCS	$[-100, 100]$	f_7	Quartic noise	UCS	$[-1.28, 1.28]$
f_2	Schwefel's 2.22	UCN	$[-10, 10]$	f_8	Rastrigin	MCS	$[-5.12, 5.12]$
f_3	Schwefel's 1.20	UCN	$[-100, 100]$	f_9	Ackley	MCN	$[-32, 32]$
f_5	Schwefel's 2.21	UCS	$[-100, 100]$	f_{10}	Griewank	MCN	$[-600, 600]$
f_6	Rosenbrock	UCN	$[-30, 30]$	f_{11}	Penalized	MCN	$[-50, 50]$
f_7	Step	UDS	$[-100, 100]$	f_{12}	Generalized penalized	MCN	$[-50, 50]$

In this experimental study, some of the common parameters; maximum number of iterations, upper and lower bounds (according to function) of search space were same in all the metaheuristics. Whereas, the distinguishing parameters related to each specific metaheuristic were set to the values suggested in related literature; FA [12], BA [11], BSO [7], BBO [8], AMO [4], EHO [9], KHA [1], GWO [6], CSO [5], CSA [2]. Stopping criteria was maximum number of iterations, which was 2000 iterations. Since, metaheuristics vary in performing number of objective function evaluations per iteration, it is better to evaluate this criteria as total number of function evaluations in all iterations. Thus, in 2000 iterations, the selected metaheuristics performed the number of function evaluations as: FA = 4871036, BA = 200100, BSO = 200100, BBO = 200100, AMO = 200050, EHO = 200100, KHA = 202100, GWO = 200000, CSO = 100050, and CSA = 100050 function evaluations.

4 Results Analysis

In order to observe two main cornerstones of the selected algorithm: exploration and exploitation, the benchmark functions are grouped into unimodal (f_1-f_7) for testing exploitative capability of the algorithms; on the other hand, multimodal group of functions (f_8-f_{12}) for observing the explorative capability. As mentioned earlier, these algorithms are run 30 times with 30 dimensions of each test function. The mean of best values found in all runs and standard deviation are taken into observations. For analyzing the robustness of the selected algorithms on each test function, the mean values are ranked from smallest to

Table 2. Performance of metaheuristic algorithms on unimodal functions. M = Mean, SD = Standard deviation, R = Rank, A.R = Average rank, O.R = Overall rank

		f_1	f_2	f_3	f_4	f_5	f_6	f_7	A.R	O.R
FA	M	4.24E−36	2.61E−18	2.61E−18	9.60E−18	14.6282	0	0.0003	2.86	1
	SD	4.73E−37	1.65E−19	1.65E−19	6.56E−19	2.3731	0	0.0001		
	R	5	4	2	2	3	1	3		
BA	M	1.11E−05	0.0151	321.2773	0.0014	0.8739	1.03E−05	0.0044	5.71	6
	SD	5.20E−07	0.0049	844.9859	0.0001	0.7707	1.40E−06	0.0050		
	R	8	8	8	6	1	4	5		
BSO	M	1.07E−42	0.0005	0.8353	0.0715	28.2020	3.80E−33	0.0106	5.57	5
	SD	1.76E−43	0.0015	0.5508	0.0418	0.7104	7.31E−33	0.0047		
	R	4	7	5	7	7	3	6		
BBO	M	0.3180	0.2726	1206.5509	13.2955	52.4394	37.1817	0.1271	9.86	10
	SD	0.0198	0.0523	297.1841	1.7200	29.3771	13.6204	0.0403		
	R	10	9	10	10	10	10	10		
AMO	M	9.76E−69	1.27E−36	0.0116	6.75E−17	14.4019	0	0.0024	3	3
	SD	1.13E−68	9.72E−37	0.0095	1.21E−16	0.7659	0	0.0006		
	R	3	3	4	3	2	2	4		
EHO	M	2.73E−12	1.34E−05	6.38E−09	0.0001	28.7371	3.0372	4.84E−06	5.43	4
	SD	5.32E−13	3.50E−06	1.53E−09	4.50E−05	0.0086	0.1147	4.37E−06		
	R	7	5	3	4	9	9	1		
KHA	M	1.96E−14	7.07E−05	78.7188	0.0013	27.9238	0.0058	0.0130	6	7
	SD	3.76E−14	6.15E−05	37.0537	0.0012	0.8590	0.0035	0.0053		
	R	6	6	7	5	6	5	7		
GWO	M	2.08E−176	5.45E−100	3.22E−56	3.06E−44	25.8277	0.1327	9.10E−05	2.43	1
	SD	0	1.58E−99	1.75E−55	5.99E−44	0.7499	0.1703	5.11E−05		
	R	1	1	1	1	4	7	2		
CSO	M	3.68E−97	4.06E−81	714.6842	3.9250	27.0649	2.9674	0.0153	6.14	8
	SD	1.77E−96	1.70E−80	472.6959	7.2895	0.7068	0.4553	0.0304		
	R	2	2	9	9	5	8	8		
CSA	M	0.0002	0.2990	8.9868	1.6162	28.2465	0.0937	0.0209	8	9
	SD	9.68E−05	0.1590	3.5145	0.5219	0.4931	0.0349	0.0061		
	R	9	10	6	8	8	6	9		

largest. These ranks are averaged to observe the performance of each method on each group of functions. This average rank is ranked to obtain overall rank.

The results obtained in unimodal function series are reported in Table 2. From the bird-eye-view, the overall rank shows that top three performers are GWO, FA, and AMO, consecutively. On the other hand, the least performers include CSO, SA, and BBO, respectively. The mediocre results were obtained from EHO, BSO, BA, and KHA. It is evident that the latest swarm-based algorithms have potential to be better optimizers while applied on wide variety of problems with some modification.

As per function-wise performance comparison, GWO achieved first in the rank by obtaining the best optimum values consistently on majority of the functions from f_1 to f_4. This proves that GWO has consistent convergence and it is a stable algorithm among others. On the other hand, BBO stood last performer among other nine algorithms in most of the cases such as f_1, f_3, f_4, f_5, and f_6. The most inconsistent metaheuristic in these experiments were EHO which

Table 3. Performance of metaheuristic algorithms on multimodal functions. M = Mean, SD = Standard deviation, R = Rank, A.R = Average rank, O.R = Overall rank

		f_8	f_9	f_{10}	f_{11}	f_{12}	A.R	O.R
FA	M	52.5005	1.60E−14	0.0015	0.0085	1.35E−32	4.6	3
	SD	18.7885	4.54E−15	0.0056	0.0325	5.57E−48		
	R	10	5	5	2	1		
BA	M	20.9631	0.0026	0.0130	0.0900	0.0468	7.4	8
	SD	35.8047	0.0004	0.0156	0.1973	0.1385		
	R	8	8	7	4	10		
BSO	M	35.3542	1.12E-14	0.0106	5.5028	0.0029	6.6	7
	SD	8.1199	4.23E−15	0.0083	1.6604	0.0056		
	R	9	4	6	10	4		
BBO	M	1.9646	0.1801	1.2862	0.9405	3.5477	8.2	10
	SD	0.5580	0.0302	0.0806	0.3018	1.2342		
	R	5	9	10	7	10		
AMO	M	6.03E−09	2.66E−15	0.0004	1.57E−32	1.35E−32	2.4	1
	SD	2.21E−08	4.01E−31	0.0022	1.11E−47	5.57E−48		
	R	4	1	4	1	2		
EHO	M	7.58E−10	1.20E−06	2.37E−10	0.2266	2.9470	5.6	6
	SD	8.87E−11	1.20E−07	4.09E−11	0.0383	0.0745		
	R	3	7	3	6	9		
KHA	M	8.4610	2.76E−08	0.0141	0.0428	6.55E−05	5.2	5
	SD	3.2298	2.77E−08	0.0135	0.1075	4.51E−05		
	R	6	6	8	3	3		
GWO	M	0	6.21E−15	0	0.0911	0.1365	4.2	2
	SD	0	2.41E−30	0	0.0262	0.1147		
	R	1	3	1	5	11		
CSO	M	0	3.61E−15	0	2.4553	1.6307	4.6	3
	SD	0	1.60E−15	0	8.6231	0.3227		
	R	2	2	2	9	8		
CSA	M	20.2545	0.7341	0.2530	1.7276	0.0314	7.8	9
	SD	7.5420	0.7674	0.0645	1.0978	0.0159		
	R	7	10	9	8	5		

performed the best in f_7, better in f_3 and f_4, insignificant in f_1 and f_2, and significantly inefficient in f_5 and f_6. Hence, it determines that EHO maintained irregular convergence rate throughout iterations in all the unimodal series of functions.

For the multimodal test functions f_8–f_{12}, Table 3 presents the experimental results. It is obvious from the overall rank that AMO topped among others on its explorative capability followed by GWO. Additionally, FA, CSO, and KHA also achieved significant global optimum values. Whereas, EHO, BSO, BA, CSA, and BBO were among the bottom performers in descending order due to being trapped in several of local minima in multimodal functions.

GWO and CSO obtained accurate results in f_8 and f_{10}. However, CSO was more consistent than GWO as it performed repeatedly better in most of the functions in this series. Apparently, EHO seems to be less consistent in this regard.

5   Conclusion

In this paper, we performed experimental study of the latest ten swarm-based metaheuristic algorithms on twelve well-known benchmark test functions with diverse characteristics. Based on the simulated results, it can be concluded that FA, GWO, and AMO are highly potential metaheuristic algorithms that can be used for solving complex unimodal and multimodal optimization problems. The analysis shows that these algorithms have balanced exploration and exploitation capability. Other than these algorithms, BA, KHA, BSO, and CSO also produced promising solutions in series of functions. For future work, further investigations can be carried out to comprehend differences in strategies adopted by these algorithms. Moreover, further statistical analysis such as significance of correlations among these metaheuristics will help group these methods according to different types of problems. On the benchmark test functions, a comprehensive research can be conducted to relate these benchmark test functions with real-world problems with the help of identification of some of the common characteristics.

Acknowledgments. The authors would like to thank Universiti Tun Hussein Onn Malaysia (UTHM), Malaysia for supporting this research under Postgraduate Incentive Research Grant, Vote No. U560.

References

1. Amudhavel, J., Kumarakrishnan, S., Anantharaj, B., Padmashree, D., Harinee, S., Kumar, K.P.: A novel bio-inspired krill herd optimization in wireless ad-hoc network (WANET) for effective routing. In: Proceedings of the 2015 International Conference on Advanced Research in Computer Science Engineering & Technology (ICARCSET 2015), p. 28. ACM (2015)
2. Askarzadeh, A.: A novel metaheuristic method for solving constrained engineering optimization problems: crow search algorithm. Comput. Struct. **169**, 1–12 (2016)
3. Karaboga, D.: An idea based on honey bee swarm for numerical optimization. Report, Technical report-tr06, Erciyes University, Engineering Faculty, Computer Engineering Department (2005)
4. Li, X., Zhang, J., Yin, M.: Animal migration optimization: an optimization algorithm inspired by animal migration behavior. Neural Comput. Appl. **24**(7–8), 1867–1877 (2014)

5. Meng, X., Liu, Y., Gao, X., Zhang, H.: A new bio-inspired algorithm: chicken swarm optimization. In: Tan, Y., Shi, Y., Coello, C.A.C. (eds.) ICSI 2014. LNCS, vol. 8794, pp. 86–94. Springer, Cham (2014). doi:10.1007/978-3-319-11857-4_10
6. Mirjalili, S., Mirjalili, S.M., Lewis, A.: Grey wolf optimizer. Adv. Eng. Softw. **69**, 46–61 (2014)
7. Shi, Y.: Brain storm optimization algorithm. In: Tan, Y., Shi, Y., Chai, Y., Wang, G. (eds.) ICSI 2011. LNCS, vol. 6728, pp. 303–309. Springer, Heidelberg (2011). doi:10.1007/978-3-642-21515-5_36
8. Simon, D.: Biogeography-based optimization. IEEE Trans. Evol. Comput. **12**(6), 702–713 (2008)
9. Wang, G.G., Deb, S., dos Coelho, L.S.: Elephant herding optimization. In: 2015 3rd International Symposium on Computational and Business Intelligence (ISCBI), pp. 1–5. IEEE (2015)
10. Yang, X.S.: Firefly algorithm. In: Nature-Inspired Metaheuristic Algorithms, vol. 20, pp. 79–90 (2008)
11. Yang, X.S.: A New Metaheuristic Bat-Inspired Algorithm, pp. 65–74. Springer, Heidelberg (2010)
12. Zhang, L., Liu, L., Yang, X.S., Dai, Y.: A novel hybrid firefly algorithm for global optimization. PLoS ONE **11**(9), e0163230 (2016)

A Mathematical Model of Information Theory: The Superiority of Collective Knowledge and Intelligence

Pedro G. Guillén$^{(\boxtimes)}$ ⓘD

Knowdle Foundation & Research Institute, Málaga, Spain
pedrogguillen@gmail.com

Abstract. The mathematical model described here is an evolution of the one constructed within the studies [1–3] by De Santos and Villa, among others. This paper aims to formalize the concept of knowledge and its properties as a basis for creating generalized economic value functions [4], focusing on the business models of the current technological sector as the application environment. The main conclusions reached are focused on the improvement of the value of information and knowledge under the assumption of collective cooperation amongst information system agents, as well as the properties of the knowledge space derived from the model.

Keywords: Knowledge · Collective · Intelligence · Wisdom · Structure · Value · Economy · Swarm · Topology

1 Algebraic Structure of Knowledge

Definition 1: Let L be a natural language. The Universe of data (also called Universal Library) is defined as all possible combinations of characters of L of arbitrary length that have meaning, that is to say, whose Boolean function of L is not null [2]. In the future, this set will be denoted by D, and a subset $d \subseteq D$ will be called data. It will be assumed that the universe of data is dense, and any combination of elements of the universe of data is an element of the universe of data.

Definition 2: The set of Universe of Contexts or Universal Context is defined as a graph with global structure of thesaurus, and locally orientable. The set will be denoted by C, and a subset $c \subseteq C$ will be called context.

Proposition 1: Whatever the L language on which the previous structures D and C are defined, and whatever the structure of C, we have $C \subset D$.

Proof: Trivial.

Then the σ-algebras of the respective sets D and C are defined, and they will be called D and C. The choice of σ-algebra is arbitrary; However, due to the structure of natural language and the results described in [1, 2], we can assume the choice of a "reasonable" non-trivial algebra.

© Springer International Publishing AG 2017
Y. Tan et al. (Eds.): ICSI 2017, Part I, LNCS 10385, pp. 12–21, 2017.
DOI: 10.1007/978-3-319-61824-1_2

Definition 3: The following set is defined as the Universe of Knowledge of the natural language L:

$$\Omega := D \times C = \{(d,c) : d \in D, c \in C\}$$

A compact subset $\Delta \subseteq \Omega$ will be called knowledge domain or simply domain, and an element $\omega \in \Omega$ of the set will be called contextualized data.

In the model, we will call the subsets that are unions of path-connected components natural knowledge or simply knowledge. The motivation to conjecture this structure for natural knowledge comes from studies on cognitive structure and brain processes [5, 6], as well as studies of the Mathematical Theory of Information [7–9]. These studies present the natural structure in which an agent of an information system processes all the knowledge it acquires. Knowledge will be the basis for the topology of the Ω space.

Proposition 2: If any knowledge domain $\Delta \subseteq \Omega$ is defined, there is always an embedding function between the domain and the set that does not depend on the structure that is given to the space.

Proof: Is immediate, since having $\Delta \subseteq \Omega$, it is verified that the restriction is always defined and is always an embedding. Note that $C \subset D$.

Once the spaces (D, D) and (C, C) are defined, it is possible to define in the set Ω the inherited σ-algebra $\Xi = D \times C$. Because of its algebraic properties, the fact that D, C have a structure of σ-algebra implies that Ξ has a structure of σ-algebra. The reason for presenting these sets by means of this algebraic structure is that a large number of useful operators can be built on it, as is defined below.

2   The Dynamic System

We define the continuous variable $t \in [0, \infty]$ which will be called time, and which must be given a starting point, t_0, in the model. Without loss of generality, we will assign $t_0 = 0$.

Analogously to the initial hypotheses, the sets of the Knowledge Universe are defined as functions of time:

$$D(t), C(t)$$

That enables the definition of a dynamic system in the σ-algebra of this universe, the Cartesian product:

$$\Omega(t) := D(t) \times C(t) = \{(d(t), c(t)) : d(t) \in D(t), c(t) \in C(t)\}$$

$$A(\Omega)(t) = A(t)$$

Regarding the model associated with the dynamic system that has been constructed, some additional assumptions must be made, because as the time variable takes a range of values or others, the set changes in a certain way.

Now the value function can be defined, which in the previous section was a vector associated with an element of σ-algebra. In the resulting dynamic system, the value function is a vector of n functions that depend of t.

$$V_t : A(t) \rightarrow K^n(t)$$

$$V_t(A(t)) := (m_1(A(t)), m_2(A(t)), \ldots, m_n(A(t)))$$

Or simplifying:

$$V_t : A(t) \rightarrow K^n(t)$$

$$V_t(t) := (m_1(t), m_2(t), \ldots, m_n(t))$$

3 The Value of Knowledge

Once the universe of knowledge is defined, the next step is to construct a function that associates each element of our σ-algebra with a numerical value. To this end the following results are introduced.

Definition 4: Any function of the form:

$$f(t) : A(\Omega)(t) \rightarrow K$$

that is a finite and bounded measure, and that it is dependent on a law of supply and demand, is defined as a value function.

Proposition 3: The value function is continuous.

Proof: Demonstration is trivial by construction, since an open set of the topology in the image space is the union of elementary images of the function, and its inverse image is precisely the union of basic openings of the topology of $\Omega(t)$. Therefore, the function is continuous.

Lemma 1: $(\Omega(t), T(A))$ is a Hausdorff topological space.

Proof: Take two elements of the space. There are two possible cases: that the points are in different components by paths or that they are in the same. In the first case, we have already found two open disjoints sets of the topology that contain both points and we are finished. In case of both points are in the sameset connected by paths, we consider all the possible paths that join both points. As the whole of the universe of data is dense, given two elements we can always find one that is in the middle, and therefore we can separate each path into two open disjoint paths, being the union of the halves the disjoint open sets containing each of the two starting points.

Lemma 2: Let A_i be a family of knowledge and f a function of value on $\Omega(t)$. Then it verifies that a $j \in N$ exists such that, if $i > j$:

$$\left|\frac{\partial f(\cap A_i)}{\partial t}\right| < 0$$

Proof: By Definition 4 we have that if f is a value function then it is dependent on a law of supply and demand. As by Definition 3 knowledge are sets belonging to agents of the information system, we then have the greater the number of agents possessing a certain knowledge (the intersection of all Ai) the greater the supply, and therefore the value of that set tends to go down in time for a sufficiently large index.

Corollary 1: A direct consequence of the previous lemma is that the temporary derivative of sufficiently infrequent knowledge value is inversely proportional to the temporal derivative of frequent knowledge. Thus, for a sufficiently large index under the hypotheses of Lemma 2:

$$\left|\frac{\partial f(\cap A_i)}{\partial t}\right| = -k\left|\frac{\partial f(\Omega \setminus \cap A_i)}{\partial t}\right|$$

is verified for some $k > 0$.

Proof: Analogous to the previous case.

In general, the intention is not to establish a numerical value for each element of the σ-algebra, but a vector with n coordinates. For this, we have only to define n different measures on the given σ-algebra that gives to each set a value from the desired point of view, defining a value function as the composition of a finite set of measures defined on the universe of knowledge.

4   Collective Knowledge

One of the fundamental results of the model that has been described is the benefit of collective versus individual knowledge. Intuitively, we can think of profit as a positive difference of value, defined in the model by the homonymous function.

In the model, the differential is the derivative of the value function regarding the time variable, that is, in the growth of the functions

$$A : t \rightarrow A(t)$$

$$f_t : A(t) \rightarrow \mathbb{R}^n(t)$$

Theorem 1: Let Ai (t) be a family of knowledge and let f be a function of value. Then, for a sufficiently large index i:

$$\left|\frac{\partial f(\cup A_i)}{\partial t}\right| \geq \sum_{j=1}^{i} \left|\frac{\partial f(A_j)}{\partial t}\right|$$

Proof: The problem will be reduced to the case of two sets of knowledge. The general case is completely analogous. We know that any value function is a measure. To demonstrate the statement, we take two of the fundamental properties of our n-dimensional measure, namely

$$\forall A, B \in A | A \subseteq B \rightarrow f(A) \leq f(B)$$

$$f(A \cup B) = f(A) + f(B) - f(A \cap B)$$

Taking these properties into account, together with the linearity properties of the derivative:

$$\left|\frac{\partial V_t(A_1 \cup A_2)}{\partial t}\right| \geq \left|\frac{\partial V(A_1)}{\partial t}\right| + \left|\frac{\partial V(A_2)}{\partial t}\right| - \left|\frac{\partial V_t(A_1 \cap A_2)}{\partial t}\right|$$

is verified.

However, we had by Lemma 1 that for a large enough family of knowledge (we suppose the result true for $i = 2$)

$$\left|\frac{\partial (\cap A_i)}{\partial t}\right| < 0$$

Therefore, it is verified that

$$\left|\frac{\partial V_t(A_1 \cup A_2)}{\partial t}\right| \geq \left|\frac{\partial V(A_1)}{\partial t}\right| + \left|\frac{\partial V(A_2)}{\partial t}\right| - \left|\frac{\partial V_t(A_1 \cap A_2)}{\partial t}\right| \geq$$

$$\left|\frac{\partial V(A_1)}{\partial t}\right| + \left|\frac{\partial V(A_2)}{\partial t}\right|$$

That proves the result. Basically, the interpretation of this theorem is that the knowledge which agents bring exclusively to others who do not possess it, generates more valuable knowledge than that is common to all agents.

5   Maximal Knowledge: The Hypersurface

Proposition 4: The image of $\Omega(t)$ by k value functions is a hypersurface of dimension k.

Proof: By hypothesis, $\Omega(t)$ is a union of sets connected by paths. Thus, locally it is a set connected by paths. Applying Lemma 1, we have that the space $(\Omega(t), T(A))$ is a

Hausdorff space, and because it is locally connected by paths it is also locally arc-connected. Therefore, by taking a point in space $\omega \in \Omega(t)$, and an open one containing it, there are $I \subseteq R$ paths that are homeomorphic to R, and whose union builds a local homeomorphism between a subspace of R^m and $\Omega(t)$. Therefore, $\Omega(t)$ is a hypersurface and since the k value functions are continuous, and it has to bethat $m = k$.

Hereinafter, the notation $f(\Omega(t)) = H(t)$ will be used.

Proposition 5: Value functions form a vector space of dimension n > k.

Proof: Amongst the properties of measures is that the sum of measures is a measure that preserves algebraic properties, and the same thing happens with the product of scalars. Therefore, any function dependent on a law of supply and demand continues to be such after an operation with another function or with a scalar.

Corollary 2: H (t) is contained in a vector space of infinite dimension.

Proof: The result is immediate, since the product of measurable functions is measurable and gives rise to a higher degree function that is not expressible as an element of the starting vector space. By repeating the process indefinitely, we obtain a vector space of infinite dimension containing $H(t)$.

From now on, this space will be called parameterization space.

Proposition 6: Once a domain of knowledge is established, there is knowledge that has extreme values (maximum or minimum).

Proof: By definition, the constraint of $H(t)$ to a domain is a compact set, since the domains are compact and the value function is continuous. Applying the Weierstrass theorem, a continuous function defined in a compact has at least one extreme value, and the result is concluded.

6 Through Knowledge: Intelligence

Once the existence of maximal knowledge in the σ-algebra/parametrization/hypersurface structure is guaranteed, it is logical to ask what structure has the maximal knowledge set or if the value of knowledge can be treated as knowledge in itself to generalize in a higher abstraction the concept of value function.

Definition 5: An intelligence function is defined as a function dependent on t of the form:

$$I(t) = \int_{H(t)} F(x)dx$$

Being F a vector field function, such that the nature of the intelligence function is equivalent to the concept of divergence in fluid mechanics [10], i.e., that F would be a function that would represent the velocity field of a fluid and the function of

Intelligence measures how intelligence flows and expands through the hypersurface of knowledge.

Being a primitive of a function, it is evident that functions of intelligence are differentiable. Assuming the conditions in the previous section, and applying the fact that in this case H (t) is a compact set, both the hypersurface and differential maximal curves at the points of the hypersurface border can be calculated.

In this case the maximal curves should represent the optimal curves of knowledge through a given local domain, and provide, through the differential of the hypersurface itself (which is differentiable) the most valuable knowledge in the future, under a Domain and given time series. The formalization of the calculation would be given by the formula:

$$I_\gamma(t) = \int_{H(t)} F(\gamma(x)) \cdot |\gamma'(x)| dx$$

The usefulness of frontier points is given by how they will relate knowledge of a given domain with other knowledge outside. Therefore, we must analyze intelligence functions from a temporal perspective, obtaining a formula dependent on the parameter selected for the auxiliary function:

$$I(x) = \int_{H(t)} F(x) dt$$

That is why the definition and choice of the auxiliary function is so important when modeling these behaviors in a dynamic system.

7   Collective Intelligence

Definition 7: For each agent i of the information system, an intelligence function with an auxiliary function is defined as:

$$I_i(t) = \int_{H(t)} F_i(x) dx$$

Proposition 7: Any agent of the system can know all the intelligence functions of all the other agents setting a sufficiently restricted domain.

Proof: Since any function of intelligence is obviously differentiable (since it is a primitive) and locally bijective, we can apply the inverse function theorem to obtain, given the starting hypersurface and results in a local domain, a local inverse, and with it an explicit local equation of $I_i(t)$. Thus, the result is concluded.

It now makes sense to define the concept of Collective Intelligence.

Definition 8: In the above conditions, the function:

$$I_c(t) = sup \int_{H(t)} |\{F_i(x)\}| dx$$

x defined as Collective Intelligence. The reason for taking the absolute value of the auxiliary function is that the positive and negative values of the field integral cancel out, and it is necessary to consider the global flow through the hypersurface, regardless of its sign.

That is, the function of intelligence that has as auxiliary function the supreme function of all the intelligence functions of the system agents. The most important property that Cognitive Intelligence verifies is the following inequality:

Theorem 2: For any set of agents, with any set of intelligence functions and any domain:

$$|I_i(t)| \leq |I_c(t)|$$

is verified.

Proof: Because of the supreme function's properties, the following has to be

$$|I_i(t)| = \left| \int_{H(t)} F_i(x) dx \right| \leq sup|I_i(t)|$$

And now, applying a measure theory result [15]:

$$sup|I_i(t)| = sup \left| \int_{H(t)} F_i(x) dx \right| \leq sup \int_{H(t)} |\{F_i(x)\}| dx = |I_c(t)|$$

thus proving our own result.

The interpretation of the previous theorem is: Collective Intelligence always obtains superior performance of the knowledge processed through value criteria than any function of intelligence of a single agent (or of a restricted part of the set of agents).

8   Extending Intelligence: Wisdom

The last level of cognitive inference of the mathematical model is a result that applies to the functions of intelligence to obtain a generalization of the model into non-parameterized places, even places of the parameterization space that are not an image of any element of the σ-algebra.

Theorem 3: Every intelligence function defined in a domain, in particular Collective Intelligence, supports a continuous extension to all the parameterization space containing H(t).

Proof: To begin with, an intelligence function is a linear form with respect to its auxiliary function, since in the space of all the continuous functions defined on a compact set (the fixed domain), the integral defined in that compact is a linear form.

In these conditions, it makes sense to apply the Hahn-Banach Theorem [11], which states that a linear form defined in a vector subspace on a field and bounded by a sublinear function (which it fulfills, since the functions are bounded by construction) then there exists a linear extension $\hat{I} : P \rightarrow \mathbb{R}$ from I to the whole space P, i.e. there is a linear function $\hat{I}$ such that:

$$\hat{I}(x) = I(x) \forall x \in P$$

This concludes the result.

Corollary 3: It is possible to know properties and values associated with knowledge or elements of space $\Omega(t)$ that do not yet exist.

Proof: In the case that $F(x)$ is the identity function and $x = t$, the function to be extended by Theorem 3 would be the hypersurface itself as a function of time, since

$$I(x) = I(t) = \int_{H(t)} 1.dt = H(t)$$

Therefore, cognitive and intelligence factors of elements of (t) that do not yet exist could be known.

Definition 9: A wisdom function is defined as the Hahn - Banach extension of an intelligence function.

9 Conclusions

Given the results of the model described, it makes sense to establish some conjectures that arise.

Conjecture 1: Is it possible to define an explicit method to obtain a function of wisdom from a given intelligence function?

It must be said that in the proof of Hahn Banach's Theorem the extension $\hat{I}$ in general is not unique and the proof, which uses the Zorn lemma, gives no method to find it. In fact, even if it were possible to find explicitly a function of wisdom from a given intelligence function, it would be highly improbable to find a general method.

Conjecture 2: Is it possible to obtain values of non-measurable sets?

This conjecture corresponds to a vacuum in model theory, related to non-measurable sets. For any non-trivial measure there are sets within the σ-algebra that can not be measured under any measure construction. These sets, such as those described in

Vitali's Theorem [12, 13], can generate distortions in the valuation of certain transformations, as with Banach-Tarski Paradox. Fortunately, these situations only occur when the elements of σ-algebra are not explicitly constructed, which in our model does not occur, since a set of Vitalican not be constructed explicitly. However, it makes sense to wonder if these sets can receive values by a function of wisdom that extends the hypersurface of values to non-measurable sets.

Conjecture 3: In an environment of a massive amount of system agents, can Collective Intelligence be subject to indetermination?

The motivation of this conjecture comes from the existence of artificial agents. Thus, an agent that is not limited to making changes per second in intelligence functions can lead the formulation of Collective Intelligence to include two different expressions of the same intelligence function, and even two simultaneous expressions of Collective Intelligence, which could have unpredictable consequences for the model in a hypothetical practical implementation [14].

References

1. Guillén, P.G.: Topological proof of the computability of the algorithm based on the morphosyntactic distance. In: Conference Series of the 9th Annual International Conference on Computer Science and Information Systems, COM2013-0595 (2013)
2. Serradilla, F., Villa, E., De Santos, A., Guillén, P.G.: Semantic construction of an univocal language. ITHEA: Inf. Theor. Appl. **19**(3), 211 (2012)
3. De Santos, A., Villa, E., Serradilla, F., Guillén, P.G.: Construction of morphosyntactic distance on semantic structures. ITHEA: Inf. Theor. Appl. **19**(4), 336 (2012)
4. Menger, C.: Principles of Economics. Ludwig von Mises Institute, Auburn (1976). 120 p.
5. Frankland, P.W., Bontempi, B.: The organization of recent and remote memories. Nat. Rev. Neurosci. **6**(2), 119–130 (2005)
6. Rugg, M., Yonelinas, A.P.: Human recognition memory: a cognitive neuroscience perspective. Trends Cogn. Sci. **7**(7), 313–319 (2003)
7. Shannon, C.E., Weaver, W.: The Mathematical Theory of Communication. University of Illinois Press, Urbana (1964). 125 p.
8. Moreiro González, J.A.: Aplicaciones al análisis automático del contenido provenientes de la teoría matemática de la información. Anales de documentación **5**, 273–286 (2002)
9. Chung, F.: Graph theory in the information age. Not. AMS **57**(6), 726–732 (2010)
10. Bourbaki, N.: Integration I. Chap. III–IV. Springer, Heidelberg (2004)
11. Narici, L., Beckenstein, E.: The Hahn-Banach theorem: the life and times. Topol. Appl. **77**, 193–211 (1997)
12. Foreman, M., Wehrung, F.: The Hahn-Banach theorem implies the existence of a non-Lebesgue measurable set. Fundamenta Mathematicae **138**, 13–19 (1991)
13. Herrlich, H.: Axiom of Choice. Springer, Heidelberg (2006). 120 p.
14. Bosyk, G.M.: Más allá de Heisenberg. Relaciones de incerteza tipo Landau-Pollak y tipo entrópicas (2014). 113 p.
15. Hawkins, T.: The Lebesgue's Theory of Integration. University of Wisconsin Press, Madison (1970)

Modelling and Verification Analysis of the Predator-Prey System via a First Order Logic Approach

Zvi Retchkiman Konigsberg[(✉)]

Centro de Investigacion en Computacion, Instituto Politecnico Nacional,
Mexico, CDMX, Mexico
mzvi@cic.ipn.mx

Abstract. Consider the interaction of populations, in which there are exactly two species, one of which the *predators* eat the *preys* thereby affecting each other. In the study of this interaction Lotka-Volterra models have been used. This paper proposes a formal modelling and verification analysis methodology, which consists in representing the interaction behavior by means of a formula of the first order logic. Then, using the concept of logic implication, and transforming this logical implication relation into a set of clauses, called Skolem standard form, qualitative methods for verification (satisfiability) as well as performance issues, for some queries, are applied.

Keywords: Predator-prey system · First order logic · Model · Verification · Unsatisfiability · Refutation methods

1 Introduction

Consider the interaction of populations, in which there are exactly two species, one of which the *predators* eat the *preys* thereby affecting each other. Such pairs exist throughout nature: fish and sharks, lions and gazelles, birds and insects, to mention some. In the study of this interaction Lotka-Volterra models have been used [1]. This paper proposes a well defined syntax modeling and verification analysis methodology which consists in representing the predator-prey interaction system as a formula of the first order logic, (where the quantifiers are chosen according to a video showing how a group of lions attack a zebra). Then, using the concept of logic implication, and transforming this logical implication relation into a set of clauses, called Skolem standard form, qualitative methods for verification (satisfiability) as well as performance issues, for some queries, are addressed. The method of Putnam-Davis based on Herbrand theorem for testing the unsatisfiability of a set of ground clauses as well as the resolution principle due to Robinson, which can be applied directly to any set of clauses (not necessarily ground clauses), are invoked. The paper is organized as follows. In Sect. 2, a first order background summary is given. In Sect. 3, the Putnam-Davis rules and the resolution principle for unsatisfiability, are recalled. In Sect. 4, the predator-prey problem is addressed. Finally, the paper ends with some conclusions.

© Springer International Publishing AG 2017
Y. Tan et al. (Eds.): ICSI 2017, Part I, LNCS 10385, pp. 22–30, 2017.
DOI: 10.1007/978-3-319-61824-1_3

2   First Order Logic Background

This section presents a summary of the first order logic theory. The reader interested in more details is encouraged to see [2,3].

Definition 1. *A first-order language $\mathcal{L}$ is an infinite collection of distinct symbols, no one of which is properly contained in another, separated into the following categories: parentheses, connectives, quantifiers, variables, equality symbol, constant symbols, function symbols and predicate symbols.*

Definition 2. *Terms are defined recursively as follows: (i) A constant is a term. (ii) A variable is a term. (iii) If f is an nth-place function symbol, and $t_1, t_2, \ldots, t_n$ are terms, then $f(t_1, t_2, \ldots, t_n)$ is a term. (iv) All terms are generated by applying the above rules.*

Definition 3. *If P is an nth-place predicate symbol, and $t_1, t_2, \ldots, t_n$ are terms, then $p(t_1, t_2, \ldots, t_n)$ is an atom. No other expressions can be atoms.*

Definition 4. *An occurrence of a variable in a formula is bound if and only if the occurrence is within the scope of a quantifier employing the variable, or is the occurrence in that quantifier. An occurrence of a variable in a formula is free if and only if this occurrence of the variable is not bound.*

Definition 5. *A variable is free in a formula if at least one occurrence of it is free in the formula. A variable is bound in a formula if at least one occurrence of it is bound.*

Definition 6. *Well-formed formulas, or formulas for short, in the first-order logic are defined recursively as follows: (i) An atom is a formula. (ii) If F and G are formulas then, $\sim (F), (F \vee G), (F \wedge G),$ and $(F \leftrightarrow G)$ are formulas. (iii) If F is a formula and x is a free variable in F, then $(\forall x)F$ and $(\exists x)F$ are formulas. (iv) Formulas are generated only by a finite number of applications of (i), (ii), and (iii).*

Definition 7. *An interpretation I of a formula F in the first-order logic consists of a nonempty domain D, and an assignment of "values" to each constant, function symbol, and predicate symbol occurring in F as follows: (1) To each constant, we assign an element in D, (2) To each nth-place function symbol, we assign a mapping from D^n to D, (3) To each nth-place predicate symbol, we assign a mapping from D^n to T, F, where T means true and F means false.*

Remark 1. Sometimes to emphasize the domain D, we speak of an interpretation of the formula over D. When we evaluate the truth value of a formula in an interpretation over the domain D, $(\forall x)$ will be interpreted as "for all elements in D," and $(\exists x)$ as "there is an element in D. For every interpretation of a formula over a domain D, the formula can be evaluated to T or F according to the following rules: (1) If the truth values of formulas G and H are evaluated, then the truth values of the formulas $\sim (F), (F \vee G), (F \wedge G), (F \rightarrow G),$ and

$(F \leftrightarrow G)$ are evaluated according to the well known formulas of propositional calculus, (2) $(\forall x)G$ is evaluated to T if the truth value of G is evaluated to T for every $d \in D$; otherwise, it is evaluated to F, (3) $(\exists x)G$ is evaluated to T if the truth value of G is T for at least one $d \in D$; otherwise, it is evaluated to F. We note that any formula containing free variables cannot be evaluated.

Definition 8. *A formula G is consistent (satisfiable) if and only if there exists an interpretation I such that G is evaluated to T in I. If a formula G is T in an interpretation I, we say that I is a model of G and I satisfies G.*

Definition 9. *A formula G is inconsistent (unsatisfiable) if and only if there-exists no interpretation I that satisfies G.*

Definition 10. *A formula G is valid if and only if every interpretation of G satisfies it.*

Definition 11. *A formula G is a logical implication of formulas $F_1, F_2, \ldots, F_n$ if and only if for every interpretation I, if $F_1, F_2, \ldots, F_n$ is true in I, G is also true in I.*

The following characterization of logical implication plays a very important role as will be shown in the rest of the paper.

Theorem 1. *Given formulas $F_1, F_2, \ldots, F_n$ and a formula G, G is a logical implication of $F_1, F_2, \ldots, F_n$ if and only if the formula $((F_1 \wedge F_2 \wedge \ldots, \wedge F_n) \rightarrow G)$ is valid if and only if the formula $(F_1 \wedge F_2 \wedge \ldots \wedge F_n \wedge \sim (G))$ is inconsistent.*

Definition 12. *A formula F in the first-order logic is said to be in a prenex normal if and only if is in the form of $(Q_1 x_1)(Q_2 x_2) \ldots (Q_n x_n)(M)$ where every $Q_i x_i, i = 1, 2, \ldots, n$ is either $\forall x_i$ or $\exists x_i$, and M is a formula containing no quantifiers. $(Q_1 x_1)(Q_2 x_2) \ldots (Q_n x_n)$ is called the prefix and M is called the matrix of the formula F.*

Next, Given a formula F, the following procedure transforms F into a prenex normal form. (1) Eliminate $\rightarrow$ and $\leftrightarrow$, (2) Move $\sim$, (3) Rename variables and (4) Pull quantifiers. (details are provided in [3].)

Let a formula F be already in a prenex normal form i.e., $(Q_1 x_1) \ldots (Q_n x_n)(M)$, where M is in a conjunctive normal form CNF (a finite conjunction of clauses, see next definition). Suppose Q_i is an existential quantifier in the prefix. If no universal quantifier appears before Q_i, we choose a new constant c different from other constants occurring in M, replace all x_i appearing in M by c and delete $Q_i x_i$ from the prefix. If $(Q_1 x_1)(Q_2 x_2) \ldots (Q_k x_k)$ $(1 \leq k < i)$ are all the universal quantifiers appearing before $Q_i x_i$, we choose a new k-place function symbol f different from other function symbols in M, replace all x_i in M by $f(x_1, x_2, \ldots, x_k)$ delete $Q_i x_i$ from the prefix. After the above process is applied to all the existential quantifiers in the prefix, the last formula we obtain is called a universal form, or Skolem standard form, of the formula F. The constants and functions used to replace the existential variables are called Skolem functions.

Remark 2. It is important to point out that universal forms are not unique.

Definition 13. *A clause is a finite disjunction of zero or more literals (atoms or negation of atoms).*

When it is convenient, we shall regard a set of literals as synonymous with a clause. A clause consisting of r literals is called an r-literal clause. A one-literal clause is called a unit clause. When a clause contains no literal, we call it the empty clause, denoted by $\square$. Since the empty clause has no literal that can be satisfied by an interpretation, the empty clause is always false. The importance of transforming a formula F in to its universal form results evident, thanks to the next result.

Theorem 2. *Let S be a set of clauses that represents a universal form of a formula F. Then F is inconsistent if and only if S is inconsistent.*

By definition, a set S of clauses is unsatisfiable if and only if it is false under all interpretations over all domains. Since it is inconvenient and impossible to consider all interpretations over all domains, it would be nice if we could fix on one special domain H such that S is unsatisfiable if and only if S is false under all the interpretations over this domain. Fortunately, there does exist such a domain, which is called the Herbrand universe of S, defined as follows.

Definition 14. *Let H_0 be the set of constants appearing in S. If no constant appears then, H_0 is to consist of a single constant, say $H_0 = a$. For $i = 0, 1, 2, \ldots$ let H_{i+1} be the union of H_i, and the set of all terms of the form $f(t_1, t_2, \ldots, t_n)$ for all n-place functions f occurring in S, where $t_j = 1, 2, \ldots n$ are members of the set H_i. Then each H_i is called the i-level constant set of S, and H_∞, is called the Herbrand universe of S.*

Definition 15. *Let S be a set of clauses. The set of ground atoms of the form $P(t_1, t_2, \ldots, t_n)$ for all n-place predicates P occurring in S, where $t_1, t_2, \ldots, t_n$ are elements of the Herbrand universe of S, is called the atom set, or the Herbrand base of S. A ground instance of a clause C of a set S of clauses is a clause obtained by replacing variables in C by members of the Herbrand universe of S.*

We have seen that the problem of logical implication is reducible to the problem of satisfiability, which in turn is reducible to the problem of satisfiability of universal sentences. Next, Herbrand's theorem is presented, which states that to test whether a set S of clauses is unsatisfiable, we need consider only interpretations over the Herbrand universe of S. This can be used together with algorithms for unsatisfiability (Davis Putnam rules discussed in Sect. 3) to develop procedures for this purpose.

Theorem 3. *Let a formula F be already in a prenex normal form i.e., $(Q_1 x_1)$ $(Q_2 x_2) \ldots (Q_n x_n)(M)$, where M is in a conjunctive normal form CNF and contains no quantifiers, i.e., is universal. Let H_∞ be the Herbrand universe of S (with S the set of clauses that represents the universal form of F). Then F is unsatisfiable if and only there is a finite unsatisfiable set $\acute{S}$ of ground instances of clauses of S.*

Remark 3. Herbrand's theorem suggests a refutation procedure: that is, given an unsatisfiable set S of clauses to prove, if there is a mechanical procedure that can successively generate sub-sets $S_1, S_2 \ldots$ of ground instances of clauses in S and can successively test $S_1, S_2 \ldots$ for unsatisfiability, then, as guaranteed by Herbrand's theorem, this procedure can detect a finite n such that S_n is unsatisfiable, otherwise it will continue forever i.e., it is undecidable.

3   Unsatisfiability Methods

3.1   Davis and Putnam Rules

Davis and Putnam introduced a method for testing the unsatisfiability of a set of ground clauses, therefore it is immediately applicable to a set of clauses S considering interpretations over the Herbrand universe. Their method consists of the following rules: (1) Delete all the ground clauses from S that are tautologies. The remaining set $\acute{S}$ is unsatisfiable if and only if S is, (2) If there is a unit ground clause L in S, obtain $\acute{S}$ from S by deleting those ground clauses in S containing L. If $\acute{S}$ is empty then, S is satisfiable, otherwise obtain a set $\acute{\acute{S}}$ by deleting $\sim (L)$ from $\acute{S}$. $\acute{\acute{S}}$ is unsatisfiable if and only if S is, (3) A literal L in a ground clause of S is said to be pure in S if and only if the literal $\sim (L)$ does not appear in any ground clause in S. If a literal L is pure in S, delete all the ground clauses containing L. The remaining set $\acute{S}$ is unsatisfiable if and only if S is, (4) If the set S can be written as: $(A_1 \vee L) \wedge (A_2 \vee L) \ldots (A_m \vee L) \wedge (B_1 \vee \sim L) \wedge (B_2 \vee \sim L) \ldots (B_m \vee \sim L) \wedge R$ where A_i, B_i and R are free of L and $\sim L$ then, obtain the sets $S_1 = A_1 \wedge A_2 \ldots A_m \wedge R$ and $S_2 = B_1 \wedge B_2 \ldots B_m \wedge R$. S is unsatisfiable if and only if both, $S_1 \cup S_2$ are.

3.2   The Resolution Principle

The procedure introduced by Davis and Putnam relies on Herbrand's theorem which has one major drawback: It requires the generation of sets $S_1, S_2 \ldots$ of ground instances of clauses. For most cases, this sequence grows exponentially. We shall next introduce the resolution principle due to Robinson, a more efficient method than Davis and Putnam procedure. It can be applied directly to any set S of clauses (not necessarily ground clauses) to test the unsatisfiability of S. Resolution is a sound and complete algorithm i.e., a formula in clausal form is unsatisfiable if and only if the algorithm reports that it is unsatisfiable. Therefore it provides a consistent methodology free of contradictions. However, it is not a decision procedure because the algorithm may not terminate.

Definition 16. *A substitution is a finite set of the form $\{t_1/v_1, t_2/v_2, \ldots, t_n/v_n\}$, where every v_i is a variable, every t_i, is a term different from v_i. When the t_i are ground terms, the substitution is called a ground substitution. The substitution that consists of no elements is called the empty substitution and is denoted by ϵ.*

Definition 17. *Let $\theta = \{t_1/x_1, \ldots, t_n/x_n\}$ and $\lambda = \{u_1/y_1, \ldots, u_m/y_m\}$ be two substitutions. Then the composition of θ and λ is the substitution, denoted by $\theta \circ \lambda$, that is obtained from the set $\{t_1\lambda/x_1, \ldots, t_n\lambda/x_n, u_1/y_1, \ldots, u_m/y_m\}$ by deleting any element $t_j\lambda/x_j$ for which $t_j\lambda = x_j$, and any element u_i/y_i such that y_i is among $x_1, x_2, \ldots, x_n$.*

Definition 18. *A substitution θ is called a unifier for a set $E_1, E_2, \ldots, E_n$ if and only if $E_1\theta = E_2\theta = \ldots, E_n\theta$ The set $\{E_1, E_2, \ldots, E_n\}$ is said to be unifiable if there is a unifier for it.*

Definition 19. *A unifier σ for a set $E_1, E_2, \ldots, E_n$ of expressions is a most general unifier if and only if for each unifier θ for the set there is a substitution λ such that $\theta = \sigma \circ \lambda$.*

Definition 20. *If two or more literals (with the same sign) of a clause C have a most general unifier σ, then $C\sigma$ is called a factor of the clause C. If $C\sigma$ is a unit clause, it is called a unit factor of C.*

Definition 21. *Let C_1 and C_2 be two clauses (called parent clauses) with no variables in common. Let L_1 and L_2 be two literals in C_1 and C_2, respectively. If L_1 and $\sim (L_2)$ have a most general unifier σ, then the clause $(C_1\sigma - L_1\sigma) \cup (C_2\sigma - L_2\sigma)$ is called a binary resolvent of C_1 and C_2. The literals L_1 and L_2 are called the literals resolved upon.*

Definition 22. *A resolvent of (parent) clauses C_1 and C_2 is one of the following binary resolvents: (1) a binary resolvent of C_1 and C_2, (2) a binary resolvent of C_1 and a factor of C_2, (3) a binary resolvent of a factor of C_1 and C_2, (4) a binary resolvent of a factor of C_1 and a factor of C_2.*

Definition 23. *Given a set S of clauses, a deduction of C from S is a finite sequence of clauses $C_1, C_2, \ldots, C_n$ such that each C_i, either is a clause in S or a resolvent of clauses preceding C_i,, and $C_k = C$. A deduction of $\square$ from S is called a refutation, or a proof of S.*

The following result, called lifting lemma, plays a key role in the proof of the soundness and completeness theorem for the resolution procedure.

Lemma 1. *If $\acute{C}_1$ and $\acute{C}_2$ are instances of C_1 and C_2, respectively, and if $\acute{C}$ is a resolvent of $\acute{C}_1$ and $\acute{C}_2$ then there is a resolvent C of C_1 and C_2 such that $\acute{C}$ is an instance of C.*

The main result of this subsection, the soundness and completeness theorem for the resolution procedure, is next presented.

Theorem 4. *A set S of clauses is unsatisfiable if and only if there is a deduction of the empty clause $\square$ from S.*

Theorem 5. *The set of unsatisfiable sentences is undecidable.*

4   Predator-Prey System

Consider the interaction of populations, in which there are exactly two species, one of which the *predators* eats the other the *preys* thereby affecting each other's growth rates. Such pairs exist throughout nature: fish and sharks, lions and gazelles, birds and insects, to mention some. It is assumed that, the predator species is totally dependent on a single prey species as its only food supply, the prey has unlimited food supply, and that there is no threat to the pray other than the specific predator. The predator-prey system behavior is described as follows: (1) States: S: preys are safe, D: the preys are in danger, B: the preys are being eaten, I: the predators are idle, L: the predators are in search for a prey, CL: the predators continue searching for a prey, A: the predators attack the preys, F: the predator has finished eating the prey, P: the predator dies; (2) Rules of Inference: (a) if S and L then CL, (b) if S and CL then P, (c) if D and (L or CL) then A, (d) if A then B, (e) if B then F (f) if F then I, (g) if I then L. Therefore, by associating variables to the states, we can define the following predicates: $S(x)$: x is a safe prey, $D(x)$: the prey x is in danger, $B(x, y)$: the prey x is being eaten by predator y, $I(x)$: the predator x is idle, $L(x, y)$: the predator y is in search for a prey x, $CL(x, y)$: the predator y continues searching for a prey x, $A(x, y)$: the predator ya attacks prey x, $F(x, y)$: the predator y has finished eating prey x, $P(x)$: the predator x passed away.

Remark 4. The main idea consists of: the predator-prey behavior is expressed by a formula of the first order logic, (where the quantifiers are chosen according to a video showing how a group of lions attack a zebra), some query is expressed as an additional formula. The query is assumed to be a logical implication of the predator-prey formula (see Theorem 13). Then, transforming this logical implication relation into a set of clauses by using the techniques given in Sect. 2, its validity can be checked. Even more using the resolution principle, unifications done during the procedure provide answers to some specific queries. The domain D of the interpretation will be considered to be formed by a set of predators and a set of preys.

The formula that models the predator-prey behavior turns out to be:

$$[(\forall x)(\forall y)(S(x) \wedge L(x.y) \rightarrow CL(x, y))] \wedge [(\forall x)(\forall y)(S(x) \wedge CL(x.y) \rightarrow P(y))] \wedge$$
$$[(\exists x)(\forall y)(D(x) \wedge (L(x, y) \vee CL(x, y))) \rightarrow A(x, y))] \wedge [(\exists x)(\forall y)(A(x, y) \rightarrow$$
$$B(x, y))] \wedge [(\exists x)(\forall y)(B(x, y) \rightarrow F(x, y))] \wedge [(\exists x)(\forall y)(F(x, y) \rightarrow I(y))] \wedge$$
$$[(\exists x)(\forall y)(I(y) \rightarrow L(x, y))] \tag{1}$$

We are interested in verifying, the following statements:

(S1) Claim: If D and (L or CL) then B. Specifically, we want to know if there is prey p such that the following formula is a logical implication of Eq. 1: $(\exists p)(\forall q)(D(p) \wedge (L(p, q) \vee CL(p, q))) \rightarrow B(p, q))$. The set of clauses for this case is given by:

$$S = \{(\sim S(x)\vee \sim L(x.y) \vee CL(x.y)), (\sim S(x)\vee \sim CL(x,y) \vee P(y)),$$
$$(\sim D(c_1)\vee \sim L(c_1, z)\vee A(c_1, z), (\sim D(c_2)\vee \sim CL(c_2, w)\vee A(c_2, w)), (\sim A(c_3, u)\vee$$
$$B(c_3, u)), (\sim B(c_4, v) \vee F(c_4, v)), (\sim F(c_5, r) \vee I(r)), (\sim I(s) \vee L(c_6, s)), (D(p)),$$
$$(L(p, f(p)) \vee CL(p, f(p)), (\sim B(p, f(p)))\}.$$

Then a resolution refutation proof, with its required substitutions, is as follows:

(a) $p = c_1, (\sim L(c_1, z) \vee A(c_1, z)) \rightarrow z = u, c_3 = c_1, (\sim L(c_1, z) \vee B(c_1, z)) \rightarrow$
$p = c_1, z = f(p), (\sim L(c_1, f(p))) \rightarrow p = c_1, (CL(c_1, f(p)))$.

(b) $p = c_2, (\sim CL(c_2, w) \vee A(c_2, w)) \rightarrow w = u, c_2 = c_3, (\sim CL(c_2, w) \vee$
$B(c_2, w)) \rightarrow p = c_2, z = f(p), (\sim CL(c_2, f(p))))$. Now, from the last two
equations of (a) and (b), setting $c_2 = c_1$, we get a proof of S i.e., $\square$. Therefore we can conclude that: we not only have proved that the claim is true,
but we have computed a value for $p, p = c_1 = c_2 = c_3$. Which tell us that
the same prey that has been attacked, it has to be the same that is being
eaten, and not another one, otherwise, the refutation procedure fails. This
result is consistent with reality.

(S2) Claim: if D and $(L$ or $CL)$ then I. Specifically, we want to know if
there is prey p such that the following formula is a logical implication of Eq. 1:
$(\exists p)(\forall q)(D(p) \wedge (L(p, q) \vee CL(p, q))) \rightarrow I(q))$. The set of clauses for this case is
given by:

$$S = \{(\sim S(x)\vee \sim L(x.y) \vee CL(x.y)), (\sim S(x)\vee \sim CL(x,y) \vee P(y)),$$
$$(\sim D(c_1)\vee \sim L(c_1, z)\vee A(c_1, z), (\sim D(c_2)\vee \sim CL(c_2, w)\vee A(c_2, w)), (\sim A(c_3, u)\vee$$
$$B(c_3, u)), (\sim B(c_4, v) \vee F(c_4, v)), (\sim F(c_5, r) \vee I(r)), (\sim I(s) \vee L(c_6, s)), (D(p)),$$
$$L(p, f(p)) \vee CL(p, f(p)), (\sim I(q))\}.$$

Then a resolution refutation proof, with its required substitutions, is as follows:

(a) $r = q, (\sim F(c_5, q)) \rightarrow c_5 = c_4, q = v, (\sim B(c_4, v))$ from (a) of (S1) we know
$(\sim L(c_1, z) \vee B(c_1, z))$ therefore $c_1 = c_4, v = z, (\sim L(c_1, z)) \rightarrow p = c_1, z =$
$f(p), (CL(c_1, f(p)))$.

(b) $r = q, (\sim F(c_5, q)) \rightarrow c_5 = c_4, q = v, (\sim B(c_4, v))$ from (b) of (S2) we know
$(\sim CL(c_2, w))\vee B(c_2, w))$ therefore $c_2 = c_4, v = w, (\sim CL(c_2, w))$. Now, from
the last two equations of (a) and (b), setting $c_2 = c_1, w = f(p)$, we get a
proof of S i.e., $\square$. Therefore, the claim holds for $p = c_1 = c_2 = c_3 = c_4 = c_5$,
and the same conclusion given in (S1) extrapolates for this case.

5   Conclusions

The main contribution of the paper consists in the study of the predator-pray
system by means of a formal reasoning deductive methodology based on first
order logic theory. The results obtained are consistent with how the predator-
prey system behaves in real life.

References

1. Haberman, R.: Mathematical Models in Mechanical Vibrations, Population Dynamics, and Traffic Flow. Prentice Hall, Upper Saddle River (1977)
2. Chang, C.-L., Lee, R.C.-T.: Symbolic Logic and Mechanical Theorem Proving. Academic Press, Cambridge (1973)
3. Davis, M., Sigal, R., Weyuker, E.: Computability, Complexity, and Languages, Fundamentals of Theoretical Computer Science. Academic Press, Cambridge (1983)

Flock Diameter Control in a Collision-Avoiding Cucker-Smale Flocking Model

Jing Ma$^{(\boxtimes)}$ and Edmund M-K Lai

Department of Information Technology and Software Engineering,
Auckland University of Technology, Auckland, New Zealand
`jing.ma@aut.ac.nz`

Abstract. Both the original Cucker-Smale flocking model and a more recent version with collision avoidance do not have any control over how tightly the system of agents flock, which is measured by the flock diameter. In this paper, a cohesive force is introduced to potentially reduce the flock diameter. This cohesive force is similar to the repelling force used for collision avoidance. Simulation results show that this cohesive force can reduce or control the flock diameter. Furthermore, we show that for any set of model parameters, the cohesive force coefficient is the single determining factor of this diameter. The ability of this modified collision-avoiding Cucker-Smale model to provide control of the flock diameter could have significance when applied to robotic flocks.

Keywords: Flock diameter control · Flocking · Cucker-Smale model · Collision avoidance

1 Introduction

Swarm Intelligence is inspired by biological swarms with emergent behaviours that evolve to collectively solve a problem. It has found applications in a diversity of areas [4,7,10]. One of these collective behaviours is flocking which is a phenomenon where individual autonomous agents use only limited information to self-organize into a state of motion consensus, starting from a disordered initial state [15]. In [11], three simple rules — separation, alignment and cohesion, are used to simulate flocking behaviour. The separation rule keeps the agents from colliding. The alignment rule helps them to reach a common speed and direction, while the cohesion rule keeps the flock together spatially. They have been used successfully for computer animation of flocks of birds, etc. However, this model is not amenable to mathematical analyses.

The first mathematical flocking model was proposed by Vicsek [14]. With this model, a group of self-propelled particles moves at the same speed but initially at random directions. Each particle updates its direction by averaging those of its neighbours within a certain radius. Based on the Vicsek's model,

© Springer International Publishing AG 2017
Y. Tan et al. (Eds.): ICSI 2017, Part I, LNCS 10385, pp. 31–39, 2017.
DOI: 10.1007/978-3-319-61824-1_4

Cucker and Smale [3] later proposed a flocking model governed by the following equations:

$$\begin{cases} \dot{p}_i = v_i \\ \dot{v}_i = \frac{1}{N} \sum_{j=1}^{N} \psi\left(\|p_j - p_i\|\right)(v_j - v_i) \end{cases} \tag{1}$$

for N agents where $1 \leq i \leq N$ and the position and velocity of the i-th agent are denoted by p_i and v_i respectively. The communication rate function ψ quantifies the influence between i-th and j-th agents. It is a positive decreasing function of the Euclidean distance between the agents. With

$$\psi(\|p_j - p_i\|) = \frac{1}{(1 + \|p_j - p_i\|^2)^\beta}, \tag{2}$$

it has been mathematically proven that when $\beta < 1/2$ flocking will emerge unconditionally, while for $\beta \geq 1/2$ flocking could only be guaranteed under some conditions on the initial positions and velocities of particles [3]. More recently, a generalization to the Cucker-Smale model has been proposed to ensure that the agents do not collide [1,2]. This is achieved by adding a repelling force function f such that the model equations become

$$\begin{cases} \dot{p}_i = v_i \\ \dot{v}_i = \frac{1}{N} \sum_{j=1}^{N} \psi(\|p_j - p_i\|)(v_j - v_i) \\ \qquad + \sum_{j \neq i} f(\|p_j - p_i\|^2)(p_j - p_i) \end{cases} \tag{3}$$

This is a more realistic model in practice.

The remaining issue is related to the flock diameter. This is defined the maximum distance between any two agents in the flock [6]. In previous studies such as [5,8,13], flocking is assumed to be achieved when the velocity is aligned and this diameter attains any finite value, no matter how large. Intuitively, we only use the term flocking to a group of agents that are moving reasonably close to each other. However, with the Cucker-Smale system, there is no control over the final flock diameter. In practice, for example, in the deployment of a group of autonomous robots, we often want to be able to exert some control over the flock diameter. The main aim of this paper is to propose a modified model based on (3) that will allow us more control over the flock diameter. Inspired by the idea of the repelling force for collision avoidance, we introduce a cohesive force in order to achieve this. The effectiveness of this modified model is demonstrated through computer simulation and the relationship between flock diameter and the cohesive force parameter is obtained.

The rest of this paper is organized as follows. A definition of flocking and a description of the repelling force function of collision-avoiding Cucker-Smale system are presented in Sect. 2. In Sect. 3.1, our proposed introduction of a general cohesive force function to the collision-avoiding Cucker-Smale model is discussed.

The effects of this model on the flock diameter are studied through computer simulation and the results are presented in Sect. 4. Finally, Sect. 5 concludes the paper and discuss the future work by using nonlinear control methods.

2   Preliminaries

2.1   Definition of Flocking

Flocking is said to be achieved for a group of N agents if the following two conditions are satisfied:

1. The velocity of every agent is virtually the same, i.e. for an arbitrarily small $\delta > 0$,

$$|v_i - v_j| \leq \delta \tag{4}$$

   for all $i, j \in [1, N], i \neq j$.
   An equivalent measure for velocity alignment is the average normalized velocity v_a which is defined by

$$v_a = \frac{\left|\sum_{i=1}^N v_i\right|}{\sum_{i=1}^N |v_i|} \tag{5}$$

   Using this measure, the criterion (4) can alternatively be stated as $|1 - v_a| < \delta'$ for some arbitrarily small $\delta' > 0$.
2. The distance between any two agents is bounded by ϵ. That is,

$$\sup_{1 \leq i,j \leq N} \|p_i - p_j\| < \epsilon \tag{6}$$

   ϵ is the upper bound on the distance between two agents that are furthest apart. It will be referred to as the flock diameter in this paper.

2.2   Collision Avoidance

Let $d_0 > 0$ be the minimum distance between any two particles. If $\|p_j - p_i\| < d_0$, then collision is said to have occurred. Collisions could occur between agents in the Cucker-Smale flocking system (1).

In [2], a repelling force is introduced to separate two agents that are too close to each other. It is suggested that this repelling force function $f : (d_0, \infty] \to [0, \infty)$ should have the following properties for $d_1 > d_0$:

1. $\int_{d_0}^{d_1} f(r)dr = \infty$, and
2. $\int_{d_1}^{\infty} f(r)dr < \infty$.

Incorporating this function into (1) results in (3) introduced earlier. A simple function such as $f(r) = (r - d_0)^{-\theta}$ suggested by [2] will have the required properties.

3   Flock Diameter Control

3.1   Cohesive Force

The Cucker-Smale models given by (1) and (3) do not provide any control over the final flock diameter which is the largest distance between any two agents when flocking condition 1 of Sect. 2.1 is satisfied. Inspired by the way collision avoidance was introduced to (1) through a repelling force, flock diameter could potentially be reduced or controlled by introducing a cohesive force.

This cohesive force function ϕ must be Lipschitz continuous, so that the existence theorems and equations for Cucker-Smale model can apply. Given that a repelling force is asserted when an agent moves within a radius of d_1 of another agent, the cohesive force should operate between agents that are at least at a distance of d_1 apart. So, $\phi(r) = 0$ when $r \leq d_1$. For $r > d_1$, $\phi(r)$ should be a monotonically non-decreasing function of r.

An example of cohesive force function is given by:

$$\phi(r) = \begin{cases} 0 & r \leq d_1 \\ k * \frac{1}{1+e^{-r}} & r > d_1 \end{cases} \tag{7}$$

where $1/(1 + e^{-r})$ is a sigmoid function and is Lipschitz continuous.

3.2   Modified Cucker-Smale System

Introducing this cohesive force into (3), the modified Cucker-Smale system becomes

$$\begin{cases} \dot{p}_i = v_i \\ \dot{v}_i = \frac{1}{N} \sum\limits_{j=1}^{N} \psi(\|p_j - p_i\|)(v_j - v_i) \\ \quad + \sum\limits_{j \neq i} f(\|p_j - p_i\|^2)(p_j - p_i) \\ \quad + \sum\limits_{j \neq i} \phi(\|p_j - p_i\|^2)(p_j - p_i) \end{cases} \tag{8}$$

for $1 \leq i \leq N$ where ϕ is the cohesive force function.

The second and third terms for $\dot{v}_i$ in (8) could be combined since they essentially have the same form. Hence we have

$$\begin{cases} \dot{p}_i = v_i \\ \dot{v}_i = \frac{1}{N} \sum\limits_{j=1}^{N} \psi(\|p_j - p_i\|)(v_j - v_i) \\ \quad + \sum\limits_{j \neq i} H(\|p_j - p_i\|^2)(p_j - p_i) \end{cases} \tag{9}$$

When the distance between any two particles $r = \|p_j - p_i\|$ is less than collision avoidance distance d_1, then H acts like the repelling force function f. Otherwise, it acts like the cohesive force function ϕ.

Using the examples for the repelling and cohesive forces from Sects. 2.2 and 3.1 respectively, a possible function H is shown in Fig. 1. Here, a positive value denotes an attractive force while negative values denote repulsion.

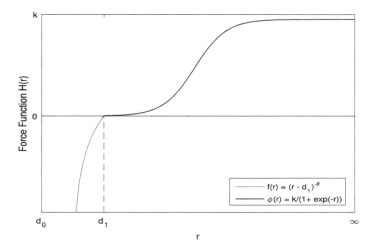

Fig. 1. Force function $H(r)$ in (9).

4   Simulation Results

The characteristics of the modified collision-avoiding Cucker-Smale system (8) given in Sect. 3.2 will now be studied using computer simulation. The aim is to evaluate how various parameters of the system affects the flock diameter. These parameters include the cohesive force coefficient k in (7), the initial field size and the collision distance d_1.

The agents are free to move in an infinitely large two-dimensional space. Thus they will not encounter any boundaries. Without loss of generality, we shall not assign units to both the distance and time. Every agent will move with the same speed of 0.5 per unit time with a uniformly random initial direction in $[0, 2\pi)$. The initial position of each agent will be randomly chosen within a circle of l units in diameter which will be referred to as the initial field size. The value of β in (2) is fixed at $1/4$ to ensure flocking occurs. The collision avoidance distance d_1 is set at 0.2. The value of d_0 is 0.01. The value of parameter θ of the repelling function is 2. The system is considered to be in a flocking state when the average velocity $v_a \geq 0.99$. In every scenario for each set of parameters, the value shown is the average over 20 independent simulation runs.

4.1   Effect of Cohesive Force Coefficient

First, we shall consider the effect of the cohesive force coefficient k in (7) on the flock diameter when flocking is achieved. We examine values of k between 0 and 3 with $k = 0$ indicating that cohesive force is not used. The size N of the flock ranges from 10 to 50 with increments of 10. The initial field size is 4.

Figure 2 shows that increasing the cohesive force coefficient leads to a substantial reduction in the flock diameter for all different values of N. Comparing between no cohesive force and $k = 3$ reveals the largest reduction in flock

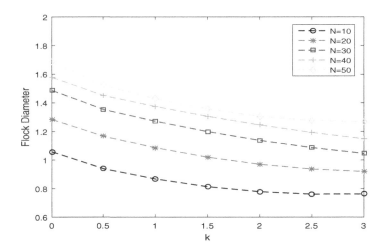

Fig. 2. Flock diameter for different cohesive force coefficients

diameter is for $N = 10$ at 29.8%. For N from 20 to 50, percentage reduction in flock diameter are 12.3%, 15.1%, 14.9% and 14.5% respectively. This shows that the cohesive force can be used to obtain a tighter group of flocking agents.

4.2   Effect of Initial Field Size

The initial field size, which reflects how closely placed the agents initially are, may have a substantial effect on the final flock diameter. In this set of simulations, we vary the intial field size while keeping other system parameters the same. Figure 3 shows that increasing the initial separation of the agents does

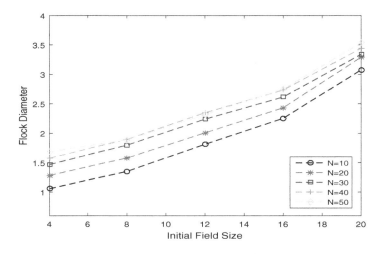

Fig. 3. Flock diameter for different initial field size without cohesive force.

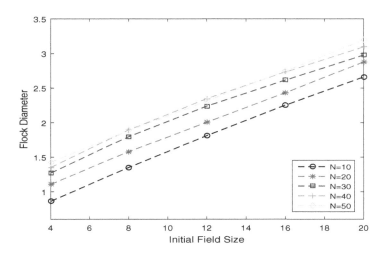

Fig. 4. Flock diameter for different initial field size with cohesive force coefficient $k = 1$.

Table 1. Flock diameter by cohesive force coefficients (k)

Initial field size	N = 10		N = 20		N = 30		N = 40		N = 50	
	k = 0	k = 1	k = 0	k = 1	k = 0	k = 1	k = 0	k = 1	k = 0	k = 1
4	1.0570	0.8634	1.2835	1.1080	1.4864	1.2703	1.5771	1.3513	1.6877	1.4280
8	1.5185	1.3644	1.7080	1.6742	1.8716	1.8059	2.0552	1.9180	2.2818	1.9707
12	2.2013	1.7812	2.4258	1.9644	2.6490	2.3418	2.7641	2.4214	2.8702	2.2588
16	2.8674	2.2808	3.0995	2.3925	3.1646	2.4683	3.2858	2.6104	3.3848	2.7377
20	3.0775	2.6609	3.2935	2.8801	3.3419	2.9797	3.4430	3.0980	3.5366	3.1976

have a substantial effect on the final flock diameter. For $N = 10$, when the initial field size is increased from 4 to 20, the flock diameter is increased almost 3 times. As N increases, the percentage increase in flock diameter is smaller. But for $N = 50$, the increase is still more than twice.

The same simulations are repeated with a cohesive force coefficient k set to 1. The results can be found in Fig. 4. The numerical values for Figs. 3 and 4 are listed in Table 1 for ease of comparison. It is interesting to note that the percentage increase in flock diameter with cohesive force is more or less the same that without cohesive force. This is true for all values of N. Thus the effect of initial field size on the final flock diameter is essentially the same for both systems.

4.3   Effect of Collision Distance

The remaining factor that could have a substantial influence on the flock diameter is the collision distance d_1. Simulation results with $k = 1$ and varying values of d_1 are shown in Fig. 5. Other parameters are the same as the previous simulations. The lowest curve shows that collision distance in 1 has approximately

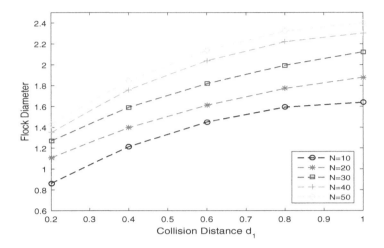

Fig. 5. Flock diameter in different collision distances

2 times impact on flock diameter when collision avoidance is 0.2 for $N = 10$. Figure 5 displays a similar tendency in the different number of groups. Consider the collision distance $d_1 = 0.2$ as the reference. When d_1 is doubled, the flock diameter increases by 130% for all values of N. For instance, with $N = 50$, the flock diameter is 1.8755 in $d_1 = 0.4$ that is 1.3 times in comparison with 1.4280 of $d_1 = 0.2$. Based on the results from Sect. 4.1, if d_1 is increased, the only way to reduce the flock diameter is by increasing the cohesive force.

5   Conclusions and Future Work

In this paper, we introduced a cohesive force into the collision-avoiding Cucker-Smale flocking model. The main purpose is to provide a way to control the diameter of the flock. Simulation results show that this cohesive force is able to reduce this diameter, and the cohesive force is obvious a nonlinear control for flocking diameter. Other factors such as the initial field size and collision distance have the same effect on the system with and without cohesive force. Thus the only significant factor in controlling the flock diameter is the cohesive force coefficient. The ability of this modified collision-avoiding Cucker-Smale model to provide control of the flock diameter could have significance when applied to robotic flocks. In the future, we are exploring ideas like [9,12] using nonlinear control for flocking diameters.

References

1. Cucker, F., Dong, J.G.: Avoiding collisions in flocks. IEEE Trans. Autom. Control **55**(5), 1238–1243 (2010)
2. Cucker, F., Dong, J.G.: A general collision-avoiding flocking framework. IEEE Trans. Autom. Control **56**(5), 1124–1129 (2011)

3. Cucker, F., Smale, S.: Emergent behaviour in flocks. IEEE Trans. Autom. Control **52**(5), 852–862 (2007)
4. Elkawkagy, M.: Improving the performance of hybrid planning. Int. J. Artif. Intell. **14**(2), 98–116 (2016)
5. Ha, S.Y., Liu, J.G., et al.: A simple proof of the cucker-smale flocking dynamics and mean-field limit. Commun. Math. Sci. **7**(2), 297–325 (2009)
6. Ha, S.Y., Tadmor, E.: From particle to kinetic and hydrodynamic descriptions of flocking. Kinet. Relat. Models **1**(3), 415–435 (2008)
7. Okubo, A.: Dynamical aspects of animal grouping: swarms, schools, flocks, and herds. Adv. Biophys. **22**, 1–94 (1986)
8. Park, J., Kim, H.J., Ha, S.Y.: Cucker-Smale flocking with inter-particle bonding forces. IEEE Trans. Autom. Control **55**(11), 2617–2623 (2010)
9. Precup, R.E., Angelov, P., Costa, B.S.J., Sayed-Mouchaweh, M.: An overview on fault diagnosis and nature-inspired optimal control of industrial process applications. Comput. Ind. **74**, 75–94 (2015)
10. Qin, Q., Cheng, S., Zhang, Q., Li, L., Shi, Y.: Biomimicry of parasitic behavior in a coevolutionary particle swarm optimization algorithm for global optimization. Appl. Soft Comput. **32**, 224–240 (2015)
11. Reynolds, C.W.: Flocks, herds and schools: a distributed behavioural model. SIG-GRAPH Comput. Graph. **21**(4), 25–34 (1987)
12. Tan, Y., Dai, H.H., Huang, D., Xu, J.X.: Unified iterative learning control schemes for nonlinear dynamic systems with nonlinear input uncertainties. Automatica **48**(12), 3173–3182 (2012)
13. Ton, T.V., Linh, N.T.H., Yagi, A.: Flocking and non-flocking behavior in a stochastic Cucker-Smale system. Anal. Appl. **12**(01), 63–73 (2014)
14. Vicsek, T., Czirók, A., Ben-Jacob, E., Cohen, I., Shochet, O.: Novel type of phase transition in a system of self-driven particles. Phys. Rev. Lett. **75**(6), 1226 (1995)
15. Vicsek, T., Zafeiris, A.: Collective motion. Phys. Rep. **517**(3), 71–140 (2012)

Building a Simulation Model for Distributed Human-Based Evolutionary Computation

Kei Ohnishi$^{(\boxtimes)}$, Junya Okano, and Mario Koeppen

Graduate School of Computer Science and System Engineering,
Kyushu Institute of Technology, 680-4 Kawazu, Iizuka, Fukuoka 820-8502, Japan
ohnishi@cse.kyuyech.ac.jp, okano@evocomp.cse.kyuyech.ac.jp,
mkoppen@ci.kyuyech.ac.jp
http://evocomp.cse.kyutech.ac.jp/

Abstract. Evolutionary computation (EC) is called "human-based EC" especially when its all main operators, which are selection, crossover, and mutation, are executed by humans. One type of human-based EC is distributed human-based EC, in which humans independently manage their solution candidates and share them by direct communication between the humans. It is expected that the EC solves problems in human organizations. However, it is not easy to conduct real experiments to investigate the effect of human behaviors on the performance of the EC because such experiments needs many cooperative people. In the paper, we, therefore, first model human behaviors and then build a simulation model including the model. The model of human behaviors focuses on physical movement and free will to decide a time of interactions with others. Furthermore, we attempt to understand the EC though simulations using the built simulation model.

Keywords: Human-based evolutionary computation · Human behaviors · Simulation model

1   Introduction

Evolutionary computation (EC) as optimization methods that is inspired by biological genetics and evolution has attracted great attentions. In case that selection, crossover, and mutation operators as main operators of EC are conducted by humans, such EC is called human-based EC [7]. There are two types of human-based EC. One is centralized human-based EC, which manages solution candidates in a centralized way such as a Web server. The other is distributed human-based EC, which forces humans to manage solution candidates by themselves and allow them to share solution candidates only through direct communication among them. We can expect that those two types of human-based EC solve problems in human organizations. However, in order to investigate how human behaviors influence the performance of human-based EC, we need great cooperations of many people to conduct human subjective experiments for the investigation.

© Springer International Publishing AG 2017
Y. Tan et al. (Eds.): ICSI 2017, Part I, LNCS 10385, pp. 40–49, 2017.
DOI: 10.1007/978-3-319-61824-1_5

Therefore, in the present paper, we make a model of human behaviors in the distributed human-based EC and build a simulation model including the model of human behaviors for simulation studies. Specifically, we model free determination of timings by humans at which they interact with others and physical movement of humans. Moreover, we attempt to understand the distributed human-based EC using the built simulation model.

The remainder of the paper is organized as follows. Section 2 describes the simulation model for the distributed human-based EC. In Sect. 3, we show and discuss the simulation results. We run the simulations using a variety of parameters values of the simulation model. Section 4 describes the related work. Finally, conclusions and future work are presented in Sect. 5.

2 Simulation Model

We assume that moving people form a mobile ad-hoc network (MANET) [1] with wireless communication devices to implement human-based EC and share solution candidates on the MANET. When using MANET as a place for sharing solution candidates in distributed human-based EC, it is basically hard to decide behaviors of humans who execute evolutionary operators such as selection, crossover, and mutation in advance. Humans have their own will. Concretely, it is hard to decide humans' fitness functions, evolutionary operators executed by humans, movement of humans, and timings at which humans create solution candidates in advance. The only one thing that an operator of distributed human-based EC can control is how to share information among humans by means of their communication devices.

However, when building a simulation model of distributed human-based EC, we have to embed the details of both what we can decide in the practical use and what we cannot decide into the simulation model. From the following section, we explain the details embedded into the simulation model.

2.1 Fitness Function and Evolutionary Operators of Humans

In the simulation model built herein, we assume that the distributed human-based EC conducts function optimization, which is real parameters optimization, and solution candidates take a form of real-valued vector. In addition, we assume that all humans who participate in the execution of the distributed human-based EC have the same fitness function. As crossover and mutation operators that all humans conduct, BLX-0.36 [3] and the uniform mutation that changes a value of parameter by a uniform random number generator are used.

However, in reality, optimization problems that humans solve are often described in a natural language and their solution candidates as well. In addition, humans would not have the same fitness function but have their own fitness functions.

2.2   Information Sharing Method Using Identifiers Representing Human Interests

Humans who participate in the distributed human-based EC share their created solution candidates among them. As a method for the sharing, we use the information sharing method with identifiers representing human interests that we previously proposed in [5].

The information sharing method sends a solution candidate that a human newly created to the surroundings by broadcasting, in which a communication device sends data to all devices within its communicable range. In addition, the method forces the communication device as a receiver to decide if it actually receives the solution candidate, and if it decides to receive it, then it again sends it to the surroundings by broadcasting. The details of this sending is described in our previous work [5].

As mentioned above, we have assumed in the simulation model that the distributed human-based EC conducts function optimization and the search space is a real-valued space. In addition, we assume that the space of human interests is the same as the search space and also that the solution candidate of each human is used as the identifier of the human representing the interests of the human. As mentioned later, a human holds the solution candidate with the best fitness value among those that the human created so far, so that the identifier of each human is equivalent to the best solution candidate for the human.

2.3   Human Movement

In our previous study [5], we considered that each human randomly moves to a position within a 1×1 square area centered on his/her current position once per unit of time. However, in the simulation model built herein, we consider that each human has a destination and go straight to the destination at a constant speed.

2.4   Timing of Human to Create a Solution Candidate

In our previous study [9], we focused on the timing of sharing a solution candidate by a human who joins the distributed human-based EC. At the timing, a human meets others and not only share his/her solution candidate with others but also create a new solution candidate by mixing his/her solution candidate and those received from others. Only when a human meets others, the human creates a new solution candidate.

Meanwhile, in the simulation model built herein, we do not use the same model used in our previous study [9] to decide a timing to share a solution candidate. Humans simply publish and forward their solution candidates to their surroundings by broadcasting right after they created them. However, we herein newly introduce a model that gives each human a timing to create a new solution candidate (in other wards, to update his/her solution candidate). We assume that humans create their solution candidates at every given cycle time. So, the simulation model built herein assigns a cycle time of creating a solution candidate to each human. The simulation model considers discrete time represented as an integer and the cycle time is determined as an integer within $[1, C]$ $(C \geq 1)$.

The cycle time of creating a solution candidate for each human is probabilistically determined. As the probability with which it becomes c for a human, $p(c)$, we use two types. One is represented by Eq. (1), which all possible cycle times appear with equal probability. The other is represented by Eq. (2).

$$p(c) = 1/C. \tag{1}$$

$$p(c) = \frac{\exp(-\alpha \times c/100)}{\sum_{k=1}^{C} \exp(-\alpha \times k/100)}. \tag{2}$$

In addition, in the simulation model, a human receives a solution candidate from other human in his/her surroundings at a time (time a) and will further forward the solution candidate to his/her surroundings at the next time (time $a+1$). Each human creates a solution candidate by applying BLX-0.36 between his/her solution candidate and each of solution candidates that he/she received during the time period from the previous time of creating a solution candidate to the present time and also by applying the uniform mutation to each new solution candidate created by BLX-0.36. Each new solution candidate is then evaluated with a fitness function of each human. Each human holds the solution candidate with the best fitness among all solution candidate created by him/her so far as his/her current solution candidate.

3   Simulation

3.1   Simulation Scenario

We consider a public space such as the large-spaced station precincts as the field of executing the distributed human-based EC. Concretely, we consider the rectangle area with the size of 20×80 shown in Fig. 1. The entrance and exit gates of the space are represented by the bold segment line with the length of 10 in Fig. 1 and there are six entrance and exit gates.

We assume two kinds of number of humans who execute the human-based EC in the field, 100 and 150. The people form MANET in the field.

All humans move by the distance of 4/3 during every unit time. Each human is given a start entrance and exit gate randomly and then given a start point on the start entrance and exit gate randomly, and then, goes straight to the end point on the end entrance and exit gate, which are also given randomly. However, the start and end entrance and exit gates are not identical. Once a human reached the end entrance and exit gate, he/she does not keep being in the field, and a new human with a randomly created solution candidate appears at a randomly given point on a randomly given entrance and exit gate.

We use several values in between 3.5 and 6.5 as a communicable range of a human. A human can communicate with others who exist within a circle of radius of his/her given communicable range.

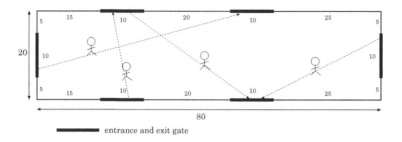

Fig. 1. The field of executing the distributed human-based EC.

We use three fitness functions represented by Eqs. (3), (4), (5). In the simulations, we set n in the three functions to be 20. The distributed human-based EC attempts to minimize the function.

$$f(\boldsymbol{x}) = 10n + \sum_{i=1}^{n}\{x_i^2 - 10\cos(2\pi x_i)\} \ (-5.12 \leq x_i \leq 5.12). \tag{3}$$

$$f(\boldsymbol{x}) = 418.9829n + \sum_{i=1}^{n}\{-x_i\sin(\sqrt{|x_i|})\} \ (-512 \leq x_i \leq 512). \tag{4}$$

$$f(\boldsymbol{x}) = \sum_{i=1}^{n-1}\{100(x_{i+1} - x_i^2)^2 + (x_i - 1)^2\} \ (-2.408 \leq x_i \leq 2.408). \tag{5}$$

We use three values, 10, 20, 30 for the parameter of d [5], which decides the probability of receiving a solution candidate that was published and forwarded by someone. The maximum value of cycle time at which a human creates a solution candidate, C, is set to be 10. In addition, we use three kinds of values of the parameter of Eq. (2), α, which are 1, 10, 15. Eq. (2) determines the probability of assigning the specific cycle time, c, to a human. The crossover and mutation rates are 1.0 and 0.01, respectively. The maximum number of solution candidates that a human can receive from others during every unit time is 60. That is to say, a human can create new 60 solution candidates at most by the application of the crossover and mutation during a unit time.

3.2   Simulation Results

Figure 2 shows the time-variations of the best fitness value when several values were used as the communicable ranges and Eq. (2) was used for determining the cycle time of creating a solution candidate. The number of humans was 100 and $d = 10$. Figure 3 shows the time-variations of the best fitness value when several values were used as the number of humans in the field of executing the distributed human-based EC and Eq. (2) was used for determining the cycle time of creating a solution candidate. The communicable range was 5 and $d = 10$. Figure 4 shows the time-variations of the best fitness value when several values were used as the parameter α of Eq. (2) for the determination of the cycle time. The number of

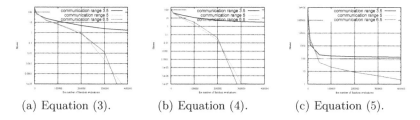

(a) Equation (3).   (b) Equation (4).   (c) Equation (5).

Fig. 2. The time-variation of the best fitness value when the communicable range is changed. The horizontal axis represents the number of function evaluations. The value of α is set to be 10.

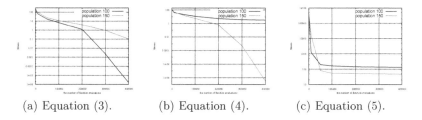

(a) Equation (3).   (b) Equation (4).   (c) Equation (5).

Fig. 3. The time-variation of the best fitness value when the number of humans is changed. The horizontal axis represents the number of function evaluations. The value of α is set to be 10.

humans was 150 and the communicable range was 5 and $d = 10$. Figure 5 shows the time-variations of the best fitness value when several values were used as the parameter d that decides the probability of receiving a solution candidate forwarded by others. The number of humans was 150 and the communicable range was 5 and $\alpha = 10$. All result graphs mentioned above are drawn by setting the horizontal axis to be the number of function evaluations. Also, all results mentioned above are the average over 100 independent runs.

When the communicable range is small, the number of solution candidates that a human receives from others becomes small. Then, the number of times of creating and evaluating a solution candidate by a human at his/her cycle time of creating a solution candidate becomes small. That indicates that the load of a human becomes small and at the same time indicates that opportunities in which a human holding better solution candidate shares his/her better one with others become few. We can observe from Fig. 2 that a larger communicable range results in better search performance.

When the number of humans in the field is small, the number of solution candidates that a human receives from others becomes small. Similar to the case of the small communicable range, we can observe from Fig. 3 that a larger number of humans in the field results in better search performance.

As mentioned above, in case of the small communicable range or in case of the small number of humans in the field, it can happen that better created solution candidates are not shared among humans in the field. A mechanism

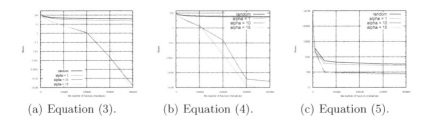

(a) Equation (3).   (b) Equation (4).   (c) Equation (5).

Fig. 4. The time-variation of the best fitness value when the value of the parameter α is changed. The horizontal axis represents the number of function evaluations. The label of "random" represents the result when using Eq. (1).

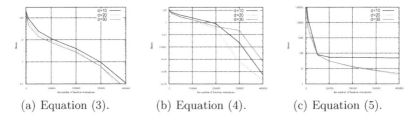

(a) Equation (3).   (b) Equation (4).   (c) Equation (5).

Fig. 5. The time-variation of the best fitness value when the value of the parameter d is changed. The horizontal axis represents the number of function evaluations.

that avoids the disappearance of better solution candidates due to non-sharing of them among humans in the field is needed. A simple solution for the problem is to place a storage to save better solution candidates in the filed and provide them from the storage at need. However, we have to decide a policy for selecting solution candidates to be saved.

The larger the value of the parameter α in Eq. (2), which represents the probability to decide the cycle time of creating a solution candidate, is, the larger the number of humans with a short cycle of time of creating a solution candidate is. As a result, the number of humans who create a solution candidate at a time increases. That indicates that the number of solution candidates forwarded to the surroundings increases, and as a result, the number of times of creating and evaluating a solution candidate by a human increases. For a human, creating and evaluating a solution candidate many times would be burden, but by doing so, local search in the search space can be sufficiently conducted. Furthermore, since humans with a randomly created solution candidate constantly appear in the field, the diversity of existing solution candidates in the field can be maintained to some extent. We can observe from Fig. 4(a), (b) that a larger value of α results in better search performance.

When the value of the parameter d [5] for determining the probability of receiving a solution candidate is large, only humans who have similar solution candidates in the search space, which are equivalent to similar identifiers representing interests, can share their solution candidates among them. That is to say, when the value of the parameter d is large, the number of times of creating

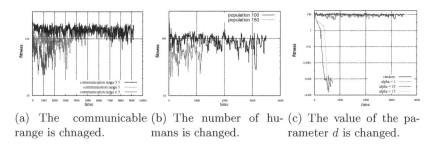

(a) The communicable range is chnaged.

(b) The number of humans is changed.

(c) The value of the parameter d is changed.

Fig. 6. The time-variation of the best fitness value when using Eq. (3) as a fitness function. The horizontal axis represents time.

and evaluating a solution candidate decreases and only local search is mainly conducted. That results in loosing the diversity of solution candidates in the field. We can observe from Fig. 5(a)–(c) that too large value of d results in worse search performance.

In the above-mentioned result graphs, we set the horizontal axis to be the number of times of evaluating a solution candidate. Here we examine the characteristics of the searching by setting the horizontal axis to be time. We used Eq. (3) as a fitness function. Figure 6(a)–(c) shows the results when varying the communicable range, the number of humans in the field, and the value of the parameter d for determining the cycle time of creating a solution candidate, respectively. In those figures, the best fitness value at each time in the field is plotted. The reason why the end time of the graph depends on the parameter settings is that the number of times of creating and evaluating a solution candidate during the unit time varies depending on the parameter settings under the fixed number of total evaluations.

We can observe from Fig. 6(a)–(c) that good results are obtained when the number of evaluations during the unit time is large. On the other hand, when the number of evaluations during the unit time is small, good results are not obtained. That would be because many good solution candidates disappear without being shared among humans in the field, as mentioned above.

We can observe from Fig. 6(a) that the best fitness value in the field does not change much when the communicable range is small. That is because the number of evaluations during a unit time is small and good solution candidates are frequently lost without being shared. On the other hand, when the communicable range is large, the improvement of the best fitness value is observed over the entire simulation time though the fitness value fluctuates. Similar to this, we can observe from Fig. 6(b) that the improvement of the best fitness value is observed when the number of humans is large. In addition, we can also observe that basically when the value of α becomes large, we can observe the improvement of the best fitness value.

In summary, we can say that facilitating solution candidates sharing even among humans who do not have very close identifiers and increasing the number of times of creating and evaluating a solution candidate during a unit time in the situation that humans with a randomly created solution candidate constantly come to the field results in that good solution candidates keep staying in the field, and consequently, better search performance is achieved. However, that forces humans to create more solution candidates during a short period of time, and it is likely that humans have heavy load. Therefore, as mentioned above, we need to consider a mechanism that selectively saves solution candidates created by humans in the field and utilizes them.

4   Related Work

The human-based genetic algorithm (human-based GA) [7], which is one type of human-based EC, was first applied to problems that are described in a natural language and require to describe solution candidates in natural language [6,7]. In this first application, the human-based GA took a centralized approach to managing communication between humans, and it used a message board on a web forum. However, as mentioned in [2], the areas of application of the human-based GA are not limited to those that involve a natural language, and in that study, an interactive GA and a human-based GA were compared in terms of their ability to solve problems that are not described in a natural language. It was shown that crossover and mutation performed by humans are useful for solving such problems.

The framework of GAs has been utilized in human organizations for solving problems and for enhancing creativity [6,8]. In [6], the components and procedures in human organizations are regarded as the fundamental genetic components (such as genes, individuals, and population) and genetic procedures (such as crossover and mutation). In [8], a new human-based genetic algorithm was proposed for solving problems of human organizations.

We proposed the use of centralized human-based EC for determining how to utilize various data available on the Internet [4]. We also proposed a method for sharing solution candidates among participants in distributed human-based EC [5], which is used in the simulation model built in this paper. In another study of ours [9], we examined the search characteristics of the particular distributed human-based EC in which a human shares his/her solution candidate with others at his/her cycle time.

5   Concluding Remarks

In the present paper we have built the realistic simulation model of the distributed human-based EC that models human behaviors. Then, we have attempted to understand the distributed human-based EC through simulations using the built model. We have observed from the simulation results that more number of times of creating (evaluating) a solution candidate during a unit time results in better search performance. If we consider creating and evaluating a solution candidate to be the load of human, the obtained result is interpreted as that the higher load a humans take, the better solutions humans obtain. So, if we execute the distributed human-based EC in the real world like in the simulation, we should introduce a mechanism that achieves better performance while relieving the human load.

In future work, we will consider such a mechanism mentioned above. In addition, we will realize the simulation model considering the situation that humans have their own fitness functions and then attempt to understand the distributed human-based EC more through simulations using the realized model.

References

1. Basagni, S., Conti, M., Giordano, S., Stojmenovic, I.: Mobile Ad Hoc Networking. Wiley-IEEE Press, New York (2004)
2. Cheng, C.D., Kosorukoff, A.: Interactive one-max problem allows to compare the performance of interactive and human-based genetic algorithms. In: Deb, K. (ed.) GECCO 2004. LNCS, vol. 3102, pp. 983–993. Springer, Heidelberg (2004). doi:10.1007/978-3-540-24854-5_98
3. Eshelman, L.J., Shaffer, D.J.: Real-coded genetic algorithms and interval-schemata. Found. Genet. Algorithms **2**, 187–202 (1993)
4. Hasebe, R., Kouda, R., Ohnishi, K., Munetomo, M.: Human-based genetic algorithm for facilitating practical use of data in the internet. In: Joint 7th International Conference on Soft Computing and Intelligent Systems and 15th International Symposium on Advanced Intelligent Systems (SCIS&ISIS2014), pp. 1327–1332 (2014)
5. Hasebe, R., Ohnishi, K., Koeppen, M.: Distributed human-based genetic algorithm utilizing a mobile ad hoc network. In: 2013 IEEE International Conference on Cybernetics (CYBCONF 2013), pp. 174–179 (2013)
6. Kosoruko, A., Goldberg, D.E.: Evolutionary computation as a form of organization. In: Genetic and Evolutionary Computation Conference (GECCO 2002), pp. 965–972 (2002)
7. Kosorukoff, A.: Human based genetic algorithm. In: 2001 IEEE International Conference on Systems, Man, and Cybernetics (SMC 2001), pp. 3464–3469 (2001)
8. Llorà, X., Ohnishi, K., Chen, Y., Goldberg, D.E., Welge, M.E.: Enhanced innovation: a fusion of chance discovery and evolutionary computation to foster creative processes and decision making. In: Deb, K. (ed.) GECCO 2004. LNCS, vol. 3103, pp. 1314–1315. Springer, Heidelberg (2004). doi:10.1007/978-3-540-24855-2_143
9. Okano, J., Hamano, K., Ohnishi, K., Koeppen, M.: Particular fine-grained parallel GA for simulation study of distributed human-based GA. In: 2014 IEEE International Conference on Systems, Man, and Cybernetics (SMC 2014), pp. 3523–3528 (2014)

Model of Interruptions in Swarm Unit

Eugene Larkin$^{(\boxtimes)}$, Alexey Ivutin, and Anna Troshina

Tula State University, Tula 300012, Russia
{elarkin,atroshina}@mail.ru, alexey.ivutin@gmail.com

Abstract. Time characteristics of algorithm interpretation by Von-Neumann computers are investigated. With use of semi-Markov process fundamental apparatus the analytical model of program runtime evaluation is worked out. It is shown that external interruptions are the result of functioning of independent random process, which develops in parallel with algorithm interpretation. For description of interaction of main program, interruption generator and interruption handler apparatus of Petri-Markov nets is used. Basic structural-parametric model of computer functioning in the presence of interruptions is worked out. It is shown that in common case Petri-Markov model is an infinite one. The recursive procedure of wandering through Petri-Markov net for case under investigation is worked out. It is shown that process of wandering through the net is not quite semi-Markov one. The method of transformation of Petri-Markov model onto strictly semi-Markov process is proposed.

Keywords: Von-Neumann computer · Interruption · Runtime · Semi-Markov process · Petri-Markov model · Wandering

1 Introduction

Interruption operational mode is of widely used in Von-Neumann computers [3,12]. In accordance with such regime, CPU switches from interpretation of main algorithm to interpretation of interruption processing algorithm when interruption signal comes. After completion of interruption handling, CPU returns to main algorithm. In practice, there are interruptions of two types: software ones and hardware ones. Software interruptions lead to common brunching of algorithms [1], so its influence on time of interpretation of common algorithm is rather routine engineering task. Hardware interruptions are generated with external computer independent unit. Due to the fact in such system, the parallel process takes place: on the one hand there is the CPU, while on the other hand, the interruption generator.

As with any parallel process in the system under investigation occurs competition for use of computer resources, in particular, a CPU time [2,5,14]. "Competition" increases time of interpretation of main algorithm. An increment of time depends on interruptions flow density, and time characteristics of interruption processing algorithm. Proper approach to evaluation of time increment in case under consideration reduced is the analysis of the "competition" between

© Springer International Publishing AG 2017
Y. Tan et al. (Eds.): ICSI 2017, Part I, LNCS 10385, pp. 50–59, 2017.
DOI: 10.1007/978-3-319-61824-1_6

current algorithm and interruption generator. A model, which allows perform such analysis, does not currently exist, which explains the necessity and urgency of the paper.

2   Time Characteristics Interruption-Free Algorithm

The process of interpretation of algorithm comes down to sequential interpretation of its operators, which unfolds in time. Natural model of the algorithm interpretation is the semi-Markov process [6,7,10] as follows

$$\mu = \{B, \boldsymbol{h}(t)\}, \qquad (1)$$

where t – is the time; $B = \{\beta_1, \ldots, \beta_j, \ldots, \beta_J, \beta_{J+1}, \ldots, \beta_M\}$ – is subset of states; $\boldsymbol{h}(t) = \lfloor h_{j,m}(t) \rfloor$ – semi-Markov matrix of size $M \times M$.

States of semi-Markov process have the next physical meaning: β_1 – is the starting state; $B \supset E = \{\beta_{J+1}, \ldots, \beta_m, \ldots, \beta_M\}$ – is the subset of absorbing states. Structure of semi-Markov process is such, that for any state $\beta_j \notin E, j \neq 1$, there is at least one way $\beta_1 \to \beta_j$ and at least one way $\beta_j \to E$. Due to the fact in the semi-Markov matrix $\boldsymbol{h}(t)$:

- first column includes zero elements only;
- rows from $(J+1)$-th till M-th includes zero elements only;
- for other rows

$$\sum_{m=1}^{M} \int_0^\infty h_{jm}(t)dt = 1, 1 \leq j \leq J. \qquad (2)$$

Thus all possible ways $\beta_1 \to E$ make a complete group of incompatible events. For such a case density of time of attainment the subset E from the state β_1 is as follows

$$f(t) = L^{-1} \left[{}^r\boldsymbol{I}_1 \sum_{k=1}^\infty \{L[\boldsymbol{h}(t)]\}^k \, {}^c\boldsymbol{I}_E \right], \qquad (3)$$

where L and L^{-1} – are direct and inverse Laplace transforms correspondingly; ${}^r\boldsymbol{I}$ – row victor first element of which is equal to one, and other elements are zeros; ${}^c\boldsymbol{I}_E$ – column vector, elements from the first till J-th of which are zeros, and other elements are equal to one.

Semi-Markov process after reduction is shown on the Fig. 1a. External interruption generator is shown on the Fig. 1b. Semi-Markov process of interruption handler is shown on the Fig. 1c. As it follows from the structure in the system it takes a couple of two-parallel processes.

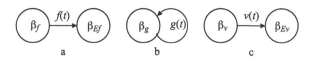

Fig. 1. Structure of simplified semi-Markov process (a), interruption generator (b) and interruption handler (c)

3 Model of Algorithm Interpretation at the Presence of Interruptions

Semi-Markov process of algorithm interpretation at the presence of interruptions is as follows

$$\tilde{\mu} = \left\{ \{\beta_g, \beta_f, \beta_v, \beta_{Ef}, \beta_{Ev}\}, \tilde{h}(t) \right\}, \tag{4}$$

where $\beta_g, \beta_f, \beta_v$ – is the set of states; β_g – is the state, which simulates the generation of one interruption; β_f – is the state, which simulates algorithm interpretation; β_v – is the state, which simulates the interruption handler; β_{Ef}, β_{Ev} – are absorbing states;

$$\tilde{h}(t) = \begin{bmatrix} \boldsymbol{h}_g(t) & \boldsymbol{0}_{12} & \boldsymbol{0}_{13} \\ \boldsymbol{0}_{21} & \boldsymbol{h}_f(t) & \boldsymbol{0}_{23} \\ \boldsymbol{0}_{31} & \boldsymbol{0}_{32} & \boldsymbol{h}_v(t) \end{bmatrix} \tag{5}$$

$$\boldsymbol{h}_g(t) = [g(t)]; \qquad \boldsymbol{h}_f(t) = \begin{bmatrix} 0 & f(t) \\ 0 & 0 \end{bmatrix}; \qquad \boldsymbol{h}_v(t) = \begin{bmatrix} 0 & v(t) \\ 0 & 0 \end{bmatrix};$$

$$\boldsymbol{0}_{12} = \boldsymbol{0}_{13} = \begin{pmatrix} 0 & 0 \end{pmatrix}; \qquad \boldsymbol{0}_{21} = \boldsymbol{0}_{31} = \begin{pmatrix} 0 \\ 0 \end{pmatrix}; \qquad \boldsymbol{0}_{23} = \boldsymbol{0}_{32} = \begin{pmatrix} 0 & 0 \\ 0 & 0 \end{pmatrix}. \tag{6}$$

All elementary processes of (4) interact between themselves. Rules of interaction are the next. Processes $\boldsymbol{h}_f(t)$ and $\boldsymbol{h}_g(t)$ begin "compete" between themselves [8,9,11]. If "wins" $\boldsymbol{h}_f(t)$, interpretation of algorithm ends, otherwise starts process $\boldsymbol{h}_v(t)$, which "compete" with the process $\boldsymbol{h}_g(t)$. If in the case "wins" $\boldsymbol{h}_g(t)$, then starts $\boldsymbol{h}_v(t)$ again, otherwise process return to "competition" of $\boldsymbol{h}_f(t)$ and $\boldsymbol{h}_g(t)$. $\boldsymbol{h}_f(t)$ starts from the state, on which it was interrupted.

Classical theory of semi-Markov process does not permit describe interaction, so for modeling Petri-Markov net (PMN) [8,9,11] apparatus was used. PMN, which describes interpretation of algorithm under interruptions, is as follows:

$$\Pi = \{A, Z, \varphi(t), \boldsymbol{\Lambda}\}, \tag{7}$$

where $A = \{{}^0 a_2, {}^1 A, \ldots, {}^k A_k, \ldots\}$ – is set of places, including ${}^0 a_2$, which simulates start of the process, and subsets ${}^k A = \{{}^k a_1, {}^k a_2, {}^k a_3\}$, $k = 1, 2, \ldots$ of places, which simulate processes under "competition" of k-th level; ${}^k a_1$ – is the place, which simulates current (main, or interruption handler); ${}^k a_2$ – is the place, which simulates functioning of interruption generator, ${}^k a_3$ – is the place, which simulates return on the previous level; $Z = \{z_1, \ldots, z_k, \ldots\}$ – is the set of transitions, which define level of interruptions handler; $\varphi(t)$ – is the matrix of densities; $\boldsymbol{\Lambda}$ – is the matrix of logical conditions of execution of switching from transitions.

The structure of PMN describes multilevel pair "competition" of three subjects: main algorithm S_f, interruption generator S_g and interruption handler S_v.

"Competition" first level is submitted with subset from the place ${}^0 a_2$ nil places ${}^1 a_3, {}^2 a_3$ i.e.

$$\Pi = \{A_1, Z_1, \varphi_1(t), \boldsymbol{\Lambda}_1\}, \tag{8}$$

where $A_1 = \{{}^0 a_2, {}^1 a_1, {}^1 a_2, {}^1 a_3, {}^2 a_1, {}^2 a_2, {}^2 a_3\}$ – is the subset of places; $Z_1 = \{z_1, z_2, z_3\}$ – is the subset of transitions; $\varphi_1(t)$ – is the 7×3 matrix of densities determining time characteristics; $\Lambda_1 = \lfloor {}^1\lambda_{ij} \rfloor$ – is the 7×3 matrix of logical conditions of switching from transitions.

$$\varphi_1(t) = \begin{bmatrix} \delta(t) & 0 & 0 \\ 0 & {}^1\varphi_1(t) & 0 \\ 0 & {}^1\varphi_2(t) & 0 \\ 0 & 0 & 0 \\ 0 & 0 & {}^2\varphi_1(t) \\ 0 & 0 & {}^2\varphi_2(t) \\ \delta(t) & 0 & 0 \end{bmatrix}; \tag{9}$$

$${}^1\lambda_{11} = {}^1\lambda_{14} = {}^1\lambda_{16} = {}^1\lambda_{17} = {}^1\lambda_{21} = {}^1\lambda_{22} = {}^1\lambda_{23} = {}^1\lambda_{27} = {}^1\lambda_{31} = {}^1\lambda_{32} =$$
$${}^1\lambda_{33} = {}^1\lambda_{34} = {}^1\lambda_{35} = {}^1\lambda_{36} = 0;$$
$${}^1\lambda_{12} = {}^1\lambda_{13} = \left({}^0a_2, z_2 \right) \vee \left({}^2a_3, z_1 \right);{}^1\lambda_{15} = \left({}^1a_2, z_2 \right);$$
$${}^1\lambda_{24} = {}^1\lambda_{25} = \left({}^1a_2, z_2 \right) \vee \left({}^3a_3, z_3 \right);{}^1\lambda_{26} = \left({}^1a_1, z_2 \right);{}^1\lambda_{37} = \left({}^2a_1, z_2 \right).$$
$$\tag{10}$$

Recursive procedure of the first level is realized with the next mode. After switching $\left({}^0a_2, z_1 \right)$ logical conditions became equal to $\lambda_{12} = 1; \lambda_{13} = 1$. In accordance with (10), switching $\left(z_1, {}^1 a_1 \right)$ and $\left(z_1, {}^1 a_2 \right)$ is allowed. After switching processes S_f and S_g, begin "compete". On the first step of the recursion

$${}^1\varphi_1(t) = f(t); \qquad {}^1\varphi_2(t) = g(t). \tag{11}$$

Weighed densities of time of switching $\left({}^1a_1, z_2 \right) / \left({}^1a_2, z_2 \right)$ the first, are as follows:

$$\begin{cases} {}^1\eta_1(t) = {}^1\varphi_1(t) \left[1 - {}^1\Phi_2(t) \right], \\ {}^1\eta_2(t) = {}^1\varphi_2(t) \left[1 - {}^1\Phi_1(t) \right]. \end{cases} \tag{12}$$

where $\cdots\Phi_{\cdots} = \int\limits_0^t \cdots\varphi_{\cdots}(\tau)d\tau.$

Probabilities and pure densities of switching $\left({}^1a_1, z_2 \right)$ and $\left({}^1a_2, z_2 \right)$ the first are as follows

$${}^1\pi_1 = \int\limits_0^\infty {}^1\eta_1(t)dt, \qquad {}^1\pi_2 = \int\limits_0^\infty {}^1\eta_2(t)dt.$$

$${}^1\psi_1(t) = \frac{{}^1\eta_1(t)}{{}^1\pi_1}, \qquad {}^1\psi_2(t) = \frac{{}^1\eta_2(t)}{{}^1\pi_2} \tag{13}$$

In the case of switching $\left({}^1a_1, z_2 \right)$ the first, PMN switches to place ${}^1a_3$, which is the analogue of absorbing state, and interpretation of main algorithm ends. In the case of switching

$$\left({}^1a_2, z_2 \right)$$

the first, PMN switches to places ${}^2a_1, {}^2a_2$, and on formula [9]

$${}^1\varphi_1(t) = \frac{1(t) \int\limits_0^\infty {}^1\varphi_2(\tau){}^1\varphi_1(t+\tau)d\tau}{\int\limits_0^\infty {}^1\Phi_2(t)d{}^1\Phi_1(t)}, \tag{14}$$

time till completion of residence in place $^1 a_1$ ($1(t)$ – is the Heaviside function). Time (14) is substituted to the process S_f.

After switching for the first time $(z_2, {}^2 a_1)$ and $(z_2, {}^2 a_2)$, in the places $^2 a_2$, begins "competition" of the interruption handling process and interruption generator, which starts newly $({}^2 \varphi_2(t) = g(t))$. "Competition" starts for the first time, thus $^2 \varphi_1(t) = v(t)$. Weighed time densities of switching $({}^2 a_1, z_3)$ and $({}^2 a_2, z_3)$ the first, probabilities and pure densities are as follows:

$$\begin{cases} {}^2\eta_1(t) = {}^2\varphi_1(t)\left[1 - {}^2\Phi_2(t)\right], \\ {}^2\eta_2(t) = {}^2\varphi_2(t)\left[1 - {}^2\Phi_1(t)\right], \end{cases}$$

$$^2\pi_1 = \int_0^\infty {}^2\eta_1(t)dt, \quad {}^2\pi_2 = \int_0^\infty {}^2\eta_2(t)dt, \tag{15}$$

$$^2\psi_1(t) = \frac{{}^2\eta_1(t)}{{}^2\pi_1}, \quad {}^2\psi_2(t) = \frac{{}^2\eta_2(t)}{{}^2\pi_2}$$

If in "competition" "wins" switching $({}^2 a_1, z_3)$, then

(a) accomplishes the substitution

$$^2\varphi_2(t) = \frac{1(t)\int_0^\infty {}^2\varphi_1(\tau)^2\varphi_2(t + \tau)d\tau}{\int_0^\infty {}^2\Phi_1(t)d^2\Phi_2(t)}, \tag{16}$$

(b) accomplish switches, at first $(z_3, {}^2 a_3)$, then $({}^2 a_3, z_1)$;
(c) accomplish pair of switches $(z_1, {}^1 a_1)$ and $(z_1, {}^1 a_2)$ and starts the "competition" in places $^1 a_1, {}^1 a_2$ with new densities $^1\varphi_1(t), {}^1\varphi_2(t)$.

If in "competition" "wins" switching $({}^2 a_2, z_3)$, then

(a) accomplishes the substitution

$$^2\varphi_1(t) = \frac{1(t)\int_0^\infty {}^2\varphi_2(\tau)^2\varphi_1(t + \tau)d\tau}{\int_0^\infty {}^2\Phi_2(t)d^2\Phi_1(t)}, \tag{17}$$

(b) PMN switches onto the next level.

k-th level is simulated with subset from the place $^{k-1} a_2$ till $^k a_3, {}^{k+1} a_3$ i.e.

$$\Pi_k = \{A_k, Z_k, \varphi_k(t), \Lambda_k\}, \tag{18}$$

where $A_k = \{{}^{k-1} a_2, {}^{k1} a_1, {}^k a_2, {}^k a_3, {}^{k+1} a_1, {}^{k+1} a_2, {}^{k+1} a_3\}$ – is the subset of places; $Z_k = \{z_k, z_{k+1}, z_{k+2}\}$ – is the subset of transitions; $\varphi_k(t) = \lfloor {}^k\varphi_{ij}(t) \rfloor$ – is the 7×3 matrix of densities; $\Lambda_k = \lfloor {}^k\lambda_{ji} \rfloor$ – is the 3×7 matrix of logical conditions;

$$\varphi_k(t) = \begin{bmatrix} {}^{k-1}\eta_2(t) & 0 & 0 \\ 0 & {}^k\varphi_1(t) & 0 \\ 0 & {}^k\varphi_2(t) & 0 \\ 0 & 0 & 0 \\ 0 & 0 & {}^{k+1}\varphi_1(t) \\ 0 & 0 & {}^{k+1}\varphi_2(t) \\ \delta(t) & 0 & 0 \end{bmatrix}; \qquad (19)$$

$${}^k\lambda_{11} = {}^k\lambda_{14} = {}^k\lambda_{16} = {}^k\lambda_{17} = {}^k\lambda_{21} = {}^k\lambda_{22} = {}^k\lambda_{23} = {}^k\lambda_{27} = {}^k\lambda_{31} = {}^k\lambda_{32} =$$
$${}^k\lambda_{33} = {}^k\lambda_{34} = {}^k\lambda_{35} = {}^k\lambda_{36} = 0;$$
$${}^k\lambda_{12} = {}^k\lambda_{13} = \left({}^{k-1}a_2, z_k\right) \vee \left({}^{k+1}a_3, z_k\right); {}^k\lambda_{15} = \left({}^k a_2, z_{k+1}\right);$$
$${}^k\lambda_{24} = {}^k\lambda_{25} = \left({}^k a_2, z_{k+1}\right) \vee \left({}^{k+2}a_3, z_{k+1}\right);$$
$${}^k\lambda_{26} = \left({}^k a_1, z_{k+1}\right); {}^k\lambda_{37} = \left({}^{k+1}a_1, z_{k+1}\right).$$
$$(20)$$

Recursive procedure on the k-th level is as follows. After switching $\left({}^{k-1}a_2, z_k\right)$ logical conditions became $\lambda_{12} = 1, \lambda_{13} = 1$. In accordance with (20) switching $\left(z_k, {}^k a_1\right)$ and $\left(z_k, {}^k a_2\right)$ may be done. After switching processes S_f and S_g begin "compete". Density ${}^k\varphi_1(t)$ is the result of previous upstream and downstream switching. If upstream switching on the k-th level take place for the first time, then

$${}^k\varphi_1(t) = f(t); \qquad {}^2\varphi_2(t) = g(t). \qquad (21)$$

Otherwise ${}^k\varphi_1(t)$ is the result of previous switching; ${}^2\varphi_2(t) = g(t)$.

Weighed time densities of switching $\left({}^k a_1, z_{k+1}\right)$ and $\left({}^k a_2, z_{k+1}\right)$ the first, probabilities and pure densities are as follows:

$$\begin{cases} {}^k\eta_1(t) = {}^k \varphi_1(t) \left\lfloor 1 - {}^k \Phi_2(t) \right\rfloor, \\ {}^k\eta_2(t) = {}^k \varphi_2(t) \left[1 - {}^k \Phi_1(t) \right], \end{cases} \qquad (22)$$

$${}^k\pi_1 = \int_0^\infty {}^k\eta_1(t)dt, \quad {}^k\pi_2 = \int_0^\infty {}^k\eta_2(t)dt, \qquad (23)$$

$${}^k\psi_1(t) = \frac{{}^k\eta_1(t)}{{}^k\pi_1}, \quad {}^k\psi_2(t) = \frac{{}^k\eta_2(t)}{{}^k\pi_2}$$

In the case of switching $\left({}^k a_1, z_{k+1}\right)$ the first, PMN returnable switches to place ${}^k a_3$, and then switches to $\left({}^k a_3, z_{k-1}\right)$.

In the case of switching $\left({}^k a_2, z_{k+1}\right)$ the first, PMN switches to places ${}^{k+1}a_1, {}^{k+1}a_2$. On formula [9]

$${}^k\varphi_1(t) = \frac{1(t) \int_0^\infty {}^k\varphi_2(\tau) {}^k\varphi_1(t + \tau)d\tau}{\int_0^\infty {}^k\Phi_2(t)d^k\Phi_1(t)}, \qquad (24)$$

time till completion of residence in place $^k a_1$ is evaluated and substituted to process S_f.

After switching $\left(z_{k+1}, ^{k+1} a_1\right)$ and $\left(z_{k+1}, ^{k+1} a_2\right)$ at places $^{k+1} a_1, ^{k+1} a_2$ begins "competition" of the processes S_v and S_g. Interruption generator restarts again $^{k+1}\varphi_2(t) = g(t)$. Process of interruption handling, if such level was passed on previous stages, may continue, thus

$$
^k\varphi_1(t) = \begin{cases} v(t), & \text{when step } \left(z_k, ^k a_1\right) \text{ is executer for the first time;} \\ ^k\varphi_1(t), & \text{depending of pre-history in all other cases.} \end{cases} \tag{25}
$$

Weighed time densities of switching $\left(^{k+1} a_1, z_{k+2}\right)$ and $\left(^{k+1} a_2, z_{k+2}\right)$ the first, probabilities and pure densities are as follows:

$$
\begin{cases} ^{k+1}\eta_1(t) = ^{k+1}\varphi_1(t)\left[1 - ^{k+1}\Phi_2(t)\right], \\ ^{k+1}\eta_2(t) = ^{k+1}\varphi_2(t)\left[1 - ^{k+1}\Phi_1(t)\right], \end{cases}
$$

$$
^{k+1}\pi_1 = \int_0^\infty {}^{k+1}\eta_1(t)dt, \quad {}^{k+1}\pi_2 = \int_0^\infty {}^{k+1}\eta_2(t)dt, \tag{26}
$$

$$
^{k+1}\psi_1(t) = \frac{^{k+1}\eta_1(t)}{^{k+1}\pi_1}, \quad {}^{k+1}\psi_2(t) = \frac{^{k+1}\eta_2(t)}{^{k+1}\pi_2}.
$$

If in "competition" "wins" switching $\left(^{K=1} a_1, z_{k+2}\right)$, then

(a) accomplishes the substitution

$$
^{k+1}\varphi_2(t) = \frac{1(t)\int_0^\infty {}^{k+1}\varphi_1(\tau)^{k+1}\varphi_2(t+\tau)d\tau}{\int_0^\infty {}^{k+1}\Phi_1(t)d^{k+1}\Phi_2(t)}, \tag{27}
$$

(b) accomplish switches, at first $\left(z_{k+2}, ^{k+1} a_3\right)$, then $\left(^{k+1} a_3, z_k\right)$;
(c) accomplish pair of switches $\left(z_k, ^k a_1\right)$ and $\left(z_k, ^k a_2\right)$ and starts the "competition" in places $^k a_1, ^k a_2$ with new densities $^k\varphi_1(t), ^k\varphi_2(t)$.

If in "competition" "wins" switching $\left(^{k+1} a_2, z_{k+2}\right)$, then

(a) accomplishes the substitution

$$
^{k+1}\varphi_1(t) = \frac{1(t)\int_0^\infty {}^{k+1}\varphi_2(\tau)^{k+1}\varphi_1(t+\tau)d\tau}{\int_0^\infty {}^{k+1}\Phi_2(t)d^{k+1}\Phi_1(t)}, \tag{28}
$$

(b) PMN switches onto the next level.

Similarly recursion may be spread onto subsequent interruption levels.

4   Interruption Model as Semi-Markov Process

Described process have the next properties:

- it is the infinite one on quantity, both the interruption levels, and returns to the same level;
- strictly, it is not quite semi-Markov, due to substitutions (14), (15), (16), (17), (24), (27), (28).

For analyses of time and stochastic characteristics PMN under investigation should be transformed to ordinary strictly semi-Markov process, with use the next method.

1. From PMN one should to extract subnet as follows

$$\boldsymbol{\Pi}_k^c = \left\{ \left\{ ^{k-1}a_2, {}^k a_1, {}^k a_2 \right\}, \{z_k, z_{k+1}\}, \boldsymbol{\varphi}_k^c, \boldsymbol{\Lambda}_k^c \right\}, k = 1, 2, \ldots, \qquad (29)$$

where $\boldsymbol{\varphi}_k^c, \boldsymbol{\Lambda}_k^c$ – are sub-matrices of current densities and logical conditions of switching.

2. One should to develop sub-graph, represented part of binary tree, as follows

$$\mu_k^c = \{B_k^c, \boldsymbol{h}_k^c(t)\}$$

$$= \left\{ \left\{ \beta_{k-1,1(2)}^c, \beta_{k,1}^c, \beta_{k,2}^c \right\}, \begin{bmatrix} 0 & {}^k\eta_1(t) & {}^k\eta_2(t) \\ 0 & 0 & 0 \\ 0 & 0 & 0 \end{bmatrix} \right\}, k = 1, 2, \ldots, \qquad (30)$$

where $\left\{ \beta_{k-1,1(2)}^c, \beta_{k,1}^c, \beta_{k,2}^c \right\}$ – are the states of semi-Markov process, describing "winning" in the "competition" interrupt handler (or main algorithm) or interrupt generator; ${}^k\eta_1(t)$ and ${}^k\eta_2(t)$ – densities calculated by (22).

Form of tree is shown on Fig. 2, where node ${}^0a_2$ of graph simulates the only starting state of semi-Markov process. Shaded nodes of graph simulate infinite quantity of absorbing states.

Due to the fact, that all prehistory was taken into account when there was recursive procedure of evaluation ${}^k\eta_1(t)$ and ${}^k\eta_2(t)$, Process is strictly semi-Markov one. For evaluation of time of reaching the subset of absorbing states from the starting state may be used any known method.

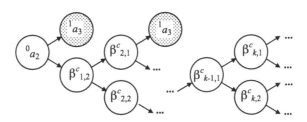

Fig. 2. Binary tree of semi-Markov process

5 Conclusion

The resulting model, based on fundamental mathematical apparatus of PMN describes classical and simplest case of the computer system with interruption. It is obviously, that in such a system both runtime, and number of interruption levels may increase till infinity. Nevertheless, model and proposed method evaluation of time intervals are the productive ones and allows estimate runtime in more complicated cases, for example, when there is a dispatching possibility in operational system [4,13]. Besides with use of proposed approach there can be solved the optimization problem, using as a criterion the time factor.

The research was carried out within the state assignment of The Ministry of Education and Science of Russian Federation (No. 2.3121.2017/PCH).

References

1. Bouchhima, A., Yoo, S., Jeraya, A.: Fast and accurate timed execution of high level embedded software using HW/SW interface simulation model. In: Proceedings of the 2004 Asia and South Pacific Design Automation Conference, pp. 469–474. IEEE Press (2004)
2. Cleaveland, R., Smolka, S.A.: Strategic directions in concurrency research. ACM Comput. Surv. (CSUR) **28**(4), 607–625 (1996)
3. Czerwinski, M., Cutrell, E., Horvitz, E.: Instant messaging and interruption: influence of task type on performance. In: OZCHI 2000 Conference Proceedings, vol. 356, pp. 361–367 (2000)
4. Duda, K.J., Cheriton, D.R.: Borrowed-Virtual-Time (BVT) scheduling: supporting latency-sensitive threads in a general-purpose scheduler. ACM SIGOPS Oper. Syst. Rev. **33**(5), 261–276 (1999)
5. Heymann, M.: Concurrency and discrete event control. IEEE Control Syst. Mag. **10**(4), 103–112 (1990)
6. Ivutin, A., Larkin, E.: Estimation of latency in embedded real-time systems. In: 2014 3rd Mediterranean Conference on Embedded Computing (MECO), pp. 236–239. IEEE (2014)
7. Ivutin, A., Larkin, E., Lutskov, Y.: Evaluation of program controlled objects states. In: 2015 4th Mediterranean Conference on Embedded Computing (MECO), pp. 250–253. IEEE (2015)
8. Ivutin, A.N., Larkin, E.V., Lutskov, Y.I., Novikov, A.S.: Simulation of concurrent process with Petri-Markov nets. Life Sci. J. **11**(11), 506–511 (2014)
9. Ivutin, A., Larkin, E.: Simulation of concurrent games. Bulletin of the South Ural state university, series: mathematical modelling. Prog. Comput. Softw. **8**(2), 43–54 (2015)
10. Korolyuk, V., Swishchuk, A.: Semi-Markov Random Evolutions, pp. 59–91. Springer, Heidelberg (1995)
11. Larkin, E.V., Kotov, V.V., Ivutin, A.N., Privalov, A.N.: Simulation of relay-races. Bull. South Ural State Univ. Ser.: Math. Model. Program. Comput. Softw. **9**(4), 117–128 (2016)
12. Regehr, J., Duongsaa, U.: Preventing interrupt overload. In: ACM SIGPLAN Notices, vol. 40, pp. 50–58. ACM (2005)

13. Squillante, M.S.: Stochastic analysis and optimization of multiserver systems. In: Ardagna, D., Zhang, L. (eds.) Run-time Models for Self-managing Systems and Applications. Autonomic Systems, pp. 1–24. Springer, Heidelberg (2010). doi:10. 1007/978-3-0346-0433-8_1
14. Valk, R.: Concurrency in communicating object petri nets. In: Agha, G.A., Cindio, F., Rozenberg, G. (eds.) Concurrent Object-Oriented Programming and Petri Nets. LNCS, vol. 2001, pp. 164–195. Springer, Heidelberg (2001). doi:10.1007/ 3-540-45397-0_5

Novel Swarm-Based Optimization Algorithms

Dolphin Pod Optimization

Andrea Serani[✉] and Matteo Diez

CNR-INSEAN, National Research Council-Marine Technology Research Institute,
Rome, Italy
andrea.serani@insean.cnr.it, matteo.diez@cnr.it

Abstract. A novel nature-inspired deterministic derivative-free global optimization method, namely the dolphin pod optimization (DPO), is presented for solving simulation-based design optimization problems with costly objective functions. DPO implements, using a deterministic approach, the global search ability provided by a cetacean intelligence metaphor. The method is intended for unconstrained single-objective minimization and is based on a simplified social model of a dolphin pod in search for food. A parametric analysis is conducted to identify the most promising DPO setup, using 100 analytical benchmark functions and three performance criteria, varying the algorithm parameters. The most promising setup is compared with a deterministic particle swarm optimization and a DIviding RECTangles algorithm, and applied to two hull-form optimization problems, showing a very promising performance.

Keywords: Dolphin pod optimization · Deterministic optimization · Global optimization · Derivative-free optimization

1 Introduction

Global optimization metaheuristics, such as particle swarm optimization (PSO) [1], have been deeply investigated in the last decades, while several new metaheuristic methods were introduced, such as firefly (FA) [2], cuckoo search (CS) [3], and bat (BA) [4] algorithms. This kind of methods are usually stochastic implying that statistically significant results can be obtained only through extensive numerical campaigns. Such an approach could be too expensive in simulation-based design optimization (SBDO) for industrial applications,

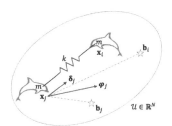

Fig. 1. Pod model

especially when CPU-time expensive computer simulations are used directly as analysis tools. For this reason, efficient deterministic global optimization algorithms are sought-after in SBDO.

The objective of the present work is to introduce and assess a novel deterministic derivative-free global optimization algorithm, based on a simplified social model of a dolphin pod in search for food: the dolphin pod optimization (DPO).

© Springer International Publishing AG 2017
Y. Tan et al. (Eds.): ICSI 2017, Part I, LNCS 10385, pp. 63–70, 2017.
DOI: 10.1007/978-3-319-61824-1_7

DPO is intended for unconstrained single-objective minimization of SBDO problems with costly objective functions and is a deterministic swarm-intelligence algorithm. To the authors' knowledge, this is a little explored field, where the global search ability of swarm-intelligence systems is implemented in a deterministic way. DPO is formulated considering the essential elements of the cetacean intelligence: congregation, self-awareness, communication, and memory. Each dolphin is subject to a pod attraction force (congregation) and an external force related to the food distribution (representing the objective function) known to each dolphin (self-awareness) and the pod (communication and memory). Based on these forces, the pod is modelled as a spring-mass system.

DPO has common features to other swarm-intelligence methods, such as PSO, central force optimization (CFO, [5]), and gravitational search algorithm (GSA, [6]), where the agent state is described by the particle system dynamics. Nonetheless, compared to PSO, CFO, and GSA, DPO presents significant differences: it is the only algorithm that is at the same time deterministic, based on agent position and directly on objective function values (absolute fitness), memory based, fully informed, and formulated by a rigorous integration of the agent dynamics, which allows to determine a set of parameters ensuring the system stability. Moreover, differently from FA, CS, and BA, DPO is based on a deterministic concept and the system state depends not only on the agent positions and comparative fitness, but also directly on the absolute fitness, though the food attraction force.

The effectiveness and efficiency of DPO depend on four main parameters: the number of dolphins interacting during the optimization, the initialization of the pod in terms of initial position and velocity, the set of coefficients controlling the pod dynamics, and finally the method to handle the box constraints. The approach for the analysis includes a parametric study using 100 analytical benchmark functions [7], with dimensionality from 2 to 50, characterized by different degrees of non-linearities and number of local minima, with full-factorial combination of: number of dolphins, initialization of the pod, 81 different coefficient sets, and the method to handle the box constraints. Three metrics are used to evaluate the algorithm performances, based on the distance between DPO-found and analytical optima. The most significant parameters for DPO are identified, based on the associated relative variability of the results [8], and the performances of the best performing (on average) DPO setup are compared with a deterministic particle swarm optimization algorithm (DPSO) [7] and the DIviding RECTangles (DIRECT) algorithm [9]. Finally, DPO is applied to two SBDO problems, pertaining to the hull-form optimization of a USS Arleigh Burke-class destroyer, namely the DTMB 5415 model, in calm water and waves. The SBDO results are finally compared to DPSO and DIRECT.

2 Dolphin Pod Optimization Algorithm

Consider an optimization problem of the type

$$
\begin{aligned}
\text{Minimize} \quad & f(\mathbf{x}) \\
\text{subject to} \quad & \mathbf{l} \leq \mathbf{x} \leq \mathbf{u},
\end{aligned}
\tag{1}
$$

where $f(\mathbf{x})$ is the objective function, $\mathbf{x} \in \mathbb{R}^N$ is the variable vector with $N \in \mathbb{N}^+$ the number of variables, and $\mathbf{l}$ and $\mathbf{u}$ are the lower and the upper bounds for $\mathbf{x}$, respectively.

Now consider a foraging pod of dolphins $\mathbf{x}_j$, exploring the variable space with the aim of finding an approximate solution for problem in Eq. 1. The pod is modelled as a dynamical system where the dynamics of the j-th individual depends on a pod attraction force $\boldsymbol{\delta}_j$ (congregation), a food attraction force $\boldsymbol{\varphi}_j$ (self-awareness, communication, and memory), as well as the hydrodynamic drag, proportional to $\dot{\mathbf{x}}_j$ (Fig. 1)

$$\ddot{\mathbf{x}}_j + \xi \dot{\mathbf{x}}_j + k\boldsymbol{\delta}_j = h\boldsymbol{\varphi}_j, \tag{2}$$

where

$$\boldsymbol{\delta}_j = \sum_{i=1}^{N_d} (\mathbf{x}_j - \mathbf{x}_i) \qquad \text{and} \qquad \boldsymbol{\varphi}_j = \sum_{i=1}^{N_d} \frac{2F(\mathbf{x}_j, \mathbf{b}_i)}{1 + \|\mathbf{x}_j - \mathbf{b}_i\|^\alpha} \mathbf{e}(\mathbf{b}_i, \mathbf{x}_j), \tag{3}$$

with

$$F(\mathbf{x}_j, \mathbf{b}_i) = \frac{f(\mathbf{x}_j) - f(\mathbf{b}_i)}{\rho}, \qquad \mathbf{e} = \frac{\mathbf{b}_i - \mathbf{x}_j}{\|\mathbf{b}_i - \mathbf{x}_j\|}. \tag{4}$$

In the above equations, ξ, k and $h \in \mathbb{R}^+$ define the pod dynamics; $N_d \in \mathbb{N}^+$ is the pod size; $\alpha \in \mathbb{R}^+$ tunes the food attraction force; $\mathbf{x}_j \in \mathbb{R}^N$ is the vector-valued position of the j-th individual; $f(\mathbf{x}) \in \mathbb{R}$ is the objective function (representing the food distribution); $\mathbf{b}_i$ is the best position ever visited by the i-th individual; $\rho = f(\mathbf{w}) - f(\mathbf{b})$ is a dynamic normalization term for f, where $\mathbf{b} = \mathrm{argmin}\{f(\mathbf{b}_j)\}$ is the best position ever visited by the pod and $\mathbf{w} = \mathrm{argmax}\{f(\mathbf{x}_j)\}$ the worst position occupied by the pod individuals at the current time instance.

Using the explicit Euler integration scheme, the DPO iteration becomes

$$\frac{\mathbf{v}_j^{n+1} - \mathbf{v}_j^n}{\Delta t} = -\xi \mathbf{v}_j^n - k\boldsymbol{\delta}_j + h\boldsymbol{\varphi}_j, \tag{5}$$

which finally yields

$$\begin{cases} \mathbf{v}_j^{n+1} = (1 - \xi\Delta t)\mathbf{v}_j^n + \Delta t(-k\boldsymbol{\delta}_j + h\boldsymbol{\varphi}_j) \\ \mathbf{x}_j^{n+1} = \mathbf{x}_j^n + \mathbf{v}_j^{n+1}\Delta t \end{cases} \tag{6}$$

where $\mathbf{x}_j^n$ and $\mathbf{v}_j^n$ represent the j-th dolphin position and velocity at the n-th iteration, respectively. Equation 6 represents a fully informed formulation, where each individual knows the story of the whole pod.

The integration step Δt (see Eq. 6) must guarantee the stability of the explicit Euler scheme, at least for the free dynamics. To this aim, consider the free dynamics of the k-th component of $\mathbf{x}$ (k-th variable), say a. Consider the dynamics of a for the j-th dolphin

$$\ddot{a}_j + \xi \dot{a}_j + k\delta_j = 0, \tag{7}$$

and finally for the entire pod

$$\left\{ \begin{matrix} \dot{a} \\ \dot{c} \end{matrix} \right\} = \begin{bmatrix} 0 & I \\ -K & -G \end{bmatrix} \left\{ \begin{matrix} a \\ c \end{matrix} \right\} = A \left\{ \begin{matrix} a \\ c \end{matrix} \right\}, \tag{8}$$

where

$$K = -k \begin{bmatrix} N_d - 1 & -1 & \cdots & -1 \\ -1 & N_d - 1 & \cdots & -1 \\ \vdots & \vdots & \ddots & \vdots \\ -1 & \cdots & -1 & N_d - 1 \end{bmatrix} \tag{9}$$

and $G = \xi I$, with I the $[N_d \times N_d]$ identity matrix. The solution of Eq. 8 is stable if $Re(\lambda) \leq 0$ where $\lambda = -\gamma \pm i\omega$ are eigenvalues of A. This yields

$$\Delta t \leq \left. \frac{2\gamma}{\gamma^2 + \omega^2} \right|_{\min} = \Delta t_{\max}. \tag{10}$$

3   DPO Setting Parameters

The DPO parameters used in the current analysis are defined in the following. Their full-factorial combination is considered for the assessment of the algorithm performance, resulting in a total of 1458 different setups.

The number of dolphins used (N_d) is defined as $N_d = 2^r N$, with $r \in \mathbb{N}[2, 4]$ therefore ranging from $4N$ to $16N$.

The initialization of dolphins' location and velocity is performed using a deterministic and homogeneous distribution, following the Hammersley sequence sampling, applied to three different sub-domains, defined as: (A) domain, (B) domain bounds, and (C) domain and bounds [7]. A non-null initial velocity is used (see Eq. 15 in [7]).

Provided that all the design variables are normalized such that the domain is confined in a unit hypercube $\mathcal{U}$ (i.e. $-0.5 \leq x \leq 0.5$), the following positions are used for the coefficients controlling the pod dynamics: $k = h = q/N_d$; $\Delta t = \Delta t_{\max}/p$; $\xi \Delta t < 1$ where q defines the weight for the attraction forces (δ_j and φ_j), and p defines the integration time step. Three values are used for each parameters: $\xi = \{0.01; 0.10; 1.00\}$, $q = \{0.10; 1.00; 10.0\}$, $p = \{2.00; 4.00; 8.00\}$, and $\alpha = \{0.50; 1.00; 2.00\}$.

The dolphins are confined within $\mathcal{U}$ using an inelastic (IW) and an elastic (EW) wall-type approach. Specifically, in the IW approach, if a dolphin is found to violate one of the bounds in the transition from two consecutive iterations, it is placed on that bound setting to zero the associated velocity component, whereas, in the EW approach, the associated velocity component is reversed.

Finally, the number of function evaluations or evaluation budget ($N_{\max}$) is assumed as $N_{\max} = 2^c N$, where $c \in \mathbb{N}[7, 12]$ and therefore ranges from $128N$ to $4096N$.

4   Performance Metrics

Three performance metrics are used to assess the algorithm performances and defined as follows [7]:

$$\Delta_x = \sqrt{\frac{1}{N}\sum_{k=1}^{N}\left(\frac{x_{k,min} - x_{k,min}^\star}{R_k}\right)^2}, \quad \Delta_f = \frac{f_{min} - f_{min}^\star}{f_{max}^\star - f_{min}^\star}, \quad \Delta_t = \sqrt{\frac{\Delta_x^2 + \Delta_f^2}{2}}.$$

(11)

Δ_x is a normalized Euclidean distance between the minimum position found by the algorithm ($\mathbf{x}_{min}$) and the analytical minimum position ($\mathbf{x}_{min}^\star$), where $R_k = |u_k - l_k|$ is the range of the k-th variable. Δ_f is the associated normalized distance in the function space, f_{min} is the minimum found by the algorithm, $f_{min}^\star$ is the analytical minimum, and $f_{max}^\star$ is the analytical maximum of the function $f(\mathbf{x})$ in the search domain. Δ_t is a combination of Δ_x and Δ_f and used for an overall assessment. Additionally, the relative variability ($\sigma_{r,s}^2$) for the performance metrics [8] is used to assess the impact of each tuning parameter s on the algorithm performance. For the numerical implementation and further details the readers is referred to [8].

5   Numerical Results

5.1   Analytical Benchmark Functions

Hundred analytical benchmark functions are used, including a wide variety of problems, such as continuous and discontinuous, differentiable and non differentiable, separable and non-separable, scalable and non-scalable, unimodal and multimodal, with $2 \le N \le 50$ (see Table A.10 and A.11 in [7]). The study is conducted, setting a part functions with less and more than ten variables. For sake of simplicity the results are shown in terms of Δ_t only. Figure 2a and b show the relative variability $\sigma_{r,s}^2$ based on Δ_t, associated to the DPO tuning parameters for $N < 10$ and $N \ge 10$ respectively. The pod initialization is the most significant parameters overall. The coefficient p (time step factor) increase its importance with the increasing of function evaluations till reach the same variability of the pod initialization, for problems with $N \ge 10$. The food attraction force coefficient (α) is the less important overall.

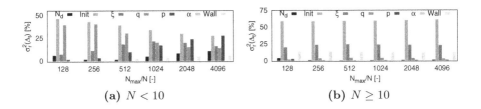

(a) $N < 10$ **(b)** $N \ge 10$

Fig. 2. DPO setting parameters relative variability $\sigma_r^2\%$ based on Δ_t

Average values of Δ_t are used to identify the best performing DPO setup, which corresponds to: $N_s = 4N$, pod initialization on domain (A), $\xi = 1.00$, $q = 1.00$, $p = 8.00$, $\alpha = 0.50$, and elastic wall-type for problems with $N < 10$; $N_s = 4N$, pod initialization on domain bounds (B) , $\xi = 0.10$, $q = 0.10$, $p = 8.00$, $\alpha = 0.50$, and elastic wall-type for problems with $N \geq 10$. Figure 3a and b show a comparison between the best DPO, DPSO, and DIRECT, for $N < 10$ and $N \geq 10$, respectively; DPO outperforms DPSO and DIRECT, on average, specially for problems with a high number of variables.

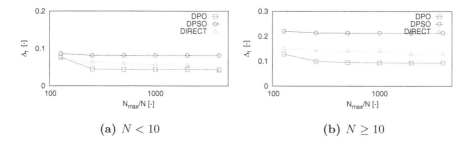

(a) $N < 10$ (b) $N \geq 10$

Fig. 3. Best DPO vs DPSO and DIRECT, based on average performance metrics

Finally three illustrative examples of algorithm convergence are shown in Fig. 4 for (a) Griewank, (b) Rastrigin, and (c) Ackley functions. For these examples, DPO convergence shows a better effectiveness and efficiency than DPSO and DIRECT.

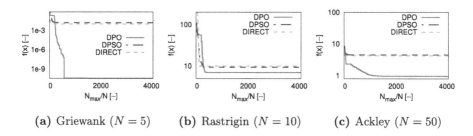

(a) Griewank $(N = 5)$ (b) Rastrigin $(N = 10)$ (c) Ackley $(N = 50)$

Fig. 4. Algorithms convergence comparison on illustrative benchmark functions

5.2   Hull-Form SBDO Problems

The optimization aims at improving separately (I) calm-water and (II) seakeeping performances. For problem I, the objective function (f) is the total resistance (normalized with the displacement) in calm water at 18 kn, whereas for problem II f is a seakeeping merit factor based on the root mean square of the bridge vertical acceleration at 30 kn in head waves (0 deg) and on the roll angle at 18 kn in stern long-crested waves (150 deg) [10]. Modification of the parent hull are performed using orthogonal functions, defined over a surface-body patches, and 90%-confidence dimensionality reduction based on Karhunen-Loève expansion [11]. A $N = 6$ design space is used and shape modification are produced directly on the computational grid. Details of the geometry modification can be found in [12,13]. Problem I is solved using the code WARP [14], whereas problem II is solved using the code SMP [15]. Details on the computational domain for the free-surface and the hull grid, numerical solvers and their validation can be found in [10]. For each problem, a budget of 4800 function evaluations is used.

DPO achieves an objective function reduction close 13% and 32%, for problem I and II respectively. Specifically, DPO, DPSO, and DIRECT have a similar objective function reduction for problem I, even if DPSO and DIRECT show a faster convergence to the global minimum than DPO (Fig. 5a). Problem II shows a greater objective function reduction by DPO, compared with DPSO and especially DIRECT (Fig. 5b), probably trapped in a local minimum.

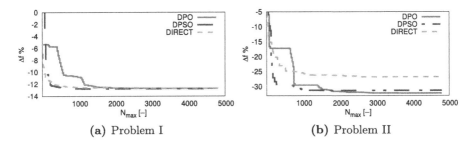

(a) Problem I   (b) Problem II

Fig. 5. Hull-form optimization, convergence history

6   Conclusions

A novel deterministic algorithm for global derivative-free optimization (dolphin pod optimization, DPO) has been presented and assessed using 100 analytical benchmark functions. All combinations of DPO parameters led to 1458 optimization runs for each benchmark function.

The coefficient q and the pod initialization are the most significant parameters for low budget of function evaluations and a number of variables $N < 10$, while for $N \geq 10$ the pod initialization is found always the most significant parameter. The best performing DPO setup has been identified based on average results and has been found slightly better than DPSO and DIRECT for benchmark functions with $N < 10$ (especially for low budget) and remarkably better for $N \geq 10$. The best performing DPO has been applied to two hull-form SBDO problems with $N = 6$. Comparable results of DPO, DPSO and DIRECT were achieved for problem I (calm-water resistance), whereas problem II (seakeeping) revealed a greater effectiveness of DPO, compared to DPSO and DIRECT.

Acknowledgements. The present research is supported by the US Office of Naval Research Global, NICOP grant N62909-15-1-2016, administered by Dr Woei-Min Lin, and by the Italian Flagship Project RITMARE. The DIRECT algorithm was taken from the DFL, Derivative-Free Library (https://www.dis.uniroma1.it/~lucidi/DFL/) administered by Dr Giampaolo Liuzzi.

References

1. Kennedy, J., Eberhart, R.: Particle swarm optimization. In: Proceedings of the Fourth IEEE Conference on Neural Networks, Piscataway, pp. 1942–1948 (1995)
2. Yang, X.S.: Nature-Inspired Metaheuristic Algorithms. Luniver Press, Bristol (2008)
3. Yang, X.S., Deb, S.: Cuckoo search via levy flights. In: Proceedings of World Congress on Nature and Biologically Inspired Computing (NaBic 2009), Coimbatore, India, pp. 210–214 (2009)
4. Yang, X.S.: A new metaheuristic bat-inspired algorithm. In: Gonzlez, J., Pelta, D., Cruz, C., Terrazas, G., Krasnogor, N. (eds.) (NICSO 2010). Studies in Computational Intelligence, vol. 284, pp. 65–74. Springer, Berlin Heidelberg (2010)
5. Formato, R.A.: Central force optimization: a new metaheuristic with applications in applied electromagnetics. Prog. Electromagn. Res. **77**, 425–491 (2007)
6. Rashedi, E., Nezamabadi-pour, H., Saryazdi, S.: GSA: a gravitational search algorithm. Inf. Sci. **179**(13), 2232–2248 (2009). Special Section on High Order Fuzzy Sets
7. Serani, A., Leotardi, C., Iemma, U., Campana, E.F., Fasano, G., Diez, M.: Parameter selection in synchronous and asynchronous deterministic particle swarm optimization for ship hydrodynamics problems. Appl. Soft Comput. **49**, 313–334 (2016)
8. Campana, E.F., Diez, M., Iemma, U., Liuzzi, G., Lucidi, S., Rinaldi, F., Serani, A.: Derivative-free global ship design optimization using global/local hybridization of the DIRECT algorithm. Optim. Eng. **17**(1), 127–156 (2015)
9. Jones, D., Perttunen, C., Stuckman, B.: Lipschitzian optimization without the Lipschitz constant. J. Optim. Theory Appl. **79**(1), 157–181 (1993)
10. Diez, M., et al.: Multi-objective hydrodynamic optimization of the DTMB 5415 for resistance and seakeeping. In: Proceedings of the 13th International Conference on Fast Sea Transportation, FAST 2015, Washington, D.C., USA (2015)
11. Diez, M., Campana, E.F., Stern, F.: Design-space dimensionality reduction in shape optimization by Karhunen-Loève expansion. Comput. Methods Appl. Mech. Eng. **283**, 1525–1544 (2015)
12. Diez, M., Serani, A., Campana, E.F., Volpi, S., Stern, F.: Design space dimensionality reduction for single- and multi-disciplinary shape optimization. In: AIAA/ISSMO Multidisciplinary Analysis and Optimization (MA&O), AVIATION 2016, Washington D.C., USA, 13–17 June 2016
13. Serani, A., Fasano, G., Liuzzi, G., Lucidi, S., Iemma, U., Campana, E.F., Stern, F., Diez, M.: Ship hydrodynamic optimization by local hybridization of deterministic derivative-free global algorithms. Appl. Ocean Res. **59**, 115–128 (2016)
14. Bassanini, P., Bulgarelli, U., Campana, E.F., Lalli, F.: The wave resistance problem in a boundary integral formulation. Surv. Math. Indus. **4**, 151–194 (1994)
15. Meyers, W.G., Baitis, A.E.: SMP84: improvements to capability and prediction accuracy of the standard ship motion program SMP81. Technical report SPD-0936-04, David Taylor Naval Ship Research and Development Center, September 1985

Teaching-Learning-Feedback-Based Optimization

Xiang Li, Kang Li[✉], and Zhile Yang

School of Electronics, Electrical Engineering and Computer Science,
Queen's University Belfast, Belfast BT9 5AH, UK
{xli25,k.li,zyang07}@qub.ac.uk

Abstract. Teaching-learning-based Optimization (TLBO) is a popular meta-heuristic optimisation method that has been used in solving a number of scientific and engineering problems. In this paper, a new variant, namely Teaching-learning-feedback-based Optimization (TLFBO) is proposed. In addition to the two phases in the canonical TLBO, an additional feedback learning phase is employed to further speed up the convergence. The teacher in the previous generation is recorded and communicates with the current teacher to provide combined feedbacks to the learners and supervise the learning direction to avoid wasting computational efforts incurred in the previous generations. Numerical experiments on 10 well-known benchmark functions are conducted to evaluate the performance of the TLFBO, and experimental results show that the proposed TLFBO has a superior and competitive capability in solving continuous optimisation problems.

Keywords: Teaching-learning-based Optimization (TLBO) · Feedback · Global optimization · Heuristic method

1 Introduction

Engineering optimization problems involving numerous design variables are often highly non-linear, non-convex, and the objective functions may be transformed to non-differentiable and discontinuous ones, which call for powerful global optimization approaches [1]. The nature-inspired heuristic optimization approaches have been intensively researched in the past decade and widely used in solving engineering problems.

Genetic Algorithms (GA) [6] was inspired from Darwin's theory of natural selection, is one of the earliest and most important population based optimization technologies. Whereas particle swarm optimization (PSO) [7] was inspired from the foraging behaviours of birds or fishes and adopts the inertia weights, social and cognitive parameters in forming a new generation. Other counterparts such as the artificial bee colony (ABC) [5] and ant colony optimization (ACO) [4] mimics the characteristics of bees and ants to maximize the likelihood in effectively and efficiently obtaining the value of the global optima. From the state-of-the-art literature reviews, the approaches can be grouped into

© Springer International Publishing AG 2017
Y. Tan et al. (Eds.): ICSI 2017, Part I, LNCS 10385, pp. 71–79, 2017.
DOI: 10.1007/978-3-319-61824-1_8

two categories [15], namely, evolutionary protocol and particle swarm intelligence [7], enlightened from different social phenomena and natural observations respectively.

Teaching-learning-based Optimization (TLBO) method is a promising variant of the population based method, which focuses on addressing the global optimization problems of continuous non-linear functions [11,12]. The TLBO method stems from the knowledge propagation among teachers and students. Two drawbacks of the TLBO were identified as follows. Firstly, the performance highly depends on the initial value and the initialization boundaries. Besides, the speed of the convergence is slower than other counterpart algorithms, e.g. PSO. Furthermore, for the majority of nature inspired algorithms, the stochastically evolved generations may suffer from the loss of the evolution clue of previous generations and the links with the past optima would be broken, which leads to a highly time consuming process to achieve the global best.

To tackle this drawback, a phase namely the feedback learning phase is integrated into the original TLBO method. Herein, the previous and current teachers form a judging panel to evaluate the learning behaviour after each iteration, give the feedback and supervise the learning of students. The core contribution of this paper is to accelerate the convergence by introducing the memory in relation with the historical generations. Furthermore, the paper represents a pilot research of improving the TLBO algorithm performance by considering the effects of the past optimum values as the genes embedded in future generation evolution.

The remainders of this paper is organised as follows. Section 2 is a brief tutorial on the TLBO algorithm. The preliminary concept of the TLFBO method is introduced in Sect. 3, which also gives an intuitive overview of the advantages intuitively. The efficiency of the proposed algorithm is evaluated on 10 well-adopted benchmark functions and the results are presented in Sect. 4, which demonstrates the effectiveness of the proposed method. Finally, Sect. 5 concludes this paper.

2   Standard TLBO Algorithm

The protocol of the TLBO method was first proposed in 2011 [11]. It involves two phases, namely the teacher and learner phases respectively. All the learners in the classroom form one population and have two chances to be trained. In the teacher phase, the most knowledgeable person in each generation is considered as the teacher, who will disseminate its optimum performance with others and affect the outcome of the learners. The output items are analogous to the scores/grades of the students. In the leaner phase, students learn from each other to further improve their personal scores/grades.

2.1   Teacher Phase

A teacher is deemed to circulate his/her knowledge to help the learners and the average grades of the class toward to his/her own level. Each student accomplishes the study upon acquiring the discrepancy between teacher and mean solution among the peers. The updating process is formulated as followings [16].

$$DM_i = rand_1(0,1) * (Teacher_i - T_F * M_i). \tag{1}$$

Where M_i is the mean value of each course in the class, and $Teacher_i$ denotes the most knowledgeable one as the teacher. Herein, due to, in the real class, the average grades can only be increased in some extent depending on the variable learning capabilities of students in the class, which is named as teaching factor T_F expressed as 1 or 2 and decides the changeable value of mean.

$$T_F = round(1 + rand_2(0,1)) \tag{2}$$

Nurturing throughout the efforts of teacher, the scores or grades of the learners are improved as

$$St_i^{new} = St_i^{current} + DM_i \tag{3}$$

where the St_i^{new} is the generated new population after the knowledge exchange among the teacher and the students. As before, the person, who is highly educated, will be voted as the new teacher for the next heuristic step.

2.2 Learner Phase

The second phase is based on learning mutually to enhance the exploitation capabilities. Students learn from the others through the group discussion, presentation as well as formal communication, etc. [11]. Each person holds one chance to share his/her knowledge with his/her peers randomly. A better performance of the class results from this phase to avoid falling into the local optima simultaneously. The modification processes will be expressed as follows.

$$St_k^{new} = \begin{cases} St_k^{current} + rand_3(0,1)(St_k - St_j) & if \ f(St_k < St_j) \\ St_k^{current} + rand_3(0,1)(St_j - St_k) & if \ f(St_j < St_k) \end{cases} \tag{4}$$

where following the teaching phase, on the basis of the fitness function or objective function, the grades of students St_k and St_j are compared randomly.

3 A Novel TLBO with Feedback Learning Phase (TLFBO) Algorithm

Analysis of the behaviours of the original TLBO method reveals two issues that emerge as the Achillest heel to restrict its performance. One is the initialization and the other is the convergence speed. Contradictory to the behaviours of the well-known PSO method, TLBO is a tardy and precise approach to search for the global optima. Though the performance is improved generation by generation during the evolution process, there exists no distinguished direction for the convergence and exploration, which is different from the PSO method. Thereby, TLBO can not maintain a persistent performance and the final results vary with different initialization values.

Inspired by the supervised learning, a Feedback phase is designed following the learning phase, which is dedicated to increase the converging speed. In this modification, the last global optima as the previous teacher works together with the newly selected teacher denoted as the current teacher to evaluate the learning results. The differences between these two teachers are recorded as the judging panel to supervise everyone in the class. This feedback learning phase is formulated as follows.

$$St_i^{new} = St_i^{current} + l_1 * w * D_{last} + l_2 * w * D_{current} \tag{5}$$

$$D_{last} = Tr_i^{last} - St_i^{current} \tag{6}$$

$$D_{current} = Tr_i^{current} - St_i^{current} \tag{7}$$

$$w = (G_{total} - G_{current}/G_{total}) \tag{8}$$

Where the value of w is the mutation weight. Its value will be decreased gradually to constrain the searching scope. l_1, l_2 are denoted as the feedback weights for the two selected teacher, and they are predefined numbers between 0 and 1, and $l_1 + l_2 = 1$. These coefficients will adjust the extent of learning to follow the best one (current teacher). G_{total} is the predefined total generations; $G_{current}$ is the current generation number. The differences $D_{last}, D_{current}$ comparing the individuals in the whole class with the two teachers are used to update the grades of the whole class.

There are two remarkable advantages of this TLFBO method. Firstly, the students are taught by accessing the information of both the past and current generations. The historical performances are recorded as the memory to help with the future evolution process. With these memories, the learning efficiency can be increased significantly as shown in the experimental section. Furthermore, it is easy to be implemented. To reflect the aforementioned memory concept, in this paper, the last optimum is utilised to help the current teacher rather than being discarded during the evolution process. This designed feedback phase help to guide the learning direction to follow the historic trajectory along the evolution process. The complete TLFBO algorithm is depicted in Fig. 1.

4   Experiments and Test Results

The performance of the TLFBO algorithm is assessed on 10 well-known benchmark functions encompassing 30 dimensions [17]. Besides, comparative study is conducted in comparison with several TLBO variants e.g. original TLBO [11], elite TLBO, [10] a modified TLBO [13], Self-learning TLBO [18] as well as several popular swarm algorithms including weighted PSO [14], PSO-CF [2], classical DE/rand//bin [3] and others emerging algorithms enumerating as GWO [9], and MFO [8].

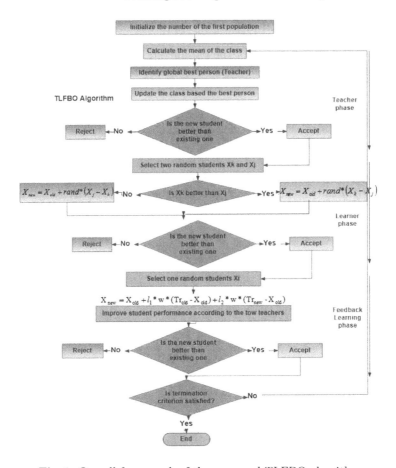

Fig. 1. Overall framework of the proposed TLFBO algorithm

4.1 Benchmark Functions

Ten benchmark functions are taken from [17] and presented as follows with the information of dimensions and boundaries.

$$
\begin{aligned}
&Sphere function(f_1) &&dimension = 30, \ [-100, 100]\\
&Schwefel's problem1.2(f_2) &&dimension = 30, \ [-100, 100]]\\
&Rosenbrock function(f_3) &&dimension = 30, \ [-30, 30]\\
&Ackley's function(f_4) &&dimension = 30, \ [-32, 32]\\
&Griewank function(f_5) &&dimension = 30, \ [-600, 600]\\
&Rastrigin function(f_6) &&dimension = 30, \ [-5.12, 5.12]\\
&Step function(f_7) &&dimension = 30, \ [-100, 100]\\
&Schwefel's problem2.21(f_8) &&dimension = 30, \ [-100, 100]]\\
&Schwefel's problem2.26(f_9) &&dimension = 30, \ [-500, 500]\\
&Quartic function(f_{10}) &&dimension = 30, \ [-1.28, 1.28]
\end{aligned}
$$

4.2   Parameters Determination

According to the rule of thumb for the parameters setting of the aforementioned algorithms, the population number is set as 30. Further details are outlined as following.

wPSO phase $c1 = 1, \ c2 = 3, \ wmax = 0.9, \ wmin = 0.4$
PSO − CF phase $c1 = c2 = 2.05, \ K = 0.729$
classical DE phase $F = 0.7, \ CR = 0.5$
elite TLBO phase   *elite number* $= 5$
self − learning TLBOphase   *weighting factor* $= 3$
TLFBO phase   *feedback wrighting* $l_1 = 0.3, \ l_2 = 0.7$

4.3   Results and Discussion

Herein, the proposed novel TLBFO algorithm was compared with other counterparts. Table 1 compares the performances in terms of the mean values and standard deviations in relation with the above 10 algorithms on the 10 remarkable benchmark functions (f_1 to f_{10}).

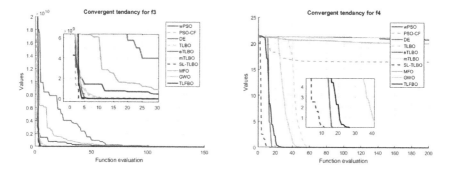

Fig. 2. Convergent tendencies of 10 algorithms for f_3 and f_4

It is shown that the value of the objective function of the proposed TLFBO algorithm is equal or better than the peers with the exception of f_4 and f_8. The proposed algorithm is further shown to converge faster for the majority of the scenarios (excepting f_9). In the cases of f_4, all the algorithms show relatively slow convergence speed, while the TLFBO performs better as shown in Fig. 2. Furthermore, for f_2, f_5, f_6, the performances were quite similar for all TLBO variants. This reveals that TLBO and its couterparts have a superior capability to find the global optima and could produce a similar performance with a less numbers of generations.

Table 1. Benchmark tests results for different algorithms

	Sphere	Schwefels problem 1.2	Rosenbrock	Ackley	Griewank
	f1	f2	f3	f4	f5
wPSO	$8.134e02 \pm 1.535e03$	$2.122e-01 \pm 2.267$	$9.340e06 \pm 3.558e07$	$2.044e01 \pm 8.943e00$	$1.188e00 \pm 3.345e-01$
PSO-CF	$1.833e03 \pm 5.046e03$	$2.534e-02 \pm 3.879e-01$	$4.649e07 \pm 2.304e08$	$1.972e01 \pm 1.157e01$	$1.427e00 \pm 9.321e-01$
DE	$7.650e01 \pm 1.568e02$	$6.002e-01 \pm 1.002e01$	$9.541e05 \pm 4.605e06$	$2.001e01 \pm 3.921e-01$	$9.978e-01 \pm 2.775e-01$
TLBO	$5.60e-45 \pm 7.755-44$	$\mathbf{0.000e00 \pm 0.000e00}$	$2.892e01 \pm 1.897e-01$	$7.751e-09 \pm 2.295e-07$	$\mathbf{0.000e00 \pm 0.000e00}$
eTLBO	$6.75e-76 \pm 9.51e-75$	$4.180e-32 \pm 8.760e-31$	$2.894e01 \pm 1.325e-01$	$1.480e-15 \pm 7.250e-15$	$\mathbf{0.000e00 \pm 0.000e00}$
mTLBO	$6.45e-91 \pm 8.27r-90$	$1.640e-33 \pm 4.850-32$	$2.893e01 \pm 2.073e-01$	$1.057e01 \pm 5.678e01$	$\mathbf{0.000e00 \pm 0.000e00}$
SL-TLBO	$3.98e-115 \pm 8.71e-114$	$\mathbf{0.000e00 \pm 0.000e00}$	$2.899e01 \pm 1.976e-01$	$\mathbf{8.880e-16 \pm 0.000e00}$	$\mathbf{0.000e00 \pm 0.000e00}$
MFO	$1.812e01 \pm 1.563e02$	$5.150e-27 \pm 4.880e-26$	$1.393e05 \pm 1.835e06$	$2.001e01 \pm 1.757e-01$	$3.3334e-01 \pm 1.174e00$
GWO	$1.91e-21 \pm 2.08e-20$	$9.628e-04 \pm 1.1183e-02$	$\mathbf{2.794e01 \pm 5.017e00}$	$7.012e00 \pm 5.432e01$	$4.760e-03 \pm 4.463e-02$
TLFBO	$\mathbf{3.76e-118 \pm 5.78e-117}$	$\mathbf{0.000e00 \pm 0.000e00}$	$2.874e01 \pm 1.670e-01$	$1.19e-09 \pm 3.52e-08$	$\mathbf{0.000e00 \pm 0.000e00}$

	Rastrigin	Step	Schwefels problem 2.21	Schwefels problem 2.26	Quartic
	f6	f7	f8	f9	f10
wPSO	$1.287e03 \pm 1.667e03$	$7.143e02 \pm 2.045e03$	$1.837e01 \pm 2.029e01$	$-4.321e03 \pm 2.680e02$	$1.452e06 \pm 7.099e06$
PSO-CF	$2.448e03 \pm 4.987e03$	$1.664e03 \pm 3.854e03$	$2.189e01 \pm 3.173e01$	$-5.223e03 \pm 2.449e02$	$7.766e06 \pm 3.426e07$
DE	$3.876e02 \pm 3.697e02$	$7.992e01 \pm 1.835e02$	$1.970e01 \pm 2.183e01$	$-5.898e03 \pm 1.752e03$	$7.429e04 \pm 2.702e05$
TLBO	$\mathbf{0.000e00 \pm 0.000e00}$	$5.784e00 \pm 3.498e00$	$1.770e-22 \pm 7.623e-22$	$-6.083e03 \pm 2.319e03$	$9.465e00 \pm 2.256e00$
eTLBO	$\mathbf{0.000e00 \pm 0.000e00}$	$6.546e00 \pm 3.897e00$	$4.400e-38 \pm 4.500e-37$	$-5.667e03 \pm 2.339e03$	$9.323e00 \pm 2.601e00$
mTLBO	$\mathbf{0.000e00 \pm 0.000e00}$	$5.878e00 \pm 3.798e00$	$2.201e-45 \pm 3.960e-44$	$-6.382e03 \pm 1.733e03$	$9.429e00 \pm 1.488e00$
SL-TLBO	$\mathbf{0.000e00 \pm 0.000e00}$	$5.194e00 \pm 4.225e00$	$\mathbf{8.680e-59 \pm 1.247e-57}$	$-6.302e03 \pm 2.588e03$	$9.549e00 \pm 2.747e00$
MFO	$2.732e02 \pm 1.211e03$	$9.316e00 \pm 2.859e01$	$4.473e01 \pm 4.683e01$	$-1.703e03 \pm 3.751e02$	$7.980e03 \pm 4.714e04$
GWO	$6.325e00 \pm 2.194e01$	$7.894e-01 \pm 1.918e01$	$1.490e-05 \pm 5.770e-05$	$-1.085e03 \pm 6.070e02$	$9.458e00 \pm 3.579e00$
TLFBO	$\mathbf{0.000e00 \pm 0.000e00}$	$\mathbf{5.171e00 \pm 4.0137e00}$	$2.023e-47 \pm 1.47e-46$	$\mathbf{-7.613e02 \pm 4.374e02}$	$\mathbf{9.304e00 \pm 2.137e00}$

5 Conclusion

A novel optimization variant TLFBO was proposed in this paper. The historical global best value was recorded (previous teacher) in the optimization processes to help supervise the learners in the new generation in cooperation with the current global best value (current teacher), which is defined as the feedback learning. The performance of the new TLFBO algorithm was tested on 10 well-known benchmark functions. The numerical results demonstrate that the new TLFBO heuristic algorithm dramatically outperforms other counterparts in terms of convergence speed and solution accuracy.

Acknowledgment. This paper was partially funded by the EPSRC under grant EP/P004636/1 and partially supported by NSFC under 61673256, and Shanghai Science Technology Commission under grant No. 14ZR1414800.

References

1. Chowdhury, P.R., Singh, Y.P., Chansarkar, R.A.: Hybridization of gradient descent algorithms with dynamic tunneling methods for global optimization. IEEE Trans. Syst. Man Cybern. - Part A: Syst. Hum. **30**(3), 384–390 (2000)
2. Clerc, M., Kennedy, J.: The particle swarm-explosion, stability, and convergence in a multidimensional complex space. IEEE Trans. Evol. Comput. **6**(1), 58–73 (2002)
3. Das, S., Suganthan, P.N.: Differential evolution: a survey of the state-of-the-art. IEEE Trans. Evol. Comput. **15**(1), 4–31 (2011)
4. Dorigo, M., Birattari, M., Stutzle, T.: Ant colony optimization. IEEE Comput. Intell. Mag. **1**(4), 28–39 (2006)
5. Gao, W.F., Huang, L.L., Liu, S.Y., Dai, C.: Artificial bee colony algorithm based on information learning. IEEE Trans. Cybern. **45**(12), 2827–2839 (2015)
6. Holland, J.H.: Adaptation in Natural and Artificial Systems: An Introductory Analysis with Applications to Biology, Control and Artificial Intelligence. MIT Press, Cambridge (1992)
7. Kennedy, J., Eberhart, R.: Particle swarm optimization. In: Proceedings of the IEEE International Conference on Neural Networks, vol. 4, pp. 1942–1948, November 1995
8. Mirjalili, S.: Moth-flame optimization algorithm: a novel nature-inspired heuristic paradigm. Knowl.-Based Syst. **89**, 228–249 (2015)
9. Mirjalili, S., Mirjalili, S.M., Lewis, A.: Grey wolf optimizer. Adv. Eng. Softw. **69**, 46–61 (2014)
10. Rao, R., Patel, V.: Comparative performance of an elitist teaching-learning-based optimization algorithm for solving unconstrained optimization problems. Int. J. Ind. Eng. Comput. **4**(1), 29–50 (2013)
11. Rao, R.V., Savsani, V.J., Vakharia, D.: Teaching-learning-based optimization: a novel method for constrained mechanical design optimization problems. Comput.-Aided Des. **43**(3), 303–315 (2011)
12. Rao, R., Savsani, V.J., Vakharia, D.: Teaching-learning-based optimization: an optimization method for continuous non-linear large scale problems. Inf. Sci. **183**(1), 1–15 (2012)

13. Satapathy, S.C., Naik, A.: Modified teaching-learning-based optimization algorithm for global numerical optimizationa comparative study. Swarm Evol. Comput. **16**, 28–37 (2014)
14. Shi, Y., Eberhart, R.C.: Parameter selection in particle swarm optimization. In: Porto, V.W., Saravanan, N., Waagen, D., Eiben, A.E. (eds.) EP 1998. LNCS, vol. 1447, pp. 591–600. Springer, Heidelberg (1998). doi:10.1007/BFb0040810
15. Wang, Z., Lu, R., Chen, D., Zou, F.: An experience information teaching-learning-based optimization for global optimization. IEEE Trans. Syst. Man Cybern.: Syst. **46**(9), 1202–1214 (2016)
16. Yang, Z., Li, K., Foley, A., Zhang, C.: A new self-learning TLBO algorithm for RBF neural modelling of batteries in electric vehicles. In: 2014 IEEE Congress on Evolutionary Computation (CEC), pp. 2685–2691 (2014)
17. Yao, X., Liu, Y., Lin, G.: Evolutionary programming made faster. IEEE Trans. Evol. Comput. **3**(2), 82–102 (1999)
18. Zhile, Y., Kang, L., Qun, N., Yusheng, X., Foley, A.: A self-learning tlbo based dynamic economic/environmental dispatch considering multiple plug-in electric vehicle loads. J. Mod. Power Syst. Clean Energy **2**(4), 298–307 (2014)

Magnetotactic Bacteria Optimization Algorithm Based on Moment Interaction Energy

Lifang Xu[1], Hongwei Mo[2(✉)], Jiao Zhao[2], Chaomin Luo[3], and Zhenzhong Chu[4]

[1] Engineering Training Center, Harbin Engineering University, Harbin, Heilongjiang 150001, China
[2] Automation College, Harbin Engineering University, Harbin, Heilongjiang 150001, China
honwei2004@126.com
[3] Department of Electrical and Computer Engineering, University of Detroit Mercy, Detroit, MI, USA
[4] College of Information Engineering, Shanghai Maritime University, Pudong, China

Abstract. In this paper, an improved magnetotactic bacteria optimization algorithm (IMBOA) is proposed to solve unconstrained optimization problems. IMBOA uses an archive to keep some better solutions in order to guide the moving of the whole population in each generation. And it uses a kind of efficient interaction energy to enhance diversity of the population for encouraging broader exploration. The proposed algorithm is compared with some relative optimization algorithms on the CEC 2013 real-parameter optimization benchmark functions. Experimental results show that the proposed algorithm IMBOA has better performance than the compared algorithms on most of the benchmark problems.

Keywords: Magnetotactic bacteria optimization algorithm · Efficient interaction energy · Diversity · Archive

1 Introduction

Swarm intelligence is a new algorithm for finding the global optimal solution, which has been widely used in various fields, and achieved good results. It is based on social organisms from insects to human. Some popular swarm intelligence algorithms, such as Particle Swarm Optimization (PSO) [1], Artificial Bee Colony (ABC) [2], Bacterial Foraging Optimization Algorithm (BFOA) [3], Firefly Algorithm [4] and Cuckoo Algorithm [5] etc. were developed based on different inspiration sources which can be seen from their names.

Magnetotactic bacteria optimization algorithm (MBOA) is also a swarm intelligence algorithm which proposed by Mo in 2012 [6]. It emulates the process of magnetotactic bacteria swimming along geomagnetic field lines of the earth to minimize

© Springer International Publishing AG 2017
Y. Tan et al. (Eds.): ICSI 2017, Part I, LNCS 10385, pp. 80–87, 2017.
DOI: 10.1007/978-3-319-61824-1_9

their magnetostatic energy [7]. In recent years, several improved MBOA have been proposed to improve the performance of MBOA [8–10].

This paper is proposed based on MBOA-BR and is organized as follows. Section 2 introduces the MBOA-BR. Section 3 describes the IMBOA. Section 4 presents the test functions and the experimental setting for each algorithm. Section 4 presents the experiment results and analysis. Conclusions are given in Sect. 5.

2   The Magnetotactic Bacteria Optimization Algorithm

Magnetotactic Bacteria Optimization Algorithm is a new optimization algorithm inspired by the biology characteristics of magnetotactic bacteria in nature. It obtains the optimal solution by regulating the moments of cells continually by the process of MTS generation, MTS regulation and MTS replacement. Assume that the cell population size is N and search space is D dimension. Then the i th cell is represented by $X_i = (x_{i1}, x_{i2}, \ldots, x_{iD})$. Before the operators of MTS generation, MTS regulation and MTS replacement, we first calculate the interaction energy.

- Interaction energy calculation
  Randomly select a cell X_r in the population. The distance $D_i = (d_{i1}, d_{i2}, \ldots, d_{iD})$ between two cells X_i^t and X_r^t is calculated using (1).

$$D_i = X_i - X_r. \tag{1}$$

The interaction energy $E_i = (e_{i1}, e_{i2}, \ldots, e_{iD})$ is calculated as follows:

$$e_{ij}(t) = \left(\frac{d_{ij}(t)}{1 + c_1 \times norm(D_i(t)) + c_2 \times d_{pq}(t)} \right)^3. \tag{2}$$

In (2), c_1, c_2 are constants and $p \in [1, N]$, $q \in [1, D]$ are randomly chosen integer indices. $norm(D_i(t))$ is the Euclidean distance between cells X_i^t and X_r.

- MTS generation
  The MTS are produced as follows:

$$v_{ij}(t) = x_{ij}(t) + r_1 \times m_{pq}(t). \tag{3}$$

$$M_i(t) = \frac{E_i(t)}{B}. \tag{4}$$

where m_{pq} are randomly selected from matrix

$$M = (M_1, \cdots M_N) = \begin{bmatrix} m_{11} & m_{21} & \cdots & m_{N1} \\ m_{12} & m_{22} & \cdots & m_{N2} \\ \vdots & \vdots & \cdots & \vdots \\ m_{1D} & m_{2D} & \cdots & m_{ND} \end{bmatrix}, \text{ and } r_1 \text{ is a random number}$$

between 0 and 1. B is a constant named magnetic field strength.

- MTS regulation
  The MTS moment of cell is regulated as follows:
  If $rand > 0.5$

$$u_{ij}(t) = v_{cbestj}(t) + r_2 \times \left(v_{cbestj}(t) - v_{kj}(t)\right). \tag{5}$$

Otherwise

$$u_{ij}(t) = v_{kj}(t) + r_2 \times \left(v_{cbestj}(t) - v_{kj}(t)\right). \tag{6}$$

In (5) and (6), v_{cbestj} stands for the j th dimension of current best cell V_{cbest} in the current generation. v_{kj} stands for the j th dimension of a cell V_k selected from current population. r_2 is a random number between 0 and 1.

- MTS replacement
  After the moment regulation, the population is evaluated according to cells' fitness. The last fifth of the cells will be replaced based on (7). m_{ab} is randomly chosen from matrix $M = (M_1, \cdots M_N)$. $rand(1, n)$ is a random vector with n dimensions, and each dimension is a random number between 0 and 1.

$$X_i(t) = m_{ab}(t) \times ((rand(1, n) - 1) \times rand(1, n)). \tag{7}$$

Other cells will not change, thus we can get (8).

$$X_i(t + 1) = U_i(t). \tag{8}$$

3 An Improved Magnetotactic Bacteria Optimization Algorithm

- External archive
  In IMBOA, we set an external archive to guide the search direction of solutions. The archive will store 2 best cells in each generation. At the end of each generation, the external archive will be updated if the new solutions are better than those in the archive. $archive_i$ denotes the ith cell stored in the external archive.
- Efficient interaction energy
  In MBOA, interaction energy affects the diversity of the whole cell population severely. By designing new interaction energy we can improve its performance. New interaction energy defined as (10):

$$d_{ij}(t) = \left|x_{ij}(t) - x_{rj}(t)\right|. \tag{9}$$

$$e_{ij}(t) = c \times \sqrt[h]{R(t)^h - \left(d_{ij}(t)\right)^h}. \tag{10}$$

$$R(t) = R_{max} - (R_{max}/Maxgeneration) \times t. \tag{11}$$

$$R_{max} = x_{maxj} - x_{minj}. \tag{12}$$

x_{maxj}, x_{minj} are the upper and lower bounds for the dimension j respectively, and
$$c = \begin{cases} +1 & x_{ij}(t) \geq 0 \\ -1 & x_{ij}(t) < 0 \end{cases}.$$
If the values of some $d_{ij}(t)$ are out of the range $[0, R(t)]$ in each generation, they will be mapped to random values in the range $[0, R(t)]$.

- MTS generation
  In the proposed algorithm parameter r_1 is not a random value between 0 and 1, but is set as (13).

$$r_1 = 2rand - 1 + (t/MaxGeneration)^2. \tag{13}$$

- MTS regulation
  In the proposed algorithm, we first compare the best solution $V_{cbest}(t+1)$ generated by MTS generation operation with the best one $X_{best}(t)$ generated in the last generation. Then we determine whether the archive is used to guide solutions moving or not. Pseudo code of this operation is as follows.

$$u_{ij}(t) = v_{ij}(t) + rand \times \left(v_{cbest}(t) - v_{ij}(t)\right). \tag{14}$$

$$u_{ij}(t) = v_{ij}(t) + rand \times \left(v_{cbest}(t) - v_{ij}(t)\right) + r_2 \times \left(\bar{x}_j - v_{ij}(t)\right). \tag{15}$$

$$\bar{x}_j = \sum_{i=1}^{k} archive_{ij}/k. \tag{16}$$

According to above description, Table 1 gives the pseudo code of the improved algorithm.

4 Experimental Setting and Results Analysis

Twenty benchmark functions from the IEEE CEC 2013 [15] are used to test the optimization performance of the IMBOA algorithm. These functions are more complex and can be used to compare the performance of different algorithms in a more systematic manner. The dimensions of benchmark functions are $D = 30$ and the population size is set as 100. The maximum number of function evaluations is $10000 \times D$. In this paper, 51 times independent experiments are conducted.

In this paper, we compare four state-of-the-art algorithms which are CoDE [16], GL-25 [17], ALC-PSO [18] and EPSDE [19]. The results of all experiments are shown in the next section.

Table 2 presents the results for all the 20 benchmark functions on 30 dimention.

Table 1. Algorithm 1

For   each solution
If $X_{best}(t) > V_{cbest}(t+1)$
MTS regulation operation is conducted according to (14)
Else
MTS regulation operation is conducted according to (15)
End
End

Table 2. The proposed algorithm (IMBOA)

Line	Pseudo code of IMBOA
1	Initialization:
2	Set cell population size N, and randomly generate the initial population in the search space.
3	Set external archive
4	The IMBOA loop:
5	While $t \leq MaxGeneration$ do
6	Keep the current best solution $X_{best}(t)$, determinate radius $R(t)$ and magnetic field $B(t)$
7	Calculate interaction distances according to (9)
8	Calculate interaction energy according to (10)
9	Obtain moments according to (4)
10	For $i = 1$ to N do
11	MTS generation according to (3) and (13)
12	End for
13	Evaluate the current population according to fitness and find the best solution
14	MTS regulation according to algorithm 1
15	Evaluate the current population according to fitness and update the external archive
16	Memorize the best solution achieved so far
17	End while

As shown in the table, in the five unimodal functions (f1 to f5), we can see that IMBOA outperforms the other four algorithms on 2 functions (f3 and f4) according to the mean preformence. CoDE performs better than IMBOA for function f2. And the performance of GL25, ALC-PSO and EPSDE are better than IMBOA for function f5. For function f1, all the algorithms attain the best performance value of zero except CoDE based on the mean values error, but the performance of all the five algorithms are identical by the Wilcoxon rank-sun text. It may be noted that IMBOA preforms better on most of the five functions particularly on the function f3. The mean values

Table 3. Experimental results on 30-dimensions of CEC2013 benchmark function.

Function		IMBOA	CoDE	GL_25	ALC-PSO	EPSDE
f_1	Mean	**0.0000e+00**	4.0904e−08 ≈	**0.0000e+00** ≈	**0.0000e+00** ≈	**0.0000e+00** ≈
	Std.	**0.0000e+00**	1.7380e−08	**0.0000e+00**	**0.0000e+00**	**0.0000e+00**
f_2	Mean	4.1687e+05	**1.6199e+05** −	3.5860e+06 +	1.3790e+07 +	6.1113e+05 +
	Std.	2.9704e+05	**2.0672e+05**	1.7720e+06	8.8582e+06	2.9014e+06
f_3	Mean	**2.2058e+04**	2.9085e+07 +	2.6280e+06 +	3.4824e+08 +	5.5938e+07 +
	Std.	**8.1071e+04**	1.3402e+07	2.4208e+06	5.5136e+08	1.8370e+08
f_4	Mean	**1.0532e+01**	1.8638e+01 +	1.0107e+03 +	4.7917e+03+	7.8170e+03 +
	Std.	**1.0204e+01**	1.2412e+01	5.5342e+02	1.8108e+03	1.6863e+04
f_5	Mean	7.3357e−06	1.5390e−05 +	**0.0000e+00** −	**0.0000e+00** −	**0.0000e+00** −
	Std.	5.0185e−06	4.4395e−06	**0.0000e+00**	**0.0000e+00**	**0.0000e+00**
f_6	Mean	2.7105e+01	1.3373e+01 −	2.5852e+01 −	8.1710e+01 +	**3.5673e−01** −
	Std.	2.4484e+00	1.9511e+00	2.0038e+01	4.1600e+01	**3.7086e−01**
f_7	Mean	**1.2340e−01**	4.0179e+01 +	1.2966e+01 +	8.5105e+01 +	2.5287e+01 +
	Std.	**2.1268e−01**	6.5980e+00	3.6632e+00	3.3072e+01	2.5118e+01
f_8	Mean	2.0948e+01	2.0955e+01 ≈	2.0962e+01 ≈	**2.0931e+01** ≈	2.0935e+01 ≈
	Std.	4.2099e−02	**3.9983e−02**	4.5752e−02	5.9860e−02	5.7660e−02
f_9	Mean	**7.7397e+00**	3.2548e+01 +	2.3505e+01 +	2.5734e+01 +	3.4578e+01 +
	Std.	**2.3345e+00**	1.6915e+00	2.9739e+00	4.1924e+00	2.3407e+00
f_{10}	Mean	**5.7958e−03**	2.2562e−01 +	1.6612e−01 +	1.7775e−01 +	4.8870e−02 +
	Std.	**8.1205e−03**	1.7583e−01	1.9755e−01	1.1746e−01	2.3342e−02
f_{11}	Mean	2.1655e+01	2.5250e+01 +	2.4767e+01 +	2.4386e+01 +	**0.0000e+00** −
	Std.	5.5382e+00	2.4168e+00	7.4443e+00	6.3782e+00	**0.0000e+00**
f_{12}	Mean	**3.3321e+01**	1.6345e+02 +	1.3729e+02 +	9.7632e+01 +	5.8921e+01 +
	Std.	**8.3205e+00**	1.3672e+01	6.5880e+01	2.2826e+01	1.1242e+01
f_{13}	Mean	**5.6574e+01**	1.7888e+02 +	1.5941e+02 +	1.5740e+02 +	7.6820e+01 +
	Std.	2.2987e+01	1.4105e+01	3.7587e+01	3.7713e+01	**1.3086e+01**
f_{14}	Mean	5.4660e+02	1.4166e+03 +	3.0633e+03 +	9.8269e+02 +	**6.7618e+01** −
	Std.	2.2835e+02	1.8431e+02	2.0263e+03	3.0638e+02	**9.7077e+01**
f_{15}	Mean	**1.8469e+03**	6.9528e+03 +	7.0457e+03 +	4.1709e+03 +	6.7747e+03 +
	Std.	5.2295e+02	**2.6879e+02**	2.9018e+02	9.9453e+02	4.0411e+02
f_{16}	Mean	**3.9978e−01**	2.4424e+00 +	2.5962e+00 +	2.2073e+00 +	2.4681e+00 +
	Std.	3.0323e−01	3.0804e−01	**2.7101e−01**	3.5923e−01	2.9840e−01
f_{17}	Mean	5.1277e+01	6.4747e+01 +	9.8616e+01 +	6.5842e+01 +	**3.0434e+01** −
	Std.	5.2967e+00	2.8797e+00	5.9337e+01	1.6186e+01	**1.5878e−04**
f_{18}	Mean	**5.9629e+01**	2.2866e+02 +	2.0386e+02 +	1.5253e+02 +	1.6218e+02 +
	Std.	**7.4254e+00**	1.0892e+01	1.0162e+01	5.0815e+01	1.2225e+01
f_{19}	Mean	2.9531e+00	8.2616e+00 +	6.6007e+00 +	3.9574e+00 +	**2.4837e+00** −
	Std.	5.5836e−01	8.2459e−01	4.9562e+00	1.3254e+00	**1.9235e−01**
f_{20}	Mean	**9.7703e+00**	1.2602e+01 +	1.1720e+01 +	1.4369e+01 +	1.3133e+01 +
	Std.	1.3857e+00	**2.2579e−01**	3.6441e−01	9.9967e−01	6.0075e−01
	+		16	16	17	12
	−		2	2	1	6
	≈		2	2	2	2

error of IMBOA is the order of 10^4, whereas other algorithms are in the range of 10^6 and 10^8. For function f4, IMBOA provides the best mean performance and it outperforms GL25, ALC-PSO and EPSDE by a significant margin. According to above observations, it is inferred that IMBOA provides better solutions for unimodal function.

In the fifteen basic multimodal functions (f6–f20), IMBOA does not performs worse than CoDE and GL25 for all basic multimodal functions except f6. IMBOA outperforms EPSDE on 9 functions (f7, f9, f10, f12, f13, f15, f16, f18 and f20). However, the performance of IMBOA is better than (or equal to) ALC-PSO for all basic multimodal functions. It is worth to note that IMBOA almost attains the best performance value of zero based on the mean value error on functions f7 and f10, and the performance of all the five algorithms are identical on function f8. EPSDE finds the best solution on function f11.

As shown in the bottom three rows of Table 3 counting the number of +, −, and ≈ results, the proposed IMBOA has the best overall performance on these benchmark functions.

5   Conclusions

In this paper, an improved magnetotactic bacteria optimization algorithm (IMBOA) is presented. It utilizes efficient interaction energy to enhance the diversity of cell population in search process. And it uses an archive to keep better solutions to direct the whole population in each generation. We also test the performance of the proposed algorithm on 20 benchmark functions provided for CEC2013 special session on real-parameter single objective optimization. Experiment results demonstrate that IMBOA performs better compared with the other algorithms for most benchmark functions.

Acknowledgements. This work is partially supported by the National Natural Science Foundation of China under Grant No. 61075113, the Excellent Youth Foundation of Heilongjiang Province of China under Grant No. JC201212.

References

1. Kennedy, J., Eberhart, R.: Particle swarm optimization. In: IEEE International Conference on Neural Networks, Piscataway, NJ, pp. 1942–1948 (1995)
2. Tereshko, V.: Reaction-diffusion model of a honeybee colony's foraging behaviour. In: Schoenauer, M., Deb, K., Rudolph, G., Yao, X., Lutton, E., Merelo, J.J., Schwefel, H.-P. (eds.) PPSN 2000. LNCS, vol. 1917, pp. 807–816. Springer, Heidelberg (2000). doi:10.1007/3-540-45356-3_79
3. Müeller, S., Marchetto, J., Airaghi, S., Koumoutsakos, P.: Optimization based on bacterial chemotaxis. IEEE Trans. Evol. Comput. **6**, 16–29 (2002)
4. Yang, X.-S.: Firefly algorithms for multimodal optimization. In: Watanabe, O., Zeugmann, T. (eds.) SAGA 2009. LNCS, vol. 5792, pp. 169–178. Springer, Heidelberg (2009). doi:10.1007/978-3-642-04944-6_14

5. Yang, X.S., Deb, S.: Cuckoo search via Lévy flights. In: Proceedings of World Congress on Nature & Biologically Inspired Computing (NaBic 2009), USA, pp. 210–214. IEEE Publications (2009)
6. Mo, H.W.: Research on magnetotactic bacteria optimization algorithm. In: The Fifth International Conference on Advanced Computational Intelligence, pp. 423–428 (2012)
7. Faivre, D., Schuler, D.: Magnetotactic bacteria and magnetosomes. Chem. Rev. **108**, 4875–4898 (2008)
8. Mo, H.W., Liu, L.L., Xu, L.F., Zhao, Y.Y.: Research on magnetotactic bacteria optimization algorithm based on the best individual. In: The Sixth International Conference on Bio-inspired Computing, Wuhan, China, pp. 318–322 (2014)
9. Mo, H.W., Liu, L.L., Xu, L.F.: A power spectrum optimization algorithm inspired by magnetotactic bacteria. Neural Comput. Appl. **25**(7), 1823–1844 (2014)
10. Mo, H.W., Liu, L.L., Zhao, J.: A new magnetotactic bacteria optimization algorithm based on moment migration. IEEE/ACM Trans. Comput. Biol. Bioinform. **14**(1), 15–26 (2017)
11. Liang, J., Qu, B.Y., Suganthan, P., Hernández-Díaz, A.: Problem definitions and evaluation criteria for the CEC 2013 special session and competition on real-parameter optimization. Computational Intelligence Laboratory, Zhengzhou University, Zhengzhou China and Technical report, Nanyang Technological University, Singapore, Technical report (2013)
12. Wang, Y., Cai, Z., Zhang, Q.: Differential evolution with composite trial vector generation strategies and control parameters. IEEE Trans. Evol. Comput. **15**(1), 55–66 (2011)
13. Garcia-Martinez, C., Lozano, M., Herrera, F., Molina, D., Sanchez, A.M.: Global and local real-coded genetic algorithms based on parent-centric crossover operators. Eur. J. Oper. Res. **185**(3), 1088–1113 (2008)
14. Chen, W.N., et al.: Particle swarm optimization with an aging leader and challengers. IEEE Trans. Evol. Comput. **17**(2), 241–258 (2013)
15. Mallipeddi, R., Suganthan, P.N., Pan, Q.K., Tasgetiren, M.F.: Differential evolution algorithm with ensemble of parameters and mutation strategies. Appl. Soft Comput. **11**(2), 1679–1696 (2011)
16. Mendes, R., Kennedy, J., Neves, J.: The fully informed particle swarm: simpler, maybe better. IEEE Trans. Evol. Comput. **8**(3), 204–210 (2004)

A Guide Sign Optimization Problem for an Added Road Based on Bird Mating Optimizer

Fang Liu, Min Huang$^{(\boxtimes)}$, Teng Zhang, and Feng Mao

Guangdong Provincial Key Laboratory of Intelligent Transportation System,
Sun Yat-sen University, Guangzhou 510006, Guangdong, China
huangm7@mail.sysu.edu.cn

Abstract. As new roads constructed frequently, it is necessary to optimize the current guide sign system. To solve the problem scientifically and systematically, this paper proposes an optimization model for current guide sign system in the situation of adding a road, which is solved by Bird Mating Optimizer (BMO). The optimization model first selects several important entrances as optimized starting nodes. Next the set of guiding routes with the best costs from these starting nodes to the targeted road are solved by BMO. The costs consider travel time costs, usage costs for vehicles and reconstruction costs for the guide signs. Then deploy guide signs in the guiding routes considering driving habits. In addition, perfect the optimized project by deploying several necessary directly-guiding items in other entrances of the targeted road. Finally, the model is applied in Huacheng Avenue in Guangzhou for comparison with current deployment. Guiding accessibility to Huacheng Avenue improves from 77.9% to 88.9%.

Keywords: Guide sign optimization · Bird mating optimizer · Guide accessibility

1 Introduction

As information technology developed, ITS (Intelligent transportation System) has become the future trend of transportation. It is investigated that 76.9% of travellers thought guide signs play an important roles in the way to their destinations [1]. Researches on intelligentizing guide signs have great significance on the improvement of traffic operation efficiency, reduction of the emission of pollutants and intelligent management of transportation infrastructures. Granular data model for guide signs achieved the first step of intelligentizing guide signs [2]. Then some scholars researched on guide sign deploying models for a new district, which achieve the intelligent deployment of guide signs [3]. Evaluation methods are established in the simulation platform to analyze the guide signs' effectiveness scientifically [4]. In the aspect of guide sign optimization, the definition of guiding accessibility is put forward and maximize guiding accessibility to improve the guide sign system [5]. New roads have close bond with guide sign system, which are always ignored by traffic managers. As a result, there are some problems in the current guide sign system such as laggard

© Springer International Publishing AG 2017
Y. Tan et al. (Eds.): ICSI 2017, Part I, LNCS 10385, pp. 88–98, 2017.
DOI: 10.1007/978-3-319-61824-1_10

updating, inaccurate or promiscuous guide information. Guide signs without timely update confuse drivers and increase the probability of traffic accidents. Motivated by above, this study proposes an optimization model for guide signs in the situation of an added road.

Bird Mating Optimizer (BMO), proposed by Askarzadeh, is one of meta-heuristic optimization approaches which imitates the bird mating behavior of different species [6]. This algorithm combines the idea of classification from swarm algorithms with the operation of mutation and mating from Evolution Algorithm (EA) and is able to provide good balance between exploration and exploitation. BMO has been proved to represent a competitive performance in comparison with other optimization algorithms in many engineering optimization problems [7, 8]. In this study, BMO is applied to solve a discrete optimization problem and verify the potential of BMO in discrete problems.

This paper is arranged in five sections; Sect. 2 describes a mathematic model of the guide sign optimization for an added road; Sect. 3 presents BMO algorithm to deal with the discrete problems which proposed in Sect. 2; the case study of Huacheng Avenue is shown in Sect. 4; finally, conclusions are stated in Sect. 5.

2 Problem Description

For the purpose of solving the guide sign optimization problem for an added road by mathematical optimization algorithm, a mathematical model of this optimization problem should be established, which consists of mathematical model of guide signs and the guide sign optimization problem.

2.1 Mathematical Model of Guide Signs

The guide sign model includes two parts: road-network model and guide sign data model. Road-network G adopts a classical Node-Arc topology road-network model [9] formulated by $G = (V, A)$. V is a set of nodes where M is the number of nodes and A is a set of arcs where a_k is an arc connected with node v_i and node v_j.

$$V = \{v_i | i = 0, 1, 2 \ldots M\} \tag{1}$$

$$A = \{a_k = \langle v_i, v_j \rangle | v_i, v_j \in V\} \tag{1 - a}$$

A guide sign system (RS) is a set of guide sign items formulated by Eq. 2 where T is the number of guide sign items. One of guide sign items includes one guide information and is the minimum unit of guide signs [10]. According to topological relationship between the located arc and the guided arc in a guide sign item, we categorize the items into two types: connectivity items(*type* = c), which means the located arc and the guided one are directly connected in road-network G, and directivity items(*type* = d), which means the located arc and the guided one are not directly connected in road-network

G. Every guide sign item's information is recorded including its type(*type*), its located arc(*larc*), its guided node(*innode*), its next arc in the guiding direction(*narc*) and its guiding arc(*garc*).

$$RS = \{rs_t | t = 1, 2, \ldots T\} \tag{2}$$

$$rs_t = \{type, larc, innode, narc, garc\} \tag{2-a}$$

$$type \in \{c, d\}, larc, narc, garc \in A, innode \in V \tag{2-b}$$

For example, the attributes of guide sign item rs_1 in Fig. 1 are shown as follows: $rs_1 = \{rs_1.type = c, rs_1.larc = a_{01}, rs_1.innode = v_1, rs_1.narc = a_{12}, rs_1.garc = a_{12}\}$.

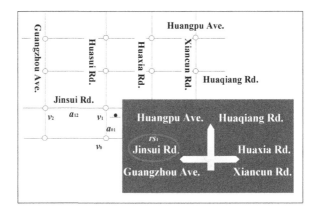

Fig. 1. Example of guide sign model

2.2   Modeling of Guide Sign Optimization Problem for an Added Road

According to the road network model, the added road F is regarded as a feature consisted of R arcs and $(R + 1)$ nodes.

$$F = \{f_r = \ <D_r, D_{r+1}> \ |f_r \in A, D_r, D_{r+1} \in V, r<R, r \in N*\} \tag{3}$$

At first, the optimization model should determine the optimized area(*Area*), optimized starting node set($\{S_p\}$). The optimized area is determined by guiding level of the targeted road F. The guiding level is an attribute to describe the importance of a feature in guide sign system. We select starting nodes in the boundary of optimized area. Generally four uniformly distributed nodes in arterial roads are selected. Terminal nodes are set as the node set of the targeted road($\{D_r\}$). The illustrations of optimized area, optimized starting nodes and terminal nodes are shown in Fig. 2.

The second step is to find out the optimal guiding routes(OR_p) from feasible solution sets $\{fr_p\}$, where fr_p represents a feasible route from the start node S_p to the target road. The quality of a route focuses on both the length and convenience of the guiding route for travellers and economic effectiveness of deploying guide signs for traffic managers. To unify dimensions of each term, we transfer all optimal object into economic costs. So, the objective function(f) is formulated as follows:

$$
\begin{aligned}
\min f &= \sum_{i=1}^{p} (a \times C_T^p + b \times C_U^p + C_R^p) \\
&= \sum_{i=1}^{p} (a \times V_{ot} \times \sum_{k} \sum_{s} \frac{Le_k(fr_p) \cdot l_k^s}{v_s} + b \\
&\quad \times \frac{\sum C}{M} \times \sum_{k} Le_k(fr_p) + w_T \times Item(fr_p) + w_D \times Sign(fr_p))
\end{aligned}
\tag{4}
$$

Where:

a: number of attracted people travelled to the target road F
   C_T^p: the travel time cost of each road user in the route fr_p
b: attracted traffic volume
   C_U^p: the use cost for one vehicle in the route fr_p
   C_R^p: the cost of optimizing guide signs in the route fr_p
V_{ot}: time value per hour
$Le_k(fr_p)$: length of arc a_k in the route fr_p

$$
l_k^s = \begin{cases} 1, & \text{guiding level of } a_k \text{ is } s \\ 0, & \text{otherwise} \end{cases}
$$

v_s: median designed speed in a road of guiding level s
$\sum C$: cost for vehicles, such as cost of acquisition, annual inspection
M: annual vehicle kilometers travelled
w_T: cost of deploying one guide sign item
w_D: cost of deploying one guide sign
$Item(fr_p)$: number of deploying guide sign items in the route fr_p
$Sign(fr_p)$: number of deploying guide signs in the route fr_p

In the objective function, $a \times C_T^p$ represents the travel time cost of travellers whose destination is the target road F. Travel time cost of a traveller is calculated by the method of time value unit, which is proved better than the method of salary pricing [11]. $b \times C_T^p$ represents the usage cost for vehicles driven towards the target road F, which covers the cost of acquisition, annual inspection, maintenance and so on [12]. C_R^p represents the cost of optimizing guide signs.

After solving the optimal guiding route set($\{OR_p\}$), guide signs are deployed in the optimal routes($\{rs_{tr}\}$). It is unnecessary to deploy guide signs on all intersections of the optimal routes. Considering the driving habits in route finding [13], some deploying rules are designed. The following rules are applied in this optimization problem:

I. A guide sign for guiding the target road should be deployed in the first arc of the route.
II. A guide sign should be deployed before left or right turn or U-turn.
III. A guide sign should be deployed after driving straight through three intersections without guiding information for the target road.

$$\{rs_{tr}|rs_{tr}.larc \in OR_p, rs_{tr}.type = \mathrm{d}\} \qquad (5)$$

$$\{rs_{ta}|rs_{ta}.innode = D_r\} \qquad (5-\mathrm{a})$$

The last step is to deploy guide signs in all terminal nodes($\{rs_{ta}\}$). The flowchart of the optimization is shown in Fig. 3. The result of the guide sign optimization problem for the targeted road F is a guide sign optimization set$\{rs_{tr}, rs_{ta}\}$.

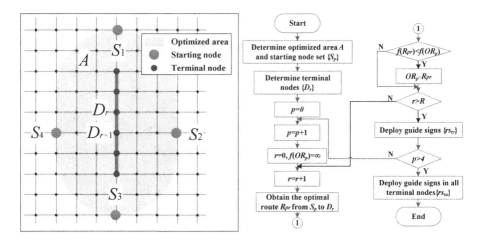

Fig. 2. Some definitions in optimization   **Fig. 3.** Flowchart of the optimization problem

3 Solution Based on BMO

Bird Mating Optimizer (BMO) is one of Evolutionary algorithms (EAs), which is inspired by mating strategies of bird species to breed broods with superior genes. The population of BMO algorithm is called society and each member in society is called a bird, representing a feasible solution in an optimization problem. All the feasible solutions are divided into female birds and male birds according to the value of objective function. The females are categorized into two groups so that the better ones are parthenogenesis and the others are polyandrous. The males with better genes are selected as monogamous and the others are chosen as polygynous. The birds with the worst genes, namely promiscuous, are removed from society and new ones are generated using a random way. And then each bird breeds brood by mutating or mating, which is detailed in Sects. 3.3 and 3.4. After breeding, each bird evaluates the quality of its brood and if the brood's genes are better than its parent bird's, the brood will stay in the society instead of its parent. Otherwise, the brood will be abandoned from the society. Repeat breeding until the predefined number of generation is met. The flowchart is shown in Fig. 4.

BMO has a great performance for solving many continuous optimization problems. The optimization problem of solving the optimal guiding routes in this study is a discrete problem. So, it is necessary to recode the feasible solution and redefine the mating and mutation modes for the discrete optimization problem.

3.1 Coding for Feasible Solution

Inspired by the definition of chromosome in GA, a sequence of nodes and their loci in the route are recorded as codes of feasible solutions in this optimization problem. For example, the guiding route in Fig. 5 can be coded as a two-dimensional array$\{(0, S_p),$ $(1, v_{14}), (2, v_{10})...(6, D_r)\}$.

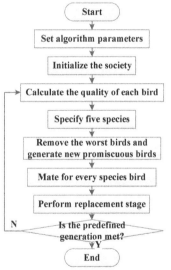

Fig. 4. The flowchart of BMO

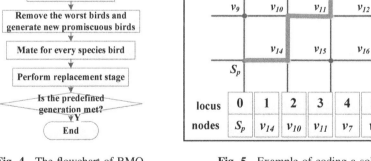

Fig. 5. Example of coding a solution

3.2 Objective Function

As is discussed in Sect. 2.2, the objective function is defined as Eq. 4. The value of objective function is related to the length of arcs and numbers of optimized guide signs in the feasible routes. In this guide sign optimization problem, the smaller value of the objective function means less cost on guide sign optimization for the target road.

3.3   Mating Mode of BMO

Considering the discontinuity of optimization problem and inspired by the idea of crossover in GA [14], we redefine the candidate mated set and mating mode. The birds which have at least one same node and different gender as the selected bird in society are able to be joined the candidate mated set for choosing. Mating mode in BMO has two patterns due to different bird species: two-parent mating and multi-parent mating.

Two-parent mating is that the two selected parents mate with each other and breed a brood. Firstly, roulette wheel approach [15] is resorted to choose a mated bird from the candidate mated set for the selected bird by Eq. 7. In this approach, the bird with better quality has a larger probability to be chosen. Next randomly choose a common node of the selected bird and the mated bird as a mating point (that is v_{10} in Fig. 6). At last the nodes of the selected bird before the mating point and the nodes of the mated bird after the mating point are recombined to generate the brood.

$$P_T = \frac{1/f(\overrightarrow{x})}{\sum\limits_{u=0}^{mt} 1/f(\overrightarrow{x_u})} \tag{7}$$

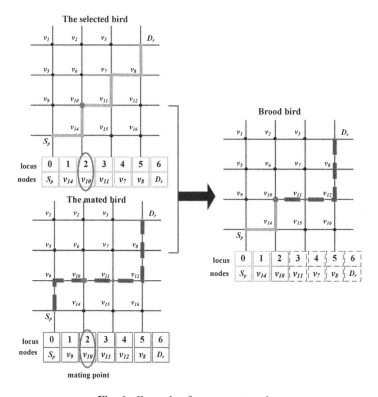

Fig. 6. Example of two-parent mating

Where: P_T is the probability of the candidate bird selection, mt is the number of birds in the candidate mated set.

Multi-parent mating is that the selected parent mate with at least two birds to breed a brood. So the brood's genes are influenced by at least three birds in society. In multi-parent mating, mated birds are chosen by an annealing function in Eq. 8. The birds with more similar quality have a larger probability to be chosen. Multi-parent mating can be regarded as a multi-step two-parent mating. Choose a bird as the first mated bird from the candidate mated set and make two-parent mating with the selected bird to generate an intermediate bird. Then the intermediate bird makes two-parent mating with other chosen birds to generate a new brood.

$$P_M = \exp(\frac{-\Delta f}{T}) \tag{8}$$

Where: P_M is the probability of the candidate bird selection, Δf is the absolute difference between the selected bird and the candidate bird. T is an adjustable parameter.

3.4 Mutation Mode of BMO

Mutation mode refers to the mutation of chromosome in GA [16]. In order to enhance diversity of the local search, two patterns of mutation are adopted: one-point mutation and two-point mutation. One-point mutation is to select a random locus as a mutation point (that is v_{10} in Fig. 7). Keep the route from the optimized starting node(S_p) to the selected mutation point and regenerate a new route from the mutation point to the destination(D_r). Two-point mutation is to select two random loci as mutation points (that is v_{10}, v_7 in Fig. 8). Regenerate a new route between these two mutation points to replace the segment in the original route.

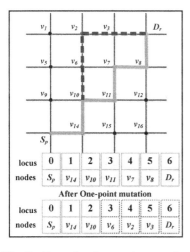

Fig. 7. Example of one-point mutation

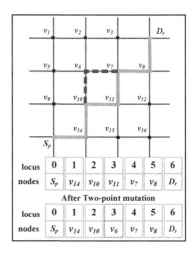

Fig. 8. Example of two-point mutation

4   Case Study

The guide sign optimization model is applied to improve the guide sign system of Huacheng Avenue in Tianhe District, Guangzhou.

A reasonable parameters setting has ability to get great balance between better result and less computing time. We have made some preliminary experiments to determine appropriate society size and number of generation for the guide sign optimization problem. The society size is set as 20 while the number of generation is set as 20. The other adjustable parameters, referred to the literature of BMO algorithm is set as follows: the number of parthenogenesis, polyandrous, monogamous, polygynous and promiscuous is respectively set at 5%, 5%, 50%, 30% and 10% of the society. The static mutation control factor is set at 0.9 while the dynamic one is set as an increasing linear function changing from 0.1 to 0.9. It is sufficient that two birds mate with polyandrous and polygynous birds [6].

As is shown in Fig. 9, the optimal routes from each starting node to the target road are worked out, whose best value of objective function are presented in Table 1, and the visualization of optimized guide signs is presented in Fig. 9. As the convergence curves of optimal routes shows in Fig. 10, convergence optimal solutions are obtained in all the four routes. And the convergence speeds up as the values of objective function decrease except that Route 3 obtains a best solution from the initial solutions. That proved that BMO algorithm can solve the discrete guide sign optimization successfully.

Guiding accessibility is a quantified index which is used to evaluate if the guide signs is able to lead drivers to their destination [17]. It is a ratio of numbers of accessibility signs and total signs guiding for the target feature. Compared with the current guide sign system, the guiding accessibility of the optimized guide signs observably improved from 77.9% to 88.9%, which is shown in Table 2. It is proved that the optimization model has ability to solve the guide sign optimization problem in the situation of adding a road.

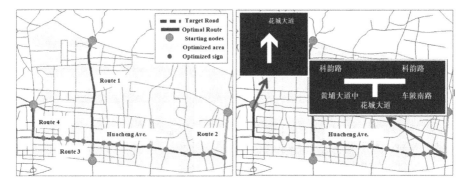

Fig. 9. Result of optimization by BMO

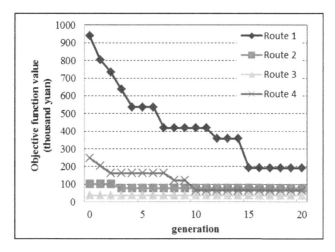

Fig. 10. Objective function value during the optimization by BMO

Table 1. Objective function values of the optimal routes (unit: thousand yuan)

Route	Travel time cost	Cost for vehicles	Cost of optimizing guide signs	Objective function values
1	1193.1	572.0	150.0	1915.1
2	522.1	250.3	5.0	777.4
3	215.8	103.5	50.0	369.3
4	377.4	180.9	100.0	658.3

Table 2. Comparison of optimization result and current situation

	Total of guide signs	Number of accessibility signs	Guiding accessibility
Current situation	68	53	77.9%
Optimization result	135	120	88.9%

5   Conclusion

In this study, an optimization model is proposed to solve the guide sign optimization problem in the situation of adding a road. At first, the model of guide signs and the guide sign optimization model for an added road are established. Then the coding and mating strategies in the Bird Mating Optimizer algorithm are redefined for the discrete optimization. At last, the model is applied to solve the guide sign optimization of Huacheng Avenue, Guangzhou. Compared with the current guide sign system, the guiding accessibility improves from 77.9% to 88.9%, which is proved the validity of the model.

Acknowledgement. This research described in this paper was supported by Science and Technology Project of Guangdong province (Nos. 2016A02022300, 2016B090918038, 2015B010110005, 2014B010118002), Science and Technology Project of Guangzhou city (201510010247).

References

1. Wang, D.M., Hu, M., Ge, L.Z., Li, Y.J.: User requirements analysis of information elements on urban road guide sign. Chin. J. Ergon. **21**(2), 26–30 (2015)
2. Huang, M., Wu, H.B., Rao, M.L., et al.: Urban road guide signs data model design and applied resreach. J. Geomat. Sci. Technol. **28**(6), 454–457 (2011)
3. Li, M., Huang, M., Niu, Z.M., et al.: Application of data model for guidance system in deploying guide signs. Appl. Res. Comput. **31**(2), 457–460 (2014)
4. Qin, L.Y., Jiang, H., Zhu, Z.L.: Analysis and simulation evaluation for traffic guide sign system of Hongqiao transport hub. Trans. Stand. **238**, 23–27 (2011)
5. Zheng, J., Huang, M., Liu, F., et al.: Artificial bee colony algorithm for guiding the accessibility optimization problem of a guide sign system. In: CICTP 2016, Shanghai, pp. 26–39 (2016)
6. Askarzadeh, A.: Bird mating optimizer: an optimization algorithm inspired by bird mating strategies. Commun. Nonlinear Sci. Numer. Simul. **19**(4), 1213–1228 (2014)
7. Li, H., Liu, J.K., Lu, Z.R.: Bird mating optimizer in structural damage identification. In: Advances in Swarm and Computational Intelligence, Beijing, pp. 49–56 (2015)
8. Huang, S.J., Tai, T.Y., Liu, X.Z., et al.: Application of bird-mating optimization to phase adjustment of open-wye/open-delta transformers in a power grid. In: IEEE International Conference on Industrial Technology (2015)
9. Song, Y., Shi, J.J., Xu, G.H.: Dynamic road-network model based on route planning system. Comput. Commun. **22**(5), 28–31 (2004)
10. Li, M.N., Huang, M., Niu, Z.M., et al.: Deployment model for urban guide signs based on road network topology. Procedia – Soci. Behav. Sci. **96**, 1631–1639 (2013)
11. Hu, Y.J.: Research of the quantifying method for determining the resident trip cost in city. J. Trans. Eng. Inf. **7**(1), 5–10 (2009)
12. Fan, X.T., Jin, W.Z.: Trip cost quantification model of urban private car. Comput. Commun. **29**(4), 24–27 (2011)
13. Li, M., Huang, M., Li, E.D.: Analytical research on influences of urban direction sign guidance system on driver behaviors. Technol. Highw. Transp. **4**, 151–155 (2015)
14. Ji, G.L.: Survey on genetic algorithm. Comput. Appl. Softw. **21**(2), 69–73 (2013)
15. Alireza, A., Alireza, R.: A new heuristic optimization algorithm for modeling of proton exchange membrane fuel cell: bird mating optimizer. Int. J. Energy Res. **37**(10), 1196–1204 (2013)
16. Xu, Q.Z., Ke, X.Z.: Genetic algorithm analysis for shortest path. Comput. Eng. Des. **29**(6), 1507–1509 (2008)
17. Li, M., Huang, M., Niu, Z.M., et al.: Analysis and evaluation for guiding accessibility considering connection signs. Trans. Syst. Eng. Inf. Technol. **14**(2), 226–231 (2014)

LGWO: An Improved Grey Wolf Optimization for Function Optimization

Jie Luo, Huiling Chen$^{(\boxtimes)}$, Kejie Wang, Changfei Tong, Jun Li,
and Zhennao Cai

College of Physics and Electronic Information,
Wenzhou University, Wenzhou, China
chenhuiling.jlu@gmail.com

Abstract. Grey wolf optimization (GWO) algorithm is a novel nature-inspired heuristic paradigm. GWO was inspired by grey wolves, which mimics the leadership hierarchy and hunting mechanism of grey wolves in nature. It has exhibited promising performance in many fields. However, GWO algorithm has the drawback of slow convergence and low precision. In order to overcome this drawback, we propose an improved version of GWO enhanced by the Lévy-flight strategy, termed as LGWO. Lévy-flight strategy was introduced into the GWO to find better solutions when the grey wolves fall into the local optimums. The effectiveness of LGWO has been rigorously evaluated against ten benchmark functions. The experimental results demonstrate that the proposed approach outperforms the other three counterparts.

Keywords: Grey wolf optimization · Function optimization · Lévy-flight

1 Introduction

In recent years, meta-heuristic optimization algorithms have become more and more popular in optimization techniques. Some of them such as Genetic Algorithm (GA) [1], Ant Colony Optimization (ACO) [2], and Particle Swarm Optimization (PSO) [3] are well-known among scientists from different fields. These optimization techniques have been applied in various fields of study. As a new member of the swarm intelligence algorithms, grey wolf optimization (GWO) [4] algorithm has been applied to many fields, such as optimal reactive power dispatch problem [5], medical diagnosis [6] and so on.

GWO algorithm is a new meta-heuristic optimization method through imitating the hunting mechanism of grey wolves in nature. In this algorithm, four types of grey wolves such as alpha, beta, delta, and omega are employed for simulating the leadership hierarchy. The fittest solution is considered to be alpha, the second and third best solutions are named as beta and delta respectively. The rest of the candidate solutions are assumed to be omega. In GWO algorithm the hunting (optimization) is guided by alpha, beta, and delta. The omega wolves follow these three wolves. The location of omega (X_i) can be calculated according the location of alpha (X_α), beta (X_β), and delta (X_δ). The inventor of this algorithm, Seyedali Mirjalili, has proved that the GWO algorithm is able to provide very competitive results compared with other state-of-the-art meta-heuristic optimization algorithms.

© Springer International Publishing AG 2017
Y. Tan et al. (Eds.): ICSI 2017, Part I, LNCS 10385, pp. 99–105, 2017.
DOI: 10.1007/978-3-319-61824-1_11

Lévy-flight was originally introduced by the French mathematician in 1937 named Paul Lévy. Lévy-flight is a statistical description of motion that extends beyond the more traditional Brownian motion discovered over one hundred years earlier. A diverse range of both natural and artificial phenomena are now being described in terms of Lévy statistics [7]. In previous works, Lévy-flight strategy has been applied in many optimization algorithms such as PSO [8], shuffled frog-leaping algorithm [9] and dragonfly algorithm [10]. Inspired from the above works, we introduce the Lévy-flight strategy into the GWO to improve the randomness, stochastic behavior, and exploration of grey wolves, and propose a Lévy-flight enhanced grey wolf optimization (LGWO) algorithm. To the best of our knowledge, it is the first time to propose to employ the Lévy-flight strategy to improve the performance of GWO algorithm. The proposed algorithm takes advantage of Lévy-flight's ability of strengthening global search and overcoming the problem of being trapped in local minima, so the proposed algorithm can lead to a faster and more robust method.

The remainder of this paper is structured as follows. In Sect. 2 the detailed implementation of the Lévy-flight grey wolf optimization algorithm is presented. Section 3 describes the experimental results. Finally, conclusions are summarized in Sect. 4.

2 Lévy-Flight Grey Wolf Optimization (LGWO)

To increase the diversity of population against premature convergence and accelerate the convergence speed as well, we propose an improved GWO algorithm, LGWO. Lévy-flight has the prominent properties to increase the diversity of population, sequentially, which can make the algorithm effectively jump out of the local optimum. We let each grey wolf perform once Lévy-flight after the position updating, which is formulated as follows.

$$X_{t+1} = X_t + \text{Lévy}(d) \times X_t \tag{1}$$

where t is the current iteration, and d is the dimension of the position vectors. The LGWO can be divided into several steps as follows.

Step 1. Population initialization.
Give the random location X_i of an individual grey wolf, where i represents the population size.
Step 2. Calculate fitness.
Calculate the fitness of grey wolves according to the object function.
Step 3. Find the alpha, beta, delta grey wolves.
Sort grey wolves according to their fitness, determine the alpha, beta and delta with the three fittest grey wolves and the corresponding location among the grey wolves.
Step 4. Update the location of all search agents.
All the moving of grey wolves are guided by alpha, beta and delta, calculate new location of each grey wolf with Eq. $X_i = (X_\alpha + X_\beta + X_\delta)/3$.

Step 5. Perform once Lévy-flight.

After search agents update their position based on the location of the alpha, beta, and delta, they will perform Lévy-flight. New position can calculated with Eq. (1).

Step 6. Iterative optimization.

Enter the iterative optimization to repeat the implementation of step 2 to 5. The circulation stops when the iterative number reaches the maximal iterative number.

The pseudo code of the LGWO is as follows.

Initialize the grey wolf population Xi (i=1,2,...,n)
Calculate the fitness of each search agent
X_α=*the best search agent*
X_β=*the second best search agent*
X_δ=*the third best search agent*
   while(t<*Max number of iterations*)
      for *each search agent*
         Update the position of the current search agent
      end for
      for *each search agent*
         Update the position of the current search agent using Lévy-flight by
equation (1)
      end for
      Calculate the fitness of all search agents
      Update X_α, X_β, X_δ
      t=t+1
   end while
   return X_α

---

3   Experimental Results and Discussions

In the experiments, the number of grey wolves was set to 30. Four models including original GWO, Lévy-flight Grey Wolf Optimization (LGWO), Particle Swarm Optimization (PSO), and Firefly Algorithm (FA) [11] were built in MATLAB R2014b. The experiments are done on Windows 10 Operation System withi5-5200U Dual-Core CPU (2.7GHZ) and 4 GB RAM.

In order to verify the effectiveness of the proposed algorithm, four methods were employed to compare against ten commonly used multidimensional functions as shown in Table 1, of which f_1–f_6 are unimodal benchmark functions and f_7–f_{10} are the multimodal benchmark functions. We test each function for 50 times, and calculated the average (Ave) and standard deviation (Std) respectively.

Table 2 shows the average and standard deviation of ten benchmark functions over different iterations. The ten benchmark functions are divided the two groups: unimodal

Table 1. Benchmark functions

Function	Range	Minimum
$f_1(x) = \sum_{i=1}^{n} x_i^2$	$[-100, 100]$	0
$f_2(x) = \sum_{i=1}^{n} |x_i| + \prod_{i=1}^{n} |x_i|$	$[-10, 10]$	0
$f_3(x) = \sum_{i=1}^{n} (\sum_{j-1}^{i} x_j)^2$	$[-100, 100]$	0
$f_4(x) = \max_i \{|x_i|, 1 \le i \le n\}$	$[-100, 100]$	0
$f_5(x) = \sum_{i=1}^{n-1} [100(x_{i+1} - x_i^2)^2 + (x_i - 1)^2]$	$[-30, 30]$	0
$f_6(x) = \sum_{i=1}^{n} ([x_i + 0.5])^2$	$[-100, 100]$	0
$f_7(x) = \sum_{i=1}^{n} -x_i \sin(\sqrt{|x_i|})$	$[-500, 500]$	$-418.9829*5$
$f_8(x) = \sum_{i=1}^{n} [x_i^2 - 10\cos(2\pi x_i) + 10]$	$[-5.12, 5.12]$	0
$f_9(x) = -20\exp(-0.2\sqrt{\frac{1}{n}\sum_{i=1}^{n} x_i^2})$ $- \exp(\frac{1}{n}\sum_{i=1}^{n} x_i^2 \cos(2\pi x_i)) + 20 + e$	$[-32, 32]$	0
$f_{10}(x) = \frac{1}{4000}\sum_{i=1}^{n} x_i^2 - \prod_{i=1}^{n} \cos(\frac{x_i}{\sqrt{i}}) + 1$	$[-600, 600]$	0

(f_1–f_6) and multi-modal (f_7–f_{10}) functions. As their names suggest, unimodal benchmark functions have single optimum, so they can benchmark the exploitation and convergence of an algorithm. In contrast, multi-modal test functions have more than one optimum. From the table we can see that the proposed LGWO not only performs better on unimodal functions than GWO, PSO and FA, but also obtains better experimental results when it is aimed at multimodal functions. It can be seen from Table 2 that the LGWO are better able to find the minimum value than the original GWO when the iterations is large. When the iterations of the benchmark function are increased, the LGWO will be more likely to go beyond the original GWO. Moreover, Table 2 also shows that the LGWO has better stability than the GWO, indicating that its probability of finding a better solution is bigger than the GWO. The superiority results suggest that the Lévy-flight strategy can make the GWO more likely to jump out of the local optimum.

Limited to the space, the example is taken only in the dimension of 10 to describe the convergence of the involved functions. Figure 1 records the convergence curve of each function for above four methods. The LGWO algorithm proposed in this paper has much higher convergence speed than GWO as these graphs shown. It can be seen from the Fig. 1 that the GWO could find better solution than the PSO and FA. However, the GWO converges so quickly that it couldn't discover good value any more. LGWO is different from the original GWO. Although the curve of the LGWO tends to be stable, it still shakes sometimes and found better value again, and gradually goes to convergence. The LGWO inherits the characteristic of the GWO to find the optimal value quickly, and it also avoids the situation of falling into a local optimum and never comes out.

Table 2. Results of testing benchmark functions

Functions	Iterations	LGWO		GWO		PSO		FA	
		Ave	Std	Ave	Std	Ave	Std	Ave	Std
f_1	200	5.49e−41	1.85e−41	1.33e−40	3.83e−40	10.3218	3.0415	115.5736	30.3075
	400	1.83e−85	9.57e−85	6.71e−81	2.41e−80	9.0246	2.5405	3.2053	0.7857
	600	1.4e−106	8.3e−106	4.4e−102	2.2e−101	8.3400	2.2393	0.0699	0.0174
	800	2.4e−117	1.5e−117	8.4e−113	3.7e−112	7.8747	2.1498	0.0035	0.0012
	1000	**1.2e−120**	**7.3e−119**	**4.0e−116**	**1.8e−115**	**7.6112**	**2.1725**	**8.6e−05**	**3.4e−05**
f_2	200	2.49e−24	4.34e−24	1.40e−23	1.38e−23	8.0192	1.3209	2.6830	0.3733
	400	2.19e−48	3.80e−48	1.73e−47	3.51e−47	7.3198	1.0921	0.4782	0.0543
	600	1.30e−60	2.82e−60	1.15e−59	2.52e−59	7.1030	1.0170	0.0984	0.0203
	800	4.36e−66	1.00e−66	3.36e−65	6.59e−65	6.9876	0.9414	0.0163	0.0043
	1000	**8.84e−68**	**2.02e−68**	**6.98e−67**	**1.36e−66**	**6.7634**	**0.8999**	**0.0027**	**6.47e−04**
f_3	200	2.18e−17	1.36e−17	1.25e−14	5.99e−14	17.3898	4.5296	210.4943	72.1206
	400	7.17e−38	3.46e−38	3.46e−31	1.44e−29	13.4786	3.7466	5.2665	1.6820
	600	1.31e−51	8.20e−51	1.55e−41	8.03e−41	12.4173	3.3408	0.1125	0.0370
	800	1.36e−59	8.25e−59	1.50e−48	8.00e−48	11.4855	2.9942	0.0080	0.0040
	1000	**3.98e−62**	**2.50e−61**	**3.08e−51**	**1.68e−50**	**10.4933**	**2.6142**	**3.08e−04**	**1.35e−04**
f_4	200	9.05e−15	2.13e−14	4.67e−15	7.65e−15	1.8764	0.2566	6.0516	0.8163
	400	1.08e−28	3.18e−28	4.76e−27	6.79e−27	1.7202	0.2525	0.9721	0.1277
	600	5.37e−35	1.21e−35	3.74e−33	6.18e−33	1.6362	0.2036	0.1482	0.0179
	800	3.63e−38	8.26e−38	2.22e−36	3.47e−36	1.5722	0.2012	0.0393	0.0076
	1000	**3.41e−39**	**7.69e−39**	**2.09e−37**	**3.25e−37**	**1.5403**	**0.2157**	**0.0061**	**0.0013**
f_5	200	6.9512	0.3782	6.8907	0.4102	2.83e + 03	1.48e + 03	3.00e + 03	1.91e + 03
	400	6.6556	0.5042	6.7009	0.5567	2.16e + 03	1.06e + 03	59.7347	75.8749
	600	6.5200	0.5420	6.6300	0.5790	1.87e + 03	918.5846	29.7589	59.8659
	800	6.4459	0.5836	6.5184	0.5875	1.58e + 03	761.0960	24.7992	56.8106
	1000	**6.3606**	**0.6087**	**6.4253**	**0.6564**	**1.46e + 03**	**672.2368**	**23.4208**	**56.2300**
f_6	200	0.0579	0.0665	0.0497	0.0551	10.2588	2.7969	120.3125	35.1500
	400	0.0290	0.0493	0.0236	0.0355	9.4074	2.4384	3.0748	0.7166
	600	0.0093	0.0122	0.0130	0.0351	8.5023	2.5372	0.0646	0.0196
	800	0.0025	0.0017	0.0073	0.0354	7.7028	2.4274	0.0035	0.0013
	1000	**2.11e-05**	**3.31e-05**	**0.0050**	**0.0354**	**7.2818**	**2.1844**	**9.45e−05**	**3.78e−05**
f_7	200	−2.0e + 03	202.7682	−2.0e + 03	199.7866	−2.4e + 03	392.3186	−2.4e + 03	253.3992
	400	−2.2e + 03	220.6800	−2.2e + 03	202.7934	−2.5e + 03	414.0335	−2.8e + 03	271.9065
	600	−2.4e + 03	241.9892	−2.4e + 03	270.8392	−2.5e + 03	413.8999	−2.8e + 03	272.3528
	800	−2.7e + 03	295.8849	−2.6e + 03	291.3653	−2.5e + 03	414.0291	−2.8e + 03	272.3546
	1000	**−2.8e + 03**	**306.5871**	**−2.7e + 03**	**321.8249**	**−2.5e + 03**	**413.9706**	**−2.8e + 03**	**272.3546**
f_8	200	1.4873	4.8465	0.5348	1.6122	73.1296	8.5777	33.9514	5.2415
	400	0.9383	3.4937	0.1421	0.7044	69.4196	8.3337	10.1710	3.8351
	600	0.6034	2.0852	0.1322	0.6544	66.5019	8.2603	7.7521	3.6072
	800	0.5266	1.8176	0.1285	0.6357	63.8435	8.3588	7.6832	3.6072
	1000	**0.4692**	**1.6174**	**0.1252**	**0.6194**	**62.4198**	**8.0957**	**7.6811**	**3.6073**
f_9	200	0.1985	1.8703	0.1770	1.2398	71.6533	8.8169	34.2501	6.1371
	400	0.1092	0.7639	0.1386	0.9802	67.8814	9.9679	9.3912	2.9315
	600	0.0580	0.1351	0.1075	0.7600	65.3662	10.0420	6.9800	2.8497
	800	0.0179	0.0351	0.0994	0.7029	62.2983	8.9925	6.9069	2.8456
	1000	**0.0039**	**0.0137**	**0.0885**	**0.6259**	**60.8011**	**8.9990**	**6.9051**	**2.8457**
f_{10}	200	0.0410	0.0795	0.0617	0.1769	0.6932	0.1238	2.0713	0.2557
	400	0.0195	0.0496	0.0451	0.0750	0.6201	0.1100	0.8524	0.0914
	600	0.0170	0.0392	0.0231	0.0694	0.5782	0.1091	0.4286	0.1249
	800	0.0144	0.0301	0.0203	0.0575	0.5582	0.1109	0.0303	0.0339
	1000	**0.0131**	**0.0248**	**0.0197**	**0.0461**	**0.5438**	**0.1012**	**0.0228**	**0.0337**

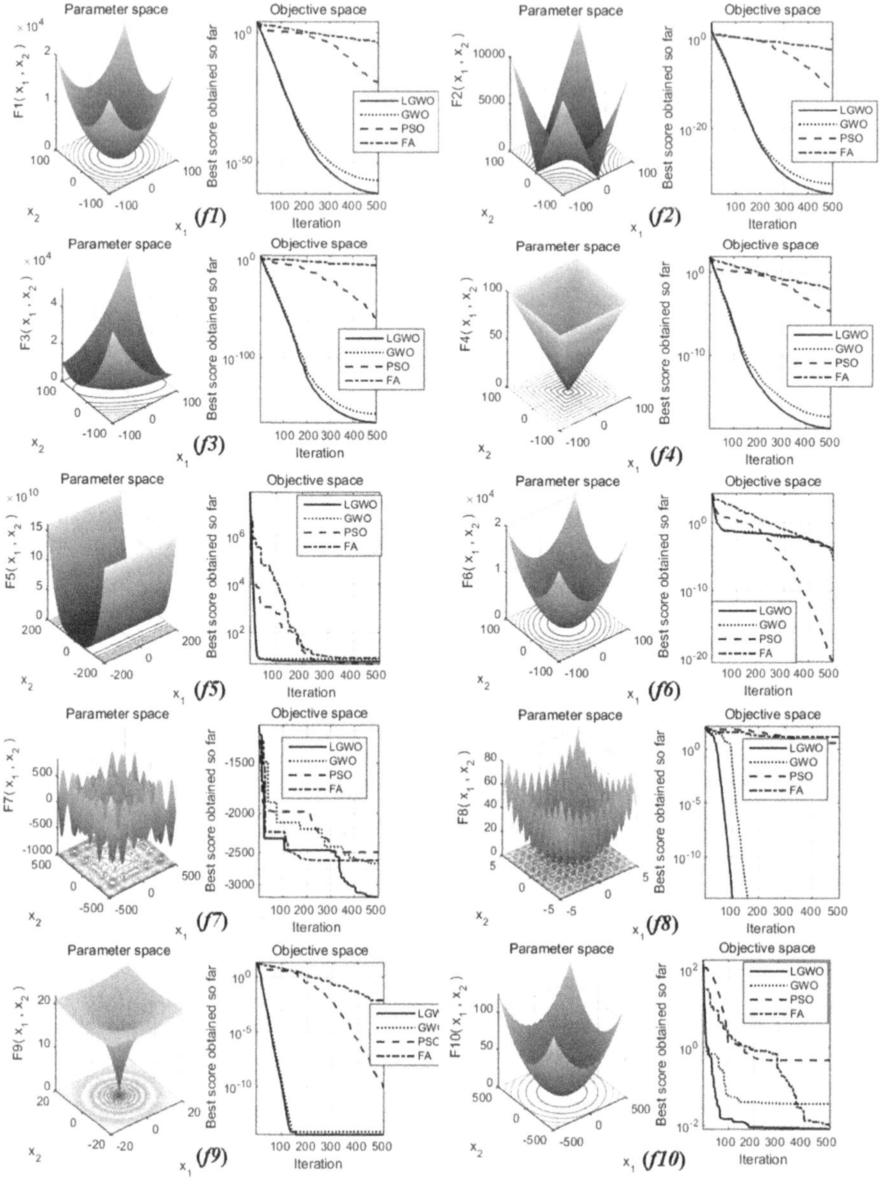

Fig. 1. Convergence curves of ten benchmark functions for four methods when Dimension = 10

4   Conclusions

This work has proposed an improved GWO, LGWO for function optimization. The main novelty of this paper lies in the proposed LGWO approach, which aims at obtaining the better solution quality and converging at a faster speed. In order to achieve this purpose, we have introduced Lévy-flight strategy into the original GWO. The empirical experiments on a set of multidimensional functions have demonstrated the superiority of the proposed method over the other three counterparts. In future works, we plan to apply the proposed method to optimize the parameters of machine learning methods such as support vector machines and kernel extreme learning machine.

Acknowledgements. This research is supported by the National Natural Science Foundation of China (NSFC) (61303113, 61402337). This research is also funded by the Zhejiang Provincial Natural Science Foundation of China (LY17F020012, LQ13G010007, LQ13F020011 and LY14F020035), the Science and Technology Plan Project of Wenzhou, China (G20140048, H20110003).

References

1. Holland, J.H.: Genetic algorithms. Sci. Am. **267**, 66–72 (1992)
2. Dorigo, M., Birattari, M., Stutzle, T.: Ant colony optimization artificial ants as a computational intelligence technique. IEEE Comput. Intell. Mag. **1**, 28–39 (2006)
3. Kennedy, J., Eberhart, R.: Particle swarm optimization. In: Proceedings of the 1995 IEEE International Conference on Neural Networks. IEEE, vol. 4, pp. 1942–1948. IEEE Press, Piscataway (1995)
4. Mirjalili, S., Mirjalili, S.M., Lewis, A.: Grey wolf optimizer. Adv. Eng. Softw. **69**, 46–61 (2014)
5. Sulaiman, M.H., Mustaffa, Z., Mohamed, M.R., Aliman, O.: Using the gray wolf optimizer for solving optimal reactive power dispatch problem. Appl. Soft Comput. J. **32**, 286–292 (2015)
6. Li, Q., Chen, H., Huang, H., Zhao, X., Cai, Z., Tong, C., Liu, W., Tian, X.: An enhanced grey wolf optimization based feature selection wrapped kernel extreme learning machine for medical diagnosis. Comput. Math. Methods Med. **2017** (2017)
7. Kamaruzaman, A.F., Zain, A.M., Yusuf, S.M., Udin, A.: Lévy flight algorithm for optimization problems—a literature review. Appl. Mech. Mater. **421**, 496–501 (2013)
8. Jensi, R., Jiji, G.W.: An enhanced particle swarm optimization with levy flight for global optimization. Appl. Soft Comput. **43**, 248–261 (2016)
9. Tang, D., Yang, J., Dong, S., Liu, Z.: A lévy flight-based shuffled frog-leaping algorithm and its applications for continuous optimization problems. Appl. Soft Comput. **49**, 641–662 (2016)
10. Mirjalili, S.: Dragonfly algorithm: a new meta-heuristic optimization technique for solving single-objective, discrete, and multi-objective problems. Neural Comput. Appl. **27**, 1053–1073 (2016)
11. Yang, X.-S.: Firefly algorithms for multimodal optimization. In: Watanabe, O., Zeugmann, T. (eds.) SAGA 2009. LNCS, vol. 5792, pp. 169–178. Springer, Heidelberg (2009). doi:10. 1007/978-3-642-04944-6_14

An Improved Monarch Butterfly Optimization with Equal Partition and F/T Mutation

Gai-Ge Wang[1(✉)], Guo-Sheng Hao[1], Shi Cheng[2], and Zhihua Cui[3]

[1] School of Computer Science and Technology,
Jiangsu Normal University, Xuzhou, Jiangsu, China
gaigewang@gmail.com, gaigewang@163.com,
guoshenghaoxz@tom.com,
[2] School of Computer Science, Shaanxi Normal University, Xi'an, China
cheng@snnu.edu.cn
[3] Complex System and Computational Intelligence Laboratory,
Taiyuan University of Science and Technology, Taiyuan 030024, Shanxi, China
zhihua.cui@hotmail.com

Abstract. In general, the population of most metaheuristic algorithms is randomly initialized at the start of search. Monarch Butterfly Optimization (MBO) with a randomly initialized population, as a kind of metaheuristic algorithm, is recently proposed by Wang *et al*. In this paper, a new population initialization strategy is proposed with the aim of improving the performance of MBO. Firstly, the whole search space is equally divided into *NP* (population size) parts at each dimension. And then, in order to add the diversity of the initialized population, two random distributions (*T* and *F* distribution) are used to mutate the equally divided population. Accordingly, five variants of MBOs are proposed with new initialization strategy. By comparing five variants of MBOs with the basic MBO algorithm, the experimental results presented clearly demonstrate five variants of MBOs have much better performance than the basic MBO algorithm.

Keywords: Benchmark · Monarch butterfly optimization · Equal partition · *F* mutation · *T* mutation

1 Introduction

In computer science, decision making, and mathematics, optimization is to minimize/maximize a function. In general, most optimization algorithms can loosely be divided into two categories: traditional mathematical programming methods and modern metaheuristic algorithms. The traditional mathematical programming methods will get the same results with the same initial conditions, but they are hard to solve modern complicated engineering problems. While, the later, modern metaheuristic algorithms can solve modern complicated problems well, though they will get different results even if they are implemented under the same conditions. Due to their excellent performance, more and more scholars pay attention to the research of metaheuristic algorithms.

© Springer International Publishing AG 2017
Y. Tan et al. (Eds.): ICSI 2017, Part I, LNCS 10385, pp. 106–115, 2017.
DOI: 10.1007/978-3-319-61824-1_12

In the past decades, many state-of-the-art metaheuristic algorithms have been proposed, such as differential evolution (DE) [1, 2], cuckoo search (CS) [3–8], particle swarm optimization (PSO) [9–12], biogeography-based optimization (BBO) [13–15], harmony search (HS) [16, 17], gravitational search algorithm (GSA) [18, 19], fireworks algorithm (FWA) [20], brain storm optimization (BSO) [21, 22], earthworm optimization algorithm (EWA) [23], elephant herding optimization (EHO) [24, 25], water wave optimization [26], ant lion optimizer (ALO) [27], multi-verse optimizer (MVO) [28], firefly algorithm (FA) [29, 30], ant colony optimization (ACO) [31], bat algorithm (BA) [32–35], grey wolf optimizer (GWO) [36], and krill herd (KH) [37–43]. For most metaheuristic algorithms, their population are initialized in a random way. In the many different random distributions, uniform is one of the most used ones.

In this paper, we will take MBO algorithm [44–46] as an example to show a new initialization strategy for metaheuristic algorithms. Firstly, the whole search space is equally divided into *NP* (population size) parts at each dimension. And then, in order to add the diversity of the initialized population, two random distributions (*T* and *F* distribution) are used to mutate the equally divided population. Accordingly, five variants of MBOs are proposed with new initialization strategy. By comparing them with the basic MBO algorithm, the experimental results presented clearly demonstrate five variants of MBOs have much better performance than the basic MBO algorithm.

The organization of this paper is outlined here. Section 2 provides the mainframe of the basic MBO algorithm. Subsequently, the improved MBO with new initialization strategy is presented in Sect. 3, and is followed by several simulation results, comparing five variants of with the basic MBO algorithm for benchmark optimization as presented in Sect. 4. Section 5 concludes our work.

2   Related Work

Though Monarch Butterfly Optimization algorithm [44] has been proposed, many scholars have worked on MBO algorithm. In this section, some of the most representative work regarding MBO are summarized and reviewed.

Ghetas *et al.* proposed an innovation method called MBHS to tackle the optimization problem [47]. In MBHS, the harmony search (HS) adds mutation operators to the process of adjusting operator to enhance the exploitation, exploration ability, and speed up the convergence rate of MBO. For the purpose to validate the performance of MBHS, 14 benchmark functions are used, and the performance is compared with well-known search algorithms.

Feng *et al.* presented a novel binary monarch butterfly optimization (BMBO) method, intended for addressing the 0–1 knapsack problem (0–1 KP) [46]. Two tuples, consisting of real-valued vectors and binary vectors, are used to represent the monarch butterfly individuals in BMBO. Three kinds of individual allocation schemes are tested in order to achieve better performance. Toward revising the infeasible solutions and optimizing the feasible ones, a novel repair operator, is employed. The comparative study of the BMBO with four algorithms clearly points toward the superiority of the former in terms of convergent capability and stability in solving the 0–1 KP.

In order to overcome this, Wang *et al.* a new version of the MBO algorithm [45, 48], by incorporating crossover operator. A variant of the original MBO, the proposed one is essentially a self-adaptive crossover (SAC) operator. A kind of greedy strategy is also utilized. The proposed methodology is essentially a new version of the original MBO, supplemented with Greedy strategy and self-adaptive Crossover operator (GCMBO). The proposed GCMBO method is benchmarked by twenty-five standard functions.

Feng *et al.* proposed a novel chaotic MBO algorithm (CMBO) [49], in which chaos theory is introduced in order to enhance its global optimization ability. In CMBO, 12 one-dimensional classical chaotic maps are used to tune two main migration processes of monarch butterflies. Meanwhile, applying Gaussian mutation operator to some worst individuals can effectively prevent premature convergence of the optimization process.

Ghanem and Jantan combined artificial bee colony (ABC) with elements from the MBO algorithm [50]. The combined method introduces a new hybrid approach by modifying the butterfly adjusting operator in MBO algorithm and uses that as a mutation operator to replace employee phase of the ABC algorithm. The new algorithm is called Hybrid ABC/MBO (HAM). The proposed algorithm was evaluated using 13 benchmark functions and compared with the nine metaheuristic methods.

Wang *et al.* proposed a discrete MBO (DMBO) [51], and firstly used to solve Chinese TSP (CTSP). Furthermore, the parametric study for one of the most parameter, butterfly adjusting rate (BAR), is also provided. The best-selected BAR is inserted into the DMBO method and then solve CTSP problem.

Feng *et al.* proposed a novel multi-strategy monarch butterfly optimization (MMBO) algorithm for the discounted 0–1 knapsack problem (DKP) [52]. In MMBO, two effective strategies, neighborhood mutation with crowding and Gaussian perturbation, are introduced into MMBO. Experimental analyses show that the first strategy can enhance the global search ability, while the second strategy can strengthen local search ability and prevent premature convergence during the evolution process. Accordingly, MBO is combined with each strategy, denoted as NCMBO and GMMBO, respectively. The experimental results on three types of large-scale DKP instances show that NCMBO, GMMBO and MMBO are all suitable for solving DKP.

3 Monarch Butterfly Optimization

By idealizing the migration behaviour of monarch butterflies, Wang *et al.* proposed a new swarm intelligence algorithm, called MBO [44].

In MBO, the number of monarch butterflies in Land 1 and Land 2 is NP_1 and NP_2. NP is population size; p is the ratio of monarch butterflies in Land 1. Subpopulation 1 and Subpopulation 2 are composed of monarch butterflies in Land 1 and Land 2, respectively. Accordingly, migration operator can be given as

$$x_{i,k}^{t+1} = x_{r_1,k}^t,\tag{1}$$

where $x_{i,k}^{t+1}$ indicates the kth element of x_i at generation $t + 1$. Similarly, $x_{r_1,k}^t$ indicates the kth element of x_{r_1}. t is the current generation number. Butterfly r_1 is randomly selected from Subpopulation 1. When $r \leq p$, $x_{i,k}^{t+1}$ is generated by Eq. (1). On the contrast, if $r > p$, $x_{i,k}^{t+1}$ is updated as follows:

$$x_{i,k}^{t+1} = x_{r_2,k}^t, \tag{2}$$

where $x_{r_2,k}^t$ indicates the kth element of x_{r_2} [44].

For all the elements in butterfly j, if $rand \leq p$, it can be updated as

$$x_{j,k}^{t+1} = x_{best,k}^t, \tag{3}$$

where $x_{j,k}^{t+1}$ indicates the kth element of x_j at generation $t + 1$. Similarly, $x_{best,k}^t$ indicates the kth element of x_{best} with the fittest solution in butterfly population. On the contrast, if $rand > p$, it can be updated as

$$x_{j,k}^{t+1} = x_{r_3,k}^t, \tag{4}$$

where $x_{r_3,k}^t$ indicates the kth element of x_{r_3}. More MBO can be found in [44].

4   Improving MBO with Equal Partition and F/T Mutation

For most metaheuristic algorithms, their individuals in initial population are randomly located in the whole search space. In this paper, we propose an enhanced initialization strategy for metaheuristic algorithms. This proposed initialization strategy is then incorporated into the basic MBO algorithm with the aim of improving the performance of MBO. The enhanced initialization strategy involves two aspects: equal partition and mutation. The main idea of the proposed initialization strategy can be given as follows.

4.1   Equal Partition

In order to fully explore the information of the whole search space, the whole search space is equally divided. In this way, all the individuals can be located in the whole search space. We will take the dth dimensional as an example to show how to partition the search space.

Suppose xid is the dth element of the ith individual for population P, and its upper and lower bound are Ld and Ud ($Ud > Ld$), respectively. Accordingly, we get a line segment **AB**. In fact, points **A** and **B** are respectively Ld and Ud, as shown in Fig. 1.

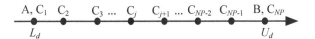

Fig. 1. Initializing population with equal partition

Suppose C1, C2, $\cdots$, Cj, C$_{j+1}$, $\cdots$, CNP are equal partition points at line segment **AB**, and their corresponding coordinates are $Y1$, $Y2$, $\cdots$, Yj, Y_{j+1}, $\cdots$, YNP, respectively. Clearly, C1 and CNP are essentially C1 and CNP, and $Y1 = Ld$, $YNP = Ud$. Because Cj ($j = 1$, $\cdots$, NP) is equal partition point at line segment **AB**, its coordinate Yj can be given as follows:

$$Y_j = L_d + (j-1) \times \frac{U_d - L_d}{NP - 1}, j = 1, \cdots, NP \tag{5}$$

Next, an integer k is randomly gotten between 1 and NP. So, we get $xid = Yk$. In order to fully explore the information of the whole search space, an integer sequence **R** = $\{R1, R2, \cdots, Rk, R_{k+1}, \cdots, RNP\}$ are generated. Accordingly, the dth element of the ith individual xik can be given below.

$$x_{ik} = Y_{R_k} \tag{6}$$

Until now, the dth element in xi (xid) for population P have been initialized. All the other elements in xi for population P can be initialized like xid. Accordingly, all the other individuals in population P will be initialized as xi. After equal partition, we get initial population P_1.

4.2 Mutation

As mentioned before, population P1 is generated by equal partition. In order to add the diversity of the population, randomness can be introduced into the generated population, and this introduced randomness can be considered as mutation operator for individuals. For xid, its responding mutation value x'_{id} can be given as

$$x'_{id} = Mu(x_{id}) \tag{7}$$

Function $Mu(xid)$ can be any random distribution. In this paper, we use T and F distribution to mutate the population. Through the mutation operator, as shown in Eq. (7), we get a new population P_2.

Up to now, we get two initial populations P_1 and P_2. Therefore, three choices for initial population can be provided: P_1, P_2, and $\{P_1 \cup P_2\}$. Accordingly, for MBO, three improved MBO algorithms can be generated. We will take T distribution as an example. Three generated MBO algorithms can be given as MBOEP, MBOTM, MBOEPTM. In MBOEP, the population P_1 generated by equal partition is considered as the initial population; in MBOTM, the mutation population P_2 generated by mutation operator as shown in Eq. (7) is considered as the initial population; while in MBOEPTM, the NP best individuals are selected from $\{P_1 \cup P_2\}$ and considered as the initial population. In short, we can use EP, TM, and EPTM short for MBOEP, MBOTM, and MBOEPTM, respectively. Similarly, function Mu is generated by F distribution, we can get another three improved MBO algorithms, which are MBOEP (EP), MBOFM (FM), and MBOEPFM (EPFM). Variants of MBOs will be verified in the next section.

5 Simulation Results

In this section, variants of MBOs (MBO, MBOEP, MBOFM, MBOEPFM, MBOTM, and MBOEPTM), are verified by twelve benchmarks, as shown in Table 1. Both maximum generation and population size are set to 200, and the dimension of twelve benchmarks are set to 100. For other parameters, they are set as follows: butterfly adjusting rate = 5/12. In order to obtain fair results, all the implementations are conducted under the same conditions as shown in [53].

In this paper, in order to get the most representative results, fifty independent runs are implemented, and the results are recorded in Table 2. From Table 2, we can see, five variants of MBOs performs better than the basic MBO on most benchmarks. Especially, MBOEPTM has the best performance for the mean function value, while MBOTM has the smallest Std values. This indicates the function values obtained by MBOTM will fall into the smallest range of the search space. Comprehensively considering all the algorithms, MBOTM and MBOEPTM have the similar performance which perform much better than other four MBOs (MBO, MBOEP, MBOFM, MBOEPFM) Ggg.

Table 1. Benchmark functions.

No.	Name	No.	Name
F01	Ackley	F07	Levy
F02	Alpine	F08	Pathological function
F03	Brown	F09	Penalty #1
F04	Csendes	F10	Penalty #2
F05	Griewank	F11	Powell
F06	Holzman 2 function	F12	Schwefel 2.21

Furthermore, the convergence history of F05 and F12 are also provided, as shown in Figs. 2 and 3.

From Fig. 2, clearly, MBOEPTM and MBOTM (especially MBOEPTM) have much better function values than other four MBO methods (MBO, MBOEP, MBOFM, MBOEPFM). MBOEPTM has a fastest convergence speed, even it can get the best results within twenty generations.

From Fig. 3, similar to Fig. 2, MBOEPTM is well capable of finding the minimum with the fastest speed. MBOEPTM has the second best performance, which is only inferior to MBOEPTM. This indicates that our proposed initialization strategy can significantly improve the performance of the basic MBO algorithm.

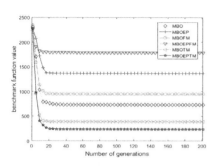

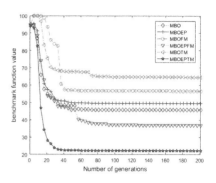

Fig. 2. Convergent process of six variants of MBOs on F05 Griewank with D = 100.

Fig. 3. Convergent process of six variants of MBOs on F12 Schwefel 2.21 with D = 100.

Table 2. Optimization results obtained by four methods with different population size and maximum generations.

	MBO	MBOEP	MBOFM	MBOEPFM	MBOTM	MBOEPTM
F01	17.83 ± 1.57	17.01 ± 3.70	**12.30 ± 6.36**	17.67 ± 2.71	19.62 ± **0.34**	15.05 ± 5.95
F02	138.36 ± 85.90	120.49 ± 123.53	100.20 ± 89.97	172.79 ± 71.46	103.24 ± **40.83**	**74.80** ± 56.49
F03	5.8E9 ± 1.4E10	3.4E12 ± 5.2E12	7.8E10 ± 1.8E11	3.9E8 ± 7.0E8	180.26 ± 365.43	**63.41** ± **43.90**
F04	48.08 ± 38.52	7.30 ± 8.15	7.70 ± 9.38	8.45 ± 6.72	5.81 ± 5.45	**0.15 ± 0.24**
F05	731.40 ± 777.32	1.4E3 ± 1.2E3	948.03 ± 597.68	1.8E3 ± 680.46	391.60 ± 526.35	**239.86 ± 393.65**
F06	3.1E6 ± 3.4E6	5.3E6 ± 2.6E6	3.2E6 ± 3.6E6	3.0E6 ± 2.7E6	**1.3E6** ± 1.6E6	1.2E6 ± **1.4E6**
F07	324.14 ± 389.35	777.65 ± 335.73	722.26 ± 379.74	466.92 ± 382.39	**237.32 ± 191.90**	258.34 ± 255.94
F08	32.47 ± 4.95	27.94 ± 5.52	25.24 ± 10.18	**18.84** ± 9.21	30.13 ± **2.85**	25.83 ± 7.92
F09	5.9E8 ± 1.1E9	1.5E9 ± 1.5E9	2.0E9 ± 1.5E9	4.7E8 ± 1.0E9	**2.3E8 ± 3.4E8**	2.8E8 ± 4.6E8
F10	9.4E8 ± 2.2E9	2.1E9 ± 2.6E9	1.5E9 ± 2.4E9	1.4E9 ± 1.9E9	**5.7E8 ± 1.4E9**	1.1E9 ± 1.7E9
F11	4.0E4 ± 2.7E4	2.4E4 ± 1.8E4	1.9E4 ± 2.9E4	1.9E4 ± 3.2E4	6.2E3 ± **8.3E3**	**1.2E4** ± 1.01E4
F12	45.68 ± 27.09	49.43 ± **22.02**	56.54 ± 35.90	37.01 ± 28.34	64.47 ± 33.70	**22.19** ± 29.42

6   Discussion and Conclusions

In this paper, a new initialization strategy is proposed in order to improve the performance of the metaheuristic algorithms. The newly proposed initialization strategy includes two parts: equal partition and mutation. MBO has been taken as an example to verify our proposed population initialization strategy. In this initialization population strategy, firstly, the whole search space is equally divided into *NP* (population size) parts at each dimension. And then, in order to add the diversity of the initialized population, two random distributions (*T* and *F* distribution) are used to mutate the equally divided population. Accordingly, five variants of MBOs are proposed with new initialization. By comparing with the basic MBO algorithm on twelve benchmark functions, the experimental results presented clearly demonstrate five variants of MBOs have much better performance than the basic MBO algorithm.

Acknowledgements. This work was supported by the Natural Science Foundation of Jiangsu Province (No. BK20150239) and National Natural Science Foundation of China (No. 61503165 and No. 61673196).

References

1. Storn, R., Price, K.: Differential evolution-a simple and efficient heuristic for global optimization over continuous spaces. J. Global Optim. **11**, 341–359 (1997)
2. Wang, G.-G., Gandomi, A.H., Alavi, A.H., Hao, G.-S.: Hybrid krill herd algorithm with differential evolution for global numerical optimization. Neural Comput. Appl. **25**, 297–308 (2014)
3. Yang, X.-S., Deb, S.: Cuckoo search via Lévy flights. In: Proceeding of World Congress on Nature & Biologically Inspired Computing (NaBIC 2009), pp. 210–214. IEEE Publications (2009)
4. Li, X., Yin, M.: Modified cuckoo search algorithm with self adaptive parameter method. Inf. Sci. **298**, 80–97 (2015)
5. Wang, G.-G., Deb, S., Gandomi, A.H., Zhang, Z., Alavi, A.H.: Chaotic cuckoo search. Soft. Comput. **20**, 3349–3362 (2016)
6. Wang, G.-G., Gandomi, A.H., Yang, X.-S., Alavi, A.H.: A new hybrid method based on krill herd and cuckoo search for global optimization tasks. Int. J. Bio-Inspired Comput. **8**, 286–299 (2016)
7. Wang, G.-G., Gandomi, A.H., Zhao, X., Chu, H.E.: Hybridizing harmony search algorithm with cuckoo search for global numerical optimization. Soft. Comput. **20**, 273–285 (2016)
8. Wang, G., Guo, L., Duan, H., Liu, L., Wang, H., Wang, J.: A hybrid meta-heuristic DE/CS algorithm for UCAV path planning. J. Inform. Comput. Sci. **9**, 4811–4818 (2012)
9. Kennedy, J., Eberhart, R.: Particle swarm optimization. In: Proceeding of the IEEE International Conference on Neural Networks, vol. 4, pp. 1942–1948. IEEE, Perth (1995)
10. Shieh, H.-L., Kuo, C.-C., Chiang, C.-M.: Modified particle swarm optimization algorithm with simulated annealing behavior and its numerical verification. Appl. Math. Comput. **218**, 4365–4383 (2011)
11. Mirjalili, S., Lewis, A.: S-shaped versus V-shaped transfer functions for binary particle swarm optimization. Swarm Evol. Comput. **9**, 1–14 (2013)
12. Wang, G.-G., Gandomi, A.H., Yang, X.-S., Alavi, A.H.: A novel improved accelerated particle swarm optimization algorithm for global numerical optimization. Eng. Comput. **31**, 1198–1220 (2014)
13. Simon, D.: Biogeography-based optimization. IEEE Trans. Evolut. Comput. **12**, 702–713 (2008)
14. Zheng, Y.-J., Ling, H.-F., Xue, J.-Y.: Ecogeography-based optimization: enhancing biogeography-based optimization with ecogeographic barriers and differentiations. Comput. Oper. Res. **50**, 115–127 (2014)
15. Duan, H., Zhao, W., Wang, G., Feng, X.: Test-sheet composition using analytic hierarchy process and hybrid metaheuristic algorithm TS/BBO. Math. Probl. Eng. **2012**, 1–22 (2012)
16. Geem, Z.W., Kim, J.H., Loganathan, G.V.: A new heuristic optimization algorithm: harmony search. Simulation **76**, 60–68 (2001)
17. Wang, G., Guo, L., Duan, H., Wang, H., Liu, L., Shao, M.: Hybridizing harmony search with biogeography based optimization for global numerical optimization. J. Comput. Theor. Nanos. **10**, 2318–2328 (2013)

18. Rashedi, E., Nezamabadi-pour, H., Saryazdi, S.: GSA: a gravitational search algorithm. Inf. Sci. **179**, 2232–2248 (2009)
19. Yin, M., Hu, Y., Yang, F., Li, X., Gu, W.: A novel hybrid K-harmonic means and gravitational search algorithm approach for clustering. Expert Syst. Appl. **38**, 9319–9324 (2011)
20. Tan, Y., Zhu, Y.: Fireworks algorithm for optimization. In: Tan, Y., Shi, Y., Tan, K. (eds.) Advances in Swarm Intelligence, vol. 6145, pp. 355–364. Springer, Heidelberg (2010)
21. Shi, Y.: An optimization algorithm based on brainstorming process. Int. J. Swarm Intell. Res. **2**, 35–62 (2011)
22. Shi, Y., Xue, J., Wu, Y.: Multi-objective optimization based on brain storm optimization algorithm. Int. J. Swarm Intell. Res. **4**, 1–21 (2013)
23. Wang, G.-G., Deb, S., Coelho, L.D.S.: Earthworm optimization algorithm: a bio-inspired metaheuristic algorithm for global optimization problems. Int. J. Bio-Inspired Comput. (2015)
24. Wang, G.-G., Deb, S., Coelho, L.D.S.: Elephant herding optimization. In: 2015 3rd International Symposium on Computational and Business Intelligence (ISCBI 2015), pp. 1–5. IEEE (2015)
25. Wang, G.-G., Deb, S., Gao, X.-Z., Coelho, L.D.S.: A new metaheuristic optimization algorithm motivated by elephant herding behavior. Int. J. Bio-Inspired Comput. **8**, 394–409 (2016)
26. Zheng, Y.-J.: Water wave optimization: a new nature-inspired metaheuristic. Comput. Oper. Res. **55**, 1–11 (2015)
27. Mirjalili, S.: The ant lion optimizer. Adv. Eng. Softw. **83**, 80–98 (2015)
28. Mirjalili, S., Mirjalili, S.M., Hatamlou, A.: Multi-verse optimizer: a nature-inspired algorithm for global optimization. Neural Comput. Appl. **27**, 495–513 (2016)
29. Yang, X.S.: Firefly algorithm, stochastic test functions and design optimisation. Int. J. Bio-Inspired Comput. **2**, 78–84 (2010)
30. Guo, L., Wang, G.-G., Wang, H., Wang, D.: An effective hybrid firefly algorithm with harmony search for global numerical optimization. Sci. World J. **2013**, 1–10 (2013)
31. Dorigo, M., Maniezzo, V., Colorni, A.: Ant system: optimization by a colony of cooperating agents. IEEE Trans. Syst. Man Cybern. B Cybern. **26**, 29–41 (1996)
32. Yang, X.-S.: Nature-Inspired Metaheuristic Algorithms. Luniver Press, Frome (2010)
33. Wang, G., Guo, L.: A novel hybrid bat algorithm with harmony search for global numerical optimization. J. Appl. Math. **2013**, 1–21 (2013)
34. Wang, G.-G., Chu, H.E., Mirjalili, S.: Three-dimensional path planning for UCAV using an improved bat algorithm. Aerosp. Sci. Technol. **49**, 231–238 (2016)
35. Xue, F., Cai, Y., Cao, Y., Cui, Z., Li, F.: Optimal parameter settings for bat algorithm. Int. J. Bio-Inspired Comput. **7**, 125–128 (2015)
36. Mirjalili, S., Mirjalili, S.M., Lewis, A.: Grey wolf optimizer. Adv. Eng. Softw. **69**, 46–61 (2014)
37. Gandomi, A.H., Alavi, A.H.: Krill herd: a new bio-inspired optimization algorithm. Commun. Nonlinear Sci. Numer. Simul. **17**, 4831–4845 (2012)
38. Wang, G.-G., Gandomi, A.H., Alavi, A.H.: Stud krill herd algorithm. Neurocomputing **128**, 363–370 (2014)
39. Wang, G.-G., Gandomi, A.H., Alavi, A.H.: An effective krill herd algorithm with migration operator in biogeography-based optimization. Appl. Math. Model. **38**, 2454–2462 (2014)
40. Wang, G.-G., Gandomi, A.H., Alavi, A.H., Deb, S.: A hybrid method based on krill herd and quantum-behaved particle swarm optimization. Neural Comput. Appl. **27**, 989–1006 (2016)
41. Wang, G.-G., Deb, S., Gandomi, A.H., Alavi, A.H.: Opposition-based krill herd algorithm with Cauchy mutation and position clamping. Neurocomputing **177**, 147–157 (2016)

42. Wang, G.-G., Gandomi, A.H., Alavi, A.H., Deb, S.: A multi-stage krill herd algorithm for global numerical optimization. Int. J. Artif. Intell. Tools **25**, 1550030 (2016)
43. Wang, G.-G., Gandomi, A.H., Alavi, A.H., Gong, D.: A comprehensive review of krill herd algorithm: variants, hybrids and applications. Artif. Intell. Rev. (2017)
44. Wang, G.-G., Deb, S., Cui, Z.: Monarch butterfly optimization. Neural Comput. Appl. (2015)
45. Wang, G.-G., Zhao, X., Deb, S.: A novel monarch butterfly optimization with greedy strategy and self-adaptive crossover operator. In: 2015 2nd International Conference on Soft Computing & Machine Intelligence (ISCMI 2015), pp. 45–50. IEEE (2015)
46. Feng, Y., Wang, G.-G., Deb, S., Lu, M., Zhao, X.: Solving 0–1 knapsack problem by a novel binary monarch butterfly optimization. Neural Comput. Appl. (2015)
47. Ghetas, M., Yong, C.H., Sumari, P.: Harmony-based monarch butterfly optimization algorithm. In: 2015 IEEE International Conference on Control System, Computing and Engineering (ICCSCE), pp. 156–161. IEEE (2015)
48. Wang, G.-G., Deb, S., Zhao, X., Cui, Z.: A new monarch butterfly optimization with an improved crossover operator. Oper. Res.: Int. J. (2016)
49. Feng, Y., Yang, J., Wu, C., Lu, M., Zhao, X.-J.: Solving 0–1 knapsack problems by chaotic monarch butterfly optimization algorithm. Memetic Comput. (2016)
50. Ghanem, W.A.H.M., Jantan, A.: Hybridizing artificial bee colony with monarch butterfly optimization for numerical optimization problems. Neural Comput. Appl. (2016)
51. Wang, G.-G., Hao, G.-S., Cheng, S., Qin, Q.: A discrete monarch butterfly optimization for Chinese TSP problem. In: Tan, Y., Shi, Y., Niu, B. (eds.) Advances in Swarm Intelligence: 7th International Conference, ICSI 2016, Bali, Indonesia, June 25-30, 2016, Proceedings, Part I, vol. 9712, pp. 165–173. Springer International Publishing, Cham (2016)
52. Feng, Y., Wang, G.-G., Li, W., Li, N.: Multi-strategy monarch butterfly optimization algorithm for discounted {0–1} knapsack problem. Neural Comput. Appl. (2017)
53. Wang, G., Guo, L., Wang, H., Duan, H., Liu, L., Li, J.: Incorporating mutation scheme into krill herd algorithm for global numerical optimization. Neural Comput. Appl. **24**, 853–871 (2014)

Particle Swarm Optimization

A Scalability Analysis of Particle Swarm Optimization Roaming Behaviour

Jacomine Grobler[1]([⊠]) and Andries P. Engelbrecht[2]

[1] Department of Industrial and Systems Engineering, University of Pretoria,
Pretoria, South Africa
jacomine.grobler@gmail.com
[2] Department of Computer Science, University of Pretoria,
Pretoria, South Africa
engel@cs.up.ac.za

Abstract. This paper investigates the effect of problem size on the roaming behaviour of particles in the particle swarm optimization (PSO) algorithm. Both the extent and impact of the roaming behaviour in the absence of boundary constraints is investigated, as well as the PSO algorithm's ability to find good solutions outside of the area in which particles are initialized. Four basic PSO variations and a diverse set of real parameter benchmark problems were used as basis for the investigation. Problem size was found to have a significant impact on algorithm performance and roaming behaviour. The larger the problem is that is being considered, the more important it is to address roaming behaviour.

1 Introduction

The tendency of particles in a particle swarm optimization (PSO) algorithm to roam beyond the boundary constraints of a search space has already been identified and analyzed [1,2]. Various strategies for constraining particles within the feasible bounds of a search space have also been developed [3–5]. This roaming behaviour can, however, be both disadvantageous, due to the computational inefficiency of searching infeasible areas, but advantageous when the feasible bounds of the problem are unknown and particles are initialized in an area not containing the optimal solution.

This paper describes an investigation into the roaming behaviour of particles in the PSO algorithm. The specific focus of the investigation is how this roaming behaviour affects algorithm performance as problem size increases. Both the extent of the roaming behaviour in the absence of boundary constraints is investigated, as well as the PSO algorithm's ability to find good solutions outside of the area in which particles are initialized. Four basic PSO variations, namely the *gbest* PSO [6], the constriction PSO (CPSO) [7,8], the barebones PSO (BBPSO) [10], and the guaranteed convergence PSO (GCPSO) [9], was used as basis for the investigation. The roaming behaviour of these algorithms were analyzed over a varied set of real parameter benchmark problems. A positive correlation was observed between the extent of roaming and problem size,

© Springer International Publishing AG 2017
Y. Tan et al. (Eds.): ICSI 2017, Part I, LNCS 10385, pp. 119–130, 2017.
DOI: 10.1007/978-3-319-61824-1_13

which both negatively affects algorithm performance. This paper is considered significant because, to the best of the authors' knowledge, it documents the first investigation into the effect of scalability on particle roaming behaviour in an attempt to investigate why PSO scales badly as dimensions increase.

The rest of the paper is organized as follows: Sect. 2 provides background information on the PSO algorithms utilized in this study. Section 3 describes the experimental setup and results obtained. Finally, the paper is concluded in Sect. 4.

2　Particle Swarm Optimization Algorithms

The four PSO algorithms used as basis for this study were selected due to their popularity in the PSO community and to maintain continuity between this paper and the previous work of Engelbrecht [1,2]. The *gbest* PSO, CPSO, BBPSO and the GCPSO are summarized throughout the rest of this section.

2.1　The *Gbest* PSO Algorithm

The *gbest* [6] model calculates the velocity of particle i in dimension j at time $t + 1$ using

$$v_{ij}(t+1) = wv_{ij}(t) + c_1 r_{1j}(t)[\hat{x}_{ij}(t) - x_{ij}(t)] + c_2 r_{2j}(t)[x_j^*(t) - x_{ij}(t)] \quad (1)$$

where $v_{ij}(t)$ represents the velocity of particle i in dimension j at time t, c_1 and c_2 are the cognitive and social acceleration constants, $\hat{x}_{ij}(t)$ and $x_{ij}(t)$ respectively denotes the personal best position (*pbest*) and the position of particle i in dimension j at time t. $x_j^*(t)$ denotes the global best position (*gbest*) in dimension j, w refers to the inertia weight, and $r_{1j}(t)$ and $r_{2j}(t)$ are sampled from a uniform random distribution, $U(0,1)$. The displacement of particle i at time t is simply derived from the calculation of $v_{ij}(t+1)$ in Eq. (1) and is given as

$$x_{ij}(t+1) = x_{ij}(t) + v_{ij}(t+1) \quad (2)$$

2.2　The Constriction PSO Algorithm

The CPSO algorithm was developed by Clerc [7,8] as an alternative approach to balancing the exploration-exploitation trade-off of the PSO algorithm. The velocity of particle i is calculated as

$$v_{ij}(t+1) = \chi[v_{ij}(t) + \phi_1(\hat{x}_{ij}(t) - x_{ij}(t)) + \phi_2(x_j^*(t) - x_{ij}(t))] \quad (3)$$

where χ is defined as

$$\chi = \frac{2\kappa}{|2 - \phi - \sqrt{\phi(\phi - 4)}|} \quad (4)$$

and $\phi = \phi_1 + \phi_2$, $\phi_1 = c_1 r_1$, $\phi_2 = c_2 r_2$, $\phi \geq 4$ and $\kappa \in [0,1]$.

2.3   The Barebones PSO Algorithm

In the BBPSO algorithm [10], the velocity of particle i is sampled from a Gaussian distribution as follows:

$$v_{ij}(t+1) \sim N(\frac{\hat{x}_{ij}(t) + x_j^*(t)}{2}, \sigma),$$

(5)

where $\sigma = |\hat{x}_{ij}(t) - x_j^*(t)|$ and $x_{ij}(t+1) = v_{ij}(t+1)$.

2.4   The Guaranteed Convergence PSO Algorithm

Unfortunately, it has been shown that the basic PSO swarm can stagnate on a solution which is not necessarily a local optimum [9]. The GCPSO algorithm [9] has been shown to address this PSO issue and requires that different velocity and displacement updates, defined as

$$v_{\tau j}(t+1) = -x_{\tau j}(t) + x_j^*(t) + wv_{\tau j}(t) + \rho(t)(1 - 2r_j(t))$$

(6)

and

$$x_{\tau j}(t+1) = x_j^*(t) + wv_{\tau j}(t) + \rho(t)(1 - 2r_j(t)),$$

(7)

are applied to the global best particle, where $\rho(t)$ is a time-dependent scaling factor, $r_j(t)$ is sampled from a uniform random distribution, $U(0,1)$, and all other particles are updated by means of Eqs. (1) and (2).

3   Empirical Evaluation of PSO Roaming Behaviour

The roaming behaviour of the four basic PSO variations were evaluated on the 2015 IEEE Congress of Evolutionary Computation benchmark problem set [11] in 10, 30, 50 and 100 dimensions. Initial algorithm control parameters were selected as specified in [2]. The number of particles in a swarm, n_s, was set to 30, the maximum number of iterations, I_{max}, was $10000n_x$ (as specified in the benchmark problem set definition [11]), where n_x denotes the number of dimensions of each problem, w was set to 0.729844 and c_1 and c_2 was set to 1.49618 for the *gbest* PSO, BBPSO and GCPSO algorithms. In the CPSO algorithm, c_1 and c_2 were set to 2.05 and κ was set to 1 to be equivalent to the parameters of the other three PSO algorithms.

Two sets of experiments were conducted and are described within the next two subsections. The first experiment investigated the extent of particle roaming outside of the feasible bounds of the problem, while the second focused on how this behaviour could be beneficial for problems where the optimal solution lies outside of the area in which the swarm is initialized.

3.1   Investigation of Particle Roaming Behaviour Outside
of the Feasible Search Space

To investigate the extent and effect of particle roaming, particles were initialized randomly in the feasible search space. However, during the search no boundary constraint mechanism was applied and particles were allowed to roam outside of the feasible search space. The performance of each algorithm on each benchmark problem-dimension combination was recorded over 30 independent simulation runs. The solution quality (measured as the difference between the optimal solution and the fitness function value obtained) and the population diversity of each algorithm, were recorded at the end of each optimization run. Diversity is calculated as

$$Diversity = \frac{1}{n_s} \sum_{i=1}^{n_s} \sqrt{\sum_{j=1}^{n_x} (x_{ij}(t) - \overline{x}_j(t))^2} \tag{8}$$

where $\overline{x}_j(t)$ is the mean of the jth dimension of all particles in the swarm at time t [12].

Finally, the percentage of time per optimization run that at least 90% of the particle positions violate the boundary constraints, τ_{90}, the percentage of time at which at least 90% of the *pbest* positions violate the boundary constraints, $\tau Pbest_{90}$, and the number of iterations at which the *gbest* position of the swarm violates the boundary constraints, $\tau Gbest$, were also recorded. A particle was considered to have violated the boundary constraint if at least one of its dimensions fell outside of the feasible space. The results of the roaming behaviour of the four PSO algorithms are presented in Tables 1, 2, 3 and 4. The notation, μ and σ, denote the mean and standard deviation associated with the corresponding metric.

For each of the four PSO variations, each dimension-problem combination was compared by means of statistical tests with every other dimension-problem combination. For every comparison, a Mann-Whitney U test at 95% significance was performed (using the two sets of 30 data points of the two dimension-problem combinations under comparison) and if the first combination statistically significantly outperformed the second combination, a win was recorded. If no statistical difference could be observed a draw was recorded. If the second combination outperformed the first combination, a loss was recorded for the first combination. As an example, (38-52-0) in row 1 column 1, indicates that when the results of the *gbest* PSO on the 10-dimensional problems were compared to the results obtained by the PSO on all other dimensions, 38 of the comparisons showed that statistically significantly better performance was obtained by the PSO algorithm on the 10 dimension problems. The number of statistically similar comparisons are 52 and the 0 losses indicate that PSO never performed worse on a 10-dimensional problem when compared to the other dimensions.

From the results it can be clearly seen that as the problem size increase, solution quality deteriorates. This can be largely attributed to the fact that the larger the problem, the larger the percentage of time that is spent by the particles outside of the feasible search space. It was observed that as problem size increases, τ_{90}, $\tau Pbest_{90}$ and $\tau Gbest$ also increases. This conclusion is logical since the more time is wasted in searching in infeasible areas of the search space, the less effective an algorithm will be. Visual checks of the optimization runs indicated that for most of the 10-dimensional problems the particles roam outside

Table 1. PSO roaming behaviour on the CEC 2015 benchmark problem set.

Problem	Dimension	Error		Diversity		τ_{90}		$\tau Pbest_{90}$		$\tau Gbest$	
		μ	σ	μ	σ	μ	σ	μ	σ	μ	σ
1	10	17062	34094	33.09	24.68	37.88	37.05	37.02	37.42	40.32	36.19
	30	$4.2246e+05$	$2.5631e+05$	165.69	74.2	89.91	25.67	89.9	25.67	89.96	25.53
	50	$1.3762e+06$	$7.3448e+05$	159.54	68.58	99.99	0	99.99	0.01	99.99	0.01
	100	$4.4965e+06$	$2.5712e+06$	184.34	42.33	100	0	99.99	0.01	100	0
2	10	9579.5	10693	0.2	0.13	3.37	18.18	3.31	18.11	3.43	18.22
	30	8301	15174	0.11	0.11	13.57	34.31	13.45	34.3	13.62	34.35
	50	15669	17224	0.17	0.14	13.74	34.26	13.7	34.28	13.79	34.31
	100	56452	$2.3234e+05$	0.07	0.08	20.55	40.1	20.54	40.09	20.57	40.11
3	10	20.12	0.14	163.82	104.14	88	30.22	86.76	29.84	93.02	25.29
	30	20.65	0.26	921.92	2211	99.98	0.01	99.83	0.17	99.96	0.06
	50	20.57	0.42	674.29	939.83	99.99	0	99.93	0.06	99.98	0.02
	100	21.07	0.4	1584.6	1468.7	100	0	99.98	0.06	99.99	0.01
4	10	15.25	7.03	0.7	1.67	0.04	0.06	0	0	0.22	0.38
	30	119.43	32.97	0	0	73.22	44.35	72.84	44.16	73.44	44.49
	50	277.76	62.17	0	0	80.09	40.32	80.02	40.4	80.13	40.31
	100	873.33	155.21	0	0	99.99	0.02	99.98	0.07	100	0.01
5	10	498.59	245.62	12.95	35.38	82.94	36.04	82.22	35.67	83.95	35.87
	30	3243	637.71	31.77	51.77	96.64	18.16	96.44	18.22	96.6	18.19
	50	6164.9	825.41	42.65	82.29	96.68	18.16	96.62	18.17	96.61	18.23
	100	14787	1560.4	81.96	130.47	100	0	99.97	0.03	99.99	0.02
6	10	3845.5	5000.6	14.23	31.67	93.75	21.56	93.55	21.12	94.07	21.6
	30	$1.0839e+05$	66003	450.47	396.68	99.98	0.01	99.97	0.03	99.99	0.01
	50	$2.8431e+05$	$1.5845e+05$	1747.6	1069.3	99.99	0	99.98	0.01	99.99	0.01
	100	$1.7075e+06$	$8.7922e+05$	6457.3	2645.8	100	0	100	0	100	0
7	10	1.14	0.52	$1.8013e+06$	$9.8361e+06$	85.8	32.66	85.58	32.72	86.1	32.7
	30	7.37	1.92	$2.8552e+40$	$1.5639e+41$	99.97	0.02	99.96	0.03	99.97	0.03
	50	18.06	3.26	$1.6288e+73$	$8.9211e+73$	99.99	0	99.99	0.01	99.99	0
	100	53.39	22.25	$4.1647e+115$	$2.253e+116$	100	0	100	0	100	0
8	10	2976.9	3058.9	$6.935e+09$	$3.2649e+10$	78.46	28.39	73.83	32.24	83.05	25.43
	30	42001	38152	$3.9474e+41$	$1.7239e+42$	99.97	0.05	99.94	0.08	99.97	0.05
	50	$1.3486e+05$	84915	$6.1947e+41$	$3.393e+61$	99.99	0	99.98	0.01	99.99	0.01
	100	$6.7941e+05$	$3.7091e+05$	$1.4317e+80$	$7.8416e+80$	100	0	99.99	0	100	0
9	10	100.34	0.28	3.65	2.83	28.41	43.77	28.2	43.77	30.36	46.05
	30	141.03	86.54	1.09	1.07	17.06	37.57	17.03	37.57	17.13	37.65
	50	150.08	135.75	1.27	1.4	17.73	37.39	17.72	37.35	17.83	37.39
	100	331.75	445.99	0.54	0.61	93.32	24.98	93.3	24.98	93.41	24.99
10	10	1874.3	1740.8	23.7	70.46	42.11	43.29	41.43	43.63	46.64	41.63
	30	45190	38342	1896.9	2489.3	99.98	0.01	99.96	0.03	99.98	0.02
	50	3027.1	1567.9	7647.8	3040.7	99.99	0	99.98	0.02	99.99	0.01
	100	22089	13354	8570.6	2990	100	0	99.99	0	100	0
11	10	311.15	79.5	10.22	10.32	91.57	20.71	90.27	21.18	92.92	20.39
	30	869.84	173.42	34.59	152.94	99.98	0	99.98	0.02	99.99	0
	50	1418.4	244.43	3.46	18.98	99.99	0	99.99	0.01	99.99	0.01
	100	3176	166.26	0	0	100	0	100	0	100	0
12	10	102.7	0.96	2.76	2.03	0.07	0.1	0.01	0.03	0.22	0.34
	30	141.96	44.93	51.1	104.68	73.73	43.4	72.92	44.36	76.8	42.68
	50	197.03	16.29	287.91	786.72	99.99	0	99.99	0	99.99	0
	100	200	0	203.36	309.63	100	0	100	0	100	0
13	10	0	0	37.98	62.97	99.94	0.01	99.94	0.01	99.97	0.01
	30	0	0	87.63	135.42	99.99	0.01	99.98	0	99.99	0
	50	0	0	59.52	89.61	99.99	0	99.99	0	99.99	0
	100	0	0	115.99	173.98	100	0	100	0	100	0
14	10	1660.5	2573.8	29.7	40.28	99.86	0.24	99.8	0.29	99.91	0.2
	30	14487	10839	642.04	869.64	99.98	0	99.98	0	99.99	0
	50	32791	31840	698.38	780.03	99.99	0	99.99	0	99.99	0
	100	29060	51386	3085.8	1662.7	100	0	100	0	100	0
15	10	100	0	0	0	0.05	0.05	0	0.01	0.25	0.22
	30	100	0	0	0	0.75	0.48	0.54	0.51	0.88	0.47
	50	108.53	3.48	1.68	0.95	17.2	34.1	17.2	34.52	17.63	34.6
	100	135.01	6.39	0.45	0.39	59.3	47.84	59.27	47.82	59.51	47.8

Table 2. CPSO roaming behaviour on the CEC 2015 benchmark problem set.

Problem	Dimension	Error μ	σ	Diversity μ	σ	τ_{90} μ	σ	$\tau Pbest_{90}$ μ	σ	$\tau Gbest$ μ	σ
1	10	17669	23822	37.96	27.79	40.94	41.78	40.26		42.76	40.79
	30	$4.9832e+05$	$4.1417e+05$	161.19	55.24	95.42	15.66	95.54		95.54	15.65
	50	$1.2455e+06$	$7.5288e+05$	145.45	48.92	99.87	0.68	99.86		99.87	0.68
	100	$3.9156e+06$	$1.8412e+06$	183.84	40.89	100	0	99.99		100	0
2	10	13511	14867	0.27	0.19	16.64	37.96	16.53		16.71	37.79
	30	10677	13704	0.15	0.1	26.82	44.72	26.72		26.87	44.76
	50	14509	16543	0.14	0.09	7.13	25.07	7.07		7.16	25.08
	100	9061.5	13505	0.06	0.05	17.34	37.37	17.33		17.36	37.37
3	10	20.13	0.12	421.98	721.37	79.85	32.59	71.17		91.1	24.87
	30	20.51	0.38	417.55	418.06	99.09	4.77	99.52		99.9	0.39
	50	20.71	0.46	577.5	303.06	99.99	0	99.94		99.98	0.03
	100	20.93	0.44	1091.7	808.3	100	0	99.98		99.99	0.03
4	10	14.33	6.57	0.15	0.61	8.9	25.73	8.31		9.66	26.35
	30	129.38	36.07	0	0	66.79	47.62	66.59		66.82	47.64
	50	305.25	68.52	0	0	93.25	25.3	93.14		93.28	25.31
	100	892.17	121.01	0	0	100	0	99.99		100	0
5	10	411.4	220.09	14.92	16.48	85.05	34.04	83.9		85.78	34.2
	30	3330.1	631.06	18.72	24.39	96.66	18.13	96.5		96.66	18.09
	50	6422.5	809.44	22.07	40.04	99.99	0.01	99.93		99.97	0.04
	100	15000	1458.8	55.77	102.26	100	0	99.97		99.99	0.03
6	10	2959	3614.5	10.69	15.9	92.94	23.52	92.32		93.36	23.34
	30	91820	95278	536.13	442.93	99.98	0.01	99.97		99.99	0.01
	50	$3.5433e+05$	$2.1517e+05$	1868.5	946.87	99.99	0	99.99		99.99	0.01
	100	$1.9347e+06$	$9.7443e+05$	6235.9	1990.7	100	0	99.99		100	0
7	10	1.2	0.35	$4.2757e+14$	$2.3419e+15$	67.23	46.65	66.92		67.57	46.44
	30	7.88	1.86	$1.0194e+45$	$5.5837e+45$	99.98	0.01	99.96		99.98	0.02
	50	22.06	17.08	$2.0865e+73$	$1.1304e+74$	99.99	0	99.99		99.99	0
	100	65.63	28.93	$1.5733e+121$	$8.4558e+121$	100	0	100		100	0
8	10	2224.7	2897.8	$1.8593e+10$	$9.1177e+10$	77.22	30.8	72.5		80.58	29.79
	30	43409	44200	$4.6417e+50$	$2.5423e+51$	99.98	0.01	99.94		99.98	0.02
	50	$1.4159e+05$	86677	$4.2033e+58$	$1.5987e+59$	99.99	0	99.98		99.99	0.01
	100	$7.0798e+05$	$2.9761e+05$	$3.5087e+81$	$1.9218e+82$	100	0	99.99		100	0
9	10	104.38	22.33	4.67	3.83	20.65	37.42	20.26		23.48	39.74
	30	137.47	78.6	1.24	1.55	17.14	37.67	17.1		17.16	37.66
	50	138.28	120.49	1.09	0.8	7.47	25.15	7.45		7.51	25.14
	100	181.42	256.69	0.71	0.56	99.77	1.14	99.77		99.85	0.77
10	10	1917.5	1798.5	10.13	7.23	37.87	42.63	37.19		41.97	41.49
	30	74317	90309	2039.7	2631.8	99.98	0.01	99.94		99.97	0.04
	50	3899.9	3021.6	6328.4	2218.6	99.99	0	99.98		99.99	0.01
	100	24407	17471	7437.1	3055.2	100	0	99.99		100	0
11	10	296.39	115.07	17.28	20.35	86.91	29.85	85.76		87.99	30.03
	30	866.71	88.61	12.75	69.81	99.98	0	99.98		99.99	0
	50	1463.5	125.86	7.93	43.41	99.99	0	99.99		99.99	0
	100	3156.4	175.93	4.53	24.8	100	0	100		100	0
12	10	102.6	1.07	4.57	4.5	6.24	20.75	4.28		7.99	25.73
	30	142.31	44.66	94.66	422.92	66.83	47.52	66.77		67.01	47.38
	50	191.16	26.97	157.05	629.46	99.99	0.03	99.99		99.99	0
	100	123.89	3.1	4.89	2.43	100	0	100		100	0
13	10	0	0	25.49	44.18	99.94	0	99.94		99.97	0.01
	30	0	0	51.44	69.36	99.99	0.01	99.98		99.99	0
	50	0	0	70.01	113.49	99.99	0	99.99		99.99	0
	100	460.26	6.89	66.16	113.55	100	0	100		100	0
14	10	1556.7	2326.6	40.3	51.38	99.88	0.1	99.84		99.93	0.08
	30	20154	8132.5	214.7	559.95	99.98	0.01	99.98		99.99	0
	50	36369	31740	630	743.78	99.99	0	99.99		99.99	0
	100	41637	58767	2563.4	1745.6	100	0	100		100	0
15	10	100	0	0	0	0.06	0.08	0.01		0.3	0.23
	30	100	0	0	0	0.69	0.46	0.49		0.84	0.46
	50	108.14	3.49	1.28	0.92	17.58	37.25	17.44		17.8	37.23
	100	160.05	22.69	0.43	0.52	58.88	48.11	58.82		59.02	48.09

Table 3. BBPSO roaming behaviour on the CEC 2015 benchmark problem set.

Problem	Dimension	Error		Diversity		τ_{90}		$\tau Pbest_{90}$		$\tau Gbest$	
		μ	σ	μ	σ	μ	σ	μ	σ	μ	σ
1	10	80950	$1.0379e+05$	76.07	58.58	68.73	38	68.97	38.17	70.45	36.72
	30	$2.3797e+06$	$2.0464e+06$	346.1	200.79	99.59	2.21	99.5	2.19	99.58	2.17
	50	$4.3121e+06$	$1.9701e+06$	312.83	104.89	99.99	0	99.87	0.1	99.98	0.01
	100	$1.2729e+07$	$4.347e+06$	439.83	100.25	100	0	99.34	0.65	99.96	0.02
2	10	14942	15580	0.12	0.11	23.28	42.76	23.15	42.65	23.38	42.83
	30	13110	17475	0.06	0.05	30.38	46.17	30.13	46.17	30.39	46.2
	50	15859	20248	0.04	0.02	10.99	30.06	10.58	30.03	10.96	30.07
	100	10053	12839	0.01	0.01	21.57	39.86	20.72	39.84	21.44	39.84
3	10	20.34	0.09	2385.6	4490.1	99.86	0.24	99.5	0.63	99.88	0.2
	30	20.96	0.03	6017	5918.4	99.99	0	99.89	0.07	99.9	0.26
	50	21.13	0.03	10715	8898.7	99.99	0	99.95	0.03	99.97	0.04
	100	21.32	0.03	29678	34575	100	0	99.97	0.04	99.99	0
4	10	15.22	8.32	3.14	5.87	10.81	30.13	10.12	29.78	11.29	30.25
	30	130.04	38.43	0.05	0.26	83.31	37.49	82.83	37.56	83.4	37.4
	50	297.59	88.83	0.07	0.4	93.42	24.9	92.93	24.88	93.35	24.9
	100	1024.4	213.49	0	0	100	0	98.07	1.42	99.81	0.1
5	10	393.05	189.2	60.37	39.91	91.74	23.87	92.15	22.05	92.66	22.75
	30	4019.3	1780.9	511.35	452.1	99.99	0	99.86	0.11	99.95	0.07
	50	8203.5	2993.7	799.93	613.24	99.99	0.04	99.9	0.07	99.9	0.29
	100	20882	8391.4	1611.8	1099.1	100	0	99.92	0.1	99.95	0.1
6	10	4380.7	7801.8	54.38	77.26	93.89	23.06	93.8	23.08	93.89	23.09
	30	$2.6208e+05$	$2.9964e+05$	1450.8	903.88	99.99	0	99.94	0.02	99.98	0.02
	50	$1.0019e+06$	$6.8224e+05$	5331.7	2750.8	99.99	0	99.81	0.65	99.98	0.01
	100	$4.9145e+06$	$2.1017e+06$	32859	13435	100	0	99.95	0.04	99.99	0.01
7	10	1.52	0.22	$9.0603e+42$	$4.9625e+43$	99.92	0.07	99.82	0.07	99.93	0.05
	30	19.01	17.74	Inf	NaN	99.99	0	99.94	0.02	99.98	0.01
	50	67.54	45.54	$1.0303e+10$	$3.8756e+10$	99.99	0	99.94	0.03	99.98	0.01
	100	144.84	46.06	$1.2941e+17$	$6.6078e+17$	100	0	99.91	0.04	99.98	0.01
8	10	2910.9	3533.7	$1.4246e+22$	$7.7989e+22$	99.09	4.02	98.52	6.01	99.31	3.02
	30	87787	68899	$1.1637e+113$	$6.3738e+113$	99.99	0	99.91	0.04	99.97	0.02
	50	$3.7305e+05$	$2.7849e+05$	$2.401e+134$	$1.3151e+135$	99.99	0	99.89	0.08	99.98	0.02
	100	$1.9521e+06$	$1.0707e+06$	Inf	NaN	100	0	99.91	0.06	99.99	0.01
9	10	100.26	0.21	9.98	3.89	21.5	36.26	21.53	36.75	29.31	43.37
	30	132.04	76.65	7.45	4.24	14.04	34.29	13.99	34.28	14.06	34.27
	50	145.54	127.07	14.71	6.99	14.86	33.79	14.91	34	14.98	33.97
	100	302.42	437.84	24.04	13.76	84.92	29.06	84.59	30.92	90.12	24.62
10	10	1839.2	1661.3	130.23	130.64	68.59	42.41	67.12	44.01	72.97	39.17
	30	$3.3715e+05$	$3.8031e+05$	7368.1	2457.8	99.99	0	99.93	0.04	99.98	0.02
	50	$1.7784e+05$	$1.6072e+05$	9924.6	1690.7	99.99	0	99.93	0.04	99.98	0.02
	100	$4.4356e+05$	$1.4941e+06$	18177	1987.9	100	0	99.95	0.03	99.99	0.01
11	10	341.96	98.68	46.79	46.56	93.71	18.33	94.26	18.15	95.55	18.17
	30	900.17	81.15	438.79	615.32	99.99	0	99.97	0.01	99.99	0
	50	1445.1	183.82	1158.7	1339	99.99	0	99.98	0	99.99	0
	100	3350.7	351.28	2978.5	2135.8	100	0	99.99	0	100	0
12	10	106.08	17.77	46.42	222.09	8.25	25.36	8.26	25.29	9.95	28.46
	30	193.84	23.43	1178.6	822.56	96.68	18.05	96.66	18.05	96.69	18.05
	50	200	0	2299.1	1310.3	99.99	0	99.99	0	99.99	0
	100	200	0	4823.8	2094.3	100	0	100	0	100	0
13	10	0	0	232.28	218.44	99.95	0.02	99.94	0.01	99.97	0
	30	0	0	867.22	550.97	99.99	0	99.98	0	99.99	0
	50	0	0	1663.7	973.26	99.99	0	99.99	0	99.99	0
	100	0	0	2540.3	1465.2	100	0	99.99	0	100	0
14	10	692.05	1236.8	233.17	168.19	99.87	0.25	99.72	0.39	99.87	0.25
	30	9841.9	11611	1494.5	1181.3	99.99	0	99.99	0.01	99.99	0
	50	26441	31411	923.5	785.71	99.99	0	99.99	0	99.99	0
	100	45499	58057	3596.6	2652.6	100	0	100	0	100	0
15	10	100	0	0	0	0.16	0.21	0.03	0.1	0.32	0.31
	30	100	0	0.42	2.28	1.29	0.66	0.77	0.71	1.44	0.68
	50	100.94	0.86	10.32	10.91	2.71	1.1	1.73	1.25	2.86	1.06
	100	114.81	4.38	76.38	46.02	17.82	32.55	16.21	32.44	18.16	32.57

Table 4. GCPSO roaming behaviour on the CEC 2015 benchmark problem set.

Problem	Dimension	Error		Diversity		$\tau 90$		$\tau Pbest_{90}$		$\tau Gbest$	
		μ	σ	μ	σ	μ	σ	μ	σ	μ	σ
1	10	16595	15953	15.12	11.44	45.04	43.76	44.47	43.47	46.42	43.65
	30	$4.0601e+05$	$2.1156e+05$	116.96	34.34	89.99	26.11	89.99	26.07	90.01	26.05
	50	$1.1714e+06$	$5.5314e+05$	129.03	49.78	99.99	0.03	99.99	0.01	99.99	0.01
	100	$2.398e+06$	$5.9582e+05$	185.08	30.83	100	0	99.99	0.01	100	0.01
2	10	13471	18121	0.12	0.1	23.23	42.74	23.15	42.68	23.34	42.91
	30	11436	14033	0.05	0.04	26.72	44.72	26.63	44.72	26.76	44.76
	50	14812	23510	0.02	0.02	3.59	18.21	3.56	18.21	3.62	18.2
	100	6956.4	7584.4	0	0	7.08	25.26	7.06	25.26	7.13	25.24
3	10	20	0	132.52	65.96	81.12	35.41	77.74	38.34	86.57	34.42
	30	20	0	212.9	187.22	89.4	30.35	89.8	30.25	89.9	30.48
	50	20	0	305.98	223.76	93.41	25.07	93.46	24.62	93.32	25.37
	100	20.01	0.07	469.77	359.02	96.76	17.72	96.76	17.61	96.66	18.26
4	10	18.57	7.7	0.59	1.19	12.77	33.09	12.39	32.31	13.33	34.27
	30	167.95	42.66	0	0	66.48	47.59	66.25	47.51	66.7	47.62
	50	422.72	78.08	0	0	89.81	30.22	89.69	30.34	89.86	30.16
	100	1084.2	146.29	0	0	93.34	25.22	93.33	25.22	93.35	25.21
5	10	510.52	198.91	7.53	9.88	55.56	49.38	54.62	48.78	55.9	49.66
	30	3540.8	642.03	11.55	33.48	83.38	37.65	83.25	37.66	83.36	37.69
	50	6637.3	766.5	11.34	23.11	93.34	25.28	93.28	25.3	93.29	25.36
	100	15283	1276.2	14.64	21.02	99.99	0.04	99.96	0.05	99.96	0.1
6	10	3373.6	3156	4.58	7.34	88.33	30.34	88.19	30.15	89.71	30.24
	30	54712	55002	447.46	456.72	99.96	0.1	99.95	0.11	99.97	0.1
	50	$1.2871e+05$	65643	898.11	575.13	99.99	0.01	99.98	0.01	99.99	0.02
	100	$5.2331e+05$	$2.5866e+05$	1050.9	848.63	100	0	100	0	100	0
7	10	1.19	0.41	3.61	5.33	74.11	41.99	73.91	41.9	74.63	42.15
	30	8.33	1.95	$2.7435e+26$	$1.0982e+27$	99.97	0.03	99.95	0.06	99.97	0.03
	50	21.51	14.41	$6.6735e+33$	$3.6536e+34$	99.99	0	99.99	0.01	99.99	0.01
	100	96.3	42.01	$1.499e+13$	$8.2104e+13$	100	0	100	0	100	0
8	10	2698.3	3494.3	$2.8048e+09$	$1.4755e+10$	69.16	38.59	66.98	39.19	71.76	38.14
	30	27797	19340	$2.1453e+21$	$1.175e+22$	99.94	0.19	99.91	0.23	99.94	0.18
	50	$1.1006e+05$	94116	$2.0351e+06$	$1.0269e+07$	99.99	0.03	99.96	0.06	99.98	0.03
	100	$2.1423e+05$	89228	1212	950.85	100	0	99.99	0.02	100	0
9	10	100.35	0.27	4.46	4.93	28.24	43.39	27.19	43.56	31.07	44.57
	30	150.06	106.36	1.01	0.84	17.01	37.73	16.98	37.72	17.05	37.72
	50	173.34	175.75	1.54	2.55	17.16	37.58	17.14	37.55	17.2	37.62
	100	440.81	616.86	1.39	1.47	96.48	18.07	96.48	18.08	96.61	18.08
10	10	2545.1	3899.8	10.56	19.89	42.92	44.75	42.89	44.4	44.59	44.36
	30	31542	40952	388	1084	99.88	0.26	99.84	0.3	99.91	0.21
	50	3222.6	4196.3	1936.1	1747.5	99.99	0.01	99.98	0.01	99.99	0.02
	100	7641.6	6612.7	3566.9	2463.2	100	0	99.99	0	100	0
11	10	317.07	52.06	8.23	10.41	96.63	3.48	96.03	3.97	97.93	2.5
	30	878	236.71	7.46	17.73	98.55	5.82	93.43	24.95	99.04	3.67
	50	1507.4	139.07	0	0	99.99	0	99.99	0	99.99	0
	100	3378.5	153.95	7.75	42.45	100	0	100	0	100	0
12	10	103.18	1.22	3.22	3.18	3.19	17.24	3.05	16.69	3.41	18.21
	30	139.49	43.53	76.98	287.18	66.91	47.01	66.2	47.09	67.34	46.99
	50	113.19	1.38	9.78	7.16	98.46	8.32	98.65	7.24	99.99	0.02
	100	120.98	1.43	11.03	7.32	100	0	100	0	100	0
13	10	0	0	31.47	69.25	99.94	0.01	99.93	0.01	99.96	0.01
	30	0	0	61.86	114.68	99.98	0	99.98	0	99.99	0
	50	219.74	5.09	88.44	188.21	99.99	0	99.99	0	99.99	0
	100	464.36	5.86	49.45	108.64	100	0	100	0	100	0
14	10	1812	2763	25.83	40.22	96.42	18.21	96.35	18.2	96.48	18.22
	30	19435	9900.3	359.75	738.69	99.98	0.01	99.98	0	99.99	0
	50	38080	30751	468.6	679.7	99.99	0	99.99	0	99.99	0
	100	24358	47790	2712.6	1273.1	100	0	99.99	0	100	0
15	10	100	0	0	0	0.08	0.1	0	0.01	0.3	0.25
	30	100	0	0	0	0.48	0.36	0.37	0.37	0.55	0.37
	50	100.13	0.71	0	0	0.73	0.43	0.61	0.44	0.87	0.42
	100	118.41	6.69	5.7	6.81	2.64	6.64	2.57	6.94	4.54	13.49

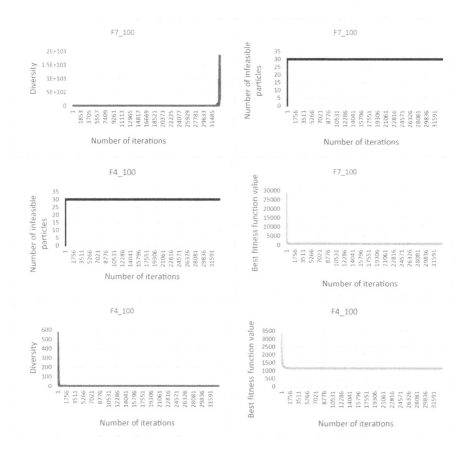

Fig. 1. Plots of key performance metrics for functions 4 and 7 in 100 dimensions (F4_100 and F7_100).

Table 5. Hypotheses analysis of the effect of problem size on various performance metrics.

Dimensions	10	30	50	100	10	30	50	100
	Error				*Diversity*			
PSO	38-52-0	22-54-14	12-54-24	1-54-35	30-53-7	19-60-11	8-61-21	7-58-25
CPSO	39-51-0	22-54-14	13-52-25	2-51-37	25-52-13	18-64-8	10-63-17	7-61-22
BBPSO	38-52-0	22-56-12	11-55-24	0-55-35	35-49-6	25-50-15	15-51-24	7-46-37
GCPSO	36-54-0	21-54-15	14-54-22	2-52-36	22-58-10	16-61-13	11-62-17	10-61-19
TOTAL	151-209-0	87-218-55	50-165-95	5-212-143	112-212-36	78-235-47	44-177-79	22-226-103
	τ_{90}				$\tau Pbest_{90}$			
PSO	43-45-2	31-45-14	16-45-29	0-45-45	43-45-2	31-45-14	16-45-29	0-45-45
CPSO	42-45-3	28-46-16	18-46-26	1-45-44	42-45-3	28-45-17	19-45-26	1-45-44
BBPSO	41-46-3	27-46-17	19-45-26	2-45-43	41-46-3	18-55-17	17-46-27	6-47-37
GCPSO	3-51-36	28-47-15	17-46-27	2-45-43	40-45-5	30-46-14	17-46-27	2-45-43
TOTAL	129-187-44	114-184-62	71-182-108	5-180-135	166-181-13	107-191-62	66-190-104	9-182-169
	$\tau Gbest$				*Error* (**Section 3.2**)			
PSO	43-47-0	30-46-14	15-46-29	0-45-45	43-47-0	24-49-17	12-54-24	1-50-39
CPSO	42-47-1	27-47-16	17-47-26	1-45-44	43-47-0	23-50-17	10-56-24	1-53-36
BBPSO	41-47-2	18-55-17	14-55-21	3-51-36	45-45-0	26-47-17	15-48-27	1-46-43
GCPSO	41-47-2	28-47-15	16-47-27	2-45-43	39-49-2	22-49-19	16-49-25	6-47-37
TOTAL	167-188-5	103-195-62	62-195-103	6-186-168	209-188-2	95-195-70	53-207-100	9-196-155

Table 6. Algorithm comparison on the CEC 2015 benchmark problem set.

Problem	Dimension	PSO		CPSO		BBPSO		GCPSO	
		μ	σ	μ	σ	μ	σ	μ	σ
1	10	$1.90E+07$	$4.85E+07$	$3.30E+07$	$1.25E+08$	$1.01E+06$	$2.75E+06$	11917	7746.3
	30	$1.75E+08$	$2.15E+08$	$2.37E+08$	$3.50E+08$	$6.11E+07$	$1.04E+08$	$1.59E+05$	75939
	50	$4.82E+08$	$7.60E+08$	$4.58E+08$	$6.67E+08$	$1.35E+08$	$1.67E+08$	$9.84E+05$	$4.33E+05$
	100	$5.24E+08$	$5.61E+08$	$5.15E+08$	$5.53E+08$	$7.91E+08$	$5.09E+08$	$3.13E+06$	$6.33E+05$
2	10	$3.99E+09$	$7.64E+09$	$4.89E+09$	$1.28E+10$	$2.41E+08$	$5.86E+08$	20750	14026
	30	$2.38E+10$	$1.64E+10$	$2.58E+10$	$1.75E+10$	$1.06E+10$	$1.04E+10$	11464	4911
	50	$5.09E+10$	$2.60E+10$	$5.97E+10$	$3.88E+10$	$2.64E+10$	$1.34E+10$	24633	19052
	100	$1.27E+11$	$4.86E+10$	$1.27E+11$	$4.93E+10$	$1.16E+11$	$2.92E+10$	7626.8	4671.4
3	10	20.01	0.02	20.01	0.03	20.15	0.13	20	0
	30	20.04	0.07	20.05	0.07	20.59	0.32	20.02	0.04
	50	20.05	0.09	20.09	0.23	21	0.28	20.01	0.02
	100	20.15	0.3	20.13	0.29	21.2	0.32	20.02	0.02
4	10	62.97	35.06	72.25	30.76	25.22	14.12	59.01	26.82
	30	237.64	71.09	299.23	85.51	158.39	37.04	306.23	79.24
	50	635	135.52	693.86	116.26	378.9	88.27	723.42	139.66
	100	1682.9	231.53	1680.5	228	1158.1	185.54	1837.5	288.63
5	10	783.71	252.87	778.43	258.23	586.31	185.93	830.18	256.41
	30	3926.9	462.64	4146.3	592.58	3917.8	486.59	4260	703.33
	50	6957.6	731.51	6787.4	923.93	6727.5	727.59	7535.9	766.15
	100	16424	1241.6	16460	1308.4	16609	1869.8	16749	1082
6	10	$1.13E+06$	$5.96E+06$	$1.49E+05$	$3.66E+05$	7692.6	3294.4	6088.1	3018.8
	30	$2.03E+06$	$5.16E+06$	$4.16E+06$	$7.73E+06$	$2.35E+06$	$5.48E+06$	90466	34628
	50	$1.20E+07$	$1.76E+07$	$3.46E+07$	$5.18E+07$	$7.54E+06$	$8.86E+06$	$1.64E+06$	90896
	100	$2.07E+07$	$2.48E+07$	$2.20E+07$	$3.41E+07$	$2.23E+07$	$1.96E+07$	$3.21E+05$	$1.11E+05$
7	10	16.95	43.66	4.82	8.06	2.44	1.43	3.11	1.62
	30	100.24	93.47	150.27	172.58	33.56	56.92	10.73	3.21
	50	352.33	404.69	341.02	294.17	126.29	88	47.31	12.52
	100	1823.4	1455.8	1792.2	1447.8	1077.4	566.27	135.55	32.27
8	10	3256.7	2208	2864.9	2004	2790.1	2061.8	2805.7	968.58
	30	81713	88759	$2.34E+05$	$5.53E+05$	$2.03E+05$	$2.50E+05$	16297	8318.1
	50	$3.15E+06$	$4.66E+06$	$6.01E+06$	$7.68E+06$	$3.59E+06$	$4.28E+06$	58562	30376
	100	$6.49E+06$	$6.47E+06$	$6.53E+06$	$6.32E+06$	$1.01E+07$	$1.06E+07$	$2.13E+05$	79213
9	10	103.69	9.48	104.9	8.69	100.47	0.61	100.34	0.22
	30	268.16	95.3	226.18	82.04	141.74	32.04	103.36	0.43
	50	367.6	126.91	473.21	143.23	232.22	75.05	105.81	0.64
	100	1649.6	633.3	1640.6	636.54	944.31	502.76	1136.7	1126
10	10	1455.9	1483	1880.6	2042.9	3955.6	7556.5	2604.8	2263.8
	30	$8.43E+06$	$1.79E+06$	$1.98E+06$	$6.64E+06$	$1.88E+06$	$2.91E+06$	29735	7986.5
	50	$4.67E+06$	$7.40E+06$	$3.95E+06$	$5.12E+06$	$5.10E+06$	$5.65E+06$	3523.1	1005.9
	100	$3.02E+08$	$3.48E+08$	$2.93E+08$	$3.42E+08$	$1.50E+08$	$1.87E+08$	6315.4	1252.5
11	10	300.87	1.27	300.97	1.29	300.8	0.54	300.2	0.13
	30	529.78	373.32	492.5	325.63	538.08	322.37	363.71	237.3
	50	475.26	277.15	464.61	329.44	447.73	301.8	417.99	435.54
	100	1414.8	1460.3	1505	1569.6	2200.6	1262.6	1034	1472.2
12	10	110.56	13.78	114.32	15.94	103.86	1.46	104.84	2.64
	30	197.68	10.21	198.18	11.5	194.25	23.58	197.43	17.02
	50	201.07	3.72	196.03	11.79	196.88	13.73	176.48	40.29
	100	204.02	2.05	204.04	2.05	200.76	0.21	201.41	0.73
13	10	1442	1697.3	1028.3	1524.6	0.05	0.12	1679	1639.5
	30	8926.4	3310.5	8358.4	3613.4	815.48	2533	10548	2487.7
	50	1219.3	3771.2	889.55	3374.2	0.21	0.18	3132.2	6150.2
	100	2.05	1.84	2.01	1.82	0.17	0.07	0.41	0.38
14	10	3232.7	992.83	3013.6	264.91	2969.8	43.06	2954.8	20.36
	30	47943	11495	46934	11531	36370	3804.1	32697	1923.6
	50	$1.09E+05$	14434	$1.08E+05$	16146	89281	10152	71650	11005
	100	$2.85E+05$	38901	$2.85E+05$	38899	$1.89E+05$	23271	$1.12E+05$	10945
15	10	2025.2	7016.5	175.66	136.09	104.16	10.14	100	0
	30	17813	33628	4509	10919	295.26	501.69	100	0
	50	40903	65381	57339	81774	5810.7	14819	103.98	2.76
	100	80868	$1.58E+05$	79145	$1.48E+05$	62728	54958	115.52	2.76

of the search space, but are eventually pulled back into the search space as better solutions are found within the constrained area. However, as the problem size increases, the fraction of runs where the particles are pulled back into the search space dramatically decreases. Finally, diversity and roaming behaviour is highly problem dependent. There is evidence of runs where the particles remain outside of the feasible search space due to high levels of diversity still present at the end of the optimization run and other runs where particles have converged to a local optimum outside of the feasible search space. See Fig. 1 for two examples.

3.2 Investigation of Particle Roaming Behaviour to Find Feasible Solutions Outside of the Initialization Space

This second investigation focused on how roaming behaviour could be beneficial for problems where the optimal solution lies outside of the area in which the swarm is initialized. The CEC 2015 problems and the four PSO variations were again used for evaluation purposes. All particles were initialized in the range $[50, 100]^{n_x}$, which translates to 25% of the problem domain and all particles were bound in the range $[-100, 100]^{n_x}$ (the problem domain) to ensure that no iterations are wasted on searching infeasible areas of the search space. As soon as a particle became infeasible, the position of the infeasible dimension was set to the boundary constraint value and the associated dimension of the particle velocity was set to zero. The purpose of the exercise was to determine the influence of problem size on algorithm performance when the optimal solution does not necessarily lie within the initialization space. The results are indicated in Tables 5 and 6.

It is again evident that algorithm performance degrades as problem dimensions increase. This was the case for 83% of the problems tested. This conclusion is further supported by the 209 times that the PSO algorithms statistically significantly outperformed the other dimension-algorithm combinations for the 10 dimensional case. Overall, the GCPSO algorithm was the best performing PSO variation and BBPSO the second best. BBPSO performed especially well on problems 4, 5, 12 and 13. In summary, the PSO algorithms were able to find good solutions outside of the initialization ranges due to roaming, but became less efficient as the dimensions increased.

4   Conclusion

This paper described an investigation into the roaming behaviour of particles in the PSO algorithm and how this behaviour is influenced by different sized problems. Both the extent of the roaming behaviour in the absence of boundary constraints, as well as the PSO algorithm's ability to find good solutions outside of the initialization area, were investigated. Four basic PSO variations were used as basis for the investigation. Problem size was found to have a major influence on roaming behaviour and algorithm performance. As problem size increased, the extent of roaming also increased and the algorithm performance decreased. Effectively addressing roaming behaviour thus becomes even more important when larger problems need to be solved.

Future research opportunities exist in analysing the roaming behaviour of other evolutionary algorithms such as differential evolution or genetic algorithms and more in depth analysis of the drivers and consequences of roaming behaviour.

References

1. Engelbrecht, A.P.: Particle swarm optimization: velocity initialization. In: IEEE Congress on Evolutionary Computation (2012)
2. Engelbrecht, A.P.: Roaming behavior of unconstrained particles. In: BRICS Congress on Computational Intelligence (2014)
3. Cheng, S., Shi, Y., Qin, Q.: Experimental study on boundary constraints handling in particle swarm optimization: from population diversity perspective. Int. J. Swarm Intell. Res. **2**(3), 29–43 (2011)
4. Chu, W., Gao, X., Sorooshian, S.: Handling boundary constraints for particle swarm optimization in high-dimensional search space. Inform. Sci. **181**(20), 4569–4581 (2011)
5. Xie, X.F., Bi, D.C.: Handling boundary constraints for numerical optimization by particle swarm flying in periodic search space. In: IEEE Congress on Evolutionary Computation, pp. 2307–2311 (2004)
6. Kennedy, J., Eberhart, R.: Particle swarm optimization. In: IEEE International Confererence on Neural Networks, pp. 1942–1948 (1995)
7. Clerc, M.: The swarm and the queen: towards a deterministic and adaptive particle swarm optimization. In: IEEE Congress on Evolutionary Computation, pp. 1951–1957 (1999)
8. Clerc, M., Kennedy, J.: The particle swarm - explosion, stability and convergence in a multidimensional complex space. IEEE Trans. Evolut. Comput. **6**(1), 58–73 (2002)
9. Van den Bergh, F., Engelbrecht, A.P.: A new locally convergent particle swarm optimiser. In: IEEE International Conference on Systems, Man and Cybernetics, pp. 6–12 (2002)
10. Kennedy, J.: Bare bones particle swarms. In: IEEE Swarm Intelligence Symposium, pp. 80–87 (2002)
11. Liang, J.J., Qu, B.Y., Suganthan, P.N., Chen, Q.: Problem definitions and evaluation criteria for the CEC 2015 competition on learning-based real-parameter single objective optimization. Technical report 201411A, Computational Intelligence Laboratory, Zhengzhou University, Zhengzhou China and Nanyang Technological University, Singapore (2014)
12. Vesterstrom, J.S., Riget, J., Krink, T.: Division of labor in particle swarm optimisation. In: Congress on Evolutionary Computation, pp. 1570–1575 (2002)

The Analysis of Strategy for the Boundary Restriction in Particle Swarm Optimization Algorithm

Qianlin Zhou, Hui Lu$^{(\boxtimes)}$, Jinhua Shi, Kefei Mao, and Xiaonan Ji

School of Electronic and Information Engineering,
Beihang University, Beijing 100191, China
mluhui@vip.163.com

Abstract. Particle swarm optimization has been applied to solve many optimization problems because of its simplicity and fast convergence performance. In order to avoid precocious convergence and further improve the ability of exploration and exploitation, many researchers modify the parameters and the topological structure of the algorithm. However, the boundary restriction strategy to prevent the particles from flying beyond the search space is rarely discussed. In this paper, we investigate the problems of the strategy that putting the particles beyond the search space on the boundary. The strategy may cause PSO to get stuck in the local optimal solutions and even the results cannot reflect the real performance of PSO. In addition, we also compare the strategy with the random updating strategy. The experiment results prove that the strategy that putting the particles beyond the search space on the boundary is unreasonable, and the random updating strategy is more effective.

Keywords: Particle swarm optimization (PSO) · Boundary restriction strategy · Random updating strategy

1 Introduction

Particle swarm optimization [1] is one kind of popular intelligence algorithms. It searches the optimal solution through the cooperation among particle swarm. Compared with other swarm intelligent algorithms, PSO is simple, fast and easy to code. Therefore, the algorithm has attracted extensive attention since it was put forward.

However, PSO is easy to get stuck in the local optimal solutions due to the lack of variance mechanism of the particle. In addition, the lower degree of convergence is also a disadvantage of PSO. In order to improve the performance of the exploration and the exploitation, many researchers modify PSO in the algorithm parameters, the topological structure and the algorithm mechanism.

For the improvement of the parameters, Shi and Eberhart introduced inertia weight into the original particle swarm optimization [2]. They also proposed the linearly decreasing inertia weight to improve the performance of PSO [3]. The nonlinear inertia weight [4] proposed by Amitava and Patrick also has a good balance between exploration and exploitation. Chen et al. [5] proposed IPSO with a self-adaptive adjustment inertia-weighted strategy factor. In addition, Ratnaweera et al. [6] proposed

© Springer International Publishing AG 2017
Y. Tan et al. (Eds.): ICSI 2017, Part I, LNCS 10385, pp. 131–139, 2017.
DOI: 10.1007/978-3-319-61824-1_14

the time-varying acceleration coefficients PSO. For the topological structure of PSO, Suganthan [7] introduced a variable neighborhood operator based on the research of the local version of PSO. Mendes et al. [8] further investigated the information flow of the particle swarm and proposed a series of topological structures. In addition, hybrid PSO has also been widely studied. Lvbjerg et al. [9] proposed hybrid particle swarm optimization algorithm through incorporating the reproduction of standard GA into the PSO. Higashi and Iba [10] proposed PSO with Gaussian mutation combining the idea of the particle swarm with concepts from evolutionary algorithms. Feng and Pan [11] presented a modified PSO algorithm based on cache replacement algorithm.

However, the strategy to update particles beyond the search space in PSO is rarely discussed. Many researches only put particles beyond the search space on the corresponding boundary without detailed analysis. In fact, there are many problems caused by this strategy in PSO. In this paper, we find the above strategy may cause PSO to get stuck in the local optimal solutions, and even worse, it may make PSO lose its real ability for some functions. In addition, we compare the performance of the strategy with that of the random updating strategy, which means the updating position will be randomly chosen from the whole search space for the particles beyond the search space. Based on the experiment results, the random updating strategy can avoid the problems existing in the former strategy significantly improve the performance of PSO.

The rest of the paper is organized as following. Section 2 illustrates a common boundary restriction strategy to restrict particles' position and problems caused by it in PSO. In Sect. 3, we conduct detailed experiments to prove the problems and compare the performance of the boundary restriction strategy introduced in Sect. 2 with that of the random updating strategy. Finally, a short conclusion follows in Sect. 4.

2 The Boundary Restriction Strategy in PSO

We focus on the improved standard global PSO algorithm. The updating formulas for particles are as following.

$$v_i(t+1) = w \times v_i(t) + c_1 \times p\_rand() \times (pbest_i - x_i) + c_2 \times g\_rand() \times (gbest - x_i). \tag{1}$$

$$x_i(t+1) = x_i(t) + v_i(t+1). \tag{2}$$

The linearly decreasing inertia weight w is adopted and shown as the Eq. (3).

$$w(t) = w_{max} - (w_{max} - w_{min}) \times t/num_{iteration} \ (w_{max} = 0.95, w_{min} = 0.4). \tag{3}$$

Along with iterations, particles may fly beyond their search space. Velocity restriction [12] shown as following is an effective way to avoid velocity explosion.

$$\text{If } v_{ij}(t+1) > v_j^{max}(v_{ij}(t+1) < -v_j^{max}) \text{ then } v_{ij}(t+1) = v_j^{max}(v_{ij}(t+1) = -v_j^{max}). \tag{4}$$

Here, v_j^{max} is the maximum velocity of the j th dimension for all particles. v_j^{max} can be obtained by the following formula in many researches.

$$v_j^{max} = k(x_j^{max} - x_j^{min})\big/2 \, k \in (0, 1].$$

(5)

Here, x_j^{max} and x_j^{min} are the j th dimensional upper bound and lower bound of the search space respectively. k is a constant in the interval $(0, 1]$.

However, particles may still fly beyond the feasible search space. The measure to solve this problem in many researches is as following.

$$\text{If } x_{ij}(t+1) > x_j^{max}(x_{ij}(t+1) < -x_j^{max}) \text{ then } x_{ij}(t+1) = x_j^{max}(x_{ij}(t+1) = -x_j^{max}).$$

(6)

From the Eq. (6), the particle will be put on the corresponding dimensional boundary if the particle flies beyond the search space in some dimension.

However, the combination of above velocity restriction and position restriction will cause some problems. Firstly, all particles are more likely to be on the boundary of the search space. It is unfair for the whole candidate solutions. Secondly, if a particle has enough updating velocity and flies beyond the search space in all dimensions, the position of the particle will be the corresponding vertex of the search space. However, if the current *gbest* is far away from the vertex, the velocity of this particle in the next iteration will be likely more than v^{max} or less than $-v^{max}$ for all dimensions. Therefore, the velocity will be $v^{max}(-v^{max})$ and the position of this particle in the next iteration is certain. If the optimal solution is in the neighborhood of this certain position, PSO will converge quickly and obtain the optimal solution. For example, given that k is equal to 1 in the Eq. (5), the certain position where a particle at vertices will fly to in probability is the center of the search space. In fact, many benchmark functions' search space is symmetric, and the optimal solution is exactly the center. It seems that PSO has great performance to obtain the optimal solution. However, if we only move the optimal solution to another place, the performance will significantly decrease. Therefore, the performance of PSO with above boundary restriction strategy is unfaithful.

The random updating strategy for the particles beyond the search space can avoid the problems caused by the above boundary restriction strategy. It means that once any dimensional position exceeds its feasible interval, the updating position for the particle in the current iteration will be randomly selected from all the candidate solutions. The random updating strategy can make PSO have the balanced performance no matter where the optimal solution is. In addition, it can also enhance the population diversity and improve the possibility to obtain the optimal solution.

3   Experiment and Analysis

The benchmark functions that we used is shown in Table 1. The range of x_i is $[-5, 5]$ for $f_1(X) \sim f_3(X)$, $[-8, 8]$ for $f_4(X) \sim f_6(X)$, and $[-10, 10]$ for the others. All functions have the same optimal solution $\max(f_i(X)) = f_i(0, \cdots 0) = 0 \, (1 \leq i \leq 10)$, which is exactly the center of the search space. As a contrast, $f_1'(X) \sim f_{10}'(X)$ are the same as $f_1(X) \sim f_{10}(X)$ except for the range of x_i, which is $[-3,5]$ for $f_1'(X) \sim f_3'(X)$, $[-6,8]$ for $f_4'(X) \sim f_6'(X)$, and $[-6,10]$ for the others. The optimal position for these changed functions is still $(0,0)$ but no longer the center of the search space.

Table 1. The benchmark functions.

Functions	Functions
$f_1(X) = -\sum_{i=1}^{n} [x_i^2 - 10\cos(2\pi x_i) + 10]$	$f_2(X) = 0.5 - \frac{\sin^2 \sqrt{x_1^2 + x_2^2} - 0.5}{[1 + 0.001(x_1^2 + x_2^2)]^2} - 1$
$f_3(X) = \frac{1 + \cos(12\sqrt{x_1^2 + x_2^2})}{0.5(x_1^2 + x_2^2) + 2} - 1$	$f_4(X) = -[\frac{1}{4000}\sum_{i=1}^{n} x_i^2 - \prod_{i=1}^{n} \cos(\frac{x_i}{\sqrt{i}}) + 1]$
$f_5(X) = -\sum_{i=1}^{n-1} \left\{ 100[(x_{i+1} + 1) - (x_i + 1)^2]^2 + [1 - (x_i + 1)]^2 \right\}$	$f_6(X) = -[-20\exp(-0.2\sqrt{\sum_{i=1}^{n} x_i^2/n}) - \exp(\sum_{i=1}^{n} \cos(2\pi x_i)/n) + 20 + e]$
$f_7(X) = -\sum_{i=1}^{n} (\sum_{j=1}^{i} x_j)^2$	$f_8(X) = -(\sum_{i=1}^{n} |x_i| + \prod_{i=1}^{n} |x_i|)$
$f_9(X) = -\{\sin^2[3\pi(x_1 + 1)] + x_1^2[1 + \sin^2(3\pi x_2 + 3\pi)] + x_2^2[1 + \sin^2(2\pi x_2 + 2\pi)]\}$	$f_{10}(X) = -[\sum_{i=1}^{n} x_i^2 + (\sum_{i=1}^{n} 0.5ix_i)^2 + (\sum_{i=1}^{n} 0.5ix_i)^4]$

3.1 The Analysis of the Strategy that Putting the Particles Beyond the Search Space on the Boundary

All the functions that we used are two dimensional in order to clearly show the problems caused by the strategy that putting the particles beyond the search space on the boundary. In addition, the more particles and iterations mean the greater possibility for PSO to obtain the better degree of convergence. In order to avoid optimization results so close to the global optimal solution that the program cannot distinguish and lead to less obvious experiment results, we use 20 particles and 200 iterations. k is equal to 1 in the Eq. (5). PSO is executed 100 times respectively for each case and the results are shown as Table 2. The ratio means the percentage of obtaining the optimal solution in the 100 runs and the threshold to decide whether PSO obtains the optimal solution is 10^{-6}. In addition, the late five columns of Table 2 are the distribution of the difference between 0 and optimization solutions which obtain the optimal solution in the 100 runs.

From Table 2, the ratio of obtaining the optimal solution decreases obviously for $f_1'(X) \sim f_4'(X)$ when compared with $f_1(X) \sim f_4(X)$. Moreover, the distribution of the optimization solutions shows that the particles can almost obtain the accurate optimal solution $(0,0)$ for $f_1(X) \sim f_{10}(X)$, while for $f_1'(X) \sim f_{10}'(X)$ the optimization solutions are only close to $(0,0)$. We analyze all particles' positions along with the iterations to uncover the causes of the significant differences. Take $f_4(X)$ as an example, the particle

Table 2. The results of the contrast experiments.

Functions	Ratio	0	$(0, 10^{-15}]$	$(10^{-15}, 10^{-12}]$	$(10^{-12}, 10^{-9}]$	$(10^{-9}, 10^{-6}]$
$f_1(X)/f_1'(X)$	99%/94%	99/0	0/57	0/35	0/2	0/0
$f_2(X)/f_2'(X)$	94%/17%	91/0	0/10	2/6	1/0	0/1
$f_3(X)/f_3'(X)$	100%/82%	100/0	0/51	0/25	0/4	0/2
$f_4(X)/f_4'(X)$	94%/56%	93/0	1/43	0/9	0/2	0/2
$f_5(X)/f_5'(X)$	100%/100%	98/0	1/89	1/11	0/0	0/0
$f_6(X)/f_6'(X)$	100%/100%	98/0	0/0	0/0	2/47	0/53
$f_7(X)/f_7'(X)$	100%/100%	99/0	1/100	0/0	0/0	0/0
$f_8(X)/f_8'(X)$	100%/100%	99/0	0/0	0/0	0/67	1/33
$f_9(X)/f_9'(X)$	100%/100%	98/0	2/97	0/3	0/0	0/0
$f_{10}(X)/f_{10}'(X)$	100%/100%	99/0	1/100	0/0	0/0	0/0

swarm obtain the optimal solution in the 29th iteration in one experiment. We give all parameters of the first particle obtaining the optimal solution from the 25th to the 34th iteration in Table 3 to show the process how it obtains the optimal solution.

Table 3. All parameters of the particle obtaining the optimal solution in the 25th–34th iteration.

No.	Position	Velocity	$pbest$	$gbest$	$p\_rand()$	$g\_rand()$
25	(5.98798, −7.92058)	(3.97809, −7.69489)	(−0.62541, 0.72835)	(3.03260, −4.55764)	0.928459	0.187363
26	(−2.01203, 0.74742)	(−8, 8)	(−0.62541, 0.72835)	(3.03260, −4.55764)	0.699531	0.662533
27	(−0.3936, 0.69719)	(1.61842, −0.05023)	(−0.3936, 0.69719)	(3.03260, −4.55764)	0.876501	0.914099
28	(7.28305, −7.30281)	(7.67665, −8)	(−0.3936, 0.69719)	(3.07380, −4.52855)	0.253488	0.234756
29	(8, −8)	(0.81243, −1.60364)	(−0.3936, 0.69719)	(3.07380, −4.52855)	0.851735	0.085847
30	(0, 0)	(−8, 8)	(0, 0)	(0, 0)	0.370618	0.762884
31	(−6.918, 6.918)	(−6.918, 6.918)	(0, 0)	(0, 0)	0.468864	0.296021
32	(−2.29837, 2.29837)	(4.61963, −4.61963)	(0, 0)	(0, 0)	0.229370	0.816503
33	(5.70163, −5.70163)	(8, −8)	(0, 0)	(0, 0)	0.993911	0.172161
34	(−0.74339, 0.74339)	(−6.44502, 6.44502)	(0, 0)	(0, 0)	−	−

From Table 3, the particle obtains the optimal position $(0, 0)$ accurately in the 30th iteration when its position in the 29th iteration is $(8, -8)$ which is a vertex of the search space. $pbest(-0.3936, 0.69719)$ and $gbest(3.07380, -4.52855)$ are relatively far away from the current vertex $(8, -8)$ in the 29th iteration and the current velocity $(0.81243, -1.60364)$ is relatively small. Therefore, the velocity in the next iteration will be more than v^{max} (8) or less than $-v^{max}$ (−8) when $p\_rand()$ and $g\_rand()$ take the appropriate values. Based on the Eqs. (1) and (2), the velocity in the next iteration will be $(8, -8)$ and the position will be $(0, 0)$.

The above analysis provides the evidence that position restriction strategy makes PSO accurately obtain $(0, 0)$ rather than close to $(0, 0)$ for $f_1(X) \sim f_{10}(X)$. However, the center of the search space is no longer the optimal solution when we decrease the search space. Therefore, the degree of convergence will degrade for $f_1'(X) \sim f_{10}'(X)$. In addition, the ratio of obtaining the optimal solution is also increased due to the potential possibility that the particles accurately fly to the optimal solution when they are located in a vertex. Therefore, the ratios of $f_1'(X) \sim f_4'(X)$ are significantly lower than that of $f_1(X) \sim f_4(X)$. However, the optimal solution of $f_5'(X) \sim f_{10}'(X)$ is more obvious and easier to search in the search space for PSO compared with $f_1'(X) \sim f_4'(X)$. PSO can still obtain the great performance without the position restrict strategy's hole for $f_5'(X) \sim f_{10}'(X)$. Therefore, the performance is similar just observing the ratio for $f_5(X) \sim f_{10}(X)$ and $f_5'(X) \sim f_{10}'(X)$. There are also some conclusions about the phenomenon that the particle will fly to a certain position in probability when it is in a vertex of the search space.

Firstly, the phenomenon that the particle at vertices flies to the center of the search space in the next iteration is a matter of probability when k is equal to 1. $p\_rand()$ and $g\_rand()$ are chosen randomly from 0 to 1. Their values may be too small to make the velocity in the next iteration exceed the restriction of the velocity. In addition, the positions of *pbest* and *gbest* also have the crucial influence on the velocity.

Secondly, any particle which flies to the center of the search space from a vertex will not leave the diagonal of the search space in the later iterations. As shown in Table 3, the particle will always move on the diagonal of the search space after the 30th iteration. The reason is that *pbest* and *gbest* is $(0, 0)$ for this particle, and they will not change. In addition, the velocity vector of the particle is parallel to the diagonal direction.

Thirdly, the phenomenon that a particle flies to the center from a vertex will be likely occur in the early iterations. Based on the linearly decreasing inertia weight, the particles have the larger velocity in the early iterations because of the larger w. Therefore, the particle is also more likely to fly beyond the search space in the early iterations. This characteristic may lead to the premature convergence and mislead our assessment about the performance of PSO.

Fourthly, the particle will fly to other certain positions when k is not equal to 1. For example, if k is equal to 0.5 when the search space is $(-8, 8)$, the particle may fly to one of $(-4, -4)$, $(-4, 4)$, $(4, -4)$ and $(4, 4)$. If the position is exactly the optimal solution, the particle will also just move on the diagonal of the search space in the later iterations. In fact, the problem will always exist as long as the maximum velocity is a constant and the measure to restrict particles' position is shown as the Eq. (6) in PSO.

Finally, with the increasing of the functions' dimension, the occurrence probability of the problem is very low if the number of the particles and the iterations remain unchanged, because the phenomenon only occurs when the velocities of all dimensions are more than v^{max} or less than $-v^{max}$ simultaneously. However, even if the velocities in part dimensions are beyond $[-v^{max}, v^{max}]$ and the particle flies to the search space's center of the corresponding dimensions, the position may also be a good solution. For example, a 10-dimensional particle may fly from $P_1(8, -8, 8 \cdots 8, x_1, y_1, z_1)$ to $P_2(0, 0, 0 \cdots 0, x_2, y_2, z_2)$ when each dimension's search space is $[-8, 8]$. If P_2 is exactly a near-optimal solution, the particle swarm may be trapped in local optimal solution. The particles may also find better solution, like $(\Delta_1, \Delta_2, \cdots \Delta_7, x_2 + \Delta_8, y_2 + \Delta_9, z_2 + \Delta_{10})(\Delta_i(1 \leq i \leq 10)$ is a small constant), than P_2 in the neighborhood of P_2 along with the convergence of the particle swarm. Therefore, it seems that the final output of PSO may be only a local optimal solution without any problems. However, the PSO's performance has been significantly influenced by the boundary restriction strategy that putting the particles beyond the search space on the boundary.

3.2   The Analysis of the Random Updating Strategy

$f_1(X) \sim f_{10}(X)$ are still used as the benchmark functions. All parameters and measures of PSO are the same as the above experiments except for the updating strategy for the particles beyond the search space. The random updating strategy is used in the following experiments. The results of the contrast experiments are shown in Table 4.

Table 4. The results of the contrast experiments.

Functions	Ratio	0	(0, 10^-15]	(10^-15, 10^-12]	(10^-12, 10^-9]	(10^-9, 10^-6]
$f_1(X)/f_1'(X)$	96%/96%	0/0	63/57	29/37	4/1	0/1
$f_2(X)/f_2'(X)$	63%/77%	0/0	27/25	24/29	9/18	3/5
$f_3(X)/f_3'(X)$	82%/94%	0/0	58/66	16/21	8/6	0/1
$f_4(X)/f_4'(X)$	78%/93%	0/0	62/71	13/10	3/7	0/5
$f_5(X)/f_5'(X)$	100%/100%	0/0	84/85	16/15	0/0	0/0
$f_6(X)/f_6'(X)$	100%/100%	0/0	0/0	0/0	56/55	44/45
$f_7(X)/f_7'(X)$	100%/100%	0/0	100/100	0/0	0/0	0/0
$f_8(X)/f_8'(X)$	100%/100%	0/0	0/0	0/0	54/70	46/30
$f_9(X)/f_9'(X)$	100%/100%	0/0	97/97	3/3	0/0	0/0
$f_{10}(X)/f_{10}'(X)$	100%/100%	0/0	100/100	0/0	0/0	0/0

From Table 4, the ratios do not decline as the experiment results shown in Table 2 for the first four functions when the center of the search space is no longer the optimal solution. On the contrary, the ratios are improved for $f_2'(X) \sim f_4'(X)$ compared with $f_2(X) \sim f_4(X)$. Due to the smaller search space, the improved ratios meet expectations. All optimization solutions are not the real global optimal solution exactly. They are only the solutions close to the global optimal solution. In addition, the distribution of the optimization solutions is similar for each pair of functions $f_i(X)$ and $f_i'(X)(1 \leq i \leq 10)$. The differences are only caused by the randomness of PSO and the reduced search space. Therefore, the random updating strategy can avoid the problem that the particle fly to the center of the search space in probability when it is at a vertex.

Compared with the experiment results in Table 2, the ratios for $f_1'(X) \sim f_4'(X)$ are larger in Table 4. Dissymmetry search space will not make the particle fly to the global optimal solution in probability. The results in Table 2 for $f_1'(X) \sim f_{10}'(X)$ reflect the real performance of PSO. However, the strategy that putting the particles beyond the search space on the boundary will make a large number of particles locate in the boundary in the early iterations. Moreover, if the center of the search space is exactly a near-optimal solution, the strategy that putting the particles beyond the search space on the boundary may cause PSO to get stuck in the near-optimal solution. Compared with the above strategy, the random updating strategy makes full use of particles and increases the diversity of particles. Taking $f_4'(X)$ as an example, all particles' traces with the two strategies are shown in Fig. 1 (a) and (b) respectively.

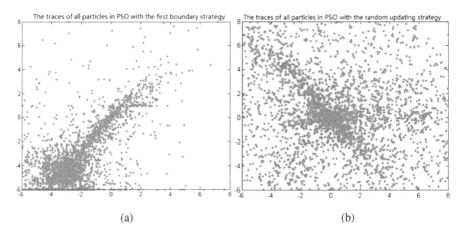

(a)   (b)

Fig. 1. The compared results of all particles' traces in PSO with the two boundary restriction strategies respectively

3.3 Summary of the Experiments

If the boundary restriction strategy that putting the particles beyond the search space on the boundary is adopted in PSO, the particle will fly to a different but certain position in the search space in probability based on the different maximum velocity when it is located at a vertex of the search space. When the certain position is exactly the global optimal solution, the performance of PSO will be incredible, though it seems that PSO has excellent performance most of the time. When the certain position is not the global optimal solution, the performance of PSO is real, but it will decline sharply for the functions with many near-optimal solutions which are close to the global optimal solution. The random updating strategy can effectively avoid the above problems. It can improve the diversity of the particles and increase the probability of obtaining the global optimal solution in the same condition compared with the former boundary restriction strategy. Therefore, we think the random updating strategy for the particles beyond the search space is a better way, and it should be widely adopted in PSO, especially for the reality optimization problems without prior knowledge of the global optimal solution.

4   Conclusion

This paper focuses on the analysis of the two different boundary restriction strategies in PSO. Combined with the experiment results, we analyze the existing problems of the strategy that putting the particles beyond the search space on the boundary and the reasons for them. In addition, we also prove the advantages of the random updating strategy. Therefore, the random updating strategy is recommended for the reality optimization problems without prior knowledge of the global optimal solution.

The future work will focus on the more effective boundary restriction strategies. In addition, the deeper analysis of the different boundary restriction strategies for the different optimization problems is also deserving of research.

Acknowledgments. This research is supported by the National Natural Science Foundation of China under Grant No. 61671041.

References

1. Kennedy, J., Eberhart, R.C.: Particle swarm optimization. In: Proceedings of the Feedback Mechanism IEEE International Conference on Neural Networks. IEEE Service Center, Piscataway, NJ, pp. 1942–1948 (1995)
2. Shi, Y.H., Eberhart, R.C.: A modified particle swarm optimizer. In: 1998 IEEE International Conference on Evolutionary Computation Proceedings. IEEE World Congress on Computational Intelligence, Anchorage, AK, pp. 69–73 (1998)
3. Shi, Y.H., Eberhart, R.C.: Parameter selection in particle swarm optimization. In: 7th International Conference, EP98 San Diego, California, USA, vol. 1447, pp. 591–600 (1998)
4. Amitava, C., Patrick, S.: Nonlinear inertia weight variation for dynamic adaptation in particle swarm optimization. Comput. Oper. Res. **33**(3), 859–871 (2006)
5. Chen H.H., Li G.Q., Liao H.l.: A self-adaptive improved particle swarm optimization algorithm and its application in available transfer capability calculation. In: 2009 Fifth International Conference on Natural Computation, Tianjin, pp. 200–205 (2009)
6. Ratnaweera, A., Halgamuge, S., Watson, H.C.: Self-organizing hierarchical particle swarm optimizer with time-varying acceleration coefficients. IEEE Trans. Evol. Comput. **8**(3), 240–255 (2004)
7. Suganthan P.N.: Particle swarm optimiser with neighbourhood operator. In: Proceedings of the 1999 Congress on Evolutionary Computation, CEC 99, Washington, DC, vol. 3, p. 1962 (1999)
8. Mendes, R., Kennedy, J., Neves, J.: The fully informed particle swarm: simpler, maybe better. IEEE Trans. Evol. Comput. **8**(3), 204–210 (2004)
9. Lvbjerg M., Rasmussen T.K., Krink T.: Hybrid particle swarm optimizer with breeding and subpopulations. In: Proceedings of the Third Genetic and Evolutionary Computation Conference, vol. 1, pp. 469–476 (2001)
10. Higashi N., Iba H.: Particle swarm optimization with Gaussian mutation. In: Proceedings of the 2003 IEEE Swarm Intelligence Symposium, SIS 2003, pp. 72–79 (2003)
11. Feng M., Pan H.: A Modified PSO Algorithm Based on Cache Replacement Algorithm. In: 2014 Tenth International Conference on Computational Intelligence and Security (CIS), Kunming, pp. 558–562 (2014)
12. Federico, M., Beata, W.: Particle swarm optimization (PSO) a tutorial. Chemometr. Intell. Lab. Syst. **149**, 153–165 (2015)

Particle Swarm Optimization with Ensemble of Inertia Weight Strategies

Muhammad Zeeshan Shirazi, Trinadh Pamulapati,
Rammohan Mallipeddi[(✉)], and Kalyana Chakravarthy Veluvolu

School of Electronics Engineering, College of IT Engineering,
Kyungpook National University, Bukgu, Daegu 41566, Republic of Korea
zeeshanshirazi85@gmail.com,
trinadhpamulapati@gmail.com,
mallipeddi.ram@gmail.com, veluvolukc@gmail.com

Abstract. Particle swarm optimization (PSO) has gained significant attention for solving numerical optimization problems in different applications. However, the performance of PSO depends on the appropriate setting of inertia weight and the optimal setting changes with generations during the evolution. Therefore, different adaptive inertia weight strategies have been proposed. However, the best inertia weight adaptive strategy depends on the nature of the optimization problem. In this paper, different inertia weight strategies such as linear, Gompertz, logarithmic and exponential decreasing inertia weights as well as chaotic and oscillating inertia weight strategies are explored. Finally, PSO with an adaptive ensemble of linear & Gompertz decreasing inertia weights is proposed and compared with other strategies on a diverse set of benchmark optimization problems with different dimensions. Additionally, the proposed method is incorporated into heterogeneous comprehensive learning PSO (HCLPSO) to demonstrate its effectiveness.

Keywords: Particle swarm optimization · Ensemble of inertia weight strategies · Gompertz decreasing inertia weight · Linear decreasing inertia weight

1 Introduction

Optimization is the process of finding the best solution in a search space from all possible solutions. In literature, a variety of population based derivate-free heuristic optimization algorithms have been proposed. Among the different optimization algorithms, Particle Swarm Optimization (PSO) is a computational method inspired by behavioral models of fish schooling, bird flocking, etc. [1, 2].

PSO has been successfully applied to optimization problems in different engineering application areas due to features such as; simplicity, ease of implementation, robustness to control parameters and high computational efficiency. However, the performance of PSO depends on the optimal setting of parameters such as inertia weight and acceleration parameters. In literature, different inertia weight strategies are adopted to improve the performance of PSO for optimization. In [3], a dynamic inertia weight strategy was explored to improve the performance of PSO. The decreasing inertia weight with linearly or Gompertz functions were also explored [4].

© Springer International Publishing AG 2017
Y. Tan et al. (Eds.): ICSI 2017, Part I, LNCS 10385, pp. 140–147, 2017.
DOI: 10.1007/978-3-319-61824-1_15

However, based on the observation that different inertia weight strategies suit different optimization problem depending on the nature of the problem, therefore we proposed an ensemble of inertia weight strategies. In [5], PSO with comprehensive learning referred to as comprehensive learning PSO (HCLPSO) was proposed. In HCLPSO [5], the exploration and exploitation are balanced by dividing the population into two subgroups and each subgroup employs linearly decreasing inertia weight along with time varying acceleration coefficients. Therefore, the proposed ensemble of inertia weight strategies is incorporated into HCLPSO to demonstrate its effectiveness. The proposed algorithm keeps track of how successful each behavior has been over a number of iterations and uses the information to select the next behavior of particle. A new self- adaptive Heterogeneous particle swarm optimizers (HPSO) was proposed to adaptively select the exploration/exploitation point of the search [6]. The particles are allowed to change their behaviors during the search in dynamic HPSOs. In [7] compared the behavior and performance of a selection of self-adaptive PSO algorithms to that of time variant algorithms. To solve the issue of parameter tuning, self-adaptive PSO algorithms proposed the adaptive nature of parameters over time [8]. In [9] proposed a HPSO algorithm with particles to follow the different search behaviors selected from a behavior pool and addressed the exploration – exploitation trade off problem. They provided a preliminary empirical analysis that much can be gained by using heterogeneous swarms the sensitivity of the PSO algorithm to the control parameters by first providing an overview of 18 inertia weight control strategies addressed in [10].

In the remainder of the paper, Sect. 2 briefly describes the working principle of standard PSO algorithm and the different strategies proposed in the literature for inertia weight adaptation. In Sect. 3, presents the proposed ensemble of inertial weight strategies. In Sect. 4, we describe our experimental set up and simulation results to highlight the effectiveness of the proposed algorithm on 11 benchmark functions. Finally, Sect. 5 presents conclusions.

2 Literature Survey

2.1 Standard PSO

PSO is a stochastic optimization technique which starts with initial random positions and velocities assigning to all the particles in the space. The movement of particles depends on their local and global best positions. The velocity and position of a particle can be updated by using the following Eqs. (1) and (2). The standard PSO employs constant acceleration coefficients (c_1, c_2) & inertia weight (w). The values of these parameters for optimal convergence in generally are 1.419 for both c_1, c_2 and 0.7 for w. However, it has been pointed out that an optimized inertia weight (w = 0.721) [11] provides a balance between global and local exploration and improves the chances to find the optimum value. The velocity update and position update equations in standard PSO are given as follows

$$\vec{v}_i(k+1) = w\vec{v}_i(k) + c_1\vec{r}_i(k)(\vec{p}_i(k) - \vec{x}_i(k)) + c_2\vec{r}_i(k)(\vec{g}(k) - \vec{x}_i(k)) \qquad (1)$$

$$\vec{x}_i(k+1) = \vec{x}_i(k) + \vec{v}_i(k+1) \qquad (2)$$

where w is the inertia weight and c_1, c_2 are the acceleration constants, i ranges from 1 to number of particles, k indicates the iteration number. $\vec{v}_i(k)$ and $\vec{x}_i(k)$ represent the velocity and position vectors, respectively at k^{th} iteration. $\vec{p}_i(k)$ denotes the personal best result of ith particle, while $\vec{g}_i(k)$ is the global best of the population. $\vec{r}_i(k)$ is a random number in the range of [0,1].

2.2 Different Inertia Weight Strategies

The performance of standard PSO is better owing to the presence of inertia weight and acceleration constants c_1 and c_2. The effectiveness of standard PSO in solving a numerical optimization problem depends on the appropriate values of inertia weight and acceleration constants. Since the optimal value of inertia weight at different iterations of the evolution process, a number of inertia weight adaptation strategies have been proposed. Table 1 shows the different inertia weight strategies to improve the performance of standard PSO. However, depending on the nature of the optimization problem a particular adaptation strategy may be apt for the problem at hand.

Table 1. Different inertia weight strategies

S. No.	Name of inertia weight	Formula of inertia weight		Reference
1	Linear decreasing inertia weight	$w(k) = w_1 + (w_2 - w_1)\frac{k}{FE}$	(3)	[12]
2	Gompertz decreasing inertia weight	$w(k) = w_1 + (w_2 - w_1)e^{-0.05u(nFE-k)}$ $$u = 10^{\left(\log\left(\frac{FE}{D}\right)-13\right)}$$	(4)	[3]
3	Chaotic inertia weight	$w(k) = (w_1 - w_2)\frac{FE-k}{FE} + 4w_2z(1-z)$		[13]
4	Oscillating inertia weight	$w(k) = 0.5\left((w_1 + w_2) + (w_1 - w_2)cos\left(\frac{22.6\pi}{FE}\right)\right)$		
5	Logarithm decreasing inertia weight	$w(k) = w_1 + (w_2 - w_1)\log\left(1 + \frac{10k}{FE}\right)$		
6	Exponent decreasing inertia weight	$w(k) = (w_1 - w_2 - 0.175)e^{\left(\frac{1}{1+\frac{5k}{FE}}\right)}$		

where $w(k)$ is inertia weight at k^{th} iteration, w_1 and w_2 are inertia weights at the start and end of a given run, respectively. Furthermore, u is the constant to adjust sharpness of the function, FE is the maximum number of function evaluations in a run, D is the dimensionality of the problem, and n is a constant to partition the Gompertz function.

3 Proposed Method

A new strategy to select dynamic inertia weight is proposed in PSO and we called it ensemble of inertia weight strategies. In the proposed method, an ensemble of linearly decreasing and Gompertz decreasing inertia weights is considered. Initially, each position vector in the population is assigned with one of the two inertia weight strategies which will be used (to update velocity and position of the particle) if the assigned strategy results in a better position of the particle then the assigned strategy is retained. If not, inertia weight strategy is changed. The above results in assigning a better inertia weight strategy to each of the particle in the population depending on its current position and velocity during each iteration. The acceleration constant c_1 is linearly decreased while c_2 linearly increased. The flowchart of the proposed ensemble method is as follows (Fig. 1).

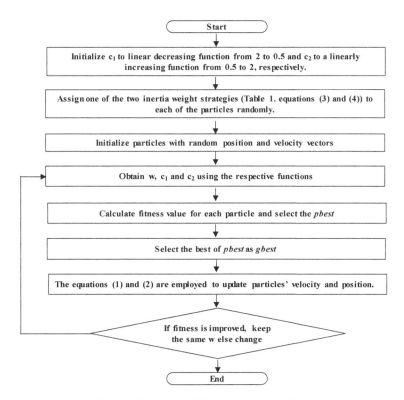

Fig. 1. Flow chart of the proposed algorithm.

4   Experimental Setup and Simulation Results

In this section, we demonstrate the performance of particle swarm optimization with different strategies along with the static and dynamic inertia weight and acceleration constants. Eleven benchmark functions (F1–F11) presented in Table 2 are used in this study for testing the performance. In Table 2, $x = [x_1, x_2, \ldots, x_D]$ represents the decisions variables and $o = [o_1, o_2, \ldots, o_D]$ is the shifted global optimum.

Table 3 shows the description and parameter settings of ten different instances employed in this study. Four different parameter settings of acceleration constants c1 and c2 were tested with dynamically changing such that c1 = c2 increasing, c1 = c2 decreasing, c1 increasing and c2 decreasing, and vice versa. Based on the optimized results, c1 decreasing and c2 increasing was used in optimization of benchmark

Table 2. Bench mark functions employed in this study

F#	Function equation	Bounds	f_{bias}
F1	$f(x) = \sum_{i=1}^{D} z_i^2 + f_{bias}, z = x - o$	$[-100, 100]^D$	−450
F2	$f(x) = \sum_{i=1}^{D} \left(\sum_{J=1}^{i} z_j \right) + f_{bias}, z = x - o$	$[-100, 100]^D$	−450
F3	$f(x) = \sum_{i=1}^{D} (10^6)^{\frac{i-1}{D-1}} z_i^2 + f_{bias}, z = x - o,$	$[-100, 100]^D$	−450
F4	$f(x) = \left(\sum_{i=1}^{D} \left(\sum_{J=1}^{i} z_j \right)^2 \right) * (1 + 0.4|N(0, 1)|) + f_{bias}, z = x - o$	$[-100, 100]^D$	−450
F5	$f(x) = \max\{|A_i x - B_i|\} + + f_{bias}, \quad A = D \times D, B_i = A_i * o$ $a_{ij} = random\ no. in\ the\ range[-500, 500]$	$[-100, 100]^D$	−310
F6	$f(x) = \sum_{i=1}^{D-1} \left(100(z_{i+1} - z_i^2)^2 + (z_i - 1)^2 \right) + f_{bias}, z = x - o$	$[-100, 100]^D$	390
F7	$f(x) = \sum_{i=1}^{D-1} \frac{z_i^2}{4000} \prod_{i=1}^{D} \cos\left(\frac{z_i}{\sqrt{i}}\right) + 1 + f_{bias}, z = x - o$	$[0, 600]^D$	−180
F8	$f(x) = -20 \exp\left\{ -0.2\sqrt{\frac{1}{D}\sum_{i=1}^{D} z_i^2} \right\} - \exp\{\frac{1}{D}\sum_{i=1}^{D} \cos(2\pi z_i)\}$ $+ 20 + e + f_{bias}, z = x - o$	$[-32, 32]^D$	−140
F9	$f(x) = \sum_{i=1}^{D} \left(z_i^2 - 10\cos(2\pi z_i) + 10 \right) + f_{bias}, z = x - o$	$[-5, 5]^D$	−330
F10	$f(x) = \sum_{i=1}^{D} \left(\sum_{k=0}^{kmax} \left(a^k \cos(2\pi b^k (z_i + 0.5)) \right) \right)$ $- D \sum_{k=0}^{kmax} \left(a^k \cos(2\pi b^k * 0.5) \right) + f_{bias}, z = x - o$	$[-0.5, 0.5]^D$	90
F11	$f(x) = \sum_{i=1}^{D-1} (A_i - B_i(x))^2 + f_{bias}$ $A_i = \sum_{i=1}^{D} (a_{ij} \sin\alpha_j + b_{ij} \cos\alpha_j), B_i(x) = \sum_{i=1}^{D} (a_{ij} \sin x_j + b_{ij} \cos x_j)$ $A\&B = D \times D, a_{ij} = random\ no. in\ the\ range[-100, 100]$ $\alpha_{ij} = random\ no. in\ the\ range[-\pi, \pi]$	$[-\pi, \pi]^D$	−460

functions. In cases 2 to 7, the acceleration constants are dynamically changing such that $c1$ is decreasing from 2 to 0.5 and $c2$ is increasing from 0.5 to 2. The proposed ensemble methods employs the linearly decreasing and Gompertz decreasing inertia strategies because these are the two better performing strategies compared to the other strategies on the benchmark problems presented in Table 2. To further demonstrate the significance of the current work, proposed ensemble of inertia weight strategies is incorporated into HCLPSO (case 9) and is referred to as AENHCLPSO (Case 10).

The maximum number of function evaluations is set to 100000 300000 for 10D, 30D problems respectively. All the experiments were run 30 times, independently on each problem. The swarm size is taken as 30 in cases 1 to 8, while in cases 9 and 10, swarm size of 20 and 40 is employed for 10D and 30D problems, respectively.

Tables 4 and 5 show the experimental results of cases 1–8 discussed above with swarm size of 30 and problem dimensions of 10, 30, and 50, respectively. From the results, it can be observed that out of first 8 cases, linear and Gompertz decreasing inertia weight strategies perform better than exponent, logarithm, chaotic and

Table 3. Description of different cases utilized in this study.

Case	Label	Description
1	SPSO	Standard PSO with acceleration constants c1, c2 = 1.496 and inertia weight w = 0.721
2	LDIW	PSO with linearly decreasing inertia weight from 0.9 to 0.4 values
3	GDIW	PSO with Gompertz decreasing inertia weight from 0.9 to 0.4 values
4	CIW	PSO with chaotic inertia weight in the interval of 0.4 and 0.9 values
5	OIW	PSO with oscillating inertia weight in the interval of 0.4 and 0.9 values
6	LOGIW	PSO with logarithmic decreasing inertia weight from 0.9 to 0.4 values
7	EXPIW	PSO with exponent decreasing inertia weight from 0.9 to 0.4 values
8	EAIW	Here, ensemble of cases 2 and 3 with adaptive inertia weight strategy is utilized
9	HCLPSO	PSO with w = 0.99–0.2, c1 = 2.5–0.5, c2 = 0.5–2.5, c = 3–1.5 with linear decreasing
10	AENHCL PSO	In AENHCLPSO case, ensemble of cases 2 and 3 with adaptive inertia weight strategy is utilized

oscillating inertia weight strategies. Based on this observation, the ensemble is formed using linear and Gompertz decreasing inertia weights. In Tables 4 and 5, it can be observed that the ensemble performs comparable to the best among the two constituent inertia weight strategies and better than the worst performing inertia weight strategy.

Table 4. Comparison between different inertia weight strategies on 10D problems size of 30

	SPSO		LDIW		GDIW		CIW		OIW		LOGIW		EXPIW		EAIW	
	MEAN	STD	MEAN	STD	MEAN	STD	MEAN	STD	MEAN	STD	MEAN	STD	MEAN	STD	MEAN	STD
F1	2.64E+01	7.34E+01	0.00E+00	0.00E+00	1.46E-03	7.98E-03	2.14E+01	1.17E+02	6.15E-03	1.56E-02	6.73E-30	3.69E-29	5.40E-13	2.78E-12	0.00E+00	0.00E+00
F2	1.12E+01	3.66E+01	6.02E-21	3.29E-20	2.07E-28	1.13E-28	3.62E-03	1.08E-02	1.29E+01	1.63E+01	2.70E-06	8.08E-06	1.81E-02	3.82E-02	2.65E-28	2.47E-28
F3	5.67E+04	2.06E+05	0.00E+00	0.00E+00	8.49E-01	2.99E+00	3.24E+02	1.69E+03	1.14E+05	2.47E+05	2.76E-01	1.51E+00	2.07E+02	4.89E+02	0.00E+00	0.00E+00
F4	2.83E+01	5.00E+01	3.56E-05	8.45E-05	1.10E+00	3.16E+01	5.73E+01	1.47E+02	3.54E+00	7.25E+00	2.77E-01	6.97E-01	1.75E+00	3.83E+01	4.35E-04	2.26E-03
F5	2.44E+02	1.34E+03	1.09E+01	5.98E+00	4.81E-06	2.63E-05	1.84E-01	9.65E-01	2.18E+01	2.44E+01	2.16E-04	5.91E+00	1.32E-01	5.31E-01	2.12E-10	8.79E-10
F6	1.04E+06	3.95E+06	1.38E+01	3.41E+01	3.20E+00	2.51E+00	1.14E+01	2.13E-01	3.36E-02	1.22E-03	4.32E+01	6.58E-01	1.88E-01	4.39E-01	3.02E+00	3.43E+00
F7	1.22E-01	7.40E-02	6.64E-02	3.74E-02	3.95E-02	2.44E-02	6.05E-02	4.46E-02	3.81E-02	3.13E-02	4.01E-02	2.50E-02	4.24E-02	2.98E-02	5.34E-02	3.61E-02
F8	4.86E+00	8.51E+00	6.66E-01	3.65E+00	8.63E-01	3.49E+00	4.18E-01	6.12E-01	2.02E+00	4.89E+00	1.05E-07	3.24E-07	2.18E+00	8.05E+00	1.40E+00	5.07E+00
F9	1.32E+01	7.66E+00	2.62E+00	1.49E+00	3.23E+00	1.50E+00	3.55E+00	2.02E+00	2.12E+00	1.32E+00	2.16E+00	1.20E+00	2.72E+00	1.73E+00	2.65E+00	1.57E+00
F10	1.08E+00	1.28E+00	2.05E-01	5.32E-01	2.17E-01	5.49E-01	4.31E-01	6.85E-01	3.63E-01	5.20E-01	1.53E-01	4.66E-01	1.83E-01	4.78E-01	2.03E-01	5.25E-01
F11	3.54E+03	7.66E+03	3.33E+02	6.04E+02	4.55E+02	7.52E+02	9.46E+02	1.81E+03	5.72E+02	6.46E+02	4.37E+02	6.10E+02	8.25E+02	1.28E+03	2.42E+02	5.11E+02

The original HCLPSO employs a linearly decreasing inertia weight strategy only. We incorporate the proposed ensemble of inertia weight strategies into original HCLPSO and the results are depicted in Table 6. From the results it can be observed that AENHCLPSO performs better or comparable to HCLPSO on most of the cases.

Table 5. Comparison between different inertia weight strategies on 30D problems size of 30

	SPSO		LDIW		GDIW		CIW		OIW		LOGIW		EXPIW		EAIW	
	MEAN	STD	MEAN	STD	MEAN	STD	MEAN	STD	MEAN	STD	MEAN	STD	MEAN	STD	MEAN	STD
F1	1.27E+03	2.89E+03	3.74E+01	2.65E+02	5.14E-28	1.52E-27	4.61E+02	4.01E+02	1.03E+02	2.67E+02	7.93E-05	1.37E-04	6.86E+01	3.09E+02	1.49E+01	8.16E+01
F2	7.78E+02	2.14E+03	1.72E+00	3.40E+00	2.62E-03	6.46E-03	2.33E+03	1.40E+03	5.76E+03	2.40E+03	1.94E+02	1.41E+02	1.42E+03	8.02E+02	5.01E-01	1.65E+00
F3	1.44E+07	3.70E+07	5.83E+02	3.19E+03	1.30E+02	4.96E+02	2.95E+06	2.11E+06	1.44E+06	1.87E+06	4.18E+04	1.18E+05	2.13E+05	3.22E+05	1.41E+02	7.39E+04
F4	4.96E+03	6.29E+03	1.18E+03	7.17E+02	6.27E+02	3.90E+02	1.38E+04	3.17E+03	8.37E+03	2.77E+03	4.62E+03	2.04E+03	1.19E+04	4.16E+03	8.73E+02	4.62E+02
F5	6.94E+03	2.12E+03	4.29E+03	9.72E+02	4.47E+03	1.14E+03	7.93E+03	2.18E+03	6.15E+03	1.84E+03	4.84E+03	1.13E+03	6.13E+03	1.15E+03	4.15E+03	1.46E+03
F6	2.39E+08	3.14E+08	9.57E+01	1.13E+02	8.62E+05	4.68E+06	2.00E+07	3.23E+07	1.14E+04	3.06E+04	2.45E+02	3.16E+02	2.82E+03	4.20E+03	4.41E+06	2.42E+07
F7	1.49E+01	2.26E-01	1.89E-01	4.77E-01	2.27E+00	1.16E-01	1.43E+01	1.43E-01	2.58E+00	3.43E+00	8.30E-01	1.77E+00	3.07E+00	9.46E+00	1.38E+00	5.07E+00
F8	1.87E+01	4.03E+00	1.94E+00	5.33E+00	5.97E+00	9.28E+00	1.14E+01	7.40E+00	8.23E+00	7.81E+00	6.97E+00	9.18E+00	9.38E+00	8.70E+00	2.66E+00	6.90E+00
F9	1.12E+02	3.86E+01	3.39E+01	8.72E+00	3.25E+01	8.18E+00	7.27E+01	1.84E+01	4.62E+01	8.71E+00	3.84E+01	8.01E+00	4.71E+01	1.29E+01	3.44E+01	7.27E+00
F10	1.26E+01	3.55E+00	3.28E+00	2.03E+00	2.97E+00	1.55E+00	1.16E+01	2.04E+00	7.51E+00	2.89E+00	3.89E+00	1.72E+00	6.08E+00	2.31E+00	2.71E+00	1.84E+00
F11	5.87E+04	5.59E+04	1.96E+04	2.25E+04	1.64E+04	1.26E+04	7.59E+04	3.67E+04	6.26E+04	3.75E+04	2.65E+04	1.75E+04	5.01E+04	2.85E+04	1.68E+04	1.14E+04

Table 6. Comparison between HCLPSO and AENHCLPSO on 10D and 30D problems

	HCLPSO Population size = 20, D = 10		AENHCLPSO Population size = 20, D = 10		HCLPSO Population size = 40, D = 30		AENHCLPSO Population size = 40, D = 30	
	MEAN	STD	MEAN	STD	MEAN	STD	MEAN	STD
F1	**0**	**0**	**0**	**0**	**0.00E+00**	**0.00E+00**	4.21E−31	2.30E−30
F2	**7.65E−15**	**1.37E−14**	5.91E−11	2.01E−10	**6.14E−05**	**4.16E−05**	7.94E−04	5.63E−04
F3	**0.00E+00**	**0.00E+00**	7.50E−26	4.04E−25	**0.00E+00**	**0.00E+00**	**0.00E+00**	**0.00E+00**
F4	8.22E−05	2.52E−04	**1.69E−05**	**4.87E−05**	6.83E+02	3.47E+02	**3.37E+02**	**1.26E+02**
F5	3.58E−01	4.22E−01	**0.00E+00**	**0.00E+00**	2.99E+03	4.78E+02	**2.14E+03**	**9.89E+02**
F6	**1.51E+00**	**2.12E+00**	9.26E+00	1.84E+01	**2.33E+00**	**2.55E+00**	4.78E+01	6.05E+01
F7	**2.07E+02**	**7.74E−14**	**2.07E+02**	5.89E−03	1.19E+03	1.52E−13	1.19E+03	2.31E−13
F8	1.54E−15	2.22E−15	**0.00E+00**	**0.00E+00**	3.26E−14	7.11E−15	**7.93E−15**	**2.59E−15**
F9	3.32E−02	1.82E−01	3.32E−02	1.82E−01	**3.32E−02**	**1.82E−01**	6.34E−01	1.29E+00
F10	**1.70E−01**	**2.31E−01**	1.06E+00	4.88E−01	**1.71E−01**	**3.18E−01**	2.25E+00	1.21E+00
F11	2.67E+05	1.45E−10	2.67E+05	1.78E−10	9.66E+05	7.11E−10	9.66E+05	7.89E+01

5 Conclusion

The performance of PSO depends on the proper selection of appropriate inertia weight strategy. Depending on the nature of problem, a particular inertia weight strategy may be apt for a given problem. Based on this observation, to effectively solve a set of optimization problems we propose an ensemble of inertia weight strategies for PSO. On most of the benchmark problem, the performance of the proposed ensemble of inertia weight strategies is better than the worst of the constituent algorithms. In addition, on most of the cases the performance of the proposed method is comparable to the best performing strategy. We further included the proposed ensemble idea in HCLPSO, an improved version of PSO algorithm to show that its effectiveness.

Acknowledgment. This research was supported by the Basic Science Research Program through the National Research Foundation of Korea (NRF) funded by the Ministry of Education, Science and Technology under the Grant NRF- 2015R1C1A1A01055669.

References

1. Eberhart, R., Kennedy, J.: A new optimizer using particle swarm theory. In: Proceedings of the Sixth International Symposium on Micro Machine and Human Science, pp. 39–43 (1995)
2. Kennedy, J., Eberhart, R.: Particle swarm optimization. In: 4th IEEE International Conference on Neural Networks (1995)
3. Gao, Y., Duan, Y.: An adaptive particle swarm optimization algorithm with new random inertia weight. In: International Conference on Intelligent Computing, pp. 342–350 (2007)
4. Xin, J., Chen, G., Hai, Y..: A particle swarm optimizer with multi-stage linearly-decreasing inertia weight. In: International Joint Conference on Computational Sciences and Optimization, pp. 505–508 (2009)
5. Lynn, N., Suganthan, P.N.: Heterogeneous comprehensive learning particle swarm optimization with enhanced exploration and exploitation. Swarm Evol. Comput. **24**, 11–24 (2015)
6. Nepomuceno, F.V., Engelbrecht, A.P.: A self-adaptive heterogeneous pso for real-parameter optimization. In: IEEE Congress on Evolutionary Computation (CEC), pp. 361–368 (2013)
7. van Zyl, E., Engelbrecht, A.: Comparison of self-adaptive particle swarm optimizers. In: IEEE Symposium on Swarm Intelligence (SIS), pp. 1–9 (2014)
8. Harrison, K.R., Engelbrecht, A.P., Ombuki-Berman, B.M.: The sad state of self-adaptive particle swarm optimizers. In: IEEE Congress on Evolutionary Computation (CEC), pp. 431–439 (2016)
9. Engelbrecht, A.P.: Heterogeneous particle swarm optimization. In: Dorigo, M., et al. (eds.) ANTS 2010. LNCS, vol. 6234, pp. 191–202. Springer, Heidelberg (2010). doi:10.1007/978-3-642-15461-4_17
10. Harrison, K.R., Engelbrecht, A.P., Ombuki-Berman, B.M.: Inertia weight control strategies for particle swarm optimization. Swarm Intell. **10**, 267–305 (2016)
11. Jiang, M., Luo, Y., Yang, S.: Stagnation analysis in particle swarm optimization. In: Proceedings of the 2007 IEEE Swarm Intelligence Symposium (2007)
12. Shi, Y., Eberhart, R.: A modified particle swarm optimizer. In: IEEE World Congress on Computational Intelligence, pp. 69–73 (1998)
13. Bansal, J.C., Singh, P., Saraswat, M., Verma, A., Jadon, S.S., Abraham, A.: Inertia weight strategies in particle swarm optimization. In: Third World Congress on Nature and Biologically Inspired Computing (NaBIC), pp. 633–640 (2011)

Hybrid Comprehensive Learning Particle Swarm Optimizer with Adaptive Starting Local Search

Yulian Cao[1], Wenfeng Li[1(✉)], and W. Art Chaovalitwongse[2]

[1] School of Logistics Engineering,
Wuhan University of Technology, Wuhan, China
{yulian,liwf}@whut.edu.cn
[2] Department of Industrial Engineering, Institute for Advanced Data Analytics,
University of Arkansas, Fayetteville, AR, USA
artchao@uark.edu

Abstract. Particle Swarm Optimization (PSO) offers efficient simultaneous global and local searches but is challenged with the problem of slow local convergence. To address this issue, a hybrid comprehensive learning PSO algorithm with adaptive starting local search (ALS-HCLPSO) is proposed. Determining when to start local search is the main of ALS-HCLPSO. A quasi-entropy index is innovatively utilized as the criterion of population diversity to depict an aggregation degree of particles and to ascertain whether the global optimum basin has been explored. This adaptive strategy ensures the proper starting of local search. The test results on eight multimodal benchmark functions demonstrate the performance superiority of ALS-HCLPSO. And comparison results on six advanced PSO variants further test the validity and superiority of ALS-HCLPSO algorithm.

Keywords: Quasi-entropy · Adaptive strategy · Population diversity · Local search

1 Introduction

Particle swarm optimization (PSO) algorithm is a population-based intelligent evolutionary algorithm, which imitates the foraging behavior of bird flocks and was first introduced by Eberhart and Kennedy [1]. PSO offers efficient simultaneous global and local searches. Many improved PSO variants are proposed, such as using dynamic clustering to maintain population diversity in [2]. As an excellent PSO algorithm, comprehensive learning PSO (CLPSO) adopts dimensional learning strategy and offers strong global search ability [3]. It takes the personal best (referred to as pbest) of all particles as candidate learning object other than its own pbest only, which maintains the population diversity without increasing computational complexity. Further research on CLPSO has been conducted and its superiority is stressed [4, 5].

However, most studies focus on the improvement of the global search, and less progress has been made to improve the local convergence rate. The traditional local search (LS) methods, such as the steepest descent method and quasi-Newton method,

© Springer International Publishing AG 2017
Y. Tan et al. (Eds.): ICSI 2017, Part I, LNCS 10385, pp. 148–157, 2017.
DOI: 10.1007/978-3-319-61824-1_16

with fast local convergence [6], are expected to compensate for this deficiency. This combination of population-based global search and individual-based local search, which is called memetic algorithm [7], can be regarded as a mode. Memetic algorithm offers an idea to enhance global and local searches simultaneously. Three major difficulties of using this hybrid mechanism should be addressed. Which particles are chosen to do LS? When to start LS? How much calculation resource will be spent for LS?

Several attempts have been made to apply the LS to PSO. The results show that the LS plays a significant role in accelerating the local search phase [8, 9]. However, there are obvious shortcomings in the existing approach. The switching strategy between the PSO algorithm and LS lacks theoretical support. The LS starts with random given number of iteration [8] and is conducted for all particles at each iteration [9]. Frequent use of LS not only waste computation resource, but also increase the risk of being trapped into a local minimum. In theory, only after the PSO performs a sufficient global search could the optimal solution be found. In addition, it wastes resources if LS starts too late. A more effective adaptive strategy is needed to convert from PSO iteration to LS. However, scarce research on adaptive starting strategy of LS has been reported.

A hybrid CLPSO algorithm with adaptive starting LS (ALS-HCLPSO) is proposed to further improve the CLPSO. The aggregation degree of selected particles is determined through a quasi-entropy index, which can be creatively employed to judge whether the particle has reached the global optimum basin. Accordingly, the key issue of adaptive starting LS is solved effectively.

The performance of ALS-HCLPSO algorithm is tested by eight multimodal functions. Comparison results on six improved PSO algorithms, including two improved CLPSO algorithms and one LS-based PSO algorithm, further verify the superiority of ALS-HCLPSO algorithm.

2   The Proposed ALS-HCLPSO Algorithm

2.1   PSO Algorithm and Local Search Method

PSO is a population based meta-heuristic algorithm. The update equation of velocity is illustrated in Eq. (1), including three parts. The first part depends on its own velocity, which is also known as inertia. The second part is self-cognition based on its own pbest. And the third one is social cognition, relating to the best solution of the whole swarm (referred to as gbest).

$$v_{id}^{t+1} = v_{id}^{t} + c_1 \cdot r_1 \cdot \left(pbest_{id}^{t} - x_{id}^{t}\right) + c_2 \cdot r_2 \cdot \left(gbest_{d}^{t} - x_{id}^{t}\right) \tag{1}$$

Numerous improvements have been made to deal with the problem of premature in PSO. In our research, the CLPSO algorithm that shows a very strong global optimization ability is chosen to be further studied. It uses pbest from all particles to maintain the population diversity, whose velocity update formula is Eq. (2). Although its global ability is enhanced, the convergence speed of its later iterations is still slow because of the inherent stochastic search mechanism [3]. We are going to utilize the traditional LS methods to speed up the local convergence of CLPSO.

$$v_{id}^{t+1} = \omega_L \cdot v_{id}^t + c \cdot r_i^d \cdot \left(p_{fi(d)d}^t - x_{id}^t \right) \tag{2}$$

Steepest descent method and quasi-Newton method are traditional LS approach. Steepest descent works in spaces of any number of dimensions, even in infinite-dimensional ones. Broyden-Fletch-Goldfarb-Shannon (BFGS) is a widely used quasi-Newton method and has super-linear convergence rate [6]. Its gradient is approximated by the finite difference method so it can be generalized to non-derivative and discontinuous problems. In this study, the steepest descent method and BFGS quasi-Newton method are introduced into CLPSO algorithm respectively.

2.2 Motivation of Adaptive Starting LS

A vital issue needs to be addressed for integrating LS into CLPSO, namely, when to start LS. In theory, if the LS is conducted before sufficient CLPSO global search, it will converge to a local minimum. Additionally, if the LS starts after CLPSO has done numerous random local search, it would waste computing resources and the advantage of the hybrid algorithm is not obvious. Therefore, it is necessary to start the LS appropriately.

In experiment, BFGS quasi-Newton coupled with CLPSO (referred to as CLbfgs), a hybrid algorithm with a fixed LS starting strategy, is applied to test the acceleration effect of LS. The maximum number of function evaluations (referred to as Max_FEs) and the number of function evaluation for LS (referred to as L_FEs) are given. LS will be triggered when CLPSO runs to the value of function evaluation at Max_FEs - L_FEs. The performance of CLbfgs with LS starting at different CLPSO iterations is analyzed by testing the 10-dimensional problem of F4 and F5 respectively. Each experiment runs 25 times independently. The convergence precision of CLbfgs and CLPSO is recorded, and the multiple of precision improvement by CLbfgs is computed. To intuitively get the difference of multiple in magnitude, the logarithmic operation is made.

Improvement multiple curves show significant peaks, where the multiple ascend first and then descend (Fig. 1). Therefore, the peak is the best opportunity to start the LS, which can maximize the benefit of the hybrid algorithm. Moreover, it can also be observed that the peak position is different for F4 and F5, and the width is not the same as well. Here, optional interval of F4 is much larger than that of F5. Additionally, because of the random initialization of PSO, even for the same function, the best time to start LS is different at each independent run. This shows that the best time to start LS is different. And it is difficult to set a fixed reasonable value for all tests in advance. Therefore, adaptive strategy is required to start the LS at an appropriate time.

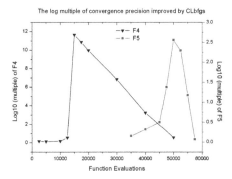

Fig. 1. Convergence precision improved by CLbfgs for F4 and F5 at different LS starting time

2.3   Adaptive Strategy Based on Quasi-Entropy

Only local optimum will be obtained if a particle for LS is not in the global optimum basin. In this part, we focus on solving how to determine the proper opportunity of starting LS. From the geometric sense, it is unable to judge whether a particle reaches the single peak region of the global optimum solution by looking at a single particle. However, if more than one top ranking particles gather together, it is believed that these particles have entered the global optimum basin. This phenomenon also indicates the end of CLPSO global search phase and the start of local search phase, and this is the best time to start the LS.

How to judge the degree of aggregation of these particles is the key for learning whether the global optimum basin is searched or not. The fitness of pbest (referred to as fit_pbest) is used to define a Quasi-Entropy (QE) index, which can characterize the population diversity. The Eq. (3) is the formula of QE at each iteration t.

$$QE(t) = -\sum_{i=1}^{N} P_i(t) \log P_i(t) \tag{3}$$

where

$$P_i(t) = \frac{f(pbest_i(t))}{\sum_{i=1}^{N} f(pbest_i(t))}$$

The proportion of the selected top ranking particles (referred to as β) are regarded as the set of Q. And $P_i(t)$ represents the proportion of fit_pbest of each particle to the sum of fit_pbest of all particles in set Q.

Taking the 10-dimensional Schwefel function (F4) for example, the performance of QE and variance of fit_pbest are compared (Fig. 2). The population size is set to be 15. The curve of QE is relatively stable first, and begins to decline at a point rapidly. The point where the slope of variance increases suddenly matches the lowest point where QE curve rapidly declines to. Common results can also be observed in other functions.

It indicates that the sharp decline of QE means the loss of population diversity, where global search has completed and the particles begin to do LS. This is the best time to start the LS.

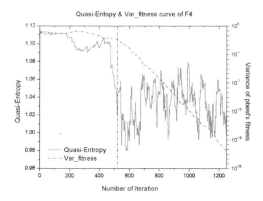

Fig. 2. Curve of quasi-entropy and variance of F4

The proposed QE index can depict the population diversity, and is more effective and timely to determine the change point compared with the variance. Therefore, we introduce the following strategy to start LS adaptively. When QE value of iteration t decreased to a certain proportion of the QE value of the initialized population, the LS will be triggered. The judgment criterion is the formula (4), where α is the decrease ratio.

$$QE(t) \leq \alpha \cdot QE(1) \tag{4}$$

2.4   Framework of ALS-HCLPSO Algorithm

The idea of ALS-HCLPSO is to combine the strong global ability of CLPSO and the fast convergence of LS. The hybrid mode of CLPSO and LS is serial. And the LS starting strategy is adaptive depending on the judgment of QE. The Fig. 3 illustrates the framework of ALS-HCLPSO.

Based on the illustration of the adaptive strategy in Sect. 2.3, once the LS starting criterion is satisfied, the hybrid algorithm will shift from CLPSO to LS. Since particles in the same unimodal region will converge to the same solution, our strategy is that only the gbest will perform LS when particles have located at the global optimum solution. This is to ensure the convergence to the global optimal solution and to maximize the optimization efficiency.

It should be noted that ALS-HCLPSO is a unified appellation. In this paper, two kinds of LS methods are utilized. Specifically, Abfgs-HCLPSO is based on the BFGS quasi-Newton method. Similarly, Asd-HCLPSO is based on the steepest descent method.

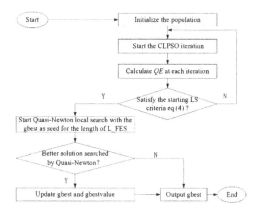

Fig. 3. Flowchart of ALS-HCLPSO algorithm

3   Numerical Experiments and Analysis

3.1   Benchmark Functions and Experimental Settings

Eight multimodal benchmark functions are selected to examine the performance of ALS-HCLPSO algorithm. Table 1 shows the property of each function. Their mathematical expression can be found in the literature [3]. Firstly, the performance of Abfgs-HCLPSO and Asd-HCLPSO is examined compared to the CLPSO algorithm. Secondly, comparison with other PSO variants is executed. The termination criterion is reaching the given Max_FEs. Since the examined functions converge at different iterations, it's not wise to set the same Max_FEs for them. The Max_FEs for F1 to F8 are 5.5E4, 3.5E4, 4E4, 2E4, 5E4, 4E4, 2E4 and 5E4, respectively. The problem dimension and population size are 10 and 40 respectively. Other parameter settings of the PSO variants are the same as their reference literatures. All experiments are run on matlab R2014a at a PC with windows10 operating system and 8 GB of memory. Each experiment runs 25 times independently.

Table 1. The search range and global optimum solution of multimodal benchmark functions

Test functions	Search range	x^*	$f(x^*)$
F1: Griewanks	$[-600, 600]^D$	$[0, 0, \ldots, 0]$	0
F2: Rastrigin	$[-5.12, 5.12]^D$	$[0, 0, \ldots, 0]$	0
F3: Noncontinuous Rastrigin	$[-5.12, 5.12]^D$	$[0, 0, \ldots, 0]$	0
F4: Schwefel	$[-500, 500]^D$	$[420.96, 420.96, \ldots, 420.96]$	0
F5: Weierstrass	$[-0.5, 0.5]^D$	$[0, 0, \ldots, 0]$	0
F6: Shifted Noncontinuous Rastrigin	$[-5.12, 5.12]^D$	$[0, 0, \ldots, 0]$	-330
F7: Shifted Schwefel	$[-500, 500]^D$	$[420.96, 420.96, \ldots, 420.96]$	-450
F8: Composition function	$[-500, 500]^D$	$[0, 0, \ldots, 0]$	0

3.2   Convergence Precision

Statistical results of convergence precision are shown in Table 2. The mean and median values of ALS-HCLPSO are better than those of CLPSO for all tested functions. Asd-HCLPSO and Abfgs-HCLPSO increase the mean accuracy by 13 and 11 orders of magnitude in F1. And they converge to the global optimum in F2 and F3, whose convergence accuracy is much higher than the e−4 of CLPSO algorithm itself. The two hybrid algorithms improve the mean accuracy by 10 orders of magnitude in F4. Similar results can also be observed in F5 to F8. It can be concluded that ALS-HCLPSO achieves several orders of magnitude higher convergence accuracy. Furthermore, the standard variance of ALS-HCLPSO is smaller than that of CLPSO algorithm (Table 1), which indicates that the performance of ALS-HCLPSO is more robust than that of CLPSO. It should be noted that the total number of function evaluation of ALS-HCLPSO algorithm is basically less than Max_FEs, because the LS starts before the number of function evaluation at Max_FEs - L_FEs in most cases. In short, the ALS-HCLPSO algorithm obtains a solution with higher precision by lower computational cost, and its performance is more stable than CLPSO.

Table 2.   Statistical results of convergence precision

Algorithm	Func.	Median	Mean	Std.	Func.	Median	Mean	Std.
CLPSO	F1	5.50E−05	2.01E−04	4.74E−04	F2	1.21E−04	2.01E−04	2.36E−04
Abfgs-HCLPSO		0.00E+00	5.33E−17	6.51E−17		0.00E+00	0.00E+00	0.00E+00
Asd-HCLPSO		6.66E−16	1.06E−15	1.06E−15		0.00E+00	0.00E+00	0.00E+00
CLPSO	F3	1.43E−04	1.83E−04	1.39E−04	F4	5.95E−02	1.00E−01	1.06E−01
Abfgs-HCLPSO		0.00E+00	0.00E+00	0.00E+00		1.27E−11	1.24E−11	1.46E−12
Asd-HCLPSO		0.00E+00	0.00E+00	0.00E+00		1.46E−11	1.42E−11	2.01E−12
CLPSO	F5	7.00E−08	9.27E−08	8.32E−08	F6	4.84E−05	1.07E−04	2.20E−04
Abfgs-HCLPSO		1.91E−11	3.55E−10	7.26E−10		5.81E−13	5.83E−13	1.48E−13
Asd-HCLPSO		4.28E−09	3.68E−08	1.07E−07		5.04E−13	5.33E−13	1.32E−13
CLPSO	F7	5.34E−03	1.21E−02	1.93E−02	F8	2.23E−02	1.05E+00	4.03E+00
Abfgs-HCLPSO		5.46E−12	3.71E−11	1.43E−10		2.76E−13	3.51E−13	1.68E−13
Asd-HCLPSO		4.55E−12	1.20E−11	2.85E−11		8.52E−15	1.66E−13	2.05E−13

3.3   Comparison with Other PSO Variants

To further evaluate the performance of ALS-HCLPSO, six PSO variants, UPSO [10], FDR-PSO [11], SRPSO [12], DNLPSO [4], HCLPSO [5] and DMS-L-PSO [8] are selected to compare with. DNLPSO and HCLPSO are also improved based on CLPSO. DMS-L-PSO is dynamic multi-swarm PSO with LS, whose LS starting strategy is different from our adaptive strategy.

The mean and variance values of the six advanced PSO variants are presented (Table 3). The results of ALS-HCLPSO and CLPSO are listed in Table 2 and not repeated here. Taking F1 for example, Abfgs-HCLPSO and Asd-HCLPSO achieve more than 10 orders of magnitudes higher accuracy compared to those of other PSO

Table 3. Convergence precision of other PSO variants

Function	Metrics	UPSO	FDR-PSO	SRPSO	DNLPSO	HCLPSO	DMS-L-PSO
F1	Mean	2.96E−02	6.01E−02	3.05E−02	4.13E−01	1.16E−02	7.69E−03
	Std.	(1.62E−02)	(3.27E−02)	(2.03E−02)	(3.54E−01)	(1.08E−02)	(7.63E−03)
F2	Mean	8.68E+00	3.02E+00	2.07E+00	1.07E+01	3.98E−02	2.61E−14
	Std.	(3.84E+00)	(1.42E+00)	(1.11E+00)	(7.18E+00)	(1.99E−01)	(4.28E−14)
F3	Mean	6.53E+00	4.44E+00	3.64E+00	1.00E+01	4.02E−02	3.30E−01
	Std.	(1.75E+00)	(1.08E+00)	(1.25E+00)	(9.25E+00)	(2.00E−01)	(5.53E−01)
F4	Mean	4.36E+02	6.85E+02	7.13E+02	9.49E+02	2.50E+01	1.42E+02
	Std.	(1.81E+02)	(2.87E+02)	(2.54E+02)	(5.44E+02)	(4.82E+01)	(1.13E+02)
F5	Mean	0.00E+00	0.00E+00	1.41E−07	9.17E−01	7.11E−16	0.00E+00
	Std.	(0.00E+00)	(0.00E+00)	(7.03E−07)	(1.26E+00)	(1.78E−15)	(0.00E+00)
F6	Mean	5.69E+00	4.12E+00	5.68E+00	8.37E+00	2.40E−01	4.83E−01
	Std.	(1.46E+00)	(1.62E+00)	(2.50E+00)	(3.90E+00)	(4.36E−01)	(7.12E−01)
F7	Mean	1.04E+02	4.61E+02	1.81E+02	3.94E+02	3.53E−03	4.74E+00
	Std.	(1.83E+02)	(2.52E+02)	(1.89E+02)	(2.84E+02)	(1.76E−02)	(2.37E+01)
F8	Mean	4.00E+00	8.00E+01	1.04E+02	1.02E+02	1.23E−17	4.39E−14
	Std.	(2.00E+01)	(9.57E+01)	(1.21E+02)	(9.37E+01)	(6.13E−17)	(2.19E−13)

variants. Similar results can also be observed on the rest of tested functions. Our proposed hybrid algorithm performs basically much better than the compared algorithms. The standard deviation values are also listed. The data on Tables 1 and 2 show that standard deviation of the two ALS-HCLPSO algorithms are consistently smaller than that of PSO variants. It means that the performance of ALS-HCLPSO is more robust.

Furthermore, the ranking analysis results based on the mean value of convergence precision among PSO variants is presented to understand the comparison intuitively (Table 4). Abfgs-HCLPSO and Asd-HCLPSO are both winners. They are superior to other comparison PSO variants. HCLPSO and DMS-L-PSO are the ranking followed algorithms, which are better than CLPSO but worse than the two ALS-HCLPSO algorithms.

Table 4. The ranking of PSO variants based on the mean value of convergence precision

Function	UPSO	FDR-PSO	SRPSO	DNLPSO	HCLPSO	DMS-L-PSO	CLPSO	Abfgs-HCLPSO	Asd-HCLPSO
F1	6	8	7	9	5	4	3	1	2
F2	8	7	6	9	5	3	4	1	1
F3	8	7	6	9	4	5	3	1	1
F4	6	7	8	9	4	5	3	1	2
F5	1	1	8	9	4	1	7	5	6
F6	8	6	7	9	4	5	3	2	1
F7	6	9	7	8	3	5	4	2	1
F8	6	7	9	8	1	2	5	4	3
Ave. Rank	6.125	6.5	7.25	8.75	3.75	3.75	4	2.125	2.125
Rank	6	7	8	9	3	3	5	1	1

Because of the similarity of each function, the convergence curve of F1 is chosen to show the optimization process over iterations intuitively (Fig. 4). DMS-L-PSO and UPSO converges fast at the beginning, however, their convergence rate starts to be slow shortly after that. Asd-HCLPSO and Abfgs-HCLPSO continue to converge to better value and is superior to other compared PSO variants. Especially, they converge to better results than CLPSO at the end of iteration.

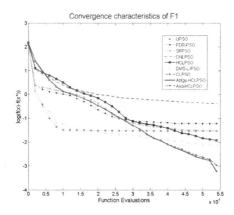

Fig. 4. Convergence curve of different PSO algorithms

4   Conclusions

This paper focuses on the adaptive strategy of the proposed ALS-HCLPSO algorithm that combines the advantage of fast local convergence of traditional LS with the strong global search ability of CLPSO. The adaptive strategy of switching from the global search of CLPSO to the LS effectively addresses the key issues of appropriate starting LS.

In this paper, two canonical LS methods are applied to ALS-HCLPSO algorithm. Experiments on eight multimodal benchmark functions verify the performance of the algorithm. The test results show that ALS-HCLPSO algorithm greatly improves CLPSO. Comparative experiments with other state-of-the-art PSO variants further validate the superiority of ALS-HCLPSO.

The proposed hybrid algorithm can be extended to other PSO algorithms with slight modifications, which will be our future research.

Acknowledgements. This work is supported by National Natural Science Foundation of China (Grants No. 61571336, No. 61603280 and No. 71672137).

References

1. Kenndy, J., Eberhart, R.C.: Particle swarm optimization. Proc. IEEE Int. Conf. Neural Netw. **4**, 1942–1948 (1995)

2. Liang, X., Li, W., Zhang, Y., et al.: An adaptive particle swarm optimization method based on clustering. Soft. Comput. **19**(2), 431–448 (2015)
3. Liang, J.J., Qin, A.K., Suganthan, P.N., et al.: Comprehensive learning particle swarm optimizer for global optimization of multimodal functions. IEEE Trans. Evol. Comput. **10**(3), 281–295 (2006)
4. Nasir, M., Das, S., Maity, D., et al.: A dynamic neighborhood learning based particle swarm optimizer for global numerical optimization. Inform. Sci. **209**, 16–36 (2012)
5. Lynn, N., Suganthan, P.N.: Heterogeneous comprehensive learning particle swarm optimization with enhanced exploration and exploitation. Swarm Evol. Comput. **24**, 11–24 (2015)
6. Luenberger, D.G., Ye, Y.: Linear and Nonlinear Programming, 4th edn. Springer, New York (2015)
7. Pablo, M.: On evolution, search, optimization, genetic algorithms and martial arts: towards memetic algorithms. Technique report, Caltech Concurrent Computation Program, USA (1989)
8. Zhao, S.Z., Liang, J.J., Suganthan, P.N., et al.: Dynamic multi-swarm particle swarm optimizer with local search for large scale global optimization. In: 2008 IEEE Congress on Evolutionary Computation, pp. 3845–3852 (2008)
9. Han, F., Liu, Q.: An improved hybrid PSO based on ARPSO and the Quasi-Newton Method. In: Tan, Y., Shi, Y., Buarque, F., Gelbukh, A., Das, S., Engelbrecht, A. (eds.) ICSI 2015. LNCS, vol. 9140, pp. 460–467. Springer, Cham (2015). doi:10.1007/978-3-319-20466-6_48
10. Parsopoulos, K.E., Vrahatis, M.N.: Unified particle swarm optimization in dynamic environments. In: Rothlauf, F., Branke, J., Cagnoni, S., Corne, D.W., Drechsler, R., Jin, Y., Machado, P., Marchiori, E., Romero, J., Smith, George D., Squillero, G. (eds.) EvoWorkshops 2005. LNCS, vol. 3449, pp. 590–599. Springer, Heidelberg (2005). doi:10.1007/978-3-540-32003-6_62
11. Peram, T., Veeramachaneni, K., Mohan, C.K.: Fitness-distance-ratio based particle swarm optimization. In: Proceedings of the IEEE Swarm Intelligence Symposium, pp. 174–181 (2003)
12. Tanweer, M.R., Suresh, S., Sundararajan, N.: Self regulating particle swarm optimization algorithm. Inform. Sci. **294**, 182–202 (2015)

A Bare Bones Particle Swarm Optimization Algorithm with Dynamic Local Search

Jia Guo[1(✉)] and Yuji Sato[2]

[1] Graduate School of Computer and Information Science, Hosei University,
Tokyo, Japan
guojia@ieee.org
[2] Faculty of Computer and Information Sciences, Hosei University,
Tokyo, Japan
yuji@hosei.ac.jp

Abstract. Swarm intelligence algorithms are wildly used in different areas. The bare bones particle swarm optimization (BBPSO) is one of them. In the BBPSO, the next position of a particle is chosen from the Gaussian distribution. However, all particles learning from the only global best particle may cause the premature convergence and rapid diversity-losing. Thus, a BBPSO with dynamic local search (DLS-BBPSO) is proposed to solve these problems. Also, because the blind setting of local group may cause the time complexity an unpredictable increase, a dynamic strategy is used in the process of local group creation to avoid this situation. Moreover, to confirm the searching ability of the proposed algorithm, a set of well-known benchmark functions are used in the experiments. Both unimodal and multimodal functions are considered to enhance the persuasion of the test. Meanwhile, the BBPSO and several other evolutionary algorithms are used as the control group. At last, the results of the experiment confirm the searching ability of the proposed algorithm in the test functions.

Keywords: Dynamic local search · Bare bones · Swarm intelligence · Diversity

1 Introduction

The swarm intelligence attracts plenty of researchers in recent years. The particle swarm optimization (PSO), proposed by Kennedy and Eberhard [1] in 1995, is one of the most famous swarm algorithms. It is inspired by the team behavior of bird flocking and fish schooling. Particles gain memories in the algorithm and each particle can remember its personal best position can be remembered by each particle. From this base, the swarm can remember the global best position. Compared with other optimization algorithms, PSO is famous for fast convergence and easy applying. However, the PSO is obsessed with the problem of the premature convergence when dealing with complex multimodal functions. So in

© Springer International Publishing AG 2017
Y. Tan et al. (Eds.): ICSI 2017, Part I, LNCS 10385, pp. 158–165, 2017.
DOI: 10.1007/978-3-319-61824-1_17

order to enhance the search ability of the PSO, Clerc and Kennedy [2] made some mathematical analysis through defining constriction coefficients.

In 2003, Kennedy [3] proposed the Bare Bones Particle Swarm Optimization (BBPSO) which canceled the velocity formula. And he employed the Gaussian distribution to choose the next position of a particle, in which, only the personal and the global best positions are considered during the process. It is the feature that makes the BBPSO faster and easier than the original PSO. In addition, different with the PSO, BBPSO is a parameter-free algorithm, which means it can be easily applied to various problems without human intervention. It is a great advantage for real-world problems. For instance, Omran proposed a clustering algorithm applying a bare bones method in unsupervised image classification [4].

Although the BBPSO has been applied in different areas, its inherent defects have not been solved. To solve these problems, the dynamic local search system makes arduous efforts to increase the diversity on the premise of not adding additional calculation into standard BBPSO. In conclusion, the rest of the paper is organized as follows: in Sect. 2 we give a brief review about the PSO and BBPSO; Sect. 3 gives an introduction about the proposed algorithm; Sect. 4 will have a clear description about the experiment to verify the issue; Sect. 5 gives the conclusion of this paper.

2   Related Work

The PSO algorithm is proposed by Kenedy and Eberhart [1]. In the PSO, each particle learns from their self-best and the global best position. In addition, to use the full information from other particles, a fully informed particle swarm (FIPS) is proposed by Mendes et al. [5]. Particles in FIPS are affected by all of their neighborhoods. After that the structure of the neighborhood group is discussed by Kennedy and Mendes [6].

The Bare bones particle swarm optimization (BBPSO) is a simple version of PSO. The next position of a particle is calculated by the Gaussian distribution. Both personal best position and swarm global best position are used in the calculation. Since the cancel of the velocity item, no parameters are needed in the algorithm, which endows the algorithm with a wilder area of application. Moreover, in order to speed up convergence, Kennedy [3] proposed a mutated version of BBPSO in which particles iterate with the following equations:

$$
\mu = \frac{p_i + gbest}{2}
$$
$$
\sigma = |p_i - gbest| \tag{1}
$$
$$
x_i(t+1) = \begin{cases} N(\mu, \sigma) & if(\varepsilon > 0.5) \\ p_i & else \end{cases}
$$

where $P = (p_1, p_2, \ldots, p_n)$ is the best position of each particle; $X = (x_1, x_2, \ldots x_i)$ is the particle swarm; $gbest$ is the best position of the whole swarm; t is the iteration time and $x_i(t)$ means the position of x_i at t time iteration; ε is a random number from 0 to 1. The parameter ε makes the mutated version able to save half of the time during the calculation. According to [3], the mutated version gives better results than some other versions of PSO in some benchmark functions.

Blackwell [7] proposed a dynamic update rule of the PSO as a second-order stochastic difference equation. The relations are applied to three particular particle swarm optimization implementations, the standard PSO of Clerc and Kennedy, a PSO with discrete recombination, and the Bare Bones swarm. Moreover, a bare bones PSO with jumps (BBPSOwj) is proposed to escape from the local minimum.

Campos et al. [8] proposed a bare bones particle swarm optimization with scale matrix adaptation (SMA-BBPSO) algorithm. In the proposed algorithm, the next position of a particle is selected from a multivariate t-distribution with a rule for adaptation of its scale matrix. Also, the Gaussian distribution acts as a particular case in the SMA-BBPSO. Furthermore, the SMA-BBPSO is explained by a theoretical analysis. Finally, the experiments with a set of benchmark functions confirm the ability of the SMA-BBPSO.

To sum up, most of the existing studies improve the performance by giving some new judgment methods or introducing some parameters to the BBPSO. Although these strategies promote some abilities of the original algorithm, the sacrifice is additional computation and inestimable increases of calculation time. Besides, in the standard BBPSO, all particles learn from the global best even if the current position is far from the global optimum. Therefore, particles may easily be trapped in local optimum if the search area includes plenty of local solutions. Hence, to balance the searching ability and the calculating speed, a simple and parameter free algorithm will be proposed in the next section.

3 Proposal of Bare Bones Particle Swarm Optimization with Dynamic Local Search

In this section, the bare bones particle swarm optimization algorithm with dynamic local search (DLS-BBPSO) will be presented. As it is introduced, the DLS-BBPSO is a parameter free algorithm for single objective function. It means that the proposed algorithm can adapt different functions without human intervention and adjustment. Unlike other variations of BBPSO, the DLS-BBPSO does no add extra calculation during the iteration process. In the following subsections, each step of the DLS-BBPSO will be presented.

3.1 Initialization

Initialization is the first step of the DLS-BBPSO. As a parameter-free algorithm, only the data connected to test functions are needed before an experiment. In particular, necessary messages are: the number of particles N; the dimension of the problem D; the fitness function F; the exploring area, R; the max iteration times T.

After all information is inputted, all particles will be randomly spread in the R. Then the first version of personal best positions, *pbest*, can be calculated by F. Meanwhile, the global best position, *gbest*, can be obtained. It can be found that no additional parameter is needed during the initialization process, which means the DLS-BBPSO can be easily applied to different problems. This is a great advantage when compared with parameter-needed algorithm in the real-world problems.

3.2 Dynamic Local Search System

In order to increase the diversity of the swarm during the iteration process, a dynamic local search (DLS) system is introduced to the standard BBPSO. In the DLS, particles are classified into different groups. Each group has only one leader and several teammates. Since all of the test functions are minimum problems, a particle with smaller fitness value is set to have a better position. At the beginning, the DLS will select a random particle to be the first "current leader", then it will select another random particle to compare with the "current leader". If the position of the selected particle is better than the "current leader", the selected particle will be a new "current leader". Otherwise, it will be a teammate of the "current leader". The DLS system will keep doing this until all particles are selected. It can be seen that the number of groups and the number of teammates of each group is not static. The process of the grouping is all random and no parameter is needed. This is the reason that the method named as "dynamic local search".

3.3 Evaluation

As a variation of standard bare bones particle swarm algorithm, DLS-BBPSO inheres the Gaussian distribution from BBPSO. The selecting mode is used both in and out groups. In particular, the next position of each leader will be selected by Gaussian distribution with a mean $(leader+gbest)/2$ and a standard deviation $|leader - gbest|$. More details are in the following equation:

$$
\begin{aligned}
u &= \frac{(pbest(leader) + gbest)}{2} \\
l &= |pbest(leader) - gbest| \\
x_t(i) &= N(u, l)
\end{aligned}
\tag{2}
$$

where $pbest(leader)$ is the personal best position best position of each leader; $gbest$ is the best position that all element has ever reached; $N(u,l)$ is a Gaussian distribution with a mean u and a standard deviation l. This equation is inherited from standard BBPSO.

Conversely, teammates in groups will not move to the global best position like the BBPSO. The next position of each teammate will be selected by the Gaussian distribution with a mean $(leader + teammate)/2$ and a standard deviation $|leader - teammate|$. More details are in the following equation:

$$u = \frac{(p(leader) + p(teammate))}{2}$$
$$l = |p(leader) - p(teammate)|$$
$$x_t(i) = N(u, l)$$

(3)

where $p(teammate)$ is the personal best position of a teammate particle and $p(leader)$ is the personal best position of the teammate's corresponding leader.

Specifically, the comparison of the evolution pattern between the DLS-BBPSO and the standard BBPSO is shown in Fig. 1. It can be observed that all particles learn from the only global best particle in BBPSO while only the leader of each group will learn from the global best in DLS-BBPSO. Besides, the iteration pattern of the DLS-BBPSO is only one possible situation. To explain the DLS-BBPSO more clearly, the pseudo-code is given in Fig. 2.

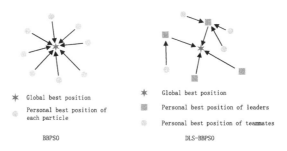

Fig. 1. The comparison of iteration pattern between the BBPSO and the DLS-BBPSO

Algorithm 1 DLS-BBPSO

Require: Max iteration time, T
Require: Fitness function, f
Require: Searching Range, R
Require: Dimension of the function, D
Require: Particle swarm $X = x_1, x_2, ...x_n$
Require: Personal best position $Pbest = p_1, p_2, ...p_n$
Require: Global best position $Gbest$
Ensure: All particles are in R
 1: **while** $t < T$ **do**
 2:    Select a random particle x_k as the first "current leader"
 3:    **while** $X \neq \varnothing$ **do**
 4:       Select a random particle x_i from X
 5:       **if** $p(x_i) < p(current\_leader))$ **then**
 6:          Choose a new position for "current leader" with equation (2)
 7:          Make $p(x_i)$ to be a new "current leader"
 8:       **else**
 9:          Make $p(x_i)$ to be a teammate of the "current leader"
10:          Choose a new position for x_i with equation (3)
11:       **end if**
12:    **end while**
13:    Update $Pbest$
14:    Update $Gbest$
15:    t=t+1
16: **end while**

Fig. 2. The pseudo-code of DLS-BBPSO

4   Experiment

4.1   Experimental Method

To verify the search ability of DLS-BBPSO, a set of comprehensive benchmark functions are chosen for the experiment. They are divided into two groups according to their properties:

(1) f_1–f_2, unimodal functions with only one global optimum in the search area;
(2) f_3–f_6, complex multimodal functions with several local minimum in the search area.

All of the 6 functions are minimum value problem. Also, the summarize of all functions are shown in Table 1. Meanwhile, a control group is set to increase the persuasive of the experiment. The setting of the group is considered in an earlier research [8]. The population of each function is 30.

Table 1. Experiment functions

Function	Search range	Reference
f_1 = Sphere function	(100, 100)	[8,9]
f_2 = Rosenbrock function	(−2.048, 2.048)	[8,9]
f_3 = Rastrigin function	(−5.12, 5.12)	[8,9]
f_4 = Ackley function	(−32.768, 32.768)	[8,9]
f_5 = Griewank function	(−600, 600)	[8,9]
f_6 = Weierstrass function	(−0.5, 0.5)	[8,9]

4.2   Experimental Results and Discussion

In order to minimize accidental errors, the empirical error is calculated from 30 independent runs while each algorithm has 1500 iteration times. The empirical error is defined as $|gbest - Optimum|$, where the $gbest$ is the global best value given by an algorithm after its last iteration, and the $Optimum$ is the theoretical optimal solution of the function. Experimental results are shown in Tables 2 and 3. In Tables 2 and 3, the experimental results are displayed, where "mean" stands for the mean empirical error of 30 independent runs; "Std. Dev." stands for the standard deviation of the 30 results. And the best results of each team are shown in **boldface**.

The statistical results in Tables 2 and 3 show that DLS-BBPSO has significant performance on the chosen functions. In the unimodal function group, the DLS-BBPSO wins a first rank and a second rank. Specifically speaking, the DLS-BBPSO gives a mean empirical error 2.14E−41 on f_1. Meanwhile, the champion on f_1, SMA-BBPSO gives a mean empirical error 2.71E−154 and the third rank algorithm PSO gives 4.34E−03. It is reasonable to believe that although the

Table 2. Comparisons of empirical error between PSO, BBPSO and FIPS

Function	PSO [2]		BBPSO [3]		FIPS [5,6]	
	Mean	Std. Dev.	Mean	Std. Dev.	Mean	Std. Dev.
f_1	**1.13E−10**	**7.94E−11**	1.58E−06	1.75E−06	1.03E−04	6.95E−05
f_2	5.26E+01	4.20E+01	9.47E+01	8.55E+01	**4.29E+01**	**4.01E+01**
f_3	8.36E+01	1.98E+01	8.28E+01	1.83E+01	**7.78E+01**	**1.63E+01**
f_4	1.90E+01	3.59E+00	**5.08E−01**	**8.70E−01**	6.86E−01	1.04E+00
f_5	**2.10E−03**	**4.16E−03**	5.23E−03	7.96E−03	5.53E−02	8.15E−02
f_6	**3.19E−03**	**4.95E−04**	5.23E−03	3.18E−03	8.01E−02	7.05E−02

Table 3. Comparisons of empirical error between BBPSOwj, SMA-BBPSO and DLS-BBPSO

Function	BBPSOwj [7]		SMA-BBPSO [8]		DLS-BBPSO	
	Mean	Std. Dev.	Mean	Std. Dev.	Mean	Std. Dev.
f_1	4.34E−03	2.37E−02	**2.42E−154**	**2.71E−154**	2.14E−41	1.17E−40
f_2	6.50E+01	4.22E+01	2.87E+01	**1.37E−02**	**2.79E+01**	8.08E−01
f_3	1.11E+01	3.45E+00	**0.00E+00**	**0.00E+00**	**0.00E+00**	**0.00E+00**
f_4	1.54E−01	3.55E−01	2.22E−15	1.81E−15	**1.72E−15**	**1.53E−15**
f_5	3.05E−02	2.84E−02	**0.00E+00**	**0.00E+00**	**0.00E+00**	**0.00E+00**
f_6	5.21E−01	1.22E−01	4.80E−12	2.39E−12	**0.00E+00**	**0.00E+00**

DLS-BBPSO is better than other algorithms in the test, it is still behind the SMA-BBPSO so far. Otherwise, on the f_2, the result of SMA-BBPSO is 2.78% larger than DLS-BBPSO's while the 3rd rank function FIPS gives a 53.76% larger result. Hence it can be conjectured that the search ability of DLS-BBPSO on unimodal functions needs to be improved.

On the other hand, in the multimodal group, the DLS-BBPSO gives excellent results. It wins on all of the four functions in the group and finds the exactly right position on three. Specifically speaking, the DLS-BBPSO gives a final result 0 on the f_3, while SMA-BBPSO gives the same answer. But the 3rd rank algorithm, BBPSOwj only gives 1.11E+01. It can be assumed that the DLS-BBPSO and the SMA-BBPSO successfully escape from the local minimum while others do not. On the f_4, the result SMA-BBPSO gives is 29.07% larger than DLS-BBPSO gives, while the BBPSOwj gives a 106.40% larger empirical error. The results of this team prove the DLS-BBPSO has better performance than other chosen algorithms but still has much room to improve. Moreover, DLS-BBPSO finds the right point on both the f_5 and the f_6 while the SMA-BBPSO only succeeds in f_5. It is a good evidence to prove that the dynamic local search system has a very strong ability to escape from the local minimum. The random selecting and dynamic grouping strategy give the swarm a powerful weapon to escape from the local minimum.

5   Conclusion

A bare bones particle swarm optimization with dynamic local search (DLS-BBPSO) is proposed in this paper. The DLS-BBPSO algorithm inherits the Gaussian distribution from the BBPSO. Apart from this, a dynamic local grouping system is used to increase the diversity of the swarm. Furthermore, no parameter is needed in DLS-BBPSO. Hence it can be fast applied to different functions.

To verify the performance of the DLS-BBPSO, both unimodal and multimodal benchmark functions are used in the experiment. Meanwhile, the standard PSO and several variants of the BBPSO are running on the same functions to contrast. Finally, the results confirmed the searching ability of the DLS-BBPSO.

Although the DLS-BBPSO shows better performance than the chosen variant of PSO and BBPSO, there still some points to be improved in the future. First, the results of the experiment on unimodal functions show that the DLS-BBPSO lacks digging ability. Hence, to balance the searching extension and depth is a good point. Moreover, applying this parameter-free algorithm to real-world functions is a topic worth to discuss.

References

1. Kennedy, J., Eberhart, R.: Particle swarm optimization. In: Proceedings of the IEEE International Conference on Neural Networks, vol. 4, pp. 1942–1948 (1995)
2. Clerc, M., Kennedy, J.: The particle swarm - explosion, stability, and convergence in a multidimensional complex space. IEEE Trans. Evol. Comput. **6**(1), 58–73 (2002)
3. Kennedy, J.: Bare bones particle swarms. In: Proceedings of the 2003 IEEE Swarm Intelligence Symposium, SIS 2003, pp. 80–87. IEEE (2003)
4. Omran, M., Al-Sharhan, S.: Barebones particle swarm methods for unsupervised image classification. In: IEEE Congress on Evolutionary Computation, pp. 3247–3252. IEEE (2007)
5. Mendes, R., Kennedy, J., Neves, J.: The fully informed particle swarm: simpler, maybe better. IEEE Trans. Evol. Comput. **8**(3), 204–210 (2004)
6. Kennedy, J., Mendes, R.: Neighborhood topologies in fully-informed and best-of-neighborhood particle swarms. IEEE Trans. Syst. Man Cybern. Part C (Appl. Rev.) **36**(4), 515–519 (2006)
7. Blackwell, T.: A study of collapse in bare bones particle swarm optimization. IEEE Trans. Evol. Comput. **16**(3), 354–372 (2012)
8. Campos, M., Krohling, R.A., Enriquez, I.: Bare bones particle swarm optimization with scale matrix adaptation. IEEE Trans. Cybern. **44**(9), 1567–1578 (2014)
9. Liang, J.J., Qin, A.K., Member, S., Suganthan, P.N., Member, S., Baskar, S.: Comprehensive learning particle swarm optimizer for global optimization of multimodal functions. IEEE Trans. Evol. Comput. **10**(3), 281–295 (2006)

Improving Multi-layer Particle Swarm Optimization Using Powell Method

Fengyang Sun, Lin Wang$^{(\boxtimes)}$, Bo Yang, Zhenxiang Chen, Jin Zhou,
Kun Tang, and Jinyan Wu

Shandong Provincial Key Laboratory of Network Based Intelligent Computing,
University of Jinan, Jinan 250022, China
wangplanet@gmail.com

Abstract. In recent years, multi-layer particle swarm optimization (MLPSO) has been applied in various global optimization problems for its superior performance. However, fast convergence speed leads the algorithm easy to converge to the local minimum. Therefore, MLPSO-Powell algorithm is proposed in this paper, selecting several swarm particles by the tournament operator in each generation to run Powell algorithm. MLPSO global searching performance with Powell local searching performance forces swarm particles to search more optima as much as possible, then it will rapidly converge as soon as it gets close to the global optimum. MLPSO-Powell enhances local search ability of PSO in dealing with multi-modal problems. The experimental results shows that the proposed approach improves performance and final results.

Keywords: Multi-layer particle swarm optimization · Powell · Particle Swarm Optimization · Tournament

1   Introduction

Particle swarm optimization (PSO) [1] can solve many complex optimization problems and has been widely used in various applications. The weakness of traditional two-search-layer PSO is that particles are very easy to fall into local optima and can't jump out.

Therefore, multi-layer particle swarm optimization (MLPSO) [2] increased layers from two to multiple, having best positions in multiple potential optimization of each layer at the same time, and the upper layer guides below. Thus, MLPSO may jump out of local optima through cross-layer collaborative behavior of the entire group. However, in this way MLPSO's fast convergence speed makes it easy converge to local optima instead. If a swarm particle accidentally flies to the inner edge of the global optimal point, unfortunately it is not the global best swarm particle of this epoch, it will be wrongly drawn away to the best swarm particle possibly. Thus it will miss out a great opportunity to approach the real global optimal solution.

© Springer International Publishing AG 2017
Y. Tan et al. (Eds.): ICSI 2017, Part I, LNCS 10385, pp. 166–173, 2017.
DOI: 10.1007/978-3-319-61824-1_18

The problem arises here is whether we can force swarm particles which are close to the inner edge of a local minimum to have the opportunity to really converge to the local minimum, thus greatly improving the possibility of finding the global optimal solution?

In order to address this problem, we propose a MLPSO-Powell algorithm, choosing a number of particles by the tournament operator in each epoch to be optimized by the Powell algorithm. In this way, the entire population is able to find more local optima, meanwhile due to the local searching performance of Powell, the swarm particles which are close to the global minimum will rapidly converge to it.

The rest of this paper is arranged as follows. Section 2 reviews the recent developments of Particle Swarm Optimization. Section 3 describes and analyzes MLPSO-Powell in detail. Section 4 overviews and discusses the results of the experiment. Section 5 gives the research conclusion.

2 Particle Swarm Optimization

2.1 Related Works

Kennedy and Eberhart [1] proposed Particle Swarm Optimization (PSO) based on birds foraging. Each particle represents one of the possible solutions. In addition to inertia from itself, each particles movement also refers to the movement of the individual itself on best historical practices to produce the migration of cognitive learning, moreover refers to the best experience of entire population as social learning. Shi and Eberhart [3] put forward an iterative linear inertia weight with reduction.

Suganthan [4] introduced a time-varying European spatial neighborhood operator: in the initial stage, the search neighborhood is defined as each particle itself, with increase of the number of iterations, neighborhood scope gradually extended to the entire population. Liang and Suganthan [5] proposed a dynamic multi-swarm PSO (DMS-PSO). Entire group is divided into many small subgroups for exchanging information. [6] used fitness-distance-ratio to select the adjacent particles. Wang et al. [7] proposed a dynamic tournament topology strategy, choosing particles from the entire swarm. Each particle is guided by several particles with better position.

Kennedy et al. [8] proposed a fully informed particle swarm (FIPS). Each particle learns best historical position of itself and the neighborhood, in addition learns the successful experience of other particles in the neighborhood. Liang et al. [9] proposed comprehensive learning particle swarm optimization (CLPSO). Every particle learns from their own or other particles at random, and its each dimension can learn from different particle. The learning strategy makes each particle have more learning objects, and fly in larger potential space, which is beneficial to global search. [10] used a phased hybrid evolution method to improve the simulation of cement hydration. Chenghui Zhang et al. studied oscillation and asymptotic behavior of a class of higher-order delay dynamic equations to improve the system stability, which is beneficial to behavior of particles [11–13]. Robinson et al. [8] integrated GA with PSO for antenna design optimization and recursive neural network design. [14] proposed a nearest neighbor partitioning method to increase the flexibility of nearest neighbors.

2.2 Challenge

Although there are many algorithms making various improvements for PSO, they still revolve around a basic characteristic of traditional PSO, which is that PSO has only two best positions. Thus, all particles within the space are very susceptible to the influence of the global best position at each epoch, which creates premature convergence. There are several algorithms, such as FIPS [8] and CLPSO [9], enhancing the awareness of particles by learning the information of other particles in the neighborhood, however it may restrict the possibility of particles exploring more potential local optima if the learning object is not actually good or in a right direction. In other words, their common characteristic is that the local searching ability is not so strong.

MLPSO-Powell algorithm is proposed in this paper on the basis of MLPSO's excellent global searching ability, selecting several particles by tournament operator in each epoch to run Powell. Due to the strong local searching ability of Powell, when a particle is close to the global optimal point, it is very likely to rapidly converge. MLPSO-Powell's details are described in the rest.

3 Methodology

3.1 Strategy of MLPSO-Powell

Multi-layer particle swarm optimization (MLPSO) [2] increased swarm layers from two to multiple. Best positions are in multiple potential optimization of each layer at the same time. The upper layer can decide more swarm particles direction and learn more comprehensive information, thus the upper layer guides below. In this way, MLPSO could jump out of local optima through cross-layer collective behavior of the entire group. The velocity updating formula of MLPSO are presented in Eq. 1,

$$v^i = \omega v^i + \sum_{j=1}^{m} \phi_j r \left(pbest_j^i - x^i \right) \tag{1}$$

where m is the number of layers, ϕ is the sum of acceleration constants, $\phi_j = \frac{\phi}{m}$ is the acceleration constant of layer j, $pbest$ is the best historical position of each particle at each layer, ω is $random\,[0,1]$ [2].

Powell belongs to the conjugate direction method. It is a effective nonlinear local search direct method designed for unconstrained optimization problem. It doesn't need to compute derivative, and uses function value only: it starts one-dimensional search from a point along each conjugate direction, and it finally can get the minimum. Nonlinear means that the initial point of the search direction is not fixed, and as the change of the initial point location the acceleration direction will also change, thus searching is not along straight line [15,16].

MLPSO-Powell absorbs the advantages of the two algorithms above. At first of each iterative epoch, it selects t particles by the tournament operator to be optimized by Powell. The tournament operators ordering criterion is average Euclidean distance between the particle and all the rest particles. These particles fitness and position are replaced by new fitness and position after Powell. The rest particles compute their fitness normally. Then the algorithm updates *pbest* or *lbest*, and updates each particles velocity and position. Then it executes next epoch until it meets the termination conditions.

3.2   Details of MLPSO-Powell

MLPSO-Powell is shown in Algorithm 1, where FE is number of fitness evaluation, MAX_FE is maximum fitness evaluation, $SP = (SP_1, SP_2, \cdots, SP_m)$ is the distribution of the population, $SPN = \prod\limits_{i=1}^{m} SP_i$ is the number of swarm particles, t is the number of swarm particles to run Powell, T is the set of index of the t swarm particles.

Algorithm 1. Algorithm of MLPSO-Powell

---

1 Initialize population;
2 $FE=0$;
3 **while** $FE < MAX\_FE$ **do**
4   Select t particles into T through Tournament operator;
5   **for** $p=1$ to SPN **do**
6   **if** $p \in T$ **then**
7   Powell(x_p);
8   Update x_p and fitness(p);
9   **end**
10   Calculate the fitness of the rest swarm particles;
11   **end**
12   Update the best position *pbest* on each layer;
13   Update velocity and position of all the swarm particles;
14     $FE++$;
15 **end**

---

4   Experiments

4.1   Benchmark Functions and Compared Algorithms

This research tries to test performance of MLPSO-Powell for various benchmark functions to simulate the manifestation of these algorithms in dealing with

Table 1. Benchmark functions

Function	Formula	Search range	Best value	Accrucy level
F_1 Quartic	$F_1(x) = \sum_{i=1}^{D} ix_i^2 + random[0,1)$	$[-1.28, 1.28]^D$	0	1.00E−06
F_2 Step	$F_2(x) = \sum_{i=1}^{D} (\lfloor x_i + 0.5 \rfloor)^2$	$[-100, 100]^D$	0	1.00E−06
F_3 Ackley	$F_3(x) = -20\exp(-0.2\sqrt{\frac{1}{D}\sum_{i=1}^{D}x_i^2}) -$ $\exp(\frac{1}{D}\sum_{i=1}^{D}\cos(2\pi x_i)) + 20 + e$	$[-32, 32]^D$	0	1.00E−02
F_4 Griewanks	$F_4(x) = \frac{1}{4000}\sum_{i=1}^{D}x_i^2 - \prod_{i=1}^{D}\cos(\frac{x_i}{\sqrt{i}}) + 1$	$[-600, 600]^D$	0	1.00E−02
F_5 Rastrigin	$F_5(x) = \sum_{i=1}^{D}(x_i^2 - 10\cos(2\pi x_i) + 10)$	$[-5, 5]^D$	0	1.00E−02
F_6 Noncontinuous Rastrigin	$F_6(x) = \sum_{i=1}^{D}(y_i^2 - 10\cos(2\pi y_i) + 10)$ $y_i = \begin{cases} x_i & |x_i| < \frac{1}{2} \\ \frac{round(2x_i)}{2} & |x_i| \geq \frac{1}{2} \end{cases}$	$[-100, 100]^D$	390	390+1.00E−2
F_7 Shifted Rotated Rastrigin	$F_7(x) = \sum_{i=1}^{D}(z_i^2 - 10\cos(2\pi z_i) + 10)$ $z = (x - o) * M$	$[-5, 5]^D$	−330	−330+1.00E−2
F_8 Shifted Expanded Griewanks plus Rosenbrocks	$F_R(x) = \sum_{i=1}^{D-1}(100(x_i^2 - x_{i+1})^2 + (x_i - 1)^2)$ $F_8(x) = F_4(F_R(z_1, z_2)) + F_4(F_R(z_2, z_3)) + ... + F_4(F_R(z_{D-1}, z_D)) + F_4(F_R(z_D, z_1))$ $z = x - o + 1$	$[-3, 1]^D$	−130	−130+1.00E−2

real complex problems. All of the 8 benchmark functions (3 unimodal and 5 muti-modal [17,18]) are tested between 10 and 20 dimensions. Table 1 shows the properties and formulas of these functions.

The experiment compared 5 evolutionary algorithms (1 genetic algorithm and 4 PSO algorithms), including proposed MLPSO-Powell. These Algorithms are genetic algorithm (GA), PSO with inertia weight (PSO-W), dynamic multi-swarm PSO (DMS-PSO), global multi-Layer PSO (Global MLPSO), global multi-layer PSO with Powell (MLPSO-Powell) (Tables 2 and 3).

In the above methods, genetic algorithm [19] is a searching optimal solution by simulating the natural evolution process. PSO-W [20] is a traditional PSO algorithm used in various applications. DMS-PSO [5] randomly redraws the entire group every few epochs. Global MLPSO [2] increases the diversity of the particles search by increasing the number of layers to improve performance. All the experimental results are performed in MATLAB 8.4 of the same machine with Intel(R) Core (TM) i5-3230M 2.60 GHz CPU, 8 G memory and Windows 7 ultimate operating system.

4.2 Convergence Analysis

To compare the convergence speed, the algorithms test the same max fitness evaluation of 200000 on 10-D and 500000 on 20-D. Each method were simulated on each benchmark function 10 times with same parameters.

Figure 1 shows the mean performance of best value of each method on each benchmark function for 10 times. Although MLPSO-Powell performs slower than

Table 2. Best, Worst, Mean, Std results on F1 to F8 for 10-D problems

	F_1	F_2	F_3	F_4	F_5	F_6	F_7	F_8	Average rank
GA									
Mean	8.97E−02	2.30E+00	7.97E−06	1.23E−02	3.98E−01	1.05E+01	−2.81E+02	−1.28E+02	
Std	3.04E−02	2.00E+00	3.15E−06	2.28E−02	5.14E−01	4.95E+00	2.44E+01	1.59E+00	
Rank	5	5	5	2	2	5	5	5	4.25
PSO-W									
Mean	5.15E−04	**0.00E+00**	2.66E−15	6.74E−02	4.97E−01	2.00E−01	−3.19E+02	**−1.29E+02**	
Std	2.53E−04	0.00E+00	0.00E+00	2.50E−02	5.24E−01	6.32E−01	4.82E+00	1.29E−01	
Rank	2	1	2	4	3	2	4	1	2.375
DMS-PSO									
Mean	1.65E−03	**0.00E+00**	1.35E−10	1.39E−01	3.34E+00	4.75E+00	−3.21E+02	**−1.29E+02**	
Std	6.34E−04	0.00E+00	8.30E−11	5.57E−02	1.86E+00	7.41E−01	9.96E−01	1.18E−01	
Rank	4	1	4	5	5	4	2	1	3.25
GMLPSO									
Mean	**1.18E−04**	**0.00E+00**	**2.31E−15**	4.48E−02	2.49E+00	3.40E+00	−3.20E+02	**−1.29E+02**	
Std	5.46E−05	0.00E+00	1.12E−15	1.86E−02	1.43E+00	1.90E+00	4.12E+00	1.39E−01	
Rank	1	1	1	3	4	3	3	1	2.125
MLPSO-Powell									
Mean	8.78E−04	**0.00E+00**	2.66E−15	**0.00E+00**	**1.99E−14**	**2.13E−15**	−3.24E+02	**−1.29E+02**	
Std	5.53E−04	0.00E+00	0.00E+00	0.00E+00	5.02E−14	5.60E−15	2.95E+00	1.77E−01	
Rank	3	1	2	1	1	1	1	1	1.375

Table 3. Best, Worst, Mean, Std results on F1 to F8 for 20-D problems

	F_1	F_2	F_3	F_4	F_5	F_6	F_7	F_8	Average rank
GA									
Mean	2.71E−01	1.00E+01	1.26E−05	1.16E−02	**2.09E+00**	1.92E+01	−5.97E+01	−1.22E+02	
Std	1.01E−01	5.98E+00	5.17E−06	1.28E−02	1.44E+00	4.66E+00	5.28E+01	3.88E+00	
Rank	5	5	5	3	1	5	5	5	4.25
PSO-W									
Mean	1.54E−03	**0.00E+00**	5.15E−15	3.81E−02	5.37E+00	**4.00E−01**	−2.41E+02	**−1.29E+02**	
Std	7.93E−04	0.00E+00	1.72E−15	2.81E−02	1.70E+00	5.16E−01	8.26E+00	3.63E−01	
Rank	2	1	2	4	3	1	3	1	2.125
DMS-PSO									
Mean	7.85E−03	**0.00E+00**	2.13E−08	4.31E−02	1.59E+01	1.61E+01	−2.34E+02	−1.27E+02	
Std	2.00E−03	0.00E+00	2.41E−08	4.73E−02	4.07E+00	2.88E+00	6.37E+00	2.51E−01	
Rank	4	1	4	5	4	4	4	4	3.75
GMLPSO									
Mean	**2.92E−04**	**0.00E+00**	**2.66E−15**	7.37E−03	1.73E+01	1.23E+01	−2.46E+02	−1.28E+02	
Std	1.22E−04	0.00E+00	0.00E+00	1.19E−02	5.73E+00	5.70E+00	7.80E+00	6.82E−01	
Rank	1	1	1	2	5	3	2	2	2.125
MLPSO-Powell									
Mean	3.87E−03	**0.00E+00**	5.86E−15	**0.00E+00**	3.50E+00	5.72E+00	**−2.99E+02**	−1.28E+02	
Std	1.90E−03	0.00E+00	1.12E−15	0.00E+00	1.84E+00	4.90E+00	7.45E+00	3.78E−01	
Rank	3	1	3	1	2	2	1	2	1.875

other algorithms sometimes, the former has an excellent search accuracy. The introduction of Powell makes particles easy to find more local optima, thus it increases the probability of finding the global optimum. In this way, the algorithm can jump out of local minima and improve the performance finally.

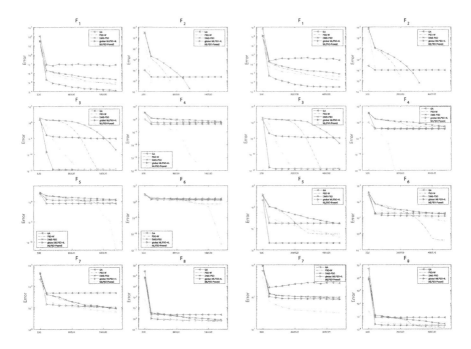

Fig. 1. The mean convergence performance on 8 benchmark functions for 10-D (left 8) and 20-D (right 8) problems

5   Conclusions

MLPSO-Powell algorithm is proposed in this article to improve the overall performance of MLPSO by Powell's local search. MLPSO-Powell strategy enhances the local search ability of the algorithm, then it can jump out of local optimum by finding as much local minima as possible.

In order to assess the performance of the proposed method, we use 8 known benchmark functions to compare different algorithms. The experimental results manifest that the proposed approach improves the accuracy and performance.

Acknowledgments. This work was supported by National Natural Science Foundation of China under Grant No. 61573166, No. 61572230, No. 61373054, No. 61472164, No. 61472163, No. 61672262, No. 61640218. Shandong Provincial Natural Science Foundation, China, under Grant ZR2015JL025. Science and technology project of Shandong Province under Grant No. 2015GGX101025. Project of Shandong Province Higher Educational Science and Technology Program under Grant No. J16LN07. Shandong Provincial Key R&D Program under Grant No. 2016ZDJS01A12.

References

1. Eberhart, R., Kennedy, J.: A new optimizer using particle swarm theory. In: Proceedings of the Sixth International Symposium on Micro Machine and Human Science, MHS 1995, pp. 39–43. IEEE, October 1995

2. Wang, L., Yang, B., Chen, Y.: Improving particle swarm optimization using multi-layer searching strategy. Inf. Sci. **274**, 70–94 (2014)
3. Shi, Y., Eberhart, R.: A modified particle swarm optimizer. In: Proceedings of the 1998 IEEE International Conference on Evolutionary Computation. IEEE World Congress on Computational Intelligence, pp. 69–73. IEEE, May 1998
4. Suganthan, P.N.: Particle swarm optimiser with neighbourhood operator. In: Proceedings of the 1999 Congress on Evolutionary Computation, CEC 1999, vol. 3, pp. 1958–1962. IEEE (1999)
5. Liang, J.J., Suganthan, P.N.: Dynamic multi-swarm particle swarm optimizer with local search. In: The 2005 IEEE Congress on Evolutionary Computation, vol. 1, pp. 522–528. IEEE, September 2005
6. Veeramachaneni, K., Peram, T., Mohan, C., Osadciw, L.A.: Optimization using particle swarms with near neighbor interactions. In: Cantú-Paz, E., et al. (eds.) GECCO 2003. LNCS, vol. 2723, pp. 110–121. Springer, Heidelberg (2003). doi:10.1007/3-540-45105-6_10
7. Wang, L., Yang, B., Orchard, J.: Particle swarm optimization using dynamic tournament topology. Appl. Soft Comput. **48**, 584–596 (2016)
8. Mendes, R., Kennedy, J., Neves, J.: The fully informed particle swarm: simpler, maybe better. IEEE Trans. Evol. Comput. **8**(3), 204–210 (2004)
9. Susilo, A.: Comprehensive learning particle swarm optimizer (CLPSO) for global optimization of multimodal functions. Undergraduate theses (2008)
10. Wang, L., Yang, B., Abraham, A.: Distilling middle-age cement hydration kinetics from observed data using phased hybrid evolution. Soft. Comput. **20**(9), 3637–3656 (2016)
11. Zhang, C., Li, T., Agarwal, R.P., Bohner, M.: Oscillation results for fourth-order nonlinear dynamic equations. Appl. Math. Lett. **25**(12), 2058–2065 (2012)
12. Zhang, C., Agarwal, R.P., Li, T.: Oscillation and asymptotic behavior of higher-order delay differential equations with p-Laplacian like operators. J. Math. Anal. Appl. **409**(2), 1093–1106 (2014)
13. Zhang, C., Agarwal, R.P., Bohner, M., Li, T.: Oscillation of fourth-order delay dynamic equations. Sci. China Math. **58**(1), 143–160 (2015)
14. Wang, L., Yang, B., Chen, Y., Zhang, X., Orchard, J.: Improving neural-network classifiers using nearest neighbor partitioning. IEEE Trans. Neural Netw. Learn. Syst. (2016). doi:10.1109/TNNLS.2016.2580570
15. Powell, M.J.: An efficient method for finding the minimum of a function of several variables without calculating derivatives. Comput. J. **7**(2), 155–162 (1964)
16. Powell, M.J.D.: A fast algorithm for nonlinearly constrained optimization calculations. In: Watson, G.A. (ed.) Numerical Analysis. LNM, vol. 630, pp. 144–157. Springer, Heidelberg (1978). doi:10.1007/BFb0067703
17. Yao, X., Liu, Y., Lin, G.: Evolutionary programming made faster. IEEE Trans. Evol. Comput. **3**(2), 82–102 (1999)
18. Suganthan, P.N., Hansen, N., Liang, J.J., Deb, K., Chen, Y.P., Auger, A., Tiwari, S.: Problem definitions and evaluation criteria for the CEC 2005 special session on real-parameter optimization. KanGAL report 2005005 (2005)
19. Holland, J.H.: Adaptation in natural and artificial systems. MIT Press (1992)
20. Yuhui, S., Eberhart, R.C.: Empirical study of particle swarm optimization (1999)

On the Improvement of PSO Scripts for Slope Stability Analysis

Zhe-Ping Shen[1] and Walter Chen[2(✉)]

[1] Department of Construction and Spatial Design, Tungnan University,
New Taipei, Taiwan
fishfishfishgoo@gmail.com
[2] Department of Civil Engineering, National Taipei University of Technology,
Taipei, Taiwan
waltchen@ntut.edu.tw

Abstract. Landslide is a frequently repeated problem in many part of the world. This study follows up on a 2015 study, which used the PSO to optimize the STABL program in the slope stability analysis and found the best result in the literature (factor of safety = 1.238). In the current study, a modification of the PSO scripts was proposed and two new parameters (cohesion intercept and frictional angle) were added to the analysis to examine the effect of soil strength parameters. Assuming a negative correlation, it was found that the cohesion intercept had a bigger influence than the frictional angle on the final outcome in the analysis of a sample homogeneous soil slope. The addition of new parameters also reduced the FS of the optimized sliding surface.

Keywords: STABL · PSO · Landslides · Slope stability

1 Introduction

Landslide is a frequently repeated problem in many part of the world. Previous study has shown that its analysis could be improved by the combined use of the PSO (Particle Swarm Optimization) and the STABL computer program developed by Purdue University [1]. When this technique was applied to a theoretical 2D soil slope, it was found that the Factor of Safety (FS) was reduced to 1.238, which was the best value when it was compared with literature data [2, 3]. In the present study, a modification of the PSO scripts was proposed, and two new parameters (cohesion intercept and frictional angle) were added to the analysis to explore the influence of soil parameters. New results were obtained and presented in 2D/3D parameter spaces.

2 PSO and Slope Stability Analysis

PSO is a unique artificial intelligence algorithm, which has the merits of simplicity and good physical intuition. It was applied to the analysis of slope stability (finding the minimum FS and the most likely sliding surface) with the following basic equations [4–6]:

© Springer International Publishing AG 2017
Y. Tan et al. (Eds.): ICSI 2017, Part I, LNCS 10385, pp. 174–179, 2017.
DOI: 10.1007/978-3-319-61824-1_19

$$v = wv + c_1 r_1 (pbest - x) + c_2 r_2 (gbest - x) \tag{1}$$

$$x = x + v \tag{2}$$

where v and x are velocity and position of individual particles, r_1 and r_2 are random numbers, w, c_1, and c_2 are parameters, and $pbest$ and $gbest$ are the best positions experienced by individual particles or the entire group. Equations (1) and (2) are the same equations presented in the original papers. Despite their simplicity, the equations were applied to the slope stability problem successfully with a special formulation of the problem, which used a mapping from the physical space of the slope to an imaginary 2-D solution space [1].

3   Improvement of PSO Scripts

STABL is a computer program that is well perceived by civil engineers working on construction and development related projects. Developed and maintained by Purdue University [7], the program has many users worldwide. In the previous study [1], PSO was added to STABL externally by scripts written in Fortran. The scripts both controlled the execution of STABL and the analysis of its results. Although the focus at that time was on the determination of trial sliding surfaces, the approach was flexible enough to allow the continual improvement and the addition of new parameters. In this study, the following new parameters were added:

> Cohesion intercept (c) in the range of $[0.52c, \ 1.48c]$
> Frictional angle (ϕ) in the range of $[0.737\phi, \ 1.263\phi]$

In addition, we assumed that c and ϕ were negatively correlated (varied inversely) in order to compare the relative influence of c and ϕ on FS. Originally, the mapping of parameters to be optimized by PSO was from the physical space of the slope to an imaginary 2-D solution space. With the addition of two new parameters c and ϕ, the problem became a mapping from the physical space of the slope to an imaginary 4-D solution space. Therefore, the dimension of the problem increased from two to four. The goal of this study was to examine if PSO still converged to the optimal solution, and which parameter (c and ϕ) had the uttermost influence on the final outcome.

4   Results and Discussion

The test case used in the present analysis is the same soil slope found in [1–3]. The slope was 5-m high with homogeneous properties: $\phi = \phi' = 10^0$, $c = c' = 9.8$ kPa, and $\gamma = 17.64$ kN/m^3. The original Fortran scripts were modified to consider the added dimensions, and the new results were plotted as shown in Figs. 1, 2, 3, 4, 5 and 6.

Figure 1 shows a comparison between the initial sliding surface (FS = 1.243) and the optimized sliding surface (FS = 0.87). The optimized sliding surface (in green) is shallower and closer to the toe of the slope. Figure 2 shows another comparison

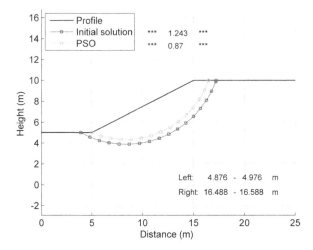

Fig. 1. A comparison between the initial sliding surface and the optimized sliding surface (using 4-parameter PSO). (Color figure online)

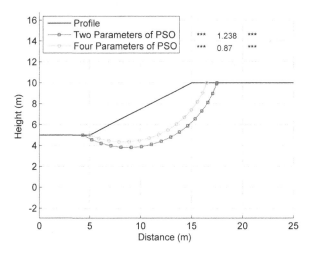

Fig. 2. A comparison between the previous optimized sliding surface (using 2-parameter PSO) and the current optimized sliding surface (using 4-parameter PSO). (Color figure online)

between the optimized sliding surface (FS = 1.238) in the previous study [1] and the optimized sliding surface (FS = 0.87) in the current study. The decrease of FS from 1.238 to 0.87 is significant, and the figure also shows a shallower sliding surface as a result of adding more parameters.

Next, we examine the movement of individual particles in the parameter space. Because it was not possible to display a 4D space directly, we presented the results in 2D (Figs. 3 and 4) and 3D spaces (Figs. 5 and 6) instead. Figure 3 shows the movement of 10 particles (200 times each) in the left boundary versus cohesion intercept space.

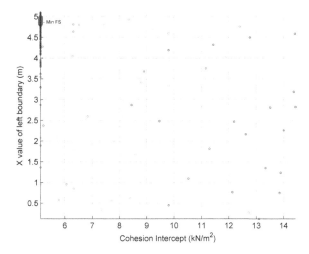

Fig. 3. The movement of particles (from blue to red) showing convergence to a small area (min FS) in the upper-left corner of the figure in the left boundary versus cohesion intercept space. (Color figure online)

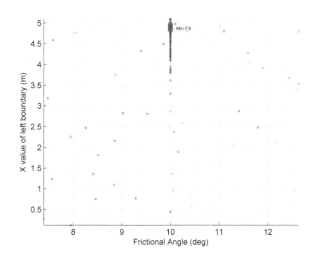

Fig. 4. The movement of particles (from blue to red) showing convergence to a small area (min FS) in the top-center of the figure in the left boundary versus frictional angle space. (Color figure online)

Color was used to denote the relative value of FS of particles. The closer to red the color is, the smaller value of FS it will represent. In contrast, the closer to blue the color is, the bigger value of FS it will symbolize. It can be seen from Fig. 3 that the particles converged to a small area (min FS) in the upper-left corner of the figure. Similar result is observed in Fig. 4, which shows the movement of 10 particles (200 times each) in the left boundary versus frictional angle space. The particles were also color-coded to

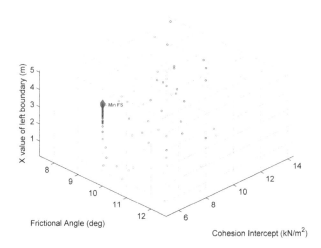

Fig. 5. The movement of particles (from blue to red) showing convergence to a small area (min FS) in the 3D parameter space of left boundary, frictional angle, and cohesion intercept. (Color figure online)

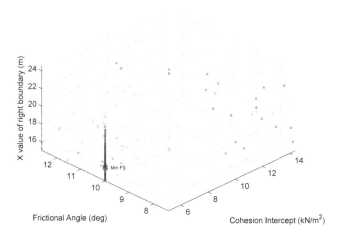

Fig. 6. The movement of particles (from blue to red) showing convergence to a small area (min FS) in the 3D parameter space of right boundary, frictional angle, and cohesion intercept. (Color figure online)

represent their FS values, and they also converged to a small area (min FS) in the top-center of the figure.

Finally, if the movement of particles are examined in 3D parameter spaces, they can be shown in Figs. 5 and 6. The three axes in Fig. 5 are left boundary, frictional angle, and cohesion intercept, whereas the three axes in Fig. 6 are right boundary, frictional angle, and cohesion intercept. Both figures show the convergence of particles to a fixed point of $\phi = 10^0$ and $c = 5.1$ kPa (9.8 * 0.52).

5 Summary and Conclusions

In this study, we proposed a modification to the PSO scripts to include more parameters (c and ϕ) in the analysis of slope stability with STABL. The obvious advantage of this approach over the previous one is that it enables a more thorough consideration of slope stability with the inclusion of soil strength parameters (and their uncertainties). The results were reasonable and satisfactory. The addition of new parameters reduced the FS of the optimized sliding surface because c and ϕ were allowed to change in the optimization process. The resulting sliding surface was shallower and closer to the toe of the slope. Since a negative correlation was assumed between c and ϕ, we found that c had a bigger influence on the final result. The particles converged to a fixed point of $\phi = 10^0$ and $c = 9.8 * 0.52 = 5.1$ kPa, which was the point of $[0.52c, \phi]$ in the parameter space. However, it is speculated that the obtained result is highly linked to the particular c and ϕ values used in the computation. If a different slope is considered (hence different c and ϕ), it is possible that a different result (ϕ more important over c) may be obtained. Therefore, the authors plan to further verify the validity of the proposed algorithm by applying the method to the real-world landslide slope discussed in [8–10] in future studies.

References

1. Chen, W.W., Shen, Z.-P., Wang, J.-A., Tsai, F.: Scripting STABL with PSO for analysis of slope stability. Neurocomputing **148**, 167–174 (2015)
2. Cheng, Y.M., Li, L., Chi, S., Wei, W.B.: Particle swarm optimization algorithm for the location of the critical non-circular failure surface in two-dimensional slope stability analysis. Comput. Geotech. **34**, 92–103 (2007)
3. Malkawi, A.I.H., Hassan, W.F., Sarma, S.K.: Global search method for locating general slip surface using Monte Carlo techniques. J. Geotech. Geoenvironmental Eng. **127**(8), 688–698 (2001)
4. Kennedy, J., Eberhart, R.: Particle swarm optimization. In: Proceedings of the IEEE International Conference on Neural Networks, pp. 1942–1948. IEEE Service Center (1995)
5. Eberhart, R., Kennedy, J.: A new optimizer using particle swarm theory. In: Proceedings of the Sixth Symposium on Micro Machine and Human Science, pp. 39–43. IEEE Service Center (1995)
6. Shi, Y., Eberhart, R.C.: Parameter selection in particle swarm optimization. In: Porto, V.W., Saravanan, N., Waagen, D., Eiben, A.E. (eds.) EP 1998. LNCS, vol. 1447, pp. 591–600. Springer, Heidelberg (1998). doi:10.1007/BFb0040810
7. Siegel, R.A.: STABL User Manual. Purdue University, West Lafayette (1975)
8. Shen, Z.-P., Chen, W.W.: Profile orientation and slope stability analysis. Sci. Program. **2016**, 1–10 (2016)
9. Shen, Z.-P., Chen, W.W.: Directional analysis of slope stability using a real example. In: Tan, Y., Shi, Y., Buarque, F., Gelbukh, A., Das, S., Engelbrecht, A. (eds.) ICSI 2015. LNCS, vol. 9140, pp. 176–182. Springer, Cham (2015). doi:10.1007/978-3-319-20466-6_19
10. Shen, Z.-P., Chen, W.W.: Slope stability analysis using multiple parallel profiles. In: Proceedings of the 11th International Conference on Natural Computation (ICNC). IEEE (2015)

A High-Dimensional Particle Swarm Optimization Based on Similarity Measurement

Jiqiang Feng[1], Guixiang Lai[1], Shi Cheng[2(✉)], Feng Zhang[3], and Yifei Sun[4]

[1] Institute of Intelligent Computing Science, Shenzhen University, Shenzhen, China
`fengjq@szu.edu.cn`
[2] School of Computer Science, Shaanxi Normal University, Xi'an, China
`cheng@snnu.edu.cn`
[3] School of Electrical Engineering, Xi'an Jiaotong University, Xi'an, China
[4] School of Physics and Information Technology, Shaanxi Normal University,
Xi'an, China

Abstract. Particle Swarm Optimization (PSO) is a kind of classical population-based intelligent optimization methods that widely used in solving various optimization problems. With the increase of the dimensions of the optimized problem, the high-dimensional particle swarm optimization becomes an urgent, practical and popular issue. Based on data similarly measurement, a high-dimensional PSO algorithm is proposed to solve the high-dimensional problems. The study primarily defines a new distance paradigm based on the existing similarity measurement of high-dimensional data. This is followed by proposes a PSO variant under the new distance paradigm, namely the LPSO algorithm, which is extended from the classical Euclidean space to the metric space. Finally, it is showed that LPSO could obtain better solution at higher convergence speed in high-dimensional search space.

Keywords: High-dimensional particle swarm optimization · Data similarity measure function · *Lclose* distance

1 Introduction

Particle Swarm Optimization (PSO) is optimized by the information interaction between individuals in the group [6,9]. Compared with other swarm intelligent optimization algorithms, PSO has the characteristics of fast convergence speed and a few parameters, thereby it is widely used for solving optimization problems, such as the optimization of neural network structure research and industrial system optimization, etc. With the complexity of the industrial production process, the dimension of optimization becomes higher and higher, even to super high-dimensional optimization problems [3]. Most of the existing PSO algorithms are optimized for 10 to 30 dimensions, and only a small part of the study is for 100-dimensional or higher optimization problems. PSO algorithms perform good for low-dimensional optimization problems, but often powerless

© Springer International Publishing AG 2017
Y. Tan et al. (Eds.): ICSI 2017, Part I, LNCS 10385, pp. 180–188, 2017.
DOI: 10.1007/978-3-319-61824-1_20

and need to be improved when dealing with variable-related high-dimensional and super-high-dimensional optimization problems. Unlike the low-dimensional optimization problem, the following issues should be considered in dealing with high-dimensional optimization problems [10]:

1. With the increase of the dimensions, the solution space of the algorithm increases sharply, and the combination of possible solutions is exponentially increasing. Thus, the optimal solution could not be found in the polynomial time. Meanwhile, the number of local minima of multimodal function increases exponentially with the increase of dimensions too.
2. The premature convergence occurs when PSO algorithm solving the high-dimensional and multimodal problems. The search direction of the PSO algorithm is determined by the vector synthesis of each dimension flight velocity, the global search ability of particles become weaker and weaker with the increase of the solution space dimensions. Thereby, weaknesses of swarm intelligent algorithms, such as premature convergence and weak in local search capability, is even more prominent in the super high-dimensional optimization.
3. In the traditional measure, the difference between the maximum and the minimum distance of the particles is approaching 0 with the increase of dimensions. Due to the characteristics of high-dimensional space, the similarity information of two data samples will be submerged by a few dimensions with large different values. Thus, in the PSO algorithm, the method of comparing different particle fitness value will lose efficacy.

For the issue 1, the hierarchical particle swarm algorithm (HPSO-RSC) is used in [4]. Based on this, an efficient packet method and coevolution framework are combined [7], which making HPSO-RSC to obtain good performance on high-dimensional optimization problems. For the issue 2, a multi-dimensional inertia weight attenuation chaotic PSO algorithm was proposed and it could enhance the group activity and local search ability in the late search period on vertical and horizontal directions.

For the issue 3, the *Lclose* distance is defined according to the existing high-dimensional spatial similarity measure function, which extends the PSO method from the classical Euclidean space to the new metric space. In the framework of classical population diversity measurement, the high-dimensional PSO in the new metric space can get a better solution at a faster convergence rate in the high-dimensional search space.

The remaining of the paper is organized as follows. Section 2 introduces the basic concepts of data similarity metric methods. Section 3 introduces the definition of LPSO. Experiments of two PSO variants (original PSO algorithm and LPSO algorithm) are conducted in Sect. 4. Finally, Sect. 5 concludes with some remarks and future research directions.

2 Similarity Measure of High-dimensional Spatial Data

In the era of big data, data attributes increases rapidly and the research space turns from low-dimensional space to high-dimensional space. Many new problems emerge while trying to utilize the similarity measure of low-dimensional space to solve high-dimensional space issues.

In the similarity algorithm of data metric, set two dots $X = (x_1, x_2, \ldots, x_n)$ and $Y = (y_1, y_2, \ldots, y_n)$ in the n dimension metric space. The traditional distance function includes Euclidean distance, Chebyshev distance, Manhattan distance and Minkowski distance [2,10]. Also, the definition of similarity between data samples is an alternative way to measure the data space. The similarity definition could affect the research results. Cosine measurement is a commonly used similarity coefficient calculation methods.

The results of traditional metric algorithm in a high-dimensional space are often meaningless due to the characteristics of high-dimensional space, such as high-dimensional data, data with large volume and sparsely distribution [11]. A conclusion on the relationship between L_k-norms and the data dimension was conducted in [1]. That is, based on the classic Euclidean distance measurement method, the relative difference between the nearest and longest distance of any two objects in space is approaching 0 with the increase of dimensions.

A method was proposed for calculating the degree of data dissipation based on the gravitational model of the earth [8]. The distance value involved in the gravitational model of the earth is changed to the information entropy between the object attributes, to avoid the calculation of distance in the high-dimensional space. The concept of projection nearest neighbor was proposed, which uses a criterion function to select the relevant dimension and only the relevant dimensions are used to calculate the similarity between the other dots and the dot itself [5]. A similarity function for high-dimensional data is given in Eq. (1), which avoids the fact that the definition of distance function in the original low-dimensional space is not applicable in the high-dimensional space. Thus, the attribute similarity and spatial similarity of high-dimensional data, and the similarity metric function of different data types are combined into a unified $HDsim(X,Y)$ [11]. The definition is given in Eq. (2).

$$Hsim(X,Y) = \frac{\sum_{i=1}^{n} \frac{1}{1+|x_i-y_i|}}{n} \tag{1}$$

$$HDsim(X,Y) = \sum_{i=1}^{n} \frac{\delta(t_x,t_y)}{n} * \frac{1}{1+\omega_i|x_i-y_i|} \tag{2}$$

A close similarity measure function to solve the failure phenomenon when using traditional similarity measure method in high-dimensional space was proposed [10]. Set $X = (x_1, x_2, \ldots, x_n)$ and $Y = (y_1, y_2, \ldots, y_n)$ are two dots in the n-dimensional space, and the proximity function is defined as shown in Eq. (3)

$$Close(X,Y) = \frac{\sum_{i=1}^{n} e^{-|x_i-y_i|}}{n} \tag{3}$$

The minimum value of the function is 0, which means the difference between X and Y in each dimension is close to infinity, and the similarities between X and Y are minimized at this time. The maximum value of the function is 1, that represents X and Y are equal in all dimensions. The X and Y in the n-dimensional space is coincident at this time, which means the maximum similarity.

3   High-Dimensional PSO Algorithm

3.1   Definition of *Lclose* Distance

It is proved that the dimensions with the similar values have a major importance in close function comparing with the traditional distance-based functions [10]. Larger amount of similar dimensions means greater similarity, which is consistent with the logical of judging the similarity of things.

Lclose **distance**: Let $X = (x_1, x_2, \ldots, x_n)$ and $Y = (y_1, y_2, \ldots, y_n)$ are two points in the n-dimensional search space. Based on the close function, *Lclose* distance is defined as:

$$Lclose(X, Y) = \frac{n}{\sum_{i=1}^{n} e^{-|x_i - y_i|}} - 1 \tag{4}$$

It usually happens that two similar particles closely at most dimensions and only varied a few dimensions in high-dimensional space. However, the large difference at a few dimensions plays the dominant role in the traditional distance similarity measure functions. For example, set n dimension space has A, B, C three dots, $A = (1, 1, 1, \ldots, 1, 1)$, $B = (1, 1, 1, \ldots, 1, 2n)$, and $C = (5, 5, 5, \ldots, 5, 5)$, respectively. The values of A and B are exactly the same on the first $(n-1)$ dimensions, but the distance on the nth dimension is $2n - 1$, A and C differs by 4 in each dimension. According to subjective logic, it should be considered A and B is more similar.

$$D_2(A, B) = \sqrt{\sum_{i=1}^{n} (a_i - b_i)^2} = 2n - 1 \quad D_2(A, C) = \sqrt{\sum_{i=1}^{n} (a_i - c_i)^2} = 4\sqrt{n}$$

It can be seen that in the traditional Euclidean space, as the dimension increases, distance between A and B are more obvious than A and C, it gains a large deviation from the subjective logic. This leads to evolutional failure of the PSO algorithm in its iterative process. Based on the definition of *Lclose* distance in Eq. (4), the AB, AC distances are calculated as follows:

$$Lclose(A, B) = \frac{n}{\sum_{i=1}^{n} e^{-|a_i - b_i|}} - 1 = \frac{n}{n - 1 + e^{1-n}} - 1$$

$$Lclose(A, C) = \frac{n}{\sum_{i=1}^{n} e^{-|a_i - c_i|}} - 1 = e^4 - 1$$

Compared with the traditional Euclidean distance, $Lclose$ distance has the following characteristics:

1. High-dimensional data distance is more coordinated with objective logic, to increase their proportions for most of the very similar dimensions, and to weaken their proportions for a few dimensions with large difference.
2. When the dimension is high, the distance value is smaller than the traditional Euclidean distance.

Under this distance paradigm, the distance between the particles is more coordinated with the evolution of PSO algorithm logic. Therefore, a new PSO algorithm, namely LPSO algorithm is designed.

3.2   PSO Algorithm Based on $Lclose$ Distance Space

In the original PSO algorithm, the particle swarm velocity and position iteration formula are:

$$v_i^{t+1} = w \times v_i^t + c_1 \times r_1 \times (pbest_i^t - x_i^t) + c_2 \times r_2 \times (gbest_i^t - x_i^t) \tag{5}$$
$$x_{i,j}^{t+1} = x_{i,j}^t + v_{i,j}^{t+1} \tag{6}$$

where $v_{i,j}$ is the velocity of each particle i in j dimension, w is the inertia weight, $pbest$ is the historical optimal position of the particle, and $gbest$ is the global optimal position of the particle swarm, c_1 and c_2 are the accelerating coefficients of individual cognition and social cognition respectively.

In this paper, based on $Lclose$ distance and the sign function, in the $Lclose$ distance space, in order to avoid the influence of the high-dimensional characteristics, which is the similarity information of the two data is submerged by a few dimensions of large different values, and the traditional PSO algorithm is difficult to achieve the search effect in high dimension distance space. So that to improve the search effect of population in high dimension space. In the $Lclose$ distance space, this paper designs the new PSO speed and position iteration update formula as follows:

$$v_{i,j}^{t+1} = w \times v_{i,j}^t + c_1 \times r_1 \times \delta(p_{i,j}^t - x_{i,j}^t) \times (\sum_{i=1}^{S} Lclose(x_{i,j}^t, p_{i,j}^t))$$
$$+ c_2 \times r_2 \times \delta(p_{i,j}^t - g_{i,j}^t) \times (Lclose(x_{i,j}^t, g_{i,j}^t)) \tag{7}$$

and

$$Lclose(X,Y) = \frac{n}{\sum_{i=1}^{n} e^{-|x_i - y_i|}} - 1, \quad \delta(a) = \begin{cases} 1, & a \geq 0 \\ 0, & a < 0 \end{cases}$$

In addition, in order to ensure that the particles remain in the search area during the iterative process, the LPSO algorithm sets a limit function on boundary constraints. When a particle exceeds the search area on a dimension, the particle will scale down on each dimension to ensure step size of the particle is not greatly changed.

4   Experimental Study

4.1   Experimental Process and Parameter Setting

In order to compare the convergence among the algorithms, the process of the simulation experiment is as follows:

1. Determining the search space and population size S, the number of iterations t, the inertia weight w and the acceleration coefficients c_1, c_2.
2. Randomly initializes the initial position and initial velocity of each particle. Calculate the historical optimal p_i of the particle and the global optimal g of the population.
3. Iteration according to Eqs. (5) and (7) respectively, update the position and velocity of particles. Updating the historical best p_i of particles and the global optimal g of the population.
4. Limiting the search area of particles. If the updated particle position exceeds the search interval, the operation is performed: $x_{i,j} = 0.8 \times x_{i,j}$.
5. Calculating the population diversity.
6. Determining whether the termination conditions are met. If yes, finish the algorithm and output the result g. If no, return to step 3.

In this paper, the parameter settings are shown in Table 1.

Table 1. Parameters setting of different PSO algorithms

Population size S	$Iteration_{\max}$	$c_1 = c_2$	w
40	2000	2.0	0.5

4.2   Experimental Results

In order to analyze the performance of the PSO algorithm and the LPSO algorithm on solving problem with different dimensions, a simple single-peak function, Sphere function, is selected in this experiment.

$$f = \sum_{i=1}^{n} x_i^2, \quad x_i \in [-100, 100] \tag{8}$$

Experimental 1: The numbers of dimensions are set to 50, 100, 200, and 500. In order to show the comparison of the convergence speed of the functions more intuitively, the experimental results are shown logarithmically as $\log(f)$. The experimental results are shown in Fig. 1. The dotted line in Fig. 1 represents the convergence result of the traditional PSO algorithm under four dimensions, and the solid line represents the LPSO algorithm. In the low dimension (100 dimensions), the traditional PSO algorithm embodies the characteristics of its fast convergence, but when the dimension reaches 500° and higher, it appears

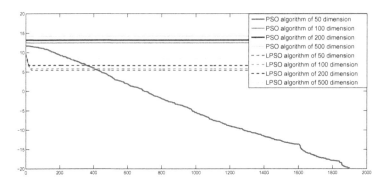

Fig. 1. PSO algorithm and LPSO algorithm in different dimensions of function fitness $(\log(f))$

various kinds of high-dimensional space Problems, the algorithm reaches to a higher degree of fitness and then stops convergence. While the LPSO algorithm is significantly less affected by dimensions. In the low dimensions, the convergence effect is slightly worse than that of the PSO algorithm. However, in the 500-fold and higher dimensions, the LPSO algorithm is less effective than the traditional PSO algorithm, and shows the superiority of the LPSO algorithm in the high-dimensional space.

The performance of various PSO algorithm could be explained via the population diversity changes during the search process. The diversity comparison is shown in Fig. 2. The decrease of diversity will lead to the decline of the global search ability of the algorithm, so it can be conducted that the LPSO algorithm has advantages in the global optimization ability in the high dimensional space.

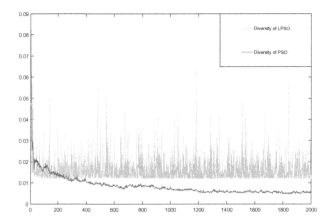

Fig. 2. Diversity comparison of LPSO and PSO algorithms

Experimental 2: The numbers of dimensions are set to 100, 500, and 1000. The results of the LPSO algorithm, PSO algorithm, and HPSO algorithm [7] are compared. In order to reflect the search performance of the proposed algorithm, the three algorithms are tested on problems with three kinds of dimensions under 20 runs respectively. The average convergence and stability result of the three experimental results are compared. The experimental results are given in Table 2.

Table 2. Experimental results of LPSO algorithm, PSO algorithm and HPSO algorithm

Dimensions	Index	PSO algorithm	HPSO algorithm	LPSO algorithm
100	Mean	**0.0146**	0.0225	50.4241
	Maximum	0.0442	0.0901	246.725
	Minimum	0.0041	0.0038	**1.287E-9**
500	Mean	279.6171	**0.4015**	0.8762
	Maximum	516.2834	0.7136	5.7592
	Minimum	102.9056	0.1520	**5.547E-4**
1000	Mean	829.8319	74.34	**12.863**
	Maximum	3.022E+3	422.2	62.9891
	Minimum	447.1972	0.523	**0.0026**

Three indicators, which include mean, maximum, and minimum, are used to measure the convergence effect in the Table 2. According to the statistical results of Table 2, it could be known that the efficiency of LPSO algorithm is far less than that of PSO algorithm and HPSO algorithm in 100-dimensional space. However, in the higher dimensional space, the performance of the PSO algorithm degrades obviously. In the 500-dimensional space, due to the high-dimensional space characteristics, the performance of the traditional PSO algorithm is not satisfied, which indicates the ineffective of the traditional PSO algorithm. Compared with the traditional algorithm, the LPSO algorithm has obvious advantages in the 500 dimensional space, but it is less than the HPSO algorithm. The maximum and minimum value of the 20 experimental results have a large standard deviation. However, in general, it is still better than the traditional PSO algorithm in the convergence performance. It could be concluded that, in the high-dimensional space the LPSO algorithm has a superiority of convergence than PSO algorithm, and this strength is more obvious with the increasing of dimensions.

5   Conclusions

Based on the close similarity measure function, *Lclose* distance is defined by the analysis of the data similarity measurement in high-dimensional space. Compared with the traditional Euclidean distance, the *Lclose* distance between particles is more coordinated with the objective logic for the high-dimensional data, which increases the specific gravity for most very similar dimensions and weakens the proportion of the dimensions with large difference in values. Then the *Lclose* distance is applied to the PSO algorithm, and the LPSO algorithm is designed. The proposed LPSO algorithm is only a preliminary research result. The efficiency of LPSO algorithm in solving high-dimensional optimization problems will be improved continuously in the following research.

Acknowledgments. This work was supported in part by the National Natural Science Foundation of China under Grant 61401283, in part by the Educational Commission of Guangdong Province, China under Grant 2014KTSCX113, and in part by the Fundamental Research Funds for the Central Universities under Grant GK201703062 and GK201603014.

References

1. Aggarwal, C.C.: On the effects of dimensionality reduction on high dimensional similarity search. In: Proceedings of the Twentieth ACM SIGMOD-SIGACT-SIGART Symposium on Principles of Database Systems, pp. 256–266 (2001)
2. Cheng, S., Shi, Y., Qin, Q.: Particle swarm optimization based semi-supervised learning on Chinese text categorization. In: Proceedings of 2012 IEEE Congress on Evolutionary Computation (CEC 2012), Brisbane, Australia, pp. 3131–3198. IEEE (2012)
3. Cheng, S., Zhang, Q., Qin, Q.: Big data analytics with swarm intelligence. Ind. Manag. Data Syst. **116**(4), 646–666 (2016)
4. Janson, S., Middendorf, M.: A hierarchical particle swarm optimizer and its adaptive variant. IEEE Trans. Syst. Man Cybern. Part B (Cybern.) **35**(6), 1272–1282 (2005)
5. Kakas, A., Moratis, P.: Argumentation based decision making for autonomous agents. In: Proceedings of the Second International Joint Conference on Autonomous Agents and Multiagent Systems (AAMAS 2003), Melbourne, Australia, pp. 883–890 (2003)
6. Kennedy, J., Eberhart, R., Shi, Y.: Swarm Intelligence. Morgan Kaufmann Publishers, San Francisco (2001)
7. Ma, Z.: Research on particle swarm optimization for high-dimensional and multi-objective optimization problems. Master's thesis, Dalian University of Technology (2014)
8. Modgil, S.: Nested argumentation and its application to decision making over actions. In: Parsons, S., Maudet, N., Moraitis, P., Rahwan, I. (eds.) ArgMAS 2005. LNCS, vol. 4049, pp. 57–73. Springer, Heidelberg (2006). doi:10.1007/11794578_4
9. Qin, Q., Cheng, S., Zhang, Q., Li, L., Shi, Y.: Particle swarm optimization with interswarm interactive learning strategy. IEEE Trans. Cybern. **46**(10), 2238–2251 (2016)
10. Shao, C.S., Lou, W., Yan, L.M.: Optimization of algorithm of similarity measurement in high-dimensional data. Comput. Technol. Dev. **21**(2), 1–4 (2011)
11. Xie, M., Guo, J., Zhang, H., Chen, K.: Research on the similarity measurement of high dimensional data. Comput. Eng. Sci. **32**(5), 92–96 (2010)

A Center Multi-swarm Cooperative Particle Swarm Optimization with Ratio and Proportion Learning

Xuemin Liu, Lili[(⊠)], and Jiaoju Ge[(⊠)]

Harbin Institute of Technology (Shenzhen), Shenzhen 518055, China
ximlli@126.com, jiaoge@hit.edu.cn

Abstract. This paper presents a center multi-swarm cooperative PSO with ratio and proportion learning (CMCPSO-RP), employing two well-known psychology theories. In the original MCPSO-CC, the convergence speed can be accelerated which comes at decreasing the diversity of sub-swarms, suffering from premature convergence. There is no mechanism to guarantee every possible region of the search space could be searched. To tackle this problem, all best particles from each sub-swarm can be collected and sent to master swarm to maintain a population of potential solutions. This process is less prone to becoming trapped in local minima, but typically has lower efficiency of iterations. To balance the ability of exploration and exploitation, a ratio and proportion learning strategy is proposed by empowering the searching particles with human-like thinking and cognitive process, inspired by Cognitive Load Theory and Human Problem Solving Theory. In our approach, a reasonable ratio design can be not only a way to exhibit a solution quality versus speed tradeoff, but also make CMCPSO-RP more in line with the laws of regular learning in nature. Application of the newly developed PSO algorithm on several benchmark optimization problems shows a marked improvement in performance over the comparison algorithms on all test functions.

Keywords: Multi-swarm · Center communication · Ratio and proportion learning

1 Introduction

Particle swarm optimization (PSO) was first developed by Kennedy and Eberhart in 1995 [1]. For general problems, it provides efficient and satisfactory solutions like other meta-heuristic methods and search algorithms [2], such as genetic algorithms (GA) [3] and differential evolution (DE) [4]. Owing to its simple concept and high efficiency, PSO has been used to solve a range of optimization problems [5], including neural network training [6], data mining [7], feature selection [8], just to name a few. Considering the fact that the original PSO works better in simple and low-dimensional search space, various attempts have been made to improve its performance from aspects of topology, parameter control mechanism, learning strategy and hybridization. For solving complex multimodal problems, numerous multi-swarm techniques, aiming to improve the population diversity, are studied, such as dynamic multi-swarm PSO

© Springer International Publishing AG 2017
Y. Tan et al. (Eds.): ICSI 2017, Part I, LNCS 10385, pp. 189–197, 2017.
DOI: 10.1007/978-3-319-61824-1_21

(DMS-PSO) [9], cooperative PSO (CPSO) [10] and multi-swarm cooperative PSO (MCPSO) [11]. Though these efforts, preventing premature convergence while retaining the fast-converging feature of PSO is still a challenging task. Hence, it is necessary to design a method that performs well in controlling the compatibility between exploration and exploitation. Note that, to our knowledge, few scholars pay attention to augment PSO from the perspective of cognitive process. Given this, Human Problem Solving Theory (HPST) [12] and the Cognitive Load Theory (CLT) [13–15] are incorporated, empowering the searching particles with human-like thinking. The goals of this work are as follows: (1) Reinterpret learning process with HPST and CLT in PSO; (2) Design a ratio and proportion learning strategy to balance convergence rate and diversity; (3) Empirically measure the effect of the learning method proposed.

The remainder of this article is organized as follows: Introduce the framework of CMCPSO-RP in Sect. 2. Then experiment setting and results are presented in Sect. 3. Followed by conclusions and future work in Sect. 4.

2   The Proposed MCPSO-RP

2.1   Reinterpret Learning Process with HPST and CLT

HPST explains the nature of problem solving [12]. When handling specified tasks, a task analysis is primary, including what processes are used, and what mechanisms perform these processes. Then seek an inventory of possible problem solving mechanisms. At last, summarize these experiences, select what can be learned from, and the relation of these to the problem-solving process.

Compared with current learning methods used in PSOs, HPST takes advantage of logic in learning and ways of thinking. As for the strategy to assign the ratio, we find the answer in CLT, stating that if interactions between many elements must be learned, then intrinsic cognitive load will be high, which will lead to a low efficiency of learning. In contrast, if elements can be learned successively rather than simultaneously, intrinsic cognitive load will be low and the learning process will be more effective [13].

2.2   CMCPSO-RP

The motivation for developing CMCPSO-RP derives from MCPSO-CC [16]. For the original MCPSO-CC, while promising, there is still room to improve its performance.

- The center communication mechanism sacrifice the diversity.
- It cannot guarantee to search every possible region of the search space.
- The particle updates its velocity with the same opportunity, which is somehow contrary to the laws of regular learning in natural ecosystems.

To address above issues, CMCPSO-RP is proposed based on HPST and CLT. According to CLT and HPS, when facing a new problem, human tend to believe themselves first and try their best to deal with the difficulty based on the existing knowledge, however, if they found it beyond their ability after attempts, they will begin to seek and learn from the neighbors immediately. This regular human problem solving psychology can be also found in other forms of life in nature, from simple cells through to mammals. Inspired by it, we can get the point that the opportunity and significance of each learning part coming from others is different.

In our approach, self-learning and social-learning come from master swarm while the center-learning and global-learning derive from slave swarms. These four parts make up the whole learning aspects which can be seen as an overall cognitive process to refine the global optimization.

(a) **Self-learning:** For the self-learning part (task analysis), a relative small proportion will be better according to HPS, especially in initial iterations, which can reduce interactivity to accelerate the convergence rate, because no good examples can be learned from.

(b) **Social-learning:** The social-learning will increase along with iterations. When best solutions begin to appear, each particle has to give up some of its experience and useless knowledge to accept the right information from neighbors to enhance the social learning.

(c) **Center-learning:** The center communication mechanism will be saved to make sure the information can be transformed among sub-swarms. CLT states that if elements can be learned successively rather than simultaneously because they do not interact, intrinsic cognitive load will be low and the learning process will be more effective.

(d) **Global-learning:** Based on CLT, it seems not hard to understand the high interactivity in global learning. However, this information transaction mechanism will lead a high cognitive load, resulting in low poor performance. We can lower the extraneous cognitive load just by reducing the information transaction between master swarm and slave swarms. As a result, a lower proportion of global learning should be made to get good performance.

To realize this mechanism, this paper proposes a modification on original MCPSO-CC velocity update equation as follows:

$$v_{i(t+1)}^M = w v_{i(t)}^M + \alpha R_1 c_1 (p_i^M - x_{i(t)}^M) + \Phi * \beta R_2 c_2 (p_g^M - x_{i(t)}^M) + (1 - \Phi) * \eta R_3 c_3 (p_g^S - x_{i(t)}^M) + \gamma R_4 c_4 (p_c^S - x_{i(t)}^M)$$

(1)

$$x_{i(t+1)}^M = x_{it}^M + v_{it}^M$$

(2)

For a minimization problem, Φ is a migration factor, given by

$$\Phi = \begin{cases} 1 \, or \, 0, & competitive \ relationship \ between \ M \ and \ S \\ between \ 0 \ and \ 1, & cooperative \ relationship \ between \ M \ and \ S \end{cases}$$

The best performed particle is found after sub-swarms updating their positions, the center particle is updated according to the following formula [22]:

$$P_c^S(t+1) = \frac{1}{N}\sum_{i=1}^{N} p_g^n(t) \tag{3}$$

The variables in Eqs. (1)–(3) are summarized in Table 1 and the pseudo code for MCPSO-CC-RP is listed in Table 2.

Table 1. Nomenclature

Variables	Descriptions	Variables	Descriptions
M	Master swarm	v	Vector of particle's velocity
S	Slave swarm	x	Vector of particle's position
w	Inertia weight	$\alpha, \beta, \eta, \gamma$	Ratio and proportion parameter, $\alpha + \beta + \eta + \gamma = 1$
R	Random number	p_i^M, p_i^S	Best previous particle in master and slave swarms
c_i	Learning factor	p_c^S	Center position of the global best particle

3 Experiment and Result

3.1 Benchmarks

In this section, six nonlinear benchmark functions are performed. All functions are designed to have minima at the origin. They are listed in Table 3.

3.2 Experimental Setting

The performance of CMCPSO-RP was compared with original MCPSO, MCPSO-CC as well as the standard PSO. The parameters setting are listed in Table 4. For fair

Table 2. Pseudocode for CMCPSO-RP

Initialization
Randomly divide the population into s=4
while (termination condition=false)
  do
    for (i=1 to number of slave particles)
      select the fittest global individual p_g^s and p_g^M, update them respectively
      select the center particle, updates its position according to Eq.(3)
  end do
  do
    for i= master swarm
      evolve the master swarm
      update the velocity and position using Eqs.(1),(2) respectively
  end do
  end

Table 3. Test functions

1. Sphere	$f_1(x) = \sum_{i=1}^{D} x_i^2$
2. Rosenbrock	$f_2(x) = \sum_{i=1}^{D} 100 \times (x_{i+1} - x_i^2)^2 + (1 - x_i)^2$
3. Rastrigrin	$f_3(x) = \sum_{i=1}^{D} (x_i^2 - 10\cos(2\pi x_i) + 10)$
4. Griewank	$f_4(x) = 1/4000 \times \sum_{i=1}^{D} x_i^2 - \prod_{i=1}^{D} \cos(x_i/\sqrt{i}) + 1$
5. Ackley	$f_5(x) = -20\exp(-0.2\sqrt{\frac{1}{30}\sum_{i=1}^{D} x_i^2}) - \exp(\frac{1}{30}\sum_{i=1}^{D}\cos 2\pi x_i) + 20 + e$
6. Quadric	$f_6(x) = \sum_{i=1}^{D} ix_i^4 + random[0,1)$

Table 4. Parameters of test functions

Test problem	Dim.	Popul. size	Search range	Initial hypercube	f_{min}
f_1	30	80	$[-100, 100]^n$	$[50, 100]^n$	0
f_2	30	80	$[-100, 100]^n$	$[15, 30]^n$	0
f_3	30	80	$[-5.12, 5.12]^n$	$[2.56, 5.12]^n$	0
f_4	30	80	$[-600, 600]^n$	$[300, 600]^n$	0
f_5	30	80	$[-32, 32]^n$	$[16, 32]^n$	0
f_6	30	80	$[-1.28, 1.28]^n$	$[-0.64, 1.28]^n$	0

comparison, in all cases the population size was set at 80 (all the sub-swarms of MCPSO, MCPSO-CC and CMCPSO-RP include the same particles) and a fixed number of maximum generations 1000 is applied to these six algorithms. A total of 50 runs for each experimental setting are conducted.

3.3   Experimental Results

Table 5 lists the experimental results (i.e. the best, worst, mean and standard deviation of the function values found in 50 runs) on Sphere, Rosenbrock, Rastrigrin, Griewank, Ackley and Quadric functions with each experimental setting.

The experimental results (i.e. the best, worst, mean and standard deviation of the function values found in 50 runs) for each algorithm on each test function are listed in Table 5. As shown in the table, the numbers in bold-face type represent the comparatively best values and all the results were reported as '0.00e+000'. Moreover, the graphs presented in Figs. 1, 2 and 3 illustrate the evolution of best fitness found by four algorithms, averaged for 50 runs for functions f_1–f_6.

Function f_1, relatively simpler, is easy to solve due to it is a unimodal problem with only a single global minimum. All algorithms converged exponentially fast toward the fitness optimum. However, as can be seen from Fig. 1, only CMCPSO-RP method achieved fast convergence rate. This implies that, for simple problems, lower the extraneous cognitive load by reducing the information transaction between master swarm and slave swarms helps MCPSO-RP solves the problem efficiently, namely, a low ratio of global learning is enough to guarantee a good result.

Table 5. Results for all algorithms on benchmark functions

		SPSO	MCPSO	MCPSO-CC	CMCPSO-RP
f_1	Best	4.20e−003	4.42e−007	1.11e−206	**2.32e−228**
	Worst	2.88e−001	9.27e−005	5.43e−178	**1.53e−214**
	Mean	4.77e−002	1.62e−005	1.09e−179	**4.55e−216**
	Std	4.79e−002	1.84e−005	0.00e+000	**0.00e+000**
f_2	Best	1.16e+002	2.08e+001	1.80e+001	**1.08e+000**
	Worst	3.68e+004	4.47e+002	**2.80e+001**	2.89e+001
	Mean	6.04e+003	6.93e+001	2.66e+001	**2.69e+001**
	Std	1.00e+004	7.32e+001	**2.00e+000**	4.37e+000
f_3	Best	2.61e+001	2.09e+001	0.00e+000	**0.00e+000**
	Worst	1.02e+002	5.07e+001	1.56e−004	**5.68e−013**
	Mean	5.82e+001	3.33e+001	3.12e−006	**6.14e−014**
	Std	1.62e+001	6.39e+000	2.21e−005	**1.20e−013**
f_4	Best	2.49e−004	0.00e+000	0.00e+000	**0.00e+000**
	Worst	1.75e−001	4.67e−003	4.55e−015	**0.00e+000**
	Mean	1.95e−002	4.60e−003	1.20e−016	**0.00e+000**
	Std	2.57e−002	8.70e−003	6.71e−016	**0.00e+000**
f_5	Best	1.48e+000	9.68e−004	8.88e−016	**8.88e−016**
	Worst	2.10e+001	2.06e+001	4.44e−015	**8.88e−016**
	Mean	1.64e+001	5.07e+000	9.59e−016	**8.88e−016**
	Std	7.13e+000	8.32e+000	5.02e−016	**0.00e+000**
f_6	Best	6.20e−003	4.35e−009	0.00e+000	**0.00e+000**
	Worst	2.14e+001	3.22e−005	9.71e−316	**3.59e−319**
	Mean	1.47e+000	9.85e−007	1.94e−317	**1.43e−320**
	Std	3.75e+000	4.54e−006	0.00e+000	**0.00e+000**

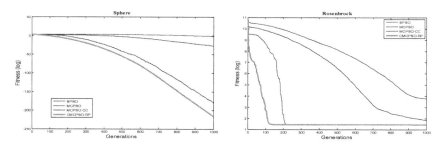

Fig. 1. Mean relative performance for Sphere and Rosenbrock function

Function f_2, a simple multimodal problem, also known as banana function, which has a long, narrow, parabolic shaped flat valley from the perceived local optima to the global optimum. For SPSO and MCPSO, to find the valley is trivial, however, convergence to the global optimum is difficult, which ultimately produced poor average best fitness for this case. In contrast, MCPSO-CC improved the average optimum solution significantly but converged slowly when compared with CMCPSO-RP, as shown in Fig. 1.

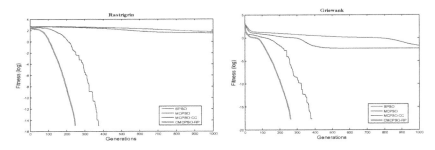

Fig. 2. Mean relative performance for Rastrigrin and Griewank function

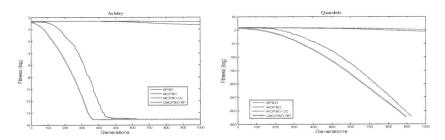

Fig. 3. Mean relative performance for Ackley and Quadric function

Function f_3–f_5 are all complex multimodal function. In this group, Rastrigin has a large number of local optima, when attempting to solve it, algorithms easily fall into a local optimum. Griewank has a $\prod_{i=1}^{D} \cos(x_i/\sqrt{i})$ component causing linkages among variables, thereby making it difficult to reach the global optimum. Ackley has one narrow global optimum basin and many minor local optima. Though these problems are difficult to deal with, MCPSO-RP gave all best optimum results with a particularly fast convergence rate due to it is capable of maintaining a larger diversity, whereas PSO and MCPSO stagnated and flatted out with no further improvement. It can be concluded, for complex multimodal problems, most schemas have to be learned simultaneously because they interact, but learning too much at one time will slow down the convergence rate, so a good ratio and proportion of each learning part can well solve this issue. That is why CMCPSO-RP converged faster than MCPSO-CC, as shown in Figs. 2 and 3.

Function f_6 is a noisy function. CMCPSO-RP had the best convergence speed, followed by MCPSO-CC, MCPSO and SPSO, see Fig. 3. It can be concluded that the introduction of ratio and proportion learning not only accelerates convergence rate but also improves the solution quality.

4   Conclusions and Future Work

Our goal in this paper is, first to reinterpret learning process in PSO with HPST, then design a new learning strategy to balance convergence rate and diversity, finally, empirically measure the effect of the learning method proposed. In this paper, an improved center multi-swarm cooperative PSO with ratio and proportion learning (CMCPSO-RP) is proposed inspired by CLT and the theory of HPST. In CMCPSO-RP, there are four learning aspects of learning: self-learning and social-learning come from master swarm while the center-learning and global-learning derive from slave swarms. According to HPS, self-learning should take a small share. As for global learning, we can reduce extraneous cognitive load by reducing the information transaction between master swarm and slave swarms according to CLT. In contrast, a higher proportion of center learning should be made to keep low information interactivity to improve the efficiency. The results are compared in terms of solution quality and convergence speed with relative previous work. The experimental results demonstrate CMCPSO-RP outperforms the comparison algorithms on all tested functions.

References

1. Eberchart, R.C., Kennedy, J.: Particle swarm optimization. In: IEEE International Conference on Neural Networks, Perth, Australia (1995)
2. Leboucher, C., Shin, H.S., Siarry, P., et al.: Convergence proof of an enhanced particle swarm optimisation method integrated with evolutionary game theory. Inf. Sci. **346**, 389–411 (2016)
3. Hassan, R., Cohanim, B., De Weck, O., et al.: A comparison of particle swarm optimization and the genetic algorithm. In: 46th AIAA/ASME/ASCE/AHS/ASC Structures, Structural Dynamics and Materials Conference, p. 1897 (2005)
4. Qin, A.K., Huang, V.L., Suganthan, P.N.: Differential evolution algorithm with strategy adaptation for global numerical optimization. IEEE Trans. Evol. Comput. **13**(2), 398–417 (2009)
5. Shi, Y.: Particle swarm optimization: developments, applications and resources. In: Proceedings of the Congress on Evolutionary Computation, vol. 1, pp. 81–86. IEEE (2001)
6. Van den Bergh, F., Engelbrecht, A.P.: Cooperative learning in neural networks using particle swarm optimizers. S. Afr. Comput. J. **2000**(26), 84–90 (2000)
7. Rana, S., Jasola, S., Kumar, R.: A review on particle swarm optimization algorithms and their applications to data clustering. Artif. Intell. Rev. **35**(3), 211–222 (2011)
8. Nguyen, H.B., Xue, B., Andreae, P.: Mutual information estimation for filter based feature selection using particle swarm optimization. In: Squillero, G., Burelli, P. (eds.) EvoApplications 2016. LNCS, vol. 9597, pp. 719–736. Springer, Cham (2016). doi:10.1007/978-3-319-31204-0_46
9. Liang, J.J., Suganthan, P.N.: Dynamic multi-swarm particle swarm optimizer with local search. In: The 2005 IEEE Congress on Evolutionary Computation, vol. 1, pp. 522–528. IEEE (2005)
10. Van den Bergh, F., Engelbrecht, A.P.: A cooperative approach to particle swarm optimization. IEEE Trans. Evol. Comput. **8**(3), 225–239 (2004)
11. Niu, B., Zhu, Y., He, X., et al.: MCPSO: a multi-swarm cooperative particle swarm optimizer. Appl. Math. Comput. **185**(2), 1050–1062 (2007)
12. Newell, A., Simon, H.A.: Human Problem Solving. Prentice-Hall, Englewood Cliffs (1972)
13. Sweller, J.: Cognitive load theory, learning difficulty, and instructional design. Learn. Instr. **4**(4), 295–312 (1994)
14. Frederiksen, N.: Implications of cognitive theory for instruction in problem solving. Rev. Educ. Res. **54**(3), 363–407 (1984)

15. Sweller, J.: Cognitive load during problem solving: effects on learning. Cogn. Sci. **12**(2), 257–285 (1988)
16. Niu, B., Li, L.: An improved MCPSO with center communication. In: International Conference on Computational Intelligence and Security, CIS 2008, vol. 2, pp. 57–61. IEEE (2008)

Applications of Particle Swarm Optimization

A Discrete Particle Swarm Algorithm
for Combinatorial Auctions

Fu-Shiung Hsieh[(⊠)]

Department of Computer Science and Information Engineering,
Chaoyang University of Technology, Taichung 41349, Taiwan
fshsieh@cyut.edu.tw

Abstract. Although combinatorial auctions make trading goods between
buyers and sellers more conveniently, the winner determination problem
(WDP) in combinatorial auctions poses a challenge due to computation complexity. In this paper, we consider combinatorial auction problem with transaction costs, supply constraints and non-negative surplus constraints. We
formulate the WDP of combinatorial auction problem as an integer programming problem formulation. To deal with computational complexity of the WDP
for combinatorial auctions, we propose an algorithm for finding solutions based
on discrete PSO (DPSO) approach. The effectiveness of the proposed algorithm
is also demonstrated by several numerical examples.

Keywords: Meta-heuristics · Combinatorial auction · Integer programming ·
Winner determination problem

1 Introduction

Auctions are a popular business model for buying and selling goods. One of the recent
trends in the development of auction mechanisms is combinatorial auctions, which
makes it possible for buyers and sellers to trade goods conveniently by placing bids on
a combination of goods rather than just individual items. However, the winner determination problem (WDP) is one of the most challenging research topics on combinatorial auctions. An excellent survey on combinatorial auctions can be found in [1, 2].
Combinatorial auctions are notoriously difficult to solve from a computational point of
view [3, 4]. The WDP can be modeled as a set packing problem (SPP) [5]. Sandholm
et al. mentions that WDP for combinatorial auction is NP-complete [6]. Many algorithms have been developed for WDP [7–10, 15].

In this paper, we propose a meta-heuristic method to solve the WDP based on Particle
swarm optimization (PSO) approach [11]. PSO is an optimization method developed
based on observations of the social behavior of animals such as bird flocking, fish
schooling and swarm theory. It has been applied successfully to nonlinear constrained
optimization problems [12]. In [13], Kennedy and Eberhart also proposed a reworking of
the algorithm to operate on discrete binary variables. In [13], trajectories of particles are
changes in the probability that a decision variable will take on a zero or one value. In this
paper, we apply the method proposed in [13] to develop a discrete Particle swarm
optimization (DPSO) algorithm for solving the WDP in combinatorial auctions. We also
compare our algorithm with Genetic Algorithm by examples to study its effectiveness.

© Springer International Publishing AG 2017
Y. Tan et al. (Eds.): ICSI 2017, Part I, LNCS 10385, pp. 201–208, 2017.
DOI: 10.1007/978-3-319-61824-1_22

The remainder of this paper is organized as follows. In Sect. 2, we first formulate the WDP for combinatorial auctions. We briefly introduce the Particle swarm approach and propose a DPSO algorithm in Sect. 3 and Sect. 4, respectively. We present our numerical results in Sect. 5 and conclude this paper in Sect. 6.

2 Problem Formulation

We first formulate the combinatorial auction problem as an integer programming problem. We then develop a meta-heuristic algorithm for it. In a combinatorial auction, there are a set of buyers, a seller and a mediator for trading goods between the buyers and seller. Buyers and the seller submit bids to the mediator. We use i to represent the seller. Let N be the number of potential buyers in a combinatorial auction. Each $n \in \{1, 2, 3, \ldots, N\}$ represents a buyer. A buyer may place a number of bids. Let H_n be the number of bids placed by buyer $n \in \{1, 2, 3, \ldots, N\}$. Let K denote the number of different types of items in the combinatorial auction. To represent the $h - th$ bid placed by buyer n, let d_{nhk} denote the buyer- n's desired units of the $k - th$ items in the $h - th$ bid and let p_{bnh} denote the price of the $h - th$ bid submitted by buyer n, where $k \in \{1, 2, 3, \ldots, K\}$ and $h \in \{1, 2, 3, \ldots, H_n\}$. We use $BB_{nh} = (d_{nh1}, d_{nh2}, d_{nh3}, \ldots, d_{nhK}, p_{bnh})$ to represent the $h - th$ bid submitted by buyer n. Similarly, we use $SB_i = (q_{i1}, q_{i2}, q_{i3}, \ldots, q_{iK}, p_{si})$ to represent the $j - th$ bid submitted by seller i, where q_{ik} is a nonnegative integer that denotes the quantity of the $k - th$ items in the $j - th$ bid submitted by seller i and p_{si} is the price of the $j - th$ bid submitted by seller i. Let T_s be the transaction cost coefficient for a seller and let T_b be the transaction cost coefficient for a buyer. The surplus of a combinatorial auction is the difference between winning buyers' total bid price and the seller' total bid price. A combinatorial auction problem aims to maximize the surplus. To formulate the problem, the following notations are defined.

Notations:

K: the number of different types of items in the combinatorial double auction.
N: the number of potential buyers in a combinatorial auction. Each $n \in \{1, 2, 3, \ldots, N\}$ represents a buyer.
H_n: the number of bids placed by buyer $n \in \{1, 2, 3, \ldots, N\}$.
d_{nhk}:   the buyer-n's desired units of the $k - th$ items in the $h - th$ bid, where $k \in \{1, 2, 3, \ldots, K\}$.
h: the $h - th$ bid created by a buyer in a combinatorial double auction.
p_s: the price of the bid submitted by the seller.
q_k: a nonnegative integer that denotes the quantity of the $k - th$ items in the bid submitted by the seller.
x: if the bid placed by the seller is a winning bid, $x = 1$, otherwise $x = 0$.
p_{bnh}:   the price of the $h - th$ bid submitted by buyer n.
y_{nh}:   if the $h - th$ bid placed by buyer n is a winning bid, $y_{nh} = 1$, otherwise $y_{nh} = 0$.
T_s: the transaction cost coefficient for a seller.
T_b: the transaction cost coefficient for a buyer.

The WDP is formulated as follows. There are several constraints for the WDP in combinatorial auctions, including the supply/demand constraints in (1) and non-negative surplus constraints in (2). Note that (1) means that, for each type of item, the total amount of goods supplied by the sellers' winning bids must be greater than or equal to the demands of the buyers' winning bids.

Winner Determination Problem (WDP):

$$\max \left[\left(\sum_{n=1}^{N} \sum_{h=1}^{H} y_{nh}(p_{bnh} + (T_b p_{bnh})) \right) - x(p_s - (T_s p_s)) \right]$$

$$s.t. \quad xq_k \geq \sum_{n=1}^{N} \sum_{h=1}^{H} y_{nh} d_{nhk} \ \forall k \in \{1, 2, \ldots, K\} \tag{1}$$

$$\sum_{n=1}^{N} \sum_{h=1}^{H} y_{nh} p_{bnh} \geq x p_s \tag{2}$$

$$x \in \{0, 1\}, y_{nh} \in \{0, 1\} \quad \forall n, h$$

3 Particle Swarm Approach

In this section, we present our discrete particle swarm algorithm for the WDP in combinatorial auction. A brief introduction to the particle swarm optimization method is given first.

With the standard particle swarm optimization, each particle of the swarm adjusts its trajectory according to its own flying experience and the flying experiences of other particles within its topological neighborhood in a D-dimensional space S. The velocity and position of particle i are represented as $v_i = (v_{i1}, v_{i2}, v_{i3}, \ldots, v_{iD})$ and $p_i = (p_{i1}, p_{i2}, p_{i3}, \ldots, p_{iD})$, respectively. Its best historical position is recorded as $pbest_i = (p_{i1}, p_{i2}, p_{i3}, \ldots, p_{iD})$, which is also called p_{best}. The best historical position that the entire swarm has passed is denoted as $g_{best} = (p_{g1}, p_{g2}, p_{g3}, \ldots, p_{gD})$, which is also called g_{best}. The velocity and position of particle i on dimension d, where $d \in \{1, 2, 3, \ldots, D\}$, in iteration $t + 1$ are updated as follows:

$$v_{id}^{t+1} = \omega v_{id}^{t} + c_1 r_1 (pbest_{id}^{t} - p_{id}^{t}) + c_2 r_2 (gbest_d^{t} - x_{id}^{t})$$

$$p_{id}^{t+1} = p_{id}^{t} + v_{id}^{t}$$

where ω is a parameter called the inertia weight, c_1 and c_2 are positive constants referred to as cognitive and social parameters, respectively, and r_1 and r_2 are random numbers generated from a uniform distribution in the region of $[0, 1]$.

In the WDP, decision variables x_{ij} and y_{nh} can be represented by a binary vector z. The dimension of z is $D = 1 + \sum_{n=1}^{N} H_n$. Each element in vector z is either 0 or 1, where 1 denotes that bid is accepted and 0 denotes that bid is not accepted.

For our problem, the fitness function is $F(x,y)$, which can be described as follows:

$$F(x,y) = \max \left[\left(\sum_{n=1}^{N} \sum_{h=1}^{H} y_{nh} (p_{bnh} + (T_b p_{bnh})) \right) - x(p_s - (T_s p_s)) \right]$$

Note that that the problem formulated in Sect. 2 is an optimization problem with binary decision variables and constraints. In this paper, we adopt a method based on biasing feasible over infeasible solutions [14]. Let $S_f = \{(x,y)|(x,y)$ is a solution in the current population, (x,y) satisfies constraints (1) $\sim$ (2).$\}$ is the set of all feasible solutions in the current population. Let $S_{f\min} = \min\limits_{(x,y)\in S_f} F(x,y)$, the object function value of the worst feasible solution in the current population. The fitness function $F_1(x,y)$ is defined as follows:

$$F_1(x,y) = \begin{cases} F(x,y) & \text{if } (x,y) \text{ satisfies constraints } (1) \sim (2) \\ U_1(x,y) & \text{otherwise} \end{cases},$$

where

$$U_1(x,y) = S_{f\min} + \sum_{k=1}^{K} ((\min(xq_k \sum_{j=1}^{J} - \sum_{n=1}^{N} \sum_{h=1}^{H} y_{nh} d_{nhk}), 0.0))$$

$$+ \min(\sum_{n=1}^{N} \sum_{h=1}^{H} y_{nh} p_{bnh} - xp_s, 0.0)$$

4 Discrete Particle Swarm Algorithm

In this paper, we apply DPSO algorithm to solve the integer programming problem with binary decision variables. To describe the DPSO algorithm, we define the required notations as follows.

M: the number of particles in the population.

S^t: the set of all particles at time t.

N_y: the number of bids placed by all buyers, $N_x = \sum\limits_{n=1}^{N} H_n$.

D: the dimension of each particle.

Z_m^t: the position of particle m at time t, where $m \in \{1, 2, \ldots, M\}$, and $Z_m^t = (x_m^t, y_m^t)$, where x_m^t is the position (one-dimensional vector) corresponding to decision variables x and y_m^t(N_y-dimensional vector)is the position (a vector) corresponding to decision variables y.

PZ_m^t: the personal best of particle m at time t, where $m \in \{1, 2, \ldots, M\}$, and $PZ_m^t = (Px_m^t, Py_m^t)$, where Px_m^t is the personal best position (one dimensional vector) corresponding to decision variables x and Py_m^t is the personal best position (N_y-dimensional vector) corresponding to decision variables y.

GZ^t: the global best at time t, and $GZ^t = (Gx^t, Gy^t)$, where Gx^t is the global best position (one dimensional vector) corresponding to decision variables x and Gy^t is the global best position (a N_y-dimensional vector) corresponding to decision variables y.

c_1:   a non-negative real parameter less than 1.
c_2:   a non-negative real parameter less than 1.
r_1,   a random variable with uniform distribution $U(0,1)$.
r_2,   a random variable with uniform distribution $U(0,1)$.
$V_{\max}$:   A maximum value of velocity.
$s(vx_{mn}^t)$:   the probability of the bit x_{mn}^t.
$s(vy_{mn}^t)$:   the probability of the bit y_{mn}^t.

The DPSO algorithm proposed in this paper is as follows:
Discrete Particle Swarm Optimization (DPSO) Algorithm

Input: $M, N, I, K, BB_{nh}, SB_i, s_{ik}, T_b, T_s$

Output: GZ^t

Step 0: $t \leftarrow 0$

   Generate Z_m^t for each particle $m \in \{1,2,...,M\}$ in the initial population of swarm, S^t

Step 1: While (stopping criteria is not satisfied)

   $t \leftarrow t+1$

   Evaluate each particle Z_m^t in S^t according to the fitness function $F_1(x, y)$

   Determine the personal best of each particle in S^t

   Determine the global best of swarm, G^t

   For each $m \in \{1,2,...,M\}$

      Calculate the velocity of particle i as follows

      Generate r_1, a random variable with uniform distribution $U(0,1)$

      Generate r_2, a random variable with uniform distribution $U(0,1)$

      $$vx_{mn}^t = vx_{mn}^{t-1} + c_1 r_1 (Px_{mn}^t - x_{mn}^t) + c_2 r_2 (Gx_{mn}^t - x_{mn}^t)$$
      $$vx_{mn}^t \in [-V_{\max}, V_{\max}]$$
      $$vy_m^t = vy_{mn}^{t-1} + c_1 r_1 (Py_{mn}^t - y_{mn}^t) + c_2 r_2 (Gy_{mn}^t - y_{mn}^t)$$
      $$vy_{mn}^t \in [-V_{\max}, V_{\max}]$$

      Apply a sigmoid limiting transformation to vx_{mn}^t to obtain $s(vx_{mn}^t)$.

      Apply a sigmoid limiting transformation to vy_{mn}^t to obtain $s(vy_{mn}^t)$.

      Update particle m as follows

      Generate $rsid$, a random variable with uniform distribution $U(0,1)$

      $$x_{mn}^t = \begin{cases} 1 & rsid < s(vx_{mn}^t) \\ 0 & otherwise \end{cases}$$

      Generate $rsid$, a random variable with uniform distribution $U(0,1)$

      $$y_{mn}^t = \begin{cases} 1 & rsid < s(vy_{mn}^t) \\ 0 & otherwise \end{cases}$$

   End While

5 Numerical Results

We conduct several numerical experiments to illustrate the effectiveness of the proposed meta-heuristic algorithm for combinatorial auctions.

Example: Suppose there are one seller and five buyers that will trade goods in a combinatorial auction. There are three types of items (goods). The bids placed by the seller and buyers are shown in Tables 1 and 2, respectively. For this combinatorial double auction problem, we have

$N = 5$, $K = 3$, $T_s = 0.5$, $T_b = 0.3$, $q_1 = 1$, $q_2 = 5$, $q_3 = 0$, $p_s = 168$, $d_{111} = 0$, $d_{112} = 3$, $d_{113} = 0$, $p_{b11} = 484$, $d_{211} = 1$, $d_{212} = 2$, $d_{213} = 0$, $p_{b21} = 472$, $d_{311} = 1$, $d_{312} = 5$, $d_{313} = 0$, $p_{b31} = 574$, $d_{411} = 0$, $d_{412} = 5$, $d_{413} = 0$, $p_{b41} = 545$, $d_{511} = 0$, $d_{512} = 3$, $d_{513} = 0$, $p_{b51} = 484$.

Table 1. The bid placed by the seller.

$k = 1$	$k = 2$	$k = 3$	Price
1	5	0	168

Table 2. The bids placed by five buyers.

Buyer	$k = 1$	$k = 2$	$k = 3$	Price
$n = 1$	0	3	0	484
$n = 2$	1	2	0	472
$n = 3$	1	5	0	574
$n = 4$	0	5	0	545
$n = 5$	4	3	0	484

The parameters for our PSO algorithm are: $M = 10$, $w = 0.4$, $c1 = 0.4$, $c2 = 0.6$, Vmax = 4.

The solutions found by applying our DPSO algorithm for this example are as follows.

$x = 1$, $y_{11} = 1$, $y_{21} = 1$, $y_{31} = 0$, $y_{41} = 0$, $y_{51} = 0$. It is an optimal solution.

In addition to the example above, we also compare the performance of our proposed algorithm with other approach by conduct several experiments. In all these experiments, the number of particles used in our DPSO algorithm is either 10 or 20 and parameters w, c1 and c2 are arbitrarily chosen from [0.3 0.8]. The results are shown in Table 3. It indicates PSO outperforms GA in efficiency and performance.

Table 3. Results for several cases.

Case	N	I	K	GA	CPU time (GA)	PSO	CPU time (PSO)
1	5	1	3	1316.4	8547 ms	1316.4	563 ms
2	8	5	4	847.0	31687 ms	1256.2	1453 ms
3	10	1	5	2542.2	2593 ms	2542.2	1171 ms
4	12	1	6	1155.6	11797 ms	1637.9	1860 ms
5	14	1	7	1619.2	60641 ms	1914.0	1797 ms

6 Conclusions

We formulate the WDP of combinatorial auctions as an integer programming problem. The problem is to determine the winners to maximize the surplus of combinatorial auctions. Due to computational complexity, it is hard to develop a computationally efficient method to find an exact optimal solution for the WDP of combinatorial auctions. To reduce computational complexity, we adopt a meta-heuristic approach and develop a solution algorithm based on discrete Particle swarm optimization method. We conduct experiments to study the performance and computational efficiency of our proposed algorithm. To study the computational efficiency of our proposed algorithm, we conduct the experiments to compare the computational time of our algorithm with Genetic Algorithms. Although our algorithm does not guarantee generation of optimal solutions, the numerical results indicate that our proposed algorithm is significantly more effective in comparison with Genetic Algorithms.

Acknowledgments. This paper is supported in part by Ministry of Science and Technology, Taiwan, under Grant MOST-105-2410-H-324-005.

References

1. de Vries, S., Vohra, R.V.: Combinatorial Auctions: a survey. Inf. J. Comput. **3**, 284–309 (2003)
2. Pekeč, A., Rothkopf, M.H.: Combinatorial auction design. Manag. Sci. **49**, 1485–1503 (2003)
3. Rothkopf, M., Pekeč, A., Harstad, R.: Computationally manageable combinational auctions. Manag. Sci. **44**, 1131–1147 (1998)
4. Xia, M., Stallaert, J., Whinston, A.B.: Solving the combinatorial double auction problem. Eur. J. Oper. Res. **164**, 239–251 (2005)
5. Vemuganti, R.R.: Applications of set covering, set packing and set partitioning models: a survey. In: Du, D.-Z. (ed.) Handbook of Combinatorial Optimization, vol. 1, pp. 573–746. Kluwer Academic Publishers, Dordrecht (1998)
6. Sandholm, T.: Algorithm for optimal winner determination in combinatorial auctions. Artif. Intell. **135**, 1–54 (2002)
7. Hsieh, F.S.: Combinatorial reverse auction based on revelation of Lagrangian multipliers. Decis. Support Syst. **48**, 323–330 (2010)

8. Jones, J.L., Koehler, G.J.: Combinatorial auctions using rule-based bids. Decis. Support Syst. **34**, 59–74 (2002)
9. Hsieh, F.S., Lin, J.B.: Assessing the benefits of group-buying based combinatorial reverse auctions. Electron. Commer. Res. Appl. **11**, 407–419 (2012)
10. Hsieh, F.S., Lin, J.B.: Virtual enterprises partner selection based on re-verse auction. Int. J. Adv. Manuf. Technol. **62**, 847–859 (2012)
11. Kennedy, J., Eberhart, R.C.: Particle swarm optimization. In: Proceedings of IEEE International Conference on Neural Networks, Piscataway, NJ, pp. 1942–1948 (1995)
12. El-Galland, A.I., El-Hawary, M.E., Sallam, A.A.: Swarming of intelligent particles for solving the nonlinear constrained optimization problem. Eng. Intell. Syst. Electr. Eng. Commun. **9**, 155–163 (2001)
13. Kennedy, J., Eberhart, R.C.: A discrete binary version of the particle swarm algorithm. In: IEEE International Conference on Systems, Man, and Cybernetics: Computational Cybernetics and Simulation, vol. **5**, pp. 4104–4108 (1997)
14. Deb, K.: An efficient constraint handling method for genetic algorithms. Comput. Methods Appl. Mech. Eng. **186**, 311–338 (2000)
15. Hsieh, F.S., Liao, C.S.: Schemes to reward winners in combinatorial double auctions based on optimization of surplus. Electron. Commer. Res. Appl. **14**(6), 405–417 (2015)

Registration of GPS and Stereo Vision for Point Cloud Localization in Intelligent Vehicles Using Particle Swarm Optimization

Vijay John[1(⊠)], Yuquan Xu[1], Seiichi Mita[1], Qian Long[2], and Zheng Liu[3]

[1] Research Center of Smart Vehicles, Toyota Technological Institute,
Nagoya, Japan
{vijayjohn,yuquan.xu86,smita}@toyota-ti.ac.jp
[2] Chinese Academy of Sciences, Beijing, China
longqian@ynao.ac.cn
[3] University of British Columbia (Okanagan), Kelowna, BC V1V 1V7, Canada
zheng.liu@ieee.org

Abstract. In this paper, we propose an algorithm for the registration of the GPS sensor and the stereo camera for vehicle localization within 3D dense point clouds. We adopt the particle swarm optimization algorithm to perform the sensor registration and the vehicle localization. The registration of the GPS sensor and the stereo camera is performed to increase the robustness of the vehicle localization algorithm. In the standard GPS-based vehicle localization, the algorithm is affected by noisy GPS signals in certain environmental conditions. We can address this problem through the sensor fusion or registration of the GPS and stereo camera. The sensors are registered by estimating the coordinate transformation matrix. Given the registration of the two sensors, we perform the point cloud-based vehicle localization. The vision-based localization is formulated as an optimization problem, where the "optimal" transformation matrix and corresponding virtual point cloud depth image is estimated. The transformation matrix, which is optimized, corresponds to the coordinate transformation between the stereo and point cloud coordinate systems. We validate the proposed algorithm with acquired datasets, and show that the algorithm robustly localizes the vehicle.

1 Introduction

Vehicle localization in the surrounding environment, such as dense 3D point clouds, is an important research problem for ADAS applications. In the standard localization algorithm, the GPS sensor is used. However, the GPS is error prone in environments, where satellite reception is limited, resulting in localization errors [2]. Researchers address this problem by performing sensor fusion with additional sensors such as camera, LIDAR, and Inertial Navigation Systems (INS) [13]. Since precise INS and LIDAR sensors are expensive, vision and GPS-based vehicle localization is increasingly receiving attention from the research community [13].

© Springer International Publishing AG 2017
Y. Tan et al. (Eds.): ICSI 2017, Part I, LNCS 10385, pp. 209–217, 2017.
DOI: 10.1007/978-3-319-61824-1_23

In this work, we propose to perform the sensor fusion of the stereo camera and GPS for depth-based vehicle localization within 3D dense point cloud using the particle swarm optimization algorithm. In the proposed algorithm, we first register the stereo and GPS coordinate systems in an offline phase. Given the estimated stereo-GPS transformation matrix, we subsequently localize the vehicle in the dense point cloud using depth information, in an online phase, using the particle swarm optimization algorithm. In this phase, stereo vision-based depth information is used within an optimization framework to perform the localization within the dense 3-D point clouds. Given the registered GPS and stereo camera, the localization is performed by estimating the optimal transformation matrix between the stereo and the point cloud coordinate system. The optimal world-stereo transformation matrix along with the GPS-stereo transformation matrix generates an optimal virtual depth map from the point cloud data. The transformation matrices are optimised by measuring the similarity between the corresponding virtual depth maps and the stereo depth maps. To efficiently measure the similarity, we prune the depth map using v-disparity [7] and restrict the similarity measurement to the unpruned regions. The proposed algorithm is validated with an acquired dataset, where we show the robustness of the proposed algorithm.

The rest of this paper is structured as follows. In Sect. 2, we present a literature review of the state-of-the-art. Details of the proposed algorithm are described in Sect. 3. The experimental results are presented in Sect. 4. Finally, we present our observations and directions for future work in Sect. 5.

2 Literature Review

The research problem of vehicle localization, typically, uses the GPS sensor. However, the GPS-based navigation systems suffer from low accuracy and intermittent missing signals under certain conditions [2]. The signal outages or localization errors, typically, occur when the vehicles moves through tunnels and roads surrounded by tall building. Additionally, the GPS errors are also introduced when the satellites are not widely positioned [2]. Researchers addressed this issue by incorporating inertial navigation systems (INS) [5] and visual odometry (VO) [8]. However these methods, inspite of improving the localization, are affected by drift errors.

To address the issue of divergence, researchers have proposed to use environmental maps, in the form of satellite images and dense point clouds, for the vehicle localization [10,11]. Typically, monocular cameras are used for the vehicle localization. In the work by Noda et al. [11] the localization was performed by matching the speeded-up robust features (SURF) observed from the on-board camera with the ones from satellite images. Similarly in [10], detected image lanes were matched with lanes in digital maps using GPS and INS information to perform localization. To enhance the accuracy of localization, Yoneda et al. [13] utilized the LIDAR sensor to perform vehicle localization in dense 3D point clouds, and achieved good localization accuracy. However, the main drawback and limitation of the LIDAR sensor is its high cost. Thus, it is more feasible to use cheaper sensors, like stereo camera, to perform the matching in dense 3D point clouds. In spite of the recent advancements in vehicle localisation techniques, it can be seen that the problem is still not solved. In this paper, we propose to use stereo vision and GPS sensors to localize the vehicle in dense point clouds.

3   Depth-Based Vehicle Localization in Point Cloud Maps

We propose the sensor fusion of stereo and GPS sensor for vehicle localization in point cloud maps using depth information. The point cloud maps are generated as a representation of the vehicle's environment [1]. The 3D point clouds, $\mathbf{P}$, contain latitude, longitude and altitude information, where the latitude and longitude are represented in the 2D Universal Transverse Mercator (UTM) system. To perform localization, we estimate the transformation matrices between the world, vehicle and stereo coordinates. The different coordinate systems are defined as follows. Firstly, the world coordinate system is defined at the origin of point cloud map's UTM system. Secondly, the vehicle coordinate system corresponds to the GPS location in the vehicle. Finally, the stereo coordinate system is defined at the location of the left camera of the stereo pair. An illustration of the different coordinate systems are provided in Fig. 1-a.

The localization is achieved by generating virtual depth maps M_v from the point cloud $\mathbf{P}$, and measuring their similarity with the stereo depth map M_s within a particle swarm optimization (PSO). While, M_s is generated by the MPV algorithm [9], M_v are generated using a set of transformation matrices. This set of matrices correspond to the transformation between the world, vehicle and stereo coordinate systems. Apart from these transformation matrices, the stereo's intrinsic calibration parameters are also needed for M_v generation. The transformation between the vehicle and stereo coordinate (T_v^s) is fixed and corresponds to the sensor fusion of the GPS and stereo camera. The transformation between the word and vehicle (T_w^v) varies as the vehicle moves. PSO localizes the vehicle in two phases, an off-line and online phase. The intrinsic calibration T_s^i and fixed transformation T_v^s are estimated in the offline phase. In the online

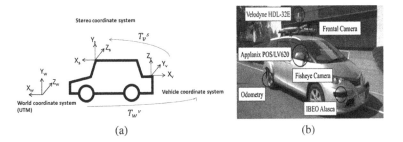

(a)   (b)

Fig. 1. (a) An illustration of the different coordinate system. (b) The intelligent vehicle used for the localization in our experiment.

phase, the varying T_w^v are estimated within an PSO-based tracker. The online tracker is initialized using the GPS-INS information. We propose a computationally effective PSO cost function, where the depth map is pruned using the u-v disparity.

3.1 Algorithm Components

Multipath Viterbi Algorithm. The MPV algorithm is a stereo matching algorithm based on the dynamic programming-based Viterbi algorithm. The structural similarity (SSIM) is used to measure the matching cost or pixel difference between the stereo images on the epipolar lines. A total variation constraint is incorporated within the Viterbi algorithm. To estimate the disparity maps, which are considered as hidden states, the Viterbi process is performed in 4 bidirectional paths. A hierarchical structure is used to merge the multiple Viterbi search paths. Additionally, an automatic rectification process is also adopted to increase the robustness of the algorithm. We refer the authors to Long et al. [9] for details of the algorithm (Fig. 2).

Depth Map Pruning. Given the estimated depth map, we perform the pruning to remove the dynamic objects in the image using the real-time curb detection algorithm proposed by Long et al. [9]. To perform the pruning, we first estimate the histogram of disparity in horizontal and vertical direction or the U-disparity (H_y) and V-disparity (H_x). Given, the U and V-disparity maps, we first detect the road surface using the V-disparity. Subsequently, we detect the curbs in the depth map. After detecting the curbs and the road, the vertical foreground objects, such as car or pedestrians, are pruned from the depth map. Examples of the pruned depth map are shown in Fig. 3. The pruned depth map reduces the computational complexity during the PSO evaluation.

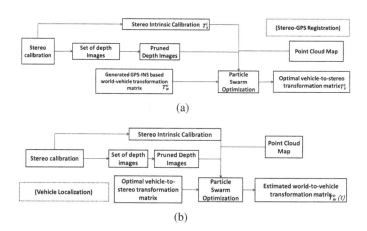

(a)

(b)

Fig. 2. A detailed overview of the (a) offline and (b) online phase of our proposed algorithm.

Fig. 3. An illustration of the depth pruning with (left) right stereo image, (middle) depth image and (right) pruned depth map.

Particle Swarm Optimization. In this paper, we use the varying inertia-based PSO proposed by Shi and Eberhart [12], which functions as global-to-local optimizer. Given a N-dimensional search space, PSO is used to identify the optimal solution using a swarm of M particles. Each particle in the PSO swarm is an N-dimensional vector ($\mathbf{x}_m = \{x_m^n\}_{n=1}^N$) representing a position. Additionally, each particle has an associated N-dimensional velocity vector, $\mathbf{v}_m = \{v_m^n\}_{n=1}^N$, to facilitate the search. The best position of each particle and its fitness function value is given by $\mathbf{p}_m = \{p_m^n\}_{n=1}^N$ and λ_m. The best particle in the swarm and its corresponding fitness function value is denoted by $\mathbf{p}_g = \{p_g^n\}_{n=1}^N$ and corresponding fitness function is stored as g respectively. The inertia weight parameter w defines the exploration of the search space. The social and cognition components of the swarm are defined by the parameters $\rho_1 = c_1 rand()$ and $\rho_2 = c_2 rand()$. In this algorithm, we used 5 PSO particles with c_1 and c_2 set at 2. A was set to 0.5 and C was set to 100 PSO iterations. A pre-defined search limits derived from the maximum possible inter-frame velocity was used.

3.2 Algorithm: Sensor Fusion

In the offline or sensor fusion phase, we first calibrate the cameras using the checkerboard. Subsequently, we estimate the fixed T_v^s using the PSO. To perform the estimation, we first acquire a dataset with our experimental vehicle known as the vehicle-stereo dataset. The vehicle-stereo dataset contains a set of stereo depth images M_s along with GPS-INS information. The GPS-INS information is used to generate the T_w^v, or world-vehicle transformation matrix. To eliminate GPS errors, the vehicle-stereo dataset is acquired in an area without any tall buildings. Given, the acquired T_w^v and the T_s^i (intrinsic), we estimate the T_v^s by generating and matching candidate virtual depth maps with the stereo depth map using the PSO. The candidate depth maps are generated from the Aisan-based point cloud data [1].

Cost Function. PSO generates the candidate matrix T_v^s or PSO particle $\mathbf{x} = \mathbf{x_v^s}'$. The candidate virtual depth map and the corresponding depth map are evaluated by the PSO-based cost function at the unpruned depth indices, given as,

$$f(\mathbf{x}') = dist(M_s, M_v(\mathbf{x}')) \tag{1}$$

where $\mathbf{x}'$ is the PSO particle which represents the vector representation of the candidate transformation matrix. $\mathbf{x}' = [e_x, e_y, e_z, \theta, t_x, t_y, t_z]$, where e_x, e_y, e_z, θ represents the axis-angle representation of the rotation matrix and t_x, t_y, t_z represents the translation parameters.

3.3   Algorithm: Online Phase

Given the estimated T_v^s and T_s^i, we localize the vehicle by estimating T_w^v at each time instant t using a PSO-based tracker. Similar to the offline phase, the unknown transformation matrix is optimized by generating and matching virtual point cloud depth images with the stereo depth image. The PSO-based tracker has two phases, an automatic initialization phase and estimation phase.

Automatic Initialization. The online PSO tracker is initialized at time instant $t = 1$ using the parameters of the T_w^v matrix or $\mathbf{x}_\mathbf{w}^\mathbf{v}(t)$ obtained from the GPS-INS. Please note that, following initialization, the PSO tracker estimates the $\mathbf{x}_\mathbf{w}^\mathbf{v}(t)$ for the remaining frames $t > 1$ *without* the GPS information.

PSO Estimation. Candidate parameters generated by the PSO algorithm, along with the previously estimated fixed transformation matrices, are used to generate the candidate virtual point cloud depth image. By measuring the similarity between the candidate and stereo depth images, the optimal $\mathbf{x}_\mathbf{w}^\mathbf{v}(t)$ transformation parameters are estimated and the vehicle is localized. The initialized PSO tracker estimates the optimal candidate for a given frame t, using the cost function Eq. 1, where $\mathbf{x} = \mathbf{x}_\mathbf{w}^\mathbf{v}(t)$.

Propagation. Once the transformation matrix is estimated for a given frame t, the PSO swarm at every subsequent frame $t + 1$ is initialized by sampling from a zero-mean Gaussian distribution centered around the previous frame's estimate $\mathbf{x}_\mathbf{w}^\mathbf{v}(t)$. The zero mean Gaussian distribution is represented by a diagonal covariance matrix with covariance value 0.01. By iteratively estimating the optimal transformation matrix and propagating the optimal candidate, the PSO tracker performs online tracking. Examples of the localization results are shown in Fig. 4.

4   Experimental Results

A comparative analysis with baseline algorithms is performed for the offline and the online phase. The Euclidean distance-based error measure between the estimated parameters and the ground truth parameters are reported. The ground truth parameter for $\mathbf{x}_\mathbf{v}^\mathbf{s}$ in the offline phase is calculated directly from the distance and orientation between the GPS and stereo camera on the experimental vehicle. The ground truth parameter for the online phase corresponds to the parameters obtained from the GPS. We implement the algorithm using Matlab and Windows (3.5 GHz Intel $i7$).

Dataset and Algorithm Parameters. The experimental vehicle was used to acquire multiple datasets with multiple sequences for the online phase.

Fig. 4. Localization results. The left column is the left stereo image, the middle column is the disparity. The right column is the optimal virtual depth map.

The online phase was evaluated with three datasets. The first dataset contains 4 sequences with total of 300 frames. The second dataset contains 5 sequences with total of 1200 frames, while the third dataset contains 3 sequences with a total of 1200 frames. The dataset for the offline phase contains 15 stereo depth maps and corresponding T_w^v parameters derived from the GPS sensor.

4.1 Offline Phase

We validate the offline phase by performing a comparative analysis with baseline optimization algorithms, the genetic algorithm (GA) and the simulated annealing algorithm (SA) [6]. The number of generations and populations in GA along with the number of iterations in SA was kept similar to the total number of PSO evaluations. Similarly, the search limits were uniform across the algorithms. The results as shown in Table 1, show that the PSO algorithm is better than the baseline algorithm.

Table 1. Error mean and variance over 3 trials on the offline dataset.

Algorithms	PSO	GA	SA
Error	0.62 ± 0.08	1.52±0.6	1.32 ± 0.5
Time taken (min)	35	37.25	69.75

Table 2. Error mean and variance for the localization experiment

Dataset	PSO	PF	APF
1	0.63 ± 0.17	0.89 ± 0.27	1.57 ± 0.7
2	0.69 ± 0.10	1.3 ± 0.53	2.04 ± 1.0
3	0.89 ± 0.06	1.1 ± 0.3	2.3 ± 1.6

4.2   Online Phase

In this section, we validate the PSO tracker by performing a comparative analysis with the widely used particle filter [3] and the APF [4]. The number of particles and iterations of both the particle filter (PF) and APF were kept the same as the PSO. The PF contains 250 particles, while the APF contains 125 particles and 2 layers. In the experiment, the GPS was used only for the initialization, and subsequent localization is only stereo-based. The results tabulated in Table 2, show that the performance of the PSO is better than the baseline algorithms. The low accuracy of the PF can be attributed to the divergence error.

5   Conclusion and Future Works

In this paper, we proposed a sensor fusion and vehicle localization algorithm using particle swarm optimization. To perform the localization, we matched the stereo depth map with virtual point cloud depth maps within a particle swarm optimization framework. Virtual depth maps are generated using a series of transformation matrices. The fixed transformation matrix between the GPS and calibrated stereo camera is estimated in an offline phase. Subsequently, the vehicle is localized in the online phase using a particle swarm optimization based tracking algorithm. We increase the computational efficiency by pruning the depth map for the evaluation. We performed a comparative evaluation with state-of-the-art techniques. Based on our results, we demonstrated the improved performance of the proposed algorithm. In the future work, we will implement the tracker phase on the GPU to achieve real-time performance.

References

1. Aisan Technology Co. Ltd. (2013). http://www.whatmms.com/
2. Colombo, O.: Ephemeris errors of GPS satellites. Bull. Godsique **60**(1), 64–84 (1986)
3. Dailey, M., Parnichkun, M.: Simultaneous localization and mapping with stereo vision. In: International Conference on Control, Automation, Robotics and Vision (2006)
4. Deutscher, J., Blake, A., Reid, I.: Articulated body motion capture by annealed particle filtering. In: CVPR (2000)
5. Farrell, J., Barth, M.: The Global Positioning System and Inertial Navigation. McGraw-Hill, New York (1999)
6. Franconi, L., Jennison, C.: Comparison of a genetic algorithm and simulated annealing in an application to statistical image reconstruction. Stat. Comput. **7**(3), 193–207 (1997)
7. Labayrade, R., Aubert, D., Tarel, J.P.: Real time obstacle detection in stereovision on non flat road geometry through "v-disparity" representation. In: Intelligent Vehicle Symposium (2002)
8. Kneip, L., Chli, M., Siegwart, R.: Robust real-time visual odometry with a single camera and an IMU. In: British Machine Vision Conference (2011)

9. Long, Q., Xie, Q., Mita, S., Tehrani, H., Ishimaru, K., Guo, C.: Real-time dense disparity estimation based on multi-path viterbi for intelligent vehicle applications. In: British Machine Vision Conference (2014)
10. Mattern, N., Schubert, R., Wanielik, G.: High accurate vehicle localization using digital maps and coherency images. In: IVS (2010)
11. Noda, M., Takahashi, T., Deguchi, D., Ide, I., Murase, H., Kojima, Y., Naito, T.: Vehicle ego-localization by matching in-vehicle camera images to an aerial image. In: Koch, R., Huang, F. (eds.) ACCV 2010. LNCS, vol. 6469, pp. 163–173. Springer, Heidelberg (2011). doi:10.1007/978-3-642-22819-3_17
12. Shi, Y., Eberhart, R.: A modified particle swarm optimizer. In: International Conference on Evolutionary Computation, pp. 69–73 (1998)
13. Yoneda, K., Tehrani, H., Ogawa, T., Hukuyama, N., Mita, S.: Lidar scan feature for localization with highly precise 3-D map. In: Intelligent Vehicles Symposium (2014)

Immersed Tunnel Element Translation Control Under Current Flow Based on Particle Swarm Optimization

Li Jun-jun[1], Xu Bo-wei[2(✉)], and Fan Qin-Qin[2]

[1] Merchant Marine College, Shanghai Maritime University, Shanghai 201306,
People's Republic of China
[2] Logistics Research Center, Shanghai Maritime University, Shanghai 201306,
People's Republic of China
xubowei138@126.com

Abstract. Translation is normally the main working mode of tunnel element transportation, which is one of key techniques in immersed tunnel. An approach which can decide the magnitudes and directions of towing forces is presented in no-power immersed tunnel element's translation control under current flow. A particle swarm-based control method, which exploits a sort of linear weighted logarithmic function to avoid weak subgoals, is utilized. In simulation, the performance of the particle swarm-based control method is evaluated in the case of Hong Kong-Zhuhai-Macao Bridge project.

Keywords: No-power immersed tunnel element · Translation control · Resultant force · Resultant moment · Particle swarm optimization

1 Introduction

The importance of immersed tunnel in the construction of roads and railways crossing a body of shallow water has been well documented during the past several decades [1]. An immersed tunnel is a kind of underwater tunnel composed of prefabricated elsewhere elements in a manageable length. Conventional towage, in which several tugs assist the tunnel element transporting from the flooded casting basin or dock to the tunnel trench, is normally used. Straight movement (moving forward or backward) and transverse movement (moving left or right), which are collectively called "translation" in this work, are normally the main working modes of immersed tunnel element floating. Therefore, it's necessary to study the modeling and optimization for immersed tunnel element translation.

In this study, the authors investigate the applications of particle swarm-based control into immersed tunnel element translation optimization under current flow. Firstly, immersed tunnel element translation is mathematically described. Secondly, immersed tunnel element translation control model is built. Thirdly, particle swarm-based control method with a kind of linear weighted logarithmic objective function is presented. Lastly, the case of Hong Kong-Zhuhai-Macao Bridge project is demonstrated to test the proposed approach.

© Springer International Publishing AG 2017
Y. Tan et al. (Eds.): ICSI 2017, Part I, LNCS 10385, pp. 218–224, 2017.
DOI: 10.1007/978-3-319-61824-1_24

2 Translation Control of Tunnel Element

2.1 Translation Velocity

For the convenience of analysis, the tunnel element with pontoons is illustrated in Fig. 1. The element center is the origin of coordinate, the element's axis is x-axis, and y-axis is perpendicular to the element's axis.

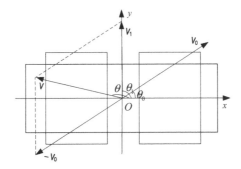

Fig. 1. Immersed tunnel element and two floating pontoons

The current velocity is set as V_0. The element velocity relative to ground is set as V_1, while the element velocity relative to current flow is set as V. The angle between V_0 and the positive direction of x-axis is θ_0; the angle between V_1 and the positive direction of x-axis is θ_1 (in actual towing process, $\theta_1 \in \{0, \pi/2, \pi, 3\pi/2\}$); the angle between V and the positive direction of x-axis is θ. The relation among V_0, V_1, and V is shown in Fig. 1. Components of V in the positive directions of x and y axis are V_x and V_y respectively, then $V_x = V_1 \cos\theta_1 + (-V_0)\cos\theta_0 = V_1\cos\theta_1 - V_0\cos\theta_0$, $V_y = V_1\sin\theta_1 + (-V_0)\sin\theta_0 = V_1\sin\theta_1 - V_0\sin\theta_0$.

2.2 Resistance of Immersed Tunnel Element Translation

According to "Guidelines for Marine Towage (2011)" [2], the marine towage resistance R_T of the tunnel element can be calculated by: $R_T = 1.15(R_f + R_B)$. R_f and R_B are friction resistance and residual resistance of tunnel element respectively. $R_f = 1.67A_{e,1}|V|^{1.83} \times 10^{-3}$ [2], $R_B = 0.62C_bA_{e,2}V^2$ [3]. The towage resistance of one floating pontoon is: $R_{TP} = 1.15(R_{fp} + R_{Bp})$. Then, the total towage resistance is:

$$R'_T = R_T + 2R_{Tp} = 1.15[1.67(A_{e,1} + 2A_{p,1})|V|^{1.83} \times 10^{-3} + 0.62C_b(A_{e,2} + 2A_{p,2})V^2] \tag{1}$$

Where, R_{fp} and R_{Bp} are friction resistance and residual resistance of the floating pontoon respectively; $A_{e,1}$ and $A_{p,1}$ are the wetted surface area under water line of tunnel element and floating pontoon respectively; C_b is block coefficient; $A_{e,2}$ and $A_{p,2}$

are the submerged part of transverse section area in tunnel element and floating pontoon respectively. Units of R_T, R_f, R_B, R_{Tp}, R_{fp}, R_{Bp} and R'_T are all "kN".

If the tunnel element is parallel to current flow, $A_{e,1} = L(B + 2d)$, $A_{e,2} = B{\cdot}d$, and $A_{p,2} = B_p{\cdot}d_p$, $A_{p,1} = L_p(B_p + 4d_p)$. If the tunnel element is perpendicular to current flow, $A_{e,1} = B(L + 2d)$, $A_{p,1} = B_p (L_p + 2d\ _p)$, $A_{e,2} = L{\cdot}d$, and $A_{p,2} = L_p{\cdot}d_p$. Where L, B and d are the length, width and draft of the tunnel element respectively; L_p, B_p, B_p' and d_p are the length, width under water, width above water and draft of the floating pontoon respectively.

The element's movement is parallel or perpendicular to current flow. Resistances R'_{Tx} and R'_{Ty} in the directions of x and y axis are calculated according to V_x and V_y respectively. R'_{Tx} and R'_{Ty} are all non-negative real numbers, then $|f_x| = R'_{Tx}$, and $|f_y| = R'_{Ty}$. Here the symbols f_x and f_y are magnitudes of f_x and f_y, which are components of f in the positive directions of x and y axis respectively.

2.3   Towing Force

The number of tugs working collaboratively is N. It is set that towing point of the ith tug (G_i) at the tunnel element is A_i ($i = 1, 2, …, N$). The coordinate of A_i is (x_i, y_i). The towing force of tug G_i is F_i. The angle of x-axis's positive direction counter clockwise to F_i is set as α_i, which is called angle of towing force F_i. The scopes of towing forces' magnitudes and angles are: $F_i \in [0, F_i^{\max}]$ and $\alpha_i \in [\alpha_i^{\min}, \alpha_i^{\max}]$. Where, $i = 1, 2, …, N$, $F_i^{\max}$ is the maximum towing force of the tug G_i. To make the tunnel element moving stably, it should be $F = -f$, $T = 0$.

3   Translation Control Model

F_x and V_x are in the same direction, while F_y and V_y are in the same direction. Then $F_x = |f_x| \cdot \text{sgn}(V_x) = R'_{Tx} \cdot \text{sgn}(V_x)$, $F_y = |f_y| \cdot \text{sgn}(V_y) = R'_{Ty} \cdot \text{sgn}(V_y)$. Where, "sgn" is sign function. Therefore, $\sum_{i=1}^{N} F_i \cos \alpha_i = F_x = R'_{Tx} \cdot \text{sgn}(V_x)$, $\sum_{i=1}^{N} F_i \sin \alpha_i = F_y = R'_{Ty} \cdot \text{sgn}(V_y)$.

All tug moments' magnitudes in four quadrants are unifiedly represented in: $T_i = \text{sgn}(y_i) \cdot F_i \sqrt{x_i^2 + y_i^2} \cos(\alpha_i + \beta_i)$, while the resultant moment' magnitude of N tugs is shown in: $T = \sum_{i=1}^{N} \text{sgn}(y_i) \cdot F_i \sqrt{x_i^2 + y_i^2} \cos(\alpha_i + \beta_i)$. Where, $\beta_i = \arctan(x_i/y_i)$. For the convenience of analysis, a variable L_i' is set as: $L_i' = \text{sgn}(y_i) \cdot \sqrt{x_i^2 + y_i^2} \cos(\alpha_i + \beta_i)$. According to Sect. 2.3, the resultant moment' magnitude will meet $T = \sum_{i=1}^{N} F_i L_i' = 0$.

Equations (4) and (5) are set as objective functions of the immersed tunnel element translation control.

$$\max g_1 = V_1 \tag{2}$$

$$\max g_2 = \prod_{i=1}^{N} (F_i^{\max} - F_i) \tag{3}$$

F_i^{sc} is set as the minimum surplus towing force for the ith tug. Then, $F_i \in [0, F_i^{\max} - F_i^{sc}]$. Constraints (in Sects. 2.3 and 3) of resultant force, resultant moment, towing forces' magnitudes and angles are restrictions of the immersed tunnel element translation control.

4 Particle Swarm-Based Translation Control Optimization of Tunnel Element

In order to treat the two subgoals equally, normalization is done firstly: $h_1 = (g_1 - g_1^{\min})/(g_1^{\max} - g_1^{\min})$, $h_2 = (g_2 - g_2^{\min})/(g_2^{\max} - g_2^{\min})$. For the purpose of avoiding weak subgoals in the optimization results, natural logarithms of h_1 and h_2 are used to replace h_1 and h_2. A sort of linear weighted logarithmic function is exploited here:

$$\max h = \lambda_1 \ln h_1 + \lambda_2 \ln h_2 \tag{4}$$

Where λ_1 and λ_2 are weighting coefficients, $\lambda_1 + \lambda_2 = 1$, $0 < \lambda_1, \lambda_2 < 1$. Through this linear weighted logarithmic function, the phenomena that one subgoal is too strong while the other subgoal is too weak can be avoided.

In translation control problem of immersed tunnel element, there are $2 * N$ decision variables: $F_1 \sim F_N$, $\alpha_1 \sim \alpha_N$. It's easy to calculate resistance by velocity, while it is difficult to calculate velocity by resistance. So decision variables of particle swarm-based control method in this work are set as "$V_1, F_1, \cdots, F_{N-3}, \alpha_1, \cdots, \alpha_N$", not "$F_1, \cdots, F_N, \alpha_1, \cdots, \alpha_N$". F_{N-2}, F_{N-1} and F_N are calculated according to Eq. (8). Where $f_y' = R_{Ty}' \cdot \text{sgn}(V_y) - \sum_{i=1}^{N-3} F_i \sin \alpha_i$, $f_x' = R_{Tx}' \cdot \text{sgn}(V_x) - \sum_{i=1}^{N-3} F_i \cos \alpha_i$, and $T' = -\sum_{i=1}^{N-3} T_i$.

$$F_{N-2} \sin \alpha_{N-2} + F_{N-1} \sin \alpha_{N-1} + F_N \sin \alpha_N = f_y' \tag{5}$$

$$F_{N-2} \cos \alpha_{N-2} + F_{N-1} \cos \alpha_{N-1} + F_N \cos \alpha_N = f_x' \tag{6}$$

$$F_{N-2} L_{N-2}' + F_{N-1} L_{N-1}' + F_N L_N' = T' \tag{7}$$

According to Kramer rule, F_{N-2}, F_{N-1} and F_N can be obtained by Eq. (8). Where

$$\Delta = \begin{vmatrix} \sin \alpha_{N-2} & \sin \alpha_{N-1} & \sin \alpha_N \\ \cos \alpha_{N-2} & \cos \alpha_{N-1} & \cos \alpha_N \\ L_{N-2}' & L_{N-1}' & L_N' \end{vmatrix}, \Delta_1 = \begin{vmatrix} f_y' & \sin \alpha_{N-1} & \sin \alpha_N \\ f_x' & \cos \alpha_{N-1} & \cos \alpha_N \\ T' & L_{N-1}' & L_N' \end{vmatrix},$$

$$\Delta_2 = \begin{vmatrix} \sin \alpha_{N-2} & f'_y & \sin \alpha_N \\ \cos \alpha_{N-2} & f'_x & \cos \alpha_N \\ L'_{N-2} & T' & L'_N \end{vmatrix}, \text{ and } \Delta_3 = \begin{vmatrix} \sin \alpha_{N-2} & \sin \alpha_{N-1} & f'_y \\ \cos \alpha_{N-2} & \cos \alpha_{N-1} & f'_x \\ L'_{N-2} & L'_{N-1} & T' \end{vmatrix}.$$

$$(F_{N-2}, F_{N-1}, F_N) = (\frac{\Delta_1}{\Delta}, \frac{\Delta_2}{\Delta}, \frac{\Delta_3}{\Delta}) \tag{8}$$

$g_2^{min} = 0$, $g_2^{max} = \prod_{i=1}^{N} F_i^{max}$, and $g_1^{min} = 0$. But g_1^{max} is unable to be obtained directly.

Here, PSO is used to seek g_1^{max}. And then, Eq. (7) is used as fitness function to optimize tunnel element translation problem.

5 Simulation

The performance of the proposed model and algorithm is validated through translation control simulation of immersed tunnel element in Hong Kong-Zhuhai-Macao Bridge project, and compared with traditional empirical method. The length, width and draft of the immersed tunnel element are: $L = 180$ m, $B = 37.95$ m and $d = 11.1$ m respectively. The length and draft of the floating pontoon are: $L_p = 40.2$ m and $d_p = 6.2$ m respectively. The width of one side float is 7.2 m, so the width (under water) of one pontoon, B_p, is 14.4 m. The width above water of the floating pontoon is $B_p' = 56.4$ m. Block coefficient $C_b = 1$, current velocity $V_0 = 2$.

Take Rongshutou sea-route as an example. The direction of immersed tunnel element moves is $12°$, $\theta_1 = 0°$. The flow directions of tide rise and retreat are $355°$ and $175°$ respectively. Therefore, $\theta_0 = 17°$ in tide rise while $\theta_0 = 197°$ in tide retreat.

The number of tugs N is 6. Their performance is shown in Tab. 1. The sketch map of towing forces' directions and the coordinates of towing points A_1–A_6 are shown in Fig. 2. The ranges of α_1–α_4 are $[-75°, 165°]$, $[-165°, 75°]$, $[35°, 255°]$, $[105°, 325°]$, $[15°, 345°]$, $[15°, 345°]$ respectively. Besides, $F_i^{sc} = 10t(i = 1, 2, \cdots, N)$.

Table 1. Performance of tugs

Tug	Power (Hp)	Main engine speed (r/min)	F_i^{max} (ton)	Remark
G_1, G_2	6800	750	56	Azimuth drive
G_3, G_4	5200	750	50	Azimuth drive
G_5, G_6	4000	750	43	Azimuth drive

The number of particles M is 20, iterations N_t is 500. In order to ensure the convergent trajectories of particles [4], $\omega = 0.7298$ and $c_1 = c_2 = 1.49618$ [5].

According to Sect. 4, g_1^{max} in tide rise and tide retreat are 6.650 knots and 2.721 knots respectively, which are calculated first by particle swarm-based control method. The sketch maps of tunnel element towing are drawn taking two situations for example. In Fig. 3, $V_1 = 5.474$ knots in tide rise. While in Fig. 4, $V_1 = 1.453$ knots in tide retreat. To compare the empirical method with particle swarm-based control method,

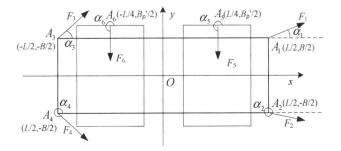

Fig. 2. Sketch map of towing forces' directions and towing points' coordinates.

the schemes acquired by these two methods are adjacent to each other. In Figs. 3 and 4, ↑ denotes north while * shows towing point of each towing force. The length of directed line is proportional to the magnitude of each towing force.

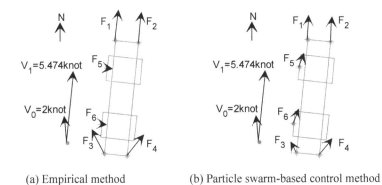

(a) Empirical method   (b) Particle swarm-based control method

Fig. 3. Sketch map of tunnel element towing in tide rise.

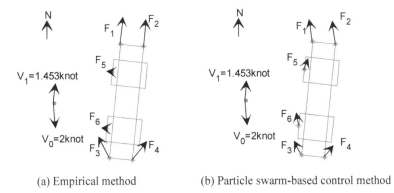

(a) Empirical method   (b) Particle swarm-based control method

Fig. 4. Sketch map of tunnel element towing in tide retreat.

In Figs. 3 and 4, the towing forces of tugs G_1–G_4 are greater in empirical method, which makes their surplus towing forces less than those of particle swarm-based control method. The minimum and mean surplus towing force of Empirical method (EM) and Particle swarm-based control method (PSM) are shown in Table 2. It can be seen that the surplus towing forces in PSM is obviously higher than that of EM.

Table 2. Surplus towing force comparison

V_1 (knot)	Minimum (ton)			Mean(ton)		
	EM	PSM	Percentage increased	EM	PSM	Percentage increased
5.474	10.757	18.995	76.58%	18.732	22.161	18.31%
1.453	15.286	21.569	41.10%	21.751	24.745	13.76%

6 Conclusion

This work mainly discusses the translation control and optimization of immersed tunnel element under current flow. It proposes a particle swarm-based control method which is validated through translation control simulation of immersed tunnel element in Hong Kong-Zhuhai-Macao Bridge project. The simulation results indicate that particle swarm-based control method is better than empirical method.

Acknowledgment. This work is supported by the Ministry of education of Humanities and Social Science project(Nos. 15YJC630145, 15YJC630059), Natural Science Foundation supported by Shanghai Science and Technology Committee (No. 15ZR1420200), National Natural Science Foundation of China (No. 61603244). Here we would like to express our gratitude to them.

References

1. Li, W., Fang, Y.G., Mo, H.H., Gu, R.G., Chen, J.S., Wang, Y.Z., Feng, D.L.: Model test of immersed tube tunnel foundation treated by sand-flow method. Tunn. Undergr. Space Technol. **40**, 102–108 (2014)
2. China Classification Society: China Classification Society. Sea towage Guide (2011). [EB/OL]. (in Chinese). http://www.ccs.org.cn/ccsewwms2007/displayNews.do?id=ff80808137de736 a0137e39c4039000b. Accessed 13 June 2012
3. Shen, P.G.: Estimation of towing resistance. Marine Technol. **5**, 9–12 (2011). (in Chinese)
4. Van den Bergh, F., Engelbrecht, A.P.: A study of particle swarm optimization particle trajectories. Inf. Sci. **176**, 937–971 (2006)
5. Shi, Y.H., Eberhart, R.: A modified particle swarm optimizer. In: Proceedings of the IEEE International Conference on Evolutionary Computation, Anchorage, pp. 69–73 (1998)

Solving Inverse Kinematics with Vector Evaluated Particle Swarm Optimization

Zühnja Riekert and Mardé Helbig[✉]

University of Pretoria, Pretoria, South Africa
u12040593@tuks.co.za, mhelbig@cs.up.ac.za

Abstract. Inverse kinematics (IK) is an optimization problem solving the path or trajectory a multi-jointed body should take for an extremity to reach a specified target location. When also considering the flow of movement, IK becomes a multi-objective optimization problem (MOP). This study proposes the use of the vector evaluated particle swarm optimization (VEPSO) algorithm to solve IK. A 3D character arm, with 7 degrees of freedom, is used during experimentation. VEPSO's results are compared to single-objective optimizers, as well as an optimizer that uses weighted aggregation to solve MOPs. Results show that the weighted aggregation approach can outperform IK-VEPSO if the correct weight combination (that is problem dependent) has been selected. However, IK-VEPSO produces a set of possible solutions.

1 Introduction

Mimicking human movement is the main focus in multiple fields, including film animation [14], game graphics [33] and humanoid robotics [30]. In robotics, Rokbani et al. [23,24] suggest that robots are built more human-like to enable them to assist humans in their environment. For example, legs can help them to climb stairs and arms can enable them to reach for and pick up objects.

While these actions come naturally to humans, it can be challenging to mimic in robotics and animation fields. Figure 1 shows an example of a typical inverse kinematics (IK) problem, where a 3D character's hand has to be animated moving towards the position of the red block. To achieve this, one first has to determine how each joint (shoulder, elbow, wrist) should be rotated for the end-effector (hand) to reach its goal. This can be done through various techniques, including trial and error, the use of an IK algorithm or motion capture. The character can then be animated using a process called inbetweening [28].

From the example shown in Fig. 1, trial and error entails the animator to shift each bone manually until the hand reaches its target. In this way a character's pose can be fine tuned to what the animator needs. However, as stated in [2,29], it is a tedious, time-consuming process.

Motion capture involves recording an actor's movements to create easy natural looking animations [7]. The drawback is that it requires additional resources, such as cameras, actors and motion capturing software. Therefore, using an algorithm to calculate the required rotation set saves human effort and cost.

© Springer International Publishing AG 2017
Y. Tan et al. (Eds.): ICSI 2017, Part I, LNCS 10385, pp. 225–237, 2017.
DOI: 10.1007/978-3-319-61824-1_25

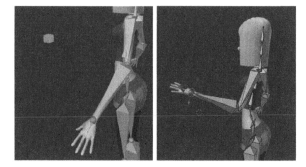

Fig. 1. 3D character model's initial and end poses for animation. The 3D model used, was downloaded with consent from [22]. (Color figure online)

Jacobian matrix-based methods [34] involve vector calculus computations that generally produce good IK solutions [4]. However, flaws of these approaches include possible singularities [3], possible continuous oscillations [4], heavy computation [11] and with high degrees of freedom (DOF) the Jacobian can become very complex to compute [4].

Learning algorithms used to solve IK, such as Neural Networks and Neurofuzzy algorithms [1,27], are computationally expensive and require suitable training sets to output adequate rotation sets [26]. Therefore, the focus of IK research, for the most part, shifted to Heuristic IK solvers. Heuristic algorithms that have been used as IK solvers include particle swarm optimization (PSO) [11], ant colony optimization [17], genetic algorithm (GAs) [35], artificial bee colony [5] and the firefly algorithm [26]. GA and PSO are both relatively easy to implement [19]. However, according to Parsopoulos PSO outperforms GA on IK [19].

Movements performed by characters in animation should not only allow for task completion, but should also look natural. People can quickly pick up on abnormalities in motion [29]. However, even making something appear too realistic can have a negative impact (the uncanny valley effect [18]). As shown in Fig. 2 multiple rotation sets could lead the end-effector to its goal, but not all of them look natural. According to [32], humans tend to minimize movement when they reach for objects. Thus, two objectives should be solved, namely minimizing the distance between the end-effector and the goal, and minimizing the total amount of movement. These two objectives lead to IK being defined as a multi-objective optimization problem (MOP). Yang et al. [35] used a PSO with a weighted aggregation fitness function to solve two IK objectives simultaneously. This approach, however, requires the search space and problem specific weights to be determined beforehand [19]. It also results in only one solution [19].

This study proposes the use of the vector evaluated genetic algorithm (VEPSO) algorithm to produce a set of IK solutions. The algorithm, IK-VEPSO, is compared against inertia weight particle swarm optimization (IWPSO) algorithms, as well as IWPSOs with a conuentional weighted aggregation (CWA) fitness function.

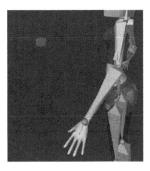 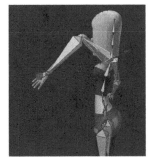 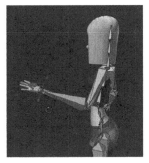

Fig. 2. Multiple rotation sets could lead the hand to the goal location. The 3D model used, was downloaded with consent from [22].

2   Background

This section provides background information regarding IK, multi-objective optimization (MOO) and the algorithms that were used in this study.

2.1   Inverse Kinematics

An end-effector refers to the extremity used to interact with the environment. On a leg, it is the foot, on a finger, the finger tip and on some mechanical structure it is its last body on a chain of connected, jointed bodies.

In animation, the bones of a rigged character follow a hierarchical chain. Moving a parent bone will move children bones as well. However, moving a child bone will not affect its parent's position. An example is shown in Fig. 3 where rotating an upper arm will move the entire rigged arm, but rotating a forearm will not change the bicep bone's position.

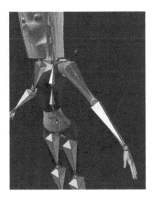  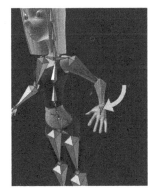

Fig. 3. Manually shifting the bones of a rigged character.

Forward kinematics (FK), as explained by Rokbani et al. [26], entails that the joint parameters of all the end-effector's hierarchical chained bodies are known to calculate the new position of the end-effector. Therefore, a FK solution will always exist [35]. In contrast to FK, with IK the desired end-effector's position is known, but the joint angles required to move the end-effector to its goal have to be determined [26]. A solution may not always exist: the target could be placed too far away to reach [4] or a linked multi-body structure simply cannot "bend" in the required way to reach the target position.

2.2   Multi-objective Optimization

MOO attempts to solve multiple objective functions simultaneously, where at least two objectives are in conflict with one another. Therefore, improving a solution towards one objective, weakens the solution with regards to another objective [8]. Since these objectives are in conflict with one another [8,15,20], a single solution does not exist. Therefore, the goal of a MOO algorithm is to find the set of optimal trade-off solutions. In the objective space this set is referred to as the Pareto front [20].

2.3   Particle Swarm Optimization Algorithms

PSO was first introduced by Kennedy and Eberhart in 1995 [13]. The idea was inspired from the behaviour of flocking birds and was initially influenced by Heppner and Grenander's work [10,21].

Standard Gbest Particle Swarm Optimization: In the standard *gbest* PSO multiple particles are initialized over a search space. The particles move around until some stopping condition is satisfied. Each particle's position represents a possible solution. A solution is ranked according to a fitness function. The swarm keeps track of its overall global best fitness, while each particle keeps track of its own personal best fitness achieved. A particle moves through the search space by constantly updating its velocity and position using (1) and (2), respectively.

$$v_{ij}(t+1) = v_{ij}(t) + c_1 r_{1j}(t)[y_{ij}(t) - x_{ij}(t)] + c_2 r_{2j}(t)[\hat{y}_j(t) - x_{ij}(t)] \quad (1)$$

$$\boldsymbol{x}_i(t+1) = \boldsymbol{x}_i(t) + \boldsymbol{v}_i(t+1) \quad (2)$$

Symbols used in (2) denote the following: t is the time step, i is the particle, $\boldsymbol{x}_i(t+1)$ is the updated position vector of particle i, $\boldsymbol{x}_i(t)$ is the current position vector of particle i and $\boldsymbol{v}_i(t+1)$ is the updated velocity vector of particle i. Symbols not present in (2), but in (1) denote the following: j is the dimension of the vector, c_1 is the cognitive component (positive), c_2 is social component (positive), r_{1j} and r_{2j} are uniform random numbers, $y_{ij}(t)$ is particle i's personal best position so far and $\hat{y}_j(t)$ is global best position of the swarm so far.

The movement of a particle takes its own knowledge, as well as the knowledge from its swarm, into account. Possible stopping conditions include stopping when a maximum number of iterations is reached, it is detected that particles stopped moving or the fitness of some particle is smaller than some predefined error (ϵ).

Inertia Weight Particle Swarm Optimization: IWPSO is a variant of PSO where an inertia weight is added to the velocity function:

$$v_{ij}(t+1) = wv_{ij}(t) + c_1 r_{1j}(t)[y_{ij}(t) - x_{ij}(t)] + c_2 r_{2j}(t)[\hat{y}_j(t) - x_{ij}(t)] \qquad (3)$$

where w denotes inertia weight and controls how much the newly calculated velocity is influenced by the current velocity.

Conuentional Weighted Aggregation: Multiple objectives can be combined into a single scalar fitness function using weighted aggregation. CWA implements weighted aggregation, where the weights are non-negative fixed values [12]:

$$F = \sum_{i=1}^{k} w_i f_i(x)_i \qquad (4)$$

where w_i refers to the weight of objective function $f_i(x)$. The weights usually sum up to 1 [19]. With this approach, a single run returns a single solution. To obtain multiple different solutions, it has to be run multiple times [12].

Vector Evaluated Particle Swarm Optimization: VEPSO is a PSO-based multi-swarm MOO algorithm proposed by Parsoulos and Vrahatis [19]. In VEPSO each objective is solved by a PSO [15]. Particle velocities are updated based on shared information between the PSOs (sub-swarms). After each iteration the archive, that holds non-dominated solutions, is updated. If a new non-dominated solution is added to the archive, and this new solution dominates any solutions in the archive, the dominated solutions are removed from the archive. If the archive is full, a solution is removed from a less-dense section of the Pareto front. The goal of VEPSO is to produce a Pareto front of solutions, where the set of solutions are as accurate, as well as diverse, as possible.

Sub-swarms in VEPSO share information by using each other's *gbest* value in the velocity update calculation (Eq. (1)). Which PSO's *gbest* value is used is determined by a knowledge transfer strategy (KTS). Various KTSs are described in [8]. VEPSO stagnation can be avoided by choosing an appropriate KTS [16].

3   Solving Inverse Kinematics Using PSO

This section discusses how the algorithms were adapted to solve IK.

3.1   PSO for Inverse Kinematics

A PSO used to solve IK is often referred to as an IK-PSO. If an IWPSO is used, it is often referred to as an IK-IWPSO. In an IK-PSO each particle represents a complete trajectory towards the target. Each particle's position represents a set of joint rotations (θ), i.e. the rotations of each joint around each axis that the joint is permitted to rotate around. After the set of rotations are found, these rotations can then be used in FK to determine the end-effector's position after the trajectory.

The search space of an IK-PSO represents the allowed movement of each joint in the multi-jointed body (rotation boundaries). For example, hinge joints can rotate around only one specific axis. In addition to restricting the search space to the rotation boundaries of joints, the search space can be further reduced to avoid collisions [29].

The main goal of an IK-PSO is to produce a trajectory that minimizes the distance between the end-effector and the target location [25]. This difference, F_{error}, is defined as:

$$F_{error} = ||Xt - Xj_n|| \tag{5}$$

where the fitness function, F_{error}, represents the Euclidean distance between the target (Xt) and the end-effector (Xj_n) [25]. Xj represents the end position of a body and n represents the total amount of bodies on a chain of jointed bodies.

3.2   PSO with a CWA Fitness Function for Inverse Kinematics

Multiple approaches exist to minimize energy and smooth over movement, including the "minimum jerk model" proposed by Flash et al. [9] and the "minimum torque-change model" [31]. Total displacement can be defined as follows:

$$F_{displacement} = \sum_{n=1}^{k}(q_{n,f} - q_{n,i})^2 \tag{6}$$

where $q_f = \{q_{f,1}, \ldots, q_{f,k}\}$ represents all the final rotations and $q_i = \{q_{i,1}, \ldots, q_{i,k}\}$ represents all the initial rotations.

Equations (5) and (6) can be aggregated with fixed weights to produce Eq. (7) [35]:

$$F = w_e F_{error} + w_d F_{displacement} \tag{7}$$

The values of the weights $(w_e$ and $w_d)$ are search space dependant [19] and have to be defined beforehand. Eq. (7) can be used instead of Eq. (5) as a fitness function in the setup described in Sect. 3.1 to produce an IK trajectory per run that uses minimal movement.

Note that Eqs. (5) and (6) are in conflict with each other, since Eq. (6) aims to minimize movement, while Eq. (5) requires movement to achieve its goal.

When assigning weights it should however be taken into consideration that it is more important for the end-effector to actually reach its target than to have minimal movement with a trajectory that does not reach the target.

3.3   VEPSO for Inverse Kinematics

VEPSO is extended in this study to solve IK. VEPSO solving IK, IK-VEPSO, has two sub-swarms, S_1 and S_2, that are initialized. S_1 uses Eq. (5) and S_2 uses Eq. (6) as their fitness function respectively.

Knowledge Transfer Strategy: A ring KTS is used to transfer knowledge between swarms, i.e. S_1 uses S_2's *gbest* and S_2 uses S_1's *gbest* to update their particles' velocities [6].

Pareto Dominance: The approach used to determine when one solution would dominate another is presented in Algorithm 1. Algorithm 1 follows the dominance rule as described in [12] and uses Eqs. (5) and (6) as the only two fitness functions.

Algorithm 1. Determining whether solution P_1 dominates solution P_2

---

1   **if** $P_1 F_{error} <= P_2 F_{error}$ *and* $P_1 F_{displacement} <= P_2 F_{displacement}$ **then**
2   **if** P_1's $F_{error} < P_2 F_{error}$ *or* $P_1 F_{displacement} < P_2 F_{displacement}$ **then**
3   return *true*
4   **end if**
5   **end if**
6   return *false*

---

4 Experimental Approach

This section discusses the algorithms, general configurations and algorithm specific configurations that were used for the experiments.

4.1 Algorithms

The following algorithms were used in the study: IK-PSO, IK-IWPSO, IK-IWPSO with a CWA fitness function and an IK-VEPSO. Three variations of IK-IWPSO with a CWA fitness function were used, where each variation had different weight values. Al of these algorithms were applied to solve IK for a 3D character model's 7 DOF arm. As illustrated in Fig. 1, the goal was to calculate the rotation configuration to move an arm from its initial position to a position that allows its hand to reach the centre of the red block.

4.2 General Configurations

In all simulations 15 particles were used per swarm [25]. Velocities of particles were initialized to zero [8] and particle positions were initialized to random positions within the search space. Each particle's position represented a set of rotations. Each set of rotations contained the combined rotation configurations of all 3 major joints present in an arm (shoulder, elbow and wrist). The 7 elements that were present in a set are listed in Table 1. Boundaries of the search space were roughly selected based on possible arm movements and adjusted to reduce collisions.

Table 1. Elements in particle rotation set of a 7 DOF arm

Element	Joint	Axis	Boundary (degrees)
$q_{1,f}$	Shoulder	X-axis	$[-130, 45]$
$q_{2,f}$	Shoulder	Y-axis	$[-61.4, 61.4]$
$q_{3,f}$	Shoulder	Z-axis	$[-130.5, 34.6]$
$q_{4,f}$	Elbow	X-axis	$[0, 130]$
$q_{5,f}$	Elbow	Y-axis	$[-45, 0]$
$q_{6,f}$	Wrist	X-axis	$[-30, 30]$
$q_{7,f}$	Wrist	Z-axis	$[-20, 45]$

4.3 Algorithm Specific Configurations

This section discusses the algorithm specific configurations of IK-PSO, IK-IWPSO, IK-PSO with CWA fitness function and VEPSO.

IK-PSO: A cognitive weight (c_1) of 1.4047 and a social weight (c_2) of 1.494 was used [25]. After each run, the total displacement (Eq. (6)) of each solution was recorded.

IK-IWPSO: An inertia weight of 0.729 (w), a cognitive weight of 1.494 (c_1) and a social weight (c_2) of 1.494 was used [25]. Similar to IK-PSO, the total displacement was recorded.

IK-IWPSO with CWA Fitness Function: Values produced from Eq. (6) are considerably larger than those produced by Eq. (5). With $w_e = 1$ and $w_d = 1$ (6), $F_{displacement}$ will have a very large influence over a particle's fitness. A w_d of $1/13.86$ was used to scale down Eq. (6) so that it can approximately have the same influence as Eq. (5). The value $1/13.86$ was calculated by considering the minimum and maximum possible values that can be produced by Eqs. (5) and (6) in the predefined search space. Since minimizing distance towards the goal is more important than minimizing movement (refer to Sect. 3.2), displacement weights of $1/(13.86 * 2)$ and $1/(13.86 * 10)$ were also tested.

4.4 VEPSO

Two IWPSOs were used as sub-swarms, both with inertia weights (w) of 0.729, cognitive weights (c_1) of 1.494 and a social weights (c_2) of 1.494. The archive was maintained according to [12] and had a fixed size of 10.

4.5 Experimental Approach

All algorithms were run 50 times. A run was terminated after 300 iterations or if a solution was found with a fitness less than 0.0001.

5   Results

The IK-VEPSO values presented in Table 2 were calculated by using the archive solutions with the best distance error fitness (as that is the most important objective).

Table 2. Distance error and rotation displacement results

	Distance error		Rotational displacement	
	Avg	StdDev	Avg	StdDev
IK-PSO	1.96484E−2	9.99659E−3	4.076534	2.231010
IK-IWPSO	**7.78310E−5**	**2.08119E−5**	3.32692	1.71769
IK-IWPSO with CWA				
$w_e = 1$ and $w_d = 1/13.86$	2.73026E−3	1.85584E−2	**1.15214**	**0.19602**
$w_e = 1$ and $w_d = 1/(13.86 * 2)$	5.06802E−3	2.44249E−2	1.42592	0.41638
$w_e = 1$ and $w_d = 1/(13.86 * 10)$	2.70516E−3	1.83785E−2	2.61991	1.16565
IK-VEPSO	2.02625E−2	1.36096E−2	**1.07691**	**0.26468**

From Table 2 it can be seen that IK-PSO performed reasonably well with regards to the distance error. However, it could not find good values for the rotational displacement, where both the average value and standard deviation values were high. IK-IWPSO obtained the best distance error values, but did not obtain very good displacement values. Both IK-PSO and IK-IWPSO only considered distance error during the search process, which may lead to unnatural poses.

IK-IWPSO with CWA and IK-VEPSO considered both distance error and displacement during the search process. The first IK-IWPSO with CWA configuration obtained the second best displacement average value and the best displacement standard deviation. IK-VEPSO obtained the best displacement average value, with the second best displacement standard deviation value. However, its distance error values were not as good as those of IK-IWPSO. It should be noted however, that IK-VEPSO did find a spread of trade-off solutions (refer to Fig. 4). These various solutions provide various trade-off values for the two objective functions.

Table 3 presents the last iteration when *gbest* was updated. The stopping condition of the algorithms was 300 iterations or when a solution with a fitness of less than 0.0001 was found. IK-IWPSO had the lowest average and standard deviation, which is confirmed by the fact that it obtained fitness levels of less than 0.0001 (refer to Table 2). IK-VEPSO had an average value of 127.22 and a very high standard deviation of 81.99. If one considers the fact that its distance error average was higher than 0.0001 (refer to Table 2), and that it therefore probably ran for 300 iterations, this may indicate that IK-VEPSO stagnated during the runs.

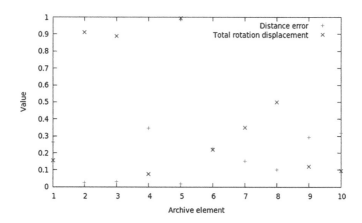

Fig. 4. Distance error and total rotation displacement of each element in archive

Table 3. Last *gbest* update iteration number

	Average	Standard deviation
IK-PSO	104.74	80.68
IK-IWPSO	**89.82**	**17.28547**
IK-IWPSO with CWA		
$w_e = 1$ and $w_d = 1/13.86$	158.1	38.97
$w_e = 1$ and $w_d = 1/(13.86 * 2)$	148.08	55.36
$w_e = 1$ and $w_d = 1/(13.86 * 10)$	95.82	29.47
IK-VEPSO	127.22	81.99

Furthermore, it can be seen from Tables 2 and 3 that the weight values used for IK-IWPSO with CWA played an important role in the performance of the algorithm. Therefore, good weight values are required for the algorithm to perform well. However, the weight values are problem-dependent. IK-VEPSO eliminates the problem of selecting weight values.

6   Conclusion

IWPSO can render solutions relatively fast and can provide excellent IK solutions, but the total movement or displacement is not considered. Both IK-IWPSO with a CWA fitness function and IK-VEPSO render solutions with better movement values than IWPSO, since the total movement used is considered. IK-IWPSO outperforms IK-VEPSO if the correct weight combination is chosen. IK-VEPSO renders a wide spread of vastly different IK solutions that an IK-IWPSO with a CWA fitness function cannot obtain. However, IK-VEPSO is prone to stagnation.

Future work will include investigating approaches to prevent stagnation of IK-VEPSO. In addition, IK-VEPSO will be compared against other MOO algorithms on the IK problem.

References

1. Bingul, Z., Ertunc, H.M., Oysu, C.: Comparison of inverse kinematics solutions using neural network for 6R robot manipulator with offset. In: Proceedings of ICSC Congress on Computational Intelligence Methods and Applications (2005)
2. Boulic, R., Huang, Z., Thalmann, D.: A comparison of design strategies for 3D human motions. In: Varghese, K., Pfleger, S. (eds.) Human Comfort and Security of Information Systems, pp. 306–319. Springer, Heidelberg (1997). doi:10.1007/978-3-642-60665-6_28
3. Boulic, R., Mas, R.: Hierarchical kinematics behaviors for complex articulated figures. Interactive Computer Animation. Prentice Hall, Upper Saddle River (1996)
4. Buss, S.R.: Introduction to inverse kinematics with jacobian transpose, pseudoinverse and damped least squares methods. IEEE J. Robot. Autom. 17(1–19), 16 (2004)
5. Çavdar, T., Mohammad, M., Milani, R.A.: A new heuristic approach for inverse kinematics of robot arms. Adv. Sci. Lett. 19(1), 329–333 (2013)
6. Cortes, O.A.C., Rau-Chaplin, A., Wilson, D., Cook, I., Gaiser-Porter, J.: A study of VEPSO approaches for multiobjective real world applications. In: Proceedings of the International Conference on Data Analytics, pp. 42–48 (August 2014)
7. Deutscher, J., Davison, A., Reid, I.: Automatic partitioning of high dimensional search spaces associated with articulated body motion capture. In: Proceedings of the IEEE Computer Society Conference on Computer Vision and Pattern Recognition. vol. 2, pp. 669–676 (2001)
8. Dibblee, D., Maltese, J., Ombuki-Berman, B.M., Engelbrecht, A.P.: Vector-evaluated particle swarm optimization with local search. In: Proceedings of the IEEE Congress on Evolutionary Computation, pp. 187–195 (May 2015)
9. Flash, T., Hogan, N.: The coordination of arm movements: an experimentally confirmed mathematical model. Neuroscience 5(7), 1688–1703 (1985)
10. Heppner, F., Grenander, U.: A stochastic nonlinear model for coordinated bird flocks. In: Krasner, E. (ed.) The ubiquity of chaos, pp. 233–238. AAAS Publications (1990)
11. Huang, H.C., Chen, C.P., Wang, P.R.: Particle swarm optimization for solving the inverse kinematics of 7-DOF robotic manipulators. In: Proceedings of the IEEE International Conference on Systems, Man, and Cybernetics, pp. 3105–3110 (October 2012)
12. Jin, Y., Olhofer, M., Sendhoff, B.: Dynamic weighted aggregation for evolutionary multi-objective optimization: why does it work and how?. In: Proceedings of the Annual Conference on Genetic and Evolutionary Computation, pp. 1042–1049. San Francisco, USA (July 2001)
13. Kennedy, J., Eberhart, R.: Particle swarm optimization. In: Proceedings of the IEEE International Conference on Neural Networks, vol. 4, pp. 1942–1948. IEEE (November 1995)
14. Lasseter, J.: Principles of traditional animation applied to 3D computer animation. Seminal Graphics, pp. 263–272. ACM, New York (1998)

15. Maltese, J., Ombuki-Berman, B., Engelbrecht, A.: Co-operative vector-evaluated particle swarm optimization for multi-objective optimization. In: Proceedings of the IEEE Symposium Series on Computational Intelligence, pp. 1294–1301 (December 2015)
16. Matthysen, W., Engelbrecht, A., Malan, K.: Analysis of stagnation behavior of vector evaluated particle swarm optimization. In: Proceedings of the IEEE Symposium on Swarm Intelligence, pp. 155–163 (April 2013)
17. Mohamad, M.M., Taylor, N.K., Dunnigan, M.W.: Articulated robot motion planning using ant colony optimisation. In: Proceedings of the International IEEE Conference Intelligent Systems, pp. 690–695 (September 2006)
18. Mori, M., MacDorman, K.F., Kageki, N.: The uncanny valley [from the field]. IEEE Robot. Autom. Mag. **19**(2), 98–100 (2012)
19. Parsopoulos, K.E., Vrahatis, M.N.: Particle swarm optimization method in multiobjective problems. In: Proceedings of the ACM Symposium on Applied Computing SAC 2002, pp. 603–607. ACM, New York (2002)
20. Parsopoulos, K.E., Tasoulis, D.K., Vrahatis, M.N., et al.: Multiobjective optimization using parallel vector evaluated particle swarm optimization. In: Proceedings of the IASTED International Conference on Artificial Intelligence and Applications, vol. 2, pp. 823–828 (2004)
21. Poli, R., Kennedy, J., Blackwell, T.: Particle swarm optimization. Swarm Intell. **1**(1), 33–57 (2007)
22. Repository, B.D.M.: Teenage girl model. https://www.blender-models.com/model-downloads/humans/id/teenage-girl-model, Accessed 9 Jan 2017
23. Rokbani, N., Alimi, M.A., Ammar, B.: Architectural proposal for a robotized intelligent humanoid, IZiman. In: Proceedings of the IEEE International Conference on Automation and Logistics, pp. 1941–1946 (August 2007)
24. Rokbani, N., Benbousaada, E., Ammar, B., Alimi, A.M.: Biped robot control using particle swarm optimization. In: Proceedings of the IEEE International Conference on Systems, Man and Cybernetics, pp. 506–512 (October 2010)
25. Rokbani, N., Alimi, A.M.: Inverse kinematics using particle swarm optimization, a statistical analysis. Proc. Eng. **64**, 1602–1611 (2013)
26. Rokbani, N., Casals, A., Alimi, A.M.: IK-FA, a new heuristic inverse kinematics solver using firefly algorithm. In: Azar, A.T., Vaidyanathan, S. (eds.) Computational Intelligence Applications in Modeling and Control, pp. 369–395. Springer International Publishing, Cham (2015). doi:10.1007/978-3-319-11017-2_15
27. Rutkowski, L., Przybyl, A., Cpalka, K.: Novel online speed profile generation for industrial machine tool based on flexible neuro-fuzzy approximation. IEEE Trans. Industr. Electron. **59**(2), 1238–1247 (2012)
28. Sakchaicharoenkul, T.: MCFI-based animation tweening algorithm for 2D parametric motion flow/optical flow. MG&V **15**(1), 29–49 (2006)
29. Tanskanen, E.: Transition synthesis for skeletal animations using optimization and simulated physics. G2 pro gradu, diplomity, Aalto University, 03 November 2014
30. Tevatia, G., Schaal, S.: Inverse kinematics for humanoid robots. In: Proceedings of the IEEE International Conference on Robotics and Automation, vol. 1, pp. 294–299 (2000)
31. Uno, Y., Kawato, M., Suzuki, R.: Formation of optimum trajectory in control of arm movement: minimum torque-change model. Japan IEICE Technical Report MBE86-79, pp. 9–16 (1987)
32. Uno, Y., Kawato, M., Suzuki, R.: Formation and control of optimal trajectory in human multijoint arm movement. Biol. Cybern. **61**(2), 89–101 (1989)

33. Wampler, K., Andersen, E., Herbst, E., Lee, Y., Popović, Z.: Character animation in two-player adversarial games. ACM Trans. Graph. **29**(3), 1–13 (2010)
34. Whitney, D.E.: Resolved motion rate control of manipulators and human prostheses. IEEE Trans. Man-Mach. Syst. **10**(2), 47–53 (1969)
35. Yang, Y., Peng, G., Wang, Y., Zhang, H.: A new solution for inverse kinematics of 7-DOF manipulator based on genetic algorithm. In: Proceedings of the IEEE International Conference on Automation and Logistics, pp. 1947–1951, August 2007

Particle Swarm Optimization for the Machine Repair Problem with Working Breakdowns

Kuo-Hsiung Wang[1] and Cheng-Dar Liou[2(✉)]

[1] Department of Computer Science and Information Management,
Providence University, Taichung 43301, Taiwan
khwang@pu.edu.tw
[2] Department of Business Administration, National Formosa University,
64, Wunhua Rd., Huwei 63201, Yunlin County, Taiwan, ROC
cdliou@nfu.edu.tw

Abstract. This paper studies the M/M/1 machine repair problem using a single service station subject to working breakdowns. This service station can be in working breakdown state only when at least one failed machine exists in the system. The matrix-analytic method is used to compute the steady-state probabilities for the number of failed machines in the system. A cost model is constructed to simultaneously determine the optimal values for the number of operating machines and two variable service rates to minimize the total expected cost per machine per unit time. The particle swarm optimization (PSO) algorithm is implemented to search for the optimal minimum value until the system availability constraint is satisfied.

Keywords: Machine repair problem · Working breakdown · Particle swarm optimization

1 Introduction

This paper deals with the machine repair problem (MRP) with M operating machines and a single non-reliable service station subject to working breakdowns. A "non-reliable" service station indicates that the service station may be subject to unpredictable breakdowns while servicing a customer. A service station operated by a team of technicians can be considered as a server in the queuing system. The team of technicians cannot always maintain the same service rate because some of them could be temporarily absent due to various reasons such as sickness, joining a training course, supporting other departments, adjusting manpower in off-peak time and so on. In this situation the service station still provides service but with a lower service rate. This is called the "working breakdown" period. The working breakdown concept is different from the working vacation concept by Servi and Finn [1]. It was first proposed by Kalidass and Kasturi [2]. They presented the steady-state analysis of an M/M/1 infinite queue with working breakdowns. In this study the service station is assumed to provide service either at a fast rate (when the service station is working) or at a lower rate (when the service station is subject to working breakdown). This model could appear in practical MRP. For example, there are M coloring machines driven and controlled by a series of printed circuit boards (PCBs) in a 24H

© Springer International Publishing AG 2017
Y. Tan et al. (Eds.): ICSI 2017, Part I, LNCS 10385, pp. 238–245, 2017.
DOI: 10.1007/978-3-319-61824-1_26

textile factory. When one of the machine PCBs fails the service station (server) operated by a team of technicians would start to repair that failed machine. Because the coloring machines are operated 24H and must satisfy various textile product color requirements, the coloring machines could frequently fail to work. When the repair facility becomes empty, the repairmen may be on stand-by until there is at least one failed machine waiting in the queue. Sometimes, due to various reasons such as attending a training course, meetings, inspecting the equipment, and other duties, some technicians could be temporarily absent. In this situation the repair station cannot provide the usual service rate (i.e. working breakdown).

Exact steady-state MRP solutions are obtained for (1) the M/M/1 model with a single service station subject to breakdowns by Wang [3]; (2) the M/E$_k$/1 model with a non-reliable service station by Wang and Kuo [4]; (3) the M/M/R model with spares and server breakdowns by Wang [5]; (4) the M/M/R model with spares operating under variable service rates by Wang and Sivazlian [6]; and (5) the M/M/R model with balking, reneging, and server breakdowns by Ke and Wang [7]. The literature on the single-server queuing models with working breakdowns starts from Kalidass and Kasturi [2]. Kim and Lee [8] presented the system size distribution and the sojourn time distribution of the M/G/1 queuing system with disasters and working breakdowns. In their study, the cold standby substitute server is considered to provide a lower service rate (i.e. working breakdown) than that of the main server. Lately, Liou [9] investigated an M/M/1 queue with an unreliable server subject to working breakdowns and impatient customers. However, existing research work regarding MRP does not include the M/M/1 MRP with working breakdowns. This motivates us to investigate the M/M/1 MRP with a single non-reliable server subject to working breakdowns.

The particle swarm optimization (PSO) algorithm was developed by Kennedy and Eberhart [10]. The huge power of this technique is its ability to satisfy a performance criterion without any prior knowledge of the candidate configurations, and the facility for searching for the global optimum result. Therefore, the PSO algorithm has high potential for analyzing complex MRP optimization problems. Excellent literature reviews can be found in the work of Clerc [11] and Alrashidi and EL-Hawary, [12].

The purpose of this paper is threefold. The first is to present a matrix-analytic method for developing steady-state solutions for an M/M/1 MRP with a single non-reliable service station subject to working breakdowns. The second is to construct the expected cost function per machine per unit time to determine the joint optimum number of machines and service rates at minimum cost until the system availability constraint is satisfied. The third is to use the PSO algorithm for searching the optimal solution of the cost minimization problem.

2   Steady-State Equations

An M/M/1 MRP with a non-reliable service station subject to working breakdowns is considered. The steady-state equations are set up first and then the matrix-analytic method with an efficient MAPLE program to calculate the steady-state probability.

2.1 Steady-State Equations

The system states are presented using pairs $\{(i,n)|i = 0, 1 \,; n = 0, 1, 2. . .M\}$, where $i = 0$ denotes that the service station is in the working period, $i = 1$ denotes that the service station is in the working breakdown period, and n is the number of failed machines in the system. The steady-state probabilities are defined as follows:

$P_0(n) \equiv$ Probability that there are n failed machines in the system when the service station is in the working period;

$P_1(n) \equiv$ Probability that there are n failed machines in the system when the service station is in the working breakdown period,

where $n = 0, 1, 2, ..., M$

The steady-state equations for an M/M/1 MRP with a non-reliable service station subject to working breakdowns are established.

The steady-state equations for $P_0(n)$ and $P_1(n)$ relating to Fig. 1 is given by:

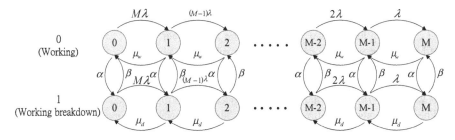

Fig. 1. State-transition rate diagram for the M/M/1 MRP.

$$(M\lambda + \alpha)P_0(0) = \mu_w P_0(1) + \beta P_1(0), \tag{1}$$

$$[(M - n)\lambda + \mu_w + \alpha]P_0(n) = (M - n + 1)\lambda P_0(n - 1) + \mu_w P_0(n+1) + \beta P_1(n), \; 1 \le n \le M - 1 \tag{2}$$

$$(\mu_w + \alpha)P_0(M) = \lambda P_0(M - 1) + \beta P_1(M), \tag{3}$$

$$(M\lambda + \beta)P_1(0) = \mu_d P_1(1) + \alpha P_0(0), \tag{4}$$

$$[(M - n)\lambda + \mu_d + \beta]P_1(n) = (M - n + 1)\lambda P_1(n - 1) + \mu_d P_1(n+1) + \alpha P_0(n), \; 1 \le n \le M - 1 \tag{5}$$

$$(\mu_d + \beta)P_1(M) = \lambda P_1(M - 1) + \alpha P_0(M). \tag{6}$$

2.2   Matrix-Analytic Method

A matrix-analytical method is used to analyze the problem further as it suffers from extreme difficulty in solving (1)–(6) in a recursive manner to develop the closed-form expressions for the steady-state probabilities $P_0(n)$ and $P_1(n)$, where $n = 0, 1, 2, ..., M$. The matrix-analytical method to simplify stationary probabilities computation is implemented in the following.

The corresponding transition rate matrix $\mathbf{Q}$ of this Markov chain has the block-diagonal form:

$$\mathbf{Q} = \begin{bmatrix} D_1 & C_0 & & & & & \\ A_1 & B_1 & C_1 & & & & \\ & A_1 & B_2 & C_2 & & & \\ & & A_1 & B_3 & C_3 & & \\ & & & \ddots & \ddots & \ddots & \\ & & & & A_1 & B_{M-1} & C_{M-1} \\ & & & & & A_1 & B_M \end{bmatrix}.$$

The rate matrix $\mathbf{Q}$ of this state process is similar to the quasi birth and death type, and this class of Markov processes has been extensively studied by Neuts [13]. Each element of the matrix $\mathbf{Q}$ is listed in the following:

$$D_1 = \begin{bmatrix} -(M\lambda + \alpha) & \alpha \\ \beta & -(M\lambda + \beta) \end{bmatrix},$$

$$A_1 = \begin{bmatrix} \mu_w & 0 \\ 0 & \mu_d \end{bmatrix}, \quad 1 \leq n \leq M$$

$$B_n = \begin{bmatrix} -[(M-n)\lambda + \mu_w + \alpha] & \alpha \\ \beta & -[(M-n)\lambda + \mu_d + \beta] \end{bmatrix}, \text{ for } 1 \leq n \leq M$$

$$C_n = \begin{bmatrix} (M-n)\lambda & 0 \\ 0 & (M-n)\lambda \end{bmatrix}, 0 \leq n \leq M-1$$

where D_1, A_1, B_n and C_n are square matrices of order 2.

Let $\mathbf{P}$ be the corresponding steady-state probability vector of $\mathbf{Q}$. By partitioning the vector $\mathbf{P}$ as $\mathbf{P} = \{P_0, P_1, P_2, ..., P_{M-1}, P_M\}$, where $P_n = \{P_0(n), P_1(n)\}, (0 \leq n \leq M)$ is a row vector of dimension 2. By solving the steady-state equations $\mathbf{PQ} = \mathbf{0}$, it follows that

$$P_0 D_1 + P_1 A_1 = 0,$$
$$P_0 C_0 + P_1 B_1 + P_2 A_1 = 0$$

$$P_{n-1} C_{n-1} + P_n B_n + P_{n+1} A_1 = 0, \text{ for } 2 \leq n \leq M-1$$
$$P_{M-1} C_{M-1} + P_M B_M = 0.$$

Thus, we obtain:

$$P_M = -P_{M-1}C_{M-1}B_M^{-1} = P_{M-1}X_M, \text{ where } X_M = -C_{M-1}B_M^{-1} \qquad (7)$$

$$P_n = P_{n-1}X_n, 2 \le n \le M - 1, \qquad (8)$$

$$P_1 = -P_0C_0(B_1 + X_2A_1)^{-1}, \qquad (9)$$

$$P_0[D_1 - C_0(B_1 + X_2A_1)^{-1}A_1] = 0, \qquad (10)$$

where $X_n = -C_{n-1}(B_n + X_{n+1}A_1)^{-1}$, $2 \le n \le M - 1$ are square matrices of order 2.

Equation (10) determines P_0 up to a multiplicative constant. The other Eqs. (7)–(9) determine $P_M, P_{M-1}, \ldots, P_2, P_1$, up to the same constant, which is uniquely determined by the following normalizing equation

$$P_0(0) + \sum_{n=1}^{M} P_n\mathbf{e} = 1,$$

where $\mathbf{e}$ is a column vector with each component equal to one. The terms $P_0(0)$, P_n and $P_j(n)$ for $j = 0, 1$ and $1 \le n \le M$ can be solved using the MAPLE computer software. Note that the main advantage for the matrix-analytical method is that it can be written as an efficient subroutine, making it easier and faster to solve a system of linear equations. Computations for huge scale MRP with N operating machines, for example $N = 300$, are entirely straightforward.

3   Cost Analysis

We develop the total expected cost function per machine per unit time for the M/M/1 MRP with a non-reliable server subject to working breakdowns, in which three decision variables M, μ_w, and μ_d are considered. The discrete variable M is a natural number, and the two continuous variables μ_w (the service rate in the working period) and μ_d (the service rate in the working breakdown period) are positive numbers. Our main objective is to determine the optimum number of operating machines M, say M^* and the optimum repair rate value (μ_w, μ_d), say (μ_w^*, μ_d^*) simultaneously so that the cost function is minimized. We also suppose that the service station can be in working breakdown at any time with breakdown rate α. Whenever the station is in the working breakdown state, it is immediately repaired at a repair rate β.

3.1   Cost Function

Let A_v denote the probability that at least one machine is operating, and A_0 represent the minimum fraction of one machine operating. We select the following cost elements:
$C_0 \equiv$ cost per unit time per failed machine in the system when the service station is the working period,

$C_1 \equiv$ cost per unit time per failed machine in the system when the service station is in the working breakdown period,

$C_2 \equiv$ fixed cost for fast service rate,

$C_3 \equiv$ fixed cost for slow service rate.

$$E[N_0] = \sum_{n=1}^{M} nP_0(n), \tag{11}$$

$$E[N_1] = \sum_{n=1}^{M} nP_1(n), \tag{12}$$

Using the definitions of these cost elements, the total expected cost function per machine per unit time is given by

$$F(M, \mu_w, \mu_d) = \frac{C_0 E[N_0] + C_1 E[N_1] + C_2 \mu_w + C_3 \mu_d}{M} \tag{13}$$

The cost minimization problem can be presented mathematically as

$$\underset{M, \mu_w, \mu_d}{Minimize} \ F(M, \ \mu_w, \ \mu_d)$$

Subject to: $A_v \geq A_0$.

The cost parameters in (13) are assumed to be linear in the expected number of indicated quantities, and it would be a hard task to develop analytical results for the optimum value (M^*, μ_w^*, μ_d^*) because the expected cost function is non-linear and complex. We use the PSO algorithm to search for the optimal value (M^*, μ_w^*, μ_d^*).

4 Numerical Results

To evaluate the PSO algorithm performance for the cost minimization problem, we tested the problems shown in Table 1. For each of the tested examples, the number of operating machines M was set in the range of 3 to 15 and both μ_w and μ_d are continuous at an interval of 0.01 to 10 and $\mu_w > \mu_d$. Although the PSO algorithm seems to be sensitive to the tuning of some parameters, according to the experiences of many experiments, the following PSO parameters can be used (see Shi and Eberhart [14]; Yoshida, et al. [15]).

PSO algorithm **Parameters**

- population size = 100;
- generations = 200;
- inertia weight factor is set to vary linearly from 0.9 to 0.4;
- the limit of change in velocity for each member in an individual was as $V_{\max} = 4.0$,
- acceleration constant $c_1 = 2.0$ and $c_2 = 2.0$.

Table 1 shows the numerical results for 100 independent experiments for each example using the PSO algorithm, where $C_0 = \$100/day$, $C_1 = \$150/day$, $C_2 = \$50/day$, $C_3 = \$15/day$ are assumed. Note that, for the convenience of comparison, the mean and maximum ratios were utilized. The solution ratio produced by the search method is calculated using V/V^*, where V is the solution generated by the PSO algorithm and V^* is the minimum solution among 100 independent experiments.

From Table 1, we observe that

(1) the mean values (V/V^*) for the PSO algorithm vary from 1.001–1.010. This implies that PSO algorithm is robust for all test instances.
(2) the max values (V/V^*) for the PSO algorithm vary from 1.005–1.019 for. This implies that the PSO algorithm search quality is very good.
(3) the average CPU time per run for the experimental solutions in Table 1 is about 8.75s for the PSO algorithm. This implies that the PSO algorithm can solve the test examples within a reasonable time.

Table 1. PSO in searching the global best solution. $(0.01 \leq \mu_w \leq 10, 0.01 \leq \mu_d \leq 10)$

(λ, α, β)	M^*	Av^*	μ_w^*	μ_d^*	$F(M^*, \mu_w^*, \mu_d^*)$	Mean	Max
(0.5, 0.1, 0.2)	15	1.0	7.083	10.000	52.262	1.003	1.011
(1.0, 0.1, 0.2)	12	1.0	7.815	10.000	83.762	1.001	1.007
(1.5, 0.1, 0.2)	8	1.0	0.010	10.000	100.671	1.004	1.011
(0.5, 0.2, 0.2)	15	1.0	6.081	10.000	51.111	1.010	1.017
(0.5, 0.3, 0.2)	14	1.0	4.554	10.000	49.624	1.007	1.019
(0.5, 0.4, 0.2)	12	1.0	2.100	10.000	47.459	1.007	1.016
(0.5, 0.1, 0.4)	15	1.0	7.72 1	8.152	52.232	1.002	1.010
(0.5, 0 1, 0 6)	15	1.0	8.018	6.589	51.604	1.002	1.009
(0.5, 0.1, 0.8)	15	1.0	8.229	4.879	50.807	1.001	1.005

5 Conclusions

This search modeled the M/M/1 MRP using a single service station subject to working breakdowns. The steady-state results were computed numerically using the matrix-analytical technique. We developed the expected cost function per machine per unit time and formulated an optimization problem to search for the minimum cost. The PSO algorithm was implemented to determine the optimal values for M^*, μ_w^*, and μ_d^* simultaneously, which minimizes the expected cost function. The PSO algorithm is applied to analyze complex MRP optimization problems including economic performance. In future research the PSO algorithm can be used to solve the optimization problems that occur in various queuing systems.

References

1. Servi, L.D., Finn, S.G.: M/M/1 queue with working vacation (M/M/1/WV). Perform. Eval. **50**, 41–52 (2002)
2. Kalidass, K., Kasturi, R.: A queue with working breakdowns. Comput. Ind. Eng. **63**, 779–783 (2012)
3. Wang, K.H.: Profit analysis of the machine repair problem with a single service station subject to breakdowns. J. Oper. Res. Soc. **41**, 1153–1160 (1990)
4. Wang, K.H., Kuo, M.Y.: Profit analysis of the $M/E_k/1$ machine repair problem with a non-reliable service station. Comput. Ind. Eng. **32**, 587–594 (1997)
5. Wang, K.H.: Profit analysis of the M/M/R machine repair problem with spares and server breakdowns. J. Oper. Res. Soc. **45**, 539–548 (1994)
6. Wang, K.H., Sivazlian, B.D.: Cost analysis of the M/M/R machine repair problem with spares operating under variable service rates. Microelectron. Reliab. **32**, 1171–1183 (1992)
7. Ke, J.C., Wang, K.H.: Cost analysis of the M/M/R machine repair problem with balking, reneging, and server breakdowns. J. Oper. Res. Soc. **50**, 275–282 (1999)
8. Kim, B.K., Lee, D.H.: The M/G/1 queue with disasters and working breakdown. Appl. Math. Model. **38**, 1788–1798 (2014)
9. Liou, C.D.: Markovian queue optimization analysis with an unreliable server subject to working breakdowns and impatient customers. Int. J. Syst. Sci. **46**(12), 2165–2182 (2015)
10. Kennedy, J., Eberhart, R.C.: Particle swarm optimization. In: Proceedings of IEEE International Conference on Neural Networks, Piscataway, NJ, pp. 1942–1948 (1995)
11. Clerc, M.: Particle Swarm Optimization (International Scientific and Technical Encyclopedia). Wiley_ISTE, London (2006)
12. Alrashidi, M.R., EL-hawary, M.E.: A survey of particle swarm optimization applications in power system operations. Electr. Power Compon. Syst. **34**, 1349–1357 (2006)
13. Neuts, M.F.: Matrix Geometric Solutions in Stochastic Models: An Algorithmic Approach. The John Hopkins University Press, Baltimore (1981)
14. Shi, Y., Eberhart, R.C.: Parameter selection in particle swarm optimization. In: Porto, V.W., Saravanan, N., Waagen, D., Eiben, A.E. (eds.) EP 1998. LNCS, vol. 1447, pp. 591–600. Springer, Heidelberg (1998). doi:10.1007/BFb0040810
15. Yoshida, H., Kawata, K., Fukuyama, Y., Nakanishi, Y.: A particle swarm optimization for reactive power and voltage control considering voltage security assessment. IEEE Trans. Power Syst. **15**, 1232–1239 (2000)

Intelligent Behavioral Design of Non-player Characters in a FPS Video Game Through PSO

Guillermo Díaz[1] and Andrés Iglesias[2,3(✉)]

[1] Master Program in Creation of Video Games, University Pompeu Fabra,
Balmes Building, Balmes 132-134, 08008 Barcelona, Spain
[2] Department of Information Science, Faculty of Sciences, Toho University,
2-2-1 Miyama, Funabashi 274-8510, Japan
[3] Department of Applied Mathematics and Computational Sciences,
University of Cantabria, Avda. de Los Castros, s/n, E-39005 Santander, Spain
iglesias@unican.es

Abstract. Although barely explored so far, swarm intelligence can arguably have a profound impact on video games; for instance, as a simple yet effective approach for the realistic intelligent behavior of Non-Player Characters (NPCs). In this context, we describe the application of particle swarm optimization to the behavioral design of NPCs in a first-person shooter video game. The feasibility and performance of our method is analyzed through some computer experiments. They show that the proposed approach performs very well and can be successfully used in a fully automatic (i.e., without any human player) and efficient way.

Keywords: Swarm intelligence · Particle swarm optimization · Video game · Non-player characters · Intelligent behavioral design

1 Introduction

Nowadays, the most classical application of artificial intelligence (AI) in video games is the behavioral animation of their virtual characters, particularly the NPCs [6–12]. They are virtual characters not controlled by the player, so their AI must be fully specified by the computer. A classical example of NPCs appears in first-person shooter (FPS) games, where the human player assumes the role of a virtual character fighting against a platoon of computer-controlled enemies, the NPCs. Very often, the human player is also assisted by a small group of members of his own squad (NPCs allies). The AI of the NPCs in FPSs is very challenging for several reasons: ons is that human player and the NPCs share a seemingly identical set of skills, abilities and goals (kill the enemies). The NPCs have to be *believable*: the human player should feel that they behave as human beings as well, taking reasonable decisions most of the time but not always, to prevent excessive predictability. The NPCs of the ally squad must be coordinated with the human player to operate in a cooperative and synchronized way. They also should protect/help the human player in a natural and realistic way.

© Springer International Publishing AG 2017
Y. Tan et al. (Eds.): ICSI 2017, Part I, LNCS 10385, pp. 246–254, 2017.
DOI: 10.1007/978-3-319-61824-1_27

Finally, *playability* is a must: players should have a chance to win and the risk to lose. An adequate balance of wins and losses improves the player engagement and makes the game more believable and fun.

For many years, the AI of NPCs was based on scripts, leading to self-replicating patterns and simple and repetitive behavioral routines. Other AI approaches for video games including NPCs were given in the form of a rules-based system, where a set of rules is used to determine the behavior of the NPCs (for instance, the *Pac-Man*). For a large list of rules, the behavior of the NPCs does not become obvious to the human player, but still the system has very little intelligence within. In the 90s, more powerful AI approaches (such as finite state machines or behavior trees) were incorporated into video games. A popular approach for action planning is the HTN (hierarchical task networks) planners, based on hierarchies of tasks that can be broken down recursively. In recent years, the level of sophistication has increased and more realistic and complex behavioral systems have been defined. Regarding the evolutionary approaches, *City Conquest* uses a genetic algorithm during the design process to identify dominant strategies and evolve the game design, even before it is released. In *Darwin's Nightmare* game, evolutionary computation is applied to drive the exploration of a large combinatorial space defining the behavior and appearance of enemy crafts. Finally, PSO is applied to the problems of pathfinding and action planning in video games in [2]. In this paper we claim that swarm intelligence can arguably be one of the best approaches for the AI of NPCs. To show it, this paper applies PSO to the behavioral design of NPCs in a FPS video game.

2   Particle Swarm Optimization

The basic idea of *swarm intelligence* (SI) is that the collective intelligence of a swarm arises from local interactions among the individuals and with the environment rather than from any centralized intelligence [3,5]. This feature can be applied to the AI routines of the (generally simple) NPCs in video games.

A popular SI technique is *particle swarm optimization* (PSO) [1,4,5]. The procedure is sketched in Algorithm 1. It starts with a population (swarm) of candidate solutions called *particles* randomly distributed over the search space and provided with initial position and velocity. A fitness function is needed to evaluate the quality of a position. Particles can communicate good positions to each other, so they can adjust their own position and velocity according to this information. The swarm evolves iteratively, so that the fitness of the global best particle improves and eventually reaches the best solution. During the iterations, each particle modifies its position and velocity as:

$$v_j^{k+1} = \omega\, v_j^k + \gamma_1 u_1 [x_{gb}^k - x_j^k] + \gamma_2 u_2 [x_{jb}^k - x_j^k] \tag{1}$$

$$x_j^{k+1} = x_j^k + v_j^k \tag{2}$$

where x_j^k and v_j^k are the position and the velocity of particle j at iteration k respectively, ω is called *inertia weight* and describes how much the old velocity

Algorithm 1. Particle swarm optimization

```
 1: for each particle do
 2:     Initialize particle                              //Initialization phase
 3: end for
 4: while (stop condition = false) do
 5:     for each particle do
 6:         Fitness ← fitness of the particle
 7:         if (Fitness is better than BestIndividualFitness) then
 8:             BestIndividualFitness ← Fitness      //update the memory individual fitness
 9:             BestIndividualPosition ← Position    //update the memory individual solution
10:         end if
11:         if (Fitness is better than BestGlobalFitness) then
12:             BestGlobalFitness ← Fitness          //update the current best global fitness
13:             BestGlobalPosition ← Position        //update the current best global solution
14:         end if
15:     end for
16:     for each particle do
17:         Update velocity                              //eqn. (1)
18:         position ← position + velocity               //eqn. (2)
19:     end for
20: end while
21: return bestGlobalPosition
```

---

will affect the new one, and coefficients γ_1 and γ_2 are constant values called *learning factors*, accounting for the "social" component (memory of the swarm best particle position at time k, x_{gb}^k) and the "cognitive" component (memory of the best historical position of particle j, x_{jb}^k). Two random numbers, u_1 and u_2, with uniform distribution on $[0, 1]$ are included to promote diversity. This procedure is iterated until a stopping condition is met or the solution is found.

3 Video Game and NPCs Design and Implementation

Our benchmark is based on a FPS developed on the game engine *Unreal 4*. In the game, the human player takes the role of a special command member who, along with other NPC team members, is assigned to difficult missions (terrorist group deactivation, hostage rescue, etc.) on different environments (indoor spaces, urban environments, natural landscapes) with no information about the enemies and the environment. Many AI tasks are required to complete the missions successfully. For instance, the player's character has to interact with the other squad members (autonomous NPCs driven by their own AI) in a cooperative and synchronized way. This introduces realistic temporal and spatial constraints. Furthermore, our NPCs exhibit a more sophisticated behavior and a wider variety of skills than in other types of video games: they can explore the environment, read and navigate maps, cooperate between them for synchronized attacks, develop advanced action planning (e.g. espionage, stealth, simulation,

Fig. 1. Top view of the game level. Red circle marks the current enemy position. (Color figure online)

counter-attack, or defensive strategies) and so on. For limitations of space, in this paper we restrict to a single environment and mission, described in next paragraphs.

To test the application of PSO to this problem, a $5,500$ square meters maze-like square public office level was created (see its top view in Fig. 1). The level has been carefully designed to represent a difficult scenario for the NPCs: it contains several tricky challenges for the characters, such as a large number of dead ends and bottlenecks such as those typically found in labyrinths. To make things more interesting and challenging, the level is fully furnished; the office rooms and corridors are filled with many objects. Some represent static obstacles the characters have to avoid, others encourage the NPCs to stay in front, obstructing the normal flow of motion, and others promote further interaction with the NPCs. Finally, in this mission, the area has been affected by some explosions, so there are several objects lying on the ground and other obstacles to avoid.

A setting where a command tries to find someone moving around some place is an ideal environment to analyze the potential of SI for dynamic goals. To this aim, we consider a particular mission: the command agents should find and capture alive an armed terrorist hidden inside this complex environment. The terrorist is an enemy NPC driven by its own AI and initialized at the center of the top side in Fig. 1. It moves throughout this environment autonomously, trying to hide and escape from the command. The command is comprised by four squads, each consisting of a variable number of members ranging from 3 to 24 initialized in the center of the bottom side in Fig. 1. Given our goals, *there is no human player in this mission*; it would interfere the simulations drastically, modifying the sequence of events and invalidating our results. *The full responsibility to accomplish the mission falls on the behavioral AI of the command NPCs.*

Fig. 2. (left) Squad ally NPCs trying to find and capture alive the enemy NPC using their PSO-based behavioral AI; (right) Enemy surrendering to the squad NPCs when getting ambushed and caught by surprise.

The behavioral system of the NPCs requires different context-sensitive actions to execute. Similar to some other FPS games, this is achieved by using a *Behavior Tree*. Basically, this is a logical structure commonly used in action planning that allows an AI to execute the correct task based on some values. In this way, the NPC could execute one or several actions depending on the environment context, the course of events, and other factors. For example, the NPC can run around the map while looking for any terrorist or enemy, and protect other members of the squad at the same time. Of course, several actions can be added or overlapped to build up a more complex behavior. For instance, the agents can move using PSO, look for enemies in sight, or shoot them in case they attack the agents.

The computational architecture of the NPC behavioral system of the enemies and command members is similar; their AI is however pretty different, as they have different goals and actions to perform. The most relevant difference concerns the inclusion of the PSO method in the AI of the command NPCs. To this aim, the command agents are considered particles of the swarm, provided with initial positions and velocities. Then, the PSO is run until the agents reach the final goal (find the enemy). The autonomous behavior of each agent and the coordination of all members of the swarm can be properly balanced through the social and the cognitive parameters γ_1 and γ_2. In this paper, we take values $\gamma_1 = \gamma_2 = 1.4$, well within the convergence area of the algorithm. This choice of the social and cognitive parameters has yielded pretty good results in our experiments. We remark, however, that changing these parameters only modifies the time taken to reach the goal, but the goal is achieved anyway. Other parameters of the PSO are the population size, p_s, and the inertia weight w. They are considered tuning parameters of the method, so they are modified during our experiments (see below for details). Of course, the enemy NPC works alone so no PSO algorithm can be applied to this character. It would be interesting, however, to see what happens if the PSO is also added to the AI of a swarm of enemy NPCs.

Extra work is also needed to set up all other aspects for a NPC in a video game, including all programming tasks related to the NPC animations, such as those shown in Fig. 2. On the left, the members of the command are moving in a formation where the agents in the front carry out searching, surveillance, and exploration tasks, while those in the rear provide protection to the squad. The enemy moves slowly and carefully, paying attention to any motion, noise or shadow, as it happens in real life. On the right, the terrorist has been caught without enough time to shoot or escape and decides to surrender. Although the graphical animations are already encoded and stored, they are activated by specific actions so they should be triggered by the AI of the NPCs.

All modeling tasks were done with *Autodesk 3D Studio Max*, *Adobe Photoshop* tool was used for texturing, and *Apple Logic Pro 9* for sound editing. The game engine *Unreal 4* has been used for the graphical environment, implement the AI algorithms, and perform the tests. An important component is the *Behavior Tree*, used to implement the behavioral routines of our NPCs. *Unreal Blueprint* visual scripting is also useful for many implementation tasks. In particular, the PSO algorithm was implemented using *Blueprint* scripting, along with some parts directly written in the programming language *Unreal* C++.

4   Experimental Results

To analyze the performance of our approach, we executed the PSO for different values of the inertia weight w (ranging from $w = 0.2$ to $w = 1.2$ with stepsize 0.2) and the population size of the swarm (from $p_s = 12$ to $p_s = 96$ with stepsize 12). This makes a total of 48 combinations. For each couple, we performed 25 independent executions and computed the CPU time taken for the swarm to find the enemy and surrender him. Table 1 shows the average time (in seconds) for the 25 executions for each pair of values (w, p_s) in the ranges indicated above.

The first and most important conclusion is that the method performs pretty well. All simulations reached the final goal, and the CPU times have been very competitive, given the large size of the environment, its difficult geometry, and the fact that the goal is highly dynamic: the enemy is constantly moving trying to escape from the command agents. Moreover, this escape strategy is not random or accidental, but driven by a smart AI. This shows that the approach works very well, and can be successfully applied to commercial video games. Second observation is that the CPU time decreases as the population size increases up to $p_s = 48$. This fact is not surprising; more agents means additional exploratory

Table 1. Average time (in seconds) on 25 executions for different values of (w, p_s).

	12 NPCs	24 NPCs	36 NPCs	48 NPCs	60 NPCs	72 NPCs	84 NPCs	96 NPCs
$\omega = 0.2$	227.1489	202.3854	179.0512	127.5967	168.5907	167.3464	153.6267	157.2611
$\omega = 0.4$	164.0612	132.1689	105.8425	87.6574	90.2230	92.6600	81.0496	106.2469
$\omega = 0.6$	126.3209	77.7963	84.5847	45.7596	79.8171	60.8325	66.3322	83.4260
$\omega = 0.8$	93.5823	60.7219	54.3361	37.4781	60.8899	63.5352	60.0020	67.6512
$\omega = 1.0$	40.8895	37.7445	37.8783	32.5637	43.1950	39.5755	43.7119	51.2208
$\omega = 1.2$	47.6974	39.3628	41.1499	36.9983	39.1020	42.7837	44.2017	46.9681

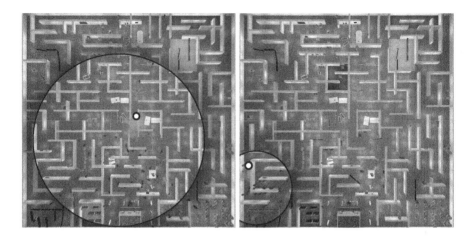

Fig. 3. Distribution of particles for a high value (left) and a small value (right) of the inertia weight in PSO. Green area marks the radius of distribution of the NPCs. White dot marks "best global position". (Color figure online)

capacity. However, this should be taken with care, because increasing the size of the swarm also increases the computational calculations. In fact, the CPU time increases again for populations larger than $p_s = 48$. In our case, the best results were obtained for a population of 48 particles; in fact, any simulation improves its results once it moves closer to that number of NPCs. However, this optimal value can vary for different configurations and game conditions. Third observation is that there is a downward trend in the average time when w gets closer to 1. In particular, higher values for w work better for dynamic goals, as the particles do not gather in any particular area of the map; instead, they move fast within a large radius around the best global position, see Fig. 3(left). Also, higher values for w work better in big scenarios, where particles should move fast through the map to approach the solution, while smaller values make the particles move together but slowly, in a small radius around the best global position, which does not work that good if the goal changes its position quickly or when the map is quite large, see Fig. 3(right). Finally, we also noticed that the initial position of the particles has a big impact on the algorithm: the more spread out the particles, the better the results. In our case, all particles begin on a small common area, thus increasing the CPU times.

5   Conclusions and Future Work

This paper reports a successful application of SI to the behavioral design of NPCs for a FPS video game. The AI of the NPCs is enriched with a powerful SI method, the popular PSO. We describe the main issues involved in this process, from the most general layer (game design) to the AI design of the NPCs. Some implementation issues are also described. Finally, several computational experiments have been carried out to assess the performance of our approach. Our results show that SI can improve the performance of the behavioral systems of the NPCs in a fully automated way, leading to realistic and autonomous intelligent behaviors from the NPCs. Furthermore, they can be achieved without the intervention of human players. Although these results come from a particular problem and video game, they can be properly adapted to other problems and games. These results validate our claims for further use of SI for video games. The ability of SI techniques to generate behaviors from models in nature make them perfect candidates to answer the need for better AI in video games.

All our simulations have been recorded frame by frame to generate videos at a frame rate of 24 FPS (all screenshots in this paper have actually been taken from the videos). All these videos along with the numerical data we collected provide an invaluable source of data and information that has still to be analyzed at full extent. This is part of our future work in the field.

Acknowledgements. This research work has been supported by Computer Science National Program, Spanish Ministry of Economy & Competitiveness, Project Ref. #TIN2012-30768, Toho University and the University of Cantabria.

References

1. Eberhart, R.C., Shi, Y.: Particle swarm optimization: developments, applications and resources. In: Proceedings of the IEEE Congress on Evolutionary Computation, CEC 2001, pp. 81–86. IEEE Computer Society Press, Los Alamitos (2001)
2. Díaz, G., Iglesias, A.: Swarm intelligence scheme for pathfinding and action planning of non-player characters on a last-generation video game. Adv. Intell. Syst. Comput. **514**, 343–353 (2017)
3. Engelbrecht, A.P.: Fundamentals of Computational Swarm Intelligence. Wiley, Chichester (2005)
4. Kennedy, J., Eberhart, R.C.: Particle swarm optimization. In: Proceedings of IEEE International Conference on Neural Networks, Perth, Australia, pp. 1942–1948. IEEE Computer Society Press, Los Alamitos (1995)
5. Kennedy, J., Eberhart, R.C., Shi, Y.: Swarm Intelligence. Morgan Kaufmann Publishers, San Francisco (2001)
6. Iglesias, A.: A new framework for intelligent semantic web services based on GAIVAs. Int. J. Inf. Technol. Web. Eng. **3**(4), 30–58 (2008)
7. Iglesias, A., Luengo, F.: Intelligent agents for virtual worlds. In: Proceedings of CW 2004, Tokyo, Japan, pp. 62–69. IEEE Computer Society Press, Los Alamitos (2004)
8. Iglesias, A., Luengo, F.: A new based-on-artificial-intelligence framework for behavioral animation of virtual actors,. In: Proceedings of CGIV 2004, Penang, Malaysia, pp. 245–250. IEEE Computer Society Press, Los Alamitos (2004)
9. Iglesias, A., Luengo, F.: New goal selection scheme for behavioral animation of intelligent virtual agents. IEICE Trans. Inf. Syst. **E88–D**(5), 865–871 (2005)
10. Iglesias, A., Luengo, F.: AI framework for decision modeling in behavioral animation of virtual avatars. In: Shi, Y., Albada, G.D., Dongarra, J., Sloot, P.M.A. (eds.) ICCS 2007. LNCS, vol. 4488, pp. 89–96. Springer, Heidelberg (2007). doi:10.1007/978-3-540-72586-2_12

11. Luengo, F., Iglesias, A.: A new architecture for simulating the behavior of virtual agents. In: Sloot, P.M.A., Abramson, D., Bogdanov, A.V., Dongarra, J.J., Zomaya, A.Y., Gorbachev, Y.E. (eds.) ICCS 2003. LNCS, vol. 2657, pp. 935–944. Springer, Heidelberg (2003). doi:10.1007/3-540-44860-8_97

12. Luengo, F., Iglesias, A.: Designing an action selection engine for behavioral animation of intelligent virtual agents. In: Gervasi, O., Gavrilova, M.L., Kumar, V., Laganà, A., Lee, H.P., Mun, Y., Taniar, D., Tan, C.J.K. (eds.) ICCSA 2005. LNCS, vol. 3482, pp. 1157–1166. Springer, Heidelberg (2005). doi:10.1007/11424857_124

Ant Colony Optimization

An Improved Ant Colony Optimization with Subpath-Based Pheromone Modification Strategy

Xiangyang Deng[1,2], Limin Zhang[1], and Jiawen Feng[1(✉)]

[1] Institute of Information Fusion, Naval Aeronautical
and Astronautical University, Yantai, Shangdong, China
xavior2012@aliyun.com, iamzlm@163.com,
fengjiawen777@163.com
[2] Institute of Electronic Engineering, Naval Engineering University,
Wuhan, China

Abstract. The performance of an ACO depends extremely on the cognition of each subpath, which is represented by the pheromone trails. This paper designs an experiment to explore a subpath's exact role in the full-path generation. It gives three factors, sequential similarity ratio (SSR), iterative best similarity ratio (IBSR) and global best similarity ratio (GBSR), to evaluate some selected subpaths called r-rank subpaths in each iteration. The result shows that r-rank subpaths keep a rather stable proportion in the found best route. And then, by counting the crossed ants of a subpath in each iteration, a subpath-based pheromone modification rule is proposed to enhance the pheromone depositing strategy. It is combined with the iteration-best pheromone update rule to solve the traveling salesman problem (TSP), and experiments show that the new ACO has a good performance and robustness.

Keywords: Ant colony optimization · Subpath-based pheromone modification strategy · Travel salesman problem · Meta-heuristic algorithm · Pheromone trails

1 Introduction

The traveling salesman problem (TSP) is a classical NP-Complete problem, and a meta-model of many problems in reality, and has received increasing attention. ACO inspired by the foraging behavior of ant colony is proposed to solve TSP firstly in 1991 [1]. After that, many improved ACO-based algorithms has been presented, Among which, literature [2] proposed a method to avoid premature convergence phenomenon, literature [3] proposed an ant system based on individual sort, literature [4] adopted an elimination cross strategy to achieve a local optimization of ant colony algorithm, literature [5] designed a random selection and perturbation strategy to improve the global exploring capability, literature [6] discussed the one with Time Windows constraints TSP problem, and proposed the ACS-TSPTW algorithm to solve it, literature [7] discussed the solution of the generalization of TSP and proposed ACO method combined with the process of mutation and local search.

ACO-based algorithms can be classified into two types by means of their pheromone updating models [8]. One of pheromone updating models associates the

© Springer International Publishing AG 2017
Y. Tan et al. (Eds.): ICSI 2017, Part I, LNCS 10385, pp. 257–265, 2017.
DOI: 10.1007/978-3-319-61824-1_28

pheromones with arcs, and the other associates pheromones with nodes. The former is most frequently used by a lot of ant algorithms, such as Ant-Q system [9] and Ant Colony System (ACS) [10], etc. And the latter is often used in solving a type of non-ergodic optimal problem, such as the pheromone mark ACO [11] and ACO applied in generalized TSP problem [12], etc. About the pheromone updating models, Literature [10] and literature [11] have made a comprehensive and detailed analysis.

In this paper, we discuss a new pheromone updating rule named as subpath-based pheromone modification rule (SPB rule), and proposed an improved ACO based on SPB rule, naturally called SPB-ACO. The SPB rule contains an iterative updating part and an adaptive part. The iterative updating part is used to updating some selected subpaths' pheromone trace, which are called r-rank subpaths chosen by counting the crossed ants of a subpath in each iteration. And the adaptive part regulates the updating strength with the iteration gradually. Then, by combining the SPB rule and the iteration-best update rule, an adaptive SPB-ACO is used to solving standard TSP. The result shows that r-rank subpaths keep a rather stable proportion in the found best route, and there play a very important role in the overall cognition of the ant colony.

2 The Basic Model of SPB-ACO

The common solution construction procedures of SPB-ACO can be described as follows: at the beginning, m ants are randomly positioned in n cities, and then each ant utilizes a pseudo-random proportional rule to select the next city continuously. After a forward city is selected, a subpath is given between the current city and the selected city, until a complete and valid route is generated. After each ant finished its tour, the pheromones on the arcs of the current best route are updated.

The main steps of the algorithm are as follows:

Step 1. Initializing ant colony.
Step 2. The ants use a pseudo-random proportional rule to select a city.
Step 3. Repeat to select city until perform a complete solution.
Step 4. According to the evaluation function to calculate the ants' fitness and get an optimal route.
Step 5. Update the pheromones according to the global pheromone update rule.
Step 6. Complete according to the algorithm termination condition or transfer to step 2.
Step 7. End the algorithm.

When an ant marked $k(1,\ldots,m)$ is in city i, it selects the next city according to the following rule:

$$j = \begin{cases} \arg\underset{l\in J_k(i)}{maax}\left[(\tau_{il})^{\alpha}\times(\eta_{il})^{\beta}\right], & if \ q \leq q_0 \\ according \ to \ formula \ (2), & otherwise \end{cases} \quad (1)$$

Where q is a random variable uniformly distributed in [0, 1], and q_0 is a given probability.

The state transition probability is calculated based on the amount of pheromones and the heuristic information on the subpaths. P_{ij}^k represents the state transition probability of ant k transferring from city i to city j:

$$p_{ij}^k = \begin{cases} \frac{\tau_{ij}^\alpha \times \eta_{ij}^\beta}{\sum_{j \in allowned} [\tau_{ij}^\alpha \times \eta_{ij}^\beta]} & if\ j \in allowned \\ 0, & otherwise \end{cases} \quad (2)$$

Where α is the pheromone heuristic factor, representing the relative importance of the trajectory, and β is the desired heuristic factor, representing the relative importance of the visibility. τ_{ij} is the pheromone density on the arc (i,j), and η_{ij} is the value of the heuristic function on the arc (i,j), and is equal to $\frac{1}{d_{ij}}$ in normal circumstances, and d_{ij} represents the distance of the arc (i,j).

After each ant has completed the tour of all the cities, according to the following rules to update the pheromone trace:

$$\begin{cases} \tau_{ij} = (1 - \rho) \times \tau_{ij} + \rho \times \Delta\tau_{ij} \\ \Delta\tau_{ij} = \frac{1}{L_{iterbest}} + \zeta \end{cases} \quad (3)$$

Where ρ is the pheromone evaporation coefficient, and $\Delta\tau_{ij}$ represents the pheromone increment on subpath (i,j) in the iteration, and $L_{iterbest}$ is the length of the best route of the iteration, and ζ is the pheromone increment of the selected subpath generated by subpath regulation strategy.

3   SPB-ACO with a Subpath-Based Pheromone Modification Strategy

Ant colony usually achieves a valid solution through iterative competition of the ant individuals from the same generation colony. They get the heritage information between different generations. It makes the experience of the previous generation of ants can guide the successive ants' searching behaviors, which makes the ants focus gradually on the optimal solution arcs with a great probability.

After a generation of ant colony building all paths, the pheromones of the subpaths belong to the iterative optimal path are updated. At the same time, the r-rank subpaths which more ants traversed are selected to deposit pheromones. The former usually called iteration-best-update rule (IB rule), and the latter is called SPB rule. The two rules together compose the subpath-based pheromone modification strategy of SPB-ACO.

3.1   Rank Subpaths and Superimpose Pheromone

In the ant colony searching process, different ants construct different solution sequences, but they always go through some same subpaths each other. In the ant colony's overall cognitive view, the more ants go through a subpath, the more important the subpath plays a role (more important role the subpath plays) in building a global optimal route. Generally, a more important subpath should be reserved, because it has a greater possibility to serve as a best route in successive iterations. Thus, after all ants complete the construction of solution sequence, the number of ants through a subpath can be counted and used to generate a new pheromone increment, which can be used to regulate the layout of pheromone.

After each iteration, rank the subpaths based on the number of r_{ij} representing the traversed ants, and update the pheromones of the first l subpaths. The amount of pheromone increment calculated in accordance with the following formula:

$$\zeta = \begin{cases} Q, & \text{if } rank \leq l \\ 0, & \text{otherwise} \end{cases} \tag{4}$$

The pheromone increment calculation process based on subpath updating rule is clearly demonstrated in the flow chart in the Fig. 1.

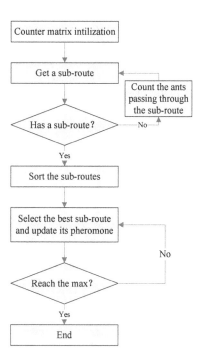

Fig. 1. The flow chart of pheromone update rule based on subpath regulation strategy

3.2 Adaptive Adjustment of Involved Subpaths

In subpath regulation strategy, the pheromones deposited on the involved subpaths have a negative effect in the late stage, and it will have an influence on the convergence. So, the involved subpaths should be gradually decreased the pheromone increment. Meanwhile, the superimposed pheromone in the late stage has two sides. On one hand it can concentrate the random search behaviors; on the other hand it has an impact on speeding the convergence maybe resulted in a premature convergence.

For a better effect of involved subpaths' pheromones on the algorithm's performance, an adaptive mechanism is established for pheromone update. In the early stage of the algorithm, through more subpaths involved in pheromone superposition, it can get better exploring ability, and makes the pheromone distribution is balanced in the solution space and let the search behaviors more randomly. In the late stage, a less amount of subpaths involved in pheromone superposition can speed the convergence. Along with the iteration, the subpath regulation strategy has a following reationship:

$$l = a \times n - iteration \qquad (5)$$

Where l is the number of the subpaths involved in the pheromone superposition in the iteration, and a represents the scale factor of the subpaths that participate in the pheromone update procedure, and n is the number of the cities, $iteration$ is the index of iteration. In order to keep the function of the subpath's pheromone in later stage, l has a minimum value.

Through the adaptive pheromone update mechanism, it can get a smooth transition of pheromone distribution in the previous and later stage of the algorithm. Moreover, the pheromone superposition of the subpaths is good for global exploring at the early stage, and the effect becomes weaker in the late stage and help to get a mall range of convergence.

4   Experimental Tests

In order to view the performance of the SPB-ACO algorithm, three tests are designed to analyze the SPB-ACO algorithm, and to compare with ACS in the following environment: MatLab 7.8.0 (R2009a), run in the Windows XP environment, computer of HP540, CPU T5470 1.6 GHz, memory 1G.

4.1   Test of SPB Rule

Expriment 1: Iteration 50 times, recording the top r subpaths by the number of ants that cross them, calculating the coincidence ratio of r subpaths in iter (>2) iteration and iter-2 iteration in each iteration, which is called SSPSR.

Expriment 2: Iteration 50 times, recording iteration-best path and the top r subpaths by the number of ants that cross them, calculating the coincidence ratio of r subpaths and iteration-best path in each iteration, which is called ISPSR.

Expriment 3: Iteration 50 times, recording global-best path and the top r subpaths by the number of ants that cross them, calculating the coincidence ratio of r subpaths and global-best path in each iteration, which is called GSPSR.

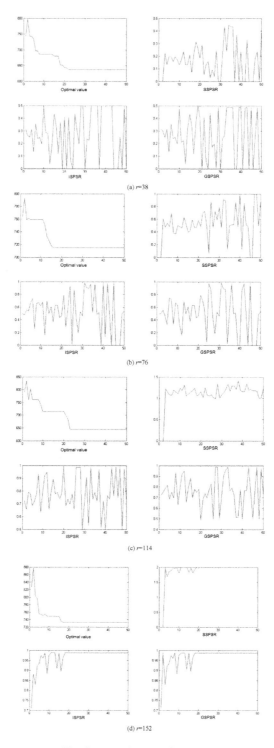

Fig. 2. Experiments of subpaths

The r is set as 38,76,114 and 152, repeat the experiments, the results are shown in Fig. 2.

As illustrated in Fig. 2, when r takes different value:

(1) In the process of convergence, SSPSR, ISPSR and GSPSR maintain relatively stable values, for example, ISPSR = 0.25 when r = 38, ISPSR = 0.5 when r = 76, ISPSR = 095 when r = 114;

(2) When the algorithm reaches the convergence state, SSPSR, ISPSR and GSPSR show very large volatility until r = 152, after that, the volatility disappears, the reason of which is that the base (denominator) is relatively big. This shows that the sub-paths obtained by counting the number of cross ants are regular in ant colony evolution. Regardless of the value of base number r, the r-subpaths are not completely contained in the iterative optimal path or the global optimal path, meanwhile, the r-subpaths could not completely contained the iterative optimal path or the global optimal path. This paper defines the r-subpaths as the r-best-subpaths and proposes an improved ant colony algorithm based on the regularity above.

During the ant colony's construction procedure, there are always some same subpaths existed in the diversity solutions of different ant individuals. From the overall cognition of the ant colony, the more ants pass through a subpath, the more important the subpath is in the process of constructing the optimal complete path. We can count the ants that cross a subpath, and then rank the subpaths. Further more, by investgating the dynamic ranking process, the role of the subpaths in the whole ant colony evolution process can be deeply explored.

4.2   Test of Comparing the SPB-ACO with ACS

ACS system will be implemented in the environment of this article, including the main parameters:

(1) Using nearest neighbor search method to construct the initial path length l_0.
(2) The initial pheromone $\tau_0 = \frac{1}{n \times l_0}$.
(3) Based on the global-best optimal update strategy.

The parameters of SPB-ACO is set as follows: $\alpha = 2$, $\beta = 1$, $\rho = 0.1$, $a = 2$, the number of ants $m = 20$, $q_0 = 0.4$, $Q = \frac{1}{L_{iterbest}}$, $iteration = 500$, τ_0 is same to ACS.

The test does 1000 iterations, and test is repeated 20 times. The standard TSP test problem eil76.tsp is used, and the best results for SPB-ACO and ACS can be seen in Fig. 3.

In the Fig. 3, it shows that the SPB-ACO algorithm can converge to a relatively optimal value, and the optimal solution is 548.13. ACS gets the optimal value 581. What's more, the SPB-ACO algorithm reaches convergence in the 75th iterations, and remains optimization characteristic in the later stage and has a wonderful performance in local optimization.

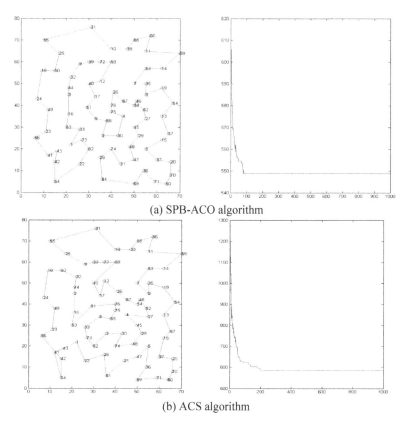

(a) SPB-ACO algorithm

(b) ACS algorithm

Fig. 3. Two results for SPB-ACO and ACS

5   Conclusion

At present, most ACO algorithms mainly pay attention to different solution structures, pheromone update strategies and the initial state of ant colony to improve the global exploring capability and convergence. However, it is difficult to achieve the two characteristics at the same time. This paper summarized the classical ant algorithms and their pheromone models, and discussed why the ant algorithms prone to premature convergence. After that, it introduced a subpath-based modification strategy to optimizing the process of constructing optimal path, and then proposed a new ACO algorithm called SPB-ACO, which provides a global pheromones update mechanism. The adaptive mechanism enabled the algorithm to obtain a compromise between the exploring ability and the exploiting ability.

However, the paper didn't establish a more precise mathematical model to describe the adaptive adjustment mechanism. So we will to do more experiments and give a more effective improvement in the near future.

References

1. Colorni, A., Dorigo, M., Maniezzo, V.: Distributed optimization by ant colonies. In: Proceedings of the 1st European Conference on Artificial Life, Paris, pp. 134–142 (1991)
2. Stützle, T., Hoos, H.H.: The MAX-MIN ant system and local search for the traveling salesman problem. In: Back, T., Michalewicz, Z., Yao, X. (eds.) Proceedings of the 1997 IEEE International Conference on Evolutionary Computation (ICEC 1997), pp. 309–314 (1997)
3. Bullnheimer, B., Hartl, R.F., Strauss, C.: A new rank-based version of the ant system: a computational study. Central Eur. J. Oper. Res. Econ. 7(1), 25–38 (1999)
4. Huang, L., Wang, K., Zhou, C., et al.: Hybrid approach based on ant algorithm for solving traveling salesman problem. J. Jilin Univ. (Sci. Ed.) 40(4), 369–373 (2002)
5. Hao, J., Shi, L., Zhou, J.: An ant system algorithm with random perturbation behavior for complex TSP problem. Syst. Eng.-Theory Pract. 9, 88–91 (2002)
6. Cheng, C.-B., Mao, C.-P.: A modfied ant colony system for solving the travelling salesman problem with time windows. Math. Comput. Model. 46, 1225–1235 (2007)
7. Yang, J., Shi, X., Marchese, M., et al.: An ant colony optimization method for generalized TSP problem. Prog. Nat. Sci. 18, 1417–1422 (2008)
8. Blum, C., Dorigo, M.: The hyper-cube framework for ant colony optimization. IEEE Trans. Syst. Man Cybern.-Part B. Also available as Technical report TR/IRIDIA/2003–03, IRIDIA, Universit Libre de Bruxelles, Belgium (2003)
9. Gambardella, L.M., Dorigo, M.: Ant-Q: a reinforcement learning approach to the traveling salesman problem. In: Proceedings of the 12th International Conference on Machine Learning, pp. 252–260 (1995)
10. Dorigo, M., Gambardella, L.M.: Ant colony system: a cooperative learning approach to the traveling salesman problem. IEEE Trans. Evol. Comput. 1(1), 53–66 (1997)
11. Deng, X., Zhang, L., Lin, H., et al.: Pheromone mark ant colony optimization with a hybrid node-based pheromone update strategy. Neurocomputing (in press). 10.1016/j.neucom.2012.12.084
12. Geng, J.Q., Weng, L.P., Liu, S.H.: An improved ant colony optimization algorithm for nonlinear resource-leveling problems. Comput. Math. Appl. 61, 2300–2305 (2011)

Decentralized Congestion Control in Random Ant Interaction Networks

Andreas Kasprzok[1]([envelope]), Beshah Ayalew[1], and Chad Lau[2]

[1] Clemson University International Center for Automotive Research,
4 Research Drive, Greenville, SC 29607, USA
akasprz@clemson.edu
[2] Harris Corporation, 2400 Palm Bay Rd NE, Mailstop HTC-5S,
Palm Bay, FL 32905, USA

Abstract. Interaction networks formed by foraging ants are among the most studied self-organizing multi-agent systems in nature that have inspired many practical applications. However, the vast majority of prior investigations assume pheromone trails or stigmergic strategies used by the ants to create foraging behaviors. We first review an ant network model where the direction and speed of each ant's correlated random walk are influenced by direct and minimalist interactions, such as antennal contact. We incorporate basic ant memory with nest and food compasses, and adopt a discrete time, non-deterministic forager recruitment strategy to regulate the foraging population. The paper's main focus is on decentralized congestion control and avoidance schemes that are activated with a quorum sensing mechanism. The model relies on individual ants' ability to estimate a perceived avoidance sector from recent interactions. Through simulation experiments it is shown that a randomized congestion avoidance scheme improves performance over alternative static schemes.

1 Background

Ants and other social insects exhibit complex problem solving capabilities in groups despite the limited capabilities of the individual insect/agent. They can share information by secreting pheromones into the environment, which can later be picked up by other ants. This mechanism of communication through the environment is referred to as stigmergy and is the focus of a multitude of algorithms and heuristics such as ant colony optimization or ACO [3]. Similar algorithms have since been applied to many optimization problems such as the traveling salesman problem (TSP) [21] to network routing and scheduling [4], protein folding [20] and others [13, 18].

Ants can also communicate directly by exchanging hydrocarbons on antennal contact. In habitats unsuited for environmental markers, ants like *Pogonomyrmex barbatus* and *Lasius Niger* instead rely on encounter rates with fellow ants for navigational purposes: by counting interactions, these ants are able to discern basic information about distance to the nest or points of interest as well

© Springer International Publishing AG 2017
Y. Tan et al. (Eds.): ICSI 2017, Part I, LNCS 10385, pp. 266–276, 2017.
DOI: 10.1007/978-3-319-61824-1_29

as initiate behavioral changes in the case of nest migration. The ant species *Temnothorax albipennis* [17] uses encounter rates to choose new nesting sites: when enough ants congregate at a certain site, increasing encounter rates above a certain threshold and signaling a consensus [8], individual ants change from nest searching to migration. This mechanism is referred to as quorum sensing [16] and is used for similar purposes in honey bees [19] and the bacteria *V. harveyi* [14]. A good discussion of the non-deterministic, yet resilient and robust characteristics of quorum sensing are summarized in [11].

In this paper, we briefly describe a microscopic model for ants that use encounter rates for the purposes of navigation and congestion avoidance. Many aspects of the model are discussed in further detail in our journal paper [12] and are inspired by the detailed descriptions of ant interaction networks by biologist Gordon [10]. Here, we specifically focus on decentralized congestion control by first implementing the quorum sensing mechanism to congestion control instead of colony migration. Each ant is considered to also estimate the perceived center of, and sector around, the congestion when detecting a concentration of ants above the quorum threshold. Three possible implementations of a decentralized congestion control scheme are proposed and evaluated. To regulate the foraging ant population, we adopt the recruitment model proposed by Gordon [15], where the rate of spawning of the ants from the nest is related to the rate of return of food bearings ants.

This work is motivated by possible applications of ant-inspired decentralized congestion control that could use minimalist direct communication without using external infrastructure or other analogues of stigmergy. Congestion control problems are prevalent in the management of transportation networks [5], data networks [7], and other multi-agent resource/information dissemination/consumption applications, such as social networks [22].

The rest of this paper is organized as follows: Sect. 2 reviews the modeling assumptions and the navigation scheme used by each ant, while Sect. 3 details the decentralized congestion control scheme. Section 4 briefly reviews the recruitment model adopted and Sect. 5 presents results and discussions focusing on variations of the congestion control scheme. Section 6 concludes the paper.

2   Navigation

The ants' (the agents') movement across a 2-dimensional plane is modeled as a correlated random walk updating at fixed time steps. During their movement, agents may physically contact other agents, which triggers an interaction during which the ants exchange whether or not they are carrying food, and, if they are, their current heading. However, the simple occurrence of the encounter/interaction is the most important information for the agents. The change in heading at each step, θ, is sampled from the von Mises distribution [1] in Eq. 1,

$$f(\theta) = M(\theta; \mu; \kappa) = \frac{1}{2\pi I_0 \kappa} e^{\kappa cos(\theta - \mu)} \tag{1}$$

where κ is the concentration parameter. It varies the dispersion of the distribution: at $\kappa = 0$, the distribution is uniform; when κ is large, the distribution becomes concentrated about its mean μ. I_0 denotes the modified Bessel equation of the first kind. This distribution is commonly applied to model a wide range of biological motions including those of ants [6]. Note that θ denotes the change in the ant's heading direction from step i to step $i + 1$. We model each ants' response to recent encounter with other ants or interactions by updating its concentration parameter with the following equation:

$$\kappa(t) = \kappa_{min} + \sum_{i=1}^{n} A e^{-T(t-t_i)} \tag{2}$$

where n is the number of interactions and t_i is the time of each interaction's occurrence. A and T are parameters, while κ_{min} denotes the minimum concentration parameter of an ant. The purpose of the latter is to retain an amount of directionality in the absence of interactions, since otherwise $\kappa = 0$ would yield a uniform distribution.

The speed of each ant is also made directly dependent on the number of recent interactions it experiences. This is modeled as follows:

$$v(t) = v + \sum_{i=1}^{n} I e^{-\lambda_I (t-t_i)} - \sum_{i=1}^{n} R e^{-\lambda_R (t-t_i)} \tag{3}$$

where v is a default or nominal speed for an ant. The two added terms represent, respectively, the impetus (I) to go faster due to more interactions, and the resistance (R) to the motion from physical impediments of more ants. The impetus parameters I and λ_I are made smaller than the resistance parameters R and λ_R in order to achieve an initial decrease in speed followed by a prolonged minute increase. The weighted sum of all (recent n) interactions experienced by each ant at time t_i is then added to the default speed v.

The goal of each foraging ant is to acquire food and return to the nest from which it is spawned. Using the mechanisms described above, the ants travel across the search space. When an ant encounters food or decides to return to the nest after a certain time of foraging unsuccessfully it does so using a *Nest Compass*. In nature, ants continuously integrate their walk in order to maintain an internal vector of the nests location [23]. We integrate this behavior into our model using a second von Mises distribution whose mean points in the direction of the nest relative to the agent's current heading. This is superimposed onto the ant's correlated random walk. The result is then normalized and used for heading decisions. While an ant is returning to the nest, interactions are used to modify the *Nest Compass* instead of the random walk: as the ant travels closer to the nest, it is more likely to encounter other ants and in return reaffirmed in its general direction. When a foraging ant encounters an ant carrying food, it sets its heading to the opposite of the encountered ant's heading before making its next decision. The assumption made is that the food source will be located in the general area opposite to where the food-carrying ant is headed, since its

Nest Compass urges it to return. Similar to the *Nest Compass*, ants who have previously found food and successfully returned it to the nest possess a *Food Compass*, which is modeled via an additional von Mises distribution added to the random walk that points to the location at which food has recently been found by the ant. With the above model components, given a certain number of food-bearing and foraging ants, if the food source is concentrated in an area, a interaction chain is eventually formed where ants influence each other in their relative directions and speeds.

3   Decentralized Congestion Avoidance Strategy

While the navigational attributes mentioned above enable the ants/agents to successfully forage in open environments, constricted scenarios such as those incorporating corridors and closed spaces greatly hinder the ants ability to forage: they get stuck in corners, are often unable to move past each other when interacting along walls, and may be unable to return to the nest depending on the geometry of the environment. Additionally, congestion in bottlenecks may cause a standstill for the colony and total loss of productivity.

We first describe how we model the behavior near walls or obstacles. When an ant contacts an obstacle other than an ant, the component of its heading perpendicular to the wall is nullified and the ant slows down, covering less distance. Additionally, we have found in computational experiments that this behavior encourages the coalescence of groups of ants along the walls; often times ants keep trying to move into opposite directions yet get stuck as they are unable to efficiently move past each other. To combat this behavior, we endow ants to avoid contact with a wall after an initial encounter with it. All directional changes that would move an ant closer to the direction of an encountered wall are avoided for a short time frame. This avoidance region is taken to be a nearly 178° sector directly perpendicular to the direction of the encountered wall. This simple behavior assumes that the wall extends a certain distance in either direction, but is found to increase the colony's productivity.

Away from obstacles, a decentralized congestion control scheme is implemented starting with the quorum sensing mechanism: when ants sense a high density of other ants around them, indicated by the concentration parameter κ reaching or exceeding a threshold value T_C, the ants change their behavior to prioritize the avoidance of further interactions [9]. To model this, we assume each ant to not only count recent interactions but also the directions from which they occurred. Then, each ant estimates the average direction to the center of the congestion using:

$$\theta_C = atan(\frac{\sum_{i=1}^{m} \sin(\theta_i)}{\sum_{i=1}^{m} \cos(\theta_i)}) \tag{4}$$

where θ_C is the heading pointing to the perceived center of congestion and θ_i the direction of contact of a given recent interaction. Higher numbers of interactions enables the ant to more precisely estimate the center of congestion, as well as the level of congestion. Once the ant estimates the direction to the center of the congestion, it creates an avoidance area/sector of a certain number of degrees equally distributed on either side of the congestion direction vector. This sector is then excluded from possible headings that the ant might choose at its next decision step, in order to avoid the congested area. The probabilities of the remaining directions are normalized and then used for further heading decisions. This behavior is illustrated in Fig. 1, where a foraging ant is avoiding a cluster of ants surrounding it.

The size of the avoidance area can be made directly dependent on the concentration parameter. This is a notion consistent our model of the formation of the ant network formed by interactions.

$$\beta_C = X_C * \kappa \tag{5}$$

where β_C is the size of the congestion area, X_C is a scaling constant, and κ is the concentration parameter. When the avoidance area is dependent on the number of recent interactions, a small amount of congestion will result in less evasive action which leads to a less drastic change in ant behavior and higher throughput in moderately traveled bottlenecks.

Instead of the static estimates of the congestion area sector offered above or the one varying with the concentration parameter, we also consider a randomized estimate. This can be done by using the von Mises distribution with a mean estimated as above and a concentration parameter κ of the random walk (indicator of encounter rates). By looking up the direction of the congestion's center as perceived by the agent from this distribution, we introduce some randomness to prevent standstills which may occur when groups of ants traveling in opposite directions encounter each other in corridors or bottlenecks. We shall refer to this case as randomized variable sector in the results below.

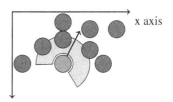

Fig. 1. An ant (center) establishes its avoidance area after several encounters with other ants

For an in-depth discussion on the decision loop of the agents, see [12].

4   Recruitment

The recruitment of foragers in the nest is abstracted using a discrete time recruitment model proposed by Prabhakar et al. [15]:

$$\alpha_n = max(\alpha_{n-1} - qD_{n-1} + cA_n - d, \underline{\alpha}), \alpha_0 = 0 \tag{6}$$

$$D_n \sim Poisson(\alpha_n) \tag{7}$$

where α_n is the rate at which ants are spawned from the nest at time n, A_n is the number of returning, food-bearing foragers at time n, and the actual number of ants departing the nest, D_n, is set to a Poisson random variable using the spawn rate as its mean. q, c, and d are parameters for the variables already discussed, while $\underline{\alpha}$ denotes the minimum spawn rate of the nest.

Ants that have previously found food and therefore possess a *Food Compass* are recruited first since according to Gordon [10], a relatively small amount of the foraging population does a majority of the work. The parameters of Eqs. 6 and 7 are tuned as to recruit ants at a rate similar to the number of ants returning plus the minimum rate. This leads to steady growth of the population until the food source is fully consumed.

5   Results and Discussion

We implemented the model described above in the modeling environment Unity t 3D using C#. Unity's built-in physics system was used for collision detection between ants as well as their navigation around each other and obstacle. Values of important parameters used for generating the simulation results presented in this section are listed in our journal paper [12]. Each experiment was executed 100 times.

We consider two geometric scenarios to primarily evaluate the decentralized congestion avoidance scheme. Scenario 1, shown in Fig. 2, depicts the nest area and food area connected by 3 paths. This was chosen to encourage interactions and create congested areas. It is a version of the double bridge experiment [2], though with two longer routes instead of one. The direct route between the two areas will be most traveled due to the ants' *Nest Compass*. We hypothesized that a correctly working congestion avoidance strategy would divert a number of ants from the shortest paths to the two alternate paths. Scenario 2 adds two additional pathways at the north and south ends of scenario 1. We run all tests in this scenario as well in order to evaluate the scalability of our strategy. The task in all experiments is to collect 180 food items arranged in a 3×3 square in the back of the food area. The results for scenario 2 were found to be in agreement with observations from scenario 1, and most of its results were therefore not included here, except where noted (discussion of Table 1), for lack of space.

Figure 3a shows the task completion times on scenario 1 when using three static sectors (with deterministically estimated perceived center, Eq. 4; and sectors of $120°$, $180°$, and $240°$), varying the congestion area/sector in relation to the concentration parameter (Eq. 5) or randomized variable sector. It can be seen that the variable and randomized sectors lead to improved task completion times by about 40 s. Furthermore, there appears to be only minute differences between the results for the experiments using static congestion avoidance sectors.

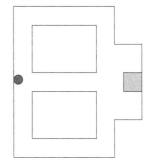

Fig. 2. Scenario 1

More information about the interactions in the network under the different congestion avoidance settings can be gleaned from Fig. 3b, which shows the

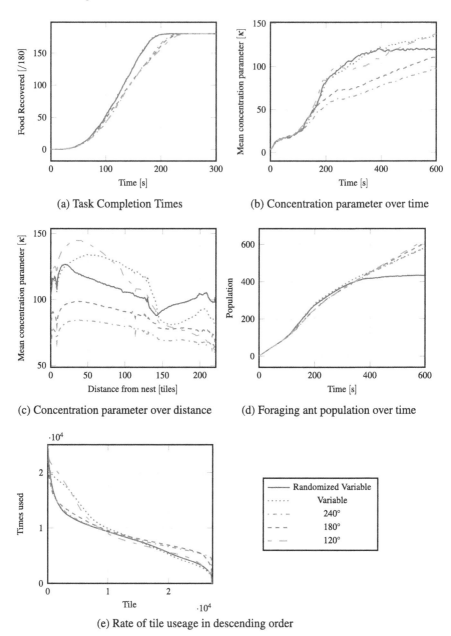

(a) Task Completion Times

(b) Concentration parameter over time

(c) Concentration parameter over distance

(d) Foraging ant population over time

(e) Rate of tile useage in descending order

Fig. 3. Experimental data comparing several congestion avoidance strategies

history of the average concentration parameter of the network during the experiment. The concentration parameter is a direct result of the recent interactions experienced by each ant. While a wider static congestion avoidance area setting

Table 1. Mean Avoidance Sector sizes for both scenarios and congestion avoidance strategies

Scenario	Avoidance strategy	Mean avoidance sector ($^\circ$)
1	Variable	243.3
	Randomized variable	232.6
2	Variable	229.5
	Randomized variable	214.0

results in less interactions/second on average, the ants in these experiments are unable to resolve the congestion created by the constricted pathways: the rate of interaction increases even after task completion (240 s, in Fig. 3a). Varying the avoidance area using κ shows a reduced trend, yet the number of interactions is still steadily increasing with time. However, randomizing the perceived center of congestion seems to allow the ants to move past each other in the congested pathways and reduce both the number of interactions which is reflected in the concentration parameter.

Table 1 references the average sizes of congestion avoidance sectors for both scenarios and the two congestion avoidance strategies using a variable sector size. The size of the congestion avoidance sector for each ant in these experiments was set to $2*\kappa$, or $T_C = 2$. While the difference between the variable and randomized variable sectors are small relative to how differently they perform, the results do support the observations made earlier that the randomized variable strategy leads to less interactions and consequently smaller concentration parameters, thereby reducing the size of the congestion avoidance sector and increasing the efficiency of the network via a faster task completion time as in Fig. 3a. Additionally, the increased number of pathways in scenario 2 compared to scenario 1 leads to less interactions as the ants are able to efficiently spread out across the increased search space in scenario 2.

Another way to evaluate the effectiveness of each congestion avoidance strategy or setting might be to measure how well it diffuses ants across the environment. To this effect, Fig. 3c illustrates the average κ for the agents with respect to their distance from the nest. What this data shows is that a high congestion avoidance angle is beneficial in avoiding interactions, lowering the concentration parameter and indicating even diffusion. The randomized variable strategy performs poorly by comparison, yet we can conclude that even distribution does not necessarily indicate higher throughput, supported by Fig. 3b. Instead the higher, static congestion avoidance sectors simply leads to a standstill, restricting individual ants' mobility and confining them to their immediate areas.

Figure 3d illustrates the foraging population of the simulated ant colony over the length of the experiment. Ants are spawned at the minimum rate according to the recruitment algorithm until the first ants return with food around 100 s into the simulation, at which point additional ants are spawned to keep the population growing at a steady pace while food is available. Once the food source has been consumed, the population should stabilize as foraging ants will return to the nest after their 180 s timeout. The population only stabilizes in the case of the randomized variable congestion avoidance area, indicating that in the other cases ants are unable to return to the nest, an observation that is also indicated by Fig. 3b. This confirms our conjecture above that the randomized variable sector gives a realistic decentralized congestion control scheme for the random interaction network created by the foraging ants.

A visual comparison of this trend is shown in Fig. 4, comparing coverage in scenario 1 between a static 120° avoidance sector and a randomized variable congestion avoidance sector. Note how most of the activity in (a) is confined to the starting area while (b) shows a more even distribution of activity. This visual approximation is supported by Fig. 3e, which shows that congestion avoidance sectors of size 180° and 240° as well as the randomized variable sector lead to more even tile coverage compared to the smaller 120° sector and the variable sector size.

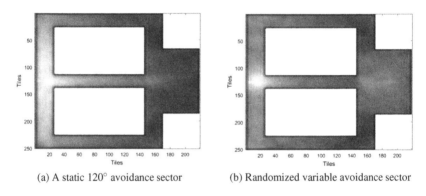

(a) A static 120° avoidance sector   (b) Randomized variable avoidance sector

Fig. 4. Examples of cumulative coverage maps for different congestion avoidance strategies

6   Conclusions

In this paper, we first reviewed the construction of a model for ants that do not use stigmergy, but instead rely on encounter rates for forming an interaction network that results in colony behavior. We then proposed and evaluated three versions of decentralized congestion control/avoidance strategies. All versions are based on endowing each ant to estimate a perceived center of congestion from its recent interactions. The main observation made from our many experiments, not all included herein, is that the randomized variable sector decentralized congestion control approach is shown to be effective in alleviating congested areas compared to its static counterparts. This model seems consistent with biologists' observations of the generally random ant decisions that lead to the formation of the overall interaction network. We are pursuing applications of these observations to practical decentralized radio networks and inter-vehicular networks in intelligent transportation systems. Scenarios where roadside units, cell towers, or other infrastructure (analogues to stigmergy) are not readily available or not yet implemented are especially attractive applications of the decentralized congestion control strategy.

References

1. Codling, E.A.: Biased random walks in biology. Ph.D. thesis, The University of Leeds (2003)
2. Deneubourg, J.L., Goss, S.: Collective patterns and decision-making. Ethol. Ecol. Evol. **1**(4), 295–311 (1989)
3. Dorigo, M.: Optimization, learning and natural algorithms. Ph.D. thesis, Politecnico di Milano, Italy (1992)
4. Dorigo, M., Stützle, T.: Ant Colony Optimization. Bradford Book, Bingley (2004)
5. Downs, A.: The law of peak-hour expressway congestion. Traffic Q. **16**(3), 393–409 (1962)
6. Fink, G.A., Haack, J.N., McKinnon, A.D., Fulp, E.W.: Defense on the move: ant-based cyber defense. IEEE Secur. Priv. **12**(2), 36–43 (2014)
7. Floyd, S., Jacobson, V.: Random early detection gateways for congestion avoidance. IEEE/ACM Trans. Netw. **1**(4), 397–413 (1993)
8. Franks, N.R., Stuttard, J.P., Doran, C., Esposito, J.C., Master, M.C., Sendova-Franks, A.B., Masuda, N., Britton, N.F.: How ants use quorum sensing to estimate the average quality of a fluctuating resource. Sci. Rep. **5** (2015). Article no. 11890. doi:10.1038/srep11890
9. Gordon, D.M., Paul, R.E., Thorpe, K.: What is the function of encounter patterns in ant colonies? Anim. Behav. **45**(6), 1083–1100 (1993)
10. Gordon, D.: Ant Encounters: Interaction Networks and Colony Behavior. Primers in Complex Systems. Princeton University Press, Princeton (2010)
11. Hamar, J., Dove, R.: Quorum sensing in multi-agent systems. INSIGHT **15**(2), 35–37 (2012)
12. Kasprzok, A., Ayalew, B., Lau, C.: A microscopic model for multi-agent interaction networks without stigmergy. Swarm Intelligence (2016, under review)
13. Leitão, P., Barbosa, J., Trentesaux, D.: Bio-inspired multi-agent systems for reconfigurable manufacturing systems. Eng. Appl. Artif. Intell. **25**(5), 934–944 (2012)
14. Liu, X., Zhou, P., Wang, R.: Small RNA-mediated switch-like regulation in bacterial quorum sensing. IET Syst. Biol. **7**(5), 182–187 (2013)
15. Prabhakar, B., Dektar, K.N., Gordon, D.M.: Anternet: the regulation of harvester ant foraging and internet congestion control. In: 2012 50th Annual Allerton Conference on Communication, Control, and Computing (Allerton), pp. 1355–1359 (2012). iD: 1
16. Pratt, S.C.: Quorum sensing by encounter rates in the ant temnothorax albipennis. Behav. Ecol. **16**(2), 488–496 (2005)
17. Pratt, S.C., Mallon, E.B., Sumpter, D.J., Franks, N.R.: Quorum sensing, recruitment, and collective decision-making during colony emigration by the ant leptothorax albipennis. Behav. Ecol. Sociobiol. **52**(2), 117–127 (2002)
18. Şahin, E.: Swarm robotics: from sources of inspiration to domains of application. In: Şahin, E., Spears, W.M. (eds.) SR 2004. LNCS, vol. 3342, pp. 10–20. Springer, Heidelberg (2005). doi:10.1007/978-3-540-30552-1_2
19. Seeley, T.D., Visscher, P.K.: Quorum sensing during nest-site selection by honeybee swarms. Behav. Ecol. Sociobiol. **56**(6), 594–601 (2004)

20. Shmygelska, A., Aguirre-Hernández, R., Hoos, H.H.: An ant colony optimization algorithm for the 2D HP protein folding problem. In: Dorigo, M., Caro, G., Sampels, M. (eds.) ANTS 2002. LNCS, vol. 2463, pp. 40–52. Springer, Heidelberg (2002). doi:10.1007/3-540-45724-0_4
21. Stützle, T., Dorigo, M.: ACO algorithms for the traveling salesman problem. In: Evolutionary Algorithms in Engineering and Computer Science, pp. 163–183 (1999). ISBN: 0471999024
22. Tian, Y.P., Liu, C.L.: Consensus of multi-agent systems with diverse input and communication delays. IEEE Trans. Autom. Control **53**(9), 2122–2128 (2008)
23. Wohlgemuth, S., Ronacher, B., Wehner, R.: Ant odometry in the third dimension. Nature **411**(6839), 795–798 (2001)

An Energy-Saving Routing Strategy Based on Ant Colony Optimization in Wireless Sensor Networks

Wei Qu[✉] and Xiaowei Wang

Shenyang Normal University, Shenyang 110034, China
quweineu@163.com

Abstract. Focus on the problem of finding the optimal path in wireless sensor networks (WSN), considering energy saving requirement, an energy-saving routing strategy based on ant colony optimization (DERS-ACO) is proposed. Our strategy designs the optimization rule of dynamic state transformation, and introduces the mechanism of rewards and penalties which further saves the search time and increase the probability of optimal path search, and prolongs lifetime of network greatly. Simulation showed that the searching probability of a global for the optimal solution is increased, and the global optimal solution is obtained quickly and effectively, furthermore the energy consumption of the nodes is saved, which will prolong the lifetime of network greatly.

Keywords: Wireless sensor networks (WSN) · Ant colony optimization (ACO) · Optimization rule of dynamic state transformation · Mechanisms of rewards and penalties

1 Introduction

Wireless Sensor Networks (WSN) is widely used in military and civilian fields [1–3]. WSN works in unattended monitoring environment generally, the network node battery replacement cost is higher, so the routing design focus more on minimizes the energy consumption of nodes [4–6]. How to find the shortest path for communication for nodes in a shortest time is our first priority to consider. The ant colony algorithm (ACO), which was first proposed by M Dora-go in 1991, is a typical method to solve this problem, and it has been broadly studied in recent years [7–9]. In wireless sensor networks, single nodes do not need to have all the information about the location of all other nodes, and only need to save the nearby nodes' in the network. However, the algorithm of ACO also has some drawbacks, such as application of convergence too fast and easily fall into local optimal solution, resulting in the whole optimal solution is ignored. In this paper, an energy-saving routing strategy based on ACO (DERS-ACO) is proposed. Through the carefully design of dynamic state transfer rule, the search probability of the best path node is realizable, and the network surviving time is prolonged.

© Springer International Publishing AG 2017
Y. Tan et al. (Eds.): ICSI 2017, Part I, LNCS 10385, pp. 277–284, 2017.
DOI: 10.1007/978-3-319-61824-1_30

2 The Energy-Saving Routing Strategy Based on ACO

2.1 State Transition Optimization Rules

Early Probability Model of DERS-ACO. Let bi(t) represent the number of ants on node i at time t, $\tau_{ij}(t)$ represents the Information on path from node i to j at time t, and m represents the total number of ants in ant colony. According to the information on each path, ant k (k = 1, 2, 3, ..., m) will decide its transfer direction during the movement. Then it will record the node that ant k have just passed by $Tabu_k$ (k = 1, 2, 3, ..., m), and the set will dynamically adjust with the evolutionary process of $Tabu_k$. $P_{ij}^K(t)$ represents state transition probabilities that ant k transfers from node i to j at time t.

$$P_{ij}^K(t) = \frac{[\tau_{ij}(t)]^{\alpha}[\eta_{ik}(t)]^{\beta}}{\sum [\tau_{is}(t)]^{\alpha}[\eta_{is}(t)]^{\beta}} \tag{1}$$

$$\eta_{ik}(t) = \frac{1}{d_{ij}} \tag{2}$$

In formula (1), information factor is α, which reflects the relative importance of pheromone accumulated during ant colony search process. If it is bigger, it is more likely that ants will choose the path previously traveled, and the collaborative between ants is stronger. Expect heuristic factor is β, and it reflects the relative importance of heuristic information in guiding the ant colony search process. In formula (2), $\eta_{ik}(t)$ is heuristic function, and d_{ij} is distance between two neighbor nodes i and j. To ant k, if d_{ij} is smaller, $\eta_{ik}(t)$ is bigger, and $P_{ij}^K(t)$ is bigger. Obviously, the heuristic function represents the desired degree that ant transfers from node i to node j.

The information on path from node i to j at time $t + t_n$ can be adjusted according to the formula (3).

$$\begin{cases} \tau(t+t_n) = (1-\rho) \times \tau(t) + \Delta\tau_{ij}(t) \\ \Delta\tau_{ij}(t) = \sum_{k=1}^{m} \Delta\tau_{ij,k}(t) \end{cases} \tag{3}$$

$$\Delta\tau_{ij}(t) = \frac{Q(t)}{L_k} \tag{4}$$

ρ represents pheromone evaporation coefficient, and 1-ρ represents pheromone residual factor. To avoid the unlimited accumulation of information, ρ must be in the range of [0,1). $\Delta\tau_{ij}(t)$ represents the pheromone increment on path from node i to j in this circle, and set its initial value is 0, that is $\Delta\tau_{ij}(t) = 0$. $\Delta\tau_{ij,k}(t)$ represents the amount of information when ant k stays on the path rom node i to j at time t in this circle. In formula (4), Q represents pheromone strength, and to some extent it affects the convergence speed. L_k represents the total length that Ant k traverses in this cycle.

Late Probability Adjustment Rules of DERS-ACO. State transition optimization rules design balance parameter q_0 first in order to balance the relative importance between "exploring" new paths and "using" the information that already exist. When $q > q_0$ explore the new path, and when $q < q_0$ use the information that already exists (q_0 is a random number between 0 and 1). The design of q_0 is important, and the value of the state transition optimization rules designed q_0 will dynamically adjust based on the condition parameters in the iterative process. The designing of adjusting in state transition optimization rules are as follows: Choose pheromones larger side when algorithm starts running. After N_C times iteration, in order to prevent falling into local optimal solution, change the value of q_0 and design a random function $rad()$ to generate a random number satisfying $rand(min(p_{ij}) \leq rad \leq max(p_{ij}))$. When $p_{ij} \geq rad$, node i choose the next hop node j in order to open up a new path, expand the search space and prevent the algorithm falling into local optimization. Finally in order to make the algorithm converges to the global optimal solution, change the value of q_0 again to be more likely to choose pheromones largest side, until the running is over. In terms of α and β, the state transition optimization rules designed the dynamic adjustment. It can targeted solve the problem that algorithm is easy to fall into local optimal solution. The state transition probabilities show that ants individual looking for the next node based on the value of pheromone concentration and distance between nodes. Then α and β respectively represent the relative emphasis of pheromone and distance during the search. The state transition optimization rules design the dynamic adjustment mechanism of α and β based on the stage where the Iteration is. Adjusted as follows:

(1) Initialize the value of q_0, α and β according to the stage where the iteration is. Use random function $rad()$ to generate a random number and assign to q, and $0 \leq q \leq 1$. Initialization model is as formula (5).

$$q_0 = \begin{cases} U & (0 < N_C < N_0) \\ U_0 & (N_0 < N_C < N_1) \\ U_1 & (N_1 < N_C < N_{c\_max}) \end{cases} \tag{5}$$

$$P_{ij}^K(t) \begin{cases} \dfrac{\tau_{ij}^\alpha(t)\eta_{ik}^\beta(t)}{\sum \tau_{is}^\alpha(t)\eta_{is}^\beta(t)} & q \leq q_0 \\ \dfrac{1}{L_n} & else \end{cases} \tag{6}$$

N_0 represents the iteration in first phase, N_1 represents the iteration in second phase, and N_c represents the iteration in local phase. $N_{c\_}$max is the maximum number of iterations, and U, U_0 and U_1 are the value of varying q_0.

(2) When $q > q_0$, node i will randomly choose the next hop node in candidate node set $j[]$ in order to open up a new path. Otherwise, it will be decided according to state transition probabilities and pheromone of the next hop. State transition probability adjustment model is shown in formula (6), L_n represents the number of neighbor node of node n.

2.2 The Reward and Punishment Mechanism

(1) When the current iteration $t == 1$, pheromone concentration updates normally, and the iteration of $\Delta\tau_{ij}(t)$ between nodes rewards on the base of the original. That is:

$$\Delta\tau_{ij}(t) = Q/L(ant) \tag{7}$$

$L(ant)$ represents the length of total path that the local ant has passed in an iteration.

(2) When the current iteration $t \geq 2$, record the length of shortest path L_{all_best} in all iterations from the beginning to last time, and reward or punish for the concentration of pheromones according to the length of the path. If the path length of current meets $L_{now} \leq L_{ave}[Nc]$ ($L_{ave}[Nc]$ is the avenue length of all the path in local iteration), $\Delta\tau_{ij}(t)$ between nodes will be rewarded based on the original according to formula (7). If path length of current meets $L_{now} > L_{ave}[Nc]$, $\Delta\tau_{ij}(t)$ between nodes will be punished based on the original according to formula (8). If path length of current meets $L_{now} \leq L_{all\_best}$, then $\Delta\tau_{ij}(t)$ between nodes will be rewarded based on the original according to formula (9).

$$\Delta\tau_{ij}(t) = -C_k/L(ant) \tag{8}$$

$$\Delta\tau_{ij}(t) = (N_C \times C_{best})/L(ant) \tag{9}$$

C_k represents the penalty parameter, and C_{best} represents the reward parameter.

2.3 Implementation Steps of DERS-ACO

(1) Confirm the starting node S (non-beacon nodes), and find a beacon node D which is the nearest node to S according to hop matrix from beacon nodes to unknown nodes, and make S_hop represents the hop between them. The first node in Tabu [](Tabu[] is Tabu list) assigned to S. And Tabu[] records the node number that m ants sequentially passed.
(2) The initial value of N_c is set to 1, and take N_c times iteration.
(3) When the ant m begin routing (the initial value of m is 1), the initial value of hop Tc that Tabu list has passed at present is 1.
(4) Starting from node S, puts the nodes that have passed at present into visited[], and in the beginning set visited[] to the subscript of nodes from column one to column Tc in row m in Tabu[] at present.
(5) In the nodes within communication range of S, put the index subscript of nodes that the hop between beacon nodes D and itself is smaller than S_hop into J_tmp[].
(6) Choose the node index that has maximum difference with hop S, that is to say, put the index subscript of nodes nearest beacon nodes D into j[] as an index of pending hop.

(7) Adjust the probability model in late stage of DEAS-ACO. And according to formula (5), determine the number of iterations at this time is in the early, mid or end stage. Initialize the value of q_0, α and β according to the stage where the iteration is. Use random function rad () to generate a random number and assign to q, and $0 \leq q \leq 1$.

(8) If $q > q_0$, the node state transition probability p is all the same, and it will randomly choose the next hop node in j[]. Otherwise, execute according to the state transition adjustment probabilities formula (6), scilicet the next hop is decided by pheromone concentration. Put the next hop that is chosen into to_visit, and put to_visit into row m column Tc++ in Tabu[].

(9) Determine whether the node of to_visit is beacon node D or not. If it is, put m to m + 1, and return to Step 3. But if not, put S to to_visit, and return to Step 4.

(10) After the ants have searching path over, calculate the total length L(ant) of path of each ant.

(11) Record the length of best path in this iteration, and put it into L_best[]. Record the best path in this iteration, and put it into R_best[]. Record the average path length in this iteration, and put it into L_ave[].

(12) Reset pheromones variety matrix $\Delta\tau_{ij}(t)$ in this iteration.

(13) In the process of searching, compare the path that is found to the average path length in this iteration and the past optimal solution. If it is better than the past optimal solution, take different rewards. But if not, punish. If $N_c == 1$, the pheromone concentration will normally update based on formula (7). But if not, record the shortest path length L_{all_best} in all iterations from the beginning to the last, and reward or punish pheromone concentration according to length of the path. If the path length of local ant meets $L_{now} \leq L_{ave}[Nc]$ ($L_{ave}[Nc]$ is the average length of all the path in local iteration), reward the parameter of $\Delta\tau_{ij}(t)$ between nodes according to formula (7); if the path length of local meets $L_{now} > L_{ave}[Nc]$, punish $\Delta\tau_{ij}(t)$ between nodes on the original according to formula (8). If the path length of local meets $L_{now} \leq L_{all\_best}$, reward $\Delta\tau_{ij}(t)$ between nodes on the bass of original according to formula (9).

(14) According to pheromones updating formula update pheromone matrix $\Delta\tau_{ij}(t)$.

(15) Clear Tabu list Tabu[]. Put $N_c + 1$ to N_c.

(16) Determine whether N_c is lager than N_c_max. If it is, finish the iteration, and if not, return to Step 2.

3   Simulation

Using MATLAB simulation platform, test and verify the node path selection performance based on the designed strategy of DERS-ACO and the algorithm of ACO.

We can see from Table 1 that the final best path length obtained by the strategy of DERS-ACO in every randomized trial remains stable, and they keep on the value of 49.4713. This is mainly due to that the next generation is influenced seriously by the previous generation with the algorithm of AOC, so it is easy to fall into local optimum.

3.1   The Iterations Performance of First Time to Find the Best Path

Table 1. The final best path length performance

	DERS-ACO	ACO
Random value1	49.4713	49.4713
Random value2	49.4713	49.4713
Random value3	49.4713	49.4713
Random value4	49.4713	49.4713
Random value5	49.4713	49.4713
Random value6	49.4713	49.5390
Random value7	49.4713	49.4713
Random value8	49.4713	49.5390
Random value9	49.4713	49.4713
Random value10	49.4713	49.5390
Average value	49.4713	49.4916

In Table 2 the minimum iterations of first time to find the best path with the strategy of DERS-ACO in every randomized experiment keep stable near the value of 3, and the volatile is less. However, the average minimum iterations of the best path found by the algorithm of ACO have greater volatility, and there will be no regular pattern. It can clearly be seen that the strategy of DERS-ACO is stable that will be beneficial to prolong the lifetime of the network.

Table 2. The table of iterations performance of first time to find the best path

	DERS-ACO	ACO
Random value1	4	1
Random value2	4	1
Random value3	1	1
Random value4	3	47
Random value5	2	4
Random value6	3	3
Random value7	3	13
Random value8	1	8
Random value9	4	2
Random value10	5	23
Average value	3	10.3

3.2   The Average Length of the Best Path Performance During N_c Times Iterations

We can see from Fig. 1 that there is a lot difference between the average length of best path and the average length of final best path during N_c times iterations, and the difference gap with the strategy of DERS-ACO is acquire to 13% of that of the algorithm of ACO.

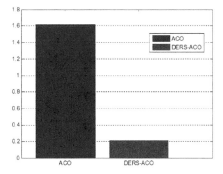

Fig. 1. The performance of difference between best paths to the average length during Nc times iterations and final best path

3.3   The Performance of the Path Avenue Length and the Best Path Length that Every Ant Has Passed During 100 Times Iterations

In Fig. 2, there is a period of time that the routing result keep higher than the other, and are mainly concentrated in the middle part of the optimization during times iterations in finding the best iteration path with the strategy of DERS-ACO. This is mainly due to the designing of the new path of exploration added in the middle of routing in the strategy of DERS-ACO.

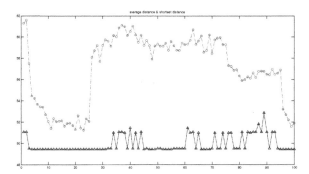

Fig. 2. The performance of the path avenue length and the best path length that every ant has passed during 100 times iterations with the strategy of DERS-ACO.

4   Conclusions

In this paper, focus on the problem of finding optimization goal of avoiding falling into local optimal solution and speeding up the search, the problem of finding the optimal path in the wireless sensor networks is studied, and an energy-saving routing strategy based on ant colony optimization (DERS-ACO) is proposes. The theoretical analysis and experimental data show that this strategy effectively expands the search capabilities of the new node, increases the searching probability of the new node, and achieves a rapid and effective global searching through dynamic state transition rule designing. Furthermore, by the designing of the mechanism of rewards and penalties, the strategy will save time while further increasing the probability of the best path search, which effectively reduce the energy consumption and prolong the lifetime of the network.

Acknowledgements. This work is supported by shenyang normal university science and technology research project of 2016 funding, No: XNL2016010, and liaoning province education science project of 2016 funding, No: JG16DB406.

References

1. Qu, W., Lin, H., Wang, J.K.: A dynamic energy-efficient routing scheme in Wireless Sensor Networks. In: ICIC-EL, vol. 8, no. 11, pp. 3113–3119 (2014)
2. Karimi, M., Naji, H.R.: Optimize cluster-head selection in wireless sensor networks using Genetic Algorithm and Harmony Search Algorithm. In: 20th Iranian Conference on Electrical Engineering, pp. 706–710 (2012)
3. Zhang, G.Y., Tang, B., Sun, J.G., Li, J.N.: Ant colony routing strategy based on distribution uniformity degree for contentcentric network. J. Commun. **36**(6), 2015126-1–2015126-12
4. Qu, D.P., Wang, X.W., Hang, M.: An aware ant routing algorithm in mobile peer-to-peer networks. Chin. J. Comput. **36**(7), 1456–1464 (2013)
5. Al-ali, R., Rana, O., Walker, D.W., et al.: G-QoSM: grid service discovery using QoS properties. Comput. Inform. **21**(4), 363–382 (2012)
6. Amaldi, E., Capone, M., Filippini, I.: Design of wireless sensor networks for mobile target detection. IEEE-ACM Trans. Netw. **20**(3), 784–797 (2012)
7. Karaboga, D., Okdem, S., Ozturk, C.: Cluster based wireless sensor network routing using artificial bee colony algorithm. Wirel. Netw. **18**(7), 847–860 (2012)
8. Okdem, S., Ozturk, C., Karaboga, D.: A comparative study on differential evolution based routing implementations for wireless sensor networks. In: Innovations in Intelligent Systems and Applications (INISTA), pp. 1–5 (2012)
9. Colorni, A., Dorigo, M., Maniezzo, V., et al.: Distributed optimization by ant colonies. In: Proceedings of European Conference on Artificial Life, Paris, pp. 134–142 (1991)

Pheromone Inspired Morphogenic Distributed Control for Self-organization of Autonomous Aerial Robots

Kiwon Yeom[(⊠)]

Department of Intelligent Robotics, Youngsan University,
288 Joonam-ro, Yangsan, South Korea
pragman@gmail.com

Abstract. A central issue in distributed formation of swarm is enabling robots with only a local view of their environment to take actions that advance global system objectives (emergence of collective behavior). This paper describes a bio-inspired control algorithm using pheromone for coordinating a swarm of identical flying robots to spatially self-organize into arbitrary shapes using local communication maintaining a certain level of density.

1 Introduction

Flying robots can fly over difficult terrain such as flooded or debris areas [1]. Rather than relying on positioning sensors which depend on the environment and are costly, flying robots rely on proprioceptive sensors and local communication with neighbors (see Fig. 1).

In this paper, we focus on a control algorithm in which flying robots self-organize and self-sustain arbitrary 2D formations. An approach to form an arbitrary shape in two dimensional space called the 'ShapeBugs' is depicted by [2]. It only requires robots to have the ability of local communication, two approximate measures for relative distance and motion, and a global compass.

In this work, we assume that our flying robots are equipped with imperfect proprioceptive sensors and a short-range wireless communication device with which robots can exchange information with only nearby neighbors. Briefly, our algorithm works as follows: first, flying robots initially wander with no information about their own coordinates or their environment. However, they have a programmed internal knowledge of the desired shape to be formed. As non-seeded robots move, they continually perform local trilaterations to figure out their location by continuous local communication. At the same time, robots maintain a certain density level among themselves using pheromones and flocking-rule-based distance measurements [3]. This enables flying robots to disperse within the specific shape and fill it efficiently.

© Springer International Publishing AG 2017
Y. Tan et al. (Eds.): ICSI 2017, Part I, LNCS 10385, pp. 285–292, 2017.
DOI: 10.1007/978-3-319-61824-1_31

Fig. 1. Artistic view of the use of a swarm of UAVs for establishing communication networks between users located on ground

2   Related Work

Several map-based UAV applications are proposed in [4,5]. In map-based applications, UAVs know their absolute position which can be shared locally or globally within the swarm.

However, obtaining and maintaining relative or global position information is challenging for UAVs or mobile robot systems. A possible advance is to adopt a global positioning system (GPS). However, GPS is not reliable and rarely possible in cluttered areas [6]. Alternatively, wireless technologies can be used to estimate the range or angle between robots of the swarm. In this case, beacon robots can be used for a reference position to other moving robots. However, depositing beacons is generally not practical for swarm systems in unknown environments [7].

Our system attempts to achieve connected arbitrary formation using a decentralized local coordinate system of robots with relative distance and density model.

3   Aerial Robot Model

Each robot has several simple equipment such as distance and obstacle-detection sensors, a magnetic compass, wireless communication, etc. (see Table 1).

Table 1. Flying robot model

Distance sensor	Provides estimated distance of each neighbor
Detection sensor	Detects obstacles in direct proximity to robot
Wireless comm.	Allows robots to communicate with each other
Locomotion	Moves robots in the world but has error
Internal shape map	Is specified by user as a target shape

We assume that robots are able to move in 2D continuous space, all robots execute the same program, and robots can interact only with other nearby robots by measuring distance and message exchange. The robots' communication architecture is based on a simple IR ring architecture because we assume that robots can interact only with nearby neighbors. The robots have omnidirectional transmission, and directional reception. This means that when a robot receives a transmission, it knows roughly which direction the transmission came from (see Fig. 2).

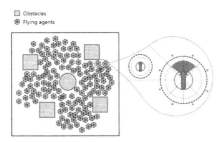

Fig. 2. Aerial robot model inspired from capabilities of real UAVs

4   Self-organizing Formation Algorithm

In our self-organizing formation algorithm, each aerial robot has a shared map of the shape to be constructed and this should be overlaid on the robot's learned coordinate system. Initially, aerial robots are randomly scattered into the simulation world without any information about the environment. Then robots begin to execute their programmed process to decide their position using only data from their proximity sensors (i.e., distance and density) and their wireless communication link with nearby neighbors.

4.1   Robot Transition Cycle

In our model, robots have a simple transition cycle model as shown in Fig. 3(a).

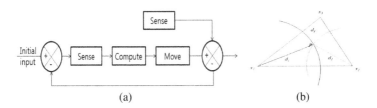

Fig. 3. (a) Robot's process cycle. (b) Robot's trilateration.

The second sense step is necessary because robots should compare the data before and after movement to determine distance and orientation from positioning error. After each transition of robot, the time until the next transition is set randomly from $[T_{min}, T_{max})$, where T_{min} is 0 to the time for computation (see Fig. 3(b)) and T_{max} is approximately the time of wait or move. Robot can move only one unit or 0 unit during *Move* and *Sense* transition process. Therefore, if robot does not move, the robot's *Move* does not have any time code.

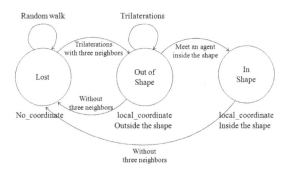

Fig. 4. An example of robot's state machine.

In this model, robots have three computational states such as *lost, out of shape*, and *in shape* (see Fig. 4 as describe in [2]. Although [2] is robust, it does not provide any stable state because there is no definition of simulation complete state.

Unlike [2], our model uses 8 IR sensors to approximately sense the direction of the referenced robot. As mentioned earlier, this enables for robot to easily and fast approach towards inside the target shape. Once robots acquire a shared coordinate system, they begin to fill a formation shape by each robot diffuses pheromone with repulse range R_{rep} (see Fig. 5(b)). The pheromone emission mechanism allows the formation shape to be robust against robots' death, while robots evenly spread out throughout the shape.

4.2 Hybrid Robot Positioning Algorithm

Trilateration allows an robot to find its perceived position (x_p, y_p) on the connected coordinate system (see Fig. 3(b)). It is also used subsequently to adjust its position.

Let the positions of the three fixed anchors be defined by the vectors $\boldsymbol{x}_1, \boldsymbol{x}_2,$ and $\boldsymbol{x}_3 \in \mathbb{R}^2$. Further, let $\boldsymbol{x}_p \in \mathbb{R}^2$ be the position vector to be determined. Consider three circles, centered at each anchor, having radii of d_i meters, equal to the distances between $\boldsymbol{x}_p$ and each anchor $\boldsymbol{x}_i$. These geometric constraints can be expressed by the following system of equations:

$$\|\boldsymbol{x}_p - \boldsymbol{x}_1\|^2 = d_1^2 \tag{1}$$
$$\|\boldsymbol{x}_p - \boldsymbol{x}_2\|^2 = d_2^2 \tag{2}$$
$$\|\boldsymbol{x}_p - \boldsymbol{x}_3\|^2 = d_3^2 \tag{3}$$

Generally, the best fit for $\boldsymbol{x}_p$ can be regarded as the point that minimizes the difference between the estimated distance (ζ) and the calculated distance from $\boldsymbol{x}_p(x_p, y_p)$ to the neighbors reported coordinate system. That is,

$$\underset{(x_p, y_p)}{\mathrm{argmin}} \sum_i |\sqrt{(\boldsymbol{x}_i - \boldsymbol{x}_p)^2} - \zeta_i|. \tag{4}$$

From this information, we can learn that this problem is closely related to the *sum-minimization problems* that arise in least squares and maximum-likelihood estimation. Therefore we suggest simple way of search of local minima (see Eqs. 5 and 6). However, in this paper, we do not consider finding any optimal or global solution but a local minimum, because it requires a lot of computational resources and it is not suitable for a small and inexpensive device. For simplification, formula 4 can be rewritten in the form of a sum as follows:

$$Q(w) = \sum_i^n Q_i(w) \tag{5}$$

where the parameter w is to be estimated and where typically each summand function $Q_i()$ is associated with the i-th observation in the data set. We perform Eq. 6 to minimize the above function:

$$w := w - \alpha \nabla Q(w) = w - \alpha \sum_{i=1}^n \nabla Q_i(w) \tag{6}$$

where α is a step size.

4.3 Flocking Movement Control

For simplification we assume that robots can continuously walk in a random zig-zag path (see Fig. 5(a)).

When robots are inside the shape, they are considered as part of the swarm that comprises it. Once robots have acquired a coincident coordinate system in the shape, they should not take any steps so that place them outside of the shape. Then robots attempt to fill a formation shape. In this work, we achieve this control by modeling virtual pheromones in a closed container. Robots react to different densities of neighbors around them, moving away from areas of high density towards those of low density [8]. They finally settle into an equilibrium state of constant density level throughout the shape over time (see pseudo code in Fig. 6).

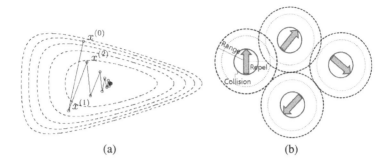

(a)   (b)

Fig. 5. (a) Robot's random walk in zig-zag path. (b) Pheromone robot's influence ranges.

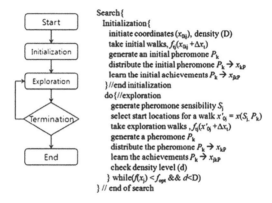

Fig. 6. An example of robot's behavior for searching a target using pheromone.

4.4 Pheromone Model for Density Control

Pheromone model is inspired by following factors: (1) biological discoveries about how cells self-organize into global patterns, and (2) distributed control systems for self-reconfigurable robot [9,10]. Pheromone provides the common mechanism that makes it possible for robots to communicate without identifiers or addresses. The basic idea of pheromone is that a swarm is a network of robots that can dynamically communicate in the network. Robot will react to pheromone according to their local topology and state information.

Let $robot_i$ and $NetEdge_i$ denote the number of robots and the number of networked edges, respectively. Then the dynamic network can be mathematically written as follows:

$$DN = (robot_i, NetEdge_i) \qquad (7)$$

Note that both $robot_i$ and $NetEdge_i$ can be dynamically changed because robots can autonomously join, leave, or be failed and died.

The diffusion and dissipation of pheromone of a given robot is denoted by $P(x, y)$, where x and y are 2D space. We simply introduce the mechanism of diffusion and dissipation of pheromone as follows:

$$\frac{\partial P}{\partial t} = \left(\alpha \frac{\partial^2 C}{\partial x^2} + \beta \frac{\partial^2 P}{\partial y^2} \right) - \Delta E \tag{8}$$

The first term on the right is for diffusion, and α and β represent the rate of diffusion in x and y directions, respectively. The second term is for dissipation and the constant δ is the rate of dissipation. Equation 8 can be considered as a part of environment function which responsible for the implementation of the dynamic communication and other effects.

4.5   Density Control and Error Correction

The density control is based on Payton approach [8], but is also similar in nature to the flocking rules proposed by Reynolds [9]. Our density control is to equalize overall density of robots at any situation. To this end, as shown Fig. 5(b), the robot has three different influence ranges. Each robot has a varying repellant (or repulsive power) that has a maximum value near center which is described as Collision area and a minimum value around Range zone (see Fig. 5(b)) regarding any adjacent robot. The robot's movement vector is weighted inversely by distance. Therefore, if any two robot are close, they push away one another. This allows robots to disperse evenly at any density.

5   Discussion

In the proposed approach, when an robot moves, it should move to another place without negatively affecting the stability of the coordinate system for adjacent robots. To demonstrate that robot movement does not negatively affect the stability, we examined the following experiment.

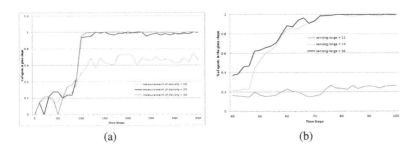

(a)   (b)

Fig. 7. (a) Percentage of robots in the shape with different measurement of density. (b) Percentage of robots in the shape with different angle of sensors.

First, we set 1000 robots in a given 100×100 world. After 150 steps, these robots are to converge on a consistent coordinate system. Then, we assign each robot a probability to move randomly with a probability. After 220 steps, the robots are no longer allowed to move (see Fig. 7(b)). For every 10 steps, the consistency is recorded. This experiment is repeated 5 times and the result is shown in Fig. 7. The consistency is sum of the difference of the actual distance between the two robots and the distance between their locations in the coordinate system. In other words, it is computed as $\Sigma^{all}_{Robotsi,j}(Distance_{i,j} - \sqrt{(x_i - x_j)^2 + (y_i - y_j)^2})$.

6   Conclusion and Future Work

We show that robots can self-organize into arbitrary user-specified shapes and maintain well the formed architecture by continuous trilateration-based on a consensus coordinate system and a virtual pheromone-based density model. We also provide several quantitative evaluations to describe the effectiveness of the proposed control algorithm in terms of percentage of robots in shapes and the variance of learned coordinate systems according to robots' movement sensor errors.

Acknowledgments. This work was supported by a 2017 research grant from Youngsan University, Republic of Korea.

References

1. Basu, P., Redi, J., Shurbanov, V.: Coordinated flocking of UAVs for improved connectivity of mobile ground nodes. In: IEEE Military Communications Conference, MILCOM 2004, vol. 3, pp. 1628–1634. IEEE (2004)
2. Cheng, J., Cheng, W., Nagpal, R.: Robust and self-repairing formation control for swarms of mobile agents. In: Proceedings of the National Conference on Artificial Intelligence, vol. 20, p. 59. AAAI Press, MIT Press, Menlo Park, Cambridge, London (1999, 2005)
3. Elston, J., Frew, E.: Hierarchical distributed control for search and tracking by heterogeneous aerial robot networks. In: IEEE International Conference on Robotics and Automation, ICRA 2008, pp. 170–175. IEEE (2008)
4. Kadrovach, B., Lamont, G.: Design and analysis of swarm-based sensor systems. In: Proceedings of the 44th IEEE 2001 Midwest Symposium on Circuits and Systems, MWSCAS 2001, vol. 1, pp. 487–490. IEEE (2001)
5. Kovacina, M., Palmer, D., Yang, G., Vaidyanathan, R.: Multi-agent control algorithms for chemical cloud detection and mapping using unmanned air vehicles. In: IEEE/RSJ International Conference on Intelligent Robots and Systems, vol. 3, pp. 2782–2788. IEEE (2002)
6. Siegwart, R., Nourbakhsh, I.: Introduction to Autonomous Mobile Robots. The MIT Press, Cambridge (2004)
7. Hu, L., Evans, D.: Localization for mobile sensor networks. In: Proceedings of the 10th Annual International Conference on Mobile Computing and Networking, pp. 45–57. ACM (2004)
8. Payton, D., Daily, M., Estowski, R., Howard, M., Lee, C.: Pheromone robotics. Auton. Robots **11**(3), 319–324 (2001)
9. Reynolds, C.: Flocks, herds and schools: a distributed behavioral model. In: ACM SIGGRAPH Computer Graphics, vol. 21, pp. 25–34. ACM (1987)
10. Rubenstein, M., Shen, W.: A scalable and distributed approach for self-assembly and self-healing of a differentiated shape. In: IEEE/RSJ International Conference on Intelligent Robots and Systems, IROS 2008, pp. 1397–1402. IEEE (2008)

Solving the Selective Pickup and Delivery Problem Using Max-Min Ant System

Rung-Tzuo Liaw, Yu-Wei Chang, and Chuan-Kang Ting[✉]

Department of Computer Science and Information Engineering,
National Chung Cheng University, Chiayi 621, Taiwan
{lrt101p,cyw104m,ckting}@cs.ccu.edu.tw

Abstract. The pickup and delivery problem (PDP) is relevant to many real-world problems, e.g., logistic and transportation problems. The problem is to find the shortest route to gain commodities from the pickup nodes and supply them to the delivery nodes. The amount of commodities of pickup nodes and delivery nodes is usually assumed to be in equilibrium; thus, all pickup nodes have to be visited for collecting all commodities required. However, some real-world applications, such as rental bikes and wholesaling business, need only to gain sufficient commodities from certain pickup nodes. A variant of PDP, namely the selective pickup and delivery problem (SPDP), is formulated to address the above scenarios. The major difference of SPDP from PDP lies in the requirement of visiting all pickup nodes. The SPDP relaxes this requirement to achieve more efficient transportation. The goal of the SPDP is to seek the shortest path that satisfies the load constraint to supply the commodities demanded by all delivery nodes with some pickup nodes. This study proposes a max-min ant system (MMAS) to solve the SPDP. The ants aim to construct the shortest route for the SPDP considering the number of selected pickup nodes and all delivery nodes. This study conducts experiments to examine the performance of the proposed MMAS, in comparison with genetic algorithm and memetic algorithm. The experimental results validate the effectiveness and efficiency of the proposed MMAS in route length and convergence speed for the SPDP.

1 Introduction

The pickup and delivery problem (PDP) is a combinatorial optimization problem applicable to many real-world cases, e.g., logistics and robotics [1,2]. This problem consists of pickup nodes and delivery nodes, where the former supplies a number of commodities for the latter to demand. The aim of the PDP is to find the shortest route to satisfy the requirement of all nodes. This problem has been proved to be NP-hard [3]. According to different requirements for the pickup and delivery nodes, the variants of PDP can be classified into one-to-one, one-to-many-to-one, and many-to-many schemes [4]. However, some real-world applications focus on supplying the demands of all delivery nodes, e.g., rental bikes and wholesaling business. These applications seek the shortest path without

Y. Tan et al. (Eds.): ICSI 2017, Part I, LNCS 10385, pp. 293–300, 2017.
DOI: 10.1007/978-3-319-61824-1_32

visiting all pickup nodes. Such relaxation of constraint allows gathering sufficient commodities from only few pickup nodes and can reduce the transportation cost for supplying the demands of delivery nodes.

Ting and Liao [5] formulated this variant of PDP as the selective pickup and delivery problem (SPDP) and proved its NP-hardness. The SPDP enables the selectivity of pickup nodes and additionally considers the vehicle capacity in the PDP. More specifically, the SPDP aims for the shortest route that can satisfy all delivery requests with the supplements from some pickup nodes under the constraint on the vehicle loading. They also applied the SPDP formulation to a real-world logistic problem for the public bike-rental service in Kaohsiung city in Taiwan. According to the classification of Berbeglia et al. [4], the SPDP belongs to the many-to-many scheme, which has no limitation on the source and destination of commodities.

This study designs a max-min ant system (MMAS) for the SPDP in view of its high capability and successes in dealing with routing and combinatorial optimization problems [6,7]. In the MMAS, a route is constructed by a population of ants with three considerations. First, the route construction of ants is affected by accumulation and evaporation of pheromone along the paths. Second, the number of nodes visited by an ant is varying due to the selectivity of pickup nodes. Third, the ants should follow the vehicle loading constraint, i.e., the loading at each node should be nonnegative and not exceed the capacity. Such constraint prohibits the circumstance of insufficient or excessive commodities. This study carries out experiments to examine the performance of the MMAS in comparison with genetic algorithm and memetic algorithm for the SPDP.

The rest of this paper is organized as follows. Section 2 introduces the SPDP formulation and its related studies. Section 3 describes the proposed MMAS. Section 4 presents the experimental results and discussions. The conclusions are drawn in Sect. 5.

2   Problem Formulation

The formulation of SPDP enables the selectivity of pickup nodes by relaxing the equilibrium of commodities supplied by pickup nodes and demanded by delivery nodes. More precisely, the SPDP assumes the amount of supplies from pickup nodes surpasses that of demands by delivery nodes. Given a complete graph $G = (V, E)$ with $V = \{v_0, \ldots, v_n\}$ and $E = \{(v_i, v_j) | v_i, v_j \in V, v_i \neq v_j\}$, each edge (v_i, v_j) is assigned a cost c_{ij} and each node v_i has a demand d_i. According to their demands, the nodes are classified into three types: pickup nodes $V^+ = \{v_i | v_i \in V, d_i > 0\}$, delivery nodes $V^- = \{v_i | v_i \in V, d_i < 0\}$, and the depot v_0 with demand $d_0 = 0$.

The SPDP aims to find the shortest route that can satisfy *all* delivery nodes with the commodities supplied from *some* pickup nodes. In the transportation process, the vehicle loading at each node should be nonnegative and no larger than the capacity Q. Let m be the number of nodes visited and $v_{(i)}$ denote the ith node visited. The SPDP selects the pickup nodes and searches for a visiting order $\boldsymbol{p} = \left(v_0, v_{(1)}, \ldots, v_{(m)}\right)$ with the minimum cost. Formally, the SPDP is formulated as follows.

$$\min \sum_{v_i, v_j \in V} c_{ij} x_{ij} \tag{1}$$

$$\text{s.t.} \sum_{v_i \in V} x_{ij} = \sum_{v_i \in V} x_{ji} \leq 1, \forall v_j \in V^+ \tag{2}$$

$$\sum_{v_i \in V} x_{ij} = \sum_{v_i \in V} x_{ji} = 1, \forall v_j \in V^- \cup \{v_0\} \tag{3}$$

$$\sum_{v_i, v_j \in S} x_{ij} \leq |S| - 1, \forall S \subseteq V \backslash \{v_0\} \tag{4}$$

$$0 \leq \ell_{(t)} \leq Q, \forall t \in \{1, .., m\} \tag{5}$$

$$x_{ij} \in \{0, 1\} \tag{6}$$

where the decision variable x_{ij} determines the inclusion of edge (v_i, v_j) in the route:

$$x_{ij} = \begin{cases} 1 & \text{vehicle travels from } v_i \text{ to } v_j, \\ 0 & \text{otherwise.} \end{cases}$$

The objective function (1) is to minimize the transportation cost, e.g., the route length, under the constraints (2)–(5). The constraints (2) and (3) concern the visits of pickup and delivery nodes. Restated, they ensure that the degree of pickup nodes should be less than or equal to one, and the degree of delivery nodes should be equal to one. The sub-tour elimination constraint (4) limits that any subset S of nodes excluding the depot v_0 has at most $|S| - 1$ edges to prevent sub-tours. The load constraint (5) guarantees the vehicle loading $\ell_{(t)}$ at each node $v_{(t)}$ is within the range $[0, Q]$, where $\ell_{(t)} = \ell_{(t-1)} + d_{(t)}$.

Ting and Liao presented the SPDP [8,9] and proved its NP-hardness [5]. The SPDP is more difficult than the PDP because the former introduces new constraints to the latter, making it harder to gain feasible solutions. To resolve this problem, they proposed a memetic algorithm (MA) based on the genetic algorithm (GA) scheme and a modified 2-opt operator. The experimental results show that the solution quality and convergence speed of the MA outperforms GA and tabu search. In addition, they applied the MA to deal with a real-world SPDP for the public bike-rental service in Kaohsiung.

The evolutionary algorithms, such as GA and MA, yield good solution quality for the SPDP; nevertheless, owing to the representation of chromosomes, these methods require a penalty function or a repair strategy for dealing with the infeasible routes. This study proposes an MMAS, which is capable of finding the feasible solutions for the SPDP. More details about the proposed MMAS are described below.

3   Methodology

This study develops an MMAS to solve the SPDP. The MMAS follows the paradigm of ant colony optimization (ACO), which has shown its effectiveness in tackling various combinatorial optimization and routing problems [10–14]. The MMAS constructs the routes for the SPDP by manipulating the ants, led by the distribution of pheromone. The following subsections introduce the major procedures in the MMAS for the SPDP: route construction and pheromone distribution.

3.1 Route Construction

The route construction involves the ants determining the visiting order of nodes, including the selected pickup nodes and all delivery nodes. Starting from the depot v_0, ants move to the remaining nodes according to the transition probability, which considers the global pheromone trail $\tau_{ij}(t)$ and the local information η_{ij}. In this study, η is defined as the reciprocal of cost c_{ij}, i.e., $\eta_{ij} = 1/c_{ij}$. Let $p_{ij}^k(t)$ be the transition probability from node v_i to node v_j at the t-th iteration. Stützle and Hoos [7] proposed computing the transition probability by

$$p_{ij}^k(t) = \begin{cases} \frac{[\tau_{ij}(t)]^\alpha [\eta_{ij}]^\beta}{\sum_{l \in N_i^k} [\tau_{il}(t)]^\alpha [\eta_{il}]^\beta} & \text{if } j \in N_i^k, \\ 0 & \text{otherwise.} \end{cases}$$

The arguments α and β control the balance between the global pheromone trails and the local heuristics. In other words, the setting of $\alpha > \beta$ intensifies the preference for the edges with high pheromone; by contrast, the setting of $\beta > \alpha$ tends to visit the edges with low cost. The MMAS maintains a set N_i^k of nodes that are feasible for visiting and vehicle loading constraints. The ants are manipulated by the above rules to iteratively construct legal routes for the SPDP.

3.2 Pheromone Distribution

The distribution of pheromone is essential to the performance of MMAS because it directly influences the selection of nodes to visit. The pheromone distribution in the MMAS is controlled by two factors: evaporation and deposition. After all ants constructed their routes, the pheromone on the edges is updated according to the following equation:

$$\tau_{ij}(t+1) = (1 - \rho)\tau_{ij}(t) + \Delta\tau_{ij}^{\text{best}},$$

where $\rho \in (0, 1]$ stands for the evaporation rate and the $\Delta\tau_{ij}^{\text{best}}$ is the deposition of pheromone

$$\Delta\tau_{ij}^{\text{best}} = \begin{cases} 1/f\left(\boldsymbol{p}^{\text{best}}\right) & \forall (v_i, v_j) \in \boldsymbol{p}^{\text{best}}, \\ 0 & \text{otherwise.} \end{cases}$$

The $f\left(\boldsymbol{p}^{\text{best}}\right)$ represents the length of the iteration-best route $\boldsymbol{p}^{\text{ib}}$ or the global-best route $\boldsymbol{p}^{\text{gb}}$. The study [7] suggested that considering only global-best route $\boldsymbol{p}^{\text{gb}}$ intensifies exploitation and thus results in premature convergence. Therefore, this study takes both iteration-best route $\boldsymbol{p}^{\text{ib}}$ and the global-best route $\boldsymbol{p}^{\text{gb}}$ into account in the update of pheromone. Specifically, the global-best ant $\boldsymbol{p}^{\text{gb}}$ is adopted as the $\boldsymbol{p}^{\text{best}}$ every fixed number of generations and the iteration-best route $\boldsymbol{p}^{\text{ib}}$ is applied in the other generations.

The MMAS further considers the background and peak concentration of pheromone, i.e., the minimum pheromone trail $\tau_{ij}^{\min}$ and the maximum pheromone trail $\tau_{ij}^{\max}$. The distribution of each pheromone trail τ_{ij} is bounded by the minimum and maximum pheromone trails to $\tau_{ij} \in \left[\tau_{ij}^{\min}, \tau_{ij}^{\max} \right]$ with

$$\tau_{ij}^{\max}(t) = \frac{1}{\rho \times f(\boldsymbol{p}^{\text{gb}})}$$

$$\tau_{ij}^{\min}(t) = \frac{\tau^{\max}(t)}{\mu \times g_\tau}$$

where μ is the number of ants in the MMAS, and the coefficient g_τ resolves on the ratio of maximum and minimum pheromone trails.

4   Experimental Results

In this study, we carried out several experiments to examine the effectiveness and efficiency of the proposed MMAS on the SPDP, compared to GA and MA. The experiments adopt the SPDP instances [15]. The notation X(Y)_Z denotes the name of the original PDP instance, the number of nodes, and the number of pickup nodes, respectively. The performance of test algorithms is evaluated on the test instances of two sizes, i.e., 91 and 181 nodes. In addition, we investigate their performance on the instances with different gains γ in the demand of pickup nodes, where the vehicle capacity $Q = 400$ for $\gamma \in \{100, 200\}$ and $Q = 600$ for $\gamma = 400$. Table 1 lists the parameter setting for GA, MA, and the proposed MMAS. The MA follows the evolutionary procedure of GA with fixed-length representation and applies the 2-opt local search [15]. The proposed MMAS uses varying-length representation. Each test algorithm runs 30 trials considering the stochastic nature of evolutionary algorithms.

We first examine the solution quality of the MMAS in comparison with GA and MA. Table 2 lists the route lengths obtained from the three test algorithms on the two sizes of SPDP instances with different γ values. The results show that MMAS and MA outperform GA in terms of route length. The MA performs best among the three test algorithms on the small instance n100mosA(91), revealing the utility of 2-opt local search in improving the routes. On the other hand, the results show that MMAS surpasses GA and MA in terms of route length on the large instance n200mosA(181). Table 2 further presents the t-test results. With confidence level $\alpha = 0.01$, the t-test results show no statistical significance between MA and MMAS on the small instance n100mosA(91), reflecting that the route lengths obtained from the proposed MMAS are comparable to those attained by MA. Furthermore, the results show that MMAS can achieve significantly shorter routes than MA does on the large instances n200mosA(181).

Table 1. Parameter setting for the test algorithms

	GA	MA	MMAS
Representation	Fixed-length	Fixed-length	Varying-length
Population size	500	500	50
Selection	2-tournament	2-tournament	-
Crossover	Order crossover $(p_c = 1.0)$	Order crossover $(p_c = 1.0)$	-
Mutation	Bit-flip $(p_m = 1/\lvert V^+ \rvert)$ + inversion $(p_m = 1.0)$	Bit-flip $(p_m = 1/\lvert V^+ \rvert)$ + inversion $(p_m = 1.0)$	-
Survival	$(\mu + \lambda)$	$(\mu + \lambda)$	-
Local search	-	2-opt	-
Intensification	-	-	$\alpha = 1,\ \beta = 5$
Evaporation rate	-	-	$\rho = 0.02$
Termination	10,000 generations $(5.0 \times 10^6$ evaluations)	3,000 generations $(1.5 \times 10^5$ evaluations)	2,000 generations $(1.0 \times 10^5$ evaluations)

Table 2. The average length and the number of selected pickup nodes in the routes obtained from GA, MA, and the proposed MMAS for different gains γ. The number after the slash indicates the average number of selected pickup nodes. The boldface marks the shortest route among the test algorithms. The p-values account for the results of t-test on the route lengths of MA and MMAS. The bold p-values signify the statistical significance with confidence level $\alpha = 0.01$.

Instance	γ	Length/#Nodes			p-value
		GA	MA	MMAS	MA vs. MMAS
n100mosA(91)_42	100	5442/2.33	5274/2.17	**5249/2.70**	1.64E−01
	200	5430/2.00	5268/2.00	**5237/2.00**	1.07E−01
	400	5366/1.00	**5170/1.00**	5193/1.00	4.34E−02
n200mosA(181)_94	100	8558/5.07	7646/5.03	**7494/5.00**	**4.38E−08**
	200	8780/3.07	7715/3.00	**7468/3.00**	**2.01E−11**
	400	8770/2.00	7581/2.00	**7479/2.00**	**6.41E−04**

Figure 1 compares the convergence of MA and the proposed MMAS. According to the results, MMAS converges much faster than MA does on n100mosA(91), where their difference in solution quality is statistically insignificant. As the size of instance increases, MMAS has faster convergence speed as well as significantly better solution quality than MA does. These satisfactory results indicate the effectiveness and efficiency of the proposed MMAS.

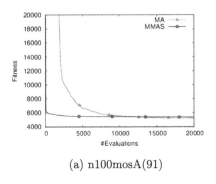

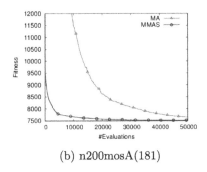

(a) n100mosA(91)   (b) n200mosA(181)

Fig. 1. The variation of mean best fitness for MA and MMAS against the number of evaluations on the four SPDP instances with gain $\gamma = 100$.

5   Conclusions

The SPDP renders a significant PDP variant pertinent to real-world logistic and transportation applications. The SPDP aims to find the shortest route satisfying all requests of delivery nodes with the supplements from some pickup nodes. A key feature of the SPDP is that it relaxes the constraint of visiting all pickup nodes and thus enables the selectivity of pickup nodes for shorter routes. To resolve the SPDP, this study proposes an MMAS, which has two major advantages:

- The route is constructed by progressively selecting the pickup nodes and delivery nodes through ants.
- The constructed route can always satisfy the constraint on the vehicle load in the SPDP.

This study carries out experiments to examine the effectiveness and efficiency of the proposed MMAS on the SPDP. Regarding solution quality, the experimental results show that the proposed MMAS is superior to GA on all test instances. In addition, MMAS performs comparably to MA on a small instance and outperforms MA on a large instance. As for convergence speed, MMAS converges much faster than MA does on all test instances. These satisfactory outcomes validate the advantage of MMAS over GA and MA in terms of solution quality and convergence speed on the SPDP.

Some directions remain for future work. First, the integration of MMAS with local enhancement operator is promising to enhance the performance. Second, extension of the MMAS to tackle the multi-objective SPDP is worthy of further study.

References

1. Parragh, S., Doerner, K., Hartl, R.: A survey on pickup and delivery problems. Part I: Transportation between customers and depot. J. für Betriebswirtschaft **58**, 21–51 (2008)

2. Parragh, S., Doerner, K., Hartl, R.: A survey on pickup and delivery problems. Part II: Transportation between pickup and delivery locations. J. für Betriebswirtschaft **58**, 81–117 (2008)

3. Savelsbergh, M., Sol, M.: The general pickup and delivery problem. Trans. Sci. **29**(1), 17–29 (1995)

4. Berbeglia, G., Cordeau, J., Gribkovskaia, I., Laporte, G.: Static pickup and delivery problems: a classification scheme and survey. Top **15**(1), 1–31 (2007)

5. Ting, C.K., Liao, X.L.: The selective pickup and delivery problem: formulation and a memetic algorithm. Int. J. Prod. Econ. **141**(1), 199–211 (2013)

6. Stützle, T., Hoos, H.: The max-min ant system and local search for combinatorial optimization problems. In: Voß, S., Martello, S., Osman, I.H., Roucairol, C. (eds.) Meta-heuristics, pp. 313–329. Springer, Heidelberg (1999)

7. Stützle, T., Hoos, H.: Max-min ant system. Future Gener. Comput. Syst. **16**(8), 889–914 (2000)

8. Liao, X.L., Ting, C.K.: An evolutionary approach for the selective pickup and delivery problem. In: Proceedings of the IEEE Congress on Evolutionary Computation, pp. 1–8 (2010)

9. Liao, X.L., Ting, C.K.: Evolutionary algorithms using adaptive mutation for the selective pickup and delivery problem. In: Proceedings of the IEEE Congress on Evolutionary Computation, pp. 1–8 (2012)

10. Pérez Cáceres, L., López-Ibáñez, M., Stützle, T.: Ant colony optimization on a budget of 1000. In: Dorigo, M., Birattari, M., Garnier, S., Hamann, H., Montes de Oca, M., Solnon, C., Stützle, T. (eds.) ANTS 2014. LNCS, vol. 8667, pp. 50–61. Springer, Cham (2014). doi:10.1007/978-3-319-09952-1_5

11. Dorigo, M., Birattari, M., Stützle, T.: Ant colony optimization. IEEE Comput. Intell. Mag. **1**(4), 28–39 (2006)

12. Dorigo, M., Blum, C.: Ant colony optimization theory: a survey. Theoret. Comput. Sci. **344**(2–3), 243–278 (2005)

13. Maur, M.: Adaptive ant colony optimization for the traveling salesman problem. Master's thesis. Technical University of Darmstadt (2009)

14. Mou, L., Dai, X.: A novel ant colony system for solving the one-commodity traveling salesman problem with selective pickup and delivery. In: Proceedings of the International Conference on Natural Computation, pp. 1096–1101 (2012)

15. Liao, X.L., Ting, C.K.: Solving the biobjective selective pickup and delivery problem with memetic algorithm. In: Proceedings of the IEEE Workshop on Computational Intelligence in Production and Logistics Systems, pp. 107–114 (2013)

An Improved Ant-Driven Approach to Navigation and Map Building

Chaomin Luo[1][✉], Furao Shen[2], Hongwei Mo[3], and Zhenzhong Chu[4]

[1] Department of Electrical and Computer Engineering,
University of Detroit Mercy, Michigan, USA
luoch@udmercy.edu
[2] Department of Computer Science and Technology,
Nanjing University, Nanjing, China
[3] Automation College, Harbin Engineering University, Harbin, China
[4] College of Information Engineering, Shanghai Maritime University,
Shanghai, China

Abstract. An improved ant-type approach, ant colony optimization (ACO) model, integrated with a heading direction scheme (HDS) to real-time collision-free navigation and mapping of an autonomous robot is proposed in this paper. The developed HDS-based ACO model for concurrent map building and safety-aware navigation is capable of remedying the shortcoming of risky distance from obstacles in combination with the Dynamic Window Approach (DWA) algorithm as a local navigator. Its effectiveness and efficiency of the developed real-time hybrid map building and safety-aware navigation of an autonomous robot have been successfully validated by simulated experiments and comparison studies.

Keywords: ACO · Motion planning · HDS · DWA · Local navigation · Grid-based map

1 Introduction

An improved ant-driven model integrated with a Dynamic Window Approach (DWA) algorithm to real-time collision-free navigation and mapping with safety consideration is proposed in this paper. Concurrent navigation and map building of an autonomous robot under unknown environments is one of the challenges in robotics motion planning.

There have been a large number of models proposed for autonomous robot navigation with obstacle avoidance such as fuzzy logic method [8], neural networks [6,9,10], machine leaning [7], and Ant Colony Optimization (ACO) [1,5], etc. Vasak and Hvizdos [8] developed a robot navigation method deployed RFID tags with the sonar through a fuzzy logic model.

Neural networks (NNs) approach plays an increasingly crucial role on robot mapping and navigation [6,9,10]. NNs algorithm aims to navigate an autonomous robot in light of sensor information in association with a fuzzy logic

© Springer International Publishing AG 2017
Y. Tan et al. (Eds.): ICSI 2017, Part I, LNCS 10385, pp. 301–309, 2017.
DOI: 10.1007/978-3-319-61824-1_33

controller [8]. Yi *et al.* [10] applied a bio-inspired neural network model to the 3D path planning and task assignment under uncertain circumstances. However, the map building and local navigator have not been integrated yet. Yang and Luo [9] proposed a new bio-inspired neuro-dynamics model for robot complete coverage motion planning under dynamically varying environments. However, the proposed model lacks the local reactive navigation and map building components under unknown environments [6]. Mo *et al.* proposed a machine-learning based Imitation Learning (IL) algorithm for robot navigation [7]. Luo *et al.* [5] proposed an ACO-based multi-goal navigation and mapping approach of an intelligent robot. However, the proposed model lacks local navigation.

In this paper, a two-level hybrid real-time mapping and navigation model of an autonomous robot is proposed. Top level is an improved ACO approach that plans a global trajectory. Bottom level utilizes DWA-based algorithm in light of the sensor information to direct a robot locally to autonomously traverse from one waypoint to another. In addition, in order to generate safer, more reasonable collision-free trajectories, novel heuristic algorithms are designed to optimize the trajectory. The developed real-time ACO model for concurrent map building and safety-aware navigation is capable of remedying the shortcoming of trajectories generated with risky distance from obstacles in combination with the DWA algorithm as a local navigator.

2   The HDS-Based ACO Algorithm

Ants in the ACO algorithm as intelligent agents in the navigation, which traverse from one waypoint to another directed by pheromone trails with an a *priori* available heuristic information. The field of "ant-like algorithms" creates models derived from the behavior observation of real ants, in which these models as a source of biological inspiration are utilized to design novel algorithms for the solution of optimizations.

Ant pheromone strength $\tau_{ij}(t)$, is a sort of numerical information defined with each arc (i, j) that is updated in the ACO algorithm, in which t is the iteration counter. The agent is initially arranged in a waypoint. At each iteration stage, a probabilistic action choice rule is applied to an agent, k (an autonomous mobile robot could be an agent). The probability $p_{ij}^k(t)$ of an agent k, currently at waypoint i, which moves to waypoint j at the tth iteration of the algorithm, is obtained as follows in Eq. (1) [2].

$$p_{ij}^k(t) = \frac{[\tau_{ij}(t)]^\alpha \times [\vartheta_{ij}]^\beta}{\sum_{l \in \aleph_i^k} [\tau_{il}(t)]^\alpha \times [\vartheta_{il}]^\beta} \quad \text{if} \quad j \in \aleph_i^k , \tag{1}$$

where $\aleph_i^k$ is the feasible adjacent waypoint of the agent k, the set of waypoints which the agent k has not yet visited. The $\vartheta_{ij} = 1/d_{ij}$ is an a *priori* available heuristic value, and d_{ij} is the distance between two waypoints. Parameter α represents importance factor of the pheromone matching a classical stochastic greedy algorithm. Parameter β is an importance factor of the heuristics function. Parameters α and β determine the relative influence of the pheromone trail and the heuristic information [2].

At each iteration stage, $\Delta\tau_{ij}^k(t)$, the amount of pheromone agent k leaves on the arcs it has visited is dynamically updated by decreasing the pheromone strength on all arcs by a constant factor before enabling each agent to supplement pheromone on the arcs. The pheromone strength τ_{ij} is dynamically updated as follows in Eq. (2) when $0 < \rho < 1$.

$$\begin{cases} \tau_{ij}(t+1) & = (1-\rho) \cdot \tau_{ij}(t) + \Delta\tau_{ij} \\ \Delta\tau_{ij} & = \sum_{k=1}^{n} \Delta\tau_{ij}^k \end{cases} \tag{2}$$

Algorithm 1. ACO algorithm for Navigation with HDS

1: **procedure** ACO WITH HDS(x, y) ▷ Find out the trajectory (x, y coordinates)
2:    Set parameters, initialize pheromone trails
3:    **while** (termination condition not met) **do**   ▷ reach the goal?
4:        ConstructSolutions
5:        ApplyLocalSearch
6:        **HeadingDirectionScheme**
7:        UpdateTrails
8:    **end while**
9:    **return** *The trajectory (x, y coordinates)*   ▷ reach the goal!
10: **end procedure**

The amount of pheromone $\Delta\tau_{ij}^k(t)$ in ant cycle system mode is defined as [2]:

$$\Delta\tau_{ij}^k(t) = \begin{cases} \frac{Q}{L^k(t)} & \text{if arc } (i, j) \text{ is used by the robot } k, \\ 0 & \text{otherwise.} \end{cases} \tag{3}$$

$L^k(t)$ is the length of the kth robot's tour in this paper. The ACO algorithm for motion planning with a heading direction scheme is summarized as Algorithm 1. Note that there is a heading direction scheme (HDS) heuristic embedded in this ACO algorithm. It will be depicted in the section later.

3   Map Building and DWA-Driven Local Navigation

The proposed robot navigation system is comprised of two layers. One is an ACO global path planner whereas the other is a DWA-based local navigator. Both effectiveness and efficiency motivate ACO to be adapted to navigation and mapping as the global plan planner. The objective of the local navigator is to generate velocity commands for the autonomous mobile robot to move towards a target. In order to assure the obstacle avoidance, Fox *et al.* [3] first successfully proposed a DWA algorithm method for the local navigation. The DWA is an obstacle-avoidance approach with synchro-drives, which considers the constraints imposed by limited velocities and accelerations, derived directly from the motion dynamics of synchro-drive autonomous mobile robots.

Unlike vector field histogram (VFH) method, the DWA considers the dynamic and kinematic constraints of a mobile robot [3]. In a nutshell, only a short time interval is periodically taken into account when calculating the next steering command to decrement the tremendous complexity of the robot motion planning. Let $x(t)$ and $y(t)$ be the coordinate at time t of a mobile robot in a global coordinate system. Let $\theta(t)$ denote the heading direction (orientation). Let the triplet $x, y, \theta(t)$ denote the kinematic configuration of the robot. The translational velocity and rotational velocity of the robot at time t are described by $v(t)$, and $\omega(t)$, respectively. The robot is assumed to move with a constant velocity (v, ω) during each control loop, in which the pair (v, ω) is considered admissible, if the robot is able to stop before it reaches the closest obstacle on the corresponding curvature. Let the term $d(v, \omega)$ represent the distance to the closest obstacle on the corresponding curvature with regard to a velocity (v, ω).

The maximal admissible velocity, over a given curvature, depends really on the distance to the next obstacle over the curvature. The accelerations are denoted to be $\dot{v}_d$ and $\dot{\omega}_d$ for breakage [3]. The set V_a of admissible velocities is expressed as (4). Therefore, the robot is allowed to stop with obstacle avoidance as the V_a is the set of velocities (v, ω) in Fig. 1(a).

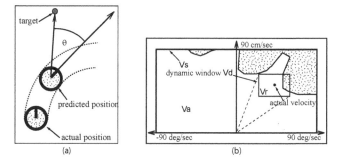

Fig. 1. The illustration of DWA algorithm (a) angle the robot's direction to the target; (b) dynamic window (redrawn from [3]).

$$V_a = \{v, \omega \mid v \leq \sqrt{2 \cdot d(v, \omega \cdot \dot{v}_d)} \wedge \omega \leq \sqrt{2 \cdot d(v, \omega \cdot \dot{\omega}_d)}\} \tag{4}$$

The dynamic window is centered on the actual velocity. All curvatures outside the dynamic window are not reachable in the next time interval, which are not taken into account for the obstacle avoidance. At the time interval t, the accelerations $\dot{v}$ and $\dot{\omega}$ will be exerted. Let (v_a, w_a) be the actual velocity, and then the dynamic window V_d is described as (5)

$$V_d = \{(v, \omega) \mid v \in [v_a - \dot{v} \cdot t, v_a + \dot{v} \cdot t] \wedge \omega \in [w_a - \dot{\omega} \cdot t, w_a + \dot{\omega} \cdot t]\} \tag{5}$$

The area V_r within the dynamic window is generated through the restrictions imposed on the search space for the velocities depicted in Fig. 3, in which V_s is assumed to be the space of possible velocities. The area V_r is defined as the intersection of the restricted areas as expressed in Eq. (6) (see Fig. 1(b)).

$$V_r = V_a \cap V_d \cap V_s \tag{6}$$

After the generated search space V_r is determined, a velocity is selected from V_r. The maximum of the objective function $G(v, \omega)$ is calculated over V_r as Eq. (7) to integrate the criteria target heading, velocity and clearance.

$$G(v, \omega) = \rho \left[\alpha \cdot H(v, \omega) + \beta \cdot D(v, \omega) + \gamma \cdot Vel(v, \omega) \right], \tag{7}$$

where ρ is a factor that smooths the weighted sum of three elements of $G(v, \omega)$. The target heading function, represented by $H(v, \omega)$ is utilized to measure the alignment of the robot with the target direction, calculated by $180 - \theta$, where θ is the angle between the target point and the heading direction of a mobile robot shown in Fig. 1(b). The function velocity $Vel(v, \omega)$ evaluates the movement of the robot on the corresponding route that is a projection on the translational velocity v. The function $D(v, \omega)$ depicts the distance to the nearest obstacle intersected with the curvature.

A $2D$ grid-based map filled with equally-sized cells, marked as either occupied or free, is constructed as the mobile robot moves. The initial value is zero, which indicates that the cell is neither occupied nor unoccupied. Concurrent map building and navigation are the essence of successful navigation under unknown environments.

4   Heading Direction Motion Based Navigation

A heading direction motion planning scheme, called heading direction scheme (HDS), is developed to acquire a more reasonable and safer trajectory in the vicinity of obstacles. The fundamental principle of this scheme is to adjust the heading direction of an autonomous robot to traverse along a safe route while carrying out the ACO navigation algorithm (illustrated in Fig. 2).

The HDS scheme is designed to check the next motion of the robot. In the vicinity of obstacles, the robot is especially guided to next cell horizontally or vertically, instead of, diagonally. By means of the environmental information provided by the LIDAR sensors, the adjacent cells to obstacles on the built map are to be *known* thus accessible. For instance, an agent (ant) is located at the center of the cells next to the eight adjacent cells including free space, and obstacles shown in Fig. 2. The next position the robot will move to is likely *a,* *b, c* or *d*. Based on the defined HDS scheme, in this kind of structure, positions indicated by red cells *d*, is unfavorable. Therefore, it is directed to the accessible cells such as *a, b* and *c* in Fig. 2. The robot shall not move to the positions located in the cells that are not preferable to be entered. Consequently, safer and more reasonable trajectory is generated in light of both the ACO algorithm and HDS scheme. A robot initiates from the initial position S to the target T in a test scenario of workspace shown in produced safe and reasonable trajectory is illustrated in Fig. 3(a). Based on the developed heading direction scheme, the produced safe and reasonable trajectory is illustrated in Fig. 3(a).

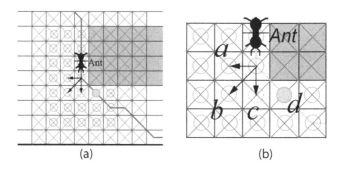

Fig. 2. Developed HDS scheme (a) the generated motion; (b) HDS in the cell structure. (Color figure online)

5 Simulation and Comparison Studies

5.1 The Hybrid Model in a Bar-Like Environment

The developed hybrid approach is applied to a test scenario with a bar-like obstacle. The developed HDS scheme associated with the ACO and DWA navigation directs an autonomous robot in the test scenario exactly identical as the one in Fig. 5 of [1]. It is now illustrated in Fig. 3(b), in order to compare our model with theirs. An ACO approach to the robot path planning in search of the final destination is developed by Chia *et al.* [1].

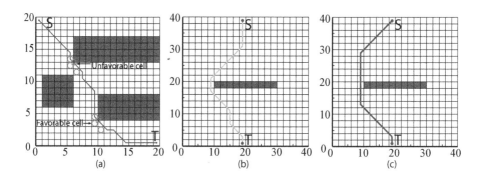

Fig. 3. Illustration of the heading direction motion scheme (a); Planned trajectory in a bar-like test scenario (b) by Chia's model of Fig. 5 (redrawn from [1]); (c) by our proposed model.

The workspace is defined with a size of 40×40, topologically organized as a cell-based map. The parameters of the ACO algorithm are selected as follows: $\alpha = 1$; $\rho = 0.3$ and $\beta = 5$. Initially, the initial position is located at $S(19, 39)$ and the robot drives toward the target at $T(19, 1)$. The trajectory planned by the model of Chia *et al.* [1] is illustrated in Fig. 3(b), whereas the trajectory produced through our ACO algorithm with HDS scheme is shown in Fig. 3(c).

It is found that the generated trajectory has the safer distance from the obstacles shown in Fig. 3 in comparison with the one generated by the model of Chia *et al.* [1]. The trajectory length, and number of turns as well as steps to complete the navigation mission are summarized in Table 1 Although the trajectory length and steps for both models are equal, with regard of the number of turns, ours is significantly better than theirs.

Table 1. Comparison of path length, steps and turns

Model	Length	Steps	Turns
Chia et al.'s model	23.14	19	6
Ours with HDS	23.14	19	3

5.2   The Hybrid Model in a Room-Like Environment

The proposed model is applied to a test scenario with populated obstacles in comparison with an identical case as Fig. 1 of [4] which is now illustrated in Fig. 4(a). Garcia *et al.* developed a hybrid model consisting of an ACO algorithm and a fuzzy logic approach [4], in which the decision-making of navigation is impacted by the distance between the source and target nodes.

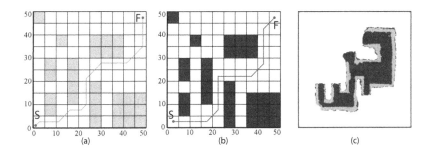

Fig. 4. Planned trajectory in a room-like test scenario (a) by Garcia *et al.*'s model of Fig. 1 (redrawn from [4]); (b) by our proposed model; (c) mapping and navigation by our proposed model.

The test scenario is a 50×50 topologically organized workspace with a grid-based map. The parameters of the ACO are chosen same as the case above with $S(1, 1)$ and $F(49, 49)$. The navigation of robot via Garcia *et al.*'s model is shown in Fig. 4(a) while the trajectory generated through our model is depicted in Fig. 4(b). The trajectory length and number of turns, steps by the proposed model and the model of Garcia *et al.* were calculated in Table 2. It reveals that the length of the trajectory by the developed ACO model is 5.17% shorter than that of their model. The number of turns of the proposed model is only 2/3 of that of their model, *i.e.*, 33.33% better than that of theirs. Additionally, the number of steps to direct the robot from the initial point to the target by the proposed ACO with HDS model is 31.82% less than that of their model. With ACO and DWA based concurrent navigation and mapping from the initial position $S(1,1)$ to the final position $F(49, 49)$, The final map built when the robot achieves the final point is shown in Fig. 4(c).

Table 2. Comparison of path length, steps and turns

Model	Length	Steps	Turns
Garcia *et al.*'s model	87	22	9
Ours with HDS	82.5	15	6

6 Conclusion

A real-time ant-driven model for map building and safety-aware navigation is developed to remedy the shortcoming of trajectories generated with risky distance from obstacles in combination with the Dynamic Window Approach algorithm as a local navigator. An efficient heading-enabled ACO algorithm is created for the real-time concurrent mapping and navigation. Its effectiveness and efficiency of the developed model have been successfully validated by simulated experiments and comparison studies.

References

1. Chia, S.H., Su, K.L., Guo, J.H., Chung, C.Y.: Ant colony system based mobile robot path planning. In: Proceedings of 2010 Fourth International Conference on Genetic and Evolutionary Computing, pp. 210–213 (2010)
2. Dorigo, M., Stutzle, T.: Ant Colony Optimization. The MIT Press, San Francisco (2004)
3. Fox, D., Burgard, W., Thrun, S.: The dynamic window approach to collision avoidance. IEEE Robot. Autom. Mag. **4**(1), 23–33 (1997)
4. Garcia, M.A.P., Montiel, O., Castillo, O., Lveda, R.S., Melin, P.: Path planning for autonomous mobile robot navigation with ant colony optimization and fuzzy cost function evaluation. Appl. Soft Comput. **9**(1), 1102–1110 (2009)
5. Luo, C., Mo, H., Shen, F., Zhao, W.: Multi-goal motion planning of an autonomous robot in unknown environments by an ant colony optimization approach. In: Proceeings of the Sixth International Conference on Swarm Intelligence, pp. 519–527 (2016)
6. Luo, C., Yang, S.X.: A bioinspired neural network for real-time concurrent map building and complete coverage robot navigation in unknown environments. IEEE Trans. Neural Netw. **19**(7), 1279–1298 (2008)

7. Mo, H., Luo, C., Liu, K.: Robot indoor navigation based on computer vision and machine learning. In: Proceedings of the Sixth International Conference on Swarm Intelligence, pp. 528–534 (2016)
8. Vasak, J., Hvizdos, J.: Vehicle navigation by fuzzy cognitive maps using sonar and RFID technologies. In: Proceedings of IEEE 14th International Symposium on Applied Machine Intelligence and Informatics, pp. 75–80 (2016)
9. Yang, S.X., Luo, C.: A neural network approach to complete coverage path planning. IEEE Trans. Syst. Man Cybern. **34**(1), 718–725 (2004)
10. Yi, X., Zhu, A., Yang, S.X., Luo, C.: A bio-inspired approach to task assignment of swarm robots in 3-D dynamic environments. IEEE Trans. Cybern. **47**(4), 974–983 (2017)

Artificial Bee Colony Algorithms

A Multi-cores Parallel Artificial Bee Colony Optimization Algorithm Based on Fork/Join Framework

Jiuyuan Huo[1,2(✉)] and Liqun Liu[3]

[1] School of Electronic and Information Engineering,
Lanzhou Jiaotong University, Lanzhou 730070, People's Republic of China
huojy@mail.lzjtu.cn
[2] Gansu Data Engineering and Technology Research Center for Resources
and Environment, Lanzhou, China
[3] College of Information Science and Technology,
Gansu Agricultural University, Lanzhou, China
liulq@gsau.edu.cn

Abstract. There are lots of computationally intensive tasks in optimization process of Artificial Bee Colony (ABC) algorithm, which requires large CPU processing time. To improve optimization precision and performance of the ABC algorithm, a parallel Multi-cores Parallel ABC algorithm (MPABC) was proposed based on the Fork/Join framework. The algorithm is to introduce the multi-populations' parallel operation to guarantee population's diversity, improve the global convergence ability and avoid falling into the local optimum. The performance of the original serial ABC algorithm and the MPABC algorithm was analyzed and compared based on four benchmark objective functions. The results show that the MPABC algorithm can achieve the speedup of 3.795 and the efficiency of 94.87% in solving complex problems. It can make full use of multi-core resources, improve the solution's quality and efficiency, and have the advantages of low parallel cost and simple realizing process.

Keywords: Parallel · Artificial bee colony algorithm · Fork/join framework

1 Introduction

For the Bio-inspired algorithms such as the Genetic Algorithm (GA) [1], Particle Swarm Optimization (PSO) [2], Differential Evolution (DE) algorithm [3] and Artificial Bee Colony (ABC) algorithm [4], increasing the number of populations is an effective solution under the premise of constant population size for improving computational efficiency and accuracy. However, in the serial computing environment, these methods will lead to large number of computing-intensive tasks in the optimization process which require lots of processing time [5].

With the popularization of multi-cores processors and in-depth study of parallel computing mechanisms, parallel computing has gradually become an important way to improve the computational efficiency [6]. Compared with the serial computing, multi-cores could effectively implement the thread-level parallelism, and improve computational efficiency [7].

© Springer International Publishing AG 2017
Y. Tan et al. (Eds.): ICSI 2017, Part I, LNCS 10385, pp. 313–319, 2017.
DOI: 10.1007/978-3-319-61824-1_34

The population-based meta-heuristics algorithms such as the GA, PSO, and ABC algorithms can explore efficiently from the parallel concepts to speed up their search process. A parallel master-slave model of the cooperative micro-particle swarm optimization approach based on the decomposition of the original search space in subspaces of smaller dimension was introduced [8]. A new technique has been described for using processors in parallel PSO to improve the performance of the algorithm [9]. For the ABC algorithms, there are several liberations discussed the parallelization. A parallel Artificial Bee Colony algorithm was applied for solving vehicle routing problem efficiently and results proved this approach to be superior [10]. In [7], three parallel models of the Artificial Bee Colony Algorithm were compared, and the trade-offs were analyzed between quality of solution and processing time.

In summary, the researches of parallel artificial bee colony algorithm are still in the initial stage. Therefore, a Multi-cores Parallel Artificial Colony (MPABC) algorithm was proposed to improve the running efficiency and optimization precision by utilizing the Fork/Join framework.

2   Java Fork/Join Framework

Fork/Join framework is a classic multi-core parallel framework that based on the divide-and-conquer strategy, and can take full advantage of multi-core CPU for massively parallel computing [11]. In the Fork/Join framework, a complex computing task is organized into a virtual tree structure. The basic idea is to use a recursive method based on threshold value to divide (Fork) a complex task into a plurality of smaller-scale, independent subtasks for solving, then the final result of the complex task could be gotten through the merge (Join) result of all the sub-tasks. Thus, this method is often referred to as the Fork/Join framework.

3   Multi-cores Parallel Artificial Bee Colony (MPABC) Algorithm

The framework of the multicore parallel Artificial Bee Colony (MPABC) algorithm is divided into upper layer and lower layer. The upper layer is implemented based on the coarse-grained model to ensure the diversity of the population. Its idea is the independent parallel run of multiple populations. It is characterized by divide-and-conquer strategy that divide population into multiple sub-populations, each sub-population using separate threads simultaneously for evolving to prevent being trapped of local optimization. In the lower layer, the Fork/Join framework is took as the parallel computing tool to realize the computing tasks of distribution and stealing. Figure 1 depicts the MPABC model with four hives at the upper level (ellipses with solid lines) and Fork/Join threads in each hive at the lower level.

The approach that repeatedly runs of the swarm optimization algorithm is better to get a more optimal solution. Each swarm independent runs on one core, and they do not make communication between multiple threads. Inside of the swarm, the Fork/Join

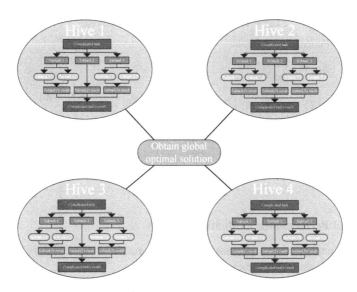

Fig. 1. The parallel framework of the MPABC algorithm

framework is used for the fine-grained parallel computing to improve performance. With the combination parallelization of fine-grained and coarse-grained, the running speed could increase to almost equivalent as they perform in the system kernel.

4   Experimental Analyses

The Experiments in this paper is to analyze and compares the performance of the original serial ABC algorithm with the parallel MPABC algorithm based on four standard benchmarking functions. The Java Fork/Join library was adopted to provide powerful and efficient parallel computing. All of our tests have been performed on an Intel (R) Core i7-4720HQ @ 2.6 GHz 4 cores with 8 GB of RAM with Microsoft Windows 8.

4.1   Benchmark Functions

The well-defined benchmark functions which based on standard mathematical functions can be used as objective functions to measure and test the performance of algorithms. The original serial ABC algorithm and the MPABC algorithm were applied to four well-known benchmark functions which were exten-sively used in the literature [12]. For ease of comparison, some functions were adjusted to make theoretical optimal value to zero. The Parameter ranges, formulations and global optimum values of these functions are given in Table 1.

Table 1. Numerical benchmark functions

Function	Function expression	Ranges	Minimum
Sphere	$f_1(\vec{X}) = \sum\limits_{i=1}^{D} X_i^2$	$[-100, 100]$	0
Griewank	$f_2(\vec{X}) = \frac{1}{4000}\sum\limits_{i=1}^{D}(X_i - 100)^2 - \prod\limits_{i=1}^{D}\cos(\frac{X_i-100}{\sqrt{i}}) + 1$	$[-600, 600]$	0
Rastrigin	$f_3(\vec{X}) = \sum\limits_{i=1}^{D}(X_i^2 - 10\cos(2\pi X_i) + 10)$	$[-5.12, 5.12]$	0
Rosenbrock	$f_4(\vec{X}) = \sum\limits_{i=1}^{D-1} 100(X_{i+1} - X_i^2)^2 + (X_i - 1)^2$	$[-50, 50]$	0

4.2   Parallel Performance Evaluation Indexes

The main evaluation indexes to measure the performance of parallel algorithms are the speedup S_p and parallel efficiency E_p [13].

(1) Speedup S_p: For a solving problem, T_s is the computation time that the sequential algorithm spends in the worst case, T_p is the computation time that a parallel algorithm spends for solving this problem in the worst case, p is the number of processors, and then the speedup of the parallel algorithm is defined as

$$S_p = T_s/T_p \tag{1}$$

As it can be seen, $1 \leq S_p \leq p$. When $S_p = p$, the parallel algorithm is the optimal parallel algorithm.

(2) Efficiency (Normalized speedup): Efficiency of parallel algorithm can be defined as the value that speedup divides the number of processors.

$$E_p = S_p/p \tag{2}$$

E_p reflects the processors' utilization in the execution of the algorithm. Since $1 \leq S_p \leq p$, $1 \leq E_p \leq 1$. If a parallel algorithm's efficiency is equal to 1, then it means that in the execution process of the algorithm each processor has been fully utilized. However, under normal circumstances, it is impossible to reach 1.

4.3   Experimental Results Analysis

In this section, experiments were taken on the four benchmark functions to compare the performance of the ABC algorithm and MPABC algorithm. By setting up the number of running cores, we evaluated the MPABC algorithm in single-core, dual-core, four-core, and eight-core environment respectively.

For all benchmark functions, the experimental parameters are set as follows: the number of subpopulations n is 4, the number of individuals in the subpopulation

NP is 200, *and D* is the dimension of the problem to be resolved. *D* was arbitrarily set to 200 to turns the problem solving process extremely difficult to all algorithms. It will help us in evaluating the behavior of the parallel models, regarding processing time and quality of solutions. The Limit value is set to 0.25 * NP * D [14]. And the maximum number of iterations of algorithms is set to 5000 and the termination condition is reached the maximum number.

Under the same condition, each of the algorithms was repeated 30 times for each test functions independently to statistically analyze the mean and standard deviation of the experiment results. The results of execution time (sec), Speedup and Efficiency for given number of cores of the two algorithms are shown in Table 2. In which, *Mean* and *SD* are the average value of mean and standard deviation of the statistical experimental data of the 30 times independent experiments. *Mean* shows precision that algorithm can achieve in a given function evaluation times, which reflecting the convergence accuracy. *SD* reflects the stability and robustness of the algorithm. For clarity, the results of the best algorithm are marked in boldface, respectively; if not all algorithms produce identical results.

For the time-consuming tasks, such as the Rosenbrock function, the ABC serial computation takes about 211 s, while the 4-cores MPABC parallel computation takes only 51 s, the speedup is 3.795, and the parallel efficiency is 94.87%. Therefore, the MPABC algorithm could make full use of the CPU multi-cores' parallel resources, greatly reduce the run time, and improve the operation efficiency. But for the simple

Table 2. Exectution time (sec), Speedup and Efficiency for given number of cores

Function	Number of cores	ABC serial runs Execution time(sec)		MPABC Parallel runs (sec) Execution time(sec)		Speedup (*Sp*)	Efficiency (*Ep*)
		Mean	SD	Mean	SD		
Sphere	1	35.662	1.284	–	–	–	–
	2			63.278	1.4550	0.5636	**28.18%**
	4			**32.420**	**0.1300**	1.100	27.50%
	8			32.541	0.1624	1.096	13.70%
Griewank	1	114.885	3.088	–	–	–	–
	2			81.970	1.245	1.402	70.08%
	4			**35.889**	**0.7617**	**3.201**	**80.83%**
	8			47.408	1.1524	2.545	31.82%
Rastrigin	1	110.129	4.536	–	–	–	–
	2			70.746	2.4805	1.557	77.85%
	4			**33.016**	**0.2498**	**3.336**	**83.39%**
	8			45.133	0.7117	2.440	30.50%
Rosenbrock	1	210.690	7.115	–	–	–	–
	2			143.497	4.034	1.468	73.41%
	4			**55.518**	**0.661**	**3.795**	**94.87%**
	8			55.536	1.023	3.794	47.42%

tasks, such as Sphere function, the parallelization does not enhance the computational efficiency, but consumes lots of computing time because of resources consumption and threads switch.

5   Conclusions

In this paper, a new parallel artificial bee colony (MPABC) algorithm based on the Fork/Join framework was proposed to improve optimization precision and performance of the ABC algorithm. Based on the benchmark functions' tests, it showed that the MPABC algorithm performs well in all performance indicators than the original serial ABC algorithm and realizes the significant increase of the computational efficiency on the ordinary multi-cores machine.

Acknowledgement. This work is supported by National Nature Science Foundation of China (Grant No. 61462058), Gansu Province Science and Technology Program (No. 1606RJZA004) and Gansu Data Engineering and Technology Research Center for Resources and Environment.

References

1. Tavares, L.G., Lopes, H.S., Lima, C.R.E.: A study of topology in insular parallel genetic algorithms. In: World Congress on Nature and Biologically Inspired Computing (NaBIC), pp. 632–635. IEEE (2009)
2. Kennedy J., Eberhart R.: Particle swarm optimization. In: Proceedings of IEEE International Conference on Neural Networks, pp. 1942–1948 (1995)
3. Fan, Q.Q., Yan, X.F.: Self-adaptive differential evolution algorithm with discrete mutation control parameters. Expert Syst. Appl. **42**, 1551–1572 (2015)
4. Karaboga, D.: An idea based on honey bee swarm for numerical optimization. Technical report-tr06, Vol. 200. Erciyes university, engineering faculty, computer engineering department (2005)
5. Zhu, X.P., Zhang, C., Yin, J.X.: Optimization of water diversion based on reservoir operating rules: analysis of the Biliu River reservoir. China J. Hydrol. Eng. **19**, 411–421 (2014)
6. Tu, K.Y., Liang, Z.C.: Parallel computation models of particle swarm optimization implemented by multiple threads. Expert Syst. Appl. **38**, 5858–5866 (2011)
7. Parpinelli, R.S., Benitez, C.M.V., Lopes, H.S.: Parallel approaches for the artificial bee colony algorithm. In: Panigrahi, B.K., Shi, Y., Lim, M.-H. (eds.) Handbook of Swarm Intelligence, pp. 329–345. Springer, Heidelberg (2011)
8. Konstantinos, E.P.: Parallel cooperative micro-particle swarm optimization: a master-slave model. Appl. Soft Comput. **12**, 3552–3579 (2012)
9. Gardner, M., McNabb, A., Seppi, K.: A speculative approach to parallelization in particle swarm optimization. Swarm Intell. **6**(2), 77–116 (2012)
10. Akancha, T., Afshar, M.A.: Implementation of parallel artificial bee colony algorithm on vehicle routing problem. Int. J. Adv. Res. Sci. Eng. (IJARSE) **2**(5), 122–130 (2013)
11. Lea, D.: A Java fork/join framework. In: Proceedings of the ACM 2000 Conference on Java Grande, pp. 36–43. ACM, June 2000

12. Gao, W., Liu, S., Huang, L.: A global best artificial bee colony algorithm for global optimization. J. Comput. Appl. Math. **236**(11), 2741–2753 (2012)
13. Alba, E., Luque, G.: Evaluation of parallel metaheuristics. In: Parallel Problem Solving from Nature (PPSN-EMAA 2006). LNCS, vol. 4193, pp. 9–14 (2006)
14. Karaboga, D., Akay, B., Ozturk, C.: Artificial bee colony (ABC) optimization algorithm for training feed-forward neural networks. In: Torra, V., Narukawa, Y., Yoshida, Y. (eds.) MDAI 2007. LNCS, vol. 4617, pp. 318–329. Springer, Heidelberg (2007). doi:10.1007/978-3-540-73729-2_30

Identification of Common Structural Motifs in RNA Sequences Using Artificial Bee Colony Algorithm for Optimization

L.S. Suma[1(✉)] and S.S. Vinod Chandra[2]

[1] College of Engineering Attingal, Thiruvananthapuram, Kerala, India
sumals.1@gmail.com
[2] University of Kerala, Thiruvananthapuram, India
vinod@keralauniversity.ac.in

Abstract. RNA molecules folded into secondary structure are found to have structure related functionalities. Efficient computational techniques are required for common structural motif identification due to its relevance in the study of various functional aspects. In this work we focus on finding the most frequent descriptor motif inherent in given set of RNA sequences. Our approach uses an efficient computational method incorporating Nature inspired optimization algorithm. The motif skeletons are obtained by applying context free grammar defined for the descriptor motif. Then swarm intelligence based Artificial Bee Colony optimization algorithm is applied to derive the common motif with minimum and maximum length values of each motif element. Optimization process is done based on the objective function defined with the frequency of occurrence as major criterion. This method is able to generate correct motif structures in Signal Recognition Particle data set. The resultant motif is compared with the common motifs generated by other evolutionary methods.

1 Introduction

RNA Genetics is a well studied area in Molecular Biology, driven by both structural and functional aspects. Involvement of RNA molecules in various cellular activities and the importance of its structure in the functionalities has motivated the research in this field. RNA molecules consist of conserved structural subunits that lead to the formation of secondary structure. Secondary structure prediction plays an important role in molecular level analysis [1]. While analyzing the secondary structure of RNA molecules, recurring local patterns are found in related ones. Digging out such structural motifs from a set of input sequences is an important problem. RNA motifs are grouped as secondary structure motifs, tertiary structure motifs, functional motifs and binding motifs. MicroRNAs are found to play critical role in gene expression regulation activities [2]. Distinguishing pre-miRNA from other stem loop hairpins is a widely addressed problem in microbiology [3].

There exist a number of mechanisms that addressed the problem of identifying secondary structural motifs from a set of input sequences. CMfinder is a framework that

© Springer International Publishing AG 2017
Y. Tan et al. (Eds.): ICSI 2017, Part I, LNCS 10385, pp. 320–327, 2017.
DOI: 10.1007/978-3-319-61824-1_35

predict RNA motifs in unaligned sequences [4]. The method is based on Expectation-Maximization algorithm that uses covariance model for motif description. A stochastic algorithm RNApromo finds local motifs with the details of known RNA secondary structure [5]. Candidate structures that occur in maximum number of the given inputs are taken to start the algorithm. RNAmine makes use of stem properties in the graph. It reframes motif finding as pattern extraction/graph mining problem [6].

Evolutionary computation is an efficient methodology inspired by biological evolution process and it is used to identify motifs from the huge search space comprising of Signal Recognition Particle dataset [7]. Genetic Algorithm and genetic programming methods are also applied to find common motifs in a set of RNA sequences [8, 9]. GeRNAMo is another technique that uses genetic programming algorithm and represents individuals in solution space as tree structures [10].

Nature inspired algorithms are found to be powerful in solving non linear problems with less computational effort. Particle Swarm Optimization is used to solve the motif finding problem in genomic sequences [11]. Artificial Bee Colony (ABC) algorithm is one of the most efficient Nature inspired algorithms that is suited for numerical optimization [12]. The principle of honey bee foraging is successfully applied on the classical optimization problem of transportation [13]. As ABC algorithm is more powerful, it is applied in solving descriptor based motif finding problem.

2 Background

RNA molecules consist of nucleotides Adenine, Uracil, Guanine & Cytocine. The pairing nature of complementary bases (A-U&G-C) leads to the formation of secondary structure. Another form of RNA structure is the tertiary structure described by intramolecular interactions. Conserved structural patterns that are recurrently present in RNA molecules are known as motifs. The commonly occurring structural motifs in RNA secondary structure include double helices, internal loops, hairpin loops or junctions.

An important aspect to be considered in motif identification process is the mode of representing the motif structures. A widely used motif representation scheme comprises 3 basic elements ss, h5 and h3 to characterize single stranded, 5' end of helix and 3' end of helix respectively. h5(4)ss(2)h3(4)ss(3) is an example representation for the compressed notation (((((..)))))... usually referred as dot bracket notation. Here the numerals in parantheses describe the number of corresponding elements.

Suppose, the same pattern exists in another RNA molecule but with slightly different motif element lengths. If h5(6)ss(3)h3(6)ss(1) is the second pattern, then the same motif is said to be present in multiple sequences and hence it is said to be a common motif. To represent such a common motif, the varying length of motif elements is also to be considered. This is done with minimum and maximum length values. The common motif in this example is represented as h5(4,6)ss(2,3)h3(4,6)ss (1,3). Here, the h5 helix in different motifs will contain 4 to 6 elements and thus min & max values in the common motif becomes (4,6).

3   Methodology

A set of related RNA sequences from SRP dataset is taken as input and tried to identify the most commonly occurring motifs by incorporating optimization process on motif length part. Our methodology consists of a preprocessing phase and an optimization phase. In the initial phase, the subsequences of varying length(input by the user) are generated. It is followed by predicting secondary structure, modeling the descriptor motif and generating the motif skeletons. In the later phase, these are converted into the common motif format by including the range of length component and finally optimization is performed to derive the common motifs. Figure 1 shows the overall processing done for motif generation.

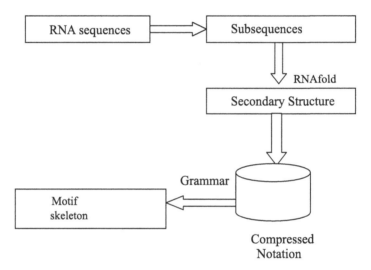

Fig. 1. Steps in motif generation process.

3.1   Motif Generation Phase

In order to generate the motif skeletons, the input RNA sequences must be processed and represented in a suitable format. Secondary structure need to be predicted for each subsequence of the given input sequences. RNAfold, a part of ViennaRNA Package 2.0 is a minimum free energy based framework and is used here for generating the secondary structure [14]. A sample secondary structure produced by RNAfold is as follows:

Input (test_sequence)   : GGGCUAUUAGCGUAGCCC
Output   : (((((((....)))))))

This output in turn is represented in the compressed notation h5(7)ss(4)h3(7). The file that stores the entire compressed notation subsequences will serve as the database for the upcoming optimization process.

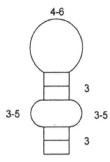

Fig. 2. Fogel's motif descriptor for SRP data

Identifying common motifs from a huge set of subsequences is a difficult task. Figure 2 shows Fogel's descriptor motif with two stems which is used in this work. From the very large search space of secondary structures, we identify only sequences that have the given motif structure. Context Free Grammar is used for modeling the motif structure. CFG is a powerful modeling technique that is used in many scientific applcations. As Context Free Grammar can easily represent the nested pairing nature of RNA structure, it is successfully applied for modeling of homologous RNA sequences [15]. To model the putative motif structure, the following grammar is used.

$$
\begin{aligned}
M &\longrightarrow \text{Yh5Xh3Y/h5Xh3Y/Yh5Xh3/h5Xh3} \\
X &\longrightarrow \text{ssM1ss/ss/null} \\
M1 &\longrightarrow \text{h5Xh3/ss/null} \\
Y &\longrightarrow \text{ss}
\end{aligned}
$$

Here, the X production is applied corresponding to the occurrence of stem in the input sequence. The subsequences are parsed by the grammar and the motif of the desired structure is generated. The motif skeleton generated by the Grammar is then converted into common motif format as described. Now we got the desired motif, but what remains is the correct length value identification of motif elements. It is done in the optimization phase.

3.2 Optimization Phase-Artificial Bee Colony Algorithm

Nature Inspired Computing provides efficient metaheuristic algorithms that are successfully applied in a variety of problems. Various species in nature-ants, birds, fish etc. has inspired scientific community to adopt their nature of work to solve problems. Artificial Bee Colony algorithm is one among this class, utilizing the honey foraging activities of bees. The basic algorithm considers three agents corresponding to 3 types of bees involved in the work. They are Employees, Onlookers & Scouts. The food sources in the algorithm correspond to the solutions of the application. The principle of optimization lies in the fact that quality of selected solution corresponds to the nectar amount (quality) of the food source.

Each cycle of ABC algorithm includes actions of employee, onlooker and scout bees. Each food source is assigned with an Employee bee. It tries to generate new food source using the previous information stored in the memory. After comparing the quality of these two solutions, it stores the best food source. Once this information is received from all the employee bees associated with N/2 food sources, onlooker bees start selecting the best one. The information is supplied to onlooker bees through woggle dance. Then scout bees are sent to explore the possible food sources. The ABC algorithm that optimizes our problem is as follows.

Step 1: Initializing length of D elements (ss, h5, h3) in N motif vectors with values in range (lb, ub)

$$z(i) = lb_i + (ub_i - lb_i) * r. \tag{1}$$

Step 2: For each employed bee, take each of these motif vectors

  (i) Perform searching of motif with length values of V in subsequence database
  (ii) Evaluate its quality by computing the objective function and update the position of food source by

$$x'(i) = x(i) + r * \left[x_j(i) - x_k(i) \right]. \tag{2}$$

  k varies from 1 to no: of solutions, $k \neq j$, r ranges between 0 & 1.

  (iii) Calculate the fitness value of new solution and compare with the fitness of the previous one.

Step 3: The probability is calculated for each employee bee by

$$P = f(x) / \sum f(x_k). \tag{3}$$

Step 4: Each onlooker bee repeats same calculations in step 3.
         Based on the probability of employed bees, onlooker bees select food source.

Step 5: After N cycles, if a particular food source is not found improved, then that source is abandoned. If one such source is found, scout bee replaces it with a random one.

As per the obtained motif structure, the parameters are chosen. The optimization algorithm starts with initialization of solution space. Here, the motif skeletons obtained constitute the particles in the solution space. These particles are initialized with length values in (lb, ub) range. After analyzing the features of the descriptor motif inherent in the dataset, suitable values have been chosen so as to facilitate the initialization of solution space. Now, each solution is to be evaluated based on its quality. Selection of objective function is so crucial that it decides the efficiency of the algorithm. Here, as we require the most frequently occurring motif in the sequences, objective function is taken as F = $\sum$ count(motif, seq)/N. After running the algorithm for a given number of cycles, optimal solution corresponding to the final objective function value is memorized.

4 Results and Discussion

We used the Signal Recognition Particle data set which is used in other methods. The data set contains 5 sequences with gi numbers 38795, 42758, 150042, 177793 and 216348.

In the pre processing steps, we generated the subsequences with length in the range (15, 50). This data set is expected to contain the descriptor motif. The problem dependent parameter, no: of stems is taken as 2. On parsing through the grammar, we have obtained 3 motif skeletons, of which we chose the most frequently occurring one. As per the obtained motif structure, the following parameters were chosen (Table 1). The comparison of resultant motifs is shown in Table 2.

After the 30th iteration of ABC algorithm, the value of objective function obtained is 0.998526. Consequently, the corresponding vector V is taken as the resultant length vector of the common motif and the common SRP motif is h5(3,4)ss(3,5)h5(2,3)ss(4,6) h3(2,3)ss(4,5)h3(2,4).

Table 1. Parameters chosen for ABC algorithm

Parameter	Type	Value
Dimension D	Problem specific	14
No: of solutions	General	30
No: of employed bee	N/2	15
Lower bound	Problem specific	2
Upper bound	Problem specific	6

Table 2. Comparing resultant motifs

Method	Motif
Fogel	h5(3,3)ss(3,5)h5(3,3)ss(4,6)h3(3,3)ss(3,5)h3(3,3)
GeRNAMo	h5(3,4)ss(4,5)h5(3,4)ss(4,5)h3(3,4)ss(4,5)h3(3,4)
GPRM	h5(3,5)ss(0,3)h5(7,9)ss(0,3)h3(3,5)ss(5,6)h3(7,8)
Our approach	h5(3,4)ss(3,5)h5(2,3)ss(4,6)h3(2,3)ss(4,5)h3(2,4)

On comparing we can see that the result obtained is very much similar to Fogel's motif as well as GeRNAMo. Figure 3 shows that minimum length values of motif elements differ from GeRNAMo by only one unit. For GPRM motif, the length of stem elements (h5 & h3) vary from the other motif structures. The comparison of maximum length values is shown in Fig. 4. As all the component length values coincide with

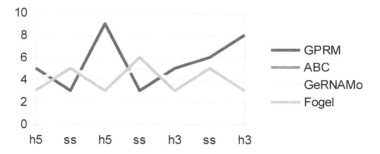

Fig. 3. Comparing minimum length values

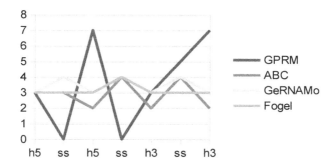

Fig. 4. Comparing maximum length values

other methods, line of ABC is not visible. This shows the accuracy of length values of the obtained motif. Moreover, ABC algorithm can solve this optimization problem in linear time.

5   Conclusion

RNA common motif identification is a well addressed problem for which we successfully applied ABC algorithm in optimizing the motif components. The powerful technique, CFG is also associated to this process for deriving the motif skeletons. Presently this approach can find out common motifs in SRP data. By designing suitable grammar, motifs can be derived from any data set. As motif identification is a combinatorial optimization problem, global optimization techniques can be used effectively. Protein motif identification is another important problem in bioinformatics. We will try to make use of variants of ABC algorithm in finding such common motifs. Hence we conclude that this global optimization method can be used for other computational problems in the field of Computational Biology.

References

1. Zuker, M., Stiegler, P.: Optimal computer folding of large RNA sequences using thermodynamics and auxiliary information. Nucleic Acids Res. **9**, 133–148 (1981)
2. Chandra, S.S.V., Reshmi, G.: A Pre-microRNA classifier by structural and thermodynamic motifs. IEEE World Congress on Nature and Biologically Inspired Computing (2009)
3. Reshmi, G., Chandra, S.S, Babu, V.J., Babu, P.S., Santhi, W.S., Ramachandran, S., Lakshmi, S., Nair, A.S., Pillai, M.R.: Identification and analysis of novel micro RNAs from fragile sites of human cervical cancer: computational and experimental approach. Genomics **97**, 333–340 (2011)
4. Yao, Z., Weinberg, Z., Ruzzo, W.L.: CMfinder—a covariance model based RNA motif finding algorithm. Bioinformatics **22**, 445–452 (2006)
5. Rabani, M., Kertesz, M., Segal, E.: Computational prediction of RNA structural motifs involved in posttranscriptional regulatory processes. Proc. Natl. Acad. Sci. **105**, 14885–14890 (2008)
6. Hamada, M., Tsuda, K., Kudo, T., Kin, T., Asai, K.: Mining frequent stem patterns from unaligned RNA sequences. Bioinformatics **22**, 2480–2487 (2006)
7. Fogel, G.B., William Porto, V., Weekes, D.G., Fogel, D.B., Griffey, R.H., McNeil, J.A., Lesnik, E., Ecker, D.J., Sampath, R.: Discovery of RNA structural elements using evolutionary computation. Nucleic Acids Res. **30**(23), 5310–5317 (2002)
8. Chen, J.H., Le, S.-Y., Maizel, J.V.: Prediction of common secondary structures of RNAs: a genetic algorithm approach. Nucleic Acids Res. **28**, 991–999 (2000)
9. Hu, Y.-J.: GPRM: a genetic programming approach to finding common RNA secondary structure elements. Nucleic Acids Res. **31**, 3446–3449 (2003)
10. Michal, S., Ivry, T., Schalit-Cohen, O., Sipper, M., Barash, D.: Finding a common motif of RNA sequences using genetic programming: the GeRNAMo system. IEEE/ACM Trans. Comput. Biol. Bioinf. **4**, 596–610 (2007)
11. Preeja, V., Abdul, Nazeer, K.A., Vinod Chandra, S.S.: Common structural motif identification in genomic sequences. In: IEEE-ICDSE, pp. 37–41 (2012)
12. Karaboga, D., Basturk, B.: A powerful and efficient algorithm for numerical function optimization: Artificial Bee Colony (ABC) algorithm. J. Glob. Optim. **39**, 459–471 (2007)
13. Saritha, R., Vinod Chandra, S.S.: A novel algorithm based on honey bee foraging principle for transportation problems. In: ACCIS Proceedings of Elsevier (2014)
14. Lorenz, R., Bernhart, S.H., zu Siederdissen, C.H., Tafer, H., Flamm, C., Stadler, P.F., Hofacker, I.L.: ViennaRNA package 2.0. Algorithms Mol. Biol. **6** (2011)
15. Knudsen, B., Hein, J.: Pfold: RNA secondary structure prediction using stochastic CFG. Bioinformatics **31**, 3423–3428 (2003)

A Mixed Artificial Bee Colony Algorithm for the Time-of-Use Pricing Optimization

Huiyan Yang, Xianneng Li[✉], and Guangfei Yang

Faculty of Management and Economics, Dalian University of Technology,
Dalian, China
xianneng@dlut.edu.cn

Abstract. Demand side management (DSM) is proposed to solve the contradiction between supply and demand of electricity market. To avoid the peak load, time-of-use (TOU) pricing strategy plays an important role in DSM to affect the behavior of using electricity by the users. In this paper, we proposed a mixed artificial bee colony (mABC) algorithm to TOU pricing optimization. Different from traditional research which optimizes the time division and electricity price separately, we consider these two factors together and optimize them simultaneously through the proposed mABC. The experimental results on a real-world scenario show the superiority of the mABC over traditional state-of-the-art methods.

Keywords: Demand side management · Swarm intelligence · Artificial bee colony · Mixed optimization · Time-of-use pricing

1 Introduction

Demand side management (DSM) is proposed to solve the supply and demand contradiction in electricity market [17]. The contradiction mainly refers to the unbalance between the stability of generation capacity and the volatility of demand which varies in the course of a day and different seasons [14]. There have been numerous DSM techniques used in the literature [7,12], within which the time-of-use (TOU) pricing is one of the most unique strategies to promote the overall electricity system efficiency, security and sustainability based on the existing electrical infrastructure [5,18]. It guides the users to choose the time of electricity consumption rationally by reasonable electricity price system and improves the electricity consumption structure by shifting the peak load to valley time, so as to achieve the peak load shifting effect [2–4,19]. The main feature of the TOU pricing strategy is to *assign different electricity prices to different time divisions based on the concrete electricity load.* For the valley load, a lower price is used to encourage the customers to use the electricity, while a higher price is levied to punish the usage of electricity in the peak load. Accordingly, the load pressure of electricity providers is relaxed and the cost of power generation

© Springer International Publishing AG 2017
Y. Tan et al. (Eds.): ICSI 2017, Part I, LNCS 10385, pp. 328–336, 2017.
DOI: 10.1007/978-3-319-61824-1_36

is reduced. In other words, TOU pricing strategy works as a soft solution to improve the profits of both energy providers and users without the construction of new electrical infrastructure.

In order to achieve the peak-valley load shifting, the TOU pricing strategy is required to deal with two considerable targets.

- **Factorization of time divisions** (for a day, 24 h) with respect to the overall electricity load. This target is to factorize each time period into 3 categories of electricity load, i.e., *valley* load, *flat* load or *peak* load.
- **Determination of electricity prices.** This target is to determine what electricity price is to be levied for the 3 categories of electricity load.

Currently, there have been various studies focusing on the TOU pricing optimization to realize the electricity load shifting effect [1,10,13,16], however, within which most of them focus on optimizing the electricity prices by a fixed, predefined factorization of time divisions. In this paper, we adopt a heuristic-based swarm intelligence method named artificial bee colony (ABC) [6,8] to address the problem. ABC is originated for solving the continuous optimization problems, while the discrete variables, i.e., Factorization of time divisions, cannot be handled. Accordingly, we develop an extended version named mixed ABC (mABC), which can optimize the discrete variable (Factorization of time divisions) and continuous variables (Determination of electricity prices) simultaneously.

The remaining paper is organized as follows. Section 2 describes the TOU pricing model. In Sect. 3, the details of mABC are presented. Section 4 conducts the experiments on a real-world scenario. Finally, we conclude the paper.

2   Time-of-Use (TOU) Pricing Model

To construct the TOU pricing model, 2 assumptions are made first. First, the amount of daily electricity consumption remains unchanged before and after the implementation of TOU pricing strategy. Second, the users' behavior of using electricity is only influenced by the electricity prices, regardless the impact of other potential factors. The notations used are presented as follows.

- Q_v, Q_f and Q_p: the amounts of power consumption for valley load, flat load and peak load before TOU pricing strategy.
- Q'_v, Q'_f and Q'_p: the amounts of power consumption for valley load, flat load and peak load after TOU pricing strategy.
- Q_{min} and Q_{max}: the maximal and minimal power load in a day.
- p_v, p_f and p_p: the electricity prices for valley load, flat load and peak load.
- p_0: the electricity price before TOU pricing strategy.
- M_0, M_{TOU}: the profits gained by the electricity provider before and after TOU pricing strategy.
- M': The saved generation cost of electricity by TOU pricing strategy.

Afterwards, we hold the following 3 equations.

$$Q_v + Q_f + Q_p = Q'_v + Q'_f + Q'_p. \tag{1}$$

$$M_0 = p_0 \left(Q_v + Q_f + Q_p \right). \tag{2}$$

$$M_{TOU} = p_v Q'_v + p_f Q'_f + p_p Q'_p. \tag{3}$$

The prime target of implementing TOU pricing strategy is to benefit both of the electricity providers and the end-users. Therefore, we have the following constraints: *For the electricity providers*: $M_{TOU} \geq M_0 - M'$; *For the end-users*: $M_{TOU} \leq M_0$. It is clear that the target of TOU pricing strategy is to reduce M' as much as possible. The realistic solutions to realize the objective are (1) minimize the peak load and (2) maximize the valley load (can be transferred to minimize the difference of peak-valley load). We assume that the end-users' reaction function between the electricity consumption and prices is represented by $f(p)$, and $Q = f(p)$. Accordingly, we formulate the TOU pricing model as follows.

$$\text{Minimize } 0.5 Q_{max} + 0.5 (Q_{max} - Q_{min}), \tag{4}$$

$$\text{subject to } M_0 - M' \leq M_{TOU} \leq M_0 \tag{5}$$

3   Mixed Artificial Bee Colony (mABC) Algorithm

3.1   Optimization Targets

To address the above TOU pricing model, two considerable targets are to be optimized, that is, **factorization of time divisions** and **determination of electricity prices**.

For the first issue, we divide a day (24 h) into 24 time periods, while we need to determine that for each time period, what category of electricity prices is levied, i.e., p_v, p_f or p_p. Therefore, there are 24 discrete variables to be optimized, where each variable can be assigned $\{valley, flat, peak\}$.

For the second target, we need to determine what electricity price is to be levied for the 3 categories of electricity load, i.e., p_v, p_f or p_p. Based on the previous research, we directly assign $p_f = p_0$ for the flat load, which is predefined. Therefore, there are 2 continuous variables to be optimized, i.e., p_v and p_p.

3.2   Mixed Artificial Bee Colony (mABC) in Details

In order to solve this problem, we propose a mixed artificial bee colony (mABC) algorithm. mABC is originated from ABC [6,8,9] which have been successfully applied to solve the continuous optimization problems. mABC extends traditional ABC to solve the mixed optimization problems, where the 24 discrete variables and 2 continuous variables of TOU pricing model are optimized simultaneously.

$H0$	$H1$	$H2$	...	$H22$	$H23$	p_v	p_p

Fig. 1. Individual structure of mABC

```
1  initialization;
2  repeat
3  |    send the employed bees;
4  |    send the onlooker bees;
5  |    if abandoned solution is found then
6  |    |    send the scout bees;
7  |    end
8  until the optimal solution is found or MFEs is reached;
```

Algorithm 1. Basic algorithm of mABC

Individual Structure: To optimize the discrete and continuous variables simultaneously, the individual structure of mABC is coded as Fig. 1.

Algorithmic Description: The basic algorithm of mABC is preserved as the traditional ABC, which is shown in Algorithm 1. It maintains a population of individuals. The population size is defined by the number of individuals SN. Three groups of artificial bees are developed and sent to the individuals, including the employed bees, onlooker bees and scout bees. They play the essential role to modify the existing individuals to find better ones until reaching the terminal condition, e.g., the maximal number of fitness evaluations ($MFEs$).

Employed Bees: The employed bees focus on searching around the current population to seek better ones. There are total SN employed bees, each of which corresponds to a unique individual. The search procedure of each individual X_i is performed in Algorithm 2. To generate V_i, we use the following strategy.

1. If the selected jth variable is *discrete*, v_i^j is randomly assigned from $\{0, 1, 2\}$ which is different from x_i^j.
2. If the selected jth variable is *continuous*, the following equation is used.

$$v_i^j = x_i^j + rand(-1, 1) \times \left(x_i^j - x_k^j \right).\tag{6}$$

Here, $k \in \{1, 2, \ldots, SN\}$ is a randomly selected neighbour individual. $rand(-1, 1)$ is a random value ranging in $[-1, 1]$.

Onlooker Bees: Afterwards, the onlooker bees (with size SN) are sent. Roulette-wheel selection is applied to guide each onlooker bee to select an individual, that, individuals with higher quality tend to have higher probability to be selected. After selection, Algorithm 2 is applied to seek a better candidate.

Scout Bees: If an individual can not be improved even that it has been searched by the employed/onlooker bees many times, i.e., *limits* times, it is considered as an abandoned solution that will be removed from the population. A scout bee will be sent to randomly generate a new individual to the population.

---

1 randomly select the jth variable, $j \in \{1, 2, ..., D\}$;

2 create offspring solution V_i, whose variables are equal to X_i;

3 change a new variable value v_i^j in V_i to replace the original value x_i^j;

4 if the V_i is better than X_i, $X_i \leftarrow V_i$; otherwise X_i will be remained;

Algorithm 2. Search procedure of each individual X_i

4 Experiments

4.1 Experimental Scenario

The data used in the experiment is from a real-world scenario of China [13]. The distribution of electricity load is described in Table 1, and the end-users' reaction data between electricity prices and consumption is shown in Table 2.

The electricity price before TOU pricing strategy, i.e., p_0, is 0.6357 RMB/KWh. Based on the realistic consideration, the electricity prices hold the following constraints.

$$\begin{cases} 0.3p_0 \leq p_v \leq 0.8p_0; \\ p_f = p_0; \\ 1.2p_0 \leq p_p \leq 1.8p_0. \end{cases} \tag{7}$$

4.2 Experimental Results and Analysis

As described in Sect. 2, the first issue to deal with the TOU pricing model is to estimate an appropriate end-users' reaction function between the electricity prices and consumption, i.e., $Q = f(p)$. Traditional research generally

Table 1. The electricity load data of a day (unit: 10^4 KWh)

Period	Electricity load	Period	Electricity load	Period	Electricity load
00:00–01:00	172.05	08:00–09:00	241.56	16:00–17:00	245.43
01:00–02:00	179.28	09:00–10:00	247.12	17:00–18:00	247.45
02:00–03:00	175.08	10:00–11:00	253.34	18:00–19:00	261.93
03:00–04:00	165.32	11:00–12:00	225.57	19:00–20:00	260.92
04:00–05:00	182.99	12:00–13:00	229.77	20:00–21:00	245.43
05:00–06:00	188.53	13:00–14:00	233.17	21:00–22:00	229.28
06:00–07:00	200.66	14:00–15:00	230.97	22:00–23:00	195.78
07:00–08:00	216.65	15:00–16:00	230.97	23:00–24:00	186.35

Table 2. The end-users' reaction data between electricity prices and consumption

Price (per unit value)	0.333	0.4	0.5	0.667	1	1.5	2	2.5	3
Consumption (per unit value)	1.2124	1.1970	1.1376	1.0897	1.0140	0.8603	0.7496	0.6217	0.5261

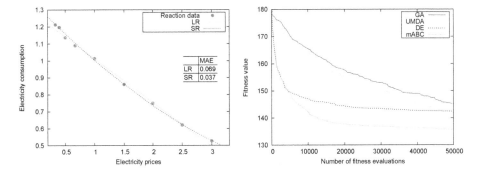

Fig. 2. The end-users' reaction function   **Fig. 3.** Performance of fitness values

applies linear regression (LR) to solve the issue, while we apply the genetic programming-based symbolic regression (SR) approach [15,20] in the paper. The reaction functions obtained by LR and SR are shown in Eq. 8, as well as the graphic illustration in Fig. 2. The mean absolute error (MAE) results show that the quadratic obtained by SR is more accurate estimation than that of LR.

$$\begin{cases} LR : Q = -0.258p + 1.27; \\ SR : Q = 0.0296p^2 - 0.356p + 1.33. \end{cases} \tag{8}$$

Afterwards, we apply mABC to optimize the factorization of time divisions and determination of electricity prices simultaneously (The source codes refer to the acknowledgements). The parameters of mABC is defined by: $SN = 10$, $limit = 260$ and $MFEs = 200,000$. The experimental results are the average values over 30 independent runs. In order to evaluate the performance of mABC, three state-of-the-art evolutionary algorithms are compared, i.e., genetic algorithm (GA), univariate marginal distribution algorithm (UMDA) [11] and a mixed version of differential evolution (DE) which treats the discrete variables as mABC.

Convergence Performance: The convergence curves of the compared algorithms are shown in Fig. 3. It is clear that mABC outperforms the other three compared algorithms. DE ranks in the second place, and GA shows the worst results. The detailed results of fitness values are shown in Table 3. Meantime, we conduct student t-test (two-tailed, paired, with confidence level 95%) to validate the statistical significance. The t-test results (P values) show that mABC outperforms the other algorithms with statistical significance.

Optimal TOU Pricing Strategy: The optimal TOU pricing strategy obtained by the studied algorithms are shown in Table 4. Three groups of results are listed, i.e., factorization of time divisions, determination of electricity prices and maximal/minimal electricity load. In 9 out of 24 time periods, the four algorithms recommend different time factorization, resulting in different final performance. As for the proposed mABC, the optimal time division scheme is: Valley hours: 00:00–06:00, 22:00–24:00 (8 h); Flat hours: 06:00–08:00, 11:00–16:00, 21:00–22:00

Table 3. Performance of fitness values and statistical test

	GA	UMDA	DE	mABC
Mean	145.2310	142.3732	142.4420	**134.3569**
Std. dev.	5.060984	6.579869	1.740538	**0.802585**
P value	**3.72E-18**	**6.01E-12**	**1.49E-07**	—

Table 4. Performance of the optimal TOU pricing strategy

Period	Factorization of time divisions					Determination of electricity prices			
	GA	UMDA	DE	mABC		GA	UMDA	DE	mABC
00:00-01:00	valley	valley	valley	valley	p_v	$0.3p_0$	$0.52162p_0$	$0.463791p_0$	$0.427678p_0$
01:00-02:00	valley	valley	valley	valley	P_p	$1.536556p_0$	$1.390754p_0$	$1.411189p_0$	$1.427157p_0$
02:00-03:00	valley	valley	valley	valley					
03:00-04:00	valley	valley	valley	valley					
04:00-05:00	valley	valley	valley	valley					
05:00-06:00	valley	valley	valley	valley		Maximal/minimal electricity load (unit: 10^4KWh)			
06:00-07:00	flat	valley	valley	flat		GA	UMDA	DE	mABC
07:00-08:00	flat	flat	flat	flat	Q_{max}	239.9997	234.6684	235.0245	**231.7929**
08:00-09:00	peak	flat	flat	peak	Q_{min}	192.6046	190.5077	193.6323	**195.7299**
09:00-10:00	flat	flat	peak	peak	Diff.	47.39502	44.16071	41.39224	**36.06298**
10:00-11:00	peak	peak	peak	peak					
11:00-12:00	flat	flat	flat	flat					
12:00-13:00	peak	flat	peak	flat					
13:00-14:00	peak	flat	flat	flat					
14:00-15:00	flat	flat	flat	flat					
15:00-16:00	flat	flat	peak	flat					
16:00-17:00	peak	flat	flat	peak					
17:00-18:00	flat	flat	flat	peak					
18:00-19:00	peak	peak	peak	peak					
19:00-20:00	peak	peak	peak	peak					
20:00-21:00	flat	peak	flat	peak					
21:00-22:00	flat	flat	flat	flat					
22:00-23:00	valley	valley	valley	valley					
23:00-24:00	valley	valley	valley	valley					

(8 h); Peak hours: 08:00–11:00, 16:00–21:00 (8 h), which is much more reasonable than the other algorithms'. The maximal/minimal electricity load shows that mABC results in much more smaller Q_{max} and the difference of Q_{max} and Q_{min}, where both are the optimization targets of the TOU pricing model. Eventually, by using the optimal TOU pricing strategy of mABC, the electricity load is adjusted to a more smooth curve, such that the supply-demand relationship is much more relaxed.

5   Conclusion and Future Work

In this paper, we proposed a mixed artificial bee colony (mABC) algorithm to Time-of-Use (TOU) pricing optimization. We consider the time division and electricity price together and optimize them simultaneously through the proposed mABC. The experimental results on a real-world scenario of China show the superiority of the mABC over traditional state-of-the-art algorithms, i.e., GA, UMDA and DE. By using the proposed mABC, a more supply-demand friendly electricity load is achieved to benefit both of the electricity providers and the end-users. In the future, we will consider to explore this approach to a more complex smart grid scenario, and develop enhanced optimization algorithms.

Acknowledgments. This work is supported by the National Natural Science Foundation of China (71601028, 71671024, 71421001, 71431002). The source codes are available at http://faculty.dlut.edu.cn/li/en/article/960204/list/.

References

1. Arteconi, A., Hewitt, N., Polonara, F.: State of the art of thermal storage for demand-side management. Appl. Energy **93**, 371–389 (2012)
2. Celebi, E., Fuller, J.D.: Time-of-use pricing in electricity markets under different market structures. IEEE Trans. Power Syst. **27**(3), 1170–1181 (2012)
3. Datchanamoorthy, S., Kumar, S., Ozturk, Y., Lee, G.: Optimal time-of-use pricing for residential load control. In: 2011 IEEE International Conference on Smart Grid Communications (SmartGridComm). pp. 375–380, IEEE (2011)
4. Gellings, C.W.: The concept of demand-side management for electric utilities. Proc. IEEE **73**(10), 1468–1470 (1985)
5. Hatami, A., Seifi, H., Sheikh-El-Eslami, M.K.: A stochastic-based decision-making framework for an electricity retailer: time-of-use pricing and electricity portfolio optimization. IEEE Trans. Power Syst. **26**(4), 1808–1816 (2011)
6. Karaboga, D., Basturk, B.: On the performance of artificial bee colony (ABC) algorithm. Appl. Soft Comput. **8**(1), 687–697 (2008)
7. Kurucz, C.N., Brandt, D., Sim, S.: A linear programming model for reducing system peak through customer load control programs. IEEE Trans. Power Syst. **11**(4), 1817–1824 (1996)
8. Li, X., Yang, G.: Artificial bee colony algorithm with memory. Appl. Soft Comput. **41**, 362–372 (2016)
9. Li, X., Yang, G., Kıran, M.S.: Search experience-based search adaptation in artificial bee colony algorithm. In: 2016 IEEE Congress on Evolutionary Computation (CEC), pp. 2524–2531. IEEE (2016)
10. Min, L., Yanhui, Z., Bailin, H.: Peak and valley periods division and time-of-use price based on power generation cost. East China Electr. Power **33**(12), 90–91 (2005)
11. Mühlenbein, H., Paaß, G.: From recombination of genes to the estimation of distributions I. Binary parameters. In: Voigt, H.-M., Ebeling, W., Rechenberg, I., Schwefel, H.-P. (eds.) PPSN 1996. LNCS, vol. 1141, pp. 178–187. Springer, Heidelberg (1996). doi:10.1007/3-540-61723-X_982
12. Ng, K.H., Sheble, G.B.: Direct load control-a profit-based load management using linear programming. IEEE Trans. Power Syst. **13**(2), 688–694 (1998)
13. Ning, N.: Research on Time-of-Use Price Optimization of DSM Based on Genetic Algorithm. Tianjin University, Tianjin (2012)
14. Rabl, V.A., Gellings, C.W.: The concept of demand-side management. In: De Almeida, A.T., Rosenfeld, A.H. (eds.) Demand-Side Management and Electricity End-Use Efficiency. NATO ASI Series, vol. 149, pp. 99–112. Springer, Netherlands (1988). doi:10.1007/978-94-009-1403-2_5

15. Schmidt, M., Lipson, H.: Distilling free-form natural laws from experimental data. Science **324**(5923), 81–85 (2009)
16. Sheen, J.N., Chen, C.S., Yang, J.K.: Time-of-use pricing for load management programs in taiwan power company. IEEE Trans. Power Syst. **9**(1), 388–396 (1994)
17. Strbac, G.: Demand side management: benefits and challenges. Energy policy **36**(12), 4419–4426 (2008)
18. Torriti, J.: Price-based demand side management: assessing the impacts of time-of-use tariffs on residential electricity demand and peak shifting in Northern Italy. Energy **44**(1), 576–583 (2012)
19. Wei, D., Jiahai, Y., Zhaoguang, H.: Time-of-use price decision model considering users reaction and satisfaction index. Autom. Electr. Power Syst. **29**(20), 10–14 (2005)
20. Yang, G., Li, X., Wang, J., Lian, L., Ma, T.: Modeling oil production based on symbolic regression. Energy Policy **82**, 48–61 (2015)

Optimization of Office-Space Allocation Problem Using Artificial Bee Colony Algorithm

Asaju La'aro Bolaji[1(✉)], Ikechi Michael[2], and Peter Bamidele Shola[3]

[1] Department of Computer Science, Faculty of Pure and Applied Sciences,
Federal University Wukari, Wukari, Taraba State, Nigeria
lbasaju@fuwukari.edu.ng
[2] VConnect Global Services Limited, No. 60, Marine View Plaza, 7th Floor,
Apongbon, Lagos 100221, Nigeria
mykehell123@gmail.com
[3] Department of Computer Science, University of Ilorin, Ilorin, Nigeria
shola.bp@unilorin.edu.ng

Abstract. Office-space allocation (OFA) problem is a class of complex optimization problems that distributes a set of limited entities to a set of resources subject to satisfying set of constraints. Due to the complexity of OFA, numerous metaheuristic-based techniques have been proposed. Artificial Bee Colony (ABC) algorithm is a swarm intelligence, metaheuristic techniques that have been utilized successfully to solve several formulations of university timetabling problems. This paper presents an adaptation of ABC algorithm for solving OFA problem. The adaptation process involves integration of three neighbourhood operators with the components of the ABC algorithm in order to cope with rugged search space of the OFA. The benchmark instances established by University of Nottingham namely Nothingham dataset is used in the evaluation of the proposed ABC algorithm. Interestingly, the ABC is able to produced high quality solution by obtaining two new results, one best results and competitive results in comparison with the state-of-the-art methods.

Keywords: Artificial Bee Colony · Timetabling problem · Office space allocation · Nature-inspired computing

1 Introduction

Typically, the office-space allocation (OFA) is a class of complex combinatorial optimization problems that has been widely investigated by the researchers in the domain of timetabling and operations research over the last few decades due to its practical utility to many organization across the globe. The problem involves assignment of a set of limited spaces to a set of resources in such a way that all resources are scheduled to the require spaces for optimal utilization subject to set of constraints satisfaction. In OFA, the classification of constraints is divided into two: hard and soft. Note that it is compulsory for the hard constraints to be

© Springer International Publishing AG 2017
Y. Tan et al. (Eds.): ICSI 2017, Part I, LNCS 10385, pp. 337–346, 2017.
DOI: 10.1007/978-3-319-61824-1_37

satisfied in an OFA problem for the solution to be *feasible*, while the satisfactions of the soft constraints is required but not mandatory. Basically, the quality of the OFA solution is determined by the satisfactions of number of soft constraints.

The introduction of several algorithmic approaches for tackling the OFA have been proposed by the workers in the domain of operational research and Artificial Intelligence over past few decades [1]. The techniques utilized for the OFA under mathematical programming approaches are [2–4]. While those that employed the usage metaheuristic approaches for the OFA are classified into local search-based and population-based approaches. Some examples of local search-based approaches which are utilized for the OFA include hill climbing [5,6], simulated annealing [5,7] and tabu search [6]. Furthermore, those used population-based approaches for the OFA are evolutionary algorithm [5,8], harmony search algorithm (HSA) [9]. Similarly, the usage of hyper-heuristic and hybrid metaheuristic approaches have also been reported [10–12]. While the success of the metaheuristic algorithms when applied independently to optimization problems have been reported, substantial research efforts emanated from hybridization especially for large scale scheduling/timetabling problems. Few example of hybridization between population-based and local search-based that algorithms applied to timetabling appeared in [13–15].

Karaboga in [16] proposed Artificial Bee Colony (ABC) algorithm as a class of swarm intelligence algorithm which imitates foraging behaviour of artificial bee in their colony. Due to the simplicity of this algorithm, it has been successfully adopted to tackle several NP-hard and complex optimization problems [17,18]. Studies have shown that ABC algorithms have been adapted and hybridized for the different formulations of timetabling problems, however non of these studies have investigated the ABC for the OFA. Therefore, the contribution of this paper is to investigates the performance ABC algorithm on OFA problem. The proposed ABC algorithm for the OFA is evaluated using the benchmark instances from the University of Nottingham, and one instance from the University of Wolverhampton. Experimentally, the ABC algorithm shows successful performance by achieving two new results, one best results and a comparable results in the remaining benchmark instances.

The remaining part of the paper is organized as follows: Sect. 2 provides brief formulation of the OFA while the main concept of artificial bee colony is presented in Sect. 3. Section 4 introduces the proposed ABC approach for the OFA problem. The computational experiments, results and discussions are given in Sect. 5, and finally conclusion and future work are presented in Sect. 6.

2   Office-Space Allocation Problem (OFA)

The formulation OFA involves a set of l entities, with dimensions $d_1, d_2, \ldots, d_l$, and a set of m rooms with capacities $c_1, c_2, \ldots, c_m$. The solution to the OFA problem is represented by a two dimensional matrix X of $[x_{i,j}]$ values, in which $x_{i,j} = 1$ if the entity j (e_j) is assigned to room i (r_i). The main objective OFA is to generates a feasible solution with the best quality. The different constraints of the OFA problem considered in this study are:

- *No sharing-* In this constraint, one particular entity should not share the room with another entity.
- *Be located in* - a particular entity should be assigned to a specific room.
- *Be adjacent to* - a particular entity should be assigned to room adjacent to another entity.
- *Be away* - a particular entity should not be allocated close to another entity.
- *Be together with* - two particular entity should be assigned to the same room.
- *Be grouped with* - A group of entities should be assigned close to each other.

The solution to the OFA is evaluated based on the fitness cost $f(x)$ as shown in Eq. (1)

$$\min f(x) = f_1(x) + f_2(x). \tag{1}$$

subject to

$$\sum_{i=0}^{l-1} \sum_{j=0}^{m-1} x_{i,j} = 1. \tag{2}$$

where $f_1(x)$ represents the space misuse function and $f_2(x)$ is used to compute the violation of the soft constraints.

$$f_1(x) = \sum_{i=0}^{l-1} WP_i + \sum_{i=0}^{l-1} OP_i. \tag{3}$$

$$f_2(x) = \sum_{r=0}^{k-1} SCP_r. \tag{4}$$

where both WP_i and OP_i are the amount of space wasted or overused for each room i; SCP_r represent the penalty for violating the rth soft constraint. For each room only i one of WP_i or OP_i has a value greater than zero, and the amount of overused for each room i is computed as shown in Eqs. 5 and 6.

$$WP_i = \max(0, c_i - \sum_{j=0}^{m-1} x_{i,j} \cdot w_j). \tag{5}$$

$$OP_i = \max(0, 2(\sum_{j=0}^{m-1} x_{i,j} \cdot w_j - c_i)). \tag{6}$$

3   Artificial Bee Colony Algorithm

In the recent time, Karaboga proposed artificial bee colony (ABC) for tackling numerical optimization [16]. It is a typical swarm intelligence population-based algorithm that simulates intelligent foraging behaviour of honey bees in their colonies. The colony of ABC consists of three categories of bees namely: employed, onlooker and scout bees. Employed bees occupy the first half of the colony while the onlooker bees take over the remaining half. During the search process, each employed bee explores its neighbourhood for the new food sources since information about the food sources exist in their memory, and shares the information of their food sources with the onlooker bees. Based on information provided by the employed bees, onlooker bees decides to select and further explores the neighbourhood of those food sources with good information in order to generate new ones. Any employed bee whose food source is exhausted and abandon by the onlooker bee will automatically turns to scout. Then scout bee randomly explore the search space in order to discover the new food source. The key procedural steps of ABC as an iterative improvement process are given as follows:

- Generate the initial population of the food sources randomly.
- REPEAT
  - Send the employed bees onto the food sources and calculate the fitness cost.
  - Evaluate the probability values for the food sources.
  - Send the onlooker bees onto the food sources depending on probability and calculate the fitness cost.
  - Abandon the exploitation process, if the sources are exhausted by the bees.
  - Send the scouts into the search area for discovering new food sources randomly.
  - Memorize the best food source found so far.
- UNTIL (requirements are met).

4 The Proposed ABC for Office-Space Allocation Problem (OFA)

In this section, the continuous nature of the ABC algorithm is modified with the introduction of three different neighourbood structures to cope with the solution search space of the OFA problem. The five steps of the ABC algorithm as adapted to the OFA is provided in the next subsections. Note that the feasibility of the solutions is preserved through out the study.

4.1 Initialize the ABC and OFA Parameters

The three control parameters of ABC algorithm which are required for solving the OFA are initialized in this step. The parameters are solution number (SN) which refers to the number of food sources in the population and similar to the population size in GA; maximum cycle number (MCN) is the maximum number of iterations; and *limit* which is normally used to diversify the search and it is responsible for the abandonment of solution, if there is no improvement for certain number of iterations. Similarly, the OFA variables like set of the entities, set of rooms, room capacity, the size of capacity required by each entity and set of constraints (i.e. hard and soft) are also extracted from the dataset.

4.2 Initialize the Food Source Memory

The food source memory (FSM) is represented by memory location which consists of sets of feasible solutions (i.e. food sources) as determined by SN as shown in Eq. 7. This step generates the initial feasible food source (i.e. solution vectors) using the peckish heuristic strategy introduced in [19] and these food sources are sorted in ascending order in the FSM in accordance with the objective function values of the food source, that is $f(x_1) \leq f(x_2) \ldots f(x_{SN})$. Algorithm 1 shows the pseudocode of the peckish heuristic strategy where each food source satisfies all hard constraints and all resources are allocated in the suitable rooms.

$$\textbf{FSM} = \begin{bmatrix} x_1^1 & x_1^2 & \cdots & x_1^N \\ x_2^1 & x_2^2 & \cdots & x_2^N \\ \vdots & \vdots & \ddots & \vdots \\ x_{SN}^1 & x_{SN}^2 & \cdots & x_{SN}^N \end{bmatrix} \begin{bmatrix} f(x_1) \\ f(x_2) \\ \vdots \\ f(x_{SN}) \end{bmatrix} . \tag{7}$$

Algorithm 1. Peckish heuristic strategy

---

    while all entities are not assigned **do**
      $K = N/3$
      Select an unassigned entity j randomly
      Select a number of K rooms which satisfy $^1/_2 \times w_j \leq w_j \leq {}^3/_2 \times w_j$ randomly
      Select the best room from K rooms with the minimum penalty
      Assign the entity j to the best room
    end while

4.3 Send the Employed Bee to the Food Sources

In this step, the employed bee operator selects feasible OFA solutions sequentially from the FSM and randomly explores their neighbourhoods using the three neighbourhood structures to produce a new set of neighbouring solutions (i.e. food sources). The neighbourhood structures utilized by employed bee are:

1. **Neighbourhood Relocate (NR):** In this neighbourhood, the entity of the allocation x_i' is moved from the current room to another room selected randomly.
2. **Neighbourhood Swap (NS):** the room of the allocation x_i' is swapped with the room of the resource x_j' selected randomly.
3. **Neighbourhood Interchange (NI):** In this neighbouhood, all entities of the two selected rooms are randomly interchanged. For example, entities assigned to room A are randomly reassigned to the room B.

The fitness of each neighbouring food source is calculated, if it is better, then it replaces the original food source food source in FSM. This process is implemented for all solutions in FSM. The detailed algorithm of this process can be found in our previous paper [20].

4.4   Send the Onlooker Bee to the Food Sources

In this step, the information of exploited food sources (i.e. OFA solutions) are shared with onlookers by the employed bees. Subsequently, onlooker bees based on shared information (i.e. proportional selection probability shown in Eq. (6)) decides to exploits the food sources randomly using the set of three neighbourhood structures discussed above. Note that the solution with highest probability has higher chanced of being chosen and perturbed to its neighbourhood using the same strategy of the employed bee. The fitness cost of the new food source is evaluated and if better, then the current food source is replaced by the new neighbouring one.

$$p_j = \frac{f(\boldsymbol{x}_j)}{\sum_{k=1}^{SN} f(\boldsymbol{x}_k)}.$$

Note that the $\sum_{i=1}^{SN} p_i$ is unity.

4.5   Send the Scout to Search for Possible New Food Sources

The scout bee as a colony explorer commences its operation once a solution is abandoned by the onlooker i.e. if the improvement of a solution in the FSM has been stopped for certain number of iterations as determined by *limit*. Then a new solution is randomly generated by the scout bee to replaces the abandoned one in FSM. Furthermore, the ABC algorithm memorizes the best food source x_{best} in FSM.

4.6   Stopping Condition

The search process of Steps 3 to 5 are repeated until a stop criterion is met as originally determined by the MCN value.

5   Computational Experiments, Results and Discussions

The proposed ABC for the OFA is coded in Microsoft Visual Basic.NET on Windows 8 platform on Intel core(TM) i3-4005u CPU @1.70 GHz and 4 GB RAM and the results all instances are obtained within computational time of 149 s. The performance of the proposed method is tested using datasets established by University of Nottingham, and University of Wolverhampton. The characteristics of these datasets is provided in Table 1.

5.1   Experimental Design

In this section, experimental design showing the influence of parameters on the performance of the proposed ABC is carried out. A series of experiments comprising 9 convergence cases of different settings of ABC parameters that are utilized for tackling the OFA problem. Table 2 shows the different experimental cases and values employed to study those two ABC parameters: SN and Limit.

Table 1. Characteristics of OFA dataset

Instances	NOTT 1		NOTT 1A		NOTT 1B		NOTT 1C		NOTT 1D		NOTT 1E		Wolver 1	
Entities	158		142		104		94		56		86		115	
Rooms	131		115		77		94		56		59		115	
Constraints	H	S	H	S	H	S	H	S	H	S	H	S	H	S
Allocation	0	35	0	30	0	9	0	35	0	9	0	26	0	0
Same room	0	20	0	15	0	20	0	0	0	0	0	30	0	0
Not sharing	100	0	90	0	34	0	46	0	46	0	38	0	115	0
Adjacency	5	15	5	10	3	6	3	15	3	6	1	5	0	0
Grouped by	0	10	0	7	0	64	0	37	0	24	0	0	0	0
Away from	6	14	0	12	0	0	0	8	0	0	4	6	0	0

Table 2. ABC experimental cases

Cases	Case 1	Case 2	Case 3	Case 4	Case 5	Case 6	Case 7	Case 8	Case 9
SN	10	10	10	20	20	20	30	30	30
Limit	100	500	1000	100	500	1000	100	500	1000

5.2 Experimental Results

Experimental results of 9 convergence cases defined previously are summarized in Table 3 for the datasets under consideration. Note that the numbers in the table refers to objective cost value (lowest is best). For each instance of both datasets, the best, ave and worst result of 10 runs are recorded and MCN is fixed at 10000 cycle. The best result among all cases in each instance is highlighted in bold. Similarly, Fig. 1 shows the boxplots that illustrate the distribution of solution quality for all experimental cases for the NOTT 1 instances. It can be seen from Fig. 1 that the gaps between the best, average, and worst solution qualities are very close, which demonstrates that the proposed ABC has the good capability of exploring the search.

As shown in Table 3, it can be seen that increase in SN has significant impact on the performance of the ABC for the OFA, however as SN increases from 20 to 30, the difference is not much. For example, Case 5 with SN = 20 achieved best results in five instances. This is followed by Cases 6 and 8 where both achieved best results in three instances. It is worthy to mention that all convergence cases achieved new results in the smallest instance (i.e. WOLVER 1). Apparently, Case 4 is better than remaining cases where obtained best results in two instances. Summarily, Case 5 with SN = 20 and Limit = 500 has the potential to enhance the search capability and therefore obtained impressive results.

5.3 Comparison with Other Approaches

The best experimental results obtained by the ABC using the both datasets are compared with other techniques that worked on the same datasets. These techniques are IPM-OFA [4], OFA-MP [4], HSA-OFA [9], SA-OFA [21], and

Table 3. Experimental results of the ABC convergence cases

Instance		Case 1	Case 2	Case 3	Case 4	Case 5	Case 6	Case 7	Case 8	Case 9
NOTT 1	Best	428.35	470.70	428.30	445.95	**425.50**	447.25	474.50	439.30	462.15
	Avg	491.74	508.96	465.24	487.06	484.34	469.02	519.65	490.39	490.88
	Worst	555.50	604.85	503.40	531.20	545.65	488.70	575.75	524.15	547.55
NOTT 1A	Best	482.75	464.50	471.55	502.15	**437.05**	471.15	474.40	466.60	453.75
	Avg	532.48	486.43	509.31	531.58	480.38	510.60	499.16	501.95	489.07
	Worst	564.55	522.70	536.75	567.35	520.00	536.95	510.40	574.30	518.40
NOTT 1B	Best	387.50	389.35	403.25	372.35	**356.60**	378.60	410.50	376.00	386.15
	Avg	409.57	408.56	412.35	416.30	413.24	377.26	425.18	394.22	395.06
	Worst	429.65	432.50	423.15	451.35	409.70	400.00	438.65	409.80	428.75
NOTT 1C	Best	360.05	349.20	339.20	365.70	**324.20**	349.95	344.20	329.20	368.55
	Avg	384.99	383.72	373.26	389.71	391.60	342.71	366.41	348.63	371.18
	Worst	427.75	418.05	398.55	408.60	420.75	361.80	418.65	378.55	388.55
NOTT 1D	Best	360.00	349.15	344.15	370.00	340.65	**334.15**	354.15	**334.15**	337.65
	Avg	373.05	359.06	358.53	379.76	354.01	349.02	370.94	352.32	345.15
	Worst	385.00	370.00	373.50	393.50	363.50	359.15	382.90	370.00	355.65
NOTT 1E	Best	152.20	157.70	157.70	**147.70**	157.70	**147.70**	152.20	**147.70**	157.70
	Avg	166.70	171.70	162.60	162.60	161.70	158.60	161.50	157.50	164.60
	Worst	188.20	197.70	172.20	167.70	167.70	167.70	177.70	167.70	177.70
WOLVER 1	Best	**634.19**	**634.19**	**634.19**	**634.19**	**634.19**	**634.19**	**634.19**	**634.19**	**634.19**
	Avg	634.19	634.19	634.19	634.19	634.19	634.19	634.19	634.19	634.19
	Worst	634.19	634.19	634.19	634.19	634.19	634.19	634.19	634.19	634.19

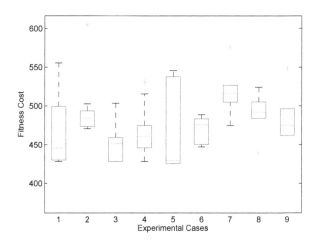

Fig. 1. Boxplot showing the effect of varying parameters on ABC algorithm

HMHPB [21]. Note that the best results obtained by the different methods are presented in bold. Interestingly, the performance ABC is better than other existing methods by achieving high quality solutions in both datasets, where the ABC obtained new results in two instances (i.e. NOTT 1 and NOTT 1E) of the University of Nottingham dataset and had comparable performance in the remaining instances of the dataset. Similarly, the proposed ABC achieved best result in one instance of the University of Wolverhampton (i.e. WOLVER 1) as achieved by IMP-OFA and HSA-OFA methods (Table 4).

Table 4. The best results achieved by the ABC and other comparative methods

Instance	ABC	IPM-OFA	OFA-MP	HSA-OFA	SA-OFA	HMHPB
NOTT 1	**425.50**	-	-	539.35	543.70	482.20
NOTT 1A	437.05	**378.88**	-	-	-	-
NOTT 1B	356.60	246.18	**243.28**	-	-	417.10
NOTT 1C	324.20	**305.73**	**305.73**	-	-	315.40
NOTT 1D	334.15	202.70	202.73	**200.10**	-	-
NOTT 1E	**147.70**	177.70	177.70	-	-	
WOLVER 1	**634.19**	634.20	**634.19**	**634.19**	-	-

6   Conclusion

In this paper an Artificial Bee Colony Algorithm (ABC) is presented for Office-space allocation (OFA) using the datasets published by University of Notting-ham and University of Wolvehampton respectively. The initial feasible solutions are generated by the proposed techniques using Peckish heuristic strategies [19]. ABC as an iterative improvement algorithm optimize the food sources with aid of three operators: employed, onlooker and scout bees. The best food source generated at each cycle is memorized. This search process is repeated until a maximum cycle number (MCN) is achieved. Experimental results produced by the proposed technique are compared with those generated by other existing techniques, which shows the promising performance of the techniques. Interestingly, the proposed ABC produced two new results one best result and comparable results on the remaining instance from the dataset of the University of Notting-ham. Further improvement could be tailored towards enhancing the exploitation capability of the proposed ABC when employed to solve the OFA. Therefore, our future work will focus on the enhance of this technique.

References

1. McCollum, B.: A perspective on bridging the gap between theory and practice in university timetabling. In: Burke, E.K., Rudová, H. (eds.) PATAT 2006. LNCS, vol. 3867, pp. 3–23. Springer, Heidelberg (2007). doi:10.1007/978-3-540-77345-0_1
2. Ritzman, L., Bradford, J., Jacobs, R.: A multiple objective approach to space planning for academic facilities. Manag. Sci. **25**(9), 895–906 (1979)
3. Benjamin, C.O., Ehie, I.C., Omurtag, Y.: Planning facilities at the university of missouri-rolla. Interfaces **22**(4), 95–105 (1992)
4. Ülker, Ö., Landa-Silva, D.: A 0/1 integer programming model for the office space allocation problem. Electron. Notes Discrete Math. **36**, 575–582 (2010)
5. Burke, E.K., Cowling, P., Landa Silva, J.D., McCollum, B.: Three methods to automate the space allocation process in UK universities. In: Burke, E., Erben, W. (eds.) PATAT 2000. LNCS, vol. 2079, pp. 254–273. Springer, Heidelberg (2001). doi:10.1007/3-540-44629-X_16

6. Lopes, R., Girimonte, D.: The office-space-allocation problem in strongly hierarchized organizations. In: Cowling, P., Merz, P. (eds.) EvoCOP 2010. LNCS, vol. 6022, pp. 143–153. Springer, Heidelberg (2010). doi:10.1007/978-3-642-12139-5_13
7. Kirkpatrick, S., Gelatt, C.D., Vecchi, M.P., et al.: Optimization by simulated annealing. Science **220**(4598), 671–680 (1983)
8. Ülker, Ö., Landa-Silva, D.: Evolutionary local search for solving the office space allocation problem. In: 2012 IEEE Congress on Evolutionary Computation (CEC), pp. 1–8. IEEE (2012)
9. Awadallah, M.A., Khader, A.T., Al-Betar, M.A., Woon, P.C.: Office-space-allocation problem using harmony search algorithm. In: Neural Information Processing, Springer 365–374(2012)
10. Burke, E.K., Silva, J.D.L., Soubeiga, E.: Multi-objective hyper-heuristic approaches for space allocation and timetabling. In: Ibaraki, T., Nonobe, K., Yagiura, M. (eds.) Metaheuristics: Progress as Real Problem Solvers, vol. 32, pp. 129–158. Springer US, New York (2005). doi:10.1007/0-387-25383-1_6
11. Burke, E., Cowling, P., Silva, J.L.: Hybrid population-based metaheuristic approaches for the space allocation problem. In: Proceedings of the 2001 Congress on Evolutionary Computation, 2001, vol. 1, 232–239. IEEE (2001)
12. Burke, E., Cowling, P., Landa Silva, J., Petrovic, S.: Combining hybrid metaheuristics and populations for the multiobjective optimisation of space allocation problems. In: Proceedings of the 2001 Genetic and Evolutionary Computation Conference (GECCO 2001), 1252–1259 (2001)
13. Bolaji, A.L., Khader, A.T., Al-Betar, M.A., Awadallah, M.A.: University course timetabling using hybridized artificial bee colony with hill climbing optimizer. J. Comput. Sci. **5**(5), 809–818 (2014)
14. Bolaji, A.L., Khader, A.T., Al-Betar, M.A., Awadallah, M.A.: A hybrid nature-inspired artificial bee colony algorithm for uncapacitated examination timetabling problems. J. Intell. Syst. **24**(1), 37–54 (2015)
15. Awadallah, M.A., Bolaji, A.L., Al-Betar, M.A.: A hybrid artificial bee colony for a nurse rostering problem. Appl. Soft Comput. **35**, 726–739 (2015)
16. Karaboga, D.: An idea based on honey bee swarm for numerical optimization. Technical report TR06, Erciyes University Press, Erciyes (2005)
17. Bolaji, A.L., Khader, A.T., Al-Betar, M.A., Awadallah, M.A.: Artificial bee colony, its variants and applications: a survey. J. Theor. Appl. Inform. Technol. (JATIT) **47**(2), 434–459 (2013)
18. Karaboga, D., Gorkemli, B., Ozturk, C., Karaboga, N.: A comprehensive survey: artificial bee colony (ABC) algorithm and applications. Artif. Intell. Rev. **42**, 21–57 (2012)
19. Corne, D., Ross, P.: Peckish initialisation strategies for evolutionary timetabling. In: Burke, E., Ross, P. (eds.) PATAT 1995. LNCS, vol. 1153, pp. 227–240. Springer, Heidelberg (1996). doi:10.1007/3-540-61794-9_62
20. Bolaji, A.L., Khader, A.T., Al-Betar, M.A., Awadallah, M.A.: Artificial bee colony algorithm for solving educational timetabling problems. Int. J. Natural Comput. Res. **3**(2), 1–21 (2012)
21. Landa-Silva, D., Burke, E.K.: Asynchronous cooperative local search for the office-space-allocation problem. INFORMS J. Comput. **19**(4), 575–587 (2007)

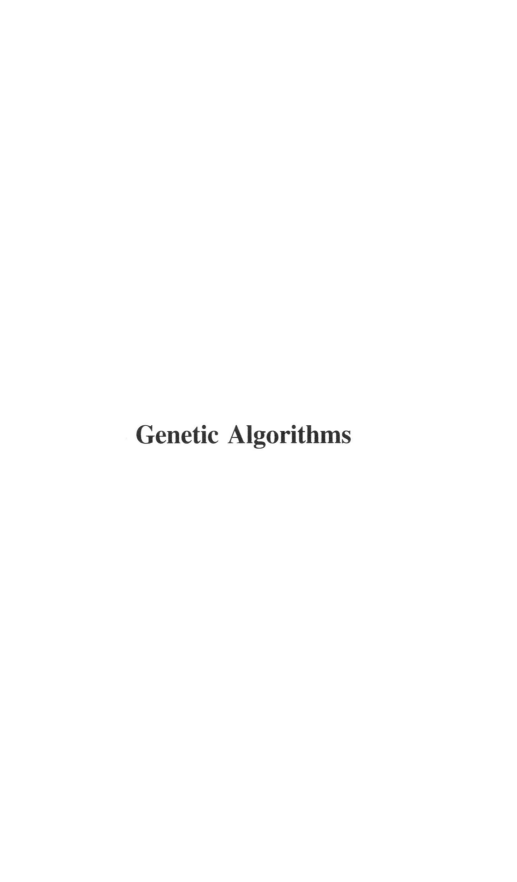

Genetic Algorithms

Enhancing Exploration and Exploitation of NSGA-II with GP and PDL

Peter David Shannon[1,2(✉)], Chrystopher L. Nehaniv[2], and Somnuk Phon-Amnuaisuk[1]

[1] School of Computing and Informatics, Universiti of Teknologi Brunei, Bandar Seri Begawan, Brunei
{Peter.Shannon, Somnuk.Phonamnuaisuk}@utb.edu.bn
[2] School of Computer Science, University of Hertfordshire, Hatfield, UK
C.L.Nehaniv@herts.ac.uk

Abstract. In this paper, we show that NSGA-II can be applied to GP and the Process Description Language (PDL) and describe two modifications to NSGA-II. The first modification removes individuals which have the same behaviour from GP populations. It selects for de-duplication by taking the result of each objective fitness function together to make a comparison. NSGA-II is designed to expand its Pareto front of solutions by favouring individuals who have the highest or lowest value (boundary points) in a front, for any objective. The second modification enhances exploitation by preferring individuals who occupy an extreme position for most objective fitness functions. The results show, for the first time, that NSGA-II can be used with PDL and GP to successfully solve a robot control problem and that the suggested modifications offer significant improvements over an algorithm used previously with GP and PDL and unmodified NSGA-II for our test problem.

Keywords: Genetic programming · Process description language · Exploration and exploitation · NSGA-II

1 Introduction

Non-dominated Sorting Genetic Algorithm II (NSGA-II) is one of the most commonly used evolutionary algorithms (EA); we are using it as a basis for further research and to replace a framework used previously [1]. Here we present some of our findings which may be of general interest to the community on using NSGA-II with Process Definition Language (PDL)—a formal representation of behavioural robot control programs—and Genetic Programming (GP).

Previous work using GP and PDL [1] used a generational model and a form of aggregated fitness function with different stages of learning (Staged Learning). Shannon and Nehaniv showed that it was possible to use GP and PDL to solve a small robot controller problem: to parallel park a car—a more sophisticated version of Koza's backup tractor-trailer truck problem [2]. Here we adapt NSGA-II to replace the EA used previously and, while we present different experiments, the simulation and test problem is otherwise similar.

© Springer International Publishing AG 2017
Y. Tan et al. (Eds.): ICSI 2017, Part I, LNCS 10385, pp. 349–361, 2017.
DOI: 10.1007/978-3-319-61824-1_38

PDL is used here as part of an investigation into exploiting its innate modularity and if it is generally a good representation to use with GP robot control problems. It is highly amenable to analysis, i.e., the effect of each submodule in a PDL program can be determined and assessed over a program's lifetime and even used as a historical account of activities [3, 4]. Such submodule analysis is much more difficult with typical tree or graph representations. We do not report here on our progress exploiting these properties, this report is about foundational work: creating an improved framework for automatically creating PDL programs with GP based on NSGA-II and two types of modification we found interesting.

We modified NSGA-II in two ways: firstly, phenotypic de-duplication of the population, and secondly, introducing a preference for individuals who occupy a greater number of single objective boundary points. GA and GP are very similar and NSGA-II can be use with GP without altering core concepts. However, an issue we find with GP is its inherently redundant representation: one program can be represented in an infinite number of ways. Enhancing exploration of the search of GA by removing duplicates from the population has been attempted several times [5–7]. The strategies employed normally operate on individuals' genotype to find similarity, often use hamming distance to compare GA bit-strings. The relative simplicity and efficiency of comparing hamming distance makes a lot of sense for GA but an analogous syntax tree comparison for GP is a lot less compelling. Not only is the operation is more complex but more importantly the veracity of the comparison is less clear. Instead we compare the phenotype of individuals, by creating n-tuples of ordered SO fitness values and compare those. Clearly with this method there is no way to differentiate between two individuals who have similar structure and similar behaviour with other individuals who have very different structure but similar behaviour. We do, however, sidestep very difficult problems, such as accounting for non-functional or equivalent but different code in our individuals, during comparison.

In order to search the solution space more thoroughly, when choosing between two candidate individuals who are in the same Pareto front, NSGA-II will always prefer individuals which, for any single objective, scored highest or lowest: these individuals are the boundary points. We modified the algorithm to differentiate between the number of boundary points individuals occupy. NSGA-II would not differentiate: when comparing two individuals α, who occupies 1 point, and δ, who occupies 2 points, neither is preferred; with our modification δ is preferred.

The first contribution this paper makes is to show that search can be enhanced by removing duplicate phenotypic individuals and preferring individuals who occupy more than one objective boundary points (see Sect. 3.2.) Previous experiments on de-duplication of the population with GA and GP have used the genotype as the basis of comparison, we use its phenotype (see Sect. 3.1, Canonical Non-dominated Sorting.) The second contribution is demonstrating GP and PDL can be successfully used in conjunction with NSGA-II.

In the rest the paper we will: describe PDL, its use with GP and our test problem; the adjustments to NSGA-II just mentioned; the results of the experiments; observations and analysis; and conclude by assessing this work and suggest improvements we might make to this work in future.

2 Problem Formulation

Steels, describes PDL as: "Control programs viewed as dynamical systems which establish a continuous relation between the time varying data coming out of sensors and a stream of values going to the actuators." [8] This report describes a slightly simplified form of PDL as used with GP before [1]. Typically, each actuator is associated with a quantity; the quantity's value is used to set the value of the actuator. We are using PDL to search for robot control programs which can parallel park a car; the robot controllers manipulate two quantities: steering and velocity. Each behaviour in our implementation of PDL is a small syntax tree, e.g., for a very simple behaviour, such as 1 + 5, the root would be an addition operator with two operands, 1 and 5. The relationship between quantities, behaviours, operators, terminals and actuators is illustrated in Fig. 1.

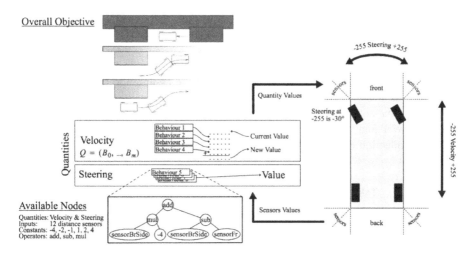

Fig. 1. Input, output and structure of a PDL program using the experiment presented here as an example.

A PDL program P could be formally defined as a tuple of quantities where quantity Q could be partially defined as a tuple of behaviours:

$$P = (Q_0, \ldots, Q_n). \tag{1}$$

$$Q = (B_0, \ldots, B_m). \tag{2}$$

Each quantity also has a value, which changes each time-step.

The value of Q, v_Q, can be determined by summing the changes in v_Q for all previous time-steps. If t stands for the current time-step and t-1 the previous time-step we can define v_Q as:

$$v_Q = \sum_{i=0}^{t-1} \delta(Q_i). \tag{3}$$

Here $\delta(Q_i)$ is the change in v_Q from time-step i-1 to time-step i. $\delta(Q)$ is determined by evaluating of each of Q's behaviours and summing the results. Each of Q's behaviour, $B_{0\ldots m}$, is a syntax tree and is evaluated (as a computer program); accessing sensor information, which depends on the current state of the simulation to determine its influence (which is a number):

$$\delta(Q) = \sum_{j=0}^{m} eval(B_j). \tag{4}$$

2.1 Representation of a PDL Genetic Program

Most discussions of genetic operators are based on the assumption that programs are represented as trees. In this case programs are represented as several collections of syntax trees, one collection for each quantity. Typical GP operators are used on syntax trees (crossover and mutation). We augment these with operators which work on a quantity's collection of trees (Table 1).

Table 1. Parameters for selecting a reproduction method.

Method	Rate
Crossover	25%
Crossover-copy (copies a whole syntax tree)	25%
Mutation	15%
Mutation-add (adds a randomly created syntax tree)	5%
Mutation-subtract (deletes a randomly selected syntax tree)	30%

Syntax tree crossover and mutation (as well as the object storing the syntax tree) are implemented using functions from DEAP [9]: they are what is normally thought of as typical GP operators. A behaviour can be copied from one quantity to another, or removed, without any concern that the resulting PDL program will crash. This feature facilitates the new forms of crossover and mutation. These new operators are designed to afford manipulations at a modular level. In this way, we might consider each behaviour to be a module and each quantity a super-module. Normally, a programmer would not consider a single line of code as a module but that is what we mean, in effect, when we refer to modularity with PDL.

```
Steering
  add(-2.0, sensorFl)
  add(mul(sensorFrSide, -4.0), mul(sensorBrSide, sensorFr))
  sub(-4.0, add(sensorBrBack, -2.0))
  add(sensorBrSide, velocity)
  add(-2.0, sensorFl)
  add(-2.0, sensorBlSide)
Velocity
  mul(-1.0, sensorFrFront)
  mul( sub(sub(4.0, sensorFlSide), add(velocity, sensorFr)),
    sub(add(sensorBrBack, velocity), add(sensorBl, sensorBl)))
  sub(sensorFr, velocity)
  mul(
    add(
      mul(mul(2.0, 0.0), mul(velocity, sensorFr)),
      sub(mul(-2.0, sensorBrBack), sub(-1.0, velocity))),
    mul(
      add(sub(sensorBlSide, sensorFlSide), add(sensorBr, velocity)),
      sub(add(sensorBl, sensorFrFront), mul(sensorBrBack, -4.0)))))
  sub(
    add(
      sub(sub(sensorFrFront, sensorFrSide), add(1, sensorBr)),
      add(add(sensorFrSide, -1.0), mul(-1.0, steering))),
    sub(
      add(mul(-2.0, velocity), add(-1.0, -4.0)),
      mul(add(1, velocity), mul(sensorFrFront, sensorBlSide))))
  mul(sensorBl, sensorFr)
  mul(-1.0, sensorFrFront)
```

Fig. 2. An example of a controller program found by the EA.

An example PDL program is shown in Fig. 2. The two quantities, steering and velocity, are shown one above the other with their respective behaviours listed underneath, each on a separate line and indented when a syntax tree is too long or complicated to put on one line.

2.2 Designing Objective Functions

The problem is represented in a 3D virtual environment and with a physics simulation. Our objective is to find a program to parallel park a robot toy car between two blocks. The assessment of success is broken down into four tests carried out after a fixed number of time-steps:

1. Distance to the centre of the parking spot from the centre of the car.
2. Rotation off parallel to the blocks.
3. Disruption to the blocks. The blocks (mass and friction parameters) are configured so that the slightest collision between them and the car will cause the blocks to drift away.
4. Movement. It should be stationary.

The experiments use fitness functions with very little *a priori knowledge* [10] incorporated—it makes no assumptions about how the problem should be solved—and is assessed at the end of the simulation after a fixed number of time-steps. The terminal set is semantically very low level, limited to distances sensors, the values of quantities, numerical constants and the operators multiply, add and subtract.

The fitness objective 1–4 are not treated equally. Each criterion is important but, in the case of parking a real car, if it was parallel to the curb, did not crash into anything but stopped ten feet from the parking spot, it would not be badly parked it would be stopped in the middle of the road. Objective 2–4 only have meaning if in the context of the parking spot, failure to get to the parking spot renders them almost meaningless. The only input to the robot is the distance sensors so the controllers are highly context sensitive. To reflect this objective 2–4 were compounded (multiplied) with objective 1. Our implementation of NSGA-II works with objective 2–4 which depend on the first objective.

Aggregate Fitness. While the EA we are using, NSGA-II, is Pareto based we find it useful when analysing results to refer to an aggregated score so each individual can be compared with only one number. For example, in Sect. 2.3, the aggregated fitness score is used to reject bad initial controllers for the first generation. The aggregate fitness is calculated by taking the mean of the normalised objective fitness scores.

2.3   Initialisation, Termination and Parameters

When a run is started, the initial generation is populated with new individuals. One to four behaviours are created for each quantity of an individual, the number of behaviours is selected at random. New behaviours are created using the using the ramped half-and-half method and are limited to a maximum depth of 4. Each new individual is evaluated and if a minimum normalised aggregated fitness of 0.1 is not achieved it is discarded and another individual will be created to replace it until the initial generation is complete.

3   Modified NSGA-II

A common approach to scoring *multi-objective* (MO) problems is to create a weighted aggregate fitness. Instead, NSGA-II selects individuals using two mechanisms. The first mechanism is ranking them using a Pareto analysis, assigning each individual to a front. The second mechanism is a density estimation tool, termed crowding distance. NSGA-II's algorithm for finalising the group of individuals which are selected from for binary tournaments to decide which reproduce, is illustrated in Fig. 3. Deb et al. [12] described how individuals are scored and compared in terms of two properties assigned to each individual. We add a third property using the same notation, $i_{canonical}$:

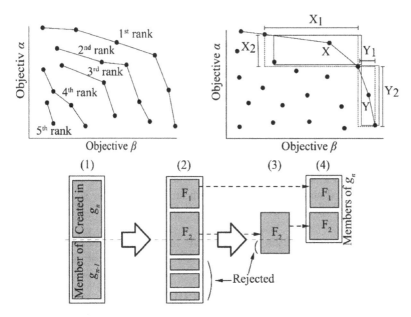

Fig. 3. Top pane: left, ranks of Pareto fronts; right, crowding distances for points X and Y. Bottom pane: non-dominated crowd distance sorting decides which individuals become members of the set which might reproduce, the dotted line indicates how large the final set of members will be; (1) prospective members are a combination of members of the previous generation (g_{n-1}) and individuals created in this generation (g_n), (2) individuals are sorted into non-dominated, crowd distance sorted, fronts, (3) fronts are added to the final selection until the selection is full, (4) if too many individuals would be added, the final front is truncated. Adapted from [11, 12].

- Rank (i_{rank}) is the front assigned to an individual during non-dominated sorting. The rank value, of individual i, is referred to as i_{rank}.
- Crowding distance $(i_{distance})$ signifies how unique an individual is in a front, the larger the value the more unusual its MO fitness should be in its front. The crowding distance value, of individual i, is referred to as $i_{distance}$.
- Canonical $(i_{canonical})$ indicates if an individual represents a unique class of behaviour within certain bounds. If several individuals have the same outcome or behaviour (their MO fitness is the same) then only one will be marked as canonical. It is assigned during non-dominated sorting and in one variation we tested indicates that no other individuals which have been added to a front have the same MO fitness scores. Comparison of MO fitness is achieved by creating an n-tuple of normalised objective fitness scores rounded to 3 decimal places and then compared position by position. This attribute does not indicate the individual is strictly canonical, merely it is the only individual which has a specific behaviour (indicated by n-tuple of fitness characteristics) which we wish to propagate. We are declaring it to be the true representation of a specific behaviour without considering how natural that representation is.

Rank, crowding distance and, the new property we add, canonical are used to compare individuals in binary tournaments for reproduction.

3.1   Modifications to Enhancing Exploration

We explored two variations on the same modification. The first, CNS, removes duplicates from the entire population. The second, CCO, only remove duplicates from a front and is a much less aggressive application of the same idea.

Canonical Non-dominated Sorting (CNS). We modified NSGA-II to adjust the non-dominated sorting procedure to enhance diversity and exploration. This was done by recording the MO fitness characteristics of each individual which are found to be non-dominated as they are added to a front. If no individuals have been added to any front, in this generation, with the same MO characteristics then it is considered canonical. In this variation, only individuals which are canonical will be added to a front. Non-canonical individuals will remain in the last front once sorting has finished and the entire front may be discarded.

This method of sorting ensures that only individuals which have different behaviour will reproduce, we expect this will increase the time spent exploring and delaying convergence of the population.

Canonical Crowd-comparison Operator (CCO). This variation on CNS and adjusts the crowd comparison operator to consult $i_{canonical}$ during tournament selection, rather than discarding non-canonical individuals during sorting. It de-duplicates the population less aggressively than CNS. CCO first consults a lookup for the two individuals being compared and if one the individuals is canonical will favour it.

The procedure to create the lookup is as follows:

- Flatten the non-dominated fronts into a list, in which order is maintained, so that individuals in the first rank are first, the second rank are second and so on.
- Iterate over each individual in the list: if the MO fitness characteristics of an individual do not occur in the set of preceding individuals—in the flattened list— then set $i_{canonical}$ to true, otherwise false.

$$i_{canonical} = if \notin P_f. \tag{5}$$

A more formally definition for $i_{canonical}$ is given in (5). The modified crowd-comparison operator is shown in (6). Partial ordering is now determined by first checking if either, but not both, i or j is considered canonical. Otherwise it is the same NSGA-II's crowd comparison operator.

$$i \prec_n j \, if((i_{canonical} \, and \, not(j_{canonical})) \\ or((i_{canonical} = j_{canonical}) \\ and((i_{rank} < j_{rank}) \\ or((i_{rank} = j_{rank}) \, and \\ (i_{distance} > j_{distance}))) \tag{6}$$

3.2 Enhancing Exploitation Using a Modified to Promotion of Crowding Distance Outliers (MCDO)

The modifications presented so far will encourage more diversity and, we expect, the EA to explore the solution space further. This modification is intended increase exploitation instead and can be used with the other proposed modification or otherwise unmodified NSGA-II. With unmodified NSGA-II, individuals in fronts which are in the most extreme positions are awarded infinite crowding distance; the EA should respond to this by broadening its search, which is desirable. This method of increasing explorations has interesting features, some of which are negative:

- An individual with an extreme characteristic may be part of a cluster of similar individuals and not truly represent interestingly unusual class of solution. We simply do not look to see with NSGA-II.
- Each time an objective is added two more individuals in a front are awarded an infinite crowding distance and its power to discern valuable solutions diminishes. If two individuals are selected for a tournament from the same front—even with only 3 objectives and therefore 6 extreme positions to occupy and chances to be awarded infinite crowding distance—discerning unusual solutions will be unlikely.
- A positive feature of awarding crowding distance this way is that a MO problems may have some relatively easy to achieve objectives and some hard to achieve objectives. In this way, an individual distinguished in a difficult to achieve objective will be recognised equally to one who is distinguished in an easier objective.

To address some of these point, this modification simply awards a share of infinity based on how many objectives there are and how many extreme positions an individual occupies. In practice ∞ is represented by a large number. The MCDO shared crowding distance for individual i, which occupies at least one extreme position, is defined as ∞ divided by the number of objectives n, multiplied by the number of extreme positions it holds, e_i.

$$c_i = \frac{\infty}{n} e_i. \tag{7}$$

4 Experiment Design

We designed two experiments to test how well the 2 modifications, in various combinations, perform compared to two controls NSGA-II and Staged Learning from [1]. In total 7 combinations were tested: CNS, CCO, NSGA-II & MCDO, CNS & MCDO, CCO & MCDO, NSGA-II and Staged Learning. The experiments were designed to test, and hopefully reject, our null hypothesis: *each of the different combinations of modified NSGA-II find good controller programs equally well as NSGA-II and Staged Learning do.*

Population size has a complex relationship with how long is spent exploring and how long is spent exploiting what has been found. The modifications we designed should affect exploitation and exploration, so we choose to run two experiments, one

Table 2. Parameters for the two experiments; each experiment has 7 samples—one for each algorithm—and a sample size of 40.

	Individuals per generation	Generations per run	Run repetitions
Experiment 1	25	90	40
Experiment 2	100	45	40

with more of individuals per generation and one with a less per generation. The smaller populations were allowed to continue longer as convergence would be slower (Table 2).

We used ANOVA to determine if the algorithms produced significantly different results and, in the case that they did, a post-hoc pairwise t-test, with the Holm method to correct for multiple testing, a significance level of $p < 0.05$ and Family Wise Error Rate (FWER) = 0.05. Each run contributed its best individual (the one with the highest aggregate fitness) to the sample for analysis; the sample size is therefore 40 (equal to the number of runs).

5 Results

For Experiment 1 we found that there is a significant difference in how well the different versions of NSGA-II find good programs at the $p < 0.05$ level [$F(6, 273) = 5.302$, $p = 0.0$]. The post-hoc test revealed that every algorithm was significantly better than Staged Learning; due to limited space, we omit the details but this outcome can be clearly seen in Fig. 4. No other significant differences in the group samples were found.

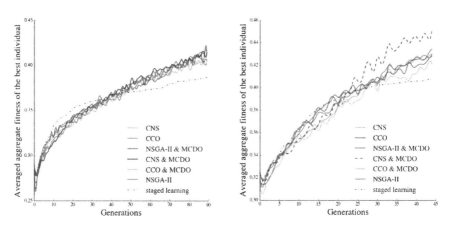

Fig. 4. The averaged performance of each group at each generation. Left, in Experiment 1, Staged Learning is significantly worse in later generations. Right, in Experiment 2, again staged learning is significantly worse. More interestingly a clear trend develops, in later generations, of a widening gap in performance between CNS & MCDO and NSGA-II. The graphs are un-smoothed except by linear interpolation between points.

Table 3. Means and standard deviations for experiment 2.

	μ	Standard deviation
CNS	0.437	0.037
CCO	0.435	0.055
NSGA-II & MCDO	0.442	0.05
CNS & MCDO	0.458	0.046
CCO & MCDO	0.425	0.033
NSGA-II	0.44	0.05
Staged Learning	0.407	0.049

Table 4. A subset of pairwise comparisons, where p < 0.2, for experiment 2.

		Statistic	P-value	Reject
CNS	Staged Learning	4.710	0.006	True
CNS & MCDO	CNS	2.023	0.05	False
CCO & MCDO	CNS	−1.361	0.181	False
CCO	CNS & MCDO	−1.862	0.07	False
CCO	Staged Learning	4.261	0.0001	True
NSGA-II & MCDO	Staged Learning	4.332	0.0001	True
NSGA-II & MCDO	CNS & MCDO	−1.555	0.128	False
CNS & MCDO	CCO & MCDO	3.947	0.0003	True
CNS & MCDO	NSGA-II	0.091	0.091	False
CNS & MCDO	Staged Learning	5.330	0	True
CCO & MCDO	Staged Learning	3.506	0.001	True
NSGA-II	Staged Learning	3.805	0.008	True

For Experiment 2 we found once more that there is a significant difference in how well the different versions of NSGA-II find good programs at the $p < 0.05$ level [$F_{(6, 273)} = 8.661$, $p = 0.0$]. The post-hoc test revealed that every algorithm significantly outperformed Staged Learning for this experiment too. More interestingly, while we cannot note significantly improved results demonstrated for CNS & MCDO over NSGA-II based on the post-hoc analysis alone, the graph in Fig. 4 shows a clear and consistently widening performance gap, generation on generation, in the second half of the experiment. Considering the graph together with the weaker but positive confidence level $p < 0.1$ found with the pairwise t test, we reject the null hypothesis: CNS & MCDO is better at finding good controller programs than NSGA-II and Staged Learning.

6   Conclusion and Future Work

Our work improves on the initial study in [1] using a widely accepted and well understood EA. We show, for the first time, that NSGA-II can be used with PDL and GP to successfully solve a robot control problem. We also found that CNS & MCDO are superior to unmodified NSGA-II for this problem. CNS can be applied to NSGA-II used with GP and MCDO more generally with any form of GA. However, we only noted significant improvement when both were used together and only demonstrated this for on a single problem. Further, we note earlier generations appear to be negatively affected by these approaches. However, performance was only noticeably worse in the earliest generations, before the 15th generation in Experiment 2 (see Fig. 4.) This, we surmise, is an effect de-duplication has on slowing convergence of the population, increasing the time spent exploring the solution space.

The logically affect of CNS is to increase diversity and for MCDO to favour individuals who are best (and worst) at the most objectives. These can be thought of as enhancing exploration and exploitation. However, here we have only shown that they improve searching when used together, not how their interactions work. An analysis of the genealogical origins and mechanisms of creation of the best individuals as a run matures may shed some light on these workings and how these modifications might be used most effectively.

Our next research priority is to create new testbed problems and gauge the generality of using GP and PDL to search for robot controllers.

References

1. Shannon, P., Nehaniv, C.L.: Evolving robot controllers in PDL using genetic programming. In: IEEE SSCI 2011-Symposium Series on Computational Intelligence-IEEE ALIFE 2011: 2011 IEEE Symposium on Artificial Life, pp. 92–99. IEEE, Paris (2011)
2. Koza, J.R.: A genetic approach to finding a controller to back up a tractor-trailer truck. In: American Control Conference, pp. 2307–2311. IEEE, Chicago (1992)
3. Nehaniv, C., Dautenhahn, K.: Embodiment and memories-algebras of time and history for autobiographic agents. In: Trappl, R. (ed.) Cybernetics and Systems, vol. 2, pp. 651–656. Austrian Society for Cybernetic Studies, Vienna, Austria (1998)
4. Nehaniv, C.L., Dautenhahn, K.: Semigroup expansions for autobiographic agents. In: First Symposium on Algebra, Languages and Computation, pp. 77–84. University of Aizu, Japan (1998)
5. Mauldin, M.L.: Maintaining diversity in genetic search. In: AAAI, pp. 247–250 (1984)
6. Shimodaira, H.: DCGA: a diversity control oriented genetic algorithm. In: Proceedings of the Ninth IEEE International Conference on Tools with Artificial Intelligence 1997, pp. 367–374. IEEE (1997)
7. Sangkawelert, N., Chaiyaratana, N.: Diversity control in a multi-objective genetic algorithm. In: The 2003 Congress on Evolutionary Computation, CEC 2003, vol. 4, pp. 2704–2711. IEEE (2003)
8. Steels, L.: Mathematical analysis of behavior systems. In: Proceedings From Perception to Action Conference 1994, pp. 88–95. IEEE (1994)
9. Fortin, F.A., Rainville, F.M.D., Gardner, M.A., Parizeau, M., Gagné, C.: DEAP: evolutionary algorithms made easy. J. Mach. Learn. Res. 13, 2171–2175 (2012)
10. Nelson, A.L., Barlow, G.J., Doitsidis, L.: Fitness functions in evolutionary robotics: a survey and analysis. Robot. Auton. Syst. 57(4), 345–370 (2009). Elsevier

11. Olson-Manning, C.F., Wagner, M.R., Mitchell-Olds, T.: Adaptive evolution: evaluating empirical support for theoretical predictions. Nature Rev. Genet. **13**(12), 867–877 (2012). Nature Publishing Group
12. Deb, K., Pratap, A., Agarwal, S., Meyarivan, T.A.M.T.: A fast and elitist multiobjective genetic algorithm: NSGA-II. IEEE Trans. Evol. Comput. **6**(2), 182–197 (2002). IEEE

A Novel Strategy to Control Population Diversity and Convergence for Genetic Algorithm

Dongyang Li[1], Weian Guo[2(⊠)], Yanfen Mao[2], Lei Wang[1], and Qidi Wu[1]

[1] School of Infomation and Electronical Information,
Tongji University, Shanghai 201804, China
[2] Sino-German College of Applied Sciences, Tongji University,
Shanghai 201804, China
guoweian@163.com

Abstract. Genetic algorithm (GA), an efficient evolutionary algorithm inspired from the science of genetics, attracts the worldwide attention for several decades. This paper tries to strengthen the search ability of the population in GA in the way of improving the distance among individuals by introducing a new solution updating strategy based on the theory of Cooperative Game. The simulation is done using fourteen benchmark functions, and the results demonstrate that this modified genetic algorithm works efficiently.

Keywords: GA · Solution distance · Cooperative game · Solution updating strategy

1 Introduction

Genetic algorithm (GA), a powerful heuristic algorithm which is established by mimicking the processes of inheritance evolution of biological, was first proposed by Holland in 1975 [1]. Since then, a number of studies have been done and prove that GA exhibits excellent optimization performance, not only on various kinds of numerical benchmarks, but also on high-dimensional and multi-objective optimization problems. In original GA, the algorithm has already outperformed many other evolutionary algorithms for some benchmarks. However, it shows a poor performance on the multi-peak functions usually. In another word, GA tends to a fast convergence to a local minimum. Therefore, it need to be improved in many of its aspects.

Since the advent of GA, many studies have been done to improve the searching ability of GA by improving the diversity of its population. For the multi-mode resource-constrained project scheduling problem, Peteghem and Vanhoucke utilize a strategy of two separate populations to enhance the seeking ability of GA and achieve good results [2]. When solving the problem of flexible job-shop scheduling, Zhang not only modifies the mutation and crossover methods, but also proposes a new population initialization strategy based on the global selection and local selection [3]. Vidal and Crainic equip GA with adaptive diversity management by introducing the efficient local

© Springer International Publishing AG 2017
Y. Tan et al. (Eds.): ICSI 2017, Part I, LNCS 10385, pp. 362–369, 2017.
DOI: 10.1007/978-3-319-61824-1_39

search based improvement procedures and diversity management approaches [4]. In the study of Castro and Soma [5], the set of feasible solutions should be made a division into regions in order to diversify the search that is used on a GA variation. An adaptive method of maintaining variable population size is proposed by Arabas and Michalewicz [6], which is used to improve the diversity of GA. Liu and Zhong [7] turn out that the population of GA which is initialized by ACO instead of randomly generated will strengthen the global search ability of GA and can be used for QoS-aware service composition problem perfectly. In [8], Tsai and Huang make an improvement of GA by consisting of two parallel EGAs along with a migration operator, their results demonstrate that this modified GA takes advantages of better population diversity, inhibiting premature convergence, and keeping parallelism in comparison with conventional Gas. In addition, Wang [9] uses ACO to improve GA to solve the problem of Job-Shop Scheduling. Tang and Pan [10] extend the search ability and convergence of original GA by incorporating an infeasible solution repairing procedure and a local optimization procedure.

Obviously, all these studies mentioned above are just trying to improving the search ability of GA by increasing the diversity of the population. The reason is that in original GA, the regular design that higher crossover probability and lower mutation probability will lead the population to poor diversity and premature convergence. In this paper, in order to improve the searching ability of GA and optimize the population of GA, we propose a new solution updating strategy to improve the search ability of GA by utilizing the theory of Cooperative Game to increase the distance between individuals in the population.

The rest of this paper is organized as follows. In Sect. 2, a brief overview of GA will be introduced. In Sect. 3, the new solution updating strategy will be presented. The simulation results will be compared and discussed in Sect. 4. We end this paper with conclusion in Sect. 5.

2   Genetic Algorithm

2.1   The Principle of Genetic Algorithm

Genetic algorithm (GA) is just established based on the evolution mechanism of natural selection and natural inheritance. The optimization problem is mapped to an evolution process, and it is considered to be the natural environment, a potential solution is seen as a chromosome which is called as an individual, and a set of solutions is defined as a species which we call it population. First, GA generates a population randomly and arranges the fitness values for individuals based on the fitness function set according to the optimization problem. Then a new population will be produced by selecting enough individuals to crosses with others and mutates. Note that, an individual with a higher fitness value should be selected with a higher probability than an individual with a lower fitness value. Through repeating this operation, potential solutions will be more and more close to the optimal solution of the problem, and the approximate optimal solution can be obtained finally.

2.2   Implementation Steps of GA

After a large efforts of scholars, a number of different genetic algorithms have been proposed. The GA proposed by Holland in 1975 is seen as the standard GA (SGA). The main steps are as follows:

(1) Encoding. Encoding is a bridge between problems and algorithms.
(2) Generate an initial population.
(3) Fitness evaluation. A fitness function ought to be defined reasonably to reflect the adaption ability to the environment of individuals.
(4) Selection operator. Selection operator is a reflection of "survival of the fittest". Generally, the probability of being selected of an individual is proportional to its fitness.
(5) Crossover operator. Crossover operator is an important operator of GA. The single point crossover strategy is adopted in SGA.
(6) Mutation operator. A random bit of the chromosome may be changed based on a mutation probability Pm.
(7) Ending criteria. An ending principle must be set to terminate GA.

Where $MaxGen$ is the maximum of evolution algebra, N is the size of the Population while Pc and Pm is the crossover probability and mutation probability respectively.

The ability of dealing with many complex problem of GA is strong has been proved for many years. However, due to its very small probability of mutation, its global exploration too poor to solve the problem with a large solution space or with a number of local minimums. Hence the main drawback of GA is tend to a premature convergence.

3   An Improved GA Based on Cooperative Game

As discussed above, GA does not have a suitable strategy to solve the problem with a large solution space or too many local minimums. In this section, we will provide an brief introduction of Cooperative Game and propose a new solution updating strategy to improve GA.

3.1   Cooperative Game

Cooperative game is first proposed by Rowland [11] in 1944. The focus of Cooperative game is how people share the benefits through cooperation, that is, the issue of benefits distribution. Cooperation and compromise is the core of cooperative game. The purpose of Cooperative game is the interests of both sides in the game have increased, or at least one's benefits of the increase, while the other one is not compromised, and the interests of the whole society will get an increase finally. More details about cooperative game can be found in [11].

A hot topic in GA is the outstanding genes in the parents must be remained to the offspring, while each individual of the offspring also should be better to keep a proper distance to others during the progression of evolution. This can be seen as a cooperative game between individuals of which the goal is to make sure the population has both proper distance among solutions and convergence. Based on this idea, this paper proposed a new solution updating strategy to increase the distance among individuals without breaking its convergence.

3.2   CGGA

As mentioned above, we should give a new solution updating strategy to increase the distance among solutions of population but cannot make GA be a random search algorithm. And this strategy based on cooperative game is introduced as follows.

(1) The *population* is seen as a game alliance S and the benefits of it is defined as $v(S)$, which is evaluated by the rate of increase shown as formula (1).

$$v(S) = w_1 \cdot \frac{D(P_1) - D(P)}{D(P)} + \frac{sum(fit) - sum(fit_1)}{sum(fit)} \tag{1}$$

Where w_1 is the weight of distance between solutions which is calculated by formula (2), *fit* is the sum of fitness of the parents, while fit_1 is the sum of fitness of the offspring, $D(P)$ and $D(P_1)$ represent the distance among individuals of parents and offspring respectively, which is defined as formula (3).

$$w_1 = \frac{popsize}{sum(fit)} \cdot N(MinFit)/20 \tag{2}$$

$$D(P) = Var(Fit) \tag{3}$$

Here, *N(MinFit)* is the number of consecutive equal values of *MinFit* up to the current generation, and $Var(\cdot)$ is the variance function which is used to evaluate the variance of *Fit* of P.

(2) Every two individuals selected to do the crossover and mutation operation will begin a cooperative game and aim to increase $v(S)$ without harming the benefits of each other. Note that a margin is set as 1.2 to avoid a too low probability of evolving. The individual *individual* that has executed the crossover and mutation will be noted as *individual_i*. The benefits of an individual is defined as in formula (4).

$$v(i) = w_1 \cdot \frac{D(P_{1c}) - D(P)}{D(P)} + \frac{sum(fit_i) - sum(fit_{ic})}{sum(fit_i)} \tag{4}$$

Where P_{ic} represents the population that replaces individual Pi with using *individual_i*, fit_i and fit_{ic} represent the fitness of *individual_i* and *individual_{ic}*.

In summary, the idea of the new solution updating strategy is every two selected individuals ready to evolve should try to improve $v(S)$ and cannot reduce the interests of its partner.

4 Simulation and Discussion

In this section, we make a simulation experiment using fourteen benchmark functions. And the results will be shown in this part.

The fourteen benchmark function which is shown in Table 1 are selected to test the modified GA.

Table 1. Benchmark functions.

Function	Name	Domain
F1	Ackley	$-30 \leq x \leq 30$
F2	Fletcher	$-\pi \leq x \leq \pi$
F3	Griewank	$-600 \leq x \leq 600$
F4	Penalty1	$-50 \leq x \leq 50$
F5	Penalty2	$-50 \leq x \leq 50$
F6	Quartic	$-1.28 \leq x \leq 1.28$
F7	Rastrigin	$-5.12 \leq x \leq 5.12$
F8	Rosenbrock	$-2.0481 \leq x \leq 2.048$
F9	Schwefel	$-65.536 \leq x \leq 65.536$
F10	Schwefel2	$-100 \leq x \leq 100$
F11	Schwefel3	$-10 \leq x \leq 10$
F12	Schwefel4	$-512 \leq x \leq 512$
F13	Sphere	$-5.12 \leq x \leq 5.12$
F14	Step	$-200 < x < 200$

More details of these benchmarks can be find in [12–14]. Five other popular EAs are also used to compare with the new algorithm CGGA. They are GA [15, 16], PBIL [17], standard PSO [18–20], ACO [21, 22], and ES [23].

Additionally, for GA and CGGA, we chose the roulette wheel selection, single point crossover, the crossover probability = 1, and the mutation probability = 0.01. And the parameters for other EAs can be found in [24].

In this simulation, the mean optimization results and the best optimization results of each EAs are shown in Tables 2 and 3.

For all of these EAs, Table 2 shows that the mean optimization results of CGGA are better than other EAs in all cases, which is benefit from the healthy distance among individuals in the population. Besides, the best optimization results, which are shown in Table 3, indicate that in the process of the evolution, CGGA performs better in searching the minimum than others in most cases, just for Fletcher, Penalty1, and

Table 2. Mean optimization results

Function	GA	CGGA	PBIL	PSO	ACO	ES
F1	6.96E+00	**2.14E+00**	1.86E+01	1.44E+01	5.89E+00	8.87E+00
F2	2.20E+04	**1.46E+04**	3.44E+05	3.33E+05	4.30E+05	5.52E+05
F3	1.28E+00	**1.10E+00**	1.71E+02	5.17E+01	1.15E+00	9.80E+01
F4	5.40E−02	**2.12E−02**	3.92E+07	1.17E+06	9.07E+07	3.95E+07
F5	6.04E−01	**2.11E−01**	1.18E+08	8.14E+06	1.78E+08	1.13E+08
F6	1.93E−06	**1.16E−06**	1.09E+01	1.49E+00	3.40E−03	1.47E+01
F7	2.63E+01	**2.75E+00**	2.00E+02	1.37E+02	7.39E+01	2.18E+02
F8	3.88E+01	**3.71E+01**	1.29E+03	3.43E+02	8.41E+02	2.48E+03
F9	4.50E+01	**2.54E+01**	4.28E+03	3.63E+03	3.16E+01	2.94E+03
F10	2.90E+03	**6.85E+02**	9.82E+03	5.41E+03	1.96E+03	1.29E+04
F11	3.89E+00	**8.95E−01**	5.37E+01	2.73E+01	2.20E+01	7.54E+01
F12	1.86E+01	**5.97E+00**	5.90E+01	3.52E+01	2.02E+01	2.10E+01
F13	9.23E−02	**8.06E−03**	5.31E+01	1.52E+01	8.02E+00	6.77E+01
F14	1.30E+01	**5.05E+00**	1.91E+04	5.67E+03	2.11E+01	1.59E+04

Table 3. Best optimization results

Function	GA	CGGA	PBIL	PSO	ACO	ES
F1	3.50E+00	**9.66E−01**	1.67E+01	1.09E+01	4.05E+00	6.66E+00
F2	**1.83E+03**	4.61E+03	1.84E+05	1.87E+05	2.62E+05	1.97E+05
F3	1.06E+00	**1.05E+00**	1.02E+02	2.63E+01	1.05E+00	5.97E+01
F4	**4.10E−03**	4.27E−03	4.35E+06	5.25E+04	2.29E+01	2.66E+06
F5	1.23E−01	**1.11E−01**	2.40E+07	1.10E+06	1.35E−32	1.91E+07
F6	4.50E−07	**7.00E−08**	4.68E+00	3.80E−01	1.30E−03	4.87E+00
F7	1.14E+01	**0.00E+00**	1.74E+02	1.09E+02	5.18E+01	1.70E+02
F8	9.84E+00	**7.43E+00**	4.04E+02	1.72E+02	3.94E+02	1.03E+03
F9	**3.83E+00**	6.63E+00	3.14E+03	2.68E+03	9.85E+00	2.30E+03
F10	1.01E+03	**2.20E+02**	5.59E+03	3.09E+03	4.62E+02	7.33E+03
F11	1.40E+00	**4.00E−01**	4.39E+01	1.48E+01	5.70E+00	4.80E+01
F12	7.00E+00	**3.00E+00**	4.26E+01	2.75E+01	6.40E+00	1.29E+01
F13	1.01E−02	**0.00E+00**	2.94E+01	8.64E+00	2.73E+00	3.59E+01
F14	**1.00E+00**	**1.00E+00**	1.27E+04	2.70E+03	9.00E+01	1.02E+04

Schwefel, the best optimization results of CGGA are the second best. It shows that CGGA works more effective than all the other EAs in finding out the minima of the fourteen benchmarks mentioned above.

5 Conclusions

In this paper, we improve GA based on the theory of cooperative game, and propose a new evolutionary algorithm CGGA. With considering the distance among individuals of population, it uses the cooperation mechanism of cooperative game to equip the solutions with a strong ability of searching. In this way, the distance among individuals in the population increased. A set of fourteen benchmarks is used to test and compare the proposed algorithm with several popular EAs. The results demonstrate the proposed algorithm is effective to deal with optimization problems.

References

1. Holland, J.H.: Adaptation in Natural and Artificial Systems, pp. 211–247. MIT Press, Cambridge (1975)
2. Peteghem, V.V., Vanhoucke, M.: A genetic algorithm for the preemptive and non-preemptive multi-mode resource-constrained project scheduling problem. Eur. J. Oper. Res. **201**(2), 409–418 (2010)
3. Zhang, G., Gao, L., Shi, Y.: An effective genetic algorithm for the flexible job-shop scheduling problem. ACM Trans. Intell. Syst. Technol. **38**(4), 3563–3573 (2011)
4. Vidal, T., Crainic, T.G., Gendreau, M., et al.: A hybrid genetic algorithm with adaptive diversity management for a large class of vehicle routing problems with time-windows. Comput. Oper. Res. **40**(40), 475–489 (2013)
5. Castro, J.L.D., Soma, N.Y.: A constructive hybrid genetic algorithm for the flowshop scheduling problem. Int. J. Comput. Sci. Netw. Secur. **9**, 219–223 (2013)
6. Arabas, J., Michalewicz, Z., Mulawka, J.: GAVaPS - a genetic algorithm with varying population size. In: Proceedings of the First IEEE Conference on Evolutionary Computation, IEEE World Congress on Computational Intelligence, vol. 1, pp. 73–78. IEEE Xplore (1994)
7. Liu, H., Zhong, F., Ouyang, B., et al.: An approach for QoS-aware web service composition based on improved genetic algorithm. In: International Conference on Web Information Systems and Mining, pp. 123–128. IEEE Xplore (2010)
8. Tsai, C.C., Huang, H.C., Chan, C.K.: Parallel elite genetic algorithm and its application to global path planning for autonomous robot navigation. IEEE Trans. Industr. Electron. **58**(10), 4813–4821 (2011)
9. Wang, L., Haikun, T., Yu, G.: A hybrid genetic algorithm for job-shop scheduling problem, pp. 271–274 (2015)
10. Tang, M., Pan, S.: A hybrid genetic algorithm for the energy-efficient virtual machine placement problem in data centers. Neural Process. Lett. **41**(2), 211–221 (2015)
11. Rowland, E.: Theory of Games and Economic Behavior. Theory of games and economic behavior, pp. 2–14. Princeton University Press (1944)
12. Back, T.: Evolutionary Algorithms in Theory and Pratice. Oxford University Press, Oxford (1996)
13. Yao, X., Liu, Y., Lin, G.: Evolutionary programming made faster. IEEE Trans. Evol. Comput. **3**, 82–102 (1999)
14. Cai, Z., Wang, Y.: A multiobjective optimization-based evolutionary algorithm for constrained optimization. IEEE Trans. Evol. Comput. **10**, 658–675 (2006)
15. Michalewicz, Z.: Genetic Algorithms + Data Structures = Evolution Programs. Springer, New York (1992)
16. Goldberg, D.: Genetic Algorithms in Search, Optimization, and Machine Learning. Addison-Wesley, Reading (1989)
17. Parmee, I.: Evolutionary and Adaptive Computing in Engineering Design. Springer, New York (2001)

18. Onwubolu, G., Babu, B.: New Optimization Techniques in Engineering. Springer, Berlin (2004)
19. Eberhart, R., Shi, Y., Kennedy, J.: Swarm Intelligence. Morgan Kaufmann, San Mateo (2001)
20. Clerc, M.: Particle Swarm Optimization. ISTE Publishing, Amsterdam (2006)
21. Dorigo, M., Stutzle, T.: Ant Colony Optimization. MIT Press, Cambridge (2004)
22. Dorigo, M., Gambardella, L., Middendorf, M., Stutzle, T.: Special section on 'ant colony optimization'. IEEE Trans. Evol. Comput. **6**(4), 317–365 (2002)
23. Guo, W., Wang, L., Ge, S.S., Ren, H., Mao, Y.: Drift analysis of mutation operations for biogeography-based optimization. Soft Comput. **19**, 1881–1892 (2015)
24. Li, D., Wang, L., et al.: Particle swarm optimization-based solution updating strategy for biogeography-based optimization. In: IEEE Congress on Evolutionary Computation (CEC), pp. 455–459 (2016)

Consecutive Meals Planning by Using Permutation GA: Evaluation Function Proposal for Measuring Appearance Order of Meal's Characteristics

Tomoko Kashima[1(✉)], Yukiko Orito[2], and Hiroshi Someya[3]

[1] Kindai University, 1, Takaya Umenobe, Higashi-Hiroshima 739-2116, Japan
kashima@hiro.kindai.ac.jp
[2] Hiroshima University, 1-2-1, Kagamiyama, Higashi-Hiroshima 739-8525, Japan
[3] Tokai University, 4-1-1, Kitakaname, Hiratsuka 259-1292, Japan

Abstract. The consecutive meals planning is a combinatorial optimization problem that determines a meals plan in one period consisting of consecutive days. This paper proposes an evaluation function using a moving entropy for this problem. The function measures the appearance order of meal's characteristics on the plan. In the numerical experiments, we apply a permutation GA to the problem. We show that our meals plan is a good solution with large variation of appearance order of meal's characteristics for the consecutive meals planning.

1 Introduction

The meals planning is to determine the food combination which minimizes or maximizes a given objective function. Many researchers have dealt with caloric intake minimization problem, cost minimization problem, and other problems for their purposes [1–5]. Each of their problems is viewed as an integer or a mixed-integer linear programming problem. Hence, we can find the high accurate solution by using a simplex method, an interior point method, or some of other optimization methods.

On the other hand, an educational institution such as elementary school and preschool provides the school meal every daily lunch. In each institution, the individual nutrition managers freely decide every meal on plan by his/her own policy under some strict rules such as an amount of calories and cost. They may consecutively provide the meals in similar characteristics though each institution has to provide various meals in different characteristics such as food style, ingredient, and cooking method, for growth of an infant or a child. We have to consider the variation of appearance order of meal's characteristics on a plan as well as the variation of meal's characteristics.

In this paper, we call such a problem "**consecutive meals planning**" and propose an evaluation function which measures the variation of appearance order of meal's characteristics on a plan by using information entropies on moving intervals.

© Springer International Publishing AG 2017
Y. Tan et al. (Eds.): ICSI 2017, Part I, LNCS 10385, pp. 370–377, 2017.
DOI: 10.1007/978-3-319-61824-1_40

We apply a permutation GA to the consecutive meals planning and then show that our evaluation function works well for maximizing the variation of appearance order of meal's all characteristics in the numerical experiments.

2   Consecutive Meals Planning

Generally, one meal consists of one main dish and some side dishes. But we deal with only a main dish as a meal and call it "**Meal Class**" in this paper. In other word, we determine the main dishes plan on one period consisting of consecutive days.

We first define the following notations.

t: Time point t $(t = 1, \cdots, T)$ on a period.

$x_{i,j}$: Item j of Characteristics i on Meal Class. That is $x_{i,j}$ $(i = 1, \cdots, I$, $j = 1, \cdots, J|_i)$.

$\mathbf{X}_m$: Meal Class which is categorized by difference of characteristics. That is $\mathbf{X}_m = \{x_{1,j}, \cdots, x_{I,j}\}$ $(m = 1, \cdots, M)$.

$\mathbf{Y}(t)$: Meal Class provided at t.

$\mathbf{Y}$: Meals plan consisting of T variables. That is a solution, $\mathbf{Y} = \{\mathbf{Y}(1), \cdots, \mathbf{Y}(T)\}$.

$\alpha_{i,j}(t)$: Appearance/non-appearance of meal's characteristics $x_{i,j}$ on meals plan $\mathbf{Y}(t)$. That is represented by a binary variable 1/0.

$P_{i,j}$: Frequency rate of meal's characteristics provided on an interval.

In this paper, we classify the individual meals into Meal Classes according to its characteristics. Let i $(i = 1, \cdots, I)$ be the kind of meal's characteristics. Let $j|_i$ $(j = 1, \cdots, J|_i)$ be the kind of items of Characteristics i. Let $x_{i,j}$ be the meal's characteristics for Item j of Characteristics i. The Meal Class $\mathbf{X}_m$, combination of meal's characteristics from $x_{1,j}$ to $x_{I,j}$, is defined as,

$$\mathbf{X}_m = \{x_{1,j}, \cdots, x_{I,j}\}, \quad (m = 1, \cdots, M), \tag{1}$$

$$m = \sum_{i=1}^{I-1} \left\{ (j|_i - 1) \prod_{k=i+1}^{I} J|_k \right\} + j|_I, \quad (j|_i \in \{1, \cdots, J|_i\}),$$

$$M = \prod_{i=1}^{I} J|_i.$$

For example, we classify the meal's characteristics into three categories; "food style", "ingredient", and "cooking method." In addition, we classify the category of food style into "Japanese", "Western", and "Chinese", the category of ingredient into "Meat", "Fish", and "Egg", and the category of cooking method into "Simmer", "Fry", "Saute", and "Deep-fry", respectively. Using these settings, Meal Class is defined as,

$$\mathbf{X}_m = \{x_{1,j}, x_{2,j}, x_{3,j}\}, \quad (m = 1, \cdots, M), \tag{2}$$

$$x_{1,j} \in \begin{cases} \text{Japanese} & (j = 1) \\ \text{Western} & (j = 2) \\ \text{Chinese} & (j = 3) \end{cases}, \qquad x_{2,j} \in \begin{cases} \text{Meat} & (j = 1) \\ \text{Fish} & (j = 2) \\ \text{Egg} & (j = 3) \end{cases},$$

$$x_{3,j} \in \begin{cases} \text{Simmer} & (j = 1) \\ \text{Fry} & (j = 2) \\ \text{Saute} & (j = 3) \\ \text{Deep-fry} & (j = 4) \end{cases},$$

$$m = \sum_{i=1}^{2} \left\{ (j|_i - 1) \prod_{k=i+1}^{3} J|_k \right\} + j|_3, \qquad (j|_i \in \{1, \cdots, J|_i\}),$$

$$M = \prod_{i=1}^{3} J|_i = 3 \cdot 3 \cdot 4 = 36.$$

From the practical viewpoints, the educational institution such as elementary school and preschool has many opportunities to provide special meals on event days. For example, Japanese preschool provides a chicken dish at Christmas. In a word, the Meal Class provided at an event date is fixed on the meals plan. Hence, we assume that the event date which provides a special meal is randomly given in the period for the consecutive meals planning.

Here, let $t^{(e)}$ be the time point for event date and let $\mathbf{X}^{(e)}$ ($\mathbf{X}^{(e)} \in \{\mathbf{X}_1 \cdots, \mathbf{X}_M\}$) be the Meal Class provided at $t^{(e)}$. The meals plan for the period consisting of T time points is defined as,

$$\mathbf{Y} = \{\mathbf{Y}(1), \cdots, \mathbf{Y}(T)\}, \tag{3}$$
$$\mathbf{Y}(t) = \{y_{1,j}, \cdots, y_{I,j}\}|_t, \qquad (t = 1, \cdots, T),$$
$$s.t.\ \mathbf{Y}(t) \in \{\mathbf{X}_1, \cdots, \mathbf{X}_M\}, \quad \mathbf{Y}(t^{(e)}) = \mathbf{X}^{(e)}, \quad \mathbf{X}^{(e)} \in \{\mathbf{X}_1 \cdots, \mathbf{X}_M\}.$$

The meals plan defined by Eq. (3) is the solution for the consecutive meals planning.

3   Evaluation Function

It is well known that the information entropy is an index which measures the predictability/unpredictability of information content and is defined as the probability of observing event. We propose an evaluation function by using the information entropy.

We first define the frequency rate of each item of Characteristics i in Meal Class, $P_{i,j}$, as the probability of observing the item on a period. The information entropy of Characteristics i is maximized if all items of Characteristics i are observed the same number in the period. However, the information entropy has a problem that it cannot measure the appearance order which the event occurs in the period though it can measure the number of occurrences.

In this paper, however, the evaluation function has to measure the variation of appearance order of meal's characteristics on meals plan. For measuring the appearance order, we divide one period consisting of T time points into the short intervals consisting of the number of items for each meal's characteristics. As described in Eq. (1), the number of items of Characteristics i is $J|_i$. Here, the first interval consists of $J|_i$ time points from $t = 1$ to $t = J|_i$. Moreover, the interval moves from the first one to the last one every one time point on the period. The last interval consists of $J|_i$ time points from $t = T - J|_i + 1$ to $t = T$. Hence, we can divide the period into $T - J|_i + 1$ intervals for Characteristics i. To maximize the information entropy on each of all intervals means to maximize the variation of appearance order of meal's characteristics because the number of time points on each interval is same as the number of items of Characteristics i.

For Characteristics i, the average of entropies in all intervals is represented as H_i and the theoretical value is represented as H_i^*. We define the evaluation function, "**the total entropy of all meal's characteristics**", as follows.

$$H = \sum_{i=1}^{I} \frac{H_i}{H_i^*}, \tag{4}$$

$$H_i = -\frac{1}{T - J|_i + 1} \sum_{t=1}^{T-J|_i+1} \sum_{j=1}^{J|_i} P_{i,j}(t) \log P_{i,j}(t), \qquad (i = 1, \cdots, I),$$

$$P_{i,j}(t) = \frac{1}{J|_i} \sum_{p=0}^{J|_i-1} \alpha_{i,j}(t+p), \qquad (t = 1, \cdots, T - J|_i + 1, \ j = 1, \cdots, J|_i),$$

$$\alpha_{i,j}(t) = \begin{cases} 1 \ (y_{i,j}(t) = x_{i,j}) \\ 0 \ (y_{i,j}(t) \neq x_{i,j}) \end{cases}, \qquad (t = 1, \cdots, T).$$

4   Permutation GA

In this paper, we apply a permutation GA to optimize this problem. The permutation GA expresses a round trip route most simply.

4.1   Genetic Representation and Fitness Value

In the genetic representation of the permutation GA, we define the individual which is expressed as different from the solution.

First, the solution $\mathbf{Y}$, given by Eq. (3), is defined as the phenotype of permutation GA.

We replace the solution to the individual. We define the individual as the following genotype consisting of the integer sequence.

$$\mathbf{v} = \{v(1), \cdots, v(T)\}, \qquad v(t) = w_s \bmod M, \tag{5}$$
$$w_s \in \{1, \cdots, T\}, \qquad w_{s_1} \neq w_{s_2}, \qquad (s_1, s_2 = 1, \cdots, T, \ s_1 \neq s_2).$$

Hence, at time point t, the permutation GA has two kinds of variables, $\mathbf{Y}(t)$ of solution and $v(t)$ of individual.

We employ the total entropy of all meal's characteristics given by Eq. (4) as the fitness value. The permutation GA tries to find the optimal solution which maximizes the fitness value.

4.2 Genetic Operations

Each operation of the permutation GA is designed as follows.

Note that the Meal Classes provided at event date are randomly given in the period and are fixed on the procedure of permutation GA. They are not operated by the crossover and the mutation.

1. Initial State
   Meal Classes provided at event date are fixed in advance. On the first generation, the permutation GA generates N_p individuals in the initial parents' population. Each individual consists of T variables which are randomly selected from the integer sequence $\{1, \cdots, T\}$ without overlapping.
2. Evaluation and Selection
   The permutation GA applies the elitism and tournament selections to select the individuals to the next population.
3. Crossover
   Let P_c be the crossover rate. The permutation GA makes new $P_c \times N_o$ individuals by using the order crossover [6] for exchanging the partial structure between two individuals according to the order of the sequence.
4. Mutation
   Let P_m be the mutation rate. The permutation GA makes new $P_m \times N_o$ individuals by exchanging two variables selected at random on one individual.
5. Terminate Criterion
   The permutation GA repeats the operations of producing the offspring population and performing the selection until the maximum number of the repetitions is satisfied.

From the last population, we choose one solution whose individual has the highest fitness value of all. This solution is the optimum or quasi-optimum meals plan for our consecutive meals planning.

5 Numerical Experiments

We applied the permutation GA to optimize the consecutive meals planning.

5.1 Experimental Setting and Parameters

In the numerical experiments, the meal's characteristics are categorized into each of food style $x_{1,j}$, ingredient $x_{2,j}$, and cooking method $x_{3,j}$. The Meal Class of experiments is defined by the same condition as Eq. (2).

The setting for the consecutive meals planning is as follows.

- The number of Meal Classes: $M = 36$ (given by Eq. (2)).
- Length of period: $T = 245$ (The number of weekdays in 2016).
- The number of event dates: 24 (Two days per one month).

Hence, the consecutive meals planning is the problem of determining the solution $\mathbf{Y} = \{\mathbf{Y}(1), \cdots, \mathbf{Y}(245)\}$ such that the evaluation function is maximized.

The parameters of permutation GA are set as follows: Parents' Population Size; $N_p = 100$, Offspring's Population Size; $N_o = 200$, Crossover Rate; $P_c = 0.9$, Mutation Rate; $P_m = 0.1$, The maximum number of the repetitions; 1000, Algorithm Run; 10.

5.2 Results and Discussion

In order to demonstrate the power of our evaluation function, the total entropy of all meal's characteristics, we compare it with other two functions. We explain three models employing these three functions as follows.

- **Model 1**
   Model 1 employs our function, the total entropy of all meal's characteristics given by Eq. (4), as the fitness value of permutation GA. In the experiments, we re-define the optimization problem of Model 1.

$$\max H = \sum_{i=1}^{I} \frac{H_i}{H_i^*}. \tag{6}$$

- **Model 2**
   Model 2 employs the function, the entropy of only Meal Classes without meal's characteristics, as the fitness value of permutation GA. In the experiments, we define the optimization problem of Model 2.

$$\max L = \frac{1}{L^*} \left\{ -\frac{1}{T-M+1} \left(\sum_{t=1}^{T-M+1} \sum_{m=1}^{M} P_m(t) \log P_m(t) \right) \right\}, \tag{7}$$

$$P_m(t) = \frac{1}{M} \sum_{p=0}^{M-1} \alpha_m(t+p), \quad (t = 1, \cdots, T-M+1, m = 1, \cdots, M),$$

$$\alpha_m(t) = \begin{cases} 1 \ (\mathbf{Y}(t) = \mathbf{X}_m) \\ 0 \ (\mathbf{Y}(t) \neq \mathbf{X}_m) \end{cases}, \quad (t = 1, \cdots, T),$$

where L^* is the theoretical value of the entropy of Meal Classes.
- **Model 3**
   Model 3 employs the function, the guaranteed minimum of the variation of appearance order of all meal's characteristics, as the fitness value of permutation GA. In the experiments, we define the following maximin problem of Model 3.

$$\max \quad \min \left(\frac{H_1}{H_1^*}, \cdots, \frac{H_I}{H_I^*} \right). \tag{8}$$

As the results of Model 1, the fitness value of the optimum or quasi-optimum solution, H, is shown in Table 1. Note that we show the maximum, the minimum,

Table 1. Results of Model 1 (Fitness value: H given by Eq. (6))

	H	L	H_1	H_2	H_3
Max.	2.6464	0.8704	0.8823	0.8823	0.8904
Min.	2.6056	0.8575	0.8633	0.8650	0.8656
Avg.	2.6274	0.8634	0.8724	0.8747	0.8803
Sd.	0.0132	0.0059	0.0076	0.0063	0.0076

Table 2. Results of Model 2 (Fitness value: L given by Eq. (7))

	H	L	H_1	H_2	H_3
Max.	2.0212	0.9815	0.6806	0.6840	0.7282
Min.	1.9174	0.9774	0.5936	0.6085	0.6519
Avg.	1.9627	0.9796	0.6337	0.6411	0.6879
Sd.	0.0343	0.0011	0.0263	0.0201	0.0206

Table 3. Results of Model 3 (Fitness value: result of maximin problem given by Eq. (8))

	H	L	H_1	H_2	H_3	Maximin
Max.	2.5472	0.8766	0.8494	0.8511	0.8494	0.8488
Min.	2.5136	0.8554	0.8384	0.8390	0.8363	0.8363
Avg.	2.5378	0.8634	0.8456	0.8465	0.8457	0.8449
Sd.	0.0113	0.0076	0.0036	0.0040	0.0042	0.0038

the average, and the standard deviation of the fitness values obtained by the permutation GA 10 times. The entropy of each of meal's characteristics, H_i ($i = 1, 2, 3$), and the entropy of only Meal Classes, L, are also shown in Table 1, respectively.

As the results of Model 2, the fitness value of the optimum or quasi-optimum solution, L, is shown in Table 2. The entropy of each of meal's characteristics, H_i ($i = 1, 2, 3$), and the total entropy of all meals's characteristics, H, are also shown in Table 2, respectively.

As the results of Model 3, the fitness value of the optimum or quasi-optimum solution obtained by the maximin problem is shown in Table 3. The entropy of each of meal's characteristics, H_i ($i = 1, 2, 3$), the total entropy of all meal's characteristics, H, and the entropy of only Meal Classes, L, are also shown in Table 3, respectively.

From Tables 1 and 2, the total entropies of all meal's characteristics, Hs, obtained by Model 1 are higher than those of Model 2. On the other hand, the entropies of only Meal Classes, Ls, obtained by Model 2 are higher than those of Model 1 because Model 2 is the problem which maximizes L. However, entropies of only meal's characteristics, H_1, H_2, and H_3, are smaller than those of Model 1.

In addition, Model 3 is the problem which maximizes the guaranteed minimum of the variation of appearance order of all meal's characteristics. Thus, the entropies of only meal's characteristics, H_1, H_2, and H_3, in Table 3 are the guaranteed minimum for the consecutive meals planning. However, the H_1, H_2, Cand H_3 obtained by Model 1 in Table 1 are larger than those of Model 3 in Table 3.

Therefore, our evaluation function of Model 1, the total entropy of all meal's characteristics H, is the effective function for the consecutive meals planning with large variation of appearance order of multiple meal's characteristics.

6   Conclusion

In the consecutive meals planning, we proposed an evaluation function which measures the variation of appearance order of meal's characteristics. We applied the permutation GA to optimize the optimization problem.

In the numerical experiments, we showed that our evaluation function is effective for the consecutive meals planning with large variation of appearance order of multiple meal's characteristics.

For obtaining the optimum or the higher accurate quasi-optimum solution by using the evolutionary algorithms, however, we need to analyze the details of the landscape of solutions and the search paths of algorithm. This is our future work.

Acknowledgements. This work was supported by JSPS KAKENHI Grant Numbers #25750007 and #15K00339.

References

1. Lancaster, L.M.: The history of the application of mathematical programming to menu planning. EJOR **57**, 339–347 (1992)
2. Darmon, N., Ferguson, E., Briend, A.: Linear and nonlinear programming to optimize the nutrient density of a population's diet: an example based on diets of preschool children in rural Malawi. Am. J. Clin. Nutr. **75**(2), 245–253 (2002)
3. Salookolayi, D.D., Yansari, A.T., Nasseri, S.H.: Application of fuzzy optimization in diet formulation. J. Math. Comput. Sci. **2**(3), 459–468 (2011)
4. Cadenas, J.M., Pelta, D.A., Pelta, H.R., Verdegay, J.L.: Application of fuzzy optimization to diet problems in Argentinean farms. EJOR **158**, 218–228 (2004)
5. Kashima, T., Matsumoto, S., Ishii, H.: Evaluation of menu planning capability based on multi-dimensional 0/1 knapsack problem of nutritional management system. IAENG IJAM **39**(3), IJAM_39_3_04 (2009)
6. Davis, L.: Applying adaptive algorithms to epistatic domains. In: 9th International Joint Conference on Artificial Intelligence, pp. 162–164. Morgan Kaufmann Publishers Inc., San Francisco (1985)

Improving Jaccard Index Using Genetic Algorithms for Collaborative Filtering

Soojung Lee[⊠]

Gyeongin National University of Education,
155 Sammak-ro, Anyang 13910, Korea
sjlee@gin.ac.kr

Abstract. As data sparsity may produce unreliable recommendations in collaborative filtering-based recommender systems, it has been addressed by many researchers in related fields. Jaccard index is regarded as effective when combined with existing similarity measures to relieve data sparsity problem. However, the index only reflects how many items are co-rated by two users, without considering whether their ratings are evaluated similar or not. This paper proposes a novel improvement of Jaccard index, reflecting not only the ratio of co-rated items but also whether the ratings of each co-rated item by two users are both high, medium, or low. A genetic algorithm is employed to find the optimal weights of the levels of evaluations and the optimal boundaries between them. We conducted extensive experiments to find that the proposed index significantly outperforms Jaccard index on moderately sparse to dense datasets, in terms of both prediction and recommendation qualities.

Keywords: Similarity measure · Jaccard coefficient · Collaborative filtering · Recommender system

1   Introduction

A recommender system has received much attention as a useful tool to reduce the work of users when searching for information on Internet. This is because it is usually designed to provide only the information that might be suitable to the users, by filtering out seemingly unnecessary information. Among several types of recommender systems, collaborative filtering (CF) is most well-known as it is successfully utilized in commerce to recommend products that might be preferred by customers. Some of the practical CF systems are GroupLens, Ringo, and Amazon.com [5]. The basic principle of CF systems is to refer to other likeminded users and recommend items which have been highly rated by them. Underlying this principle is surely the assumption that users with similar preferences for the items in the past would also have similar preference for the unseen items.

A main task of CF systems is to find likeminded or similar users, since items with high ratings given by them should be recommended to the current user.

© Springer International Publishing AG 2017
Y. Tan et al. (Eds.): ICSI 2017, Part I, LNCS 10385, pp. 378–385, 2017.
DOI: 10.1007/978-3-319-61824-1_41

Determination of similar users is a critical aspect of the CF system which significantly affects its performance. In literature, similarity calculation is usually made by two kinds of approaches, correlation-based and vector cosine-based [1,5,8,12]. Examples of the former approaches are the Pearson correlation and its variants, constrained Pearson correlation and Spearman rank correlation [5,12]. Cosine-based method treats each user or item as a vector and measures an angle between the vectors.

Despite the popularity of the traditional similarity measures, their disadvantages are often disclosed when the ratings data of the two users are insufficient. This problem, known as *data sparsity*, is inherent to CF systems and may produce unreliable similar users. Drawbacks resulting from this problem associated with traditional similarity measures are well analyzed in [2,11]. While various techniques have been developed to take care of the data sparsity problem, a simpler technique which incorporates Jaccard index into the previous similarity measure draws attention of many researchers [3,4,7,8,11]. Jaccard index is a useful tool to reflect the number of common items rated by two users. It plays an important role in computing more reliable similarity and is reported to contribute to enhancing performance of CF. This paper proposes a new index based on Jaccard index, which utilizes a genetic algorithm. The proposed index reflects not only the ratio of co-rated items but also the levels of their ratings by the two users; that is, whether they are both high, medium, or low. The superiority of the proposed index is demonstrated through extensive experiments using datasets with different characteristics.

2   Related Work

Traditional similarity measures are reported unreliable for CF systems when the ratings data are sparse [2]. As a simple but efficient way to solve this problem, the number of items co-rated by two users is often taken into consideration in calculating similarity. Jamali and Ester introduced a sigmoid function which has the number of common users as input. They combined this sigmoid function with Pearson correlation to define a similarity function [6]. Hence, the resulting similarity decreases exponentially as the number of common users decreases. Ren et al. measured the degree of rating overlap and combined this heuristic factor with Pearson correlation and the cosine similarity [9]. Their new measure is proved to mitigate the sparsity problem through various experiments and to outperform the corresponding traditional measures.

Another approach to handle data sparsity is to incorporate *Jaccard index* [7] into the previous similarity measure. Jaccard index measures the ratio of the number of co-rated items by two users to the total number of items rated by them. It has been succesfully adopted in several studies to compensate for the drawbacks of previous similarity measures which are mainly caused by data sparsity or cold-start users [3,4,8,11,13]. For example, Bobadilla et al. proposed a new similarity measure that combines mean squared differences with Jaccard index [3]. The results state that their approach outperforms Pearson correlation with datasets of short-ranged ratings but not much with a dataset of rather long-ranged ratings such as FilmAffinity. In addition, their experiments only include Pearson correlation for comparison, thus the proposed measure being not fully verified over the other traditional similarity measures.

Saranya et al. presented a weighted combination of Pearson correlation and Jaccard index as similarity measure [11]. They reported that the measure achieved a little improvement in recommendation quality [11]. However, the improvement is very slight compared to Jaccard Uniform Operator Distance [13] and no comparison is made to more popular similarity measures such as Pearson correlation, cosine similarity, or mean squared differences.

Liu et al. also proposed a formulation of similarity measure that encompasses not only the local context information of user ratings, but also the global preference of user behavior. Within the formulation, PIP similarity, Jaccard index, and the mean and variance of the ratings are included in a heuristic manner [8]. They conducted experiments with very sparse datasets of MovieLens and Epinions which have a rating range of 1 to 5. In their experiments, as many as ten state-of-the-art similarity measures are compared with their proposed method. In terms of recommendation qualities, the results show that the cosine similarity performs best when the number of consulting users is few to medium in both datasets, comparable to the best with more consulting users with Epinions, and is defeated only by the adjusted cosine measure with MovieLens with more consulting users. Hence, it is rather out of expectations that such a complex and well-formulated measure proposed by [8] is mostly outperformed by the cosine and the adjusted cosine similarities.

From the above discussion, Jaccard index is known to be used as a popular component comprising a new formulation of similarity measure which was partly successful in improving CF performance using previous similarity measures. This achievement is still notable in that the index considers only relative number of common items rated by two users, without taking the ratings themselves into account. The index is formally defined as follows. Let I_u be the set of items rated by user u and $|I_u|$ be its cardinality. Then Jaccard index between users u and v is

$$Jaccard(u, v) = \frac{|I_u \cap I_v|}{|I_u \cup I_v|}.$$

3   The Proposed Index

Although Jaccard index is designed to take only the number of common ratings, it can be modified to reflect more from the common ratings between two users, as far as their similarity is concerned. For instance, consider two cases: (a) $r_{u,i} = 3$, $r_{u,j} = 3$, $r_{v,i} = 3$, $r_{v,j} = 4$ and (b) $r_{u,i} = 1$, $r_{u,j} = 1$, $r_{v,i} = 1$, $r_{v,j} = 2$, where $r_{u,i}$ is the rating given by user u to item i. Assuming that there are only two common ratings made by u and v and the rating range is [1..5] where 1 is the lowest, it may be inferred that case (b) indicates higher similarity between the users than case (a). Our idea is based on this observation and also on the work by [3]. This work discovered that users tend to give ratings higher than the median and avoid the extreme values. This result implies that two users giving a same extreme rating can be treated as more similar than those giving a more common rating.

Based on the above discussion, our proposed index is designed to be different from Jaccard index in that it considers whether the rating of a common item is normal or extreme. That is, we are interested in how many items are commonly rated with normal or extreme values. Hence, we divide the rating range of the system into three subintervals of low, medium, and high ratings. Within each subinterval Jaccard index is computed separately. Specifically, let L_{bd} and H_{bd} be boundaries of the sub-intervals, where $L_{bd} < H_{bd}$. Then the set of items rated by user u, I_u, is divided into three as follows, based on the rating values assigned by u.

$$I_{L,u} = \{i \in I_u | r_{u,i} \leq L_{bd}\}, \ I_{M,u} = \{i \in I_u | L_{bd} < r_{u,i} < H_{bd}\}, \ I_{H,u} = \{i \in I_u | r_{u,i} \geq H_{bd}\}.$$

Also let us define three types of Jaccard indexes between users u and v as follows.

$$J_L(u,v) = \frac{|I_{L,u} \cap I_{L,v}|}{|I_{L,u} \cup I_{L,v}|}, \ J_M(u,v) = \frac{|I_{M,u} \cap I_{M,v}|}{|I_{M,u} \cup I_{M,v}|}, \ J_H(u,v) = \frac{|I_{H,u} \cap I_{H,v}|}{|I_{H,u} \cup I_{H,v}|}$$

Then our metric, named as J_{LMH}, is calculated as an arithmetic average of the three Jaccard indexes as follows.

$$J_{LMH}(u,v) = \frac{1}{3}(J_L(u,v) + J_M(u,v) + J_H(u,v))$$

Given the above framework, we further refine our index by optimizing the boundaries and the weight portion of each component in J_{LMH}. The final index is presented as

$$J_{LMH,GA}(u,v) = \alpha_L J_L(u,v) + \alpha_M J_M(u,v) + \alpha_H J_H(u,v), \ \ \alpha_L + \alpha_M + \alpha_H = 1,$$

where a genetic algorithm is used to determine the optimal α's and the boundaries, L_{bd} and H_{bd}. Hence, in our genetic algorithm, the set of five weights comprises an individual of the population. We choose MAE (Mean Absolute Error) for the fitness function, a representative metric for measuring prediction quality of collaborative filtering systems. Initially, each weight is randomly created using N_b bits to represent a real number in the range R. The algorithm terminates either when it reaches a given number of generations N_{gens} or when the best fitness of solutions in the population is less than a given threshold f_{th}.

Concerning genetic operators, three typical operators, i.e., selection, crossover, and mutation, are used. At each generation, a solution with the higher fitness is selected with higher proability. Two solutions selected as such are then crossed over with crossover probability $Prob_C$ to produce two new offsprings. Each of these offsprings is to undergo the mutation step where a random bit is flipped with mutation probability $Prob_M$. The three steps of genetic operation are repeated until the number of new offsprings becomes the given number of solutions, PS. We keep the population size PS constant throughout the generations. Table 1 describes the parameters used in our genetic algorithm and their values for experiments.

Table 1. Parameter description and values for the genetic operation

Parameter	Description	Value
N_b	Number of bits composing a gene	10
PS	Population size	60
R	Real range of a gene	[0, 1]
N_{gens}	Number of generations for algorithm termination	20
f_{th}	Fitness threshold (MovieLens/BookCrossing/Jester datasets)	0.65/1.0/2.5
$Prob_C$	Crossover probability	0.85
$Prob_M$	Mutation probability	0.05

Table 2. Characteristics of the datasets

	Matrix size (users × ratings)	Rating scale	Sparsity level
MovieLens	1000 × 3952	1∼5 (integer)	0.9607
Jester	998 × 100	−10∼+10 (real)	0.2936
BookCrossing	1014 × 883	1∼10 (integer)	0.9775

4 Performance Experiments

4.1 Design of Experiments

Our experiments are conducted with the purpose of investigating how much improvement the proposed metric achieves over Jaccard index, when it is used as a sole similarity measure for collaborative systems. Hence, we experimented with three similarity measures, Jaccard index (Jaccard), J_{LMH} (JLMH), and $J_{LMH,GA}$ (JLMHGA). Each experiment result is obtained through the five-fold cross validation [5], where the ratio of training and testing data is set to 80:20.

Among the datasets used in literature, three popular datasets are selected in our experiments, each having very unique characteristics different from the others, as presented in Table 2. Sparsity level represents how sparse the dataset is. It is defined by 1-(total number of ratings/matrix size).

Performance is evaluated based on two well-known standards in related studies, prediction quality and recommendation quality. To examine prediction quality, MAE (Mean Absolute Error) is popularly used, which is the mean difference between the predicted rating of an unrated item and its corresponding real rating. Rating prediction is typically made by referring to ratings of similar users, weighted average of which is used [10]. Hence, determination of most similar users, called the nearest neighbors (topNNs), critically affects the accuracy of prediction, which emphasizes the importance of similarity measure. Recommendation quality is usually measured by precision and recall metrics or their harmonic mean F1 [12]. We presented F1 results only, due to the space constraint.

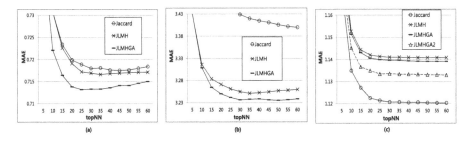

Fig. 1. Comparison of MAE performance: (a) MovieLens, (b) Jester, and (c) Book-Crossing datasets

4.2   Performance Results

Figure 1 shows MAE results with varying number of nearest neighbors (topNN) for the three different datasets. It is observed that the results of Jaccard are either far best or worst, depending on the dataset. It seems the most advantageous when the dataset is very sparse, such as BookCrossing. Although MovieLens is also as sparse as BookCrossing, Jaccard shows the worst result with slight difference on MovieLens, implying that a small difference of data sparsity can cause a significant performance difference in terms of MAE. This observation is convincing because Jaccard yields very worse MAE results on a dense dataset like Jester.

With analogous reasonings as above, JLMH results in far better MAE than Jaccard on Jester, which proves successfulness of our idea of separate application of Jaccard index to sub-ranges of ratings. This achievement is rather moderate on MovieLens and not effective on BookCrossing, all of which seem due to the dataset sparseness. By comparing the MAE results of JLMH and JLMHGA, it can be figured out that how a genetic algorithm impacts on performance. As expected, regardless of the datasets, JLMHGA yields better MAE than JLMH. However, the degree of improvement is almost ignorable on BookCrossing. Since the reason for such results on BookCrossing is presumed as data sparsity, we further experimented with two sub-intervals instead of three used for JLMHGA. This additional experiment, the legend of whose results is JLMHGA2 in the figure, differs from JLMHGA in the number of weights in the chromosome: that is, four weights rather than five. Observe that using only one boundary, JLMHGA2 notably outperforms JLMHGA. This result is because the Jaccard on each of the two sub-intevals is now more meaningful as more common number of ratings should be included in each sub-interval.

Regarding recommendation quality, Fig. 2 depicts F1 results of the measures. Similar to the behavior shown in MAE results, Jaccard leads to worst F1 performance on MovieLens and Jester, whereas it shows competitiveness with the others on BookCrossing. Therefore, the overall results of Jaccard is relatively poor in terms of F1, even though it demonstrates excellency on a very sparse dataset like BookCrossing in terms of MAE. On the contrary, JLMHGA is absolutely superior to the other two on MovieLens, which is also proved through MAE results.

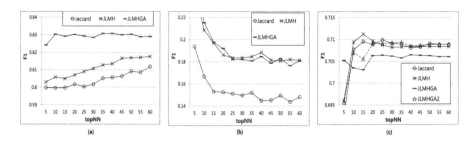

Fig. 2. Comparison of F1 performance: (a) MovieLens, (b) Jester, and (c) BookCrossing datasets

However, its effectiveness exploiting a genetic algorithm is hardly shown on the other two datasets. This rather unexpected behavior of JLMHGA, especially on Jester, is most probably due to the rough criteria for defining preference in measuring precision or recall; note that the items of ratings higher than a given threshold is considered relevant in measuring precision, thus prediction of the exact rating not so important as in MAE. Observe that the extra experiments to improve performance of JLMHGA on BookCrossing, i.e., JLMHGA2, are also effective, as shown in the figure.

To conclude, our idea of enhancing performance of Jaccard is well verified through various experiments. The idea is more successful on a denser dataset and effective in both recommendation and prediction qualities. Specifically, its improvement over Jaccard is about 19.1 to 29.8% on Jester in terms of F1 performance and about 4.7 to 7.3% on the same dataset in terms of MAE. In addition, the proposed index has a potential to further performance improvement, as demonstrated through the experiment results of JLMHGA2 on BookCrossing.

5   Conclusion

This paper proposed a novel improvement of Jaccard index, where not only the number of common items but also the levels of their ratings is considered. That is, we divide the whole rating range of the system into subintervals and examine how many items are commonly rated with normal or extreme values. The level of contribution of each subinterval to the new index as well as its boundaries is determined using a genetic algorithm. We investigated the performance of the proposed index for collaborative filtering using datasets with different characteristics. It is found that our index significantly outperformed Jaccard index on the datasets in terms of both recommendation and prediction qualities. The only exception occurs for prediction quality on a very sparse BookCrossing dataset where Jaccard index defeats all the other measures. Consequently, the superiority of the proposed index is verified on moderately sparse to dense datasets, opening a new prospect that it can be better utilized in CF systems in conjunction with other metrics for similarity computation.

References

1. Aamir, M., Bhusry, M.: Recommendation system: state of the art approach. Int. J. Comput. Appl. **120**(12), 25–32 (2015)
2. Ahn, H.J.: A new similarity measure for collaborative filtering to alleviate the new user cold-starting problem. Inf. Sci. **178**(1), 37–51 (2008)
3. Bobadilla, J., Serradilla, F., Bernal, J.: A new collaborative filtering metric that improves the behavior of recommender systems. Knowl.-Based Syst. **23**(6), 520–528 (2010)
4. Bobadilla, J., Ortega, F., Hernando, A., Bernal, J.: A collaborative filtering approach to mitigate the new user cold start problem. Knowl.-Based Syst. **26**, 225–238 (2012)
5. Bobadilla, J., Ortega, F., Hernando, A., Gutierrez, A.: Recommender systems survey. Knowl.-Based Syst. **46**, 109–132 (2013)
6. Jamali, M., Ester, M.: Trustwalker: a random walk model for combining trust-based and item-based recommendation. In: 15th ACM SIGKDD International Conference on Knowledge Discovery and Data Mining, pp. 397–406. ACM (2009)
7. Koutrica, G., Bercovitz, B., Garcia-Molina, H.: FlexRecs: expressing and combining flexible recommendations. In: The 2009 ACM SIGMOD International Conference Management of Data, pp. 745–758. ACM (2009)
8. Liu, H., Hu, Z., Mian, A., Tian, H., Zhu, X.: A new user similarity model to improve the accuracy of collaborative filtering. Knowl.-Based Syst. **56**, 156–166 (2014)
9. Ren, L., Gu, J., Xia, W.: A weighted similarity-boosted collaborative filtering approach. Energy Procedia **13**, 9060–9067 (2011)
10. Resnick, P., Lakovou, N., Sushak, M., Bergstrom, P., Riedl, J.: Grouplens: an open architecture for collaborative filtering of netnews. In: Proceedings the ACM Conference on Computer Supported Cooperative Work, pp. 175–186. ACM Press (1994)
11. Saranya, K.G., Sadasivam, G.S., Chandralekha, M.: Performance comparison of different similarity measures for collaborative filtering technique. Indian J. Sci. Technol. **9**(29) (2016)
12. Su, X., Khoshgoftaar, T.M.: A survey of collaborative filtering techniques. Adv. Artif. Intell. **2009**, 4 (2009)
13. Sun, H.-F., et al.: JacUOD: a new similarity measurement for collaborative filtering. J. Comput. Sci. Technol. **27**(6), 1252–1260 (2012)

Optimizing Least-Cost Steiner Tree in Graphs via an Encoding-Free Genetic Algorithm

Qing Liu[1,2(✉)], Rongjun Tang[1], Jingyan Kang[1], Junliang Yao[1,2],
Wenqing Wang[1,2], and Yali Wu[1,2]

[1] Faculty of Automation and Information Engineering,
Xi'an University of Technology, Xi'an, China
liuqing@xaut.edu.cn
[2] Shaanxi Key Laboratory of Complex System Control and Intelligent
Information Processing, Xi'an, China

Abstract. Most bio-inspired algorithms for solving the Steiner tree problem (STP) require the procedures of encoding and decoding. The frequent operations on both encoding and decoding inevitably result in serious time consumption and extra memory overhead, and then reduced the algorithms' practicability. If a bio-inspired algorithm is encoding-free, its practicability will be improved. Being motivated by this thinking, this article presents an encoding-free genetic algorithm in solving the STP. To verify our proposed algorithm's validity and investigate its performance, detailed simulations were carried out. Some insights in this article may also have significance for reference when solving the other problems related to the topological optimization.

Keywords: Steiner tree problem · Genetic algorithm · Encoding-free · Tree-based genotype

1 Introduction

Many practical problems like network routing [1], VLSI design [2], data-aggregation [3], boil down to the Steiner tree problem (STP) in graphs [4]. Given an undirected graph with non-negative edge costs and a subset of vertices, usually referred to as terminals, the STP in graphs requires a least-cost tree that contains all terminals but not limited to the terminals only. Since the STP is NP-complete [8], no exact algorithm can solve it within polynomial time. A large number of heuristic algorithms were thus proposed for finding the approximate solution to the STP within acceptable time, such as the well-known shortest path heuristic (SPH) [5], the distance network heuristic (DNH, also referred to as KMB, implying its three proposers' initials) [6], and the average distance heuristic (ADH) [7], etc. These heuristics run fast indeed, but their worst-case performance is poor [9], even never return the optimal solution. Given the situation that the heuristic algorithms perform poorly in the worst-case while the exact algorithms inevitably suffer from extremely serious time-overhead, increasingly more researchers tend to use the bio-inspired algorithms to obtain the optimal solution or at least the sub-optimal ones to the STP in consideration of the bio-inspired algorithms' global search ability and fast convergence.

© Springer International Publishing AG 2017
Y. Tan et al. (Eds.): ICSI 2017, Part I, LNCS 10385, pp. 386–393, 2017.
DOI: 10.1007/978-3-319-61824-1_42

So far, the bio-inspired algorithms for solving the STP are not uncommon [10–13]. Without loss of generality, these bio-inspired algorithms always rely on a population consisting of a number of candidate trees and their optimization mechanism could be summed up as "generation + evaluation". For "generation" alone, the bio-inspired algorithms generate candidate trees in different manners and thus require different encoding schemes to represent them. For instance, Fan et al. [10] adopted a binary tree-based scheme when using a hybrid genetic algorithm in optimizing multicast routing; Ma et al. [13] employed a predecessor-based tree-representation to prevent from generating loops in their research. Beyond that, as long as a certain scheme of encoding candidate trees is adopted, a corresponding operator for decoding is also required. For example, Liu et al. [1] utilized 0/1 strings as the encoding representations and thus Prim's algorithm [14] was used as the operator for decoding trees. In literature [12], Zhong et al. did the similar thing. When using the bio-inspired algorithms to solve the STP, the frequent operations on both encoding and decoding inevitably result in serious time consumption and extra memory overhead. In consideration of this, the practicability of the bio-inspired algorithms in solving the STP is thus reduced, especially in the application scenarios requiring solution speed, such as network routing optimization.

Therefore, a bio-inspired algorithm for solving the STP but not requiring the link of encoding is more needed. Actually, to propose a bio-inspired algorithm without encoding is feasible as long as we directly use the candidate trees as the genotype in the algorithms. Genetic programming (GP) pioneered by John Koza is such an instance, in which syntax tree-based genotype is employed [15]. Such a successful instance further motivates us to propose an encoding-free bio-inspired algorithm for solving the STP better. Nevertheless, how to implement the related operators acting on the genotype of uncoded trees becomes a new issue, and our research work to be presented in this paper mainly focuses on solving it. In this paper, we present an encoding-free genetic algorithm (GA) to optimize the least-cost Steiner tree for the given STP. The rest part of this paper are organized as follows. Section 2 formulates the mathematical description of the STP. The implementation of the proposed GA is to be detailed in Section 3. The simulation will be given in Section 4, followed by our conclusion as Section 5.

2 Mathematical Model of Steiner Tree Problem

Given an undirected graph G consisting of a vertex set V and an edge set E, in which each edge e is associated with a positive cost $c(e)$, the STP on such a graph G can be related as that of determining a least-cost Steiner tree spanning $|V'|$ vertices (i.e. a root and several destinations). Suppose that $T^* = (V_{T^*}, E_{T^*})$ is the least-cost Steiner tree, the optimization objective can thus be expressed as below:

$$\min \sum_{e \in T^*} c(e) \tag{1}$$

It must be noticed that, for the case of $|V'| = 2$, the STP reduces to shortest path problem (SPP); And for the case of $|V'| = |V|$, the STP reduces to the minimal spanning tree problem (MSTP). Either the SPP or the MSTP can be easily solved within polynomial time. The STP in graphs specially refers to the case of $2 < |V'| < |V|$.

3   Tree-Based Genetic Algorithm for Evolving Steiner Tree

3.1   Representation of Candidate Steiner Trees

We took the undirected graph G shown as in Fig. 1(a) as an instance to explain. For a set of specified vertices, including the vertex 1 as the root (R) and the vertices 10, 4, 5, 12, 7 as the destinations (D_1, D_2, D_3, D_4, D_5), the tree structure shown as in Fig. 1(b) is a candidate Steiner tree spanning the specified vertices. Since we are to abolish the link of encoding on purpose during optimizing, the tree structure in Fig. 1(b) can thus be directly used as the genotype in our proposed GA. And the genotype is stored as the form of its several branches. For example, the candidate Steiner tree shown in Fig. 1(b) is stored as a set of branches, i.e. $\{10{\rightarrow}3{\rightarrow}2{\rightarrow}1,\ 4{\rightarrow}11{\rightarrow}3,\ 5{\rightarrow}4,\ 12{\rightarrow}4,\ 7{\rightarrow}4\}$, which is noted as g-ST for simplicity and the set of all the vertices involved in the g-ST is noted as v-ST. For our proposed approach, the phenotype is the genotype. So the tree cost is directly used to evaluate a genotype's fitness.

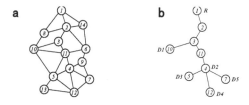

Fig. 1. An undirected graph G and a randomly generated Steiner tree spanning a set of specified vertices: (a) A undirected graph G; (b) A candidate Steiner tree

The process of randomly generating a candidate Steiner tree on a graph G can be mainly described as the following steps:

Step 1 Initialization: g-ST = $\varnothing$; root = R; destination set $D = \{D_i | i = 1 \ldots |V'|\}$; v-ST = $\varnothing$;

Step 2 Add R into the v-ST;

Step 3 Choose an element D_i from D at random to be the starting vertex of the branch to be generated, then remove D_i from D;

Step 4 Generate a random route from D_i to any one of the vertices constituting v-ST to be a branch, and then add this generated branch into the g-ST, meanwhile add the vertices included by this branch into v-ST;

Step 5 Go back to Step 3 if $D \neq \varnothing$ but terminate the process otherwise.

Through finishing the above-described process on G, various Steiner trees spanning R and D may be generated. We encapsulated such a process as an operator, referred to as RST-generator($\cdot$), for being expediently called when generating a random Steiner tree on an arbitrary undirected graph arg-G. The calling syntax is RST-generator(arg-G), where arg-G is passed to the operator as the argument.

3.2   Genetic Operators

GA evolves the optimal solution to a problem by iteratively acting the genetic operators. This subsection is to detail the specific implementation of them.

Crossover. So far, numerous crossover techniques have been widely used in problem-solving, such as single-point crossover, two-point crossover, uniform crossover, etc. However, they are not suitable to be acted on the tree-based genotype. A brand new way of implementing the crossover that adapts the tree-based genotype is thus required.

Essentially, crossover is the process that two parent solutions generate their offspring by exchanging their genetic information. For the tree-based genotype, the genetic information refers to the tree topology in fact. We thus implement the crossover operator by exchanging two parent solutions' partial topologies. To be specific, we assume that the parent1 in Fig. 2(a) and the parent2 in Fig. 2(b) are randomly selected out of the population of GA. Since the topologies of their tree-based genotypes imply their respective genetic information, the undirected graph obtained by merging the parent1 and the parent2 could thus be treated as a gene pool, as shown in Fig. 2(c). As thus, any offspring Steiner tree being generated from such a gene pool deservedly inherits the genetic information of the two parents. For instance, Fig. 2(d) and (e) separately exhibit two randomly generated offspring, of which the topologies are indeed different from but similar to that of their parents. As for the method of generating the offspring, the afore-said operator RST-generator($\cdot$) can be directly used by passing the gene pool as its argument.

Furthermore, we actually made more improvements on the crossover operator. Usually, two parents generate two offspring during the process of crossover. In our proposed approach, two parents may generate $k > 2$ offspring individuals in order to enhance the crossover operator's effectiveness on the respect of local exploitation, where k is a new parameter of our proposed GA. Nevertheless, among the k generated offspring individuals, only the best two are permitted to enter the offspring population. Therefore, the number of the generated individuals still equals to $PopSize \cdot Pc$, where Pc refers to the probability of crossover.

Mutation. Mutation alters a candidate solution's genetic information from its initial state. For our proposed approach, the genetic information refers to the topology in fact. However, to alter a Steiner tree's topology at random can result in the Steiner tree being unconnected or invalid. To avoid this risk, we implement mutation operator in a similar way to the crossover. To be specific, we beforehand generate a brand new Steiner tree at random and merge it with the one to be mutated. Namely, we built a gene pool

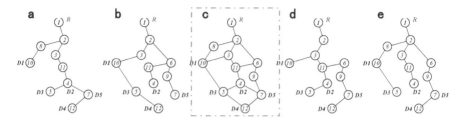

Fig. 2. Diagram of crossover operator: (a) Parent1; (b) Parent2; (c) Gene pool being composed of the topology of parent1 and that of parent2; (d) Offspring1; (e) Offspring2

containing the genetic information from the one to be mutated and the one being newly generated. As thus, any Steiner tree derived from such a gene pool can be treated as the mutant, because the mutant inherits the partial topologies of the one to be mutated and also some new topologies. As for the method of generating the mutant from the built gene pool, we still use the encapsulated RST-generator(·) by passing the built gene pool as its argument. Among the population, a total of *PopSize·Pm* candidate solutions are to be mutated, where *Pm* refers to the probability of mutation.

Sorting-Based Selection. Selection is the operator by which the superior candidate solutions survive for later breeding while the inferior ones are eliminated. We adopted the commonly used sorting-based selection to eliminate the inferior candidate solutions. Admittedly, the usage of the sorting-based selection will accelerate the decrease of the population diversity. This drawback may result in premature convergence. We thus proactively replace the repetitive individuals among the population by the newly generated ones. This practice helps to avoid the drawback of the sorting-based selection.

3.3 Primary Procedure During One Generation

Based on the previous description, this section introduces how the previously related genetic operators are organized together. During the period of one generation, the primary procedure can be depicted as the following schematic diagram in Fig. 3. The superior individuals among the original population and the offspring population get survived and then form the new generation. Evolving in this way, generation by generation, our proposed GA finally returns the optimal solution to the STP.

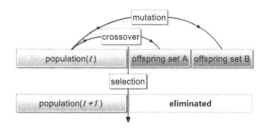

Fig. 3. Schematic diagram to the procedure of our proposed GA during one generation

4 Simulation

We implemented the proposed GA in C++ and carried out the simulation experiments on a computer with an Intel Core i7-3520 M 2.5 GHz CPU, 8 GB RAM, and Windows 7 sp1 (x64) operating system, in order to investigate its validity and performance.

4.1 Comparison with Other Bio-Inspired Algorithms

To verify our proposed GA's validity, we tested it through the task of finding the least-cost Steiner tree spanning the set of vertices {2, 4, 6, 8, 10, 12, 14, 16, 18, 20, 22, 24} on an undirected graph, as shown in Fig. 4. For comparison, some other bio-inspired algorithms based on various encoding schemes in Refs. [1, 12, 13] were also tested. The parameters of the proposed GA are set as *PopSize* = 100; *Pc* = 0.95; *Pm* = 0.05; k = 4; *MaxGeneration* = 300. As for the compared algorithms' parameters, the same settings as reported were adopted.

The final solutions obtained by the tested algorithms are compared in terms of the structure and the corresponding tree cost, as demonstrated in Fig. 5(a) and (b). Our proposed GA as well as the AFSA in literature [1] and the PSO in literature [12] found the optimal Steiner tree, whose tree cost equals to 314, while another version of the AFSA [13] only found a sub-optimal Steiner tree. Obviously, our proposed GA is valid for solving the STP. Much more importantly, unlike any of the compared algorithms, our proposed GA is encoding-free, and thus no decoding algorithm is needed. Its practicability is thus better than the compared ones in terms of time consumption, space utilization, and easy programming.

4.2 Comprehensive Investigation on Convergence

The STP's difficulty is varying when specifying different number of vertices to span, while the optimization task in the previous simulation only tests the case of 12 vertices. The influence on our proposed GA's convergence resulted by the newly introduced parameter k is also not investigated. Thus, in the subsequent part, we are to investigate our proposed GA from the two aspects through three groups of simulation experiments, in which each group investigates the convergence effectiveness when setting k as different values. The optimization tasks assigned for the three groups are to find the least-cost Steiner trees of spanning 18, 20, and 22 vertices respectively. For each optimization task as well as each k, our proposed GA was run 100 times and the convergence duration of each run was recorded.

We demonstrated the recorded convergence duration of all runs in the form of box plot as in Fig. 6, which is a convenient way of graphically depicting groups of numerical data through their quartiles. And the lines extending vertically from the boxes indicating variability outside the upper and lower quartiles. Outliers are also plotted, marked as red '+'. According to Fig. 6, we are able to conclude that our proposed GA converges faster and steadily with larger k in stastical sense, which actually benefits from the enhanced local exploitation during crossover.

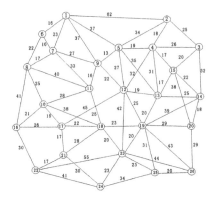

Fig. 4. Graph with 26 vertices and 65 edges

Fig. 5. Steiner trees respectively obtained by the compared AFSA [1, 13], PSO [12], and our proposed GA: (a) Steiner tree obtained by AFSA [13], tree cost = 320; (b) Steiner tree obtained by PSO [12], AFSA [1], and our proposed GA, tree cost = 314

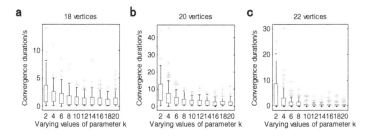

Fig. 6. Box plots of the recorded convergence duration of all runs: (a) Optimization task 1; (b) Optimization task 2; (c) Optimization task 3 (Color figure online)

5 Conclusion

The effort of our research work is to present an encoding-free genetic algorithm in solving the Steiner tree problem in graphs, such that the originally required procedures of both encoding and decoding are no longer needed, such that the practicability of the algorithm could be improved. In the proposed approach, candidate trees are directly used as the genotype. To make the related genetic operators adaptive to the tree-based genotype, we redefined the crossover and mutation based on a specially defined operator RST-generator(·), which is originally used to generate candidate Steiner trees at random when initializing the population in fact. Since all the related genetic operators are based on the RST-generator(·), our proposed encoding-free GA is very succinct and easy programming. Detailed simulations were carried out to verify the algorithm's validity and investigate its performance. Besides, some insights in this article, such as the implementation of crossover, may also have significance for reference when solving the other problems related to the topological optimization.

Acknowledgements. This research is supported in part by National Science Foundation of China (No. 61502385, No. 61401354, No. 61503299), Key Basic Research Fund of Shaanxi Province (2016JQ6015), and Scientific Research Program Funded by Shaanxi Provincial Education Department (No. 16JK1554).

References

1. Liu, Q., Odaka, T., Kuroiwa, J., et al.: An artificial fish swarm algorithm for the multicast routing problem. IEICE Trans. Commun. **E97-B**(5), 996–1011 (2014)
2. Zhou, Z., Jiang, C., Huang, L., et al.: On optimal rectilinear shortest paths and 3-Steiner tree routing in presence of obstacles. J. Softw. **14**(9), 1503–1514 (2003). (in Chinese with an English abstract)
3. Li, Z., Shi, H.: A data-aggregation algorithm based on minimum Steiner tree in wireless sensor networks. J. Northwest. Polytech. Univ. **27**(4), 558–564 (2009). (in Chinese with an English abstract)
4. Hwang, F.K., Richards, D.S., Winter, P.: The Steiner Tree Problem, vol. 53. Elsevier, Amsterdam (1992)
5. Takahashi, H., Matsuyama, A.: An approximate solution for the Steiner problem in graphs. Math. Jpn. **24**(6), 573–577 (1980)
6. Kou, L., Markowsky, G., Berman, L.: A fast algorithm for Steiner trees. Acta Informatica **15**, 141–145 (1981)
7. Rayward-Smith, V.J.: The computation of nearly minimal Steiner trees in graphs. Int. J. Math. Educ. Sci. Tech. **14**(1), 15–23 (1983)
8. Gary, M.R., Johnson, D.S.: Computers and Intractability: A Guide to the Theory of NP-completeness (1979)
9. Plesník, J.: Worst-case relative performances of heuristics for the Steiner problem in graphs. Acta Math. Univ. Comen. **LX**(2), 269–284 (1991)
10. Fan, Y., Jianjun, Yu., Fang, Z.: Hybrid genetic simulated annealing algorithm based on niching for QoS multicast routing. J. Commun. **29**(5), 65–71 (2008). (in Chinese with an English abstract)
11. Ma, X., Liu, Q.: A particle swarm optimization for Steiner tree problem. In: Proceedings of the 6th International Conference on Natural Computation (ICNC), pp. 2561–2565 (2010)
12. Zhong, W.L., Huang, J., Zhang, J.: A novel particle swarm optimization for the Steiner tree problem in graphs. In: IEEE World Congress on Evolutionary Computation, pp. 2460–2467 (2008)
13. Ma, X., Liu, Q.: An artificial fish swarm algorithm for Steiner tree problem. In: IEEE-FUZZ, pp. 59–63 (2009)
14. Prim, R.C.: Shortest connection networks and some generalizations. Bell Syst. Tech. J. **36**(6), 1389–1401 (1957)
15. Koza, J.: Genetic Programming: On the Programming of Computers by Means of Natural Selection. MIT Press, Cambridge (1992)

An Energy Minimized Solution for Solving Redundancy of Underwater Vehicle-Manipulator System Based on Genetic Algorithm

Qirong Tang[1(✉)], Le Liang[1], Yinghao Li[1], Zhenqiang Deng[1], Yinan Guo[2], and Hai Huang[3]

[1] Laboratory of Robotics and Multibody System, Tongji University,
Shanghai 201804, China
qirong.tang@outlook.com
[2] University of Mining and Technology, Xuzhou 221116, China
[3] National Key Laboratory of Science and Technology on Underwater Vehicle,
Harbin Engineering University, Harbin 150001, China

Abstract. An energy minimized method using genetic algorithm for solving redundancy of underwater vehicle-manipulator system is proposed in this paper. Energy minimization is here set up as an optimization problem. Under the constraints of the dynamic and kinematic equations, the inverse kinematic solution with the optimal index is formed by using the weight pseudoinverse matrix. Energy consumption function is chosen as the objective function, and then the energy minimized solution based on genetic algorithm for solving the redundancy of the system is performed. Two numerical examples are carried out to verify the proposed method and promising result is obtained.

Keywords: UVMS · Redundancy · Energy minimization · Genetic algorithm

1 Introduction

Underwater vehicle-manipulator system (UVMS) is kind of autonomous robot, which has the ability to complete certain intervention missions in the ocean environment [1]. With the exploration of the ocean stepping more widely, the research on UVMS is of great scientific and economic value.

The UVMS generally consists of a vehicle and a manipulator with links. It is a complex coupled redundant system in both kinematics and dynamics. Meanwhile, it becomes very complicated in the underwater environment where UVMS will be affected by restoring moments and disturbances. A solution with redundancy issue solved is to determine the motion trajectory of each degree of freedom (DOF) under the premising of completing missions [2]. The typical kinematic redundancy resolution methods, such as gradient projection method (GPM) and task-priority method [3], have been successfully applied to solve the

Y. Tan et al. (Eds.): ICSI 2017, Part I, LNCS 10385, pp. 394–401, 2017.
DOI: 10.1007/978-3-319-61824-1_43

UVMS's redundancy with some specific purposes. Nevertheless, for a UVMS, it is important to handle the redundant DOFs and at the same time to consider the limitation of energy resources.

There are some researches on the energy optimization of redundant DOFs for the UVMS. One of the most common methods is to use the weighted pseudoinverse Jacobian matrix to solve inverse kinematics [4]. This method guarantees the instantaneous kinetic energy to reach a minimum. However, the minimum of energy consumption can not be achieved because of the existence of restoring moments and disturbance. To solve the energy minimizing problem for UVMS, Sarkar in [5] used GPM to reduce the drag forces. The energy consumption is reduced by actual thruster forces which are decided by thruster dynamics. Similarly, the energy consumption is reduced by minimizing restoring moments in [6]. Those GPMs are applied to optimize the solution of redundancy and the distance between the center of gravity and the center of buoyancy is selected as the objective function. The same optimization method is also applied in [7]. Overall control effort is reduced with the reduction of gravity and buoyancy loadings on both vehicle and manipulator during end-effector motion.

The performance of the optimization using GPM strongly depends on the parameter which is very difficult to select in general. Besides, the energy gradient function of the UVMS is difficult to be obtained. Moreover, with the increase of the DOF of the UVMS, calculating the derivative of the gradient function becomes very complex. Therefore, instead of constructing the gradient function, the weight pseudoinverse matrix is used to solve the redundancy. Meanwhile, to find the solution of redundancy which minimizes the energy consumption, the global optimal solution of redundancy is studied based on genetic algorithm.

This paper is organized as follows. In Sect. 2, the research problem is described. The expression of optimization problem and its constraints are given. In Sect. 3, the dynamic and kinematic constraints of UVMS are introduced. In Sect. 4, genetic algorithm is used for energy optimization. Two numerical examples are carried out in Sect. 5 and the effectiveness of the energy minimized redundancy solving method is verified with comparisons. Conclusions are given in Sect. 6 to close this paper.

2   Problem Description

The UVMS, which is usually composed of a vehicle and a manipulator, is kinematically a redundant system. It is assumed that the vehicle is a m-DOF floating base and the manipulator is a n-DOF serial mechanism, the kinematically redundant is referred to $m + n > w$, where w is the number of workspace DOFs. Because of the existence of redundant DOFs, the system will possess infinite set of kinematical solutions when the workspace task is specified. These redundant DOFs will result different indicators, such as the shortest movement time and the lowest energy consumption. Therefore, it is very necessary to find an optimal solution for a specific situation.

For UVMS, endurance is an important performance factor. It is of great economic value to research on the redundant DOFs to carry out the trajectory which can minimize the consumption of energy while operating autonomously in the ocean environment. The solution of redundancy with energy minimized can be formulated as an optimization problem

$$min \quad \int_{t_0}^{t_f} f_e(t)dt, \tag{1}$$

$$s.t. \quad \boldsymbol{\tau}(t) = f_D(\boldsymbol{\zeta}, \dot{\boldsymbol{\zeta}}, \ddot{\boldsymbol{\zeta}}, t), \tag{2}$$

$$\boldsymbol{\zeta}(t_0) = \boldsymbol{\zeta}_0 , \quad \boldsymbol{\zeta}(t_f) = \boldsymbol{\zeta}_f, \tag{3}$$

$$\dot{\boldsymbol{\zeta}}(t_0) = \dot{\boldsymbol{\zeta}}_0 , \quad \dot{\boldsymbol{\zeta}}(t_f) = \dot{\boldsymbol{\zeta}}_f, \tag{4}$$

$$\boldsymbol{x}_E(t) = f_K(\boldsymbol{\zeta}), \tag{5}$$

where $f_e(t)$ is the energy function of the UVMS, t_0 and t_f represent initial time and final time of system motion, respectively. Here $\boldsymbol{\tau} \in \mathbb{R}^{m+n}$ represents generalized driving force which can be obtained by solving the system dynamic equations $f_D(t)$, $\boldsymbol{\zeta}, \dot{\boldsymbol{\zeta}}, \ddot{\boldsymbol{\zeta}} \in \mathbb{R}^{m+n}$ indicates the generalized position coordinates, generalized velocity and generalized acceleration, respectively. The (3) and (4) are the boundary conditions of the system task space. Function $f_K(t)$ is the kinematic constraint of the system, specifically, the kinematic equation between the generalized coordinates and the end effector of the manipulator. An energy minimization trajectory is a rational solution that satisfies the constraints and makes the objective function minimized.

3   Constraints

3.1   Dynamic Modeling of UVMS

Because of the characteristics of the underwater environment, it is necessary to consider the effect of buoyancy and restoring moments acting on the UVMS [8]. The dynamic equation of UVMS can be established based on Lagrange equation and can be written in a compact state-space form by

$$\boldsymbol{M}(\boldsymbol{\zeta})\ddot{\boldsymbol{\zeta}} + \boldsymbol{C}(\boldsymbol{\zeta}, \dot{\boldsymbol{\zeta}})\dot{\boldsymbol{\zeta}} + \boldsymbol{G}(\boldsymbol{\zeta}) = \boldsymbol{\tau}, \tag{6}$$

where $\boldsymbol{M}(\boldsymbol{\zeta}) \in \mathbb{R}^{(m+n) \times (m+n)}$ is the inertia matrix of the UVMS including added mass terms, $\boldsymbol{C}(\boldsymbol{\zeta}, \dot{\boldsymbol{\zeta}}) \in \mathbb{R}^{(m+n) \times (m+n)}$ is the matrix of Coriolis and centripetal terms, $\boldsymbol{G}(\boldsymbol{\zeta}) \in \mathbb{R}^{(m+n)}$ is the vector of gravity and buoyancy effects (restoring moment), $\boldsymbol{\tau} \in \mathbb{R}^{(m+n)}$ is the generalized driving force.

3.2   Solution for Kinematic Redundancy

The kinematic equations can be obtained by getting the derivation of kinematic constraint in (5) which is then governed by

$$\dot{\boldsymbol{x}}_E = \boldsymbol{J}_E \dot{\boldsymbol{\zeta}}, \tag{7}$$

where $J_E \in \mathbb{R}^{w \times (m+n)}$ is the Jacobian matrix defined as $J_E = \partial f_K / \partial \zeta$. The Jacobian matrix here is a transformation matrix that describes the relationship between the end effector velocity and the generalized velocity. Because of the kinematic redundancy of the system, there are infinite inverse kinematics solutions. In order to obtain the one that satisfies energy minimization, a feasible solution of redundancy is given by

$$\dot{\zeta} = \tilde{J}^{\dagger}_E \dot{x}_E, \tag{8}$$

where

$$\tilde{J}^{\dagger}_E = D^{-1} J_E^T (J D^{-1} J_E^T)^{-1}, \tag{9}$$

where matrix $D \in \mathbb{R}^{(m+n) \times (m+n)}$ is a diagonal matrix and $\tilde{J}^{\dagger}_E$ is the weight pseudoinverse matrix which depends on matrix D. The solution $\dot{\zeta}$ given in (8) minimizes $\dot{\zeta}^T D \dot{\zeta}$ over all feasible solutions. According to [8], the instantaneous kinetic energy, which is represented as $1/2 \, \dot{\zeta}^T M \dot{\zeta}$, is minimized when matrix D is replaced by inertia matrix M. However, the total energy consumption is not minimized in the motion process because of the existence of the restoring moments. Therefore, finding a suitable matrix D is the key to solve the problem. Once matrix D is determined, the solution that minimizes energy consumption can be obtained.

4   Solution Optimized by Genetic Algorithm

In order to obtain the optimal matrix D, the energy consumption function is chosen as the objective function which can be expanded as

$$E = \int_{t_0}^{t_f} |\tau(t)^T| \cdot |\dot{\zeta}(t)| dt = \int_{t_0}^{t_f} |\tau(t)^T| \cdot |D^{-1} J_E^T (J D^{-1} J_E^T)^{-1} x_E(t)| dt, \tag{10}$$

where E represents the consumption of energy during the motion process.

The expression of objective function E is very complex, so it is difficult to get the gradient. Traditional optimization methods, such as climbing hill method, are powerless in this problem. Genetic algorithm (GA) is a general optimization method based on natural selection and genetic evolution and has been successfully applied to solve optimization problems. Here GA is used to determine an optimal matrix D. The diagonal coefficients of the matrix D are used as the inputs.

Here, genetic algorithm is used to solve a constrained optimization problem. It is necessary to adjust the chromosome representation and the selection method. There will be infeasible or even illegal solutions in constrained optimized problem while using binary coding. Using order-based chromosome representation can ensure that new individuals satisfy the constraint conditions [9]. Instead of roulette wheel selection, $(\mu + \lambda)$ selection, which is considered to be the most suitable method for optimization problems, is applied because it selects the best chromosomes from the each generations. And the swap mutation, which ensures that there are no illegal solutions, is a feasible method to our problem [10].

The main procedures of this algorithm are,

(1) create initial population, coding population using order-based method
(2) evaluate the population using Eq. (10), and select optimal individuals as elite solution
(3) if the stop criterion, i.e., the maximum number of iterations is achieved or the global optimal value has not changed over 20 iterations, is reached, move to step 6)
(4) apply the mutation operation to population, elite solutions do not participate in mutation operation
(5) apply $(\mu + \lambda)$ selection and order crossover, generate new population, move to step 2)
(6) output optimal solution as the matrix D

The off-line optimization result matrix D, which can reduce the energy consumption of the system, is used as the weight matrix of the redundancy resolution.

5   Numerical Example

Numerical examples are performed to verify the proposed method. The physical data of UVMS used here for simulation is taken from a practical system. However, here it only considers in planar situation, see in Fig. 1(a) and (b). Thus, the 3-DOF vehicle and 3-DOF manipulator together result in the redundancy.

In this example, the end of manipulator is desired to move along the red dash line in Fig. 1(a), meanwhile the energy required for the motion should be minimized. The constraints of the path are set as

$$\zeta(t_0) = [0 \ 0 \ 0 \ -\pi/4 \ \pi/4 \ \pi/4]^T, \tag{11}$$

$$\dot{\zeta}(t_0) = \mathbf{0}, \quad \dot{\zeta}(t_f) = \mathbf{0}, \tag{12}$$

$$\ddot{x}_E = \begin{cases} [-0.02 \ -0.04]^T & t_0 < t \le (t_0 + t_f)/2 \\ [0.02 \ 0.04]^T & (t_0 + t_f)/2 < t < t_f \end{cases}, \tag{13}$$

where $\zeta(t_0)$ represents initial position of UVMS. Both initial and terminal velocity are set to zero. It uses SI units.

The operators and parameters of the genetic algorithm are set as follows: chromosome representation is order-based; population size is 50; initial population is selected as inertia matrix M; crossover strategy is order crossover; selection strategy is $(\mu + \lambda)$ selection; elite rate is 0.05 and mutation rate is 0.02; maximal iteration is 200 generations.

The optimization problem is solved by genetic algorithm and the optimal matrix D is obtained as $D = diag\{46.96 \ 29.43 \ 26.16 \ 18.26 \ 10.63 \ 1\}$. Using this matrix D, a solution that minimized energy consumption is obtained. To verify the effectiveness of the method, another simulation case using matrix M has been executed.

The results using matrix M are shown in Fig. 1(a), (c) and (e). Figure 1(a) shows the motion process of the trajectory of UVMS. It can be seen that

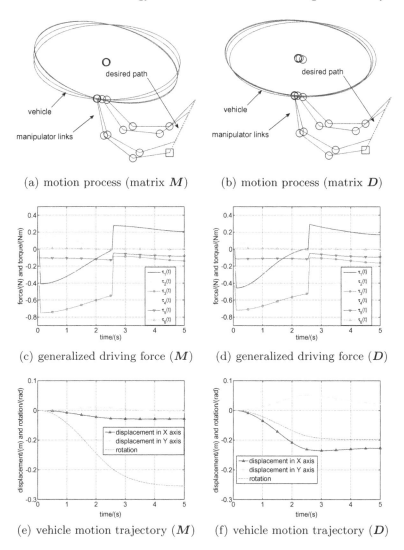

(a) motion process (matrix M)   (b) motion process (matrix D)

(c) generalized driving force (M)   (d) generalized driving force (D)

(e) vehicle motion trajectory (M)   (f) vehicle motion trajectory (D)

Fig. 1. Planning result using matrix M and matrix D (Color figure online)

the motion of the manipulator end is coincident with the desired trajectory. Figure 1(c) shows the force/torque value-time curves of the generalized driving force. The vehicle rotating torque $\tau_3(t)$ is more larger, because the clockwise rotation of the vehicle will against the restoring moments. Figure 1(e) is the curves of the vehicle displacement.

The results using matrix D are shown in Fig. 1(b), (d) and (f). In Fig. 1(b), the maximum of rotation of the vehicle becomes smaller compared to Fig. 1(a). In Fig. 1(d), it can be found that the generalized driving force of vehicle in Y-axis $\tau_2(t)$ is significantly smaller than the one in Fig. 1(c). One can find that the rotation magnitude in Fig. 1(f) of the vehicle is apparently smaller than the one in Fig. 1(e). In the case that matrix D is used, the vehicle restricts its rotational motion to reduce the energy used of counteracting the restoring moments.

Figure 2 shows the comparison of energy consumption between two examples. It can be clearly seen that the energy consumption of the path planned by using the inertia weight matrix M is higher. In the underwater environment, due to the influence of restoring moments, kinetic consumption can not represent the total energy consumption. In order to achieve the minimal energy consumption, it should limit the displacement in the direction of opposite to the restoring moment.

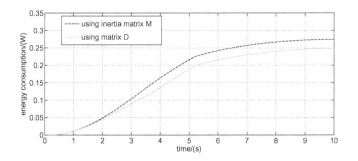

Fig. 2. Comparison of energy consumption

6   Conclusion

In this paper, an energy minimized method for solving the redundancy of UVMS using genetic algorithm is carried out. First of all, the energy minimization is described as a optimization problem. Then the dynamic and kinematic equation of UVMS are given as the constraint condition. The inverse kinematics is solved by using weight pseudoinverse matrix. Genetic algorithm is used for choosing this weight matrix. Through the comparison of two numerical examples, the effectiveness of the proposed energy optimization method is verified.

In our following work, more efficient genetic algorithm will be discussed to enhance the effectiveness of optimization. Meanwhile, it is necessary to verify our method on a real physical UVMS system.

Acknowledgments. This work is supported by the Key Basic Research Project of 'Shanghai Science and Technology Innovation Plan' (No. 15JC1403300), the National Natural Science Foundation of China (No. 61603277; No. 51579053), the State Key Laboratory of Robotics and Systems (Harbin Institute of Technology), key project (No. SKLRS-2015-ZD-03), and the SAST Project (No. 2016017). Meanwhile, this work is also partially supported by the Fundamental Research Funds for the Central Universities (No. 2014KJ032; 'Interdisciplinary Project' with No. 20153683), and 'The Youth 1000 program' project (No. 1000231901). It is also partially sponsored by 'Shanghai Pujiang Program' project (No. 15PJ1408400), the National College Students Innovation Project (No. 1000107094), as well as the project from Nuclear Power Engineering Co., Ltd. (No. 20161686). All these supports are highly appreciated.

References

1. Marani, G., Choi, S.K., Yuh, J.: Underwater autonomous manipulation for intervention missions AUVs. Ocean Eng. **36**, 15–23 (2009)
2. Mohan, S., Kim, J.: Indirect adaptive control of an autonomous underwater vehicle-manipulator system for underwater manipulation tasks. Ocean Eng. **54**, 233–243 (2012)
3. Antonelli, G., Chiaverini, S.: Task-priority redundancy resolution for underwater vehicle-manipulator systems. In: IEEE International Conference on Robotics and Automation, Leuven, Belgium, pp. 756–761 (1998)
4. Angeles, J.: On the numerical solution of the inverse kinematic problem. Int. J. Robot. Res. **4**, 21–37 (1985)
5. Sarkar, N., Podder, T.K.: Coordinated motion planning and control of autonomous underwater vehicle-manipulator systems subject to drag optimization. IEEE J. Ocean. Eng. **26**, 228–239 (2001)
6. Han, J., Chung, W.K.: Redundancy resolution for underwater vehicle-manipulator systems with minimizing restoring moments. In: 2007 IEEE/RSJ International Conference on Intelligent Robots and Systems, San Diego, USA, pp. 3522–3527 (2007)
7. Ismail, Z.H., Dunnigan, M.W.: Redundancy resolution for underwater vehicle-manipulator systems with congruent gravity and buoyancy loading optimization. In: 2009 IEEE International Conference on Robotics and Biomimetics, Guilin, China, pp. 1393–1399 (2009)
8. Ishitsuka, M., Sagara, S., Ishii, K.: Dynamics analysis and resolved acceleration control of an autonomous underwater vehicle equipped with a manipulator. In: International Symposium on Underwater Technology, Taipei, Taiwan, pp. 277–281 (2005)
9. Liao, C.C., Ting, C.K.: Extending wireless sensor network lifetime through order-based genetic algorithm. In: 2008 IEEE International Conference on Systems, Man and Cybernetics, Singapore, pp. 1434–1439 (2008)
10. Back, T.: Selective pressure in evolutionary algorithms: a characterization of selection mechanisms. In: The First IEEE Conference on Evolutionary Computation, Orlando, USA, pp. 57–62 (1994)

Study of an Improved Genetic Algorithm for Multiple Paths Automatic Software Test Case Generation

Erzhou Zhu[1(✉)], Chenglong Yao[1], Zhujuan Ma[2], and Feng Liu[1(✉)]

[1] School of Computer Science and Technology,
Anhui University, Hefei 230601, China
{ezzhu, fengliu}@ahu.edu.cn
[2] School of Economic and Technical,
Anhui Agricultural University, Hefei 230011, China

Abstract. Automatic generation of test case is an important means to improve the efficiency of software testing. As the theoretical and experimental base of the existing heuristic search algorithm, genetic algorithm shows great superiority in test case generation. However, since most of the present fitness functions are designed by a single target path, the efficiency of the generating test case is relatively low. In order to cope with this problem, this paper proposes an efficiency genetic algorithm by using a novel fitness function. By generating multiple test cases to cover multiple target paths, this algorithm needs less iterations hence exhibits higher efficiency comparing to the existing algorithms. The simulation results have also shown that the proposed algorithm is high path coverage and high efficiency.

Keywords: Software testing · Test case generation · Genetic algorithm · Multiple paths coverage

1 Introduction

Software testing is expensive, time-consuming and tedious. It is estimated that software testing requires about 50% of the total cost of software development. As an evolutionary approach for computing, Genetic algorithm (GA) has the ability to determine appropriate approx for providing solutions to optimization problems. It has been successfully applied in automatic generation of test cases. Mansour and Salame [1] combine simulated annealing algorithm and genetic algorithm to generate test case covering specific paths. Chen and Zhong [2] uses multi-population GA to generate test case for path coverage. Previous test case generators based on genetic algorithms, however, are inefficient in covering multiple target paths. As a consequence, researches are focusing on multi-path coverage test case generation. Ahmed and Hermadi [3] proposed the idea of generating test data for multiple paths, but the fitness algorithm is rather complex, resulting in a relatively long test case generation time and low efficiency. Gong and Zhang [4] used the Huffman coding method to generate multi-path test case based on genetic algorithm, however, the program under testing must be converted into binary tree before test case generation, and its time consumption is increased.

© Springer International Publishing AG 2017
Y. Tan et al. (Eds.): ICSI 2017, Part I, LNCS 10385, pp. 402–408, 2017.
DOI: 10.1007/978-3-319-61824-1_44

Based on the ideas of [3, 4], this paper presents a new GA, by redesigning the fitness function and modifying the operating operators to improve the ability of the GA's coverage and test case generation efficiency. Simulation results have verified the superiority of the proposed algorithm.

2 Algorithm Implementation

Assuming that the control flow graph (CFG) of program T contains n branches, the branch set B_s of program T can be expressed as: $B_s = <b_1, b_2, \ldots b_n>$, where b_i is a branch of T and P_i is composed of multiple b_i. All branches that T runs through x_i are denoted as $B_e(x_i)$, T executes a set of test cases N, which is a set of all the branches that N is executed on T, denoted as B_{es}.

2.1 Fitness Function Design

The fitness function is designed based on path and branch coverage. The path is encoded according to the path and branch information of the generated CFG. During the execution of an arbitrary test case, the true branch at the node is marked as 1 while the false is marked as 0. The w_{th} evolutionary population N_w is denoted as $N_w = \langle x_1, x_2, \ldots x_m \rangle$, where x_i is an individual in the population and m is the population size. With the individual x_i as the test case, the execution path $P_e(x_i)$ is obtained after executing a path in C_s. From the path encoding we can see that $P_e(x_i)$ consists of 0 and 1 strings. Let $L(x_i)$ be the number of bits in the string, that is the number of execution branches of the path. $L(x_i, j)$ is the maximum number of identical bits between the arbitrary path P_j of C_s and the execution path $P_e(x_i)$ of individual x_i. The similarity value of the k_{th} bit in the same bit of P_j and $P_e(x_i)$ is denoted as $V_k(x_i, j), V_k(x_i, j)$, is the XOR operation result of the k_{th} value in both paths. The similarity between the execution path $P_e(x_i)$ and the target path P_j is:

$$f_{sim}(P_e(x_i), P_j) = \sum_{k=0}^{L(x_i, j)} k * V_k(x_i, j). \tag{1}$$

The similarity between individual x_i and the entire set C_s is:

$$f_{sim}(P_e(x_i), C_s(w)) = (\sum_{k=0}^{m} f_{sim}(P_e(x_i), P_j))/k. \tag{2}$$

The average similarity of the w_{th} generation population N_W is:

$$f_{avgsim} = (\sum_{k=0}^{m} f_{sim}(P_e(x_i), C_s(w)))/k. \tag{3}$$

The branch coverage of individual x_i is the ratio of the specific branch $B(x_i)$ of the individual to the total number of branches $B(T)$ in the procedure under test, it can be defined as:

$$f_b(x_i) = B(x_i)/B(T). \tag{4}$$

Thus in w_{th} population, the fitness function of individual x_i can be given as:

$$f_{fit}(x_i, C_s(w)) = f_b(x_i) + \left[f_{avgsim} - f_{sim}(P_e(x_i), C_s(w))\right]^2. \tag{5}$$

The formula (5) is adopted as the fitness function of the individual x_i in this paper.

2.2 Genetic Operator Construction

Selection Operator. In each generation, individuals with higher fitness than average ones are regarded as elite individuals and are directly copied into the next generation population. The remaining individuals are selected by roulette strategy. In roulette strategy, the probability of an individual is selected depends on the proportion of the fitness of the individual to the fitness of all individuals. Taking the individual x_i of the w_{th} generation population N_W as an example, the accumulation probability of x_i can be expressed as: $P_{acc}(x_i) = \sum_{j=1}^{i} f_{fit}(x_j, C_s(W))$ thus the selection probability of individual x_i can be denoted as:

$$P(x_i) = f_{fit}(x_i, C_s(W))/ \sum_{j=1}^{N} f_{fit}(x_j, C_s(W)). \tag{6}$$

The probability interval of individual x_i is $(\sum_{j=1}^{i-1} f_{fit}(x_j, C_s(W)), \sum_{j=1}^{i} f_{fit}(x_j, C_s(W)))$, by generating a random number in interval $(0, \sum_{j=1}^{N} f_{fit}(x_j, C_s(W)))$, judging that the value belongs to which section is the individual selected in this roulette strategy.

Crossover Operator and Mutation Operator. Mutation operator simulates biological phenomenon of individual mutations to improve the diversity and avoid inbreeding through random changes. In genetic algorithm based on path coverage and branch coverage, the crossover and mutation operations are aimed at the specific value of the solution space and do not directly change the specific execution path. The binary string corresponding to the specific value is simulated as the chromosome gene of the individual in the genetic algorithm sequence. The crossover of the algorithm is that the two concrete solutions in the solution space exchange the binary code string with the crossover probability P_c according to the binary coding characteristic. The mutation of the algorithm is a specific solution to the mutation probability P_m of its binary string in a bit inversion operation. In this paper, we adopt the idea of adaptive crossover and mutation operator proposed by Srinivas and Patnaik [5]. The formulas of the adaptive crossover operator P_c and the mutation operator P_m are as follows:

$$P_c = \begin{cases} P_{cf}\left(f_{fit} - f_{avg}\right)\frac{P_{cf}-P_{cl}}{f_{max}-f_{avg}} & f_{avg} \leq f \\ P_{cf} & f < f_{avg} \end{cases}. \tag{7}$$

$$P_m = \begin{cases} P_{mf}(f - f_{avg}) \frac{P_{mf} - P_{ml}}{f_{max} - f_{avg}} & f_{avg} \leq f \\ P_{mf} & f < f_{avg} \end{cases}. \tag{8}$$

In Eq. (7), P_{cf} and P_{cl} are the crossover parameters at the time of algorithm initialization, thus the fitness interval of the crossover operator P_c is $[P_{cf}, P_{cl}]$, the same as P_m. f_{avg} is the average fitness of this generation and f_{max} is the more adaptive one of the two individuals to be crossed. By judging the relationship between individual fitness and the average fitness of the population, adaptively adjusting the crossover and mutation operations dynamically, the population can retain the high fitness and keep the species diversity of the whole population. This can effectively prevent the population from premature convergence to the local optimal solution and avoid the mutation probability becoming too large to evolve into random evolution.

3 Algorithm Steps

Steps for generating test case by using multi-path GA are described as follows:

Step1. Using static analysis to obtain T_s, C_s, B_s and other information, initialize basic parameters of the algorithm;

Step2. For each individual x_i in the population NW:
  (1) Execute x_i and collect the relevant information;
  (2) Delete the path executed in C_s, and delete the branch executed in B_s;
  (3) The fitness value and the whole space average fitness value are calculated.

Step3. If the termination condition is met, then go to Step4, otherwise go to Step7;

Step4. The residual solution is selected and the next generation solution space N_{W+1} is formed with individuals above the average fitness value;

Step5. For each individual x_i in the population N_{W+1}:
  (1) Randomly select two individuals according to P_c crossover operation;
  (2) Randomly select two individuals according to P_m mutation operation.
  Then go to Step2;

Step6. Stop evolution and output test cases.

4 Experiments and Results

Experiments are carried out on Lenovo ThinkCentre M8500t, with Intel i7 4790U CPU (3.6 GHz), 16 GB DDR3 1333 RAM and 64 bits Ubuntu 13.10 OS. The IDAPro software is used to perform the static analysis on executable software. The compilation environment for test cases is gcc4.4.7. The experiment of genetic algorithm is based on Visual Studio 2012 in C++ Genetic Algorithm Development Kit. The strategies and parameters related to GA are listed in Table 1. In this experiment, the coverage of the multi-path algorithm and the generation efficiency of the test case are tested to evaluate the superiority of our method.

Table 1. Parameters in the genetic algorithm

Parameter	Value
Selection operator	Elite and roulette selection
Crossover operator	Adaptive probabilities
Crossover rate	0.8
Mutation operator	Adaptive probabilities
Mutation rate	0.25
Gene coding	Binary coding
Population size	30
Termination generations	1000

4.1   The Coverage of the Multi-path Algorithm

In order to verify the coverage of the multi-path algorithm in this paper, we choose the three benchmark procedures, namely triangle classifier, three number sort, bubble sort. The end condition of the algorithm is to find all paths or reach the maximum number of iterations. Since each evolutionary time is only a few milliseconds, 100 total execution times are recorded as the execution time of the algorithm. Table 2 documents the basic information of the three benchmarks and the algorithm execution time and path coverage. Because there is an infeasible path in three number sorting and bubble sorting, it can be concluded that the test cases generated by this algorithm cover all the paths that can be covered.

Table 2. Basic information of the three benchmark program and its path coverage

Program	Number of theoretical paths	Branch structure	Number of components	Domain	Execution time	Path coverage
Triangle classifier	4	3 select	3	$[0,100]^3$	0.0652	100%
Three number sort	8	3 select	3	$[0,100]^3$	0.3794	87.5%
Bubble sort	8	2 loop 1 select	3	$[0,100]^3$	0.1063	87.5%

4.2   Test Case Generation Efficiency

In order to confirm the efficiency of our method, three other methods of generating test case are chosen for comparison: the single path method [6], Ahmed's method [7] and multi-path [3] methods. The single path method generates test case for a target path only once in a GA run. The fitness of an individual is the sum of the branch distance and the approach level which are not normalized. Ahmed's method generates test case for all target paths at one GA run, in which the fitness of an individual is the average of the sum of the branch distance and the approach level for all target paths. The multi-path method design the fitness function considering the matching degree of traversing path and each target path, and runs the GA once to generate the test data across all feasible paths.

The main difference between our method and the three other methods is the calculation of the fitness of an individual. Each method run 100 times independently, recording the time and number of iterations required to generate the test case, and calculating their averages. In Table 3, the fitness function is redesigned with the combination of path coverage and branch coverage, the improved selection operator and the adaptive crossover operator and mutation operator proposed by Srinivas are used on the genetic operator, which greatly reduces the number of iterations needed to achieve the optimal solution.

Table 3. Performance index value by using different methods in three benchmark program

Program	Method	Run time	Run time rate (%)	Average number of iterations	Iteration rate (%)
Triangle classifier	Single path method	0.4378	14.89	373.3	8.44
	Ahmed's method	0.2936	22.21	63.7	49.45
	Multi-path method	0.1526	42.73	34.1	92.38
	Our method	0.0652	100	31.5	100
Three number sort	Single path method	4.68	8.11	482.5	3.98
	Ahmed's method	0.825	45.99	46.9	40.94
	Multi-path method	0.3921	81.24	23.8	80.67
	Our method	0.3794	100	19.2	100
Bubble sort	Single path method	1.7642	6.03	230.4	10.98
	Ahmed's method	0.321	33.12	40.2	62.94
	Multi-path method	0.1325	80.23	32.0	79.06
	Our method	0.1063	100	25.3	100

5   Conclusion and Future Work

Finding an effective method for test case generation is one of the important contents of software testing research. In this paper, we redesigned the fitness function, and proposed a multi-path coverage algorithm based on path coverage and branch coverage, and modified the genetic operator. The proposed method is applied to three typical benchmark programs. The comparisons with single path method, Ahmed's method and multi-path method have shown that our method is superior the other ones from perspectives coverage rate and efficiency.

Acknowledgments. This paper is supported by the National Natural Science Foundation of China (Grant No. 61300169) and the Natural Science Foundation of Education Department of Anhui province (Grant No. KJ2016A257).

References

1. Mansour, N., Salame, M.: Data generation for path testing. Softw. Qual. J. **12**(2), 121–136 (2004)

2. Chen, Y., Zhong, Y.: Automatic path oriented test data generation using a multi population genetic algorithm. In: Proceedings of the 4th International Conference on Natural Computation, pp. 566–570. IPICNC, Jinan, China (2008)

3. Ahmed, M.A., Hermadi, I.: GA-based multiple paths test data generator. Comput. Oper. Res. **35**, 3107–3124 (2008)

4. Gong, D.W., Zhang, Y.: Novel evolutionary generation approach to test data for multiple paths coverage. Acta Electron. Sin. **38**(6), 1299–1304 (2010)

5. Srinivas, M., Patnaik, L.M.: Adaptive probabilities of crossover and mutation in Genetic Algorithm. IEEE Trans. Syst. Man Cybern. **24**(4), 656–667 (1994)

6. Gong, D.W., Yao, X.J., Zhang, Y.: Evolution Theory and Application for Testing Data Generation, 1st edn, pp. 8–32. Science Press, Beijing (2014)

7. Sthamer, H.H.: The automatic generation of software test data using genetic algorithms. Ph.D. thesis. University of Glamorgan, Pontyprid, Wales, UK, pp. 25–48 (1995)

Differential Evolution

An Adaptive Differential Evolution with Learning Parameters According to Groups Defined by the Rank of Objective Values

Tetsuyuki Takahama[1(✉)] and Setsuko Sakai[2]

[1] Hiroshima City University, 3-4-1 Ozuka-Higashi, Asaminami-ku 731-3194, Hiroshima, Japan
takahama@info.hiroshima-cu.ac.jp
[2] Hiroshima Shudo University, 1-1-1 Ozuka-Higashi, Asaminami-ku 731-3195, Hiroshima, Japan
setuko@shudo-u.ac.jp

Abstract. Differential Evolution (DE) has been successfully applied to various optimization problems. The performance of DE is affected by algorithm parameters such as a scaling factor F and a crossover rate CR. Many studies have been done to control the parameters adaptively. One of the most successful studies on controlling the parameters is JADE. In JADE, the values of each parameter are generated according to one probability density function (PDF) which is learned by the values in success cases where the child is better than the parent. However, search performance might be improved by learning multiple PDFs for each parameter based on some characteristics of search points. In this study, search points are divided into plural groups according to the rank of their objective values and the PDFs are learned by parameter values in success cases for each group. The advantage of JADE with the group-based learning is shown by solving thirteen benchmark problems.

Keywords: Adaptive differential evolution · Group-based learning · Differential evolution · Evolutionary algorithms

1 Introduction

Optimization problems, especially nonlinear optimization problems, are very important and frequently appear in the real world. There exist many studies on solving optimization problems using evolutionary algorithms (EAs). Differential evolution (DE) is an EA proposed by Storn and Price [9]. DE has been successfully applied to optimization problems including non-linear, non-differentiable, non-convex and multimodal functions [2,3,6]. It has been shown that DE is a very fast and robust algorithm.

The performance of DE is affected by algorithm parameters such as a scaling factor F, a crossover rate CR and population size, and by mutation strategies such as a rand strategy and a best strategy. Many studies have been done to

© Springer International Publishing AG 2017
Y. Tan et al. (Eds.): ICSI 2017, Part I, LNCS 10385, pp. 411–419, 2017.
DOI: 10.1007/978-3-319-61824-1_45

control the parameters and the strategies. One of the most successful studies on controlling the parameters is JADE (adaptive DE with optional external archive) [18]. In JADE, the values of parameters F and CR are generated according to the corresponding probability density function (PDF) and a child is created from the parent using the generated values. The values in success cases, where the child is better than the parent, are used to learn the PDFs. As for F, a location parameter of Cauchy distribution is learned, the scale parameter is fixed and values of F are generated according to the Cauchy distribution. As for CR, a mean of normal distribution is learned, the standard deviation is fixed and values of CR are generated according to the normal distribution. However, search performance might be improved by learning multiple PDFs for F and CR based on some characteristics of search points.

In this study, group-based learning of the PDFs is proposed. Search points are divided into plural groups according to the rank of their objective values. The PDFs are learned by parameter values in success cases for each group. The advantage of JADE with the group-based learning is shown by solving thirteen benchmark problems.

In Sect. 2, related works are described. DE and JADE are briefly explained in Sect. 3. In Sect. 4, JADE with the group-based learning is proposed. The experimental results are shown in Sect. 5. Finally, conclusions are described in Sect. 6.

2   Related Works

The performance of DE is affected by control parameters such as the scaling factor F, the crossover rate CR and the population size N, and by mutation strategies such as the rand strategy and the best strategy. Many researchers have been studying on controlling the parameters and the strategies.

The methods of controlling the parameters can be classified into some categories as follows:

(1) selection-based control: Strategies and parameter values are selected regardless of current search state. CoDE (composite DE) [15] generates three trial vectors using three strategies with randomly selected parameter values from parameter candidate sets and the best trial vector will head to the survivor selection.

(2) observation-based control: The current search state is observed, proper parameter values are inferred according to the observation, and parameters and/or strategies are dynamically controlled. FADE (Fuzzy Adaptive DE) [5] observes the movement of search points and the change of function values between successive generations, and controls F and CR. DESFC (DE with Speciation and Fuzzy Clustering) [10] adopts fuzzy clustering, observes partition entropy of search points, and controls CR and the mutation strategies between the rand and the species-best strategy. LMDE (DE with detecting Landscape Modality) [11,12] detects the landscape modality such as unimodal or multimodal using the change of the objective values at sampling

points which are equally spaced along a line. If the landscape is unimodal, greedy parameter settings for local search are selected. Otherwise, parameter settings for global search are selected.

(3) success-based control: It is recognized as a success case when a better search point than the parent is generated. The parameters and/or strategies are adjusted so that the values in the success cases are frequently used. It is thought that the self-adaptation, where parameters are contained in individuals and are evolved by applying evolutionary operators to the parameters, is included in this category. DESAP (DE with Self-Adapting Populations) [14] controls F, CR and N self-adaptively. SaDE (Self-adaptive DE) [7] controls the selection probability of the mutation strategies according to the success rates and controls the mean value of CR for each strategy according to the mean value in success case. jDE (self-adaptive DE algorithm) [1] controls F and CR self-adaptively. JADE (adaptive DE with optional external archive) [18] and MDE_pBX (modified DE with p-best crossover) [4] control the mean or power mean values of F and CR according to the mean values in success cases. CADE (Correlation-based Adaptive DE) [13] introduces the correlation of F and CR to JADE.

In the category (1), useful knowledge to improve the search efficiency is ignored. In the category (2), it is difficult to select proper type of observation which is independent of the optimization problem and its scale. In the category (3), when a new good search point is found near the parent, parameters are adjusted to the direction of convergence. In problems with ridge landscape or multimodal landscape, where good search points exist in small region, parameters are tuned for small success and big success will be missed. Thus, search process would be trapped at a local optimal solution. JADE adopted a weighted mean value for F, which is larger than a usual mean value, and succeeded to reduce the problem of the convergence.

In this study, we propose to improve JADE in the category (3) by introducing group-based learning according to the rank of objective values, which belongs the category (2). Thus, the proposed method is a hybrid method of the category (2) and (3).

3   Optimization by Differential Evolution

3.1   Optimization Problems

In this study, the following optimization problem with lower bound and upper bound constraints will be discussed.

$$\text{minimize } f(\boldsymbol{x}) \tag{1}$$
$$\text{subject to } l_j \le x_j \le u_j, \ j = 1, \ldots, D,$$

where $\boldsymbol{x} = (x_1, x_2, \cdots, x_D)$ is a D dimensional vector and $f(\boldsymbol{x})$ is an objective function. The function f is a nonlinear real-valued function. Values l_j and u_j are the lower bound and the upper bound of x_j, respectively.

3.2   Differential Evolution

In DE, initial individuals are randomly generated within given search space and form an initial population of size N. Each individual $\boldsymbol{x}^i, i = 1, 2, \cdots, N$ contains D genes as decision variables. At each generation, all individuals are selected as parents. Each parent is processed as follows: The mutation operation begins by choosing several individuals from the population except for the parent in the processing. The first individual is a base vector. All subsequent individuals are paired to create difference vectors. The difference vectors are scaled by a scaling factor F and added to the base vector. The resulting vector, or a mutant vector, is then recombined with the parent. The probability of recombination at an element is controlled by a crossover rate CR. This crossover operation produces a child, or a trial vector. Finally, for survivor selection, the trial vector is accepted for the next generation if the trial vector is better than the parent.

There are some variants of DE that have been proposed. The variants are classified using the notation DE/*base*/*num*/*cross* such as DE/rand/1/bin and DE/rand/1/exp.

"*base*" specifies a way of selecting an individual that will form the base vector. For example, DE/rand selects an individual for the base vector at random from the population. DE/best selects the best individual in the population.

"*num*" specifies the number of difference vectors used to perturb the base vector. In case of DE/rand/1, for example, for each parent $\boldsymbol{x}^i$, three individuals $\boldsymbol{x}^{p1}$, $\boldsymbol{x}^{p2}$ and $\boldsymbol{x}^{p3}$ are chosen randomly from the population without overlapping $\boldsymbol{x}^i$ and each other. A new vector, or a mutant vector $\boldsymbol{x}'$ is generated by the base vector $\boldsymbol{x}^{p1}$ and the difference vector $\boldsymbol{x}^{p2} - \boldsymbol{x}^{p3}$, where F is the scaling factor.

$$\boldsymbol{x}' = \boldsymbol{x}^{p1} + F(\boldsymbol{x}^{p2} - \boldsymbol{x}^{p3}) \tag{2}$$

"*cross*" specifies the type of crossover that is used to create a child. For example, 'bin' indicates that the crossover is controlled by the binomial crossover using a constant crossover rate, and 'exp' indicates that the crossover is controlled by a kind of two-point crossover using exponentially decreasing the crossover rate.

3.3   JADE

In JADE, the mean value of the scaling factor μ_F and the mean value of the crossover rate μ_{CR} are learned to define two PDFs, where initial values are $\mu_F = \mu_{CR} = 0.5$. The scaling factor F_i and the crossover rate CR_i for each individual $\boldsymbol{x}^i$ are independently generated according to the two PDFs as follows:

$$F_i \sim C(\mu_F, \sigma_F) \tag{3}$$
$$CR_i \sim N(\mu_{CR}, \sigma_{CR}^2) \tag{4}$$

where F_i is a random variable according to a Cauchy distribution $C(\mu_F, \sigma_F)$ with a location parameter μ_F and a scale parameter $\sigma_F = 0.1$. CR_i is a random variable according to a normal distribution $N(\mu_{CR}, \sigma_{CR}^2)$ of a mean μ_{CR} and a standard deviation $\sigma_{CR} = 0.1$. CR_i is truncated to $[0, 1]$ and F_i is truncated to be 1 if $F_i > 1$ or regenerated if $F_i \leq 0$. The location μ_F and the mean μ_{CR} are updated as follows:

$$\mu_F = (1-c)\mu_F + cS_{F^2}/S_F \tag{5}$$

$$\mu_{CR} = (1-c)\mu_{CR} + cS_{CR}/S_N \tag{6}$$

where S_N is the number of success cases, S_F, S_{F^2} and S_{CR} are the sum of F, F^2 and CR in success cases, respectively. A constant c is a weight of update in $(0,1]$ and the recommended value is 0.1.

JADE adopts a strategy called "current-to-pbest" where an intermediate point between a parent $\boldsymbol{x}^i$ and a randomly selected individual from top individuals is used as a base vector. A mutation vector is generated by current-to-pbest without archive as follows:

$$\boldsymbol{m} = \boldsymbol{x}^i + F_i(\boldsymbol{x}^{pbest} - \boldsymbol{x}^i) + F_i(\boldsymbol{x}^{r2} - \boldsymbol{x}^{r3}) \tag{7}$$

where $\boldsymbol{x}^{pbest}$ is a randomly selected individual from the top $100p\%$ individuals. The child $\boldsymbol{x}^{child}$ is generated from $\boldsymbol{x}^i$ and $\boldsymbol{m}$ using the binomial crossover.

In order to satisfy bound constraints, a child that is outside of the search space is moved into the inside of the search space. In JADE, each outside element of the child is set to be the middle between the corresponding boundary and the element of the parent as follows:

$$x_j^{child} = \begin{cases} \frac{1}{2}(l_j + x_j^i) & (x_j^{child} < l_j) \\ \frac{1}{2}(u_j + x_j^i) & (x_j^{child} > u_j) \end{cases} \tag{8}$$

This operation is applied when a new point is generated by JADE operations.

4   Proposed Method: Group-Based Learning

In this study, a population of individuals $\{\boldsymbol{x}^i \mid i = 1, 2, \cdots, N\}$ is divided into K groups according to a criterion, where N is the number of individuals and K is the number of groups. All individuals are sorted according to the criterion and the rank r_i $(r_i = 1, 2, \cdots, N)$ is assigned to each individual $\boldsymbol{x}^i$. In this study, the objective value of each individual is used as the criterion. The rank of the best individual, who has the best objective value, is 1. In case of $K = 2$, the individuals are divided into good individuals (group 1) and bad individuals (group 2).

The group ID of $\boldsymbol{x}^i$, $group(\boldsymbol{x}^i)$ is defined as follows:

$$group(\boldsymbol{x}^i) = \left\lceil \frac{r_i}{N} K \right\rceil \tag{9}$$

In order to realize group-based learning using parameter control of JADE, the following equations are adopted for each group $k = 1, \cdots, K$.

$$F_i \sim C(\mu_F^k, \sigma_F) \tag{10}$$

$$CR_i \sim N(\mu_{CR}^k, \sigma_{CR}^2) \tag{11}$$

$$\mu_F^k = (1-c)\mu_F^k + cS_{F^2}^k/S_F^k \tag{12}$$

$$\mu_{CR}^k = (1-c)\mu_{CR}^k + cS_{CR}^k/S_N^k \tag{13}$$

where μ_F^k is the location of Cauchy distribution for F in group k, μ_{CR}^k is the mean of normal distribution for CR in group k. S_N^k is the number of success cases in group k, where the better child than the parent is generated. S_F^k, S_{F2}^k and S_{CR}^k are the sum of F_i, F_i^2, CR_i at success cases in group k, respectively. As well as JADE, CR_i is truncated to $[0, 1]$ and F_i is truncated to be 1 if $F_i > 1$ or regenerated if $F_i \leq 0$.

5   Numerical Experiments

In this paper, well-known thirteen benchmark problems are solved by the proposed method ADEGL (Adaptive DE with Group-based Learning).

5.1   Test Problems and Experimental Conditions

The 13 scalable benchmark functions are sphere(f_1), Schwefel 2.22(f_2), Schwefel 1.2(f_3), Schwefel 2.21(f_4), Rosenbrock(f_5), step(f_6), noisy quartic(f_7), Schwefel 2.26(f_8), Rastrigin(f_9), Ackley(f_{10}), Griewank(f_{11}), and two penalized functions (f_{12} and f_{13}), respectively [17,18]. Every function has an optimal objective value 0. Some characteristics are briefly summarized as follows: Functions f_1 to f_4 are continuous unimodal functions. The function f_5 is Rosenbrock function which is unimodal for 2- and 3-dimensions but may have multiple minima in high dimension cases [8]. The function f_6 is a discontinuous step function, and f_7 is a noisy quartic function. Functions f_8 to f_{13} are multimodal functions and the number of their local minima increases exponentially with the problem dimension [16].

Experimental conditions are same as JADE as follows: Population size $N = 100$, initial mean for scaling factor $\mu_F = 0.5$ or $\mu_F^k = 0.5$ and initial mean for crossover rate $\mu_{CR} = 0.5$ or $\mu_{CR}^k = 0.5$, the pbest parameter $p = 0.05$, and the learning parameter $c = 0.1$.

Independent 50 runs are performed for 13 problems. The number of dimensions for the problems is 30 ($D = 30$). Each run stops when the number of function evaluations (FEs) exceeds the maximum number of evaluations $FE_{\max}$. In each function, different $FE_{\max}$ is adopted.

5.2   Experimental Results

Table 1 shows the experimental results on JADE, ADEGL ($K = 2$) and ADEGL ($K = 3$). The mean value and the standard deviation of best objective values in 50 runs are shown for each function. The maximum number of function evaluations is selected for each function and is shown in column labeled $FE_{\max}$. The best result among algorithms is highlighted using bold face fonts. Also, Wilcoxon signed rank test is performed and the result for each function is shown under the mean value. Symbols '+', '−' and '=' are shown when ADEGL is significantly better than JADE, is significantly worse than JADE, and is not significantly different from JADE, respectively. Symbols '++' and '−−' are shown when the significance level is 1% and '+' and '−' are shown when the significance level is 5%.

Table 1. Experimental results on 13 functions

	FE_{max}	JADE	ADEGL ($K=2$)	ADEGL ($K=3$)
f_1	150,000	9.38e−59 ± 6.5e−58	**4.32e−66 ± 1.3e−65**	3.36e−64 ± 2.2e−63
			++	++
f_2	200,000	4.19e−31 ± 2.4e−30	5.10e−32 ± 2.7e−31	**2.57e−37 ± 1.6e−36**
			=	++
f_3	500,000	**8.17e−62 ± 3.0e−61**	1.77e−59 ± 1.2e−58	2.25e−60 ± 1.5e−59
			=	=
f_4	500,000	2.01e−23 ± 9.8e−23	**1.20e−24 ± 4.3e−24**	3.70e−24 ± 1.0e−23
			+	=
f_5	300,000	5.78e−01 ± 3.5e+00	**7.97e−02 ± 5.6e−01**	7.26e−01 ± 3.5e+00
			=	=
f_6	10,000	3.02e+00 ± 1.3e+00	**1.78e+00 ± 1.2e+00**	1.98e+00 ± 1.1e+00
			++	++
f_7	300,000	**6.04e−04 ± 2.4e−04**	7.11e−04 ± 2.3e−04	6.80e−04 ± 2.2e−04
			=	=
f_8	100,000	2.37e+00 ± 1.7e+01	**2.46e−05 ± 3.1e−05**	1.18e+01 ± 3.6e+01
			++	+
f_9	100,000	1.01e−04 ± 3.9e−05	**5.64e−05 ± 2.8e−05**	5.95e−05 ± 3.0e−05
			++	++
f_{10}	50,000	9.20e−10 ± 6.4e−10	4.22e−10 ± 3.0e−10	**3.41e−10 ± 3.1e−10**
			++	++
f_{11}	50,000	**1.15e−08 ± 6.9e−08**	1.97e−04 ± 1.4e−03	3.46e−04 ± 1.7e−03
			+	=
f_{12}	50,000	2.40e−16 ± 1.6e−15	4.99e−18 ± 2.6e−17	**1.37e−18 ± 5.5e−18**
			++	++
f_{13}	50,000	1.15e−16 ± 2.2e−16	2.17e−17 ± 5.1e−17	**1.69e−17 ± 7.5e−17**
			++	++
+		—	9	8
=		—	4	5
−		—	0	0

ADEGL ($K=2$) attained best mean results in 6 functions f_1, f_4, f_5, f_6, f_8 and f_9 out of 13 functions. ADEGL ($K=3$) attained best mean results in 4 functions f_2, f_{10}, f_{12} and f_{13}. JADE attained best mean results in 3 functions f_3, f_7 and f_{11}. Also, ADEGL ($K=2$) attained significantly better results than JADE in 9 functions f_1, f_4, f_6, f_8, f_9, f_{10}, f_{11}, f_{12} and f_{13}. ADEGL ($K=3$) attained significantly better results than JADE in 8 functions f_1, f_2, f_6, f_8, f_9, f_{10}, f_{12} and f_{13}. Thus, it is thought that ADEGL ($K=2$) is the best method among 3 methods and ADEGL ($K=3$) is the second best method. JADE could not attain significantly better results than ADEGL of $K=2$ nor $K=3$.

Figure 1 shows the change of F and CR over the number of function evaluations for f_1 in case of $K = 2$. ADEGL ($K = 2$) tends to learn smaller values of F and CR than those of JADE for the best group (group 1) and larger values of F and CR than those of JADE for the worst group (group 2).

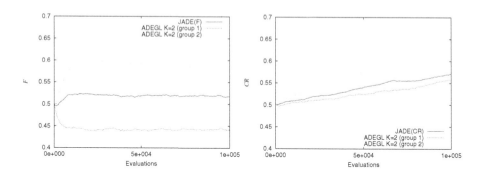

Fig. 1. The graph of F and CR in f_1

6   Conclusion

In this study, group-based learning of algorithm parameters is proposed, where individuals are divided into plural groups according to the rank of objective values and the parameters are learned for each group. DE with group learning is applied optimization of various 13 functions including unimodal functions, a function with ridge structure, multimodal functions. It is shown that the proposed method ADEGL is effective compared with JADE. Also, it is shown that parameters for good individuals are controlled to intensify convergence and parameters for bad individuals are controlled to keep divergence.

In the future, we will apply group-based learning to other adaptive optimization algorithms including differential evolution and particle swarm optimization.

Acknowledgments. This study is supported by JSPS KAKENHI Grant Numbers 26350443 and 17K00311.

References

1. Brest, J., Greiner, S., Boskovic, B., Mernik, M., Zumer, V.: Self-adapting control parameters in differential evolution: a comparative study on numerical benchmark problems. IEEE Trans. Evol. Comput. **10**(6), 646–657 (2006)
2. Chakraborty, U.K. (ed.): Advances in Differential Evolution. Springer, Heidelberg (2008)
3. Das, S., Suganthan, P.: Differential evolution: a survey of the state-of-the-art. IEEE Trans. Evol. Comput. **15**(1), 4–31 (2011)

4. Islam, S.M., Das, S., Ghosh, S., Roy, S., Suganthan, P.N.: An adaptive differential evolution algorithm with novel mutation and crossover strategies for global numerical optimization. IEEE Trans. Syst. Man Cybern. Part B: Cybern. **42**(2), 482–500 (2012)
5. Liu, J., Lampinen, J.: A fuzzy adaptive differential evolution algorithm. Soft. Comput. **9**(6), 448–462 (2005)
6. Price, K., Storn, R., Lampinen, J.A.: Differential Evolution: A Practical Approach to Global Optimization. Springer, Heidelberg (2005)
7. Qin, A., Huang, V., Suganthan, P.: Differential evolution algorithm with strategy adaptation for global numerical optimization. IEEE Trans. Evol. Comput. **13**(2), 398–417 (2009)
8. Shang, Y.W., Qiu, Y.H.: A note on the extended Rosenbrock function. Evol. Comput. **14**(1), 119–126 (2006)
9. Storn, R., Price, K.: Differential evolution - a simple and efficient heuristic for global optimization over continuous spaces. J. Global Optim. **11**, 341–359 (1997)
10. Takahama, T., Sakai, S.: Fuzzy c-means clustering and partition entropy for species-best strategy and search mode selection in nonlinear optimization by differential evolution. In: Proceedings of the 2011 IEEE International Conference on Fuzzy Systems, pp. 290–297 (2011)
11. Takahama, T., Sakai, S.: Differential evolution with dynamic strategy and parameter selection by detecting landscape modality. In: Proceedings of the 2012 IEEE Congress on Evolutionary Computation, pp. 2114–2121 (2012)
12. Takahama, T., Sakai, S.: Large scale optimization by differential evolution with landscape modality detection and a diversity archive. In: Proceedings of the 2012 IEEE Congress on Evolutionary Computation, pp. 2842–2849 (2012)
13. Takahama, T., Sakai, S.: An adaptive differential evolution considering correlation of two algorithm parameters. In: Proceedings of the Joint 7th International Conference on Soft Computing and Intelligent Systems and 15th International Symposium on Advanced Intelligent Systems (SCIS&ISIS2014), pp. 618–623 (2014)
14. Teo, J.: Exploring dynamic self-adaptive populations in differential evolution. Soft. Comput. **10**(8), 673–686 (2006)
15. Wang, Y., Cai, Z., Zhang, Q.: Differential evolution with composite trial vector generation strategies and control parameters. IEEE Trans. Evol. Comput. **15**(1), 55–66 (2011)
16. Yao, X., Liu, Y., Lin, G.: Evolutionary programming made faster. IEEE Trans. Evol. Comput. **3**, 82–102 (1999)
17. Yao, X., Liu, Y., Liang, K.H., Lin, G.: Fast evolutionary algorithms. In: Ghosh, A., Tsutsui, S. (eds.) Advances in Evolutionary Computing: Theory and Applications, pp. 45–94. Springer-Verlag New York, Inc., New York (2003). doi:10.1007/978-3-642-18965-4_2
18. Zhang, J., Sanderson, A.C.: JADE: adaptive differential evolution with optional external archive. IEEE Trans. Evol. Comput. **13**(5), 945–958 (2009)

Comparison of Differential Evolution Algorithms on the Mapping Between Problems and Penalty Parameters

Chengyong Si[1(✉)], Jianqiang Shen[1], Xuan Zou[1], and Lei Wang[2]

[1] Shanghai-Hamburg College, University of Shanghai for Science
and Technology, Shanghai 200093, China
sichengyong_sh@163.com
[2] College of Electronics and Information Engineering,
Tongji University, Shanghai 201804, China
wanglei@tongji.edu.cn

Abstract. Penalty parameters play a key role when adopting the penalty function method for solution ranking. In the previous study, a corresponding relationship between the constrained optimization problems and the penalty parameters was constructed. This paper tries to verify whether the relationship is related with the evolutionary algorithms (EAs), i.e., how the EAs influence the relationship. Two differential evolution algorithms are taken as an example. Experimental results confirm the influence and show that an improved EA will enlarge the available value of corresponding penalty parameter, especially for the intermittent relationship. The findings also prove that EA can make up the shortcoming of constraint handling techniques to some extent.

Keywords: Constrained optimization · Constraint handling techniques · Differential evolution · Penalty parameter · Algorithm selection

1 Introduction

Constrained Optimization Problems (COPs) are very common and important in the real-world applications. The COPs can be generally expressed by the following formulations:

$$
\begin{aligned}
\text{Minimize} \quad & f(\vec{x}) \\
\text{Subject to:} \quad & g_j(\vec{x}) \leq 0, \quad j = 1, \cdots, l \\
& h_j(\vec{x}) = 0, \quad j = l+1, \cdots, m
\end{aligned}
\tag{1}
$$

where $\vec{x} = (x_1, \cdots, x_n)$ is the decision variable. The decision variable is bounded by the decision space S which is defined by the constraints:

$$
L_i \leq x_i \leq U_i, \quad 1 \leq i \leq n .
\tag{2}
$$

where l is the number of inequality constraints and $m - l$ is the number of equality constraints.

© Springer International Publishing AG 2017
Y. Tan et al. (Eds.): ICSI 2017, Part I, LNCS 10385, pp. 420–428, 2017.
DOI: 10.1007/978-3-319-61824-1_46

The Evolutionary Algorithms (EAs) are essentially unconstraint search techniques [1] and play an important role in generating solutions. After solution generating, how to choose the better solutions especially for the COPs is another equivalently important issue, which leads to the development of various constrained optimization evolutionary algorithms (COEAs) [2]. The three most frequently used constraint handling techniques (CHTs) in COEAs are penalty functions, biasing feasible over infeasible solutions and multi-objective optimization.

Besides these basic CHTs, some other concepts like cooperative coevolution [3] and ensemble [4, 5] have also been proposed, which can be seen as a dynamic adjustment process. Also, some other dynamic approaches based on the three different situations in solving COPs [6] have been developed.

Among all of these aforementioned methods, the problem characteristics are rarely considered. But as Michalewicz summarized [7], it seems that all the components of Evolutionary Algorithms might be problem-specific.

Some researchers have emphasized the importance of the relationship between problem characteristics and algorithms, and have tried to realize some simple combination of algorithm variants, although the results are not very satisfactory. For example, Tsang and Kwan [8] pointed out the need to map constraint satisfaction problems to algorithms and heuristics. But they did not give an exact relationship between them. Mezura-Montes *et al.* [9] proposed a simple combination of two DE variants (i.e., DE/rand/1/bin and DE/best/1/bin) based on the empirical analysis of four DE variants. Gibbs *et al.* [10] identified the relationship between the optimal number of GA generations and the problem characteristics, through quantifying different problem characteristics of unconstrained problems.

Other methods concerning the problem characteristics were also reported [11, 12]. As presented in [11], a method to construct the relationship between problems and algorithms as well as constraint handling techniques from the qualitative and quantitative point of view was proposed. In the paper, the problem characteristics were also summarized systematically. In [12], a new framework combining promising aspects of two different Constraint Handling Techniques (CHTs) in different situations with consideration of problem characteristics was proposed.

In the previous paper [13], a corresponding relationship between the problems and the penalty parameters was confirmed through an empirical study based on different classification of benchmark functions. But whether the relationship is related with the evolutionary algorithms are not studied. In this work, we try to study this effect by comparing a new DE algorithm with the original one.

The rest of this paper is organized as follows. Section 2 briefly introduces DE. Section 3 illustrates the basic idea of this paper and the penalty parameter setting. The experimental results and analysis are presented in Sect. 4. Finally, Sect. 5 concludes this paper and provides some possible paths for future research.

2 Differential Evolution (DE)

DE, which was proposed by Storn and Price [14], is a simple and efficient EA. The mutation, crossover and selection operations are introduced in DE. The first two operations are used to generate a trial vector to compete with the target vector while the third one is used to choose the better one for the next generation. To date several variants of DE have been proposed [15].

The population of DE consists of NP n-dimensional real-valued vectors

$$\vec{x}_i = \{x_{i,1}, x_{i,2}, \ldots, x_{i,n}\}, \quad i = 1, 2, \ldots, NP. \tag{3}$$

The mutation, crossover and selection operations are defined as follows.

A. *Mutation Operation*

Taking into account each individual $\vec{x}_i$ (named a target vector), a mutant vector $\vec{v}_i = \{v_{i,1}, v_{i,2}, \ldots, v_{i,n}\}$ is defined as

$$\vec{v}_i = \vec{x}_{r1} + F \cdot (\vec{x}_{r2} - \vec{x}_{r3}). \tag{4}$$

where $r1$, $r2$ and $r3$ are randomly selected from [1, NP] and satisfying: $r1 \neq r2 \neq r3 \neq i$ and F is the scaling factor.

If $v_{i,j}$ violates the boundary constraint, it will be reset as follows [16]:

$$v_{i,j} = \begin{cases} \min\{U_j, 2L_j - v_{i,j}\}, & \text{if} \quad v_{i,j} < L_j \\ \max\{L_j, 2U_j - v_{i,j}\}, & \text{if} \quad v_{i,j} > U_j \end{cases}. \tag{5}$$

B. *Crossover Operation*

A trial vector $\vec{u}_i$ is generated through the binomial crossover operation on the target vector $\vec{x}_i$ and the mutant vector $\vec{v}_i$

$$u_{i,j} = \begin{cases} v_{i,j} & \text{if } rand_j \text{ or } j = j_{rand} \\ x_{i,j} & \text{otherwise} \end{cases}. \tag{6}$$

where $i = 1, 2, \ldots, NP$, $j = 1, 2, \ldots, n$, j_{rand} is a randomly chosen integer within the range [1, n], $rand_j$ is the jth evaluation of a uniform random number generator within [0, 1], and C_r is the crossover control parameter. The introduction of $j = j_{rand}$ can guarantee the trial vector $\vec{u}_i$ is different from its target vector $\vec{x}_i$.

C. *Selection Operation*

Selection operation is realized by comparing the trial vector $\vec{u}_i$ against the target vector $\vec{x}_i$ and the better one will be preserved for the next generation.

$$\vec{x}_i = \begin{cases} \vec{u}_i & if \ f(\vec{u}_i) \le f(\vec{x}_i) \\ \vec{x}_i & otherwise \end{cases}. \tag{7}$$

3   Comparison of Differential Evolution on the Mapping

3.1   Basic Idea

In the previous paper, the basic idea for mapping constrained optimization problems and penalty parameters are illustrated, i.e., by using different penalty parameters to solve different problems, the types of problems that the penalty parameters are good at solving can be obtained and summarized, and then the corresponding relationship can be constructed.

In this paper, the basic idea is similar (as in Fig. 1), but as the main aim is to check whether the evolutionary algorithm will influence the relationship, for page limited, the problem characteristics will not be in detail. In Fig. 1, Ori-DE means DE adopted in the previous paper [13].

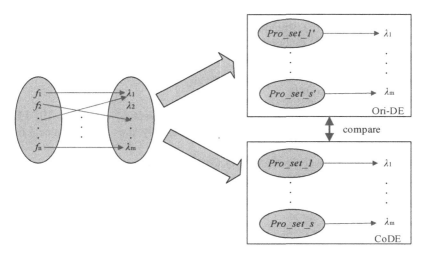

Fig. 1. Illustration of the basic idea

3.2   Penalty Parameter Setting

The penalty parameter setting is the same as in the previous paper.

As analyzed in [18], if the penalty parameter exceeds some value (λ_{max}), or is less than some value (λ_{min}), the ranking will not be influenced. Here, λ_{max} and λ_{min} are determined by the current population.

Based on this conclusion, the penalty parameter setting in this experiment is selected ranging from 0.001 to 100000. As there is no guidelines how to set the penalty parameters in such case so as to distinguish the effect better, a simple method with the same scale (i.e., 0, 0.001, 0.01, 0.1, 1, 10, 100, 1000, 10000, 100000) is adopted.

4 Experimental Study

4.1 Experimental Settings

23 benchmark functions [19] were used in our experiment.

In the previous paper, *DE/rand/1/bin* was adopted as the search algorithm. In this paper, CoDE [17] was adopted as a comparison algorithm. In CoDE, several trial vector generation strategies with a number of control parameter settings are randomly combined at each generation to create new trial vectors. The three selected trial vector generation strategies are *DE/rand/1/bin, DE/rand/2/bin, and DE/current-to-rand/1*. The three control parameter settings are $[F = 1.0, Cr = 0.1]$, $[F = 1.0, Cr = 0.9]$, and $[F = 0.8, Cr = 0.2]$. It should be pointed out that some minor changes have been made to the selection operation of CoDE, i.e., the offspring are selected from the pool which is composed by all the trial vectors and target vectors.

The parameters in the Ori-DE are set as follows: the population size (*NP*) is set to 100; the scaling factor (*F*) is randomly chosen between 0.5 and 0.6, and the crossover control parameter (*Cr*) is randomly chosen between 0.9 and 0.95.

4.2 Experimental Results

25 independent runs were performed for each test function using 5×10^5 FES at maximum, as suggested by Liang *et al.* [19]. Additionally, the tolerance value δ for the equality constraints was set to 0.0001.

Table 1 shows the succeed rate (SR) for different penalty parameters on different problems with DE and CoDE respectively.

4.3 Comparison

In this section, we compare the results by Ori-DE and CoDE, to check the effect of different EAs on the mapping between COPs and penalty parameters.

The summarized results are listed in Table 2. Here, in the first column "result", "Always" means the results will not be affected by the penalty parameters; "Continuity" means when penalty parameter is larger than certain value (the value adopted in this paper, not so exact), it can always find the optimal value; "Intermittent" means that the optimal value can be obtained only with certain penalty parameter values (also limited to the values adopted in this paper).

Table 1. Success rate for different parameters with Ori-DE and CoDE

Prob.		λ									
		0.001	0.01	0.1	0	1	10	100	1000	10000	100000
g01	Ori-DE	0	0	0	0	1	1	1	1	1	1
	CoDE	0	0	0	0	1	1	1	1	1	1
g02	Ori-DE	0	0	0	0	0.92	0.88	0.92	0.72	0.76	0.80
	CoDE	0	0	0	0	0.52	0.56	0.44	0.52	0.48	0.48
g03	Ori-DE	0	0	0	0	0	0	0	**0.28**	0	0.04
	CoDE	0	0	0	0	0	0	0	**0.24**	0	0
g04	Ori-DE	0	0	0	0	0	0	0	1	1	1
	CoDE	0	0	0	0	0	0	0	1	1	1
g05	Ori-DE	0	0	0	0	0	1	1	0.24	0	0
	CoDE	0	0	0	0	0	1	1	**1**	1	0.80
g06	Ori-DE	0	0	0	0	0	0	0	0	**0.56**	0
	CoDE	0	0	0	0	0	0	0	0	1	1
g07	Ori-DE	0	0	0	0	0	1	1	1	1	1
	CoDE	0	0	0	0	0	1	1	1	1	1
g08	Ori-DE	0.16	0.12	0.40	0.16	0.32	0.68	1	1	1	1
	CoDE	0	0	0	0	0	0	1	1	1	1
g09	Ori-DE	0	0	0	0	0	1	1	1	1	1
	CoDE	0	0	0	0	0	1	1	1	1	1
g10	Ori-DE	0	0	0	0	0	0	0	0	**0.32**	0
	CoDE	0	0	0	0	0	0	0	0	**1**	0
g11	Ori-DE	0	0	0	0	1	1	1	1	0.60	0.68
	CoDE	0	0	0	0	0	1	1	1	1	1
g12	Ori-DE	1	1	1	1	1	1	1	1	1	1
	CoDE	1	1	1	1	1	1	1	1	1	1
g13	Ori-DE	0	0	**1**	0	0.20	0	0	0	0	0
	CoDE	0	0	**1**	0	0.72	0.28	0	0	0	0
g14	Ori-DE	0	0	0	0	0	0	1	0.96	0.48	0.48
	CoDE	0	0	0	0	0	0	1	1	1	1
g15	Ori-DE	0	0	0	0	0	1	0.36	0.16	0.04	0.20
	CoDE	0	0	0	0	0	1	1	1	0.96	0.80
g16	Ori-DE	0	0	0	0	1	1	1	1	1	1
	CoDE	0	0	0	0	1	1	1	1	1	1
g17	Ori-DE	0	0	0	0	0	–	**1**	0	0	0
	CoDE	0	0	0	0	0	0	**1**	0.64	0	0
g18	Ori-DE	0	0	0	0	1	1	1	1	1	1
	CoDE	0	0	0	0	0.76	1	1	1	1	0.92
g19	Ori-DE	0	0	0	0	1	1	1	1	1	1
	CoDE	0	0	0	0	1	1	1	1	1	1
g21	Ori-DE	0	0	0	0	0	0	0	0	0	0

(*continued*)

Table 1. (*continued*)

Prob.		λ									
		0.001	0.01	0.1	0	1	10	100	1000	10000	100000
	CoDE	0	0	0	0	0	0	0	0.24	0.28	0.04
g23	Ori-DE	0	0	0	0	0	0	0	**1**	0	0
	CoDE	0	0	0	0	0	0	0	**1**	0.72	0
g24	Ori-DE	0	0	0	0	1	1	1	1	1	1
	CoDE	0	0	0	0	1	1	1	1	1	1

From Table 2, some interesting finding can be observed:

(1) For the class of "Continuity", the corresponding problems almost keep the same compared to the original results, except for g11, from $\lambda \geq 1$ to $\lambda \geq 10$. The difference lies in SR, with some getting larger, e.g., g14, and some getting smaller, e.g., g02 and g18.

(2) For the class of "Intermittent", almost all the corresponding λ changed. Some value range gets larger, e.g., g13, g17, g23, with some becoming continuity, e.g., g05, g06, and g15; some keeps the same, e.g., g03 and g10, but SR changes.

(3) For all the problems, we can see that there is no difference on these λ with SR = 1, while others change with different EAs.

(4) Good EAs can improve SR of some problems, e.g., g05, g06, g10, g14, g15, but not in all problems, e.g., g08. The consequent question is how to judge or evaluate a good evolutionary algorithm, or why some parameter settings perform better than others. This finding can help to illustrate that EAs can improve some drawbacks of CHTs to some extent. But how much can be improved is the next research topic we will do in the future.

Table 2. Comparison of results by Ori-DE and CoDE

Result	λ	Problem	
		Ori-DE	CoDE
Always	Any	g08, g12	g12
Continuity	$\lambda \geq 1$	g01, g16, g18, g19, g24, g02, g11	g01, g16, g18, g19, g24, g02
	$\lambda \geq 10$	g07, g09	g07, g09, g11, g15, g05
	$\lambda \geq 100$	g14	g08, g14
	$\lambda \geq 1000$	g04	g04
	$\lambda \geq 10000$	-	g06
Intermittent	0.1	g13	-
	10	g15	-
	100	g17	-
	1000	g03, g23	g03
	10000	g06, g10	g10
	10/100/1000	g05	-
	0.1/1	-	g13
	100/1000	-	g17
	1000/10000	-	g23

5   Conclusion

In this paper, two differential evolution algorithms are adopted to verify how the EAs influence the relationship between the constrained optimization problems and the penalty parameters obtained by the previous study. The experimental results show that the corresponding relationship will be affected by the EAs, and EAs can improve some drawbacks of CHTs to some extent.

This research can be seen as just an example toward this research direction, as only one kind of algorithm (i.e., DE) was adopted. What is the inner mechanism behind these features, and how much can the EAs improve the CHTs, as well as which factors are involved will be our future research.

Acknowledgments. This work was supported in part by the National Natural Science Foundation of China under Grants 71371142, Shanghai Young Teachers' Training Program under Grants ZZslg15087.

References

1. Mezura-Montes, E., Coello, C.A.C.: A simple multimembered evolution strategy to solve constrained optimization problems. IEEE Trans. Evol. Comput. **9**(1), 1–17 (2005)
2. Mezura-Montes, E., Coello, C.A.C.: Constraint-handling in nature-inspired numerical optimization: past, present and future. Swarm Evol. Comput. **1**(4), 173–194 (2011)
3. Li, X., Yao, X.: Cooperatively coevolving particle swarm for large scale optimization. IEEE Tran. Evol. Comput. **16**(2), 210–224 (2012)
4. Zhao, S.Z., Suganthan, P.N., Zhang, Q.: Decomposition based multiobjetive evolutionary algorithm with an ensemble of neighborhood sizes. IEEE Trans. Evol. Comput. **16**(3), 442–446 (2012)
5. Mallipeddi, R., Suganthan, P.N.: Ensemble of constraint handling techniques. IEEE Tran. Evol. Comput. **14**(4), 561–579 (2010)
6. Wang, Y., Cai, Z., Zhou, Y., Zeng, W.: An adaptive tradeoff model for constrained evolutionary optimization. IEEE Trans. Evol. Comput. **12**(1), 80–92 (2008)
7. Michalewicz, Z.: Quo Vadis, evolutionary computation? on a growing gap between theory and practice. In: Liu, J., Alippi, C., Bouchon-Meunier, B., Greenwood, Garrison W., Abbass, Hussein A. (eds.) WCCI 2012. LNCS, vol. 7311, pp. 98–121. Springer, Heidelberg (2012). doi:10.1007/978-3-642-30687-7_6
8. Tsang, E., Kwan, A.: Mapping constraint satisfaction problems to algorithms and heuristics. Technical report, CSM-198 (1993)
9. Mezura-Montes, E., Miranda-Varela, M.E., Gómez-Ramón, R.C.: Differential evolution in constrained numerical optimization: an empirical study. Inform. Sci. **180**(22), 4223–4262 (2010)
10. Gibbs, M., Maier, H., Dandy, G.: Relationship between problem characteristics and the optimal number of genetic algorithm generations. Eng. Optim. **43**(4), 349–376 (2011)
11. Si, C., Wang, L., Wu, Q.: Mapping constrained optimization problems to algorithms and constraint handling techniques. In: Proceedings of the CEC, pp. 3308–3315 (2012)
12. Si, C., Hu, J., Lan, T., Wang, L., Wu, Q.: A combined constraint handling framework: an empirical study. Memet. Comput. **9**(1), 69–88 (2017)

13. Si, C., Shen, J., Zou, X., Wang, L., Wu, Q.: Mapping constrained optimization problems to penalty parameters: an empirical study. In: Proceedings of the CEC, pp. 3073–3079 (2014)
14. Storn, R., Price, K.: Differential evolution: a simple and efficient adaptive scheme for global optimization over continuous spaces. Technical report, TR-95-012 (1995)
15. Das, S., Suganthan, P.N.: Differential evolution: a survey of the state-of-the-art. IEEE Trans. Evol. Comput. **15**(1), 4–31 (2011)
16. Menchaca-Mendez, K., Coello Coello, C.A.: Solving multiobjective optimization problems using differential evolution and a maximin selection criterion. In: Proceedings of the CEC, pp. 3143–3150 (2012)
17. Wang, Y., Cai, Z., Zhang, Q.: Differential evolution with composite trial vector generation strategies and control parameters. IEEE Trans. Evol. Comput. **15**(1), 55–66 (2011)
18. Si, C., Lan, T., Hu, J., Wang, L., Wu, Q.: On the penalty parameter of the penalty function method. Control Decis. **9**, 1707–1710 (2014)
19. Liang, J.J., Runarsson, T.P., Mezura-Montes, E., Clerc, M., Suganthan, P.N., Coello Coello, C.A., Deb, K.: Problem definitions and evaluation criteria for the CEC 2006. Technical report, Special Session on Constrained Real-Parameter Optimization (2006)

Cooperation Coevolution Differential Evolution with Gradient Descent Strategy for Large Scale

Chen Yating[⊠]

School of Date and Computer Science,
Sun Yat-Sen University, 510275 Guangzhou, China
yating.chen@foxmail.com

Abstract. In order to better solve the large scale optimization problem, we propose a cooperation coevolution differential evolutionary (CCDE) algorithm with a gradient descent strategy (GDS). The GDS based CCDE algorithm (CCDE/GDS) benefits the solution of large scale optimization problems in two critical aspects. Firstly, the optimization turned out to be far less time consuming due to that GDS is helpful for guiding the search direction on the globally best individual position. More importantly, the GDS is controlled by an elastic operator to be carried out only when the globally best individual has been trapped, making the algorithm fast respond to the large scale evolutionary environment. Secondly, GDS was reported in the literature to approximate the local best value on most object functions. Therefore, the GDS used in CCDE can promote the globally best individual position to more promising region when it is trapped into local optimum, so as to achieve high accuracy. We designed experiments on CEC2010 benchmark functions for evaluating our newly proposed algorithm, which shows that the proposed algorithm and modified framework can obtain very competitive results on the large scale optimization problem efficiently.

Keywords: Cooperation coevolution differential evolutionary · Gradient descent strategy · Large scale optimization problem

1 Introduction

In the case of numerical and combinatorial optimization, evolutionary optimization algorithms have made a huge impact [1]. However, the performance of most existing evolutionary algorithms (EA) still seriously suffered from increases on the dimensionality of the search space increases [1]. For the past few years, studies on large scale optimization have been paid much more attention because of its important effect on theoretical research and practice. Some inchoate efforts focusing large multidimensional problems led to unrealistic or even false assumptions, in which the problem is treated as a separable object. Different from those early researches, this paper will put emphasis on non-separable functions for avoiding small sub-problems brought by separable functions and making up their drawbacks for multi-dimension.

Cooperative coevolution (CC) specializes in segmentation processing large and complicated problems [2], which is usually seen as a kind of automatic implementing

Y. Tan et al. (Eds.): ICSI 2017, Part I, LNCS 10385, pp. 429–439, 2017.
DOI: 10.1007/978-3-319-61824-1_47

method for the "divide-and-conquer" strategy. A CC algorithm would break down an original problem into several sub-components under ideal conditions, which can min-imize interdependencies between different subcomponents. There are two simple decomposition strategies of algorithms in [3–6] based on the standard framework of CC: the "one-dimensional based" and "splitting-in-half". The former attempts to dis-assemble a multi-dimensional problem into several one-dimensional sub-problems. It is straightforward for separable functions but incapable of tackling non-separable prob-lems satisfactorily. The later strategy splits a multi-dimensional problem in half, which is still unclear details of interdependencies between diverse variables that can be obtained for non-separable problems. Moreover, existing CC-based algorithms do not consider more effective evolutionary optimization algorithms that have been proposed in recent years. Thus, there is still room for improving the CC algorithms for better handling large and complex problems as more advanced are adopted.

The remainder of this paper is organized as follows: Sect. 2 summarizes the background including CC as well as CCDE-G. Section 3 presents the CCDE/GDS algorithm in detail. Section 4 introduces the evaluation experiments on a diverse set of CEC2010 functions and analyses the test results in detail. And the conclusions in Sect. 5 summarize this paper and discuss future research activities based on this paper.

2 Background

2.1 Cooperative Coevolution (CC)

CC is designed as a generic framework for evolutionary optimization algorithm on large or complex problems making use of the "divide-and-conquer" strategy. The objective system is decomposed into several submodules belonging to a subpopulation which separately goes through evolution separately with only cooperation occurrence. When it comes to optimization process on multi-dimensional problems, main opera-tions of the CC framework can be described as follows:

1. Problem division: Divide a multi-dimensional objective vector into m low-dimensional child components which can be processed by traditional EA.
2. Loop initiation: Set i = 1 and recommence a new loop.
3. Child component optimization: Evolve the ith child component separately by some evolutionary algorithm under a predetermined number of fitness evolutions. If i < m then i++, and continue optimization operations.
4. Child component coadaptation: Coadaptation is crucial when capturing such interdependencies during optimization, because interdependencies may exist between subcomponents. Stop once terminal conditions are met; or else jump to Step 2 for a new loop.

To be noted, an integral loop here covers all child components and their complete evolution. Once a multi-dimensional problem is first disassembled into several low-dimensional child components, these child components would evolve cooperatively within a predetermined number of loops. The size of any one of subcomponents should be within the evolutionary algorithm's optimization ability.

2.2 Cooperation Coevolution Differential Evolution Grouping (EACC-G)

Cooperation coevolution differential evolution grouping (EACC-G) [7] was proposed with a group-based problem decomposition strategy which groups structures in dynamically changing. It can raise the possibility of optimizing interacting variables together, and maintain coadaptation among diverse groups for overall controlling. The core idea of EACC-G framework is a two-step process to divide a complete vector into child components in lower dimensions, and process all child components respectively by means of an evolutionary algorithm. Also, EACC-G employs an adaptive weighting strategy for coadaptation that assigns a weight to every child component when a loop ends, and evolves the relevant weight vector with come optimizers. In order to make the paper self-contained, the core process of EACC-G described in [7] is also briefly introduced herein as the following steps:

1. Set $i = 1$ and recommence a new loop.
2. Divide a n-dimensional vector into m child components in lower dimension stochastically.
3. Compute the ith child component separately by using a conventional EA for a predetermined ceiling of fitness evolutions. If $i < m$, let $i + +$ and re-access optimization operations.
4. Assign a weight to every child component. Perform evolutions of the weight vectors to get three key parameters of current population: the best, the worst and stochastic members.
5. Stop once terminal conditions are met; otherwise re-access Step 1 for a new loop.

The major distinctions between EACC-G and those conventional CCs, for instance in [3, 4], are a dynamically changing grouping structure for variables as well as a novel adaptive weighting which maintains coadaptation among child components at the end of each loop. After each loop, a weight vector will be build up for each of these child components. Because the determination of the optimal weight vector can be processed by evolutionary algorithms in existence, because it is a typical optimization problem in lower dimension.

Given a non-separable function, no CCs would perform well once all variables of it are heavily interdepending on each other. The EACC-G treats tightly interdependent variables as a group and optimizes them in batch, instead of one-by-one optimization strategy of other existing algorithms. In order to gain a cleaner vision of the grouping strategy's advantages, let's take the following theorem as a quick example of the probability when optimizing two interacting variables together in EACC-G, which shows how such a grouping strategy can step up the probability of obtaining the interdependencies of variables. In order to make the paper self-contained, the Theorem that given in [8] is also described as follows:

Theorem 1. The EACC-G's probability to apply two interactional variables x_i and x_j as a unitary child component for more than k loops is:

$$p_k = \sum_{r=k}^{N} \binom{N}{r} \left(\frac{1}{m}\right) \left(1 - \frac{1}{m}\right)^{N-r}, \tag{1}$$

where N is the upper limit of loops and m specifies the sum of child components [8].

Proof 1. In any one of independent loops, the EACC-G's probability to apply x_i and x_j as a unitary child component is:

$$p = \frac{\binom{m}{1}}{m^2} = \frac{1}{m}, \tag{2}$$

The times of splitting operation in EACC-G are merely in accord with the sum of loops. For example, if there is N loops, then EACC-G will obtain N executions of splitting operations. Notice that these disintegration operations are mutually independent among each other. Set p_r represent EACC-G's probability to apply x_i and x_j as a unitary child component for r loops. The value of p_r is easy to get

$$p_r = \binom{N}{r} p^r (1-p)^{N-r} = \binom{N}{r} \left(\frac{1}{m}\right)^r \left(1 - \frac{1}{m}\right)^{N-r}, \tag{3}$$

Thus,

$$p_k = \sum_{r=k}^{N} p_r = \sum_{r=k}^{N} \binom{N}{r} \left(\frac{1}{m}\right) \left(1 - \frac{1}{m}\right)^{N-r}, \tag{4}$$

The Theorem 1 is proved. That is, EACC-G's grouping strategies are valid for figuring out interdependencies of variables without profession or specialized knowledge [8].

3   Proposed CCDE/GDS Algorithm

In this section, the proposed CCDE/GDS algorithm for solving large scale optimization is described. Firstly, the differential evolution under CCDE-G framework. Then, the gradient descent strategy (GDS) is presented. Finally, the whole complete CCDE/GDS algorithm is presented in detail.

3.1   Cooperative Coevolution (CC)

Differential evolution (DE) [9, 10] is a straight forward algorithm against global optimization problems. DE process variation by employing a weighted difference vector to distinguish two objects from the others. And these mutated objects will depend on greedy selection and discrete crossover on matching objects to generate offspring. Many researches have been done on the control arguments of DE (e.g., crossover rate CR, population size NP and scaling factor F) [11, 12]. Also, DE has been combined with some other algorithms leading to various hybrid algorithms [13, 14]. First of all, we will introduce one of the latest DE variants for child components in our CC algorithm.

A self-adaptive DE algorithm (SaDE) [14] was recently designed to automatically adjust arguments during evolutionary process. And two learning strategies of DE were employed as candidates due to their good performance. Crossover rate CR and scaling factor F are automatically adjusted during the evolutionary process. And SaDE had shown a good behavior in verification test on the set of 25 benchmark functions from CEC2005 Special Session [15].

SaDE attempts to improve classical DE's performance in a different manner. SaDE focus on self-adaption of control arguments, which is different from NSDE mixing search biases of various NS operators adaptively. And the self-adaptive NSDE (SaNSDE), just as its name implies, combines the advantages of NSDE and SaDE together in a single algorithm [16]. SaNSDE and NSDE behave very similarly except where shown below:

1. Following SaDE's self-adaptive mechanism.
2. Using SaDE's dynamically adaptive strategy to adapt the value of CR.
3. Introducing SaDE's self-adaptive strategy to dynamically balancing Gaussian and Cauchy operators.

Based on SaNSDE, the corresponding cooperative coevolution, called CCDE-G, is designed by means of EACC-G's operations mentioned in Sect. 2. Moreover, another optimizer for the weighting strategy of CCDE-G is specified. We prefer to employ the classical DE in this paper for evaluating the effectiveness of SaNSDE as a child component optimizer, although SaNSDE could be used for this purpose as well. It will be less complex to evaluate its contribution of the new CC framework.

3.2   Gradient Descent Strategy (GDS)

Gradient descent (GD) method is a first-order iterative optimization algorithm. It is common practice to take steps proportional to the gradient's negative at the current point when a function is searching the local minimum by means of GD. GD is built on the observation that if the function $F(x)$ of several variables is defined and differentiable on point a's neighborhood, then $F(x)$ achieves its fastest deceleration once one starts with a in the negative gradient direction of F at a, $-\nabla F(a)$. It follows that, if

$$b=a-\gamma \nabla F(a),$$

for γ small enough, then $F(a) > F(b)$. In other words, the term $\gamma \nabla F(a)$ is subtracted from a because we want to move against the gradient, namely down toward the minimum. Based on this observation, one goes from a conjecture x_0 for a local minimum of F, and reckons the sequence $x_0, x_1, x_2, x_3, \ldots$ such that

$$x_{n+1}=x_n-\gamma_n \nabla F(x_n), n \geq 0,$$

We have

$$F(x_0) \geq F(x_1) \geq F(x_2) \geq \cdots.,$$

3.3   CCDE/GDS

The whole complete CCDE/GDS algorithm is given in Fig. 1. Round-Robin would be employed to optimize grouping once CC framework grouping is finished. The differential value between best values of two consecutive iterations has been adopted to determine whether to proceed with gradient descent algorithm. If

```
best_val⁶ - best_val⁶⁻¹ < epsilon
```

the gradient descent algorithm would be executed to update the global best_mem when searching velocity drops. The GD obtain functions' extremum value along the direction of grandian descent:

```
best_mem_i = best_mem - α *gradient
```

where gradient is the gradient magnitude of best_mem in ith iteration, and α is the velocity coefficient that might need to be adjusted based on demand of computation. To be noted, gradient of function may be unable to get. We explored gradient magnitude of best_mem by finite difference method (FDM). If the best_val obtained by GDS is better than the previous one, the best_val will be update.

```
Decompose the decision variables into group{1},
group{2} .....
for each group such as group{i}
    optimize by evolutionary algorithm
    update bestval
    if bestval_old-bestval <epsilon
        compute the gradient for variables in group{i} based on
        bestmem
        %update bestmem by gradient decent method
        bestmem{j}=bestmem{j}-lamda*gradient{j}
        update bestval according the new value of bestmem
    end if
end for
```

Fig. 1. Pseudocode of the CCDE/GDS

4   Benchmark Tests and Comparisons

4.1   Experimental Settings

In this paper, in order to investigate the advantages of CCDE/GDS, we will compare the performance of CCDE/GDS and CCDE-G proposed in [8] on CEC2010 benchmark functions. These 20 benchmark functions are divided into 5 groups as detailed in [17].

Group1: Separable functions, F1–F3;
Group2: Single-group m-nonseparable functions, F4–F8;
Group3: $\frac{n}{2m}$ -group m-nonseparable functions, F9–F13;
Group4: $\frac{n}{m}$ -group m-nonseparable functions, F14–F18;
Group5: Nonseparable functions, F19, F20.

In our algorithm, D is the dimension and set as 1000. m is introduced here for regulating the sum of variables in all groups and hence, defining the degree of separability. And m gets its value of 50 in the experimental suite. We also set population size NP = 50, the number of fitness evaluations(FEs) will be set as 3×10^6. The threshold value for gradient descent algorithm is set to 10^{-5}.

4.2 Experimental Results

We test CCDE/GDS and CCDE-G's performance on the CEC'2010 benchmark functions. The statistics over 25 independent executions of our performance tests are listed in Table 1. Mann–Whitney U test was used between CCDE/GDS and CCDE-G and expressed as p-value. in p-value, symbol '−' means that the CCDE/GDS optimization results are better than CCDE-G. And the "#" means that the results of two objects are significantly different.

As shown in Table 1, we executed CCDE/GDS optimization operations on 20 functions from test suite. Obviously, eleven of them has obtained better average optimization results than those produced from CCDE-G. And 8 best results are in

Table 1. Optimization results of CCDE/GDS and CCDE-G on CEC2010 benchmark functions

Group	Function		CCDE-G	CCDE/GDS
Group1 Separable functions	F1	Mean	3.26E−26	0.00E + 00
		Std.	1.78E−25	0.00E + 00
		p-value	−3.3E−01	−
	F2	Mean	2.39E + 02	3.27E + 02
		Std.	1.43E + 01	1.83E + 01
		p-value	3.01e−11#	−
	F3	Mean	1.31E−13	1.28E−13
		Std.	6.32E−15	1.87E−15
		p-value	**−2.35e−02#**	−
Group2 Single-group m-nonseparable	F4	Mean	4.59E + 11	7.74E + 11
		Std.	1.59E + 11	2.49E + 11
		p-value	2.88e−06#	−
	F5	Mean	2.60E + 08	2.37E + 08
		Std.	6.45E + 07	6.63E + 07
		p-value	−2.06E−01	−
	F6	Mean	2.10E + 06	2.26E + 06
		Std.	6.80E + 05	4.16E + 05
		p-value	6.46E−01	−
	F7	Mean	7.83E + 00	1.98E + 02
		Std.	4.15E + 00	7.44E + 01
		p-value	3.02e−11#	−

(continued)

Table 1. (*continued*)

Group	Function		CCDE-G	CCDE/GDS
	F8	Mean	2.08E + 07	2.56E + 07
		Std.	2.43E + 07	2.75E + 07
		p-value	1.76E−01	−
Group3 $\frac{D}{2m}$ − group m-nonseparable	F9	Mean	3.33E + 07	2.09E + 07
		Std.	2.79E + 06	2.12E + 06
		p-value	−3.02e−11#	−
	F10	Mean	1.32E + 04	5.14E + 03
		Std.	2.02E + 02	2.20E + 02
		p-value	−3.02e−11#	−
	F11	Mean	1.57E−13	1.70E−13
		Std.	1.06E−14	1.09E−14
		p-value	4.86e−05#	−
	F12	Mean	4.54E + 06	6.31E + 02
		Std.	1.83E + 05	5.67E + 01
		p-value	−3.02e−11#	−
	F13	Mean	7.81E + 02	8.19E + 02
		Std.	1.42E + 02	2.29E + 02
		p-value	7.28E−01	−
Group4 $\frac{D}{m}$ − group m-nonseparable	F14	Mean	1.13E + 08	4.22E + 07
		Std.	7.71E + 06	3.65E + 06
		p-value	−3.02e−11#	−
	F15	Mean	1.59E + 04	7.06E + 03
		Std.	2.52E + 02	3.17E + 02
		p-value	−3.02e−11#	−
	F16	Mean	1.67E−13	1.32E + 01
		Std.	1.97E−14	7.24E + 01
		p-value	6.01e−10#	−
	F17	Mean	7.63E + 06	1.71E + 03
		Std.	4.13E + 05	1.16E + 02
		p-value	−3.02e−11#	−
	F18	Mean	1.63E + 03	1.85E + 03
		Std.	3.00E + 02	5.34E + 02
		p-value	1.09E−01	−
Group5 Nonseparable functions	F19	Mean	1.85E + 07	2.30E + 05
		Std.	1.61E + 06	2.14E + 05
		p-value	−3.02e−11#	−
	F20	Mean	1.24E + 03	1.21E + 03
		Std.	9.54E + 01	8.27E + 01
		p-value	−2.77E−01	−

boldface. The 8 best results are mainly distributed from the 3rd to 5th, which means that the proposed new CCDE/GDS approach in this paper performance better on solving non-separable problems with multi-dimension and multi-group.

Figure 2 shows the evolution of CEC2010 benchmark functions. It is obvious that CCDE/GDS had better performance when compared to CCDE-G. CCDE- G/GD gained much better results than the non-GD algorithm on multi-group and nonseparable functions, which shows the superiority of our proposed algorithms for large optimization problems.

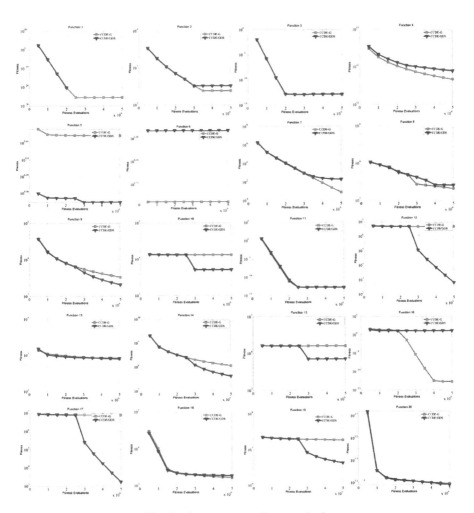

Fig. 2. Converge rate of two methods

5 Conclusion

In this paper, we attempted to develop a novel algorithm based on the GD to CCDE-G for solving large scale optimization problems. The CCDE-G method itself is suitable for tackling large multi-dimensional problems because it can decompose large and complex problems into several submodules which separately goes through evolution separately with only cooperation occurrence. Moreover, GDS can improve the bestval from a single iteration to the whole evolution, which can help to improve the performance of coevolutionary optimization significantly. Experimental tests have been conducted on 20 typical CEC'2010 benchmark functions with different characteristics. The experimental results convinced us not only the CCDE-G promising performance but also the CCDE-G/GDS outstanding performance.

References

1. Sarker, R., Mohammadian, M., Yao, X.: Evolutionary Optimization. Kluwer Academic Publishers, Norwell (2002)
2. Liu, Y., Yao, X., Zhao, Q., Higuchi, T.: Scaling up fast evolutionary programming with cooperative coevolution. In: Proceedings of the Congress on Evolutionary Computation, pp. 1101–1108 (2001)
3. Potter, M., De Jong, K.: A cooperative coevolutionary approach to function optimization. In: Proceedings of the 3th Conference on Parallel Problem Solving from Nature, pp. 249–257 (1994)
4. Shi, Y., Teng, H., Li, Z.: Cooperative coevolutionary differential evolution for function optimization. In: Proceedings of the 1st Conference on Natural Computation, pp. 1080–1088 (2005)
5. Sofge, D., De Jong, K., Schultz, A.: A blended population approach to cooperative coevolution for decomposition of complex problems. In: Proceedings of the Congress on Evolutionary Computation, vol. 1, pp. 413–418 (2002)
6. Potter, M., De Jong, K.: Cooperative coevolution: an architecture for evolving coadapted subcomponents. Evol. Comput. 8(1), 1101–1108 (2000)
7. Yang, Z., Tang, K., Xiao, X.: Large scale evolutionary optimization using cooperative coevolution. Inf. Sci. 178, 2985–2999 (2008)
8. Price, K., Storn, R., Lampinen, J.: Differential Evolution: A Practical Approach to Global Optimization. Springer, Heidelberg (2005). ISBN 3-540-20950-6
9. Storn, R., Price, K.: Differential evolution – a simple and efficient heuristic for global optimization over continuous spaces. J. Global Optim. 11(4), 341–359 (1997)
10. Gamperle, R., Muller, S.D., Koumoutsakos, P.: A Parameter Study for Differential Evolution. In: Proceedings of the WSEAS International Conference on Advances in Intelligent Systems, pp. 293–298 (2002)
11. Zaharie, D.: Critical values for the control parameters of differential evolution algorithms. In: Proceedings of the 8th International Conference on Soft Computing, pp. 62–67 (2002)
12. Sun, J., Zhang, Q., Tsang, E.: DE/EDA: a new evolutionary algorithm for global optimization. Inf. Sci. 169, 249–262 (2005)

13. Zhang, W., Xie, X.: DEPSO: hybrid particle swarm with differential evolution operator. In: Proceedings of the 2003 IEEE International Conference on Systems, Man and Cybernetics, vol. 4, pp. 3816–3821 (2003)

14. Qin, A.K., Suganthan, P.N.: Self-adaptive differential evolution algorithm for numerical optimization. In: Proceedings of the Congress on Evolutionary Computation, vol. 2, pp. 1785–1791 (2005)

15. Suganthan, P.N., Hansen, N., Liang, J.J., Deb, K., Chen, Y.P., Auger, A., Tiwari, S.: Problem Definitions and Evaluation Criteria for the CEC 2005 Special Session on Real-parameter Optimization, Technical report, Nanyang Technological University, Singapore (2005). http://www.ntu.edu.sg/home/EPNSugan

16. Yang, Z., Tang, K., Yao, X.: Self-adaptive differential evolution with neighborhood search. In: Proceedings of the 2008 Congress on Evolutionary Computation, in press

17. Tang, K., Li, X., Suganthan, P.N., Yang, Z., Weise, T.: Benchmark functions for the CEC'2010 special session and competition on large-scale global optimization, Technical report, Nature Inspired Computation and Applications Laboratory (NICAL), USTC, China (2010)

Chebyshev Inequality Based Approach to Chance Constrained Optimization Problems Using Differential Evolution

Kiyoharu Tagawa[✉] and Shohei Fujita

Kindai University, Higashi-Osaka 577-8502, Japan
tagawa@info.kindai.ac.jp

Abstract. A new approach to solve Chance Constrained Optimization Problem (CCOP) without using the Monte Carlo simulation is proposed. Specifically, the prediction interval based on Chebyshev inequality is used to estimate a stochastic function value included in CCOP from a set of samples. By using the prediction interval, CCOP is transformed into Upper-bound Constrained Optimization Problem (UCOP). The feasible solution of UCOP is proved to be feasible for CCOP. In order to solve UCOP efficiently, a modified Differential Evolution (DE) combined with three sample-saving techniques is also proposed. Through the numerical experiments, the usefulness of the proposed approach is demonstrated.

Keywords: Chebyshev inequality · Chance constraint · Optimization

1 Introduction

The Chance Constrained Optimization Problem (CCOP) is one of the possible formulations of the stochastic optimization problem. The quality of solution can be guaranteed by a specified probability in CCOP [8]. A number of real-world problems have been formulated as CCOPs [2,8]. However, CCOP is difficult to solve because its solutions have to be evaluated by using a time-consuming Monte Carlo simulation. Even though a number of Evolutionary Algorithms (EAs) have been reported for solving optimization problems under uncertainties [3], only a few EAs are dealing with CCOPs [2,4,7] insofar as we know.

This paper proposes a novel approach to solve CCOP without using the Monte Carlo simulation. The prediction interval based on Chebyshev inequality is used to estimate a stochastic function value included in CCOP from a set of samples. By using the prediction interval, CCOP is transformed into Upper-bound Constrained Optimization Problem (UCOP). It is proved that the feasible solution of UCOP is also feasible for CCOP. In order to solve UCOP efficiently, a modified Differential Evolution (DE) is proposed. That is because DE is arguably one of the most powerful stochastic real-parameter optimization algorithms in current use [9]. Since the multiple sampling of a solution is usually expensive for real-world problems, the modified DE employs three sample-saving techniques,

© Springer International Publishing AG 2017
Y. Tan et al. (Eds.): ICSI 2017, Part I, LNCS 10385, pp. 440–448, 2017.
DOI: 10.1007/978-3-319-61824-1_48

namely accumulative sampling [6], reliability relaxation [12], and upper-bound cut [13]. Finally, the usefulness of the proposed approach is demonstrated.

2   Problem Formulation of CCOP

Let $x = (x_1, \cdots, x_D) \in X$, $X = [\underline{x}_j, \overline{x}_j]^D \subseteq \Re^D$ be a vector of decision variables, or a solution. The uncertainty is given by a vector of random variables $\xi \in \Xi$ with support Ξ. Stochastic functions are defined as $g_m : X \times \Xi \to \Re$. Thereby, for a required probability level $\alpha \in (0, 1)$, CCOP is stated as

$$
\begin{bmatrix}
\min_{x \in X} & \gamma \\
\text{sub. to} & \Pr(\forall \xi \in \Xi : g_0(x, \xi) \leq \gamma) \geq 1 - \alpha \\
& \Pr(\forall \xi \in \Xi : g_m(x, \xi) \leq 0) \geq 1 - \alpha, \ m \in \mathcal{I}_M
\end{bmatrix}
\tag{1}
$$

where $\mathcal{I}_M = \{1, 2, \cdots, M\}$ is an index set of constraints.

Let $f : \Xi \to (0, \infty)$ be the Probability Density Function (PDF) of $\xi \in \Xi$. The PDF $f(\xi)$ of ξ is usually known. However, we can't derive the PDF of $g_m(x, \xi)$ from $f(\xi)$ analytically because the procedure of $g_m(x, \xi)$ is too complex or black box in many real-world problems. Therefore, the feasibility of $x \in X$ has to be verified empirically by a time-consuming Monte Carlo simulation.

3   Proposed Approach to CCOP

3.1   Chebyshev Inequality from Samples

Due to $\xi \in \Xi$ in (1), different function values $g_m(x, \xi) \in \Re$ are observed for repeated evaluations of the same solution $x \in X$. We derive the prediction interval of the random variable $g_m(x, \xi)$ from Chebyshev inequality.

Saw et al. [10] have extended Chebyshev inequality to cases where the mean and variance are not known and may not exist, but you want to use the sample mean and sample variance from N samples to bound the expected value of a new drawing from the same distribution. Let $g_m(x, \xi^n)$, $\xi^n \sim f(\xi)$ be a sample of $g_m(x, \xi)$. By using the sample mean $\overline{g}_m(x)$ and the sample variance $s_m^2(x)$ calculated from $g_m(x, \xi^n)$, $n = 1, \cdots, N$, Chebyshev inequality [10] is

$$
\Pr\left(|g_m(x, \xi) - \overline{g}_m(x)| \geq \lambda \sqrt{\frac{N+1}{N}} s_m(x)\right)
$$
$$
\leq \frac{1}{N+1} \left\lfloor \frac{(N+1)(N-1+\lambda^2)}{N\lambda^2} \right\rfloor
\tag{2}
$$

where $\lfloor \cdot \rfloor$ denotes the floor function. $\lambda > 1$ is an arbitrary real number.

The following theorem [11] provides the prediction interval of $g_m(x, \xi)$ which can be calculated from the sample mean $\overline{g}_m(x)$ and variance $s_m^2(x)$.

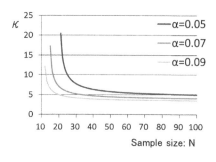

Fig. 1. Change of κ in (4) for N and α **Fig. 2.** Control of κ by rule in (11)

Theorem 1. *Let* $g_m(\boldsymbol{x}, \boldsymbol{\xi}^n)$, $n = 1, \cdots, N$ *be a set of samples and* $N \geq N_{\min}$. *From a probability level* $\alpha \in (0, 1)$, *the minimum sample size is*

$$N_{\min} = \left\lfloor \frac{1}{\alpha} + 1 \right\rfloor. \tag{3}$$

From the probability level α *and the sample size* N, κ *is defined as*

$$\kappa = \sqrt{\frac{N^2 - 1}{N\,(\alpha\,N - 1)}}. \tag{4}$$

By using $\overline{g}_m(\boldsymbol{x})$, $s_m^2(\boldsymbol{x})$, *and* κ, *the prediction interval of* $g_m(\boldsymbol{x}, \boldsymbol{\xi})$ *is*

$$\Pr([\overline{g}_m(\boldsymbol{x}) - \kappa\,s_m(\boldsymbol{x}),\ \overline{g}_m(\boldsymbol{x}) + \kappa\,s_m(\boldsymbol{x})] \ni g_m(\boldsymbol{x}, \boldsymbol{\xi}))$$
$$= \Pr([g_m^L(\boldsymbol{x}),\ g_m^U(\boldsymbol{x})] \ni g_m(\boldsymbol{x}, \boldsymbol{\xi})) \geq 1 - \alpha. \tag{5}$$

Proof. See [11]. ☐

Figure 1 shows the change of coefficient value κ in (4) that depends on both the probability level α and the sample size N. $\kappa = \sqrt{1/\alpha}$ holds for $N \to \infty$.

3.2   Upper-Bound Constrained Optimization Problem (UCOP)

As stated above, CCOP in (1) is transformed into UCOP. By using the upper-bounds $g_m^U(\boldsymbol{x})$, $m \in \{0\} \cup \mathcal{I}_M$ shown in (5), we formulate UCOP as

$$\left[\begin{array}{ll} \min_{\boldsymbol{x} \in \boldsymbol{X}} & g_0^U(\boldsymbol{x}) = \overline{g}_0(\boldsymbol{x}) + \kappa\,s_0(\boldsymbol{x}) \\ \text{sub. to} & g_m^U(\boldsymbol{x}) = \overline{g}_m(\boldsymbol{x}) + \kappa\,s_m(\boldsymbol{x}) \leq 0,\ m \in \mathcal{I}_M \end{array}\right. \tag{6}$$

where we suppose that the sample size N is larger enough than $N_{\min}$ in (3).

Theorem 2. *If* $\boldsymbol{x} \in \boldsymbol{X}$ *is a feasible solution of UCOP in (6) then* $\boldsymbol{x} \in \boldsymbol{X}$ *is also a feasible solution of CCOP in (1) under a condition* $\gamma = g_0^U(\boldsymbol{x})$.

Proof. We assume that $x \in X$ is a feasible solution of UCOP in (6). Therefore, the solution $x \in X$ provides the prediction interval as shown in (5).

Since $g_0^U(x) = \gamma$ holds,

$$\begin{aligned}
\Pr(g_0(x, \xi) \leq \gamma) &= \Pr((-\infty, \ \gamma] \ni g_0(x, \xi)) \\
&\geq \Pr([g_0^L(x), \ g_0^U(x)] \ni g_0(x, \xi)) \geq 1 - \alpha.
\end{aligned} \tag{7}$$

Since $g_m^U(x) \leq 0$, $m \in \mathcal{I}_M$ holds,

$$\begin{aligned}
\Pr(g_m(x, \xi) \leq 0) &= \Pr((-\infty, \ 0] \ni g_m(x, \xi)) \\
&\geq \Pr([g_m^L(x), \ g_m^U(x)] \ni g_m(x, \xi)) \geq 1 - \alpha.
\end{aligned} \tag{8}$$

From (7) and (8), $x \in X$ satisfies all chance constraints in (1).   □

4   Basic Differential Evolution for UCOP

We present a basic DE (DEB) for solving UCOP in (6). Like classic DE [9], DEB works by building a population $P \subset X$ of vectors which is a set of possible solutions to UCOP. The initial vectors $x_i \in P$, $i = 1, \cdots, N_P$ are generated randomly. Each $x_i \in P$ is evaluated N times. Then the upper-bounds $g_m^U(x_i)$, $m \in \{0\} \cup \mathcal{I}_M$ are calculated from samples $g_m(x_i, \xi^n)$, $n = 1, \cdots, N$.

Within a generation of a population, each vector $x_i \in P$ is assigned to the target vector in turn. A basic strategy named "DE/rand/1/bin" [9] generates a new solution called the trial vector $u \in X$ from the target vector $x_i \in P$.

A self-adapting setting of the control parameters [1] is used by DEB. In order to handle the constraints of UCOP in (6), ε-constrained method [14] is also employed. The constraint violation $\phi(x)$ of $x \in X$ is defined as

$$\phi(x) = \sum_{m \in \mathcal{I}_M} \max\{0, \ g_m^U(x)\}. \tag{9}$$

A newborn trial vector $u \in X$ is judged to be better than the target vector $x_i \in P$ if at least one of the following three criteria is satisfied: (1) $\phi(u) \leq \varepsilon(t)$ and $g_0^U(u) \leq g_0^U(x_i)$; (2) $\phi(u) \leq \varepsilon(t)$ and $\phi(x_i) > \varepsilon(t)$; (3) $\phi(u) > \varepsilon(t)$ and $\phi(u) \leq \phi(x_i)$. The value of $\varepsilon(t)$ is controlled dynamically based on the current generation t [14]. If u is better than $x_i \in P$, $x_i \in P$ is replaced by u.

5   Three Sample-Saving Techniques

5.1   Accumulative Sampling

The accumulative sampling evaluates each $x_i \in P$ in the initial population only a few times for calculating the upper-bounds $g_m^U(x_i)$, $m \in \{0\} \cup \mathcal{I}_M$. Thereafter, each $x_i \in P$ is reevaluated one time at every generation to update the upper-bounds $g_m^U(\boldsymbol{x}_i)$ until they are regarded to be converged. If $\boldsymbol{x}_i \in \boldsymbol{P}$ satisfies the following condition for three consecutive generations, we regard that the values of the upper-bounds $g_m^U(\boldsymbol{x}_i)$, $m \in \{0\} \cup \mathcal{I}_M$ have been converged.

$$\forall m \in \{0\} \cup \mathcal{I}_M : \left| \frac{g_m^U(\boldsymbol{x}_i^t) - g_m^U(\boldsymbol{x}_i^{t-1})}{g_m^U(\boldsymbol{x}_i^t)} \right| < 10^{-3} \tag{10}$$

where $\boldsymbol{x}_i \in \boldsymbol{P}$ is denoted by $\boldsymbol{x}_i^{t-1}$ and $\boldsymbol{x}_i^t$ at generation $t-1$ and t.

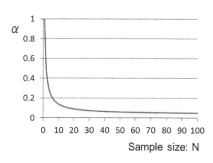

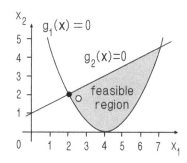

Fig. 3. Change of α for sample size N **Fig. 4.** Feasible region of DCOP in (13)

5.2 Reliability Relaxation

As a drawback of the accumulative sampling, κ in (4) has to be evaluated with a small sample size N at the beginning of search. Since the small sample size N makes κ large as shown in Fig. 1, the upper-bound is overestimated.

Let N_i^t be the sample size of vector $\boldsymbol{x}_i^t \in \boldsymbol{P}$ at generation t. Instead of (4), the proposed reliability relaxation decides κ for $\boldsymbol{x}_i^t \in \boldsymbol{P}$ as

$$\kappa = \begin{cases} \hat{\kappa}, & \text{if } N_i^t < N_{\min} \\ \min\{\hat{\kappa}, \ \kappa_i^t\}, & \text{if } N_i^t \geq N_{\min} \end{cases} \tag{11}$$

where $\hat{\kappa}$ is chosen from $\hat{\kappa} > \sqrt{1/\alpha}$ and κ_i^t is calculated as

$$\kappa_i^t = \sqrt{\frac{(N_i^t)^2 - 1}{N_i^t (\alpha N_i^t - 1)}}. \tag{12}$$

Figure 2 illustrates the value of κ controlled by the rule in (11). When $\kappa = \hat{\kappa}$ holds in (11), the value of α is changed for N. For example, Fig. 3 shows the value of α that satisfies the relation in (4), but $\hat{\kappa} = 0.5$ is used instead of κ. The reliability relaxation enlarges α over the original value specified by CCOP.

5.3   Upper-Bound Cut

The upper-bound cut (U-cut) judges hopeless trial vectors only by few samplings and discards them. When a newborn trial vector u is compared with the target vector $x_i^t \in P$, U-cut takes and examines its samples $g_m(u, \xi^n)$, $n = 1, \cdots, N_i^t$ one by one. Then u is judged to be worse than $x_i^t \in P$ and discarded if at least one of the following three criteria is satisfied along the way: (1) $\phi(x_i^t) \leq \varepsilon(t)$ and $g_0^U(x_i^t) \leq g_0(u, \xi^n)$; (2) $\phi(x_i^t) \leq \varepsilon(t)$ and $\phi(u) > \varepsilon(t)$; (3) $\phi(x_i^t) > \varepsilon(t)$ and $\phi(x_i^t) \leq \phi(u)$. The constraint violation $\phi(u)$ of the trial vector u is evaluated by (9) in which $g_m(u, \xi^n)$, $m \in \mathcal{I}_M$ is used instead of $g_m^U(u)$.

6   Numerical Experiment

6.1   Comprehensive Example of CCOP

We show the validity of the proposed approach. The test problem of CCOP is based on a Deterministic Constrained Optimization Problem (DCOP):

$$\left[\begin{array}{ll} \min_{x \in X} & g_0(x) = x_1^2 + (x_2 - 2)^2 \\ \text{sub. to} & g_1(x) = (x_1 - 4)^2 - 2\,x_2 \leq 0 \\ & g_2(x) = -x_1 + 2\,x_2 - 2 \leq 0 \end{array} \right. \tag{13}$$

where $x = (x_1, x_2)$ and $X = [-5, 10]^2 \subseteq \Re^2$.

Considering disturbance $\xi_j \in \Re$, we extended DCOP in (13) to CCOP as

$$\left[\begin{array}{ll} \min_{x \in X} & \gamma \\ \text{sub. to} & \Pr(g_0(x, \xi) = g_0(x + \xi) \leq \gamma) \geq 1 - \alpha, \\ & \Pr(g_m(x, \xi) = g_m(x + \xi) \leq 0) \geq 1 - \alpha, \; m \in \mathcal{I}_M = \{1, 2\} \end{array} \right. \tag{14}$$

where $\xi_1, \xi_2 \sim \mathcal{N}(0, 0.01^2)$, $\xi = (\xi_1, \xi_2)$, and $\alpha = 0.05$.

We transformed CCOP in (14) into UCOP as shown in (6). By using DEB, we obtained a solution $x^\circ = (2.061, 1.979)$ of UCOP with $g_0^U(x^\circ) = 4.444$.

Figure 4 illustrates the feasible region of DCOP in (13) on the decision variable space. The optimal solution $x^\star \in \Re^2$ of DCOP denoted by "•" exists on the boundary of the feasible region. On the other hand, the solution $x^\circ \in \Re^2$ of UCOP denoted by "○" exists on the inside of the feasible region of DCOP.

We will verify whether a solution $x \in \Re^2$ is feasible for CCOP in (14) or not. Let $\Pr(g_m(x, \xi) > z) = \alpha_m$ be a true probability of the complementary event of a chance constraint in (14). The value $z \in \Re$ is chosen as $z = \gamma$ for $m = 0$ and $z = 0$ for $m \in \mathcal{I}_M = \{1, 2\}$. From a huge number of samples $g_m(x, \xi^n)$, $\xi^n \sim f(\xi)$, $n = 1, \cdots, N$ generated by the Mote Carlo simulation, an empirical probability $\hat{\alpha}_m \in [0, 1]$ of the complementary event is calculated as

$$\left[\begin{array}{l} \hat{\alpha}_m = \dfrac{1}{N} \displaystyle\sum_{n=1}^{N} \ell(g_m(x, \xi^n) > z) \\[2ex] \ell(g_m(x, \xi^n) > z) = \begin{cases} 1; & \text{if } g_m(x, \xi^n) > z \\ 0; & \text{otherwise.} \end{cases} \end{array} \right. \tag{15}$$

For any $\varepsilon \in (0, 1)$ and $\delta \in (0, 1)$, if $N \geq (1/2\,\varepsilon^2) \log(2/\delta)$ samples are used to calculate $\hat{\alpha}_m$ then Chernoff bound [15] ensures $\Pr(|\alpha_m - \hat{\alpha}_m| \leq \varepsilon) \geq 1 - \delta$.

Table 1 compares the solution x° of UCOP with the solution $x^\star$ of DCOP in the empirical probabilities evaluated for the chance constraints in (14). Selecting $\varepsilon = 0.01$ and $\delta = 10^{-3}$ for N, we used $N = 2,649,159$ samples to calculate each $\hat{\alpha}_m$, $m \in \{0\} \cup \mathcal{I}_M$. From Table 1, we can confirm that x° is feasible for CCOP significantly because $\forall m \in \{0\} \cup \mathcal{I}_M : \hat{\alpha}_m \approx \alpha_m < \alpha$ holds. Even though $x^\star$ is better than x° in the objective value γ, $x^\star$ is not feasible for CCOP.

Table 1. Empirical probability evaluated for CCOP

Problem	Solution	γ	$\hat{\alpha}_0$	$\hat{\alpha}_1$	$\hat{\alpha}_2$
DCOP	$x^\star = (2.000,\ 2.000)$	4.000	0.501	0.500	0.499
UCOP	$x^\circ = (2.061,\ 1.979)$	4.444	0.000	0.000	0.000

Table 2. Comparison of modified DEs for UCOP

TC	$t_{\max} = 100$					$TN \approx 101 \times 10^3\, D$				
CCOP	PC	DEB	DEA	DEAR	DEARU	PC	DEB	DEA	DEAR	DEARU
G4	γ	-30625	-30622	-30624	-30624	γ	-30625	-30629	-30632	-30633
$D = 5$	WT	—	—	—		WT	**	**	—	
$M = 6$	TN	$505,000$	$247,247$	$169,737$	$70,843$	NS	$101\,N_P$	$170\,N_P$	$204\,N_P$	$256\,N_P$
G7	γ	35.672	36.090	35.948	35.832	γ	35.672	31.703	30.284	28.646
$D = 10$	WT	—	—	—		WT	**	**	**	
$M = 8$	TN	$1,010,000$	$582,603$	$426,259$	$120,839$	NS	$101\,N_P$	$150\,N_P$	$174\,N_P$	$419\,N_P$
G9	γ	692.87	693.68	693.11	692.95	γ	692.87	691.54	690.74	690.28
$D = 7$	WT	—	*	—		WT	**	**	**	
$M = 4$	TN	$707,000$	$388,101$	$278,065$	$104,011$	NS	$101\,N_P$	$153\,N_P$	$192\,N_P$	$258\,N_P$

6.2 Effects of Sample-Saving Techniques

In order to verify the effects of the sample-saving techniques, we compared four modified DEs: (1) DEB, (2) DEB with Accumulative sampling (DEA), (3) DEA with Reliability relaxation (DEAR), and (4) DEAR with U-cut (DEARU).

We used three test problems from [5], namely G4, G7, and G9, to compare the modified DEs. They are given by DCOPs with D decision variables and M constraints. By adding a disturbance $\xi_j \sim \mathcal{N}(0, 0.01^2)$ to decision variable x_j, $j = 1, \cdots, D$, we extended each DCOP to CCOP with $\alpha = 0.05$ in the same way with CCOP in (14). Then we transformed each CCOP into UCOP. By using the population size $N_P = 10\,D$, we applied modified DEs respectively to each UCOP 30 times. The sample size was fixed to $N = 100$ for DEB. The Termination

Condition (TC) of DEs was given by either the maximum number of generations: $t_{\max} = 100$ or the Total Number (TN) of samples: $TN \approx 101 \times 10^3\, D$.

Table 2 compares the four modified DEs in three Performance Criteria (PC): the objective value $\gamma = g_0^U(\boldsymbol{x}^\circ)$, TN spent by DE, and Number of Solutions (NS) examined by DE. The values of PC are averaged over 30 runs. Besides, by using Wilcoxon Test (WT) about γ, DEARU was compared with the others. Symbols "**" and "*" mean that there is significant difference between DEARU and an opponent respectively with risks 1[%] and 5[%]. Symbol "—" means that there is no significant difference between DEARU and an opponent.

From the results in Table 2, if TC is given by $t_{\max}$, there is not much difference between DEARU and the others in the value of γ. However, from the value of TN, we can confirm the effects of the three sample-saving techniques. Actually, DEARU finds its solution with the least TN. On the other hand, if TC is given by TN, DEARU always finds the best solution that provides the minimum γ value. From the value of NS, we can see that DEARU examines the largest number of solutions until it terminates the search of the best solution.

7   Conclusion

A new approach for solving CCOP without using the Monte Carlo simulation was proposed. By using the prediction interval of a stochastic function value, CCOP was transformed into UCOP. For solving UCOP efficiently, a modified DE combined with three sample-saving techniques was also proposed. Through the numerical experiments conducted on several test problems, we verified the usefulness of the proposed approach. In the future work, we would like to apply the proposed approach to real-world problems formulated as CCOPs.

References

1. Brest, J., Greiner, S., Bošković, B., Merink, M., Žumer, V.: Self-adapting control parameters in differential evolution: a comparative study on numerical benchmark problems. IEEE Trans. Evol. Comput. **10**(6), 646–657 (2006)
2. Jiekang, W., Jianquan, Z., Guotong, C., Hongliang, Z.: A hybrid method for optimal scheduling of short-term electric power generation of cascaded hydroelectric plants based on particle swarm optimization and chance-constrained programming. IEEE Trans. Power Syst. **23**(4), 1570–1579 (2008)
3. Jin, Y., Branke, J.: Evolutionary optimization in uncertain environments. IEEE Trans. Evol. Comput. **9**(3), 303–317 (2005)
4. Liu, B., Zhang, Q., Fernández, F.V., Gielen, G.G.E.: An efficient evolutionary algorithm for chance-constrained bi-objective stochastic optimization. IEEE Trans. Evol. Comput. **17**(6), 786–796 (2013)
5. Michalewicz, Z., Schoenauer, M.: Evolutionary algorithms for constrained parameter optimization problems. Evol. Comput. **4**(1), 1–32 (1996)
6. Park, T., Ryu, K.R.: Accumulative sampling for noisy evolutionary multi-objective optimization. In: Proceedings of the GECCO 2011, pp. 793–800 (2011)

7. Poojari, C.A., Varghese, B.: Genetic algorithm based technique for solving chance constrained problems. Eur. J. Oper. Res. **185**, 1128–1154 (2008)
8. Prékopa, A.: Stochastic Programming. Kluwer Academic Publishers, Berlin (1995)
9. Price, K.V., Storn, R.M., Lampinen, J.A.: Differential Evolution - A Practical Approach to Global Optimization. Springer, Heidelberg (2005)
10. Saw, J.G., Yang, M.C.K., Mo, T.C.: Chebyshev inequality with estimated mean and variance. Am. Stat. **38**(2), 130–132 (1984)
11. Tagawa, K.: Worst case optimization using Chebyshev inequality. In: Proceedings of BIOMA 2016, 173–185 (2016)
12. Tagawa, K., Fujita, S.: Robust optimization based on Chebyshev inequality and accumulative sampling with reliability relaxation. Inf. Process. Soc. Jpn. Trans. Math. Model. Appl. **9**(3), 75–86 (2016)
13. Tagawa, K., Harada, S.: Multi-noisy-objective optimization based on prediction of worst-case performance. In: Dediu, A.-H., Lozano, M., Martín-Vide, C. (eds.) TPNC 2014. LNCS, vol. 8890, pp. 23–34. Springer, Cham (2014). doi:10.1007/978-3-319-13749-0_3
14. Takahama, T., Sakai, S.: Constrained optimization by the ε constrained differential evolution with gradient-based mutation and feasible elites. In: Proceedings of IEEE CEC 2006, pp. 1–8 (2006)
15. Tempo, R., Calafiore, G., Dabbene, F.: Randomized Algorithms for Analysis and Control of Uncertain Systems: With Applications. Springer, Heidelberg (2012)

Solving the Distributed Two Machine Flow-Shop Scheduling Problem Using Differential Evolution

Paul Dempster[1]($\boxtimes$), Penghao Li[2], and John H. Drake[3]

[1] Artificial Intelligence and Optimisation Research Group,
School of Computer Science, University of Nottingham Ningbo China,
Ningbo 315100, China
paul.dempster@nottingham.edu.cn
[2] School of Computer Science, University of Nottingham Ningbo China,
Ningbo 315100, China
zy10611@nottingham.edu.cn
[3] Operational Research Group, Queen Mary University of London,
Mile End Road, London E1 4NS, UK
j.drake@qmul.ac.uk

Abstract. Flow-shop scheduling covers a class of widely studied optimisation problem which focus on optimally sequencing a set of jobs to be processed on a set of machines according to a given set of constraints. Recently, greater research attention has been given to distributed variants of this problem. Here we concentrate on the distributed two machine flow-shop scheduling problem (DTMFSP), a special case of classic two machine flow-shop scheduling, with the overall goal of minimising makespan. We apply Differential Evolution to solve the DTMFSP, presenting new best-known results for some benchmark instances from the literature. A comparison to previous approaches from the literature based on the Harmony Search algorithm is also given.

1 Introduction

Flow-shop scheduling is a classic optimisation problem with an extensive literature dedicated to studying a number of different variants. At its core, the aim is to generate a schedule of jobs to be completed on a set of machines, where each job has a given processing time on each machine, optimising for a particular objective. Although some variants of the problem can be solved easily, minor extensions can lead to a great increase in computational complexity. When the processing times for each job are fixed, although the two-machine case is solvable by a polynomial time algorithm [4], it becomes NP-hard when a third machine [3] or operation [5] is added.

Differential Evolution (DE) [9] is a popular population-based metaheuristic which has been particularly successful at solving continuous optimisation problems [1]. Despite the fact that DE is chiefly designed to operate in a continuous

© Springer International Publishing AG 2017
Y. Tan et al. (Eds.): ICSI 2017, Part I, LNCS 10385, pp. 449–457, 2017.
DOI: 10.1007/978-3-319-61824-1_49

environment, it has frequently been adapted to solve combinatorial optimisation problems and in particular, flow-shop scheduling problems. Onwubolu and Davendra [6] defined methods to transform solutions between real-valued vector and integer permutation representations, searching in continuous space whilst evaluating solution quality in discrete space. Qian et al. [8] combined differential evolution operating in continuous space with local search in discrete space to solve a multi-objective variant of the flow-shop scheduling problem with buffers in-between machines. Other work by Pan et al. [7] and Wang et al. [11] defined specific operators in discrete space to perform direct search on flow-shop problems using DE.

Here we focus on the distributed two machine flow-shop scheduling problem (DTMFSP). This variant of flow-shop scheduling considers a distributed set of locations containing two machines, with a set of jobs of varying lengths which must be processed on both machines at a single location. Previous work by Deng et al. [2] compared three variants of the Harmony Search algorithm over a set of benchmarks to this problem. In this paper we apply DE to the benchmark instances for the DTMFSP and compare to the previously presented Harmony Search based approaches.

2 The Distributed Two Machine Flow-Shop Scheduling Problem (DTMFSP)

The well-known two machine flow-shop scheduling problem [4] is defined as follows. Given a set of n jobs $\mathcal{J} \in \{J_1, \ldots, J_n\}$, to be processed on two machines M_1 and M_2. Each job $J_i \in \mathcal{J}$ consists of two sequential operations O_{i,M_1} and O_{i,M_2} with associated processing time p_{i,M_1} and p_{i,M_2} respectively. A schedule is a permutation $\pi = (\pi(1), \ldots, \pi(n))$ of jobs representing the order in which the jobs are processed on the two machines. Under the constraints that each machine can only process one job at a time and jobs must be processed sequentially on the two machines, using makespan as the performance criteria, the goal is to minimise the overall time taken to complete all of the jobs. This problem can be solved to optimality in polynomial-time using Johnson's rule [4].

The distributed two machine flow-shop scheduling problem (DTMFSP) is an extended version of this problem [2]. The difference between the two is that jobs in the DTMFSP can be processed in any of f identical factories, each containing a machine $M_{f,1}$ and a machine $M_{f,2}$, whereas the classical problem considers only a single factory. Additionally, jobs cannot be transferred to another factory, i.e. once a job has been processed on machine $M_{f,1}$ at a given factory, it is not possible to then process the second operation of the job on a machine in another factory $M_{g,2}$ where $g \neq f$. Rather than searching directly on the space of possible job permutations, when solving the DTMDSP search takes place over a vector of integers representing the allocated factory of each job. For each set of schedules, the overall makespan can be calculated as the longest completion time of a single factory after applying Johnson's rule to each factory.

Given a processing sequence π^k at a particular factory k, with a total of n_k jobs at each factory, the makespan of a set of schedules C_{MAX} can be calculated as:

$$C_{\pi^k(1),M_1} = p_{\pi^k(1),M_1}, k = 1, 2, \ldots, f \tag{1}$$

$$C_{\pi^k(i),M_1} = C_{\pi^k(i-1),M_1} + p_{\pi^k(i),M_1}, k = 1, 2, \ldots, f; i = 2, 3, \ldots, n_k \tag{2}$$

$$C_{\pi^k(1),M_2} = C_{\pi^k(1),M_1} + p_{\pi^k(1),M_2}, k = 1, 2, \ldots, f; \tag{3}$$

$$C_{\pi^k(i),M_2} = max\{C_{\pi^k(i),M_1}, C_{\pi^k(i-1),M_2}\} + p_{\pi^k(i),M_2}, k = 1, 2, \ldots, f; i = 2, 3, \ldots, n_k \tag{4}$$

$$C_{MAX} = max\{C_{\pi^k(n_k),M_2}\}, k = 1, 2, \ldots, f; \tag{5}$$

3   Proposed Differential Evolution-Based Approach

Differential Evolution (DE) [9] is a relatively simple evolutionary algorithm, primarily used to solve continuous optimisation problems. Through the nature-inspired concepts of selection, mutation and crossover, DE iteratively attempts to improve a set of candidate solutions to a particular problem, where a solution is represented by a vector of real values, by updating the population when better solutions are found until some termination criteria is met. DE has previously been used to solve a variety of optimisation problems, with a wide range of DE variants existing in the literature [1].

The general DE framework used in this paper is as follows. The first step is to initialise N solutions $x_1 \ldots x_N$, where N is the population size, with random values in each dimension. The algorithm then repeats the following steps until some termination criterion is met. At generation G, for each individual $x_{i,G}$ in the population, depending on the mutation strategy used, up to five solutions are chosen at random to use for mutation. The random solutions are used to calculate difference vectors to diversify the solution. A new mutant vector $v_{i,G}$ is generated using the given mutation strategy. In general a DE mutation strategy is referred to in the format: *mutation strategy/number of difference vectors*. Given x_{best} as the best solution found so far, randomly selected solutions x_{r1} to x_{r5} and $F \in [0, 2]$ as a weighting (the differential weight) used to control the influence of different vectors within the mutation, the five mutation strategies used in this paper are as follows:

$$rand/1 \qquad v_{i,G} = x_{r1,G} + F \cdot (x_{r2,G} - x_{r3,G}) \tag{6}$$

$$rand/2 \qquad v_{i,G} = x_{r1,G} + F \cdot (x_{r2,G} - x_{r3,G}) \\ + F \cdot (x_{r4,G} - x_{r5,G}) \tag{7}$$

$$best/1 \qquad v_{i,G} = x_{best,G} + F \cdot (x_{r1,G} - x_{r2,G}) \tag{8}$$

$$best/2 \qquad v_{i,G} = x_{best,G} + F \cdot (x_{r2,G} - x_{r3,G}) \\ + F \cdot (x_{r4,G} - x_{r5,G}) \tag{9}$$

$$current-to-best/1 \qquad v_{i,G} = x_{i,G} + F \cdot (x_{best,G} - x_{i,G}) \\ + F \cdot (x_{r1,G} - x_{r2,G}) \tag{10}$$

After a mutant vector $v_{i,G}$ has been obtained, crossover is performed between the original target vector $x_{i,G}$ and the mutant vector $v_{i,G}$. In each case we use binomial crossover to generate a trial vector $u_{i,G}$, with the value assigned to each dimension j of the trial vector as follows:

$$u_{j,i,G} = \begin{cases} v_{j,i,G} & \text{if } \mathrm{rand}[0,1) \leq CR \text{ or } j = j_{rand} \\ x_{j,i,G} & \text{otherwise} \end{cases} \tag{11}$$

Here $CR \in [0,1]$ is the crossover probability, controlling which of the two parents, either the target vector or the mutant vector, has the greatest influence on the trial vector generated. j_{rand} is the value of a random dimension to ensure that at least one dimension is taken from each of the parent solutions. Finally the target vector and the trial vector are compared using a fitness function g, with the best of the two kept in the population for the following generation:

$$x_{i,G+1} = \begin{cases} u_{i,G} & \text{if } g(u_{i,G}) < g(x_{i,G}) \\ x_{i,G} & \text{otherwise} \end{cases} \tag{12}$$

3.1   Encoding and Decoding Solutions to the DTMFSP

As mentioned above, given n jobs, the search process when solving the DTMFSP is over a vector of integers $S = s_1, \ldots, s_n$, with each $s \in S$ able to take a value between 1 and f corresponding to the factory that job is allocated to. As DE operates over a space of real values, a method to map a DE vector to a DTMFSP solution in discrete space is needed. Assuming that there are n jobs and f factories in the DTMFSP problem, each vector x has n dimensions with values in the range $[0,1]$. To interpret a vector as a solution to the DTMFSP problem, the range $[0,1]$ is partitioned into n equal parts, that is $[0, 1/f)$, $[1/f, 2/f)$, $\ldots$, $[(n-1)/f, n)$. For example with 4 factories and 5 jobs the vector $[0.21, 0.85, 0.42, 0.63, 0.31]$ indicates that the 1st job is assigned to 1st factory (as it is in the interval $[0, 0.25)$), the 2nd job is assigned to 4th factory (it is within the interval $[0.75, 1)$), with the remaining jobs assigned to factory 2, factory 3, and factory 2 respectively. After determining which jobs are assigned to which factory based on the DE vector, we can then determine the job sequence of each factory. For each factory, finding the minimum makespan is the classical two machine flow-shop scheduling problem, so we can use Johnson's rule to compute the optimal job sequence. The maximum makespan of this schedule can then be computed.

4   Experimental Framework and Parameter Tuning

In our experiments we use the DTMFSP benchmark instances based on a real-world scenario introduced by Deng et al. [2]. This set consists of 100 instances with 5 instances for each $f \in \{2, 3, 4, 5, 6\}$ and $n \in \{20, 50, 100, 200\}$. Here we will focus on the largest of these instances where $f = 3, 4$ and 5. As Deng et al. [2]

used $0.1 * n$ seconds CPU time as a termination criterion on different hardware, only an indirect comparison can be made to their results. For the benefit of future work in this area, we use the number of fitness evaluations as a termination criterion, allowing each run to evaluate 500,000 individuals. All experiments were performed on an Intel Core i7 CPU @ 2.40 GHz with 16 GB RAM.

The set of possible parameter settings for each DE component are given in Table 1. In order to test all of these parameter settings, we would have to test 625 (5^4) combinations. However, using an orthogonal array we are able to reduce this to 25 combinations, and use the results of these 25 experiments to derive the best set of parameters to use on the full benchmark set. Consistent with the parameter tuning performed by Deng et al. [2], we use the 11th 4-factory instance (F4_11), which has $n=100$ jobs to schedule for parameter tuning.

Table 1. Set of possible settings for each DE parameter

Parameter	Possible values
F	{0.25, 0.50, 0.75, 1.00, 1.25}
CR	{0.02, 0.04, 0.06, 0.08, 0.10}
N	{10, 25, 50, 100, 200}
Mutation strategy	{rand/1, rand/2, best/1, best/2, current-to-best/1}

After applying each of the 25 parameter combinations to F4_11, we are able to define the best set of parameters as DE_{Best}. It was found that the parameter set where $F = 0.50$, $CR = 0.02$, $N = 25$ and mutation strategy $= $ rand/2 performed best, and will be referred to as DE_{Best} herein.

5   Results

Deng et al. [2] presented three Harmony Search variants applied to the DTMFSP: classic Harmony Search (HS), 'improved' Harmony Search (IHS) and Global-best Harmony Search (GHS). Tables 2, 3 and 4 show the results of the best value obtained from 10 runs of DE_{Best} compared to the results of IHS, HS and GHS obtained by Deng et al. [2] for instances with 4, 5 and 6 factories respectively. Although only an indirect comparison can be made due to the differing termination criteria used, DE_{Best} does not run for a significantly different amount of time to the methods of Deng et al. [2], even when differences in hardware are taken into consideration. The best results obtained for each instance are highlighted bold, with new best results also highlighted with an asterisk(*). Due to space limitations, in the case of 4 factories in Table 2, we have only included those instances for which new best results are obtained. For the remaining instances of this type, DE_{Best} and IHS [2] obtain identical results.

Table 2. Best makespan obtained by DE_{Best} and Harmony Search variants over 10 runs of each instance of the Deng et al. [2] benchmark set with $f = 4$ factories

Instance	DE_{Best}	IHS	HS	GHS
F4_4	**312***	313	313	313
F4_11	**1199***	1200	1201	1200
Average (F4_1–F4_20)	**1171.0**	1171.1	1173.1	1173.0

Table 3. Best makespan obtained by DE_{Best} and Harmony Search variants over 10 runs of each instance of the Deng et al. [2] benchmark set with $f = 5$ factories

Instance	DE_{Best}	IHS	HS	GHS
F5_1	**227**	**227**	229	231
F5_2	**234**	**234**	236	235
F5_3	**243***	244	244	244
F5_4	**272**	**272**	275	275
F5_5	**234**	**234**	236	238
F5_6	**543***	545	547	546
F5_7	**537**	**537**	547	544
F5_8	**518***	519	525	522
F5_9	**534**	**534**	542	542
F5_10	466	**465**	473	469
F5_11	1008	**1007**	1024	1029
F5_12	**1044**	**1044**	1057	1060
F5_13	**982**	**982**	988	985
F5_14	**986***	988	1004	1005
F5_15	**1090***	1092	1096	1097
F5_16	**2079**	**2079**	2086	2089
F5_17	**2131**	**2131**	2140	2143
F5_18	**2087**	**2087**	2094	2095
F5_19	**2050**	**2050**	2073	2069
F5_20	2018	**2017**	2029	2029
Average	**964.2**	964.4	972.3	972.4

For the instances with $f = 4$ factories Table 2, DE_{Best} and IHS clearly outperform HS and GHS, with DE_{Best} obtaining new best solutions for instances F4_4 and F4_11. DE_{Best} also outperforms IHS on average in these instances. GHS performs particularly badly on these instances, only obtaining the same best result as the other 3 methods in 1 of the 20 instances, with HS only performing marginally better. As noted by Deng et al. [2], as the number of factories increases, the task of assigning jobs to factories assignment becomes more

Table 4. Best makespan obtained by DE_{Best} and Harmony Search variants over 10 runs of each instance of the Deng et al. [2] benchmark set with $f = 6$ factories

Instance	DE_{Best}	IHS	HS	GHS
F6_1	**173**	**173**	**173**	174
F6_2	**235**	**235**	**235**	**235**
F6_3	**192***	194	194	194
F6_4	**217***	218	219	219
F6_5	233	**232**	234	236
F6_6	**434***	438	449	443
F6_7	**423***	426	433	428
F6_8	393	**391**	406	405
F6_9	445	**442**	462	459
F6_10	460	**459**	474	465
F6_11	907	**904**	925	926
F6_12	**787**	**787**	798	796
F6_13	**867**	**867**	880	882
F6_14	877	**874**	901	900
F6_15	883	**878**	915	904
F6_16	**1782***	1786	1801	1799
F6_17	1725	**1724**	1740	1739
F6_18	1664	**1662**	1698	1689
F6_19	**1707**	**1707**	1724	1731
F6_20	1680	**1677**	1728	1715
Average	804.2	**803.7**	819.5	817.0

complex, making it more difficult to find the best combination of jobs for a particular factory. As we see in Table 3 where $f = 5$, HS and GHS are no longer able to find same best result as DE_{Best} for any of these instances. DE_{Best} is the best performing method on 17 of these 20 instances and overall on average, finding new best solutions for 5 instances. Table 4 presents the results when the number of factories $f = 6$. In this set of large instances the performance of DE_{Best} drops off somewhat, although it is able to find new best results for 5 of the 20 instances and obtains the best solution of all methods for 10 of the 20 instances. It is notable here that IHS is the best performing method when the instances get bigger, particularly for instances F6_11–F6_20 where the number of jobs is 100 and 200. One possible reason that DE_{Best} is being outperformed on these larger instances is the nature of the parameter tuning experiments performed. As the parameter tuning was done on slightly smaller instances, where $f = 4$, it could be the case that the parameters of DE_{Best} are overfitted to instances of this size and are not scaling well to the largest instances in the benchmark set.

Despite this, DE_{Best} outperforms the three Harmony Search variants presented by Deng et al. [2] on average for instances with $f = 4$ and 5, providing new best-known results for 12 of the 60 benchmark instances tested.

6   Conclusions and Future Work

In this paper we have presented experiments applying Differential Evolution (DE) to the distributed two machine flow-shop scheduling problem (DTMFSP). Although the search space of the problem is discrete, a mapping is defined from continuous to discrete space in order to apply DE to the problem indirectly. An initial set of parameter tuning experiments were performed to decide the parameters for DE before the best combination was applied to a set of benchmark instances and compared to 3 existing methods based on the Harmony Search algorithm. DE was able to outperform the Harmony Search methods in 2 of 3 sets of instances when clustered by number of factories and provide new best-known results for 12 instances of the 60 tested.

All of the DE methods tested use fixed parameters throughout a run, with the same set of parameters used across all instances of the benchmark set. It is possible that good parameter settings are dependent on the particular instance under consideration, or even the current state of the search. The literature contains existing variations of DE such as SHADE [10] and JADE [12], which control parameters including crossover probability and differential rate adaptively. Future work will apply methods such as these to the DTMFSP to see if any gain can be made by controlling parameters in a dynamic manner.

References

1. Das, S., Suganthan, P.N.: Differential evolution: a survey of the state-of-the-art. IEEE Trans. Evol. Comput. **15**(1), 4–31 (2011)
2. Deng, J., Wang, L., Shen, J., Zheng, X.: An improved harmony search algorithm for the distributed two machine flow-shop scheduling problem. In: Kim, J.H., Geem, Z.W. (eds.) Harmony Search Algorithm. AISC, vol. 382, pp. 97–108. Springer, Heidelberg (2016). doi:10.1007/978-3-662-47926-1_11
3. Garey, M.R., Johnson, D.S., Sethi, R.: The complexity of flowshop and jobshop scheduling. Math. Oper. Res. **1**(2), 117–129 (1976)
4. Johnson, S.M.: Optimal two-and three-stage production schedules with setup times included. Naval Res. Logist. Q. **1**(1), 61–68 (1954)
5. Lin, B.M., Hwang, F., Gupta, J.N.: Two-machine flowshop scheduling with three-operation jobs subject to a fixed job sequence. J. Sched., 1–10 (2017, to appear)
6. Onwubolu, G., Davendra, D.: Scheduling flow shops using differential evolution algorithm. Eur. J. Oper. Res. **171**(2), 674–692 (2006)
7. Pan, Q.K., Wang, L., Gao, L., Li, W.: An effective hybrid discrete differential evolution algorithm for the flow shop scheduling with intermediate buffers. Inf. Sci. **181**(3), 668–685 (2011)
8. Qian, B., Wang, L., Huang, D., Wang, W., Wang, X.: An effective hybrid de-based algorithm for multi-objective flow shop scheduling with limited buffers. Comput. Oper. Res. **36**(1), 209–233 (2009)

9. Storn, R., Price, K.: Differential evolution-a simple and efficient heuristic for global optimization over continuous spaces. J. Glob. Optim. **11**(4), 341–359 (1997)
10. Tanabe, R., Fukunaga, A.: Success-history based parameter adaptation for differential evolution. In: Proceedings of the IEEE Congress on Evolutionary Computation (CEC 2013), pp. 71–78. IEEE (2013)
11. Wang, L., Pan, Q.K., Suganthan, P.N., Wang, W.H., Wang, Y.M.: A novel hybrid discrete differential evolution algorithm for blocking flow shop scheduling problems. Comput. Oper. Res. **37**(3), 509–520 (2010)
12. Zhang, J., Sanderson, A.C.: JADE: adaptive differential evolution with optional external archive. IEEE Trans. Evol. Comput. **13**(5), 945–958 (2009)

A Multi-objective Differential Evolution
for QoS Multicast Routing

Wenhong Wei[1(✉)], Zhaoquan Cai[2], Yong Qin[1], Ming Tao[1],
and Lan Li[3]

[1] School of Computer, Dongguan University of Technology,
Dongguan 523808, China
weiwh@dgut.edu.cn
[2] Huizhou University, Huizhou 516007, China
[3] School of Software, Nanchang University, Nanchang 330047, China

Abstract. This paper presents a new multi-objective differential evolution
algorithm (MODEMR) to solve the QoS multicast routing problem, which is a
well-known NP-hard problem in mobile Ad Hoc networks. In the MODEMR,
the network lifetime, cost, delay, jitter and bandwidth are considered as five
objectives. Furthermore, three QoS constraints which are maximum allowed
delay, maximum allowed jitter, and minimum requested bandwidth are included.
In addition, we modify the crossover and mutation operators to build the
shortest-path multicast tree to maximize network lifetime and bandwidth, minimize cost, delay and jitter. In order to evaluate the performance and the
effectiveness of MODEMR, the experiments are conducted and compared with
other algorithms for these problems. The simulation results show that our proposed method is capable of achieving faster convergence and more preferable
for multicast routing in mobile Ad Hoc networks.

Keywords: Mobile Ad Hoc network · Multicast routing · Differential
evolution · Quality of service

1 Introduction

Differential evolution (DE), which was first proposed by Storn and Price [1], is one of
the most powerful evolutionary algorithms for global numerical optimization. The
advantages of DE are its ease of use, simple structure, speed, efficiency, and robustness
[2, 3]. Recently, DE has been successfully applied in diverse domains [4, 5]. DE also
can be used to solve multi-objective optimization problems [6, 7].

Multicast routing has drawn a lot of attention in recent years, and it is one type of
data transmission service in mobile Ad Hoc networks where the data are sent from
source node to many destination nodes through more than one path [8]. That is, it sends
the data packet only once and then it is duplicated and sent to different multicast group
members. In multicast routing, transmissions with high level of quality of service
(QoS) are fundamental guarantee of media applications, and the most important QoS
requirements are cost, delay, jitter and bandwidth [9]. However, in mobile Ad Hoc
networks, the data transmission range is influenced by the node battery power. If the

Y. Tan et al. (Eds.): ICSI 2017, Part I, LNCS 10385, pp. 458–465, 2017.
DOI: 10.1007/978-3-319-61824-1_50

battery power consumption is high in any one of the nodes, the chances of network lifetime reduction due to path breaks are also more. This may lead to packet loss during packet forwarding in the multicast network. Thus, the QoS becomes more lower and lower. That is to say, the higher node battery power, the higher is QoS.

In general, the objects in the multicast routing optimization problem include minimization of routing cost, minimization of link delay, minimization of link jitter and maximization of the bandwidth etc. [10]. It is well known that multicast routing optimization is a non-deterministic polynomial (NP) hard problem for large-scale and wide area network, so many metaheuristic algorithms and their variants are often used for solving multicast routing optimization problem [11].

It is obvious that multicast routing problem is a multi-objective optimization problem. So, in this work, the multicast routing problem with five objectives including cost, delay, jitter, bandwidth and network lifetime is studied. We adopt multi-objective differential evolution algorithm to solve these problems, and modify the crossover and mutation operators to build the shortest-path multicast tree. In addition, we also adopt non-dominant sorting and handing constraint techniques to solve the five objectives optimization and three QoS constraints problems.

The rest of this paper is organized as follows. We firstly introduce the related work in Sect. 2. Section 3 discusses the mobile Ad Hoc networks model and the problem description. The proposed algorithm (MODEMR) is then presented in detail in Sect. 4. Section 5 reports experimental results. Finally, conclusions are drawn in Sect. 6.

2 Related Work

As multicast routing problem is a NP-complete problem, several metaheuristic methods have been employed to solve this problem. These optimization methods are as follows: Genetic Algorithm (GA), Particle Swarm Optimization (PSO), Ant Colony Optimization (ACO), Artificial Neural Networks (ANN) and Tabu Search (TS) research activities. GA and its variants are the most used in multicast routing. Haghighat et al. [10] proposed a novel QoS-based multicast routing algorithm based on the genetic algorithms, and proposed connectivity matrix of edges for representation of the Steiner trees, they considered a fitness function based on the total cost, the total delay and the minimum bandwidth of a path using the penalty technique. Yen et al. [12] proposed a multi-constrained QoS multicast routing method using the genetic algorithm for mobile Ad Hoc networks, this algorithm floods data using the available resources and minimum computation time in a dynamic environment. Karthikeyan and Baskar [13] proposed an ensemble of immigrant strategies with genetic algorithm which optimizes the combined objectives of network lifetime and delay is proposed for solving multicast routing problem. The proposed algorithm ensembles random immigrant with random replacement, random immigrant with worst replacement, elitism-based immigrant and hybrid immigrant strategies. Sun et al. [14] proposed a modified quantum-behaved particle swarm optimization method for QoS multicast routing in mobile Ad Hoc networks. In the proposed method, QoS multicast routing is converted into an integer programming problem with QoS constraints and is solved by the QPSO algorithm combined with loop deletion operation. Bitam and Mellouk [15] proposed a novel bee colony optimization algorithm called bees life algorithm applied to solve the quality of service multicast routing problem for vehicular ad hoc networks as NP-Complete problem with multiple constraints. The algorithm was applied to solve QoS multicast routing problems with four objectives which were cost, delay, jitter, and bandwidth and three QoS metrics constraints.

3 Mobile Ad Hoc Networks Model and Problem Description

For a given mobile Ad Hoc networks, it can be modeled as a weighted undirected graph $G = (V, E)$, where $V = \{v_1, v_2, ..., v_p\}$ denotes the set of nodes and $E = \{e_1, e_2, ..., e_q\}$ denotes the set of links connecting these nodes. There are a weight vectors w containing QoS attributes corresponding to each edge in E, such as cost, delay, jitter and bandwidth. They represent the link transmission cost, the real delay during the link transmission, the real link transmission delay variation (jitter) and the estimated link band-width, respectively.

Suppose $s \in V$ be the source node of the multicast tree, and $M = \{d_1, d_2, ..., d_n\} \subseteq \{V - s\}$ be the set of destination nodes of the multicast tree. Let $T(s, M)$ be the multicast tree, $|V|$ be the total number of nodes, $|E|$ be the total number of links, R+ be the set of positive real numbers. So $T(s, M)$ which the source node will send the message to multiple destination nodes can be generated from the nodes of graph G through.

In the multicast tree $T(s, M)$, let $p(s, t)$ denotes the routing path of $T(s, M)$ from the source node s to destination node $t(t \in M)$. Hence, the total QoS metrics are given by the following functions:

$$Cost(T(s, M)) = \sum_{e \in T(s,M)} Cost(e) \tag{1}$$

$$Delay(p(s, t)) = \sum_{e \in p(s,t)} Delay(e) \tag{2}$$

$$Jitter(p(s, t)) = \sum_{e \in p(s,t)} Jitter(e) \tag{3}$$

$$Bandwidth(p(s, t)) = \min(Bandwidth(e)), e \in p(s, t) \tag{4}$$

In addition, let Deg_i denotes the degree of the node i, Eny_i denotes the residual battery energy of the node i, φ_{ij} denotes the flow from the node i to the node j, and θ_{ij} denotes the energy consumption of the node i to transmit a unit of datagram to the node j. Assume that the values of φ_{ij} and θ_{ij} are same in all nodes, the datagram flow rate $C_{ij} = \varphi_{ij} * \theta_{ij}$. Hence, the lifetime of the multicast tree $T(s, M)$ is given as follows:

$$Tot\_life(T(s, M)) = \min_{i \in T(s,M)} \left\{ \frac{Eny_i}{\sum_{j \in Deg_i} C_{ij}} \right\} \tag{5}$$

Total lifetime of the mobile Ad Hoc networks is determined only for the topology of the multicast tree $T(s, M)$, furthermore, from Eq. (5), the higher the Deg_i, the greater is energy consumption. As we well known, in mobile Ad Hoc networks, the Deg_i represents the set of nodes connected to node i by links. So, the higher the Deg_i, the lower is the routing cost.

Above all, for the sake of practical deployment, in this work, the two maximization objectives bandwidth and network lifetime are transferred to two minimization objectives according to add sign "−" to the objective functions. So the five objectives can be formulated as follows:

$$Min : f(T(s, M)) = (f_1, f_2, f_3, f_4, f_5) \tag{6}$$

where

$f_1 = Cost(T(s, M))$
$f_2 = Delay(T(s, M))$
$f_3 = Jitter(T(s, M))$
$f_4 = -Bandwidth(T(s, M))$
$f_5 = -Tot\_life(T(s, M))$

Subject to:

$Delay(p(s, t)) \leq QD$
$Jitter(p(s, t)) \leq QJ$
$\underset{t \in T}{Min}\{Bandwidth(p, s))\} \geq QB$

where the maximum delay and the jitter should be lower or equal to the delay threshold QD and the jitter threshold QJ respectively, and the minimum of Bandwidth($p(s, t)$) in every link in the whole multicast tree should be greater or equal to the minimum bandwidth threshold QB.

4 MODEMR Algorithm

To solve the multicast routing problem in mobile Ad Hoc networks, we design multi-objective differential evolution algorithm named as MODEMR using modified the crossover and mutation operators to build the shortest-path multicast tree. In the MODEMR, since there exist QoS constraints, constraint handling scheme is used to handle QoS constraints.

4.1 Coding Scheme

In this work, the path coding scheme is employed to represent individual. In this coding scheme, the individual is represented as single solution which is multicast tree contains the different paths from source node s to each destination node $d_i \in M$ via a set of intermediate nodes. Path is encoded as a nodes sequence, and then an individual is represented by tree structure contains its different paths.

4.2 Initialization Scheme

It is a great helpful that population evolve using a good initialization scheme, which can make the promising areas quickly be discovered. In this work, a new population initialization scheme is designed, the details are given as follows:

Firstly, for given weighted graph G including cost, delay, jitter, bandwidth and lifetime, some spanning trees are generated using the spanning tree algorithm. Suppose that the number of spanning trees is NP and the number of destination nodes is n. that is, there are NP individuals with n-dimensional parameter vectors. Then, based on each obtained spanning tree, an initial population is generated using the n paths staring from the root node s to the destination nodes $\{d_1, d_2, ..., d_n\}$ in it.

4.3 Mutation Scheme

In DE, mutation is an operator which is used to generate a mutated individual according to differential vector. In this work, we modify mutation operator, and an example for rand/1 mutation operator is given as follows:

Firstly, three paths towards three destinations are selected from three individuals, such as p_1, p_2 and p_3. For the p_3, we selected randomly an intermediate node according to the probability F from destination node to source node, and we search the same node in the p_2 from destination node to source node. If the same node is found in the p_2, the part path of the p_3 from destination node to the intermediate node will swap the same part of the p_2. If the chosen intermediate node is not found, this process is repeated until success. When this process successes, the p_1 and p_2 will become two new paths. Then, we selected randomly an intermediate node from destination node to source node in the p_2, and we search the same node in the p_1 from destination node to source node. Similarly, if the same node is found in the p_1, the part of the p_2 from destination node to the intermediate node will swap the same part of the p_1. If the chosen intermediate node is not found, this process is repeated until success. After these operators finish, p_1, p_2 and p_3 will success to mutate.

4.4 Crossover Scheme

The crossover is a binomial operator which is used to recombine the vector of two individuals to generate the trial vectors according to the probability CR. In this work, path crossover is used to substitute binomial crossover operator. In the path crossover operator, if there are two paths with the same destination nodes in two individuals, the two paths are selected and exchanged. This process is repeated until reaching all individuals with crossover probability of CR.

4.5 Selection Scheme

Before selection operator is executed, we must get the fitness of objective functions using constraint handling scheme. Then, the offspring replaces the parent immediately if the parent is dominated by the offspring. If the parent dominates the offspring, the offspring is discarded. Otherwise, when the offspring and parent are non-dominated with each other, the parent and the offspring are stored in archive together.

5 Experimental Results and Analysis

To demonstrate the effectiveness of the MODEMR, the comprehensive experiments are conducted to evaluate the performance of the MODEMR. We use Waxman's random graph generator generate a network topologies. All simulations are carried out in an Intel Core Quad CPU with 2.83 GHz and 4.00 GB RAM. The proposed algorithm was programmed in Matlab 2013b and was run on a windows 7 operating system (64 bit).

In the experiments, the proposed MODEMR is compared with other state-of-the-art EAs for multicasting rouging problem, such as EISGA [13], QPSO [14] and BLA [15]. In addition, since this is a stochastic optimization algorithm, the comparative approaches are performed on 30 independent runs for each test, and the maximum number of generations is 200.

The number of nodes in the networks was set to be 10, 20, 30, 40 and 50. The characteristics of the links are also described by a quaternary (C, D, J, B), and the QoS constraints set are described by set $\{QD, QJ, QB\}$. These constraints are as follows: (1) the cost C = rand(2, 10); (2) the delay D = 2/3 * C ms; (3) the jitter J = rand (5, 15) ms; (4) the bandwidth B = rand(50, 200) kbit/s; (5) the network lifetime T = rand(15, 30) min; (6) path maximum delay threshold at QD = rand(60, 80) ms; (7) maximum path jitter threshold QJ = rand(20, 50) ms; (8) minimum path bandwidth threshold QB = rand(100, 170) kbit/s.

To certify that the MODEMR can outperform the EISGA, QPSO, and BLA in each objective preference, the results of the average rankings of these different algorithms by the Friedman test are shown in Tables 1. It is clearly shown that the MODEMR consistently obtain best rankings in all algorithms for the five objectives. In addition, from Table 1, we also find that the QPSO and BLA are ranked 2 and 3 respectively.

Table 1. Average ranking of all the algorithms by the Friedman test for five objectives.

Algorithms	Ranking				
	Cost	Delay	Jitter	Bandwidth	Lifetime
MODEMR	**1.6(1)**	**1.6(1)**	**1.4(1)**	**1.6(1)**	**1(1)**
EISGA	4(4)	4(4)	4(4)	3.2(4)	4(4)
QPSO	1.8(2)	1.8(2)	1.6(2)	2.2(2)	2.2(2)
BLA	2.6(3)	2.6(3)	3(3)	3(3)	2.8(3)

Then, the Wilcoxon Signed Rank test is used to perform statistically significant testing between pairs of algorithms. Table 2 shows Wilcoxon Signed Rank test result on the non-dominated solutions obtained by the MODEMR and the best solutions obtained by the EISGA, QPSO, and BLA in each objective preference. It is clearly seen that the MODEMR obtain higher R+ values than R− values, which indicates that the MODEMR is significantly better than the EISGA, QPSO and BLA.

Table 2. The Wilcoxon sign rank test results for MODEMR Against EISGA, QPSO and BLA.

Algorithms	Criteria	R+	R−
MODEMR - to - EISGA	Cost	**15**	0
	Delay	**15**	0
	Jitter	**15**	0
	Bandwidth	**13.5**	1.5
	Lifetime	**15**	0
MODEMR - to - QPSO	Cost	**8**	7
	Delay	**9**	6
	Jitter	**10**	5
	Bandwidth	**13.5**	1.5
	Lifetime	**15**	0
MODEMR - to - BLA	Cost	**12**	3
	Delay	**12**	3
	Jitter	**15**	0
	Bandwidth	**13.5**	1.5
	Lifetime	**15**	0

6   Conclusion

In mobile Ad Hoc network, the multicast routing problem was studied as a multi-objective optimization problem with constraints. The objectives include transmission cost, delay, jitter, bandwidth and network lifetime. To solve multicast routing problems in Mobile Ad Hoc Network, this work proposed a QoS multicast routing algorithm based on multi-objective DE algorithm. The proposed algorithm encoded the number of the path from the source node to each destination nodes in the network as each individual vector, and its mutation and crossover operators were modified. Thus, the multicast routing problem with the constraints representing QoS requirements was solved by multi-objective DE. In order to prove the reliability and the efficiency of this proposal, the experiments are conducted. The obtained results compared with other algorithms show that the proposed algorithm is capable of achieving faster convergence and more preferable for multicast routing problem in mobile Ad Hoc network.

Acknowledgement. This work was supported by the National Nature Science Foundation of China (Nos. 61370185, 61402217), Guangdong Higher School Scientific Innovation Project (No. 2014KTSCX188), the outstanding young teacher training program of the Education Department of Guangdong Province (YQ2015158); and Guangdong Provincial Science and Technology Plan Projects (Nos. 2016A010101034, 2016A010101035). Guangdong Provincial High School of International and Hong Kong, Macao and Taiwan cooperation and innovation platform and major international cooperation projects (No. 2015KGJHZ027).

References

1. Storn, R., Price, K.: Differential evolution–a simple and efficient heuristic for global optimization over continuous spaces. J. Glob. Optim. **11**, 341–359 (1997)
2. Wei, W., Wang, J., Tao, M.: Constrained differential evolution with multiobjective sorting mutation operators for constrained optimization. Appl. Soft Comput. **33**, 207–222 (2015)
3. Zhou, X., Zhang, G., Hao, X., Yu, L.: A novel differential evolution algorithm using local abstract convex underestimate strategy for global optimization. Comput. Oper. Res. **75**, 132–149 (2016)
4. Rajesh, K., Bhuvanesh, A., Kannan, S., Thangaraj, C.: Least cost generation expansion planning with solar power plant using differential evolution algorithm. Renew. Energy **85**, 677–686 (2016)
5. Malathy, P., Shunmugalatha, A., Marimuthu, T.: Application of differential evolution for maximizing the loadability limit of transmission system during contingency. In: Pant, M., Deep, K., Bansal, J.C., Nagar, A., Das, K. (eds.) Proceedings of Fifth International Conference on Soft Computing for Problem Solving. AISC, vol. 437, pp. 51–64. Springer, Singapore (2016). doi:10.1007/978-981-10-0451-3_6
6. Wei, W., Wang, J., Tao, M., Yuan, H.: Multi-objective constrained differential evolution using generalized opposition-based learning. Comput. Res. Dev. **53**(6), 1410–1421 (2016)
7. Cheng, J., Yen, G.G., Zhang, G.: A grid-based adaptive multi-objective differential evolution algorithm. Inf. Sci. **367–368**, 890–908 (2016)
8. Liu, Y., Dong, M., Ota, K., Liu, A.: ActiveTrust: secure and trustable routing in wireless sensor networks. IEEE Trans. Inf. Forensics Secur. **11**(9), 2013–2027 (2016)
9. Tao, M., Lu, D., Yang, J.: An adaptive energy-aware multi-path routing strategy with load balance for wireless sensor networks. Wirel. Pers. Commun. **63**(4), 823–846 (2012)
10. Haghighat, A., Faez, K., Dehghan, M.: GA-based heuristic algorithms for QoS based multicast routing. Knowl. Based Syst. **16**, 305–312 (2003)
11. Koyama, A., Nishie, T., Arai, J., Barolli, L.: A GA-based QoS multicast routing algorithm for large-scale networks. Int. J. High Perform. Comput. Netw. **5**, 381–387 (2008)
12. Yen, Y., Chao, H., Chang, R., Vasilakos, A.: Flooding-limited and multi-constrained QoS multicast routing based on the genetic algorithm for MANETs. Math. Comput. Model. **53**, 2238–2250 (2011)
13. Karthikeyan, P., Baskar, S.: Genetic algorithm with ensemble of immigrant strategies for multicast routing in ad hoc networks. Soft. Comput. **19**, 489–498 (2015)
14. Sun, J., Fang, W., Wu, X., Xie, Z., Xu, W.: QoS multicast routing using a quantum-behaved particle swarm optimization algorithm. Eng. Appl. Artif. Intell. **24**, 123–131 (2011)
15. Bitam, S., Mellouk, A.: Bee life-based multi constraints multicast routing optimization for vehicular ad hoc networks. J. Netw. Comput. Appl. **36**, 981–991 (2013)

Energy-Saving Variable Bias Current Optimization for Magnetic Bearing Using Adaptive Differential Evolution

Syuan-Yi Chen[(⊠)] and Min-Han Song

Department of Electrical Engineering, National Taiwan Normal University,
Taipei 106, Taiwan
chensy@ntnu.edu.tw

Abstract. This study proposes an adaptive differential evolution (ADE)-based variable bias current control strategy to improve the energy efficiency of an active magnetic bearing (AMB) system. In the AMB system, the drive current is composed of a control current and a superimposed bias current in which the former is controlled by an external controller used to regulate the rotor position while the latter is set as a pre-designed constant used to improve the linearity and dynamic performance. Generally, the bias current causes power loss even if no force is required. In this regard, the ADE-based variable bias current control strategy is proposed to minimize the energy consumption of the AMB control system without altering the control performance. Experimental results demonstrate the high-accuracy control and significant energy saving performances of the proposed method. The energy improvements compared to baseline were 20.24% and 17.65% for the operation periods of 10 s and 50 s, respectively.

Keywords: Variable bias current optimization · Differential evolution · Magnetic bearing · Energy-saving

1 Introduction

Active magnetic bearing (AMB) uses electromagnetic forces to support and regulate a rotor to the predefined positions [1, 2]. Specific features of such bearings, such as noncontact and low friction, make AMBs appropriate for uses in some special environments such as flywheel energy storage devices [3] and spindles [4]. Most drive schemes of AMB use a constant bias current with a superimposed dynamic control current to linearize the relationship between the control current and the electromagnetic force [1, 2]. The control current is controlled by an external controller used to regulate the rotor position while the bias current is set as a pre-designed constant used to improve the control linearity. Although the bias current can improve the linearity of AMB, there are always some power losses even if no force is required [5].

In general, lowering the bias current may enhance the nonlinearities of the AMB positioning system and lead to a control singularity [6]. The existent methods for determining the variable bias current can be categorized into adaptive control [5] and switching control [6]. Both methods depend on the dynamic model and system

© Springer International Publishing AG 2017
Y. Tan et al. (Eds.): ICSI 2017, Part I, LNCS 10385, pp. 466–474, 2017.
DOI: 10.1007/978-3-319-61824-1_51

parameters heavily. For the time being, there is no research that uses evolutionary algorithm to optimize the bias current without system information dynamically.

Differential evolution (DE) uses mutation, recombination, and selection operators at each generation to move the population toward the global optimal solution. It has been successfully applied in diverse fields [7–9]. However, three crucial control parameters of DE, population size (NP), mutation factor (F), and crossover rate (CR), may significantly affect the optimization performance [8]. Therefore, many strategies and operations have been proposed to accelerate the convergence speed and enhance the optimization performance of the conventional DE algorithm [8, 9].

In this study, an energy saving variable bias current control strategy is proposed to optimize the bias current automatically for improving the energy efficiency and control performance simultaneously. The core of the strategy is based on an online DE algorithm. In this regard, no mathematical model and system parameters are required during the control process. Experimental results illustrated the validity and advantages of the proposed variable bias current optimization for the AMB positioning system.

2   AMB Positioning System

Figure 1 shows the control principle of the AMB positioning system. To regulate the position of the rotor to the zero, the currents circulated through the two electromagnets should be controlled exactly and dynamically. The deviation of the nominal air gap z_0 is denoted by variable z, which is also referred to as the rotor position. Two power amplifiers are used to transfer the voltage signals to drive currents. Moreover, the bias current i_0 is decided by a predesigned constant bias voltage v_0. On the other hand, the control current i_z, which is determined by the output of the controller v_z in real-time, is added to the bias current in the right electromagnetic coil and subtracted from the bias current in the left one. According to the dynamic adjustment of the control current i_z, the rotor can be positioned and moved to the arbitrary position.

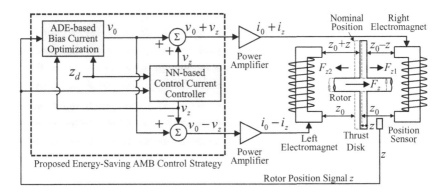

Fig. 1. Drive system of the AMB positioning system.

The nonlinear attractive electromagnetic force can be modeled as follows [1, 2]:

$$F_z = k(F_{z1} - F_{z2}) = k\left[\frac{(i_0 + i_z)^2}{(z_0 - z)^2} - \frac{(i_0 - i_z)^2}{(z_0 + z)^2}\right] \tag{1}$$

where k is the electromagnet factor; F_{z1} and F_{z2} are the forces caused by the right and left electromagnets, respectively. The dynamic model of the AMB positioning system can be derived using as follows [2]:

$$m\ddot{z} = -c\dot{z} + K_z z + K_i i_z + F_d \tag{2}$$

where m is the mass of the rotor, c is the friction factor due to the supporting bearings in radial direction, and F_d is the external disturbance. The instantaneous power loss is introduced to compare the energy efficiency of the drive system as below:

$$P = [(i_0 + i_z) + (i_0 - i_z)]R \tag{3}$$

where R is the resistance of the coil. The control current i_z can be positive, zero, or negative values depending on the rotor position while the bias current i_0 is a constant. Hence, the accumulated energy consumption E_p during the time interval T [0, T] is:

$$E_p(t) = \int_0^T P(t)dt \tag{4}$$

3   Adaptive DE Algorithm

The steps of the adopted adaptive DE (ADE) algorithm are described as follows [8]:

1. *Initialization:* The initial target vector X_i^g are generated randomly as follows:

$$X_i^g = \left[x_{i,1}^g, x_{i,2}^g, \ldots x_{i,j}^g, \ldots, x_{i,d}^g\right] \tag{5}$$

   where $i = 1, 2, \ldots, NP$, in which NP is the population size; $j = 1, 2, \ldots, d$, in which d is the dimension of the search space; and g represents the g_{th} generation.

2. *Mutation:* For the individual target vector X_i^g, three individual vectors, $X_{r1}^g, X_{r2}^g,$ and X_{r3}^g among the population are randomly selected to generate the mutant vector V_i^{g+1} according to the following mutation mechanism:

$$V_i^{g+1} = X_{r1}^g + F(X_{r2}^g - X_{r3}^g) \tag{6}$$

   where $r1 \neq r2 \neq r3 \neq i$, and F is a mutation factor. In general, the small constant mutation factor F may lead to premature convergence, whereas a large constant mutation factor may result in poor convergence efficiency. Therefore, an adaptive selection mechanism for the factor F is adopted as follows [8]:

$$F = \phi \times \zeta \times \lambda \tag{7}$$

where ϕ is a random value between $(0, 1)$, λ is an adaptive factor, and ζ is an inertia weight defined as

$$\zeta = \sqrt{\frac{g_{max} - g_{now}}{g_{max}}} \tag{8}$$

where g_{max} is the maximum number of generations, and g_{now} is the current number of generations. Moreover, the factor λ is used to balance the global exploration and local search abilities as follows:

$$\lambda = \begin{cases} \lambda_b \geq 1, & if\ q \leq q_d \\ \lambda_s < 1, & otherwise \end{cases} \tag{9}$$

where q is the improvement rate of the target vector X_i^g, q_d is the threshold of the acceptable improvement rate, λ_b and λ_s are the big and small step sizes, respectively.

3. *Recombination:* In this step, the individual target vector X_i^g is crossed over with its mutated mutant vector V_i^{g+1} to generate the trial vector U_i^{g+1} as follows:

$$u_{i,j}^{g+1} = \begin{cases} v_{i,j}^{g+1}, & if\ rand_j \leq CR \\ x_{i,j}^g, & otherwise \end{cases} \tag{10}$$

where $u_{i,j}^{g+1}$, $v_{i,j}^{g+1}$, and $x_{i,j}^g$ are the jth elements of U_i^{g+1}, V_i^{g+1}, and X_i^g, respectively; $rand_j$ is a random value between $(0, 1)$; and CR is a predesigned crossover rate.

4. *Selection:* The trial vector U_i^{g+1} is compared with the target vector X_i^g via the fitness values. The better one is then selected as a new target vector for the next generation.

$$X_i^{g+1} = \begin{cases} U_i^{g+1}, & if\ FIT(U_i^{g+1}) \geq FIT(X_i^g) \\ X_i^g, & otherwise \end{cases} \tag{11}$$

where FIT is a fitness function defined by the users. Repeat Steps 1–4 until the best fitness value is achieved or a preset count of the generation is reached.

4   Proposed Control Strategy

Selecting the bias current manually with the simultaneous considerations of the control performance and energy efficiency is difficult. In this regard, an energy saving ADE-based variable bias current control strategy is proposed. First, the rotor position was controlled by the control current through a well-designed neural network (NN) controller; moreover, the variable bias current was optimized by the ADE to minimize the power loss of the AMB. For example, NP target vectors $X_{1,1}$–$X_{4,NP}$ with one searching space dimension, which is simplified as X_1–X_{NP} here, were assumed and selected randomly in the searching range as follows:

$$X_{min} < X_i \le X_{max} \tag{12}$$

In the beginning of the control process, a predesigned bias voltage v_0 was applied directly to the AMB positioning system as shown in Fig. 2. Each trial vector X_i was then sequentially adopted as the bias voltage v_0 and learned dynamically through the ADE algorithm. The execution of the ADE algorithm in each time interval T includes N_g generation for evolution. All target vectors X_1–X_{NP} were evaluated and learned during the evolution, respectively. The positioning errors e_a and power loss P during one period are measured to evaluate the fitness value of X_i as follows:

$$FIT(X_i, I) = \frac{1}{\varepsilon + \sum\limits_{I=1}^{N_g} \left[\tau_c e_a(I) + \tau_p P(I) \right]} \tag{13}$$

where I represents the Ith iteration; τ_c and τ_p are the trade-off weights for considered positioning performance and energy efficiency respectively. The vector X_i with the highest fitness value will be selected as the bias voltage v_0 for the practical AMB positioning system in the next period so that the bias current i_0 can be adjusted dynamically by considering the control performance and energy efficiency effectively.

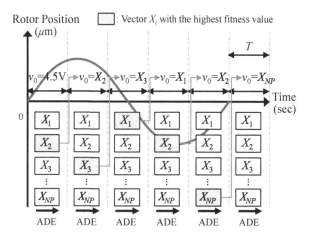

Fig. 2. Proposed energy saving variable bias current control strategy for the AMB positioning system using the ADE algorithm.

5 Experimental Results

5.1 Experimental Setup

In the experimental setup, a digital signal processor (DSP) with 14-b resolution analog-to-digital converters (ADCs) and 16-b resolution digital-to-analog converters (DACs) was used as the control core. In this study, the bias voltage v_0 was selected as 4.5 V to achieve the best transient and steady state control performance through some trials. Moreover, the scaling of the input voltage and output current of the power amplifier was 0.2 A/V. Furthermore, a low-pass filter with cutting frequency of 60 Hz was used to filter the high-frequency noise of the position signal. In the DSP, the interrupt service routine with execution frequency of 5 kHz first calculated the rotor position from the ADC interface. Subsequently, the control signal was determined according to the controller and transmitted to the power amplifier through the DAC. In addition, the maximum positioning error T_M, average positioning error T_A, and standard deviation of the positioning error T_S for the trajectory positioning are measured to compare the control performances during the control process [10].

5.2 Experimentation

The experimental setup is illustrated in Fig. 3. In the experiment, a NN controller with constant bias voltage v_0 was tested for comparing the control performance. The control voltage v_z was determined via the NN controller while the bias voltage v_0 was fixed at a constant value of 4.5 V. The control voltage v_z was changed for generating the required dynamic electromagnetic force. Though the rotor can be controlled accurately, the energy efficiency problem was not considered. The bias current i_0 circulated in the electromagnetic coils of the AMB even if no force was required. To improve the energy efficiency of the AMB positioning system, the ADE algorithm was further adopted for the optimization of the bias current. The control parameters were designed as $NP = 6$, $N_g = 50$, $CR = 0.3$, $\lambda_b = 1.5$, $\lambda_s = 0.5$, $q_d = 0.3$, $d = 1$, $\tau_c = 0.5$, $\tau_p = 0.5$, $X_{min} = 3.5$, and $X_{max} = 4.5$. The experimental results are presented in Fig. 4. As seen in Fig. 4,

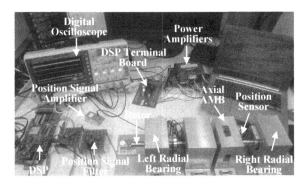

Fig. 3. Practical experimental setup of the AMB positioning system.

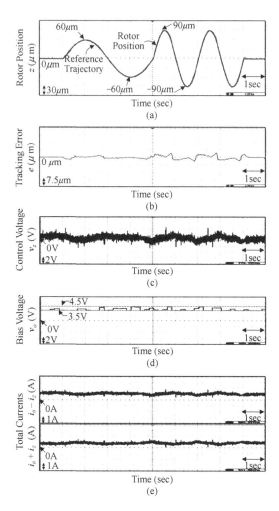

Fig. 4. Experimental results of the AMB positioning system using the NN-ADE control strategy. (a) Reference trajectory and positioning response of the rotor. (b) Positioning error e. (c) Control voltage v_z. (d) Variable bias voltage v_0. (e) Total currents $i_0 + i_z$ and $i_0 - i_z$.

the bias voltage v_0 was varied in the search range to optimize the positioning performance and energy efficiency simultaneously.

The control performance measures of the NN controller with constant bias voltages v_0 of 3.5 V and 4.5 V and NN-ADE control strategy with variable bias voltage v_0 are shown in Fig. 5. It shows that the higher bias voltage can produce stronger electromagnetic force but also consume more energy unavoidably. On the other hand, the energy consumptions of the NN controller with constant bias voltages v_0 of 3.5 V and 4.5 V and NN-ADE with variable bias voltage for operation periods of 10 s and 50 s are shown in Fig. 6. The experimental results showed that the AMB positioning system using the NN-ADE control strategy consumed lower energy than that using the NN

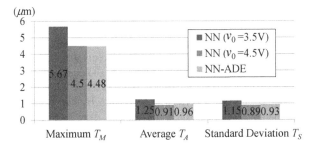

Fig. 5. Control performance measures of the NN and NN-ADE controllers.

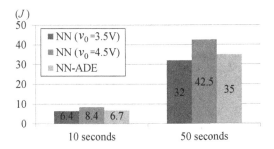

Fig. 6. Energy consumptions of the various controllers for 10 s and 50 s operation periods.

controller with constant bias voltages of 4.5 V; moreover, it exhibited higher control performance than the NN controller with a constant bias voltage 3.5 V. Compared with the NN controller with constant bias voltages v_0 of 4.5 V, the energy improvements of the proposed NN-ADE control strategy were 20.24% and 17.65% for the operation periods of 10 s and 50 s, respectively. The outstanding energy improvement achieved by the proposed NN-ADE control strategy can be verified clearly.

6   Conclusion

This study successfully demonstrated the development and application of the novel ADE-based energy saving variable bias current control strategy for an AMB positioning system considering the energy efficiency and control performance simultaneously. The theoretical bases of the proposed NN-ADE-based variable bias current control strategy were introduced to minimize the energy consumption of the AMB considering the requirements of positioning performance and robustness. The experimental results showed that not only the rotor can track the reference trajectory precisely but energy consumption can also be minimized through online bias current optimization with regard to the highly nonlinear AMB positioning system.

References

1. Schweitzer, G., Bleuler, H., Traxler, A.: Active Magnetic Bearings: Basics, Properties, and Applications of Active Magnetic Bearings. vdf Hochschulverlag, Zurich (1994)
2. Chen, S.Y., Lin, F.J.: Robust nonsingular terminal sliding-mode control for nonlinear magnetic bearing system. IEEE Trans. Control Syst. Technol. **19**(3), 636–643 (2011)
3. Mukoyama, S., Matsuoka, T., Hatakeyama, H., Kasahara, H., Furukawa, M., Nagashima, K., Ogata, M., Yamashita, T., Hasegawa, H., Yoshizawa, K., Arai, Y., Miyazaki, K., Horiuchi, S., Maeda, T., Shimizu, H.: Test of REBCO HTS magnet of magnetic bearing for flywheel storage system in solar power system. IEEE Trans. Appl. Supercond. **25**(3), 1–4 (2015)
4. Pesch, A.H., Smirnov, A., Pyrhonen, O., Sawicki, J.T.: Magnetic bearing spindle tool tracking through μ-synthesis robust control. IEEE/ASME Trans. Mechatron. **20**(3), 1448–1457 (2015)
5. Sahinkaya, M.N., Hartavi, A.E.: Variable bias current in magnetic bearings for energy optimization. IEEE Trans. Magn. **43**(3), 1052–1060 (2007)
6. Motee, N., de Queiroz, M.S.: A switching control strategy for magnetic bearings with a state-dependent bias. In: Proceeding of the 42nd IEEE Conference on Decision and Control, Maui, Hawaii USA, vol. 1, pp. 245–250, 9–12 December 2003
7. Lei, Q., Wu, M., She, J.: Online optimization of fuzzy controller for coke-oven combustion process based on dynamic just-in-time learning. IEEE Trans. Autom. Sci. Eng. **12**(4), 1535–1540 (2015)
8. Lee, W.P., Chien, C.W., Cai, W.T.: Improving the performance of differential evolution algorithm with modified mutation factor. J. Adv. Eng. **6**(4), 255–261 (2011)
9. Tang, L., Dong, Y., Liu, J.: Differential evolution with an individual-dependent mechanism. IEEE Trans. Evol. Comput. **19**(4), 560–573 (2015)
10. Lin, F.J., Chen, S.Y., Huang, M.S.: Tracking control of thrust active magnetic bearing system via Hermite polynomial-based recurrent neural network. IET Electr. Power Appl. **4**(9), 701–714 (2010)

Fireworks Algorithm

Acceleration for Fireworks Algorithm Based on Amplitude Reduction Strategy and Local Optima-Based Selection Strategy

Jun Yu[1] and Hideyuki Takagi[2(✉)]

[1] Graduate School of Design, Kyushu University, Fukuoka 815-8540, Japan
[2] Faculty of Design, Kyushu University, Fukuoka 815-8540, Japan
`takagi@design.kyushu-u.ac.jp`

Abstract. We propose two strategies for improving the performance of the Fireworks Algorithm (FWA). The first strategy is to decrease the amplitude of each firework according to the generation, where each firework has the same initial amplitude and decreases in size every generation rather than by dynamic allocation based on its fitness. The second strategy is a local optima-based selection of a firework in the next generation rather than the distance-based selection of the original FWA. We design a set of controlled experiments to evaluate these proposed strategies and run them with 20 benchmark functions in three different dimensions of 2-D, 10-D and 30-D. The experimental results demonstrate that both of the two proposed strategies can significantly improve the performance of the original FWA. The performance of the combination of the two proposed strategies can further improve that of each strategy in almost all cases.

Keywords: Fireworks algorithm · Decrement strategy · Local optima-based selection strategy

1 Introduction

Swarm intelligence has attracted the attention of many researchers because of its simplicity, robustness, parallelism and others. It simulates the mutual cooperation among simple individuals to achieve complex social behavior, such as in particle swarm optimization (PSO) [1] and ant colony optimization [2]. The fireworks algorithm (FWA) [3] is an emerging swarm intelligence algorithm proposed in 2010, which repeatedly simulates the explosion of fireworks to find the optimal solution.

Some improved versions of FWA have subsequently been proposed. For example, the enhanced fireworks algorithm (EFWA) [4] improves several operations of the original FWA and can achieve a better performance. Dynamic FWA (dyn-FWA) [5] uses a dynamic explosion amplitude for the currently best firework.

© Springer International Publishing AG 2017
Y. Tan et al. (Eds.): ICSI 2017, Part I, LNCS 10385, pp. 477–484, 2017.
DOI: 10.1007/978-3-319-61824-1_52

Although these modifications versions of FWA have improved the performance of the original FWA, there are still some limitations, and many researchers are still trying to propose new improvements.

The objective of this paper is to propose two strategies and improve the performance of the original FWA with the same cost consumption. The first approach used to attain this objective is to decrease the amplitude of each firework in accordance with the firework's generation rather than its fitness in order to achieve a good balance between exploration and exploitation. The second approach proposes a local optima-based selection strategy to keep the diversity of the population instead of the distance-based selection used in original FWA. We compare the performances of each proposal and their combination together with the original FWA.

We introduce the framework of original FWA and propose our two new strategies in Sect. 2. Then, we evaluate them compared with the original FWA using 20 benchmark functions of 3 different dimensions in Sect. 3. Finally we discuss the experimental evaluations in Sect. 4.

2   Improvements of Fireworks Algorithm

2.1   Original Fireworks Algorithm

In the real world, fireworks are launched into the sky, and many sparks are generated around the fireworks. The explosion process of a firework can be viewed as a local search around a specific point. FWA simulates the explosion process iteratively to find the optimal solution. Figure 1 demonstrates the explosive process of the FWA. Algorithm 1 shows the flowchart of FWA consisting of three operations principally: explosion, mutation and selection [3].

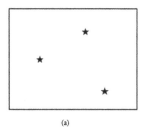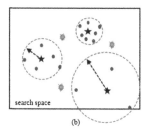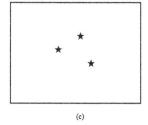

Fig. 1. Search process of FWA. (a) fireworks are generated, (b) sparks are generated around each firework, and mutation point is generated, (c) new fireworks are generated in the next generation using the (b). The (b) and (c) are iterated until a termination condition is satisfied.

2.2   Proposed Improvements

In this paper, we propose two strategies to replace the corresponding operations of the original FWA. The firework amplitudes of the original FWA are

Algorithm 1. The framework of the fireworks algorithm.

1: Initialize n fireworks randomly.
2: Evaluate the fitness of each firework.
3: **while** termination condition is not satisfied **do**
4:     Generate explosion sparks for each firework.
5:     Use Gauss mutation to obtain Gauss sparks.
6:     **if** sparks are generated outside search area **then**
7:         use a mapping rule for bringing back to the area.
8:     **end if**
9:     Evaluate the fitness of each generated sparks.
10:     Select n new fireworks for next generation.
11: **end while**
12: end of program.

automatically decided by their fitness values using the formula mentioned in the previous section. Better fireworks have relatively smaller amplitudes, while worse fireworks have relatively larger amplitudes.

The first strategy is to decrease the amplitude sizes of all fireworks from one generation to the next regardless of their fitness. We use the formula of Eq. (1) to determine the amplitude of fireworks. Figure 2 shows how the amplitude changes throughout the exploration period.

$$A_i = \begin{cases} A_{init} * (1 - \frac{FE_{cur}}{FE_{max}}) & if \quad FE_{cur} < c * FE_{max} \\ A_{init} * (1 - c) & Others \end{cases} \qquad (1)$$

where, A_{init} is the initial maximum amplitude of fireworks; FE_{cur} and FE_{max} represent the current and maximum number of fitness evaluations, respectively; and c is a constant for preventing the amplitude from becoming too small.

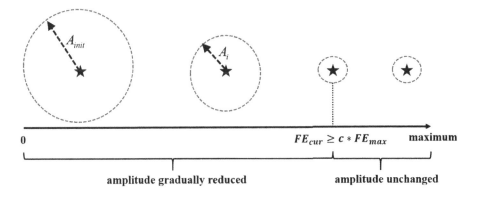

Fig. 2. Changes in amplitude throughout the exploration period

The second proposed strategy is to use a local optima-based selection of fireworks in the next generation instead of the distance-based selection used in the original FWA. Since the generated sparks can be considered as a local search around each firework, a set of a firework and its generated sparks can be considered a local subgroup. Then, we can obtain n local subgroups and a mutation subgroup consisting of all mutated sparks as the $(n+1)$-th subgroup. Because n new fireworks should be selected in the next generation, we merge the mutation subgroup and the subgroup of the worst firework into a new subgroup. The proposed local optima-based selection strategy takes the best firework or spark from each subgroup to form the next generation. Figure 3 demonstrates this selection strategy.

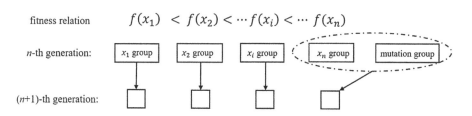

Fig. 3. The best one in each subgroup will be selected and go to next generation.

3   Experimental Evaluations

We use 20 benchmark functions from the CEC2013 benchmark test suite [6] in our evaluations. Table 1 shows their types, characteristics, variable ranges, and optimum fitness values. These landscape characteristics include shifted, rotated, global on bounds, unimodal and multi-modal. We test them with 3 dimensional settings: $D = 2$, 10 and 30.

To analyze the effect of each proposed improvement, we design the following four experiments; Experiments 1, 2, 3, and 4 are, respectively, the original FWA, the original FWA + the first proposed strategy (amplitude decrease strategy), the original FWA + the second proposed strategy (selection method of the firework in the next generation), and the original FWA + both strategies. Table 2 shows the parameter settings of the canonical FWA. The parameter settings for Experiments 2–4 are the same as the canonical FWA except the initial amplitudes; initial amplitudes A_{init} and constant c of the Eq. (1) are set as 10 and 0.95, respectively.

We evaluate convergence along the number of fitness calls instead of generations. We test each benchmark function with 30 trial runs in 3 different dimensional spaces. We apply the Friedman test and Holm's multiple comparison to the fitness values at the stop condition, i.e. maximum number of fitness calculations, for each benchmark function to check for significant difference among the methods. Table 3 shows the result of these statistical tests.

Table 1. Benchmar Function: Uni = unimodal, Multi = multimodal.

No.	Types	Characteristics	Ranges	Optimum fitness value
F_1	Uni	Sphere function	$[-100, 100]$	-1400
F_2		Rotated high conditioned elliptic function		-1300
F_3		Rotated Bent Cigar function		-1200
F_4		Rotated discus function		-1100
F_5		Different powers function		-1000
F_6	Multi	Rotated Rosenbrock's function	$[-100, 100]$	-900
F_7		Rotated Schaffers function		-800
F_8		Rotated Ackley's function		-700
F_9		Rotated Weierstrass function		-600
F_{10}		Rotated Griewank's function		-500
F_{11}		Rastrigin's function		-400
F_{12}		Rotated Rastrigin's function		-300
F_{13}		Non-continuous rotated Rastrigin's function		-200
F_{14}		Schwefel's function		-100
F_{15}		Rotated Schwefel's function		100
F_{16}		Rotated Katsuura function		200
F_{17}		Lunacek BiRastrigin function		300
F_{18}		Rotated Lunacek BiRastrigin function		400
F_{19}		Expanded Griewank's plus Rosenbrock's function		500
F_{20}		Expanded Scaffer's F_6 function		600

Table 2. Parameter setting of original FWA.

Paramaters	Values
# of fireworks for 2-D, 10-D and 30-D search	5
# of sparks m	50
# of Gauss mutation sparks,	5
constant parameters	$a = 0.04\ b = 0.8$
Maximum amplitude A_{max}	40
stop condition; MAX_{NFC}, for 2-D, 10-D, and 30-D search	4,000, 40,000, 100,000
dimensions of benchmark functions, D	2, 10, and 30
# of trial runs	30

4   Discussions

We begin our discussion with an explanation of the superiority of our proposed strategies. In the original FWA, better fireworks can obtain more resources within a small range, thus undertaking responsibility for exploitation. Exploration is achieved by worse fireworks obtaining less resources in a larger range

Table 3. Statistical test result of the Friedman test and Holm's multiple comparison for average fitness values of 30 trial runs of 4 methods. $A \gg B$ and $A > B$ mean that A is significant better than B with significant levels of 1% and 5%, respectively. $A \approx B$ means that there is no significant difference between A and B. Numbers in the table represent that 1: original FWA, 2: original FWA + proposed strategy 1, 3: original FWA + proposed strategy 2, and 4: original FWA + proposed strategies 1 and 2.

f_1	2-D	10-D	30-D
f_1	$4 \approx 2 \gg 3 \gg 1$	$4 \gg 2 \gg 3 \gg 1$	$4 \gg 2 \gg 3 \gg 1$
f_2	$4 \approx 3 \approx 2 > 1$	$4 > 2 \gg 3 \approx 1$	$4 > 2 \gg 3 \gg 1$
f_3	$3 \approx 2 \gg 1 \gg 4$	$4 \approx 2 \approx 3 \gg 1$	$4 \approx 2 > 3 \gg 1$
f_4	$3 \approx 2 \approx 4 \approx 1$	$2 \gg 1 \approx 4 \approx 3$	$2 > 4 \gg 3 > 1$
f_5	$4 \approx 2 \gg 3 \gg 1$	$4 \gg 2 \gg 3 \gg 1$	$4 \gg 2 \gg 3 \gg 1$
f_6	$4 \approx 2 \approx 3 \approx 1$	$4 > 3 \approx 2 \gg 1$	$2 \approx 3 \approx 4 \gg 1$
f_7	$3 \gg 2 \approx 4 \gg 1$	$1 \approx 2 \approx 4 \approx 3$	$4 \approx 1 \approx 3 \approx 2$
f_8	$4 > 3 \approx 1 > 2$	$4 \approx 3 \approx 2 \gg 1$	$4 \approx 2 > 1 \approx 3$
f_9	$4 \gg 3 \gg 2 \gg 1$	$3 \approx 4 \approx 1 \approx 2$	$3 \approx 4 \approx 1 \approx 2$
f_{10}	$4 \approx 2 > 3 \gg 1$	$4 > 2 \gg 3 \gg 1$	$4 \gg 2 \gg 3 \gg 1$
f_{11}	$3 \approx 4 \approx 2 \gg 1$	$3 \gg 1 \approx 4 \gg 2$	$3 \approx 1 \approx 4 \gg 2$
f_{12}	$3 \approx 2 \approx 4 \approx 1$	$1 \approx 3 \approx 2 \approx 4$	$1 \approx 4 \approx 3 \approx 2$
f_{13}	$3 \approx 2 \approx 1 \approx 4$	$3 \approx 1 \approx 2 \gg 4$	$1 \approx 3 \approx 2 \approx 4$
f_{14}	$3 \approx 1 \approx 2 \approx 4$	$3 \gg 1 \approx 4 \gg 2$	$3 > 4 \approx 1 \gg 2$
f_{15}	$1 \approx 3 \approx 2 \approx 4$	$4 \approx 2 \approx 3 \approx 1$	$4 \approx 2 \approx 3 \gg 1$
f_{16}	$4 \approx 2 \gg 3 \approx 1$	$4 \approx 2 \gg 3 > 1$	$2 \approx 4 \gg 3 \gg 1$
f_{17}	$4 \approx 2 \approx 3 \approx 1$	$4 \gg 2 \gg 3 \gg 1$	$4 \gg 2 > 3 \gg 1$
f_{18}	$4 \approx 2 \approx 3 \approx 1$	$2 \approx 4 \approx 1 \approx 3$	$2 > 4 \gg 1 \approx 3$
f_{19}	$2 \approx 4 \approx 3 \gg 1$	$4 > 2 \approx 3 \gg 1$	$4 \gg 2 \gg 3 \gg 1$
f_{20}	$2 \approx 4 \approx 3 \gg 1$	$3 \approx 1 \approx 4 \approx 2$	$1 \approx 3 \approx 2 \approx 4$

through the whole search period. However, exploration should be a task performed primarily in the early stages of search, while exploitation should be gradually emphasized along with the convergence of the population. So the first proposed strategy uses a decrement strategy to make all fireworks responsible for exploration in the early generations, with this exploration ability becoming gradually weaker as the exploitation ability becomes gradually stronger to achieve a good balance between exploration and exploitation.

We simply use the number of fitness evaluations to control the amplitude of fireworks in this paper, but this is not the unique realization of the proposed strategy 1; there must be other realizations which would allow us to improve its performance even more. For example, the amplitude can be adjusted adaptively according to optimization tasks, not just based on the number of fitness evaluations.

The distance-based selection used in the original FWA aims to preserve the diversity of fireworks, but there are still some shortcomings. This selection strategy gives higher selection probabilities to individuals located far away from other individuals. However, there is no guarantee that the fireworks selected by this original strategy have better fitness in the next generation than those in the current generation except the best individual. Further, there is also no guarantee that individuals coming from each subgroup will be selected fireworks in the next generation. If no individual from a certain subgroup is selected in the next generation, the area will not be explored in the next generation and the diversity may be lost.

The second proposed strategy can overcome these shortcomings and ensure each local optimum individual can remain in the next generation to maximize and the preserve the population diversity. This strategy may develop to become a new niche method for finding multi local or global optima at one time run.

Next, we discuss the effectiveness of our proposed strategies. To analyze their performances, Friedman test and Holm's multiple comparison test were applied at the stop condition in three different dimensions. The two strategies do not add additional fitness computation cost. Nevertheless, the statistical results in the Table 3 show that either of the two proposed strategies can improve the performance of the original FWA, and their combination can further improve performance in almost all evaluation cases.

Although combining two proposed strategies 1 and 2 with original FWA works well, it did not show clear performance for f_{11}–f_{14} in the Table 3. Figure 4

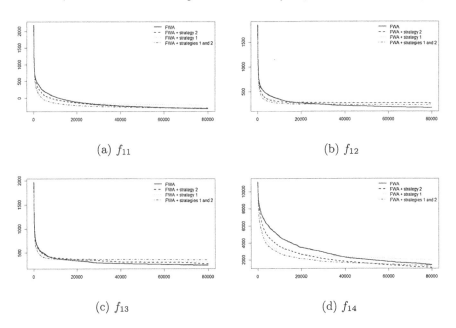

(a) f_{11}

(b) f_{12}

(c) f_{13}

(d) f_{14}

Fig. 4. Convergence curves of the original FWA, the original FWA + proposed strategy 1, the original FWA + proposed strategy 2, and the original FWA + proposed strategies 1 and 2 for 30-D f_{11}–f_{14}, respectively.

shows the average convergence curves of 4 methods for these 30-dimensional benchmark functions. These improved strategies for Rastrigin's function and Schwefel's function showed better performance in the early searching stages, while it could not keep their better performance in the later period and even became worse than the original FWA. It may be due to their many local optima; the local optima-based selection can maximize the diversity of the population, but it may reduce the convergence speed. We need further analysis of this result to understand the real reason and develop its solution.

5   Conclusion

We proposed two strategies to enhance the performance of the original FWA. The first strategy further emphasizes the balance between exploration and exploitation, and the second one selects local optimum individuals to preserve search diversity. Controlled experiments confirmed that they can improve the performance of the original FWA significantly.

In future work, we will further study these strategies and make full use of local information to obtain better performance.

Acknowledgment. This work was supported in part by Grant-in-Aid for Scientific Research (JP15K00340).

References

1. Kennedy, J., Eberhart, R.C.: Particle swarm optimization. In: IEEE International Conference on Neural Networks, Perth, Australia, vol. 4, pp. 1942–1948 (1995)
2. Dorigo, M., Maniezzo, V., Colorni, A.: Ant system: optimization by a colony of cooperating agents. IEEE Tran. Syst. Man Cybern. Part B Cybern. **26**(1), 29–41 (1996)
3. Tan, Y., Zhu, Y.: Fireworks algorithm for optimization. In: Tan, Y., Shi, Y., Tan, K.C. (eds.) ICSI 2010. LNCS, vol. 6145, pp. 355–364. Springer, Heidelberg (2010). doi:10.1007/978-3-642-13495-1_44
4. Zheng, S.Q., Janecek, A., Tan, Y.: Enhanced fireworks algorithm. In: IEEE International Conference on Evolutionary Computation, Cancun, Mexico, pp. 2069–2077 (2013)
5. Zheng, S.Q., Janecek, A., Li, J.Z., Tan, Y.: Dynamic search in fireworks algoritm. In: IEEE International Conference on Evolutionary Computation, Beijing, China, pp. 3222–3229 (2014)
6. Liang, J., Qu, B., Suganthan, P., Hernández-Díaz, A.G.: Problem definitions and evaluation criteria for the CEC 2013 special session on real-parameter optimization (2013). http://al-roomi.org/multimedia/CEC_Database/CEC2013/RealParameter Optimization/CEC2013_RealParameterOptimization_TechnicalReport.pdf

From Resampling to Non-resampling: A Fireworks Algorithm-Based Framework for Solving Noisy Optimization Problems

JunQi Zhang[1,2]([✉]), ShanWen Zhu[1,2], and MengChu Zhou[3]

[1] Department of Computer Science and Technology, Tongji University,
Shanghai, China
zhangjunqi@tongji.edu.cn, zhushanwen321@hotmail.com
[2] Key Laboratory of Embedded System and Service Computing,
Ministry of Education, Shanghai, China
[3] Department of Electrical and Computer Engineering,
New Jersey Institute of Technology, Newark, NJ 07102, USA
zhou@njit.edu

Abstract. Many resampling methods and non-resampling ones have been proposed to deal with noisy optimization problems. The former provides accurate fitness but demands more computational resources while the latter increases the diversity but may mislead the swarm. This paper proposes a fireworks algorithm (FWA) based framework to solve noisy optimization problems. It can gradually change its strategy from resampling to non-resampling during the evolutionary process. Experiments on CEC2015 benchmark functions with noises show that the algorithms based on the proposed framework outperform their original versions as well as their resampling versions.

Keywords: Fireworks algorithm · Noisy environment · Resampling · Non-resampling

1 Introduction

Noise is ubiquitous in real-world problems. Resampling methods [3,7] are commonly employed to resolve it. Equal resampling (ER) is a simple allocation method that evaluates all the solutions for the same number of times regardless their fitness values. Optimal Computing Budget Allocation (OCBA) [1] equally resamples them at first and then sequentially allocates different budget to each solution based on the first stage's solution quality and variation. Another method, called Equal Resampling top-N (ERN) [8], originates from ER but gives extra resampling resources to the most promising solutions. Besides, a powerful tool in stochastic environment, learning automaton, is introduced to alleviate the impact of noise in [12]. The studies have revealed that resampling methods may not be necessary to deal with noises. The studies [6,7] claim that particle swarm optimization (PSO) is stable and efficient in noise environment.

© Springer International Publishing AG 2017
Y. Tan et al. (Eds.): ICSI 2017, Part I, LNCS 10385, pp. 485–492, 2017.
DOI: 10.1007/978-3-319-61824-1_53

Furthermore, a non-resampling method, called evaporation [9], progressively reduces the objective value of the best position found by the swarm such that the positions with worse objective values are allowed to be considered.

From the above discussions, resampling methods provide accurate fitness value but demand more resources while non-resampling methods increase the diversity but may mislead the population to unpromising areas. This paper presents a fireworks algorithm (FWA) based framework for Noisy Optimization problems (FWANO for short), to allow gradual changes of its strategy from resampling to non-resampling along the evolution. The resampling strategy is employed to locate the true promising area in the early evolutionary stage. Then, to keep the diversity, a non-resampling strategy gradually takes charge and utilizes the unique characteristic of FWA that the information of sparks around a firework represents whether the firework is prospective. The consensus of promising sparks are adopted in FWANO instead of the best spark to neutralize noises. This feature reduces the resampling cost and obtains appreciable accuracy of the estimation, especially when sparks are close to each other in a later evolutionary stage.

The rest of the paper is arranged as follows. Section 2 reviews the background and some related works. FWANO is presented in Sect. 3. Experimental results are demonstrated in Sect. 4. Section 5 concludes the paper.

2   Background

2.1   Optimization Problems Subject to Noise

This study considers the following minimization problems:

$$\min_{x \in X} f(x), x = [x^1, x^2, ..., x^d] \tag{1}$$

where d is the number of dimensions and X is the feasible region of solution x.

Noises in optimization problems can be modeled as random variables following a Gaussian distribution [2]. The severity is described by its standard deviation σ. Besides, two types of Gaussian noises, additive and multiplicative, are presented as follows:

$$\hat{f}_+(x) = f(x) + N(0, \sigma^2) \tag{2}$$

$$\hat{f}_\times(x) = f(x) \times N(1, \sigma^2) \tag{3}$$

where $f(x)$ represents the true fitness value of a solution x. In this paper, the noisy fitness evaluation and the fitness estimation of solution x are represented as $\hat{f}(x)$ and $\tilde{f}(x)$. Obviously, $f(x) = \hat{f}(x) = \tilde{f}(x)$ when $\sigma = 0$.

2.2   Two Resampling Methods

ER first re-evaluates the fitness value of a solution for B times, and then utilizes the sample means as the estimation of true fitness:

$$\tilde{f}(x) = \frac{1}{B} \times \sum_{b=1}^{B} \hat{f}_b(x) \tag{4}$$

To improve the accuracy of promising solutions, ERN is proposed [8]. It estimates all solutions at first and then allocates extra budget B_Σ to top N solutions.

2.3   Fireworks Algorithms

The idea behind a fireworks algorithm [10] is to search around promising areas as if generating sparks around fireworks. The number of sparks and amplitude of explosions are allocated to fireworks according to their fitness values as follows:

$$s_i = M \cdot \frac{y_{max} - f(x_i) + \varepsilon}{\sum_{i=1}^{n}(y_{max} - f(x_i)) + \varepsilon} \tag{5}$$

$$A_i = \hat{A} \cdot \frac{f(x_i) - y_{min} + \varepsilon}{\sum_{i=1}^{n}(f(x_i) - y_{min}) + \varepsilon} \tag{6}$$

where x_i represents the location of firework i ($i = 1, 2, ..., n$), M and $\hat{A}$ are parameters to control the total number of sparks and amplitude of explosions, $y_{max} = \max(f(x_i))$ and $y_{min} = \min(f(x_i))$ ($i = 1, 2, ..., n$) are the maximum and minimum fitness value among n fireworks, and ε is the machine epsilon to prevent zero-division-error.

Nevertheless, conventional FWA fails in shifted functions. To solve the problem, a comprehensive study of FWA [14] reports five modifications and an enhanced fireworks algorithm (EFWA). Based on EFWA, an adaptive fireworks algorithm (AFWA) [4] is proposed to determine a proper amplitude for the firework with minimal fitness without parameters. In [13], a dynamic search fireworks algorithm (dynFWA) is proposed. It defines the concept of core firework (CF) as the firework with minimal fitness among all the fireworks, and non-core fireworks (non-CFs) as the rest of the fireworks. DynFWA applies a dynamic amplitude strategy on CF to improve its local search ability. The performances of FWA, EFWA, AFWA and dynFWA are compared in [16]. The work [15] proposes a cooperative framework for FWA (CoFFWA), which adopts an independent selection strategy to reinforce the exploitation ability of non-CFs.

2.4   CoFFWA

CoFFWA computes the number of sparks s_i and amplitude of explosion A_i as does in FWA, i.e., (5) and (6). Define x_c as the position of CF, x^* as the individual with the best fitness among all sparks and fireworks in this generation, and A_c^t as the amplitude of CF in generation t. A_c^t is updated as follows:

$$A_c^t = \begin{cases} A_c^{t-1} \times C_+ & \text{if } f(x^*) - f(x_c) < 0 \\ A_c^{t-1} \times C_- & \text{otherwise} \end{cases} \tag{7}$$

where C_+ and C_- are two parameters to increase and decrease. Then, each firework explodes according to the given s_i and A_i. In the independent selection operator, each firework chooses the best candidate among its generated sparks and itself in the current generation as the new firework. If the distance between firework i and CF is less than a threshold τ, firework i is reinitialized in the feasible space. More details can be found in [15].

3   A Fireworks Algorithm-Based Framework for Noisy Optimization

This work proposes a fireworks algorithm-based framework that changes its strategy from resampling to non-resampling gradually. Its main procedure is presented in Algorithm 1.

Algorithm 1. FWANO

1: Initialize n fireworks and evaluate their fitness $f(x_i)$ for B times
2: **while** (stopping criterion not met) **do**
3:    Update the number of resampling B (and B_Σ) using (10) and (11)
4:    Update the number of top sparks N using (12)
5:    Compute the number of sparks using (9)
6:    Compute the amplitude for each firework using (7)
7:    **for** each firework **do**
8:       Generate explosion sparks
9:       Re-evaluate generated sparks
10:       Select the new firework independently
11:       Compute the consensus for each firework using (8)
12:    **end for**
13: **end while**

3.1   Resampling Methods

At first, resampling is performed in FWANO. It re-evaluates sparks to provide accurate fitness through the sample mean as given in (4). This method ensures that FWANO finds true promising areas in noisy environments as in stationary environments. Two specific methods are introduced to FWANO and thus lead to two algorithms. The algorithm with ER, called FWANO-ER, evaluates all the sparks for B times. The one with ERN, named as FWANO-ERN, evaluates all the sparks for B times first in each generation. Then for firework i $(i = 1, 2, ..., n)$, it re-evaluates its top N_i sparks for additional B_Σ times.

3.2   The Consensus of Top Sparks

The consensus of top N_i sparks around firework i $(i = 1, 2, ..., n)$ is introduced. Note that they are the same top N_i sparks in FWANO-ERN. For firework i, FWANO sorts all its sparks by fitness in an ascending order and then takes the first N_i sparks as set T_i. In case that a firework has less sparks than N_i, bounds are set as $1 \leq N_i \leq \min(s_i, N)$, where s_i is the number of sparks from firework i, N is the maximum number of top sparks. The fitness of the consensus is calculated as follows:

$$f_c(x_i) = \frac{1}{N_i} \times \sum_{x \in T_i} \tilde{f}(x) \tag{8}$$

where $\tilde{f}(x)$ is the sample mean. Obviously, the more sparks are in T_i, the higher the diversity is. Moreover, this method consumes no additional evaluation.

$f_c(x_i)$ substitutes the fitness of the firework to calculate the number of sparks s_i in the next generation:

$$s_i = M \cdot \frac{\tilde{y}_{max} - f_c(x_i) + \varepsilon}{\sum_{i=1}^{n}(\tilde{y}_{max} - f_c(x_i)) + \varepsilon} \tag{9}$$

where $\tilde{y}_{max} = \max(f_c(x_i))$, $\tilde{y}_{min} = \min(f_c(x_i))$ $(i = 1, 2, ..., n)$ and ε is the machine epsilon to prevent any zero-division-error. Note that $f_c(x_i) = f(x_i)$ when $N = 1$.

3.3   From Resampling to Non-resampling

When sparks get closer and closer to each other, the fitness of the consensus in noisy environment gradually approaches its true value. To ensure that sparks are substantially near each other as the optimization proceeds, the dynamic amplitude strategy in (7) is adopted for all fireworks in this study. Thus, resampling in the later evolutionary stage becomes unnecessary. FWANO transforms a resampling strategy to a non-resampling one by reducing the numbers of resamplings, i.e., B and B_Σ:

$$B = \lceil \hat{B} \times \frac{E_m - E}{E_m} \rceil \tag{10}$$

$$B_\Sigma = \lceil \hat{B}_\Sigma \times \frac{E_m - E}{E_m} \rceil \tag{11}$$

while enlarging the number of top sparks N:

$$N = \lceil \hat{N} \times \frac{E}{E_m} \rceil \tag{12}$$

$\hat{B}$ and $\hat{B}_\Sigma$ are parameters controlling the largest number of resamplings, while $\hat{N}$ controls that of top sparks. E_m is the maximum number of evaluations and E is the number of used evaluations. $\lceil \ \rceil$ is a ceiling function.

4   Experiments

4.1   Experimental Setting

The test suite is CEC2015 single objective optimization benchmark [5] covering 15 functions, whose fitness values are all positive. Four types of functions

Table 1. Parameters of resampling

Algorithm	Parameter(s)
CoFFWA-ER	$B = 6$
FWANO-ER	$\hat{B} = 6$, $\hat{N} = 6$
CoFFWA-ERN	$B = 6$, $N = 2$, $B_{\Sigma} = 25$
FWANO-ERN	$\hat{B} = 6$, $\hat{N} = 6$, $\hat{B}_{\Sigma} = 10$

are employed: unimodal, multimodal, hybrid and composition functions. The dimensionality of these functions is $d = 30$ and all algorithms are run for 51 times on each function. The maximum number of evaluations is $10000 \times d$ for each run. Errors of fitness values of an algorithm are calculated as its performance. Multiplicative noises are applied to all the functions and the levels are set to $\sigma = \{0.06, 0.12, 0.18, 0.24, 0.30\}$. To keep the noisy fitness a positive value, noises are allowed to fall into $[1 - 3\sigma, 1 + 3\sigma]$.

Besides FWANO-ER and FWANO-ERN, CoFFWA and its ER (CoFFWA-ER) and ERN (CoFFWA-ERN) versions are also tested for comparison. For fairness, most of the parameters are the same in FWANO-ER and CoFFWA-ER, and so do ERN based algorithms. Common parameters for all algorithms are set as in [15]. For CoFFWA-ER, B is set to 6. Correspondingly, the maximum number of resamplings and top sparks in FWANO-ER is $\hat{B} = 6$ and $\hat{N} = 6$, respectively. For CoFFWA-ERN, $N = 2$ and $B_{\Sigma} = 25$ as suggested in [8]. But in FWANO-ERN, $\hat{N}$ is set to 6 and $\hat{B}_{\Sigma}$ is set to 10 to roughly keep the same number of resources as in CoFFWA-ERN. Details are given in Table 1.

4.2   Experimental Results

Results are partially presented in Fig. 1 for unimodal functions, Fig. 2 for multimodal functions and Fig. 3 for hybrid and composition functions. In all the following box plots, 'abcde' in bottom axis represents CoFFWA, CoFFWA-ER, CoFFWA-ERN, FWANO-ER and FWANO-ERN respectively, and numbers in the top axis represents the level of noises. In some box plots, the figure of CoFFWA exceeds the bound of plots for visibility of the results from other algorithms.

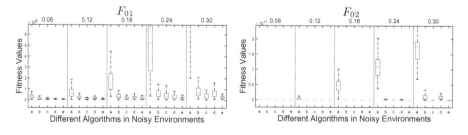

Fig. 1. The comparison of fitness values (left axis) on unimodal functions among (a) CoFFWA, (b) CoFFWA-ER, (c) CoFFWA-ERN, (d) FWANO-ER, (e) FWANO-ERN (bottom axis) under different noisy environments (top axis)

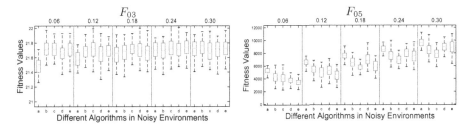

Fig. 2. The comparison of fitness values (left axis) on multimodal functions among (a) CoFFWA, (b) CoFFWA-ER, (c) CoFFWA-ERN, (d) FWANO-ER, (e) FWANO-ERN (bottom axis) under different noisy environments (top axis)

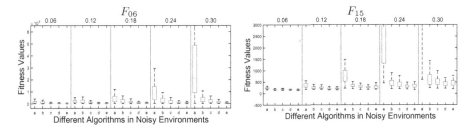

Fig. 3. The comparison of fitness values (left axis) on hybrid and composition functions among (a) CoFFWA, (b) CoFFWA-ER, (c) CoFFWA-ERN, (d) FWANO-ER, (e) FWANO-ERN (bottom axis) under different noisy environments (top axis)

The higher the noise level is, the worse the fitness estimation is. Original CoFFWA deteriorates severely, which can be observed in most functions. Comparing algorithm (a) with (b) or (c), it is obvious that resampling methods significantly improve the performance of CoFFWA. Both algorithms, FWANO-ER and FWANO-ERN, defeat their contenders. In total, the former outperforms CoFFWA-ER on 13 functions while the latter outperforms CoFFWA-ERN on 12 among total 15 functions. Those results manifest that the idea of changing its strategy from resampling to non-resampling can remarkably enhance the performance of FWA in noisy environments.

5   Conclusions

This paper presents a new FWA-based framework for optimization problems subject to noises. Its key idea is to change its strategy from resampling to non-resampling. Its application to real-world problems, e.g., [11] should be pursued.

Acknowledgments. This work is supported by China NSF under Grants No. 61572 359 and 61272271, and partly supported by the Fundamental Research Funds for the Central Universities of China (No. 0800219332).

References

1. Chen, C.H., Lin, J., Yücesan, E., Chick, S.E.: Simulation budget allocation for further enhancing the efficiency of ordinal optimization. Discrete Event Dyn. Syst. **10**(3), 251–270 (2000)
2. Jin, Y., Branke, J.: Evolutionary optimization in uncertain environments-a survey. IEEE Trans. Evol. Comput. **9**(3), 303–317 (2005)
3. Zhang, J., Xu, L., Ma, J., Zhou, M.: A learning automata-based particle swarm optimization algorithm for noisy environment. In: Proceedings of IEEE Congress on Evolutionary Computation, pp. 141–147. IEEE (2015)
4. Li, J., Zheng, S., Tan, Y.: Adaptive fireworks algorithm. In: 2014 IEEE Congress on Evolutionary Computation (CEC), pp. 3214–3221, July 2014
5. Liang, J.J., Qu, B.Y., Suganthan, P.N., Chen, Q.: Problem definitions and evaluation criteria for the CEC 2015 competition on learning-based real-parameter single objective optimization. Technical report 201411A, Zhengzhou University, Zhengzhou, China and Nanyang Technological University, Singapore, November 2014
6. Parsopoulos, K., Vrahatis, M.N.: Particle swarm optimizer in noisy and continuously changing environment. In: Hamza, M.H. (ed.) Artificial Intelligence and Soft Computing, pp. 289–294 (2001)
7. Parsopoulos, K., Vrahatis, M.N.: Particle swarm optimization for imprecise problems. In: Scattering and Biomedical Engineering: Modeling and Applications, pp. 254–264 (2002)
8. Rada-Vilela, J., Zhang, M., Johnston, M.: Resampling in particle swarm optimization. In: 2013 IEEE Congress on Evolutionary Computation (CEC), pp. 947–954 (2013)
9. Rada-Vilela, J., Zhang, M., Seah, W.: A performance study on the effects of noise and evaporation in particle swarm optimization. In: 2012 IEEE Congress on Evolutionary Computation (CEC), pp. 1–8 (2012)
10. Tan, Y., Zhu, Y.: Fireworks algorithm for optimization. In: Tan, Y., Shi, Y., Tan, K.C. (eds.) ICSI 2010. LNCS, vol. 6145, pp. 355–364. Springer, Heidelberg (2010). doi:10.1007/978-3-642-13495-1_44
11. Lu, X., Wang, L., Wang, H., Wang, X.: Kalman filtering for delayed singular systems with multiplicative noise. IEEE/CAA J. Automatica Sinica **3**(1), 51–58 (2016)
12. Zhang, J., Xu, L., Li, J., Kang, Q., Zhou, M.: Integrating particle swarm optimization with learning automata to solve optimization problems in noisy environment. In: 2014 IEEE International Conference on Systems, Man and Cybernetics (SMC), pp. 1432–1437 (2014)
13. Zheng, S., Janecek, A., Li, J., Tan, Y.: Dynamic search in fireworks algorithm. In: 2014 IEEE Congress on Evolutionary Computation (CEC), pp. 3222–3229, July 2014
14. Zheng, S., Janecek, A., Tan, Y.: Enhanced fireworks algorithm. In: 2013 IEEE Congress on Evolutionary Computation, pp. 2069–2077, June 2013
15. Zheng, S., Li, J., Janecek, A., Tan, Y.: A cooperative framework for fireworks algorithm. IEEE/ACM Trans. Comput. Biol. Bioinf. **14**(1), 27–41 (2017)
16. Zheng, S., Liu, L., Yu, C., Li, J., Tan, Y.: Fireworks algorithm and its variants for solving ICSI2014 competition problems. In: Tan, Y., Shi, Y., Coello, C.A.C. (eds.) ICSI 2014. LNCS, vol. 8795, pp. 442–451. Springer, Cham (2014). doi:10.1007/978-3-319-11897-0_50

Elite-Leading Fireworks Algorithm

Xinchao Zhao[1(✉)], Rui Li[1], Xingquan Zuo[2], and Ying Tan[3]

[1] School of Science, Beijing University of Posts and Telecommunications,
Beijing 100876, China
zhaoxc@bupt.edu.cn
[2] School of Computer Science,
Beijing University of Posts and Telecommunications,
Beijing 100876, China
[3] School of Electronics Engineering and Computer Science,
Peking University, Beijing 100871, China

Abstract. Fireworks algorithm (FWA) is effective to solve optimization problems as a swarm intelligence algorithm. In this paper, the elite-leading fireworks algorithm (ELFWA) is proposed based on dynamic search in fireworks algorithm (dynFWA), which is an important improvement of FWA. In dynFWA firework is separated to two group: core-firework (CF) and non-core fireworks (non-CFs). This paper takes some beneficial information from non-CFs to reinforce the local search effect of CF. Random reinitialization and elite-leading operator are used to maintain the diversity of the non-CFs, which play an important role in global search. Based on the CEC2015 benchmark functions suite, ELFWA has a very competitive performance when comparing with state-of-the-art fireworks algorithms, such as dynFWA, dynFWACM and eddynFWA.

Keywords: Fireworks algorithm · Elite-leading operator · Random reinitialization

1 Introduction

Both in the academic field and in the industrial world, many problems can be simplified as optimization problems. In order to solve those problems many swarm intelligence (SI) algorithms were proposed in recent years. Fireworks algorithm (FWA) [1], proposed by Tan and Zhu in 2010, is one of SI algorithms based on simulating the fireworks explosion process. The performance of twelve evolutionary algorithms are tested and compared by Bureerat in 2011 [2], in which FWA ranks at the sixth, which verifies that FWA works effectively on some optimization problems.

Due to its bloom developing, FWA has many improved variants to enrich its research field. The Enhanced Fireworks Algorithm (EFWA) [3], proposed by Zheng et al. in 2013, is an important improvement of the FWA. Five main operators of the FWA have been improved or corrected in the EFWA, which are the methods of calculating explosion amplitude, generating new explosion sparks, generating Gaussian sparks, selecting the population for the next iteration and the new mapping strategy for sparks which are out of the search space. Hence, some algorithms, like dynFWA, are

© Springer International Publishing AG 2017
Y. Tan et al. (Eds.): ICSI 2017, Part I, LNCS 10385, pp. 493–500, 2017.
DOI: 10.1007/978-3-319-61824-1_54

applied to the EFWA rather than FWA. Based on the EFWA, Zheng et al. [4] proposed Dynamic Search in Fireworks Algorithm (dynFWA) in 2014. In dynFWA, fireworks are separated into two groups. The first group consists of the firework with best fitness named core firework (CF), while the second group consists of all other fireworks named non-core fireworks (non-CFs). Compared with non-CFs, the explosion amplitude of CF is smaller, hence CF is good at local search. Non-CFs have larger explosion amplitudes which are fitter for global search. Moreover, the biggest difference between two groups is that the CF has very high probability to generate the best candidate which will be selected to the next iteration as firework.

Based on the dynFWA Yu et al. [5] proposed the dynamic fireworks algorithm with covariance mutation (dynFWACM) in 2015. It introduced the mutation operator into dynFWA, which calculates the mean value and covariance matrix of the better sparks and produces sparks according to Gaussian distribution. Zheng et al. [6] proposed an exponentially decreased dimension number strategy based dynamic search fireworks algorithm (eddynFWA) in 2015. Yu et al. [7] put forward a new FWA with differential mutation (FWA-DM) by using differential operator in 2014. Li et al. [8] proposed an adaptive fireworks algorithm (AFWA) in 2014. Zhang et al. [9] proposed an improving enhanced fireworks algorithm (IEFWA) with new Gaussian explosion and population selection strategies.

Moreover, many developments for multi-objective optimization have also been proposed. Zheng et al. [10] proposed a multi-objective fireworks optimization for variable-rate fertilization in oil crop production. Tan [11] proposed an S-metric based multi-objective fireworks algorithm in 2015. FWA has also been applied to many practical fields and problems. FWA has been used for digital filters design [12], pattern recognition [13] and so on.

2 dynFWA

In dynFWA [4] firework is separated into two groups. One group is named as Core Firework (CF) and the other is non-core fireworks (non-CFs). In each iteration, CF means the firework with the currently best fitness and non-CFs mean all the rest fireworks. The main operations of dynFWA are listed as follows.

2.1 Calculate the Numbers of Explosion Sparks

In order to take full advantage of all the fireworks, different fireworks have different numbers of sparks in dynFWA, which depends its fitness as Eq. (1).

$$S_i = M_e \cdot \frac{y_{\max} - f(X_i) + \varepsilon}{\sum_{j=1}^{N} (y_{\max} - f(X_j)) + \varepsilon} \tag{1}$$

In this equation, S_i represents the number of sparks for the firework i, M_e controls the number of sparks. $y_{\max} = max(f(X_i))$, $f(X_i)$ denotes the fitness value of the firework i. N is the number of fireworks.

2.2 Calculate the Explosion Amplitude

Fireworks with better fitness should have smaller explosion amplitudes to bias the local search. On the contrary, fireworks with worse fitness should have larger explosion amplitudes to bias the global search. So dynFWA uses Eq. (2) to calculate the explosion amplitude for non-CF.

$$A_i = \hat{A} \cdot \frac{f(X_i) - y_{\min} + \varepsilon}{\sum_{j=1}^{N} (f(X_j) - y_{\min}) + \varepsilon} \tag{2}$$

In this equation, A_i represents the explosion amplitude of the firework i, $\hat{A}$ controls the explosion amplitude of sparks. $y_{\min} = \min(f(X_i)), f(X_i)$ denotes the fitness value of the firework i. N is the number of fireworks.

But for CF, the explosion amplitude is generated with Eq. (3). Parameters C_a and C_r are used to control the amplification and reduction ratio of the exploitation amplitudes.

$$A_{CF}(t) = \begin{cases} C_a \times A_{CF}(t-1) & \text{if } f(X_{CF}(t)) < f(X_{CF}(t-1)), \\ C_r \times A_{CF}(t-1) & otherwise. \end{cases} \tag{3}$$

2.3 Generate the Explosion Sparks

After getting the information of explosion sparks and explosion amplitude, each firework explodes and creates explosion sparks with Eq. (4).

$$X_i^k = \begin{cases} X_i^k + A_i \times rand(-1, 1) & \text{if firework is non-CF} \\ X_{CF}^k + A_{CF} \times rand(-1, 1) & \text{if firework is CF} \end{cases} \tag{4}$$

X_i^k is the i-th firework and A_i is the explosion amplitude, rand $(-1,1)$ represents a random number between -1 and 1.

The location of a new spark will be mapped within the search space with Eq. (5) if it exceeds the search range in dimension k.

$$X_i^k = X_{\min}^k + rand * (X_{\max}^k - X_{\min}^k) \tag{5}$$

3 Elite-Leading Fireworks Algorithm (ELFWA)

Before proposing the Elite-leading Fireworks Algorithm (ELFWA), the difference between CF and non-CFs in the dynFWA should be introduced. The biggest difference is that CF has more chance to generate a better spark and then be selected into the next iteration. The main role of non-CFs is to keep population diversity and perform global search. But the global search is possible to be more effective if other more effective operations were used and the CF's effect is possible to be more obvious if more sparks were given. Based on this motivation, a new improvement of dynFWA, Elite-leading Fireworks Algorithm (ELFWA) is proposed. When comparing to dynFWA, the CF in ELFWA have more sparks and the non-CFs will generate none sparks. As a compensation, non-CFs will run another operation to evolve constantly for global search.

3.1 CF Operations

Some operations of CF in ELFWA will be introduced. (1) In ELFWA, CF denotes the global best solution which is the optimal solution found until now. (2) Different from dynFWA, the number of sparks in ELFWA is not alterable. In this paper the number of CF sparks is a constant which is equal to the number of non-CFs. (3) The method of calculating explosion amplitude of CF is not changed and Eq. (3) is also used. (4) The way of CF generating sparks is the same as dynFWA doing in Eq. (4). (5) CF will be updated with the best solution from non-CFs or sparks of CF at the current iteration or the global best solution in the previous iteration, depending on the fitness as Eq. (6) illustrates.

$$CF(t) = \arg\min\{f(nonCF(t)), f(SparksOfCF(t)), f(CF(t-1))\} \qquad (6)$$

3.2 Non-CFs Operations

A new strategy is proposed and utilized to the non-CFs which includes two operations, i.e., random reset operation and Elite-leading operation. Random reset operation decides whether to reset the non-CFs with a probability. If a random number r_1 is less than the given probability, non-CFs will be reinitialized with Eq. (7).

$$nonCF_i^k = X_{\min}^k + rand * (X_{\max}^k - X_{\min}^k) \qquad (7)$$

$nonCF_i^k$ represents the k-th dimension of the i-th non-CF.$X_{\max}^k$ and $X_{\min}^k$ represent the upper bound and lower bound of the k-th dimension.

After non-CFs are reinitialized, all solutions will be redistributed in the search space again and usually become worse. So in order to scan more a little more beneficial areas ELFWA uses Elite-leading operation to improve the quality of non-CFs at the successive iteration with Eq. (8).

$$nonCF_i(t) = nonCF_i(t-1) + rand(0, 1) * (gBS(t-1) - lBS(t-1)) \qquad (8)$$

In Eq. (8), gBS(t-1) indicates the global best solution and lBS(t-1) indicates the best solution in the non-CFs of the previous iteration. It will be changed when the non-CFs are reinitialized. rand(0,1) is a random number between 0 and 1. It will be found that all the non-CFs have the same directions, which is decided by gBS and lBS. This process can be regarded as the best solution of the current non-CFs leading all the non-CFs to the best one of them. Additionally, gBS is equal to the CF, so gBS is very crucial for the algorithm. That is why this research is named as Elite-leading Fireworks Algorithm.

The mapping operator of the ELFWA is changed into Eq (9).

$$X_i^k = \begin{cases} \min(X_{\max}^k, 2 * X_{\max}^k - X_i^k)) & \text{if } X_i^k > X_{\max}^k \\ \max(X_{\min}^k, 2 * X_{\min}^k - X_i^k)) & \text{if } X_i^k < X_{\min}^k \end{cases} \tag{9}$$

After mapping operation, ELFWA evaluates the quality of the explosion sparks and non-CFs. So, the framework of Elite-leading Fireworks Algorithm (ELFWA) is presented as follows.

```
Initialize N/2 fireworks as non-CFs
Evaluate non-CFs and set non-CFs' best as CF
while termination criteria are not met do
     %%CF operation:
     Calculate explosion amplitude for CF (Eq.(2));
     Generate explosion sparks (Eq.(4));
     Map sparks to search space (Eq.(9));
     %%Non-CF operation:
     If r1 < 0.05
          Reinitialize the non-CFs, evaluate the quality of
          non-CFs and reset lBS;
     End if
     Update non-CFs (Eq.(7));
     Map non-CFs back to search space (Eq.(9));
     Evaluate quality of explosion sparks, non-CFs;
     if f(nonCFs(t-1))< f(nonCFs(t))
          update explosion amplitude of non-CFs;
     Update explosion amplitude of CF (Eq.(3));
End while
Output the final solution(s) and the fitness.
```

4 Experiment and Analysis

15 benchmark functions of CEC 2015 [14] competition are used to verify the effectiveness of ELFWA. Following four state-of-the-art algorithms are compare toELFWA, EFWA [3], dynFWA [4], dynFWACM [5] and eddynFWA [6].

4.1 Experimental Setup

Several parameters in ELFWA are set as follows. The dimension of benchmark function is 30. Parameters C_a and C_r in Eq. (3) are empirically set to 0.9 and 1.1. In order to make full use of the ability of global search of CF, A_{CF} is set to the size of the space in the beginning. Both the number of non-CFs and the number of sparks are 50 in ELFWA. All the algorithms are performed 30 runs on each benchmark functions; the final mean results are recorded with 300 000 function evaluations.

4.2   Experimental Results and Analysis

The online performance comparison of five FWA algorithms is shown as Fig. 1, which clearly shows that ELFWA is effective. Especially, ELFWA is very effective for f1, f2, f6, f8, f10 and f12. For functions f3 and f5, all algorithms nearly find the same results. For the rest functions, although ELFWA does not perform best, it can find a competitive solution in short time with the less benchmark functions evaluations.

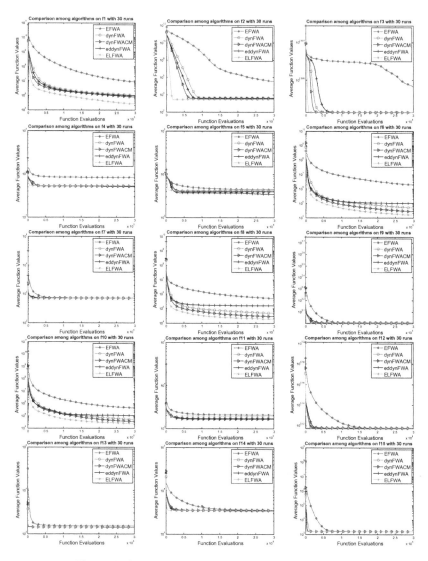

Fig. 1. Online performance comparison among algorithms

Table 1 shows that three improvements of dynFWA are better than dynFWA and all of four dynFWAs outperform EFWA. Among them, ELFWA is the slightly best one both in mean value and in the best solution, a little better than, or performs comparably with dynFWACM and eddynFWA. According to the general ranks of several FWA variants, eddynFWA, dynFWACM and ELFWA have similar ranks, however, ELFWA has the best rank. When comparing with dynFWA, ELFWA outperforms dynFWA on 10 from 15 benchmarks and performs equally on 2 functions. dynFWA slightly outperforms, however, comparably with ELFWA on other 3 functions. The value of std. also shows that the result of ELFWA is relatively stable and robust. In order to compare the difference between the existing algorithms and ELFWA, the value of student test is given. If the mean value of ELFWA is smaller than the mean value of existing algorithms and the Wilcoxon's rank-sum test under 5% significance level is true, then it is believed that the results of ELFWA are significant better than existing algorithms. In Table 1, an algorithm which is significant better than ELFWA is marked with '+', no performance significant difference is marked with '≈', significant worse than ELFWA is marked with '−'. In Table 1, ELFWA is significant better than those four algorithms at 13, 10, 8, 7 functions. The performance of ELFWA at other functions is competition with those four algorithms.

Table 1. The result of the experiment

	dynFWA			dynFWACM			eddynFWA			ELFWA	
	Mean	Std		Mean	Std		Mean	Std		Mean	Std
f1	1.03E+06	3.23E+05	−	7.71E+05	4.53E+05	−	9.52E+05	6.34E+05	−	**2.67E+05**	**1.37E+05**
f2	4.24E+03	3.93E+03	−	3.78E+03	4.02E+03	−	3.54E+03	4.06E+03	−	**2.92E+03**	**3.15E+03**
f3	3.20E+02	5.87E−06	≈	3.20E+02	1.51E−05	≈	3.20E+02	1.88E−06	≈	**3.20E+02**	**2.64E−03**
f4	5.26E+02	3.35E+01	≈	5.23E+02	3.51E+01	≈	**3.20E+02**	3.34E+01	≈	5.30E+02	**3.32E+01**
f5	4.10E+03	7.01E+02	≈	3.95E+03	6.85E+02	≈	**3.46E+03**	**6.60E+02**	≈	4.28E+03	8.64E+02
f6	4.96E+04	3.62E+04	−	2.72E+04	2.07E+04	−	1.01E+05	6.12E+04	−	**1.63E+04**	**1.02E+04**
f7	7.18E+02	1.37E+01	−	7.15E+02	**4.73E+00**	≈	**7.17E+02**	1.51E+01	≈	7.18E+02	1.27E+01
f8	4.85E+04	1.95E+04	−	2.87E+04	1.34E+04	−	**1.59E+05**	**1.03E+05**	−	1.92E+04	1.10E+04
f9	1.02E+03	6.15E+01	−	1.01E+03	3.58E+01	≈	**1.01E+03**	3.34E+01	≈	1.01E+03	**3.97E+01**
f10	4.91E+04	1.66E+04	−	3.45E+04	1.36E+04	−	1.10E+05	**6.88E+04**	−	**2.48E+04**	1.22E+04
f11	1.71E+03	2.60E+02	≈	**1.66E+03**	2.44E+02	≈	1.68E+03	**2.10E+02**	≈	1.93E+03	2.15E+02
f12	1.31E+03	1.94E+00	−	1.31E+03	1.98E+00	−	1.31E+03	1.95E+00	−	**1.31E+03**	**1.71E+00**
f13	1.43E+03	**5.79E+00**	−	1.43E+03	6.91E+00	−	**1.41E+03**	7.68E+00	≈	1.43E+03	7.44E+00
f14	3.50E+04	1.75E+03	−	3.55E+04	1.86E+03	−	**3.47E+04**	1.43E+03	≈	3.49E+04	**1.42E+03**
f15	1.60E+03	8.76E−12	−	1.60E+03	8.52E−13	≈	1.60E+03	4.53E−13	−	**1.60E+03**	9.23E−13
Rank	3.13			2.26			2.2			2.13	

5   Conclusion and Future Work

An improved variant of dynFWA is proposed based on the information borrowing and elite leading strategies in this paper. In order to explore whether strengthen the search ability of elites is effective. Based on the two groups of dynFWA, this paper uses some beneficial information of non-CFs to reinforce the effect of CF. Another strategy is used to maintain the diversity of the non-CFs and to make them play an important role in global research. The experiments show that this inspiring motivation works well. The proposed ELFWA algorithm significantly outperforms dynFWA and performs a little better than, and comparably with two recently enhanced dynFWAs, i.e., dynFWACM and eddynFWA. In the future, more useful global research operators will be considered to improve FWA algorithm and other SI algorithms.

Acknowledgments. This research is supported by National Natural Science Foundation of China (61375066, 61374204). We will express our awfully thanks to the Swarm Intelligence Research Team of BeiYou University and to the reviewers for their helpful suggestions.

References

1. Tan, Y., Zhu, Y.: Fireworks algorithm for optimization. In: Tan, Y., Shi, Y., Tan, K.C. (eds.) ICSI 2010. LNCS, vol. 6145, pp. 355–364. Springer, Heidelberg (2010). doi:10.1007/978-3-642-13495-1_44
2. Bureerat, S.: Hybrid population-based incremental learning using real codes. In: Coello, C. A.C. (ed.) LION 2011. LNCS, vol. 6683, pp. 379–391. Springer, Heidelberg (2011). doi:10.1007/978-3-642-25566-3_28
3. Zheng, S., Janecek, A., Tan, Y.: Enhanced fireworks algorithm. In: Evolutionary Computation, pp. 2069–2077. IEEE (2013)
4. Zheng, S.Q., et al.: Dynamic search in fireworks algorithm. In: Evolutionary Computation, pp. 3222–3229. IEEE (2014)
5. Yu, C., Kelley, L.C., Tan, Y.: Dynamic search fireworks algorithm with covariance mutation for solving the CEC 2015 learning based competition problems. In: Evolutionary Computation, pp. 1106–1112. IEEE (2015)
6. Zheng, S.Q., et al.: Exponentially decreased dimension number strategy based dynamic search fireworks algorithm for solving CEC2015 competition problems. In: Evolutionary Computation, pp. 1083–1090. IEEE (2015)
7. Yu, C., et al.: Fireworks algorithm with differential mutation for solving the CEC 2014 competition problems. In: Evolutionary Computation, pp. 3238–3245. IEEE (2014)
8. Li, J.Z., Zheng, S., Tan, Y.: Adaptive fireworks algorithm. In: Evolutionary Computation, pp. 3214–3221. IEEE (2014)
9. Zhang, B., Zhang, M., Zheng, Y.-J.: Improving enhanced fireworks algorithm with new gaussian explosion and population selection strategies. In: Tan, Y., Shi, Y., Coello, C.A.C. (eds.) ICSI 2014. LNCS, vol. 8794, pp. 53–63. Springer, Cham (2014). doi:10.1007/978-3-319-11857-4_7
10. Zheng, Y., Song, Q., Chen, S.Y.: Multi-objective fireworks optimization for variable-rate fertilization in oil crop production. Appl. Soft Comput. **13**(11), 4253–4263 (2013)
11. Tan, Y.: S-metric based multi-objective fireworks algorithm. In: Evolutionary Computation, pp. 1257–1264. IEEE (2015)
12. Gao, H., Diao, M.: Cultural firework algorithm and its application for digital filters design. Int. J. Model. Ident. Control **14**(4), 324–331 (2011)
13. Zheng, S.Q., Tan, Y.: A unified distance measure scheme for orientation coding in identification. In: IEEE Third International Conference on Information Science and Technology, pp. 979–985. IEEE (2013)
14. Liang, J., Qu, B., Suganthan, P., Chen, Q.: Problem definitions and evaluation criteria for the CEC 2015 competition on real-parameter single objective optimization (2014)

Guided Fireworks Algorithm Applied to the Maximal Covering Location Problem

Eva Tuba[1], Edin Dolicanin[2], and Milan Tuba[3(✉)]

[1] Faculty of Mathematics, University of Belgrade, Belgrade, Serbia
[2] Department of Technical Sciences, State University of Novi Pazar,
Novi Pazar, Serbia
[3] Graduated School of Computer Science, John Naisbitt University,
Belgrade, Serbia
tuba@ieee.org

Abstract. Maximal covering location problem is an interesting and applicable hard optimization problem. It is a facility location problem where facilities have to be placed in optimal positions to maximize coverage of the weighted demand nodes, while respecting some constraints. This hard optimization problem has been successfully tackled by swarm intelligence algorithms. In this paper we propose adjustment of the recent guided fireworks algorithm for maximal covering location problem. We tested our approach on standard benchmark data sets and compared results with other approaches from literature where our proposed algorithm proved to be more successful, both in the quality of the coverage and convergence speed.

Keywords: Maximal covering location problem · Optimization · Meta-heuristics · Fireworks algorithm · Swarm intelligence

1 Introduction

Location problems are optimization problems that are part of many different applications. They represent an interesting topic in computer science but are also applicable in real life. Location problems refer to searching for facility locations that optimize some service to demand nodes respecting given constraints. The goal is to minimize (or maximize) some characteristics, e.g. traveling cost or distance. Facility location problem was applied for optimizing carbon market trading [1], fuel station deployment [2], etc.

Covering location problems are specific location problems where each facility has some cover range and the goal is to minimize or maximize covered nodes demands. Demand nodes can be weighted, e.g. if demand nodes are buildings, weight for each building can be the number of residents, the number of apartments, etc. In this paper the maximal covering location problem (MCLP) will be

M. Tuba—This research is supported by Ministry of Education, Science and Technogical Development of Republic of Serbia, Grant No. III-44006.

Y. Tan et al. (Eds.): ICSI 2017, Part I, LNCS 10385, pp. 501–508, 2017.
DOI: 10.1007/978-3-319-61824-1_55

considered. MCLP is NP-hard problem [3] and such kind of problems for larger dimension cannot be solved in a reasonable time by deterministic algorithms. In recent years nature inspired algorithms were successfully used for finding relatively good solutions for NP-hard problems. Nowadays, one group of nature inspired algorithms, swarm intelligence algorithms, have been proven to be very successful. In this paper a novel swarm intelligence algorithm, guided fireworks algorithm, was adjusted and used to solve maximal covering location problem. The results favorably compared to other approaches from literature, both in term of coverage and convergence speed.

The rest of the paper is structured as follows. In Sect. 2 mathematical model of maximal coverage location problem is defined. In Sect. 3 the main features of guided fireworks algorithm are presented. Our proposed algorithm for solving the MCLP is explained in Sect. 4. Experimental results are shown in Sect. 5 and in Sect. 6 conclusion and propositions for further improvements are given.

2   Mathematical Formulation of the MCLP

Maximal coverage location problem refers to the task of finding optimal locations for P facilities with some covering range on a network of N demand nodes so that covered or serviced number of population-weighted demand nodes is maximal. Mathematical formulation of the problem is given as follows:

$$\text{maximize } z = \sum_{i \in I} a_i * y_i. \tag{1}$$

so that the following conditions are met:

$$\sum_{j \in N_i} x_j \geq y_i \quad \text{for all } i \in I. \tag{2}$$

$$\sum_{j \in J} x_j = P. \tag{3}$$

where a_i is the population needed to be served at demand node i, I is the set of demand nodes, J is set of facilities, P is the number of facilities that need to be located, while N_i is the set of facilities able to cover demand of node i. Variables x_j and y_i are defined by the following equations:

$$x_j = \begin{cases} 1 & \text{if a facility is allocated to site } j \\ 0 & \text{otherwise} \end{cases}. \tag{4}$$

$$y_i = \begin{cases} 1 & \text{if demand node } i \text{ is covered by a facility within distance } S \\ 0 & \text{otherwise} \end{cases}. \tag{5}$$

where distance S represents the covering range of the facility. This definition can be changed so that the MCLP is written as:

$$\text{minimize } \sum_{i=1}^{n} \overline{y_i} * a_i. \tag{6}$$

where $\overline{y_i} = 1 - y_1$. In order to use this formulation, condition from Eq. 2 is changed to

$$\sum_{j \in N_i} x_j + \overline{y_i} \geq 1 \text{ for all } i \in I. \tag{7}$$

The MCLP represents NP-hard problem. In the past years different methods for solving this problem were proposed. In [4] modified particle swarm optimization (PSO) was proposed for localization of special facilities in healthcare which is modeled by capacitated maximal covering location problem. In [5] genetic algorithm was adjusted for this problem. A hybrid PSO combined with genetic algorithm was proposed in [6]. In [7] a large-scale MCLP was studied. In order to solve the problem, a combination of greedy variable neighborhood search and fuzzy simulation was proposed.

In this paper we propose a method for solving MCLP based on adjusted swarm intelligence guided fireworks algorithm which was successfully used for various applications.

3   Guided Fireworks Algorithm

Guided fireworks algorithm (GFWA) is the latest update of the widely used fireworks algorithm. Guided FWA was proposed by Li et al. in 2016 [8]. Original fireworks algorithm (FWA) was proposed in 2010 [9] and the idea was to simulate fireworks explosions of well and badly manufactured fireworks which was used to implement exploitation and exploration.

Since its introduction, fireworks algorithm has been used as part of many different applications for solving hard optimization problems. It was used to solve constrained portfolio optimization problem in [10], while in [11] it was used for parameter tuning of local-concentration model for spam detection. In [12] FWA was used for solving RFID network planing problem, in [13] for WSN localization problem and in [14] FWA was used for finding optimal threshold values for multilevel image thresholding. Support vector machine was optimized by fireworks algorithm in [15]. In [16] guided fireworks was applied to retinal image registration problem.

Wide use of the fireworks algorithm resulted in numerous improvements and modifications. Authors he original fireworks algorithms have been collecting ideas and results and proposed several upgraded versions. In the first modification, enhanced fireworks algorithm in 2013, five modification to the initial fireworks algorithm were introduced [17]. Cooperative FWA (CoFWA) was proposed in [18] as the second version of the FWA. CoFWA enhanced the exploitation ability by introducing independent selection operator, while the exploration capacity was increased by crowdness-avoiding cooperative strategy. In [19] two methods for improving exploration were proposed and that version is known as the FWA with dynamic resource allocation (FWA-DRA). The latest version proposed by Li et al. is named guided FWA [8]. Objective function information was used to create a guiding vector with promising direction and adaptive length hence the quality of the solution as well as the convergence speed was improved.

Created vector was added to the position of the firework and an elite solution was generated (guiding spark).

Guided fireworks algorithm uses μ fireworks and for each of them the number of sparks is calculated. Fireworks and sparks represent points in d-dimensional space, where d is the dimension of the problem. For each firework explosion amplitude is calculated based on the previous solution and the best solution in generation is used to generate sparks positions.

In the GFWA the firework with the best fitness in each generation is named core firework (CF) and explosion amplitude calculation was adjusted. Also, in each generation a guiding spark (GS) is generated for each firework. The guiding spark is generated by adding a guiding vector (GV) to the firework's position. The position of the guiding spark G_i for the firework X_i is determined by the following algorithm [8]:

Algorithm 1. Generating the Guiding Spark for X_i

Require: X_i, s_{ij}, λ_i and σ

    Sort the sparks by their fitness values $f(s_{ij})$ in ascending order.

    $\Delta_i \leftarrow \frac{1}{\sigma\lambda_i}(\sum_{j=1}^{\sigma\lambda_i})s_{ij} - \sum_{j=\lambda_i-\sigma\lambda_i+1}^{\lambda_i} s_{ij}$

    $G_i \leftarrow X_i + \Delta_i$

    return G_i

The guiding vector Δ_i is the mean of $\sigma\lambda_i$ vectors which is defined by the following equation:

$$\Delta_i = \frac{1}{\sigma\lambda_i} \sum_{j=1}^{\sigma\lambda_i}(s_{ij} - s_{i,\lambda_i-j+1}). \tag{8}$$

At the end of each generation the best individual is kept as firework, and from the rest of individuals random $\mu - 1$ were chosen to be saved for the next generation.

4 The Proposed Algorithm

In this paper we propose adjusted guided fireworks algorithm for solving maximal covering location problem. In order to use the GFWA for the MCLP few adjustments were necessary.

Dimension of input vector is equal to the number of facilities that need to be set multiplied by the dimension of network. Optimal locations for P facilities need to be determined and location is defined by its coordinates. Search range was set to $[x_{i\_min}, x_{i\_max}]$ where i in the index represents dimension while min and max are minimal and maximal coordinates of dimension i of demand nodes.

Guided fireworks algorithm searches for the solution in real space. In the case of maximal covering location problem, solutions, i.e. facility locations, cannot be just any combination of numbers from the search space. In most problems facilities can be placed to one of the locations of demand nodes. In some other cases facilities cannot be placed in specific locations, e.g. mobile antennas cannot be in the center of a crossroad. In this paper we tested with standard datasets where facilities have to be in locations of the demand nodes. In order to secure that only possible solutions are used we implemented the following mechanism. After a solution is generated distances from all possible locations are calculated and the solution is placed to the nearest one.

Another adjustment was made due the fact that coordinates can be large numbers and the distance between two possible locations can also be large while the GFWA changes locations for some small vectors. In this case, a new solution that GFWA generated would be again returned to the same possible position. In order to overcome this problem, coordinates were mapped into range $[-1, 1]$ for use in the GFWA.

Since the objective function given by Eq. 6 needs to be maximized and the GFWA tries to minimize the objective function, negative value of Eq. 6 was be used.

Our proposed algorithm is summarized in Algorithm 2.

Algorithm 2. Guided fireworks algorithm for the MCLP

Randomly initialize μ fireworks in the potential space.
Evaluate the fireworks' fitness.
repeat
    For each firework calculate number of sparks (λ_i).
    For each firework calculate explosion amplitude A_i.
    For each firework, generate λ_i sparks within the amplitude A_i
    Each generated spark is moved to nearest valid location.
    For each firework, generate guiding sparks according to Algorithm 1.
    Each guided spark is moved to nearest valid location
    Evaluate all the sparks' fitness according to Eq. 6.
    Keep the best individual as a firework.
    Randomly choose other $\mu - 1$ fireworks among the rest of individuals.
until termination criteria is met.
return the position and the fitness of the best individual.

---

5 Experimental Results

The proposed algorithm was implemented in Matlab version R2016b. All experiments were performed on Intel ® Core™ i7-3770K CPU at 4 GHz, 8 GB RAM computer with Windows 10 Professional OS.

Our proposed method was tested on standard dataset given in [20]. Dataset is publicly available on http://www.lac.inpe.br/~lorena/instancias.html. Data in this dataset are real data obtained from Sao Jose dos Campos city in Brazil for purpose of finding optimal locations for internet antennas with coverage radius of 800 m. Dataset contains five instances where the number of demand nodes is changed. Each instance has a file with coordinates of N demand nodes (buildings) that are also possible locations for facilities (internet antennas) and file with weights for the nodes (population of the building). The goal is to find optimal locations for P antennas to achieve maximal weighted covering.

In order to prove the quality of our proposed method we compared our results with results in [21]. In [21] particle swarm optimization based algorithm was proposed for solving the MCLP and the results were compared with [20]. The proposed method was tested on the same dataset. In Table 1 results reported in [21] along with the results from [20] and results obtained by our proposed algorithm are presented. The first column N is the number of locations (buildings), while the second column P is the number of antennas. As in [20,21], we reported the best result in 20 runs.

Table 1. Comparison of our proposed algorithm and the methods proposed in [20,21]

N	P	GA [20]			PSO [21]			GFWA		
		Eval	Cover(%)	Time(s)	Eval	Cover(%)	Time(s)	Eval	Cover(%)	Time(s)
324	5	12,152	100	5,485	115	100	0.9	95	100	0.2
324	5	-	-	-	260	100	2.0	-	-	-
324	5	-	-	-	280	100	2.1	-	-	-
402	6	15,984	100	3,502	2,720	100	31.8	2,079	100	8.3
402	6	-	-	-	5,000	99.75	15.8	-	-	-
402	6	-	-	-	8,580	100	99.0	-	-	-
500	8	19,707	100	7,695	4,050	100	80.9	3,963	100	14.0
500	8	-	-	-	5,000	99.8	8.0	-	-	-
500	8	-	-	-	12,780	100	238.6	-	-	-

In [21] results achieved by using different number of particles were presented. Because of the different numbers of sparks generated for each firework the numbers of iterations are not comparable. Instead, we compared required objective function evaluations. Evaluation numbers for results from [21] were calculated as number of particles multiplied by number of iterations. Coverage percentage was in almost all cases 100%, thus computational time was the characteristics that could be improved. As shown in Table 1, by using GFWA the number of objective function evaluations was reduced. For dataset with 324 possible locations execution time was reduced from 0.9 s that was obtained by the proposed PSO from [21] to 0.2 s. Evaluations number is also slightly better. Other reported computational times in [21] for this instance are worse. For the instance with 402 possible locations results were improved more than in the previous case. The best result by PSO was with covering of 100% that was achieved in 2,720 evaluation and computational time was 31.8 s. Guided FWA successfully covered 100% in 2,079 evaluations while computational time was reduced almost 4 times since our proposed algorithm found the solution in 8.3 s. Finally, for the third tested instance with 500 locations, our proposed algorithm arranged facilities so the coverage was 100% in 3,963 evaluations. For this task our computational time was 14.0 s. The best results reported in [21] was for full coverage achieved in 4,050 evaluations but with the computational time of 80.9 s. Again, our computational time was significantly reduced, almost six times. Based on these results and the fact that results in [21] outperformed the results from [20] it can be concluded that our proposed algorithm outperformed both methods proposed in [20, 21].

6   Conclusion

In this paper a novel swarm intelligence algorithm, guided fireworks algorithm, was adjusted for solving maximal covering location problem. MCLP represents an important problem that can be used for various applications, thus an efficient method for solving it is needed. It has been proven that the MCLP represents NP-hard problem, so deterministic algorithms cannot solve it. In this paper we have shown that GFWA adjusted for this problem achieved excellent results. Our proposed method was compared with other approaches from literature and based on the simulation results it can be concluded that our proposed algorithm has faster convergence and significantly reduced computational time. Also, our proposed algorithm easily found maximal coverage for the largest benchmark problem which means that this algorithm is promising for larger problems. In further work, proposed algorithm can be adjusted for specific problems that will include some additional constraints.

References

1. Diabat, A., Abdallah, T., Al-Refaie, A., Svetinovic, D., Govindan, K.: Strategic closed-loop facility location problem with carbon market trading. IEEE Trans. Eng. Manag. **60**, 398–408 (2013)
2. MirHassani, S.A., Ebrazi, R.: A flexible reformulation of the refueling station location problem. Trans. Sci. **47**, 617–628 (2013)
3. Megiddo, N., Zemel, E., Hakimi, S.L.: The maximum coverage location problem. SIAM J. Algebraic Discret. Methods **4**, 253–261 (1983)
4. ElKady, S.K., Abdelsalam, H.M.: A modified particle swarm optimization algorithm for solving capacitated maximal covering location problem in healthcare systems. In: Hassanien, A.-E., Grosan, C., Fahmy Tolba, M. (eds.) Applications of Intelligent Optimization in Biology and Medicine. ISRL, vol. 96, pp. 117–133. Springer, Cham (2016). doi:10.1007/978-3-319-21212-8_5
5. Pasandideh, S.H.R., Niaki, S.T.A.: Genetic application in a facility location problem with random demand within queuing framework. J. Intell. Manuf. **23**, 651–659 (2012)
6. Hongxiang, W., Wenxian, G., Jianxin, X., Hongmei, G.: A hybrid PSO for optimizing locations of booster chlorination stations in water distribution systems. In: International Conference on Intelligent Computation Technology and Automation (ICICTA), vol. 1, pp. 126–129. IEEE (2010)

7. Davari, S., Zarandi, M.H.F., Turksen, I.B.: A greedy variable neighborhood search heuristic for the maximal covering location problem with fuzzy coverage radii. Knowl.-Based Syst. **41**, 68–76 (2013)
8. Li, J., Zheng, S., Tan, Y.: The effect of information utilization: introducing a novel guiding spark in the fireworks algorithm. IEEE Trans. Evol. Comput. **21**, 153–166 (2017)
9. Tan, Y., Zhu, Y.: Fireworks algorithm for optimization. In: Tan, Y., Shi, Y., Tan, K.C. (eds.) ICSI 2010. LNCS, vol. 6145, pp. 355–364. Springer, Heidelberg (2010). doi:10.1007/978-3-642-13495-1_44
10. Bacanin, N., Tuba, M.: Fireworks algorithm applied to constrained portfolio optimization problem. In: IEEE Congress on Evolutionary Computation (CEC), pp. 1242–1249 (2015)
11. He, W., Mi, G., Tan, Y.: Parameter optimization of local-concentration model for spam detection by using fireworks algorithm. In: Tan, Y., Shi, Y., Mo, H. (eds.) ICSI 2013. LNCS, vol. 7928, pp. 439–450. Springer, Heidelberg (2013). doi:10.1007/978-3-642-38703-6_52
12. Tuba, M., Bacanin, N., Beko, M.: Fireworks algorithm for RFID network planning problem. In: 25th International Conference Radioelektronika, pp. 440–444 (2015)
13. Tuba, E., Tuba, M., Beko, M.: Node localization in ad hoc wireless sensor networks using fireworks algorithm. In: 5th International Conference on Multimedia Computing and Systems (ICMCS), pp. 223–229. IEEE (2016)
14. Tuba, M., Bacanin, N., Alihodzic, A.: Multilevel image thresholding by fireworks algorithm. In: 25th International Conference Radioelektronika, pp. 326–330 (2015)
15. Tuba, E., Tuba, M., Beko, M.: Support vector machine parameters optimization by enhanced fireworks algorithm. In: Tan, Y., Shi, Y., Niu, B. (eds.) ICSI 2016. LNCS, vol. 9712, pp. 526–534. Springer, Cham (2016). doi:10.1007/978-3-319-41000-5_52
16. Tuba, E., Tuba, M., Dolicanin, E.: Adjusted fireworks algorithm applied to retinal image registration. Stud. Inform. Control **26**, 33–42 (2017)
17. Zheng, S., Janecek, A., Tan, Y.: Enhanced fireworks algorithm. In: IEEE Congress on Evolutionary Computation (CEC), pp. 2069–2077 (2013)
18. Zheng, S., Li, J., Janecek, A., Tan, Y.: A cooperative framework for fireworks algorithm. IEEE/ACM Trans. Comput. Biol. Bioinf. **14**, 27–41 (2017)
19. Li, J., Tan, Y.: Enhancing interaction in the fireworks algorithm by dynamic resource allocation and fitness-based crowdedness-avoiding strategy. In: IEEE Congress on Evolutionary Computation (CEC), pp. 4015–4021 (2016)
20. Arakaki, R.G.I., Lorena, L.A.N.: A constructive genetic algorithm for the maximal covering location problem. In: Proceedings of Metaheuristics International Conference, pp. 13–17 (2001)
21. Drakulic, D., Maric, M., Takaci, A.: Solving maximal covering location problem (MCLP) by using the particle swarm optimization (PSO) method. Math. Inform. Phys. **51**, 19–22 (2012)

Brain Storm Optimization Algorithm

An Improved Brain Storm Optimization with Learning Strategy

Hong Wang[1,2], Jia Liu[1], Wenjie Yi[1], Ben Niu[1,3(✉)],
and Jaejong Baek[3(✉)]

[1] College of Management, Shenzhen University, Shenzhen, China
Drniuben@gmail.com
[2] Department of Mechanical Engineering, Hong Kong Polytechnic University,
Hung Hom, Hong Kong
[3] School of Computing, Informatics and Decision Systems Engineering,
Arizona State University, Tempe, AZ 85281, USA
jbaek7@asu.edu

Abstract. Brain Storm Optimization (BSO) algorithm is a brand-new and promising swarm intelligence algorithm by mimicking human being's behavior of brainstorming. This paper presents an improved BSO, i.e., BSO with learning strategy (BSOLS). It utilizes a novel learning strategy whereby the first half individuals with better fitness values maintain their superiority by keeping away from the worst ones while other individuals with worse fitness values improve their performances by learning from the excellent ones. The improved algorithm is tested on 10 classical benchmark functions. Comparative experimental results illustrate that the proposed algorithm performs significantly better than the original BSO and standard particle swarm optimization algorithm.

Keywords: Brain Storm Optimization · Improved BSO · Learning strategy · Benchmark functions

1 Introduction

Brain Storm Optimization (BSO) is a swarm intelligence algorithm that simulates the problem-solving process of human brainstorming. The basic framework of BSO was proposed by Shi [1, 2], who designed the clustering and creating operators by mimicking brainstorming process based on Osborn's four rules in 2011.

As a young and promising algorithm, BSO can be further improved by developing various searching strategies. In [3–6], a variety of clustering methods were utilized into BSO instead of k-means clustering to reduce the computational burden of the algorithm. Zhou et al. [7] employed an adaptive step size and generated new individuals in a batch-mode. Krishnanand et al. [8] presented a hybrid algorithm combining BSO and Teaching-Learning-Based Optimization algorithm. Sun et al. [9] designed a closed-loop strategy based BSO (CLBSO) by taking advantage of feedback information and developed three versions of CLBSO. Cao et al. [10] incorporated differential evolution strategy into the creating operator of individuals and introduced a new step size control

© Springer International Publishing AG 2017
Y. Tan et al. (Eds.): ICSI 2017, Part I, LNCS 10385, pp. 511–518, 2017.
DOI: 10.1007/978-3-319-61824-1_56

method. Cheng et al. [11] removed the clustering strategy and divided the solutions into elitist and normal classes. Yadav et al. [12] modified BSO with the inclusion of a mathematical theory called fractional calculus.

In addition, the basic BSO and its variants have been applied successfully to several kinds of real-world problems. Most applications of BSOs focused on electric power systems [8], design problems in aeronautics field [9], wireless sensor networks [4] and optimization problems in finance [13]. However, in evolutionary computation research, there have always been attempts to further improve any given findings. In this paper, we present an improved variant of the BSO algorithm named BSO with learning strategy (BSOLS). In original BSO, new individuals were generated by only one or two individuals. It may trap into local optima easily. The BSOLS implement a novel learning strategy imitating human behavior of seeking benefits and avoiding weakness. The proposed algorithm is tested on 10 benchmark functions, and the results show that the BSOLS algorithm significantly improves the performance of BSO and the diversity of population.

The remaining paper is organized as follows. Section 2 will give a brief introduction of the original BSO algorithm. The proposed learning strategy and the improved algorithm are described in detail in Sect. 3. Simulations on benchmark functions and experimental results are given in Sect. 4. Finally, the conclusions and future works are made in Sect. 5.

2 Original Brain Storm Optimization Algorithm

Derived from the human brainstorming process, Shi [1] first proposed BSO algorithm and gained success. The detailed procedure of BSO can be described as follow:

Step 1: Generate n potential solutions randomly and calculate their fitness values. Step 2: Cluster n solutions into m classifications using k-means clustering method. Then select the best solution as a cluster center in each classification. Step 3: Utilize a newly generated idea in place of a randomly selected cluster center with a small probability P_1 to explore more potential solutions. Step 4: If rand $(0, 1)$ is smaller than P_2, randomly choose one cluster. Otherwise, two randomly chosen clusters are utilized to obtain X_{old}. Based on one cluster, a cluster center or a random solution is selected according to P_3. On the basis of choosing two clusters, combine two cluster centers or two random solutions from selected clusters according to P_4. The combination is defined as

$$X_{old} = rand() \cdot X_1 + (1 - rand()) \cdot X_2 \tag{1}$$

Step 5: Update X_{old} into X_{new} according to

$$X_{new} = X_{old} + \xi \cdot N(\mu, \sigma). \tag{2}$$

Where $N(\mu, \sigma)$ is the Gaussian random value with mean μ and variance σ. ξ is an alterable factor which can be expressed as:

$$\xi = \log sig(\frac{0.5T - t}{k}) \cdot rand(0, 1). \tag{3}$$

Where $logsig$ () is a logarithmic sigmoid transfer function, T and t are respectively the maximum and current iteration number, and k is for changing the slope of $logsig$ () function. Step 6: Compare the newly generated idea with the previous one and keep the better one as the next iteration of the new information.

The assigned values to the set of parameters of BSO are presented in Table 1.

Table 1. Parameter settings for original BSO

m	P_1	P_2	P_3	P_4	k	μ	σ
5	0.2	0.8	0.4	0.5	20	0	1

3 Brain Storm Optimization with Learning Strategy

In this paper, we propose a novel learning strategy inspired by the learning capacity of humans. In general, normal individuals have a willingness to make progress. However, different individuals have different learning capacity and learning strategy. Some individuals are so outstanding in special fields that what they need to pay attention to is avoiding mistakes. Others perform poorly in one area so that it is required for them to make great efforts for improvement by learning from those individuals who are excellent. Based on the above strategies, we introduce an improved BSO with learning strategy called BSOLS.

The original BSO selects only one idea or two combined idea as X_{old} to generate X_{new}, which may obtain local optima easily. In this paper, we add a learning strategy after updating operator to enhance the population diversity and to jump out of local optima.

All the updated individuals X_{new} are ranked from the best to the worst according to their fitness values to differentiate which ones are better. The top P_e % of individuals will be categorized as "elitists" while the last P_l % will be categorized as "laggards". The first half individuals have already been better so their ideas only need to keep far away from laggards' ideas. While, the second half individuals have great space to improve. So they should learn towards elitists. The learning rule is as follow:

$$X'_{new} = \begin{cases} X_{ranked} - (X_{last} - X_{ranked}) \cdot q_1 \cdot rand() \\ X_{ranked} + (X_{top} - X_{ranked}) \cdot q_2 \cdot rand() \end{cases} \text{if } i < n/2 + 1 \tag{4}$$

Where X'_{new} s is a new idea after learning operator, X_{last} is randomly selected from last P_l percentage ideas, X_{last} is randomly selected from top P_e percentage ideas, q_1 and q_2 are similar to c_2 in PSO which expresses the influence of other individuals.

According to the parameters investigation in [14], the current replacing operator makes less or even no contributions to the BSO. To simplify the algorithm, we remove the replacing operator from BSO. The procedure of the proposed BSO with learning strategy can be described as follows:

(1) The initialization, evaluating, cluster, selecting and updating operator are the same as the basic BSO. While the replacing operator in Step 3 of original BSO is omitted.
(2) Sorting. Sort fitness values for each idea X_{new} in ascending or descending order (It depends on the expected fitness value, maximum or minimum) to obtain X_{ranked}.
(3) Learning. If the index of X_{ranked} is no more than half of total individuals, execute the learning strategy of avoiding weakness. Otherwise, implement the learning strategy of seeking benefits.

The assigned values to the set of parameters of BSOLS are presented in Table 2. In BSOLS, the percentage of elitists and laggards are both set to be 0.1. q_1 and q_2 are equal to 0.13 and 0.15, respectively.

Table 2. Parameter settings for BSOLS

m	P_2	P_3	P_4	k	μ	σ	P_e	P_l	q_1	q_2
5	0.8	0.4	0.5	20	0	1	0.1	0.1	0.13	0.15

4 Benchmark Tests and Experimental Results

4.1 Test Problems

To validate the BSOLS, we use 10 benchmark functions, which have often been used to test population-based algorithms in the literature. All the 10 benchmark functions and their dynamic ranges are from [15], among which the first five functions are unimodal functions and the remaining five functions are multimodal functions. All functions are minimization problems with minimum being zero. Each benchmark function will be tested with three different dimension setting, i.e., 10, 20 and 30, respectively. The proposed BSOLS will be compared with the original BSO and the standard particle swarm optimization (SPSO). To obtain reasonable statistical results, the tested BSO algorithms for each benchmark function will be run 30 times. All the experiments are run under the MATLAB R2014a environment on the same machine with an Intel 2.2 GHz CPU, 4 GB memory. The operating system is Windows 10.

4.2 Parameter Settings

The common parameters for both BSO algorithms are set the same for the purpose of comparison, that is, the population size n is set to be 100, the maximum number of iterations is 2000. The parameters for BSO and BSOLS are given in Tables 1 and 2, respectively. In SPSO, we set c1 = c2 = 2, and inertia weight decrease from 0.9 to 0.5.

4.3 Results and Analysis

The mean fitness values, the minimum and the maximum obtained by the three algorithms are listed in Table 3. The best of all the numerical values obtained on each function are emphasized by using a bold type. Additionally, the average results obtained by the three algorithms with 30 dimensions are visually shown for the example in Fig. 1. The red dotted line, the blue line and the green line are the means of each iteration for running 30 times of BSOLS, BSO and SPSO, respectively.

Table 3. Experiment results on benchmark functions

Function	D	BSOLS			BSO			SPSO		
		Mean	Best	Worst	Mean	Best	Worst	Mean	Best	Worst
f_1	10	2.88E−54	**7.47E−77**	8.64E−53	3.55E−44	4.88E−45	6.61E−44	**2.42E−59**	4.01E−66	**6.39E−58**
Sphere	20	**3.57E−53**	**2.54E−75**	**7.47E−52**	3.11E−43	1.87E−43	4.06E−43	4.75E−26	1.95E−29	4.99E−25
	30	**4.18E−53**	**3.89E−76**	**1.25E−51**	9.58E−43	6.29E−43	1.48E−42	3.10E−14	1.92E−16	3.65E−13
f_2	10	**1.81E−31**	**1.06E−38**	**5.29E−30**	1.19E−22	7.91E−23	1.59E−22	3.06E−18	6.80E−21	4.48E−17
Schwefel's	20	**1.43E−31**	**6.77E−40**	**4.15E−30**	2.75E−04	2.66E−12	3.29E−03	5.77E−03	7.55E−04	1.53E−02
P221	30	**2.38E−33**	**4.39E−40**	**4.00E−32**	6.93E−02	9.77E−03	1.94E−01	2.58E+00	8.51E−01	6.25E+00
f_3	10	**0**	**0**	**0**	**0**	**0**	**0**	**0**	**0**	**0**
Step	20	**0**	**0**	**0**	**0**	**0**	**0**	**0**	**0**	**0**
	30	**0**	**0**	**0**	3.33E−02	**0**	2.00E+00	**0**	**0**	**0**
f_4	10	4.60E−26	**1.75E−37**	5.44E−25	4.68E−22	3.48E−22	7.29E−22	**1.03E−34**	7.81E−37	**7.81E−34**
Schwefel's	20	**3.88E−29**	**3.44E−38**	**8.97E−28**	4.02E−10	1.30E−21	1.19E−08	3.26E−17	1.13E−18	1.84E−16
P222	30	**8.63E−26**	**1.06E−37**	**1.00E−24**	1.63E−02	4.34E−17	2.79E−02	2.50E−10	2.63E−11	7.94E−10
f_5	10	**4.42E−05**	**4.42E−06**	**1.63E−04**	4.25E−04	5.92E−05	1.73E−03	1.57E−03	1.99E−04	3.78E−03
Quartic	20	**3.76E−05**	**6.26E−07**	**1.26E−04**	2.21E−03	3.38E−04	5.08E−03	7.59E−03	2.39E−03	1.32E−02
Noise	30	**4.58E−05**	**2.82E−06**	**2.50E−04**	1.13E−02	3.44E−03	2.33E−02	2.00E−02	7.08E−03	4.78E−02
f_6	10	**8.88E−16**	**8.88E−16**	**8.88E−16**	4.44E−15	4.44E−15	4.44E−15	5.39E−15	4.44E−15	7.99E−15
Ackely	20	**8.88E−16**	**8.88E−16**	**8.88E−16**	7.16E−15	4.44E−15	1.15E−14	3.89E−14	1.15E−14	1.82E−13
	30	**8.88E−16**	**8.88E−16**	**8.88E−16**	1.52E−14	4.44E−15	2.93E−14	5.36E−08	2.86E−09	1.56E−07
f_7	10	**0**	**0**	**0**	4.44E+00	1.99E+00	7.96E+00	1.03E+00	**0**	3.98E+00
Rastrigin	20	**0**	**0**	**0**	1.96E+01	1.19E+01	3.18E+01	1.07E+01	4.03E+00	1.89E+01
	30	**0**	**0**	**0**	3.59E+01	2.19E+01	5.67E+01	3.05E+01	1.49E+01	4.97E+01
f_8	10	**0**	**0**	**0**	6.12E+00	4.16E+00	1.23E+01	2.71E+00	8.34E−03	1.16E+01
Rosenbrock	20	**0**	**0**	**0**	2.26E+01	1.59E+01	9.18E+01	3.34E+01	1.35E+00	1.11E+02
	30	**0**	**0**	**0**	6.15E+01	2.60E+01	1.24E+02	4.38E+01	4.94E+00	1.20E+02
f_9	10	1.27E−04	1.27E−04	1.27E−04	1.35E+03	4.74E+02	2.47E+03	2.21E+02	**1.27E−04**	4.74E+02
Schwefel's	20	7.90E+01	2.55E−04	2.37E+03	3.32E+03	2.11E+03	4.72E+03	6.75E+02	3.55E+02	**1.30E+03**
P226	30	**3.82E−04**	**3.82E−04**	**3.82E−04**	5.17E+03	3.57E+03	8.27E+03	1.10E+03	4.74E+02	2.01E+03
f_{10}	10	**0**	**0**	**0**	2.04E+00	7.06E−01	4.29E+00	5.90E−02	2.46E−02	9.83E−02
Griewank	20	**0**	**0**	**0**	1.02E−01	**0**	1.44E+00	3.92E−02	**0**	1.20E−01
	30	**0**	**0**	**0**	1.18E−02	1.97E−13	6.63E−02	1.28E−02	7.77E−16	8.60E−02

According to Table 3, we can draw the following conclusions:

BSOLS performs significantly better than the original BSO on all the tested benchmarks, which means the learning strategy is effective in terms of the search accuracy no matter the dimensions and categories of the functions.

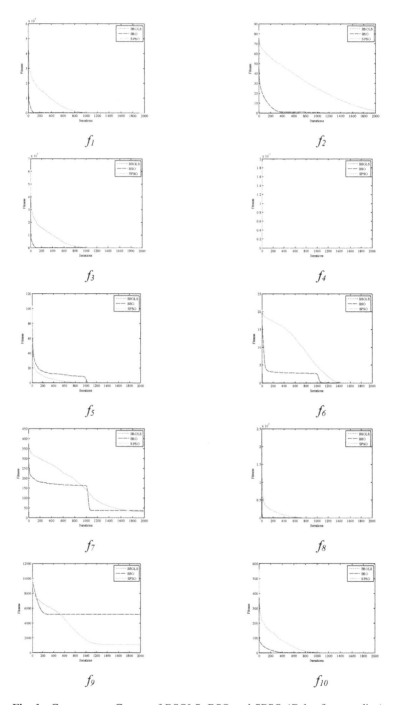

Fig. 1. Convergence Curves of BSOLS, BSO and SPSO (Color figure online)

For f_3, f_7, f_8 and f_{10}, BSOLS can find the optimum zero all the time. It means that the search stability and superiority of BSOLS in specific optimization problem is best as compared to BSO and SPSO.

Although the mean fitness values of BSOLS on f1 and f4 with 10 dimensions are slightly worse than SPSO, BSOLS still performs quite well in terms of the minimum.

We observed visually from Fig. 1 that BSOLS achieves the highest quality of the solution and the most rapid convergence rate. Altogether, whether in unimodal functions or in multimodal functions, it can be observed that BSOLS outperforms the other two algorithms.

5   Conclusions

In this paper, we propose a novel learning strategy and combine it with BSO algorithm. We apply BSOLS, the original BSO and SPSO to a set of benchmark functions to demonstrate the effect of our proposed algorithm. The results on benchmark functions show that the learning strategy significantly improves the performance of the original BSO.

In the future, BSOLS will be compared with more state-of-art algorithms and also be applied to some real-world problems. Moreover, our future work will focus on the development of modified BSO with a lighter computational burden to promote its efficiency.

Acknowledgment. This work is partially supported by The National Natural Science Foundation of China (Grants Nos. 71571120, 71271140, 71471158, 71001072, 61472257), Natural Science Foundation of Guangdong Province (2016A030310074) and Shenzhen Science and Technology Plan (CXZZ20140418182638764).

References

1. Shi, Y.: Brain storm optimization algorithm. In: Tan, Y., Shi, Y., Chai, Y., Wang, G. (eds.) ICSI 2011. LNCS, vol. 6728, pp. 303–309. Springer, Heidelberg (2011). doi:10.1007/978-3-642-21515-5_36
2. Shi, Y.: An optimization algorithm based on brainstorming process. Int. J. Swarm Intell. Res. (IJSIR) 2(4), 35–62 (2011)
3. Zhan, Z.H., Zhang, J., Shi, Y.H., Liu, H.L.: A modified brain storm optimization. In: 2012 IEEE Congress on Evolutionary Computation (CEC), pp. 1–8, June 2012
4. Chen, J., Xie, Y., Ni, J.: Brain storm optimization model based on uncertainty information. In: 2014 Tenth International Conference on Computational Intelligence and Security, pp. 99–103 (2014)
5. Zhu, H., Shi, Y.: Brain storm optimization algorithms with k-medians clustering algorithm. In: Proceedings of the Seventh International Conference on Advanced Computational Intelligence (ICACI 2015), pp. 107–110. IEEE (2015)
6. Cao, Z., Shi, Y., Rong, X., Liu, B., Du, Z., Yang, B.: Random grouping brain storm optimization algorithm with a new dynamically changing step size. In: Tan, Y., Shi, Y., Buarque, F., Gelbukh, A., Das, S., Engelbrecht, A. (eds.) ICSI 2015. LNCS, vol. 9140, pp. 357–364. Springer, Cham (2015). doi:10.1007/978-3-319-20466-6_38

7. Zhou, D., Shi, Y., Cheng, S.: Brain storm optimization algorithm with modified step-size and individual generation. In: Tan, Y., Shi, Y., Ji, Z. (eds.) ICSI 2012. LNCS, vol. 7331, pp. 243–252. Springer, Heidelberg (2012). doi:10.1007/978-3-642-30976-2_29

8. Krishnanand, K.R., Hasani, S.M.F., Panigrahi, B.K., Panda, S.K.: Optimal power flow solution using self-evolving brain–storming inclusive teaching–learning–based algorithm. In: Tan, Y., Shi, Y., Mo, H. (eds.) ICSI 2013. LNCS, vol. 7928, pp. 338–345. Springer, Heidelberg (2013). doi:10.1007/978-3-642-38703-6_40

9. Sun, C., Duan, H., Shi, Y.: Optimal satellite formation reconfiguration based on closed-loop brain storm optimization. IEEE Comput. Intell. Mag. 8(4), 39–51 (2013)

10. Cao, Z., Wang, L., Hei, X., Shi, Y., Rong, X.: An improved brain storm optimization with differential evolution strategy for applications of ANNs. Math. Probl. Eng. 2015, 1–18 (2015)

11. Shi, Y.: Brain storm optimization algorithm in objective space. In: Proceedings of 2015 IEEE Congress on Evolutionary Computation, (CEC 2015), pp. 1227–1234. IEEE, Sendai (2015)

12. Yadav, P.: Case retrieval algorithm using similarity measure and adaptive fractional brain storm optimization for health informaticians. Arab. J. Sci. Eng. 41, 1–12 (2016)

13. Niu, B., Liu, J., Liu, J., Yang, C.: Brain storm optimization for portfolio optimization. In: Tan, Y., Shi, Y., Li, L. (eds.) ICSI 2016. LNCS, vol. 9713, pp. 416–423. Springer, Heidelberg (2016). doi:10.1007/978-3-319-41009-8_45

14. Zhan, Z.H., Chen, W.N., Lin, Y., Gong, Y.J., Li, Y.L., Zhang, J.: Parameter investigation in brain storm optimization. In: Proceedings of the 2013 IEEE Symposium on Swarm Intelligence (SIS 2013), pp. 103–110 (2013)

15. Yao, X., Liu, Y., Lin, G.: Evolutionary programming made faster. IEEE Trans. Evol. Comput. 3(2), 82–102 (1999)

Difference Brain Storm Optimization for Combined Heat and Power Economic Dispatch

Yali Wu[1,2(✉)], Xinrui Wang[1,2], Yulong Fu[1,2], and Yingruo Xu[1,2]

[1] Faculty of Automation and Information Engineering,
Xi'an University of Technology, Xi'an, China
yliwu@xaut.edu.cn
[2] Shaanxi Key Laboratory of Complex System Control and Intelligent
Information Processing, Xi'an, China

Abstract. Brain Storm Optimization (BSO) is inspired by human being brain storm process. A novel Difference Brain Storm Optimization (DBSO) is proposed to solve combined heat and power economic dispatch (CHPED) problem in power plant. The difference mutation operation is adopted to replace the Gaussian mutation in the original BSO algorithm for increasing the diversity of the population and the speed of convergence. A test system with 7 units taken from the literature is simulated to verify the performance of the proposed algorithm. The results show that comparing with other intelligent optimization method, both BSO and DBSO can provide the better solution. The convergence speed of the DBSO is better than BSO algorithm.

Keywords: Brain storm algorithm · Difference brain storm optimization · Combined heat and power economic dispatch

1 Introduction

The efficiency of conventional power production unit is less than 60% in power plant. While cogeneration unit can reach about 90%, so it is now widely used in power plant. Moreover, the emission in cogeneration unit is only about 13%–18% in the environment aspects (such as CO_2, SO_2, etc.). Cogeneration units in the thermal power plant can not only supply power to the broad masses of electricity users, but also the heating for the users. In order to improve the economic efficiency of thermal power plant, economic dispatch problem which is called combined heat and power economic dispatch (CHPED) problem has been paid more attention in recent years. Its objective is to determine the best output of cogeneration unit under the constraints of operating constraints of electricity and heat demand.

Compared with conventional power plants, cogeneration unit not only the load demand of electricity and heat must be met respectively, but also the coupling relationship between them must be met in the CHP system. This makes the constraints

Y. Tan et al. (Eds.): ICSI 2017, Part I, LNCS 10385, pp. 519–527, 2017.
DOI: 10.1007/978-3-319-61824-1_57

more complex than ever. The traditional optimization algorithm is difficult to solve this problem. Many researchers proposed different intelligent optimization algorithm to solve CHPED problem. Su [1] improved genetic algorithm with an improved evolutionary direction operator and multiplier updating (IGA-MU) to avoid deforming the augmented Lagrange function. The proposed algorithm provided an efficacious approach for large-scale systems of the CHPED problem. Song [2] worked on CHPED using genetic algorithm based penalty function method. Sudhakaran and Slochanal [3] involved an integrating genetic algorithms and tabu search for combined heat and power system which consists of a conventional thermal unit, two cogeneration units and a heat-only unit. Tyagi [4] proved that particle swarm optimization algorithm can be skilled to solve CHPED problem. Mohammadi-Ivatloo [5] introduced a novel time varying acceleration coefficients particle swarm optimization (TVAC-PSO) algorithm to solve CHPED problem. The acceleration coefficients in PSO algorithm were varied adaptively during iterations to improve solution quality of original PSO and avoid premature convergence. The proposed method is applied to five test cases which are four units, five units, seven units, twenty-four units and forty-eight units. The results demonstrate the superiority of the proposed method in solving non-convex and constrained CHPED problem.

In recent years, many kinds of new type of intelligent optimization algorithm were used in CHPED problem. Basu [6] involved an opposition-based group search optimization which employs opposition-based learning for population initialization and also for iteration wise update operation. It was found that the proposed opposition-based group search optimization based approach can provide better solution. Meng [7] used crisscross optimization algorithm for solving CHPED problem. Beigvand [8] adopted gravitational search algorithm for CHPED problem.

As a novel swarm intelligence optimization algorithm, Brain Storm Optimization (BSO), which is inspired by the human brainstorming process, was presented by Shi [9] on the Second International Conference on Swarm Intelligence (ICSI11) in 2011. Zhao [10] analyzed the parameter of BSO algorithm and proposed an improved BSO which prepared with OPTICS clustering algorithm. BSO algorithms with k-medians clustering algorithms [11] and BSO algorithm in objective space [12] was put forward respectively by Shi in 2015. Xue [13] solved multi-objective optimization problems by BSO algorithm, which used non-dominated solutions to fit the true Pareto-front, the simulation results shows that it is effective to solve the multi-objective problem. More and more researchers focused their attention on the improvement of the performance on BSO algorithm. In this paper, we try to apply BSO to solve CHPED problem.

The structure of this paper is listed as follows. The mathematical model of CHPED problem is described in Sect. 2. The BSO and DBSO are proposed in Sect. 3. The DBSO algorithm is applied to CHPED problem in Sect. 4. And the simulation and result discussion is shown in Sect. 5. The conclusion is given in Sect. 6.

2   The Formulation of Combined Heat and Power Economic Dispatch Problem

2.1   The Objective Function of CHPED Problem

The CHPED system consists of power-only units, CHP units and heat-only units. The objective function of CHPED problem is minimizing the system cost. Considering valve-point effects and transmission loss, CHPED problem is a non-convex and non-linear problem. The objective function is represented as follows.

$$
\begin{aligned}
\min C_T &= \sum_{i=1}^{N_p} C_i(P_i^p) + \sum_{j=1}^{N_c} C_j(P_j^c, H_j^c) + \sum_{k=1}^{N_h} C_k(H_k^h) \\
&= \sum_{i=1}^{N_p} \left\{ \alpha_i (P_i^p)^2 + \beta_i P_i^p + \gamma_i + \left| \lambda_i \sin(\rho_i(P_i^{p\min} - P_i^p)) \right| \right\} \\
&\quad + \sum_{j=1}^{N_c} \left\{ a_j(P_j^c)^2 + b_j P_j^c + c_j + d_j(H_j^c)^2 + e_j H_j^c + f_j H_j^c P_j^c \right\} \\
&\quad + \sum_{k=1}^{N_h} \left\{ \phi_k(H_k^h) + \varphi_k H_k^h + \psi_k \right\}
\end{aligned}
\tag{1}
$$

Where C_T is the total production cost. $C_i(P_i^p)$ is the cost of power-only unit i. $C_j(P_j^c, H_j^c)$ is the cost of CHP unit j. $C_k(H_k^h)$ is the cost of heat-only unit k. N_p, N_c, N_h are the number of power-only units, CHP units and heat-only units respectively. P_i^p is power output of power-only unit i. P_j^c, H_j^c are the power output and heat output of CHP unit j respectively. H_k^h is heat output of heat-only unit k. $\alpha_i, \beta_i, \gamma_i$ are cost coefficients of power-only unit i. $a_j, b_j, c_j, d_j, e_j, f_j$ are coefficients of CHP unit j. $\phi_k, \varphi_k, \psi_k$ are coefficients of heat-only unit k.

2.2   The Constraints of CHPED Problem

The constraints of power and heat production and demand balance can be stated as follows.

$$
\sum_{i=1}^{N_p} P_i^p + \sum_{j=1}^{N_c} P_j^c = P_d + P_{loss}
\tag{2}
$$

$$
P_{loss} = \sum_{i=1}^{N_p}\sum_{m=1}^{N_p} P_i^p B_{im} P_m^p + \sum_{i=1}^{N_p}\sum_{j=1}^{N_c} P_i^p B_{ij} P_j^c + \sum_{j=1}^{N_c}\sum_{n=1}^{N_c} P_j^c B_{in} P_n^c
\tag{3}
$$

$$
\sum_{j=1}^{N_c} H_j^c + \sum_{k=1}^{N_h} H_k^h = H_d
\tag{4}
$$

Where P_d, H_d are the demand of power and heat, respectively. P_{loss} is transmission loss of power. B is the loss coefficients.

The capacity limits of power-only units, CHP units and heat-only units are presented as follows.

$$P_i^{pmin} \leq P_i^p \leq P_i^{pmax} \quad i = 1, \ldots, N_p \tag{5}$$

$$P_j^{cmin}(H_j^c) \leq P_j^c \leq P_j^{cmax}(H_j^c) \quad j = 1, \ldots, N_c \tag{6}$$

$$H_j^{cmin}(P_j^c) \leq H_j^c \leq H_j^{cmax}(P_j^c) \quad j = 1, \ldots, N_c \tag{7}$$

$$H_k^{hmin} \leq H_k^h \leq H_k^{hmax} \quad k = 1, \ldots, N_h \tag{8}$$

Where P_i^{pmin}, P_i^{pmax} are the minimum and the maximum power output of power-only unit i. $P_j^{cmin}(H_j^c)$ and $P_j^{cmax}(H_j^c)$ are the minimum and the maximum power limit of CHP unit j which are functions of generated heat (H_j^c). $H_j^{cmin}(P_j^c), H_j^{cmax}(P_j^c)$ are the minimum and the maximum heat limit of CHP unit j which are functions of generated heat (P_j^c). The feasible heat-power operation regions of CHP units [14] is depicted in Fig. 1. H_k^{hmin}, H_k^{hmax} are the minimum and the maximum heat output of heat-only unit k.

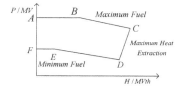

Fig. 1. The heat-power feasible regions of CHP units

3 Difference Brainstorm Optimization Algorithm

3.1 Brain Storm Optimization Algorithm

The design of BSO algorithm is based on the brainstorming process [15]. In the brainstorming process, the generation of the idea obeys the Osborn's original four rules. The people in the brainstorming group will need to be open-minded as much as possible and therefore generate more diverse ideas. Any judgment or criticism must be held back until at least the end of one round of the brainstorming process, which means no idea will be ignored. The optimization algorithm based on the idea human produce process should be superior to that of clustering behavior of animals, for human are the most intelligent animals in the world. Inspired by this process, Shi proposed BSO algorithm [9] in 2011.

3.2 Difference Brainstorm Optimization Algorithm

The Gaussian random values will be used as random values which are added to generate new individuals in classical BSO algorithm. The new individual generation in Step 6 can be represented as the Eq. (9), which is listed in the following.

$$X_{new}^d = X_{selected}^d + \xi * (\mu, \sigma) \tag{9}$$

Where $X_{selected}^d$ is the dth dimension of the individual selected to generate new individual. X_{new}^d is the dth dimension of the individual newly generated. $n(\mu, \sigma)$ is the Gaussian random function with mean μ and variance σ. The parameter ξ is a coefficient that weights the contribution of the Gaussian random value. For the simplicity and for the purpose of fine tuning, the ξ can be calculated as:

$$\xi = \log sig((0.5 * \max\_iteration - current\_iteration)/K) * rand() \tag{10}$$

Where $\log sig()$ is a logarithmic sigmoid transfer function. $\max\_iteration$ is the maximum number of iterations. $current\_iteration$ is the current iteration number. K is for changing function's slope. $rand()$ is a random value within $(0, 1)$.

As a common mutation method, the variance of Gaussian mutation is obtained by multiplying ξ and the Gaussian random function. The value of $\log sig()$ is between 0 and 1. $\xi \in (0, 1)$. As shown in Fig. 2, the Gaussian distribution follows the principle of 3σ. when μ and σ is fixed, the Gaussian function produces most values between $(\mu - 3\sigma, \mu + 3\sigma,)$ which is shown in Fig. 3. When $\mu = 0, \sigma = 1$, the range of variation is $(-3, 3)$.

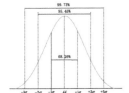

Fig. 2. Gaussian distribution

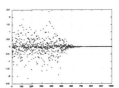

Fig. 3. The range of variation when $\mu = 0, \sigma = 1$

During the process of searching the optimal solution, all the solutions can be uniformly generated in the decision space at first. And the global search should be carried out. At the latter stage, the local variation should be optimized due to the overall rise of the solution, quality, The variance should be smaller with the stage process lasts. However, Gaussian variation is consistent, because of the range of variation is basically fixed, the individuals cannot make full use of information in the current population. In addition, the Gaussian mutation operation process is very complex.

Differential evolution (DE) mutation can be served as a useful tool in this paper to achieve the purpose. The mutation operation can be described in Eq. (11).

$$x_{newd} = \begin{cases} rand\left(L_d, H_d\right) & \text{if } rand() < P_r \\ x_{selectd} + rand(0, 1) \times (x_1 - x_2) & \text{else} \end{cases} \quad (11)$$

Where $X^d_{selected}$ is the dth dimension of the individual selected to generate new individual. X^d_{new} is the dth dimension of the individual newly generated. The upper and lower bounds of the dimension is L_d, H_d. And x_1, x_2 for the two different individuals chosen in the contemporary world.

Compared with Gaussian variation, only the random function and the four mixed operations are considered in the difference variation, so the computation complexity will be reduced significantly. And the running speed will be significantly improved. Whenever the generation of random values is based on other individuals within the contemporary population, can the information of the other individuals within the population be obtained, this make the searching efficiency higher than ever. In addition, the differential mutation can balance the local search and global search in the search process greatly which can improve the algorithm performance effectively.

4   DBSO Algorithm for Combined Heat and Power Economic Dispatch

The steps of DBSO for solve CHPED problems as follows:

1. Initialize operation.

Each individual in the population represents a solution, it can be stated as:

$$X(i) = [P^p_{i1}, \ldots, P^p_{iN_p}, P^c_{i1}, \ldots, P^c_{iN_c}, H^c_{i1}, \ldots, H^c_{iN_c}, H^h_{i1}, \ldots, H^h_{iN_k}]$$

Due to every individual satisfy the constraints, the steps of this constraints are handled as follows:

(1) The initial values of the power-only units and heat-only units are randomly generated using the formula (12). The power and the heat of CHP units are randomly generated according to the formula (13).

$$P^p_{ij} = P^{pmin}_j + r \times (P^{pmax}_j - P^{pmin}_j)\, j = 1, \ldots, N_p$$
$$H^h_{ij} = H^{hmin}_j + r \times (H^{hmax}_j - H^{hmin}_j)\, j = 1, \ldots, N_h \quad (12)$$

$$P^c_{ij} = P^{cmin}_j(\cdot) + r \times (P^{cmax}_j(\cdot) - P^{cmin}_j(\cdot))\, j = 1, \ldots, N_c$$
$$H^c_{ij} = H^{cmin}_j(\cdot) + r \times (H^{cmax}_j(\cdot) - H^{cmin}_j(\cdot))\, j = 1, \ldots, N_c \quad (13)$$

(2) The power and the heat of CHP units must satisfy the feasible regions requirement, as shown in Fig. 1, assuming BC, DC, DE that the relationship between P^c_{ij} and H^c_{ij} are $H^c_{ij} = f_1(P^c_{ij}), H^c_{ij} = f_2(P^c_{ij}), H^c_{ij} = f_3(P^c_{ij})$, respectively. The feasible regions constraints of the CHP units can be handled as the same as the formula (14).

$$
(P_{ij}^c, H_{ij}^c) = \begin{cases} (P_{ij}^c, f_1(P_{ij}^c)) & \text{if } P_C^c \le P_{ij}^c \le P_B^c \text{ and } f_1(P_{ij}^c) \le H_{ij}^c \\ (P_{ij}^c, f_2(P_{ij}^c)) & \text{if } P_E^c \le P_{ij}^c \le P_C^c \text{ and } f_2(P_{ij}^c) \le H_{ij}^c \\ (P_{ij}^c, f_2(P_{ij}^c)) & \text{if } P_D^c \le P_{ij}^c \le P_E^c \text{ and } f_2(P_{ij}^c) \le H_{ij}^c \\ (P_{ij}^c, f_3(P_{ij}^c)) & \text{if } P_D^c \le P_{ij}^c \le P_E^c \text{ and } f_3(P_{ij}^c) \ge H_{ij}^c \end{cases} \tag{14}
$$

(3) The load requirements of the power and the heat output may be satisfied, by calculating the difference (m and n) between the power and heat of individuals using Eq. (15). And then the difference m, n are distributed evenly over the power of the power-only units and the heat of the heat-only units, respectively.

$$
\begin{aligned}
m &= P_d + P_{loss} - P_{i1}^p - \ldots - P_{iN_p}^p - P_{i1}^c - \ldots P_{iN_c}^c \\
n &= H_d - H_{i1}^c - \ldots - H_{iN_c}^c - H_{i1}^h - \ldots - H_{iN_k}^h
\end{aligned} \tag{15}
$$

2. The constraints satisfaction of update operation.

When the differential mutation operation is operated, the new individual may not meet the capacity constraints. If it is exceeds the upper bound, the individual takes the upper limit of the corresponding unit; if the individual shorts of the lower bound, the individual takes the lower limit of the corresponding unit. If the power and the heat of CHP units are not satisfy the feasible regions requirement, the feasible regions constraints of the CHP units can be handled according to the formula (14). If the power and the heat output are not satisfy the load requirements, the load constraints of the power and the heat output are handled according to the third step of this constraints are handled.

In this way, the new individual which satisfies all the constraint conditions is generated. The fitness value of the new individual is calculated and the optimal individual is retained.

5   Experiments and Discussions

In this section, the CHP system with seven units [5] is considered as an example. The parameters set of DBSO algorithm are presented in Table 1.

Table 1. Parameters of DBSO algorithm

The size of population	Max-iteration	Number of clusters	P_{5a}	P_{6b}	P_{6biii}	P_{6c}
100	500	2	0.01	0.8	0.4	0.5

The power and heat generations corresponding to best cost obtained from DBSO and other algorithms is shown in Table 2. Evidently, it is seen that DBSO and BSO algorithm can provide the better solution than the other reported algorithms.

Table 2. Optimal dispatch results obtained by different algorithms

Algorithms	P_1	P_2	P_3	P_4	P_5	P_6	H_5	H_6	H_7	P_o	H_o	Cost
PSO [16]	18.463	124.26	112.78	209.816	98.814	44.011	57.924	32.76	59.316	608.143	150	10613
CPSO [5]	75	112.38	30	250	93.27	40.159	32.566	72.674	44.761	600.809	150	10325.3339
TVAC-PSO [5]	47.338	98.54	112.674	209.816	92.372	40	37.847	75	37.153	600.739	150	10100.3164
EP [16]	61.361	95.121	99.943	208.732	98.8	44	18.071	77.555	54.374	607.956	150	10390
DE [17]	44.212	98.538	112.691	209.774	98.822	44	12.538	78.348	59.114	608.037	149.999	10317
RCGA [18]	74.683	97.958	167.231	124.908	98.801	44	58.097	32.412	59.492	607.581	150	10667
AIS [16]	50.133	95.555	110.752	208.768	98.8	44	19.424	77.078	53.498	608.008	150	10355
BCO [18]	43.946	98.589	112.932	208.772	98.8	44	12.097	78.024	59.879	608.035	150	10317
TLBO [19]	45.266	98.548	112.979	209.828	94.412	40.006	25.837	74.997	49.167	600.739	150	10094.8384
OTLBO [19]	45.886	98.54	112.674	209.814	93.825	40	29.291	75	45.708	600.119	150	10094.3529
CSO [7]	45.491	98.54	112.673	209.816	94.184	40	27.179	75	47.821	600.704	150	10094.1267
BSO	45.45	98.54	112.674	209.816	94.076	40	27.822	74.999	47.18	600.555	150	10093.6666
DBSO	45.45	98.54	112.674	209.816	94.076	40	27.822	74.999	47.18	600.555	150	10093.6666

6 Conclusions

In this paper DBSO is successfully used to solve CHPED problem. DBSO use difference mutation instead of Gaussian mutation in the original BSO algorithm for improve the diversity of population and increases the speed of convergence. Compared with reported algorithms. It is found that DBSO and BSO algorithm can provide the better solution, and the convergence speed of the DBSO is better than BSO algorithm.

Acknowledgments. This paper is supported by National Youth Foundation of China with Grant Number 61503299.

References

1. Su, C.T., Chiang, C.L.: An incorporated algorithm for combined heat and power economic dispatch. Electr. Power Syst. Res. **69**(2), 187–195 (2004)
2. Song, Y.H., Xuan, Q.Y.: Combined heat and power economic dispatch using genetic algorithm based penalty function method. Electr. Mach. Power Syst. **26**(4), 363–372 (1998)
3. Sudhakaran, M., Slochanal, S.M.R.: Integrating genetic algorithms and tabu search for combined heat and power economic dispatch. In: TENCON 2003, Conference on Convergent Technologies for Asia-Pacific Region, vol. 1, pp. 67–71 (2003)
4. Tyagi, G., Pandit, M.: Combined heat and power dispatch using Particle swarm optimization. In: Electrical, Electronics and Computer Science, pp. 1–4. IEEE (2012)
5. Mohammadi-Ivatloo, B., Moradi-Dalvand, M., Rabiee, A.: Combined heat and power economic dispatch problem solution using particle swarm optimization with time varying acceleration coefficients. Electr. Power Syst. Res. **95**(1), 9–18 (2013)
6. Basu, M.: Combined heat and power economic dispatch using opposition-based group search optimization. Int. J. Electr. Power Energy Syst. **73**, 819–829 (2015)
7. Meng, A., Mei, P., Yin, H., et al.: Crisscross optimization algorithm for solving combined heat and power economic dispatch problem. Energy Convers. Manag. **105**, 1303–1317 (2015)

8. Beigvand, S.D., Abdi, H., Scala, M.L.: Combined heat and power economic dispatch problem using gravitational search algorithm. Electr. Power Syst. Res. **133**, 160–172 (2016)
9. Shi, Y.: Brain storm optimization algorithm. In: Proceedings of the Advances in Swarm Intelligence - Second International Conference, ICSI 2011, Chongqing, China, 12–15 June 2011, pp. 1–3 (2011)
10. Zhao, X.: The application of Brain storm optimization. Xi'an University of Technology (2013)
11. Zhu, H., Shi, Y.: Brain storm optimization algorithms with k-medians clustering algorithms. In: Seventh International Conference on Advanced Computational Intelligence. IEEE (2015)
12. Shi, Y.: Brain storm optimization algorithm in objective space. In: Evolutionary Computation, pp. 1–3. IEEE (2015)
13. Xue, J., Wu, Y., Shi, Y., Cheng, S.: Brain storm optimization algorithm for multi-objective optimization problems. In: Tan, Y., Shi, Y., Ji, Z. (eds.) Advances in Swarm Intelligence. LNCS, vol. 7331, pp. 513–519. Springer, Heidelberg (2012). doi:10.1007/978-3-642-30976-2_62
14. Sashirekha, A., Pasupuleti, J., Moin, N.H., et al.: Combined heat and power (CHP) economic dispatch solved using Lagrangian relaxation with surrogate subgradient multiplier updates. Int. J. Electr. Power Energy Syst. **44**(1), 421–430 (2013)
15. Smith, R.: The 7 Levels of Change, 2nd edn. Tapeslry Press, Littleton (2002)
16. Basu, M.: Artificial immune system for combined heat and power economic dispatch. Int. J. Electr. Power Energy Syst. **43**(1), 1–5 (2012)
17. Basu, M.: Combined heat and power economic dispatch by using differential evolution. Electr. Power Compon. Syst. **38**(8), 996–1004 (2010)
18. Basu, M.: Bee colony optimization for combined heat and power economic dispatch. Expert Syst. Appl. **38**(11), 13527–13531 (2011)
19. Roy, P.K., Paul, C., Sultana, S.: Oppositional teaching learning based optimization approach for combined heat and power dispatch. Int. J. Electr. Power Energy Syst. **57**(5), 392–403 (2014)

Cuckoo Search

Multiple Chaotic Cuckoo Search Algorithm

Shi Wang[1], Shuangyu Song[2], Yang Yu[2], Zhe Xu[2], Hanaki Yachi[2],
and Shangce Gao[2(✉)]

[1] College of Computer Science and Technology, Taizhou University,
Taizhou, Jiangsu, China
[2] Faculty of Engineering, University of Toyama, Toyama, Japan
`gaosc@eng.u-toyama.ac.jp`

Abstract. Cuckoo search algorithm (CSA) is a nature-inspired meta-heuristic based on the obligate brood parasitic behavior of cuckoo species, and it has shown promising performance in solving optimization problems. Chaotic mechanisms have been incorporated into CSA to utilize the dynamic properties of chaos, aiming to further improve its search performance. However, in the previously proposed chaotic cuckoo search algorithms (CCSA), only one chaotic map is utilized in a single search iteration which limited the exploitation ability of the search. In this study, we consider to utilize multiple chaotic maps simultaneously to perform the local search within the neighborhood of the global best solution found by CSA. To realize this, three kinds of multiple chaotic cuckoo search algorithms (MCCSA) are proposed by incorporating several chaotic maps into the chaotic local search parallelly, randomly or selectively. The performance of MCCSA is verified based on 48 widely used benchmark optimization functions. Experimental results reveal that MCCSAs generally perform better than CCSAs, and the MCCSA-P which parallelly utilizes chaotic maps performs the best among all 16 compared variants of CSAs.

Keywords: Cuckoo search algorithm · Chaotic local search · Neighborhood search · Optimization · Computational intelligence

1 Introduction

Optimization is basically ubiquitous, from engineering to economics, and from holiday planning to Internet routing. Because of the limited availability of capital, resources and time, the best utility of these available resources is critical. Under a variety of complex constraints, most real-world optimization is highly nonlinear and multi-modal. Many of the problem solving process is heuristic throughout human history. However, heuristic as a scientific optimization method is a common modern phenomenon. Heuristic intelligent algorithms [1] such as genetic algorithm, ant colony algorithm and particle swarm algorithm are the hot topics in the field of computational algorithm design and analysis, which are simulated by the behavior of biological or natural phenomenon.

© Springer International Publishing AG 2017
Y. Tan et al. (Eds.): ICSI 2017, Part I, LNCS 10385, pp. 531–542, 2017.
DOI: 10.1007/978-3-319-61824-1_58

Although a lot of meta-heuristic algorithms have been proposed in the literature, it is still a big demand and challenge to propose more new heuristic algorithms. The famous Free Lunch Theorem [2] proves that no meta-heuristic is best suited to solve all optimization problems. New heuristic rules can bring new inspiration, mechanism, motivation and characteristics to the problem-solving research community.

With the development of simulated biological behavior, in 2009, Yang and Deb [3] of Cambridge University proposed a novel search algorithm named cuckoo search algorithm (CSA). CSA is inspired by the behavior of cuckoo breeding. Cuckoos are charming birds that can make beautiful sounds, and their aggressive reproduction strategy make them even more fascinating. Some of the cuckoo species evolved in such a way: the female parasitic cuckoo is usually very specialized in imitating the color and pattern of the eggs of several selected host species. This reduces the chances of the eggs being abandoned and increased their fecundity. If the host bird finds that the eggs are not their own, it either throws the foreign egg away or simply gives up its nest and builds a new nest elsewhere. Parasitic cuckoos usually choose a nest, where the host bird laid its eggs not long ago. Usually, the host eggs hatch slightly later than the cuckoo eggs. If the first cuckoo hatches chicks, his first instinct is to evict the host eggs by blindly pushing the eggs out of the nest. This action leads to an increase in the food share of the cuckoo chick provided by its host bird. From the algorithmic perspective, CSA is simple, efficient and it has been successfully applied to engineering optimization and other practical problems [4].

Aiming to further enhance the search performance of CSA, chaotic mechanisms have been incorporated and several chaotic cuckoo search algorithms (CCSA) have been proposed in the literature. In [5], Ouyang et al. utilized the famous Logistic chaotic map to construct the chaotic local search and embedded it into CSA to realize a search balance between exploitation and exploration. This work was extended by Wang et al. [6,7] where 12 different chaotic maps are tested to find out which chaotic map is the most suitable for improving the performance of CSA. Their experimental results showed that the Sinusoidal chaotic map was the best choice. In contrast to the usage of chaotic map to perform local search, Wang and Zhong [8] used chaotic maps to define the scaling factor and the fraction probability in CSA, aiming to enhance the solution quality and convergence speed. Huang et al. [9] used chaotic sequences to enhance initialized host nest location, change step size of Lévy flight and reset the location of host nest beyond the boundary. Although these chaotic mechanisms have achieves some improvement for CSA, only a single chaos embedded CSA in a search iteration still suffer from the inefficient search ability due to (1) the dynamic system derived from the single chaos remains the same properties during the whole search progress, and (2) the resultant dynamic system is lack of variability. Thus, it is natural and reasonable to simultaneously combine *multiple* chaotic maps with CSA to fully utilize the distinct characteristics of each map.

Based on the above considerations, in this study we propose three kinds of multiple chaotic cuckoo search algorithms (MCCSAs) by incorporating 12 different chaotic maps into CSA parallelly, randomly, or selectively, and call them MCCSA-P, MCCSA-R or MCCSA-S respectively. As demonstrated in our previously works [10,11], chaos used to perform local search is more efficient than chaotic sequences. Thus, the chaotic maps in MCCSAs are all used to perform the exploitation around the global best solution found by CSA in an iteration. To verify the proposed algorithms, 48 widely used benchmark optimization functions including unimodal, multimodal, shifted, rotated and hybrid composition functions are adopted. Extensive simulation and statistical results suggest that MCCSA generally performs significantly better than CSA and all 12 variants of CCSAs. Moreover, MCCSA-P which utilizes all different chaotic maps parallelly in a single iteration performs the best among all compared algorithms, indicating that multiple chaotic maps can play complementary effects in the neighborhood exploitation due to their different dynamic properties.

The remainder of this paper is organized as follows. Section 2 gives a brief description of CSA. Section 3 elaborates the details of the proposed MCCSA. Section 4 gives the experimental results and discussions. Finally, some general remarks are given to conclude this paper.

2 Cuckoo Search Algorithm

CSA is a population-based optimization algorithm, and each cuckoo (i.e. solution) can be formally expressed as $X_i = \{x_i^1, x_i^2, \ldots, x_i^D\}$, $(i = 1, 2, \ldots, N)$, where N is the population size and D is the dimension of the optimization problem. There are three idealized rules of cuckoo search: (1) each cuckoo lays one egg at a time, and it is placed in a randomly chosen nest; (2) the optimal nests will be transferred to the next generation; and (3) The number of available host nests is fixed, and the alien egg can be discovered by the host with a fixed probability p_a. Based on these three rules, the pseudo-code of CSA can be illustrated in Algorithm 1.

Algorithm 1. Cuckoo Search Algorithm

---

begin
    Set the generation counter $t = 1$;
    Initialize the population of hot nests randomly;
    Set the discovery rate p_a;
    while $t < Maximum\ number\ of\ generations$ **do**
        Sort the population according to their fitness f;
        Get a cuckoo randomly X_i and replace its solution by performing Lévy
          flights;
        Choose a nest X_j among the population randomly;
        if $f(X_i) < f(X_j)$ **then**
          Replace X_j with the new solution X_i;
        A fraction (p_a) of the worse nests is deleted and new ones are generated;
        Sort the population and find and output the current best;
        $t = t + 1$;

Table 1. The definition of twelve chaotic maps.

Name	Equation	Initial z_0
Logistic map	$z_{k+1} = 4z_k(1 - z_k)$	0.152
PWLCM	$z_{k+1} = \begin{cases} z_k/0.7, & z_k \in (0, 0.7) \\ (1 - z_k)(1 - 0.7), & z_k \in [0.7, 1) \end{cases}$	0.002
Singer map	$z_{k+1} = 1.073(7.86z_k - 23.31z_k^2 + 28.75z_k^3 - 13.302875z_k^4)$	0.152
Sine map	$z_{k+1} = \sin(\pi z_k)$	0.152
Sinusoidal map	$z_{k+1} = 2.3z_k^2 \sin(\pi z_k)$	0.74
Tent map	$z_{k+1} = \begin{cases} z_k/0.4, & 0 < z_k \leq 0.4 \\ (1 - z_k)/0.6, & 0.4 < z_k \leq 1 \end{cases}$	0.152
Bernoulli map	$z_{k+1} = \begin{cases} z_k/0.6, & 0 < z_k \leq 0.6 \\ (z_k - 0.6)/0.4, & 0.6 < z_k < 1 \end{cases}$	0.152
Chebyshev map	$z_{k+1} = \cos(0.5 \cos^{-1} z_k)$	0.152
Circle map	$z_{k+1} = z_k + 0.5 - \frac{1.1}{\pi} \sin(2\pi z_k) \mathrm{mod}(1)$	0.152
Cubic map	$z_{k+1} = 2.59z_k(1 - z_k^2)$	0.242
Gaussian map	$z_{k+1} = \begin{cases} 0, & z_k = 0 \\ (1/z_k)\mathrm{mod}(1) & z_k \neq 0 \end{cases}$	0.152
ICMIC	$z_{k+1} = \sin(70/z_k)$	0.152

In Algorithm 1, the Lévy flight is used to expand the search range and increase the diversity of the population to be more likely to jump out of the local optima. For the cuckoo X_i, the Lévy flight is implemented as follows.

$$X_i := X_i + \alpha \bigoplus L\acute{e}vy(\lambda) \tag{1}$$

where the parameter $\alpha > 0$ denotes the step size. In our experiment, α is set to 0.01. The product $\bigoplus$ means entry-wise multiplications. The Lévy flight in fact provides a random walk, and the random step length is calculated by the Lévy distribution formula shown as:

$$L\acute{e}vy \sim u = t^{-\lambda} \quad (1 < \lambda \leq 3) \tag{2}$$

Essentially, a random walk process is formed by the consecutive jumps of a cuckoo. The random walk process obeys a power-law step-length distribution with a heavy tail.

3   Multiple Chaotic Cuckoo Search Algorithm

Chaos is a kind of dynamic behavior of non-liner systems. Recently, it has attracted much attention in the field of scientific identification such as chaos control, pattern recognition and optimization theory. Twelve one-dimensional chaotic maps shown in Table 1 are considered in this study. These chaotic maps display bounded dynamic instability, pseudo-random, ergodicity and non-period behavior relied on control parameters and initial value. For its ergodicity and randomness, the chaotic system changes randomly, but if it lasts long enough, it eventually passes through each state.

To improve the convergence speed and solution quality of cuckoo search, the local search procedure is considered in this study. We should pay attention to that the local search is only applied to the current global best agent X_g because (1) the area around the current global best agent may be the most promising area to find good solutions, and (2) if we perform a local search on all agents, it will take a lot of implementation time at each iteration. It can save much implementation time by apply on current global best agent. After a local search for current global best agent X_g, an agent X_g' is generated. If X_g' is better than X_g, X_g' will replace the current global worst agent X_w to survive into next iteration.

The chaotic local search that uses only one chaotic map is defined as:

$$X_g'(t) = X_g(t) + r(t)(Ub - Lb)(2c(t) - 1) \tag{3}$$

$X_g(t)$ is the current global best agent at the tth iteration. $X_g'(t)$ is a new agent that generated by the chaotic local search. Ub represent the upper bound of the search, and Lb is the lower bound. $c(t)$ is a chaotic variable produced from one of the 12 chaotic maps shown in Table 1. Additionally, it is a fact that chaotic search in a small range is more effective. $r(t)$ represents a chaotic search radius parameter which becomes gradually smaller. The equation of $r(t)$ is defined by:

$$r(t + 1) = \rho r(t) \tag{4}$$

where ρ is a shrinking parameter which is set to be 0.988 in the experiment. Initially, the search radius $r(0)$ is set to 0.03. After the local search, the agent updating procedure is implemented. The current global worst agent will be replaced by the x_g' if the fitness of x_g is improved, and the others survive to enter into the next iteration. These single chaotic map embedded cuckoo search algorithms are named as CCSA1–CCSA12, respectively.

As different chaotic maps show different and unique dynamic characteristics, different chaos therefore have different effects on different problems. In other words, the performance of CCSA is effected not only by the characteristics of the embedded chaotic map, but also by the landscape of the solved problems. To fully utilize the search dynamics of chaotic maps and their complementary ergodicities, three kinds of multiple chaotic maps embedded are proposed for CSA to construct MCCSA-P, MCCSA-R, and MCCSA-S by parallelly, randomly and selectively incorporating them, respectively.

In this study, all 12 considered chaotic maps are used in MCCSAs. In MCCSA-P, these chaotic maps are parallelly utilized to perform the chaotic local search within the neighborhood of current global best cuckoo, shown as:

$$X_{g'}^j(t) = X_g(t) + r(t)(Ub - Lb)(2c^j(t) - 1) \tag{5}$$

$$X_{g'}(t) = argmin\{f(X_{g'}^j(t))\} \tag{6}$$

where $X_{g'}^j (j = 1, 2, \ldots, 12)$ presents a candidate solution temporarily generated by the chaotic local search. $X_{g'}(t)$ is the best one of the twelve candidate solutions obtained by using different chaotic maps. If $X_{g'}(t)$ is better than $X_g(t)$, replace current worst agent $X_w(t)$ with $X_{g'}(t)$, while the others survive to enter into the next iteration.

In MCCSA-R, we randomly select a chaotic map to carry out the chaotic local search using:

$$X_{g'}(t) = X_g(t) + r(t)(Ub - Lb)(2c^R(t) - 1) \tag{7}$$

where $X_{g'}(t)$ represents the chaotic sequence generated by the Rth chaotic map randomly selected from 12 chaotic maps. j is a randomly generated from the set $\{1, 2, \ldots, 12\}$. Each chaotic map is selected with a same probability $1/12$.

In MCCSA-S, the local search is applied to the current global best agent X_g by a probability. The probability that a chaotic map is selected from 12 chaotic maps depends on the success rate and failure rate in generating improved solutions within a certain number (i.e. LP) of previous iterations. Assume that the probability of jth chaotic map be selected is P_j, $j = 1, 2, \ldots, 12$. The probabilities with respect to each chaotic map are initialized as $1/12$. At the generation t, if the new agent generated by the jth chaotic map is better than current best agent, we assign $ns_{j,t}$ a value of 1 while assign $nf_{j,t}$ a value of 0. On the contrary, we assign $ns_{j,t}$ a value of 0 while assign $nf_{j,t}$ a value of 1. We use success memory and failure memory to store these numbers within a fixed number of previous generations hereby named learning period (LP). At the generation t, the new agent generated by different chaotic map that can enter or fail to enter the next generation over the previous LP generations are stored in different columns of the memories. Once the memories overflow after $LP = 50$ generations, the earliest records stored i.e., nf_{t-LP} or ns_{t-LP} will be removed so that those numbers in the current generation can be stored in the memories. After the initial 50 generations, the probabilities of selecting different chaotic maps will be updated at each subsequent generation based on the success and failure memories. With the lapse of iteration, at the generation t, the selection probabilities are updated according to:

$$p_{j,t} = \frac{S_{j,t}}{\sum_{j=1}^{12} S_{j,t}} \tag{8}$$

$$S_{j,t} = \frac{\sum_{t-LP}^{t-1} ns_{j,t}}{\sum_{t-LP}^{t-1} ns_{j,t} + \sum_{t-LP}^{t-1} nf_{j,t}} + \varepsilon, (j = 1, 2, \ldots, 12; t > LP) \tag{9}$$

where p_j, t represents the probability of selecting the jth chaotic map at the tth iteration. S_j, t represents the success rate of the new agent generated by the jth chaotic map and successfully entering the next generation within the previous 50 generations with respect to generation t. $\varepsilon = 0.01$ is set to avoid the possible null success rates. At each generation, the probabilities of choosing each chaotic map in the candidate pool are summed to 1. S_j, t represents the success rate of the new agent generated by the jth chaotic map and successfully entering the next generation within the previous 50 generations with respect to generation t.

$\varepsilon = 0.01$ is a small constant value which is used to avoid the possible null success rates. p_j, t represents the probability of selecting the jth chaotic map at the tth iteration. To ensure that the probability of selecting all chaotic maps is always 1, we divide $S_{j,t}$ by $\sum_{j=1}^{12} S_{j,t}$ to calculate $p_{j,t}$. Obviously, the larger the success rate for the jth chaotic map within the previous 50 generations is, the larger the probability of applying it to do local search at the current generation is.

4   Experimental Results and Discussions

To evaluate the performance of these methods, we adopt 48 widely used benchmark optimization functions. F1–F25 are the CEC2005 composite benchmark functions and F26–F48 are the most commonly used benchmark numerical functions. Function F1–F5 are shifted unimodal functions; Function F6–F12 are multimodal basic functions; Function F13–F14 are multimodal expanded functions. Function F15–F25 are hybrid composition functions. Functions F26–F38 are high-dimensional problems. Functions F26–F30 are unimodal functions. Function F31 is the step function, which has one minimum and is discontinuous. Function F32 is a noisy quartic function. Functions F33–F38 are multimodal functions where the number of local minima increases exponentially with the problem dimension. Functions F39–F48 are low-dimensional functions which have only a few local minima.

In the experiment, the population size is set to 30, the dimension of all tested functions is 30, and the maximum number of iterations is 1000. In order to make a statistical analysis, all compared algorithms are implemented 30 independent times for 48 benchmark functions. We list the experimental results of CSA, CCSA1-CCSA12, MCCSA-P, MCCSA-R and MCCSA in Tables 2, 3, 4 and 5.

Table 2. Experimental results of test functions (F1–F12).

	F1	F2	F3	F4	F5	F6
CSA	1.06E-02±6.56E-04	1.73E+03±6.93E+02	1.19E+07±3.28E+06	1.74E+04±4.48E+03	5.73E+03±1.11E+03	1.27E+03±1.37E+03
CCSA1	1.21E-04±2.05E-04	5.67E+02±2.65E+02	3.85E+06±1.26E+06	1.53E+04±5.35E+03	4.91E+03±1.39E+03	1.14E+03±2.29E+03
CCSA2	7.45E-05±4.70E-05	5.57E+02±3.09E+02	4.15E+06±1.66E+06	1.34E+04±4.11E+03	5.41E+03±1.30E+03	1.13E+03±2.33E+03
CCSA3	1.66E-04±5.33E-04	5.93E+02±3.53E+02	3.86E+06±1.84E+06	1.36E+04±3.80E+03	5.58E+03±1.18E+03	1.43E+03±3.34E+03
CCSA4	8.99E-05±1.39E-04	5.74E+02±2.91E+02	**3.43E+06±1.35E+06**	1.48E+04±3.90E+03	5.22E+03±1.23E+03	1.50E+03±2.95E+03
CCSA5	8.93E-05±1.01E-04	6.76E+02±5.24E+02	3.61E+06±2.00E+06	1.34E+04±4.62E+03	5.43E+03±1.42E+03	5.11E+02±4.54E+02
CCSA6	8.38E-05±9.62E-05	6.98E+02±3.54E+02	3.56E+06±1.94E+06	1.36E+04±3.87E+03	5.73E+03±1.45E+03	6.14E+02±1.02E+03
CCSA7	6.54E-05±5.82E-05	6.57E+02±3.53E+02	3.59E+06±1.84E+06	1.65E+04±4.90E+03	5.28E+03±1.11E+03	1.26E+03±2.81E+03
CCSA8	9.02E-05±8.45E-05	5.70E+02±3.43E+02	3.72E+06±1.73E+06	1.41E+04±4.06E+03	5.10E+03±1.32E+03	1.10E+03±2.70E+03
CCSA9	1.06E-04±1.29E-04	6.05E+02±3.69E+02	3.47E+06±1.49E+06	1.33E+04±3.13E+03	5.24E+03±1.17E+03	1.13E+03±2.78E+03
CCSA10	1.02E-04±1.30E-04	6.26E+02±4.17E+02	3.93E+06±2.13E+06	1.36E+04±2.85E+03	5.32E+03±1.41E+03	5.40E+02±5.04E+02
CCSA11	5.59E-04±8.01E-04	5.89E+02±3.64E+02	4.98E+06±3.15E+06	1.48E+04±5.11E+03	4.94E+03±1.00E+03	1.27E+03±2.36E+03
CCSA12	1.05E-04±1.41E-04	5.76E+02±3.84E+02	4.21E+06±1.92E+06	1.35E+04±3.74E+03	5.26E+03±9.49E+02	6.68E+02±7.81E+02
MCCSA-P	8.78E-03±4.26E-03	1.82E+03±6.53E+02	1.07E+07±2.72E+06	1.79E+04±4.67E+03	5.62E+03±1.14E+03	8.81E+02±7.11E+02
MCCSA-R	6.80E-05±7.18E-05	**5.22E+02±2.64E+02**	3.79E+06±1.72E+06	1.40E+04±5.01E+03	**4.71E+03±1.08E+03**	**4.86E+02±4.01E+02**
MCCSA-S	**5.74E-05±5.16E-05**	5.73E+02±2.73E+02	3.62E+06±1.67E+06	**1.31E+04±4.32E+03**	4.98E+03±1.13E+03	4.99E+02±4.05E+02

	F7	F8	F9	F10	F11	F12
CSA	4.70E+03±1.55E-01	2.10E+01±7.33E-02	1.13E+02±1.63E+01	1.90E+02±3.26E+01	3.22E+01±1.45E+00	9.70E+04±2.12E+04
CCSA1	4.70E+03±7.39E+00	2.04E+01±9.80E-02	8.45E+01±2.24E+01	2.08E+02±4.23E+01	3.00E+01±2.93E+00	2.57E+04±1.78E+04
CCSA2	**4.70E+03±1.63E-12**	2.04E+01±1.37E-01	8.92E+01±2.43E+01	1.82E+02±4.58E+01	2.98E+01±2.67E+00	3.47E+04±2.19E+04
CCSA3	4.70E+03±2.65E-12	2.04E+01±9.48E-02	9.21E+01±2.08E+01	1.85E+02±4.60E+01	3.01E+01±1.88E+00	3.17E+04±2.99E+04
CCSA4	4.70E+03±8.73E+00	2.04E+01±1.45E-01	8.87E+01±2.41E+01	1.82E+02±4.59E+01	2.90E+01±3.26E+00	2.98E+04±2.45E+04
CCSA5	4.70E+03±5.54E+00	2.04E+01±1.92E-01	9.72E+01±2.00E+01	1.86E+02±4.35E+01	2.94E+01±3.05E+00	2.66E+04±1.96E+04
CCSA6	4.70E+03±2.54E-12	2.04E+01±1.71E-01	8.90E+01±2.55E+01	1.79E+02±2.68E+01	3.02E+01±2.64E+00	3.02E+04±2.57E+04
CCSA7	4.70E+03±2.06E-07	2.04E+01±6.22E-02	9.68E+01±2.56E+01	1.73E+02±3.89E+01	2.91E+01±2.23E+00	3.13E+04±2.00E+04
CCSA8	4.70E+03±5.19E+00	2.04E+01±7.33E-02	9.43E+01±2.35E+01	1.86E+02±3.59E+01	2.99E+01±2.56E+00	2.65E+04±2.03E+04
CCSA9	4.70E+03±2.45E-12	2.04E+01±1.47E-01	8.37E+01±2.36E+01	1.75E+02±3.44E+01	2.92E+01±2.14E+00	3.27E+04±2.56E+04
CCSA10	4.70E+03±8.73E+00	2.04E+01±1.57E-01	8.68E+01±2.39E+01	1.65E+02±3.07E+01	2.99E+01±1.75E+00	2.82E+04±1.94E+04
CCSA11	4.70E+03±1.56E-02	2.04E+01±1.46E-01	8.30E+01±2.14E+01	1.73E+02±5.66E+01	2.95E+01±3.44E+00	**2.24E+04±1.61E+04**
CCSA12	4.70E+03±2.19E+00	2.04E+01±2.23E-01	8.76E+01±2.42E+01	1.78E+02±2.46E+01	2.98E+01±2.44E+00	2.58E+04±1.53E+04
MCCSA-P	4.70E+03±4.77E-03	2.10E+01±1.31E-01	1.14E+02±1.85E+01	2.00E+02±3.65E+01	3.25E+01±1.42E+00	8.73E+04±1.81E+04
MCCSA-R	4.70E+03±2.97E-12	2.03E+01±9.20E-02	**8.18E+01±2.65E+01**	1.73E+02±4.24E+01	**2.90E+01±2.57E+00**	2.38E+04±1.84E+04
MCCSA-S	4.70E+03±5.38E+00	**2.03E+01±8.51E-02**	8.66E+01±1.99E+01	**1.55E+02±2.96E+01**	3.00E+01±2.12E+00	2.69E+04±1.72E+04

Table 3. Experimental results of test functions (F13–F24).

	F13	F14	F15	F16	F17	F18
CSA	9.78E+00±1.38E+00	1.34E+01±1.67E-01	4.45E+02±4.70E+01	2.59E+02±4.71E+01	3.14E+02±5.86E+01	9.12E+02±2.27E+00
CCSA1	6.37E+00±2.16E+00	1.32E+01±2.82E-01	4.08E+02±6.14E+01	2.48E+02±9.83E+01	3.06E+02±7.32E+01	9.17E+02±8.22E+00
CCSA2	6.75E+00±1.69E+00	1.32E+01±3.38E-01	4.25E+02±5.39E+01	2.15E+02±7.51E+01	3.04E+02±6.87E+01	9.15E+02±7.52E+00
CCSA3	6.94E+00±2.04E+00	1.32E+01±3.10E-01	4.44E+02±5.18E+01	2.71E+02±1.06E+02	2.99E+02±8.76E+01	9.19E+02±1.17E+01
CCSA4	6.72E+00±2.04E+00	1.32E+01±3.02E-01	4.16E+02±5.69E+01	2.62E+02±9.10E+01	3.11E+02±7.17E+01	9.18E+02±9.01E+00
CCSA5	7.03E+00±1.75E+00	1.32E+01±3.03E-01	4.06E+02±6.24E+01	2.30E+02±8.13E+01	3.07E+02±9.11E+01	9.20E+02±1.32E+01
CCSA6	6.39E+00±1.74E+00	1.32E+01±2.37E-01	4.16E+02±6.50E+01	2.13E+02±7.27E+01	3.03E+02±6.61E+01	9.16E+02±1.21E+01
CCSA7	6.94E+00±1.89E+00	1.32E+01±3.17E-01	4.19E+02±5.27E+01	2.49E+02±8.26E+01	3.01E+02±6.73E+01	9.16E+02±7.58E+00
CCSA8	6.68E+00±1.73E+00	1.32E+01±2.98E-01	4.19E+02±5.33E+01	2.50E+02±8.75E+01	2.90E+02±6.94E+01	9.17E+02±9.49E+00
CCSA9	6.35E+00±1.52E+00	1.32E+01±2.35E-01	4.08E+02±4.20E+01	2.23E+02±8.82E+01	3.29E+02±8.58E+01	9.12E+02±2.27E+01
CCSA10	6.60E+00±2.16E+00	1.32E+01±3.79E-01	4.27E+02±5.23E+01	2.48E+02±8.77E+01	3.05E+02±7.10E+01	9.15E+02±4.70E+00
CCSA11	6.50E+00±1.67E+00	1.32E+01±3.06E-01	**4.04E+02±2.66E+01**	2.24E+02±6.86E+01	3.08E+02±7.66E+01	9.14E+02±6.68E+00
CCSA12	6.73E+00±1.99E+00	1.32E+01±4.06E-01	4.25E+02±4.62E+01	2.43E+02±7.70E+01	2.91E+02±7.17E+01	9.14E+02±4.66E+00
MCCSA-P	1.11E+01±1.83E+00	1.36E+01±1.64E-01	4.44E+02±3.64E+01	2.84E+02±6.64E+01	3.41E+02±5.61E+01	**9.04E+02±6.25E+00**
MCCSA-R	6.33E+00±2.02E+00	**1.31E+01±3.18E-01**	4.05E+02±7.19E+01	2.27E+02±8.36E+01	3.19E+02±9.55E+01	9.16E+02±6.01E+00
MCCSA-S	**5.98E+00±1.67E+00**	1.32E+01±2.42E-01	4.09E+02±4.20E+01	**1.95E+02±4.64E+01**	**2.79E+02±6.58E+01**	9.15E+02±5.65E+00

	F19	F20	F21	F22	F23	F24
CSA	9.12E+00±2.52E+00	9.13E+01±2.34E+00	**5.35E+02±9.39E+01**	9.79E+02±3.27E+01	**6.26E+02±1.67E+02**	8.87E+02±2.06E+02
CCSA1	9.14E+02±4.59E+00	9.19E+02±9.31E+00	6.81E+02±2.76E+02	9.65E+02±3.45E+01	7.62E+02±2.52E+02	7.41E+02±3.42E+02
CCSA2	9.16E+02±7.38E+00	9.18E+02±1.17E+01	6.55E+02±2.55E+02	9.55E+02±3.25E+01	6.39E+02±1.57E+02	7.25E+02±3.67E+02
CCSA3	9.22E+02±1.34E+01	9.18E+02±1.21E+01	6.21E+02±2.43E+02	9.66E+02±3.94E+01	7.83E+02±2.50E+02	7.06E+02±3.67E+02
CCSA4	9.14E+02±5.32E+00	9.12E+02±2.20E+01	6.41E+02±2.56E+02	9.74E+02±4.76E+01	6.78E+02±1.99E+02	7.08E+02±3.55E+02
CCSA5	9.17E+02±7.17E+00	9.16E+02±1.02E+01	6.12E+02±2.37E+02	9.71E+02±4.54E+01	6.78E+02±2.19E+02	7.86E+02±3.61E+02
CCSA6	9.16E+02±2.95E+01	9.16E+02±1.28E+01	6.32E+02±2.57E+02	9.59E+02±4.02E+01	6.52E+02±1.66E+02	6.48E+02±3.68E+02
CCSA7	9.15E+02±5.64E+00	9.14E+02±5.91E+00	6.41E+02±2.68E+02	9.74E+02±4.31E+01	7.05E+02±2.11E+02	8.07E+02±3.22E+02
CCSA8	9.16E+02±6.78E+00	9.15E+02±6.20E+00	6.97E+02±2.95E+02	9.71E+02±4.35E+01	7.43E+02±2.18E+02	6.38E+02±3.90E+02
CCSA9	9.17E+02±9.47E+00	9.16E+02±8.37E+00	5.64E+02±1.96E+02	9.64E+02±4.08E+01	6.55E+02±1.76E+02	6.70E+02±3.84E+02
CCSA10	9.19E+02±1.08E+01	9.20E+02±1.37E+01	6.03E+02±2.20E+02	9.55E+02±3.86E+01	6.95E+02±2.05E+02	8.25E+02±2.93E+02
CCSA11	9.15E+02±7.58E+00	9.16E+02±6.43E+00	5.66E+02±1.80E+02	9.88E+02±5.62E+01	7.59E+02±2.68E+02	7.05E+02±3.81E+02
CCSA12	9.16E+02±5.86E+00	9.19E+02±1.23E+01	5.89E+02±2.17E+02	9.66E+02±5.19E+01	7.03E+02±2.10E+02	8.23E+02±3.16E+02
MCCSA-P	**9.03E+02±5.59E+00**	**9.03E+02±5.83E+00**	5.62E+02±1.65E+02	9.70E+02±3.36E+01	6.85E+02±1.92E+02	8.20E+02±2.79E+02
MCCSA-R	9.16E+02±4.04E+00	9.17E+02±6.10E+00	6.90E+02±2.88E+02	9.59E+02±5.41E+01	6.59E+02±1.83E+02	**6.33E+02±3.65E+02**
MCCSA-S	9.16E+02±6.46E+00	9.16E+02±9.26E+00	6.16E+02±2.43E+02	**9.51E+02±3.69E+01**	6.73E+02±2.04E+02	6.84E+02±3.61E+02

Table 4. Experimental results of test functions (F25–F36).

	F25	F26	F27	F28	F29	F30
CSA	1.67E+03±1.23E+01	2.88E-03±1.91E-03	7.48E-02±2.84E-02	2.64E+02±7.26E+01	8.54E+00±3.01E+00	7.00E+01±4.54E+01
CCSA1	1.67E+03±1.28E+01	2.25E-05±2.65E-05	6.95E-04±3.38E-04	7.13E+01±4.31E+01	1.21E+01±3.56E+00	7.01E+01±2.97E+01
CCSA2	1.67E+03±1.01E+01	2.42E-05±2.46E-05	8.66E-04±4.25E-04	7.27E+01±3.25E+01	1.24E+01±3.91E+00	6.80E+01±4.20E+01
CCSA3	1.67E+03±1.33E+01	2.73E-05±2.79E-05	5.34E-04±3.46E-04	7.57E+01±3.93E+01	1.24E+01±3.14E+00	6.18E+01±4.09E+01
CCSA4	1.67E+03±1.33E+01	2.81E-05±2.89E-05	6.50E-04±3.53E-04	9.09E+01±4.23E+01	1.15E+01±2.80E+00	6.65E+01±3.77E+01
CCSA5	1.67E+03±1.04E+01	2.83E-05±3.34E-05	5.50E-04±2.49E-04	9.35E+01±4.05E+01	1.26E+01±2.79E+00	6.06E+01±3.29E+01
CCSA6	1.67E+03±1.35E+01	5.50E-05±1.30E-04	8.08E-04±3.31E-04	9.37E+01±5.10E+01	1.16E+01±2.72E+00	7.52E+01±4.64E+01
CCSA7	1.67E+03±1.15E+01	2.60E-05±3.33E-05	7.44E-04±3.59E-04	7.72E+01±4.45E+01	1.17E+01±3.22E+00	8.10E+01±4.78E+01
CCSA8	1.66E+03±1.45E+01	5.71E-05±9.57E-05	5.66E-04±2.42E-04	7.90E+01±4.71E+01	1.17E+01±3.58E+00	6.53E+01±3.31E+01
CCSA9	1.66E+03±1.30E+01	3.80E-05±3.33E-05	8.36E-04±2.87E-04	7.77E+01±3.67E+01	1.17E+01±2.99E+00	7.00E+01±5.49E+01
CCSA10	1.67E+03±1.18E+01	2.15E-05±1.91E-05	8.29E-04±4.33E-04	8.73E+01±5.78E+01	1.17E+01±2.63E+00	6.44E+01±3.77E+01
CCSA11	1.67E+03±1.21E+01	1.46E-04±1.84E-04	1.61E-03±1.17E-03	9.74E+01±5.69E+01	1.17E+01±3.33E+00	6.88E+01±4.58E+01
CCSA12	1.66E+03±9.99E+00	2.79E-05±5.61E-05	5.66E-04±3.62E-04	9.60E+01±4.56E+01	1.28E+01±3.21E+00	7.13E+01±6.50E+01
MCCSA-P	**1.64E+03±8.30E+00**	**1.82E-14±3.79E-14**	**5.21E-08±4.02E-08**	**2.22E+01±8.47E+01**	**1.65E-08±1.40E-08**	**1.55E-13±2.14E-13**
MCCSA-R	1.66E+03±9.77E+00	1.90E-05±1.21E-05	5.45E-04±1.51E-04	6.86E+01±3.83E+01	1.17E+01±3.65E+00	6.08E+01±4.07E+01
MCCSA-S	1.66E+03±1.36E+01	2.14E-05±1.43E-05	4.99E-04±2.27E-04	7.30E+01±3.69E+01	1.10E+01±3.18E+00	6.21E+01±3.65E+01

	F31	F32	F33	F34	F35	F36
CSA	2.67E-01±7.85E-01	6.41E+02±3.07E-02	-8.61E+03±3.42E+02	7.58E+01±1.16E+01	2.73E+00±1.35E+00	7.90E-02±6.42E-02
CCSA1	1.67E-01±3.79E-01	7.80E-02±3.65E-02	-9.26E+03±5.74E+02	5.05E+01±1.31E+01	2.71E+00±8.97E-01	1.52E-02±2.30E-02
CCSA2	0.00E+00±0.00E+00	7.79E-02±3.59E-02	-9.41E+03±5.08E+02	5.14E+01±1.32E+01	2.35E+00±8.69E-01	1.65E-02±1.86E-02
CCSA3	0.00E+00±0.00E+00	7.49E-02±3.21E-02	-9.16E+03±6.54E+02	5.07E+01±1.59E+01	2.63E+00±8.56E-01	1.18E-02±1.24E-02
CCSA4	3.33E-02±1.83E-01	7.57E-02±3.27E-02	-9.27E+03±5.52E+02	5.37E+01±1.89E+01	2.52E+00±8.91E-01	1.13E-02±1.49E-02
CCSA5	3.33E-02±1.83E-01	7.90E-02±3.31E-02	-9.15E+03±5.99E+02	5.14E+01±1.48E+01	2.56E+00±9.06E-01	1.70E-02±1.99E-02
CCSA6	6.67E-02±2.54E-01	7.38E-02±3.19E-02	-9.43E+03±4.88E+02	5.04E+01±1.43E+01	2.32E+00±6.56E-01	1.34E-02±1.51E-02
CCSA7	6.67E-02±2.54E-01	7.54E-02±2.55E-02	-9.51E+03±5.85E+02	5.27E+01±1.34E+01	2.60E+00±6.93E-01	1.34E-02±1.64E-02
CCSA8	6.67E-02±2.54E-01	7.39E-02±2.76E-02	-9.44E+03±5.28E+02	5.22E+01±1.37E+01	2.30E+00±1.10E+00	1.11E-02±1.47E-02
CCSA9	0.00E+00±0.00E+00	6.63E-02±3.91E-02	-9.45E+03±5.42E+02	5.04E+01±1.78E+01	2.29E+00±1.00E+00	1.26E-02±1.37E-02
CCSA10	6.67E-02±2.54E-01	7.31E-02±3.35E-02	-9.49E+03±5.30E+02	5.13E+01±1.11E+01	2.33E+00±5.62E-01	1.14E-02±2.24E-02
CCSA11	1.00E-01±4.03E-01	6.28E-02±2.10E-02	-9.56E+03±7.89E+02	5.00E+01±1.70E+01	2.32E+00±9.81E-01	8.77E-03±1.05E-02
CCSA12	3.33E-02±1.83E-01	7.47E-02±2.06E-02	-9.33E+03±5.22E+02	5.07E+01±1.40E+01	2.51E+00±7.32E-01	1.46E-02±1.19E-02
MCCSA-P	**0.00E+00±0.00E+00**	**1.16E-04±1.30E-04**	**-1.24E+04±7.92E+02**	**9.22E+00±1.56E+01**	**3.10E-08±2.19E-08**	**4.66E-14±5.96E-14**
MCCSA-R	6.67E-02±2.54E-01	6.77E-02±2.93E-02	-9.56E+03±6.59E+02	4.69E+01±1.43E+01	2.24E+00±7.66E-01	6.78E-03±1.01E-02
MCCSA-S	1.13E+00±1.31E+00	6.12E-02±3.23E-02	-9.53E+03±6.62E+02	4.78E+01±1.41E+01	2.49E+00±1.10E+00	9.41E-03±9.14E-03

The results of the records are shown in the form of $Ave \pm Dev$, where Ave is the average of the final best-so-far solution of 30 runs, and Dev represents its standard deviation. The results in bold are the best result among all the compared 16 algorithms for each function. For test functions $F41$–$F48$ and F39, all 16 algorithms obtain the almost same results (i.e., the f_{min} of these test

Table 5. Experimental results of test functions (F37–F48).

	F37	F38	F39	F40	F41	F42
CSA	1.96E+00±8.35E-01	2.54E+00±4.06E+00	**9.98E-01±0.00E+00**	3.08E-04±6.60E-08	-1.03E+00±6.78E-16	**3.98E-01±0.00E+00**
CCSA1	9.58E-01±7.07E-01	6.80E+00±5.10E+00	**9.98E-01±0.00E+00**	3.07E-04±2.07E-19	-1.03E+00±6.71E-16	**3.98E-01±0.00E+00**
CCSA2	8.63E-01±9.26E-01	7.39E+00±6.31E+00	**9.98E-01±0.00E+00**	3.07E-04±8.10E-19	-1.03E+00±6.78E-16	**3.98E-01±0.00E+00**
CCSA3	9.61E-01±7.62E-01	1.01E+01±8.05E+00	**9.98E-01±0.00E+00**	**3.07E-04±1.80E-19**	-1.03E+00±6.71E-16	**3.98E-01±0.00E+00**
CCSA4	1.19E+00±9.46E-01	8.27E+00±6.70E+00	**9.98E-01±0.00E+00**	3.07E-04±7.33E-19	-1.03E+00±6.78E-16	**3.98E-01±0.00E+00**
CCSA5	1.32E+00±1.31E+00	7.75E+00±6.05E+00	**9.98E-01±0.00E+00**	3.07E-04±3.11E-19	-1.03E+00±6.78E-16	**3.98E-01±0.00E+00**
CCSA6	7.34E-01±7.97E-01	5.97E+00±6.38E+00	**9.98E-01±0.00E+00**	3.07E-04±1.24E-18	-1.03E+00±6.71E-16	**3.98E-01±0.00E+00**
CCSA7	8.85E-01±1.19E+00	5.97E+00±6.34E+00	**9.98E-01±0.00E+00**	3.07E-04±4.22E-19	-1.03E+00±6.78E-16	**3.98E-01±0.00E+00**
CCSA8	8.02E-01±9.97E-01	7.69E+00±6.05E+00	**9.98E-01±0.00E+00**	3.07E-04±2.76E-14	-1.03E+00±6.78E-16	**3.98E-01±0.00E+00**
CCSA9	9.59E-01±1.14E+00	4.72E+00±5.91E+00	**9.98E-01±0.00E+00**	3.07E-04±4.02E-19	-1.03E+00±6.71E-16	**3.98E-01±0.00E+00**
CCSA10	7.98E-01±6.79E-01	7.58E+00±6.38E+00	**9.98E-01±0.00E+00**	3.07E-04±6.64E-19	-1.03E+00±6.78E-16	**3.98E-01±0.00E+00**
CCSA11	7.95E-01±9.35E-01	6.49E+00±5.55E+00	**9.98E-01±0.00E+00**	3.07E-04±7.52E-19	-1.03E+00±6.78E-16	**3.98E-01±0.00E+00**
CCSA12	1.07E+00±1.15E+00	7.45E+00±5.11E+00	**9.98E-01±0.00E+00**	3.07E-04±3.17E-19	**-1.03E+00±6.71E-16**	**3.98E-01±0.00E+00**
MCCSA-P	**1.04E-01±5.68E-01**	1.13E-15±2.50E-15	**9.98E-01±0.00E+00**	3.07E-04±3.59E-09	-1.03E+00±6.78E-16	**3.98E-01±0.00E+00**
MCCSA-R	6.84E-01±6.92E-01	6.90E+00±5.22E+00	**9.98E-01±0.00E+00**	3.07E-04±2.19E-19	-1.03E+00±6.78E-16	**3.98E-01±0.00E+00**
MCCSA-S	6.78E-01±8.87E-01	5.29E+00±4.78E+00	**9.98E-01±0.00E+00**	3.07E-04±3.70E-19	-1.03E+00±6.78E-16	**3.98E-01±0.00E+00**

	F43	F44	F45	F46	F47	F48
CSA	3.00E+00±1.12E-15	**-3.86E+00±2.71E-15**	-3.32E+00±2.92E-15	-1.02E+01±7.34E-14	-1.04E+01±2.40E-15	-1.05E+01±2.53E-13
CCSA1	3.00E+00±1.96E-15	**-3.86E+00±2.71E-15**	-3.32E+00±1.36E-15	-1.02E+01±6.51E-15	-1.04E+01±7.38E-16	-1.05E+01±1.36E-15
CCSA2	3.00E+00±1.89E-15	**-3.86E+00±2.71E-15**	-3.32E+00±1.34E-15	-1.02E+01±6.56E-15	-1.02E+01±9.63E-01	-1.05E+01±2.14E-15
CCSA3	3.00E+00±1.97E-15	**-3.86E+00±2.71E-15**	-3.32E+00±1.34E-15	-1.02E+01±6.96E-15	-1.04E+01±8.73E-16	**-1.05E+01±8.73E-16**
CCSA4	3.00E+00±1.85E-15	**-3.86E+00±2.71E-15**	-3.32E+00±2.17E-02	-1.02E+01±6.74E-15	-1.04E+01±9.33E-16	-1.05E+01±2.36E-15
CCSA5	3.00E+00±1.88E-15	**-3.86E+00±2.71E-15**	-3.32E+00±1.34E-15	-1.02E+01±6.74E-15	-1.04E+01±8.73E-16	-1.05E+01±2.36E-15
CCSA6	3.00E+00±1.23E-15	**-3.86E+00±2.71E-15**	-3.32E+00±1.36E-15	-1.02E+01±6.68E-15	-1.04E+01±9.33E-16	-1.05E+01±2.56E-15
CCSA7	3.00E+00±1.87E-15	**-3.86E+00±2.71E-15**	**-3.32E+00±1.33E-15**	**-1.02E+01±6.45E-15**	-1.04E+01±4.66E-16	-1.05E+01±1.36E-15
CCSA8	3.00E+00±1.09E-15	**-3.86E+00±2.71E-15**	-3.31E+00±3.02E-02	-1.02E+01±6.90E-15	-1.04E+01±6.60E-16	-1.05E+01±8.73E-16
CCSA9	3.00E+00±1.08E-15	**-3.86E+00±2.71E-15**	-3.32E+00±1.34E-15	-1.02E+01±6.51E-15	-1.04E+01±8.08E-16	-1.05E+01±9.33E-16
CCSA10	3.00E+00±2.08E-15	**-3.86E+00±2.71E-15**	-3.32E+00±1.36E-15	-1.02E+01±6.51E-15	-1.04E+01±4.66E-16	-1.05E+01±1.09E-15
CCSA11	3.00E+00±1.77E-15	**-3.86E+00±2.71E-15**	-3.32E+00±1.34E-15	-1.02E+01±7.17E-15	-1.04E+01±7.38E-16	-1.05E+01±2.14E-15
CCSA12	3.00E+00±1.08E-15	**-3.86E+00±2.71E-15**	-3.32E+00±2.17E-02	-1.02E+01±6.96E-15	-1.02E+01±9.70E-01	-1.05E+01±9.33E-16
MCCSA-P	**3.00E+00±1.07E-15**	**-3.86E+00±2.71E-15**	-3.32E+00±1.91E-14	-1.02E+01±6.51E-15	**-1.04E+01±3.30E-16**	-1.05E+01±2.93E-15
MCCSA-R	3.00E+00±1.86E-15	**-3.86E+00±2.71E-15**	-3.32E+00±2.17E-02	-1.02E+01±6.74E-15	-1.04E+01±8.08E-16	-1.05E+01±9.90E-16
MCCSA-S	3.00E+00±1.98E-15	**-3.86E+00±2.71E-15**	-3.32E+00±1.36E-15	-1.02E+01±6.68E-15	-1.04E+01±8.08E-16	-1.05E+01±2.21E-15

Table 6. Performance ranking of CSA, 12 variants of CCSAs, MCCSA-P, MCCSA-R, and MCCSA-S obtained by Friedman statistical test.

	CSA	CCSA1	CCSA2	CCSA3	CCSA4	CCSA5	CCSA6	CCSA7
Average	10.85	8.58	8.58	8.54	8.5	8.53	8.31	8.54
Rank	16	14	13	12	9	10	5	11

	CCSA8	CCSA9	CCSA10	CCSA11	CCSA12	MCCSA-P	MCCSA-R	MCCSA-S
Average	8.37	8.18	8.33	8.77	8.39	7.77	7.9	7.87
Rank	7	4	6	15	8	1	3	2

functions). It is because that these functions are very simple to be solved, and the compared algorithms can find the global optimal solutions in almost all runs. In addition, it is clear that the variants of CSA combined with chaotic local search process are always outperform the traditional CSA, suggesting that the chaotic local search definitely enhances the search ability of CSA. Moreover, compared with single chaotic map embedded CCSAs, the multiple chaotic maps embedded MCCSAs perform better for 37 out of 48 tested functions, which indicates that the utilization of multiple chaotic maps is more effective than a single one. The reason is that these multiple chaotic maps might contribute to different search dynamics and thus make more fruitful results for optimization. Besides, the best performance is achieved by MCCSA-P.

To further demonstrate the effectiveness and robustness of the proposed MCCSA, the average rankings of the algorithms obtained by the Friedman test [12,13] on all tested 48 benchmark optimization functions are summarized in Table 6. The Friedman test is a nonparametric statistical test which applies the

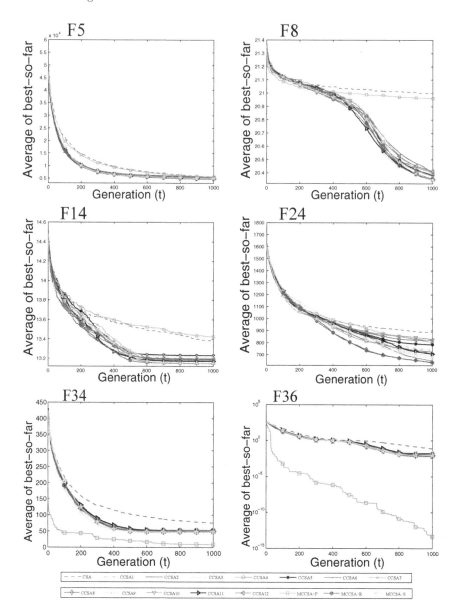

Fig. 1. Convergence graphs of F5, F8, F14, F24, F34 and F36.

post hoc method of Iman-Davenport. It can rank the algorithms for each problem separately. The best performing algorithm among all compared algorithms should have rank 1, the second best rank 2, and so on. From Table 6, it can be found that MCCSA-P gets the smallest value of 7.77, which means that it averagely performs the best for all functions. The second smallest value 7.87 is acquired by MCCSA-S, while MCCSA-R gets the third one. It is worth pointing out that CSA gets the largest ranking value, indicating that all chaotic CCSAs performs better than CSA. In addition, it is mostly desired that a general well-performing algorithm should be designed. From this practical problem-solving perspective, we can conclude that the proposed multiple chaos incorporation scheme is effective for improving the performance of CSA.

To give some insights into the search dynamics of proposed algorithms, we depict the convergence graphs for F5, F8, F14, F24, F34 and F36 in Fig. 1, where the average of best-so-far versus the iteration number are illustrated. From this figure, we also confirm that MCCSAs perform better than the others.

5   Conclusions

In this paper, we proposed a multiple chaotic cuckoo search algorithm for optimization. Taking into account the abundant searching dynamics of different chaos, the proposed MCCSA is demonstrated to be more powerful than traditional CSA and single chaotic map embedded cuckoo search algorithms through an extensive experiments. In the experiment 48 widely used benchmark functions are tested and statistical analysis is also performed. Experimental results verify the effectiveness and robustness of the proposed MCCSA. Especially, the parallelly embedding scheme for CSA is demonstrated to be the most effective based on the Friedman test.

Acknowledgment. This research was partially supported by the National Natural Science Foundation of China (Grant Nos. 11572084, 11472061, and 61472284), the project of Talent Development of Taizhou University (No. QD2016061) and JSPS KAKENHI Grant Number 17K12751, 15K00332 (Japan).

References

1. Osman, I.H., Kelly, J.P.: Meta-Heuristics: Theory and Applications. Springer Science & Business Media, Berlin (2012)
2. Wolpert, D.H., Macready, W.G.: No free lunch theorems for optimization. IEEE Trans. Evol. Comput. **1**(1), 67–82 (1997)
3. Yang, X.S., Deb, S.: Cuckoo search via lévy flights. In: Proceedings of World Congress on Nature & Biologically Inspired Computing (NaBIC 2009), pp. 210–214. IEEE (2009)
4. Fister Jr., I., Yang, X.S., Fister, D., Fister, I.: Cuckoo search a brief literature review. In: Yang, X.-S. (ed.) Cuckoo Search and Firefly Algorithm. SCI, vol. 516, pp. 49–62. Springer, Cham (2014). doi:10.1007/978-3-319-02141-6_3
5. Ouyang, A., Pan, G., Yue, G., Du, J.: Chaotic cuckoo search algorithm for high-dimensional functions. J. Comput. **9**(5), 1282–1290 (2014)
6. Wang, G.G., Deb, S., Gandomi, A.H., Zhang, Z., Alavi, A.H.: A novel cuckoo search with chaos theory and elitism scheme. In: 2014 International Conference on Soft Computing and Machine Intelligence (ISCMI), pp. 64–69. IEEE (2014)
7. Wang, G., Deb, S., Gandomi, A.H., Zhang, Z., Alavi, A.H.: Chaotic cuckoo search. Soft. Comput. **20**, 3349–3362 (2016)

8. Wang, L., Zhong, Y.: Cuckoo search algorithm with chaotic maps. Math. Probl. Eng. **2015** (2015). Article ID 715635
9. Huang, L., Ding, S., Yu, S., Wang, J., Lu, K.: Chaos-enhanced cuckoo search optimization algorithms for global optimization. Appl. Math. Model. **40**(5), 3860–3875 (2016)
10. Gao, S., Vairappan, C., Wang, Y., Cao, Q., Tang, Z.: Gravitational search algorithm combined with chaos for unconstrained numerical optimization. Appl. Math. Comput. **231**, 48–62 (2014)
11. Shen, D., Jiang, T., Chen, W., Shi, Q., Gao, S.: Improved chaotic gravitational search algorithms for global optimization. In: IEEE Congress on Evolutionary Computation (CEC), pp. 1220–1226. IEEE (2015)
12. Garcia, S., Fernandez, A., Luengo, J., Herrera, F.: Advanced nonparametric tests for multiple comparisons in the design of experiments in computational intelligence and data mining: Experimental analysis of power. Inf. Sci. **180**(10), 2044–2064 (2010)
13. Gao, S., Wang, Y., Cheng, J., Inazumi, Y., Tang, Z.: Ant colony optimization with clustering for solving the dynamic location routing problem. Appl. Math. Comput. **285**, 149–173 (2016)

Cuckoo Search Algorithm Approach for the IFS Inverse Problem of 2D Binary Fractal Images

Javier Quirce[1], Andrés Iglesias[1,2(✉)], and Akemi Gálvez[1,2]

[1] Department of Applied Mathematics and Computational Sciences,
University of Cantabria, Avenida de los Castros s/n, 39005 Santander, Spain
iglesias@unican.es
[2] Department of Information Science, Faculty of Sciences, Toho University,
Narashino Campus, 2-2-1 Miyama, Funabashi 274-8510, Japan
http://personales.unican.es/iglesias

Abstract. This paper introduces a new method to solve the IFS inverse problem for fractal images, known to be a very difficult optimization problem. Given a source binary fractal image, the method computes the IFS code of an IFS fractal whose attractor approximates the input image accurately. The proposed method is based on the cuckoo search algorithm, a powerful swarm intelligence method for continuous optimization. The good performance of the method is illustrated by its application to two examples of 2D binary fractal images.

Keywords: Swarm intelligence · Cuckoo search algorithm · Fractal images · Iterated function systems · Collage theorem

1 Introduction

Fractals have been widely used to recreate mountains, rivers, coastlines, and other natural structures. The reason is their ability generate complex shapes based on the repetition of simple pattern rules [2,5,12,14,15]. There are many methods to obtain fractal images, e.g. escape-time fractals, L-systems, recursive fractals, and so on [6,8–10]. One of the most popular is the *Iterated Function Systems* (IFS), given by a finite system of contractive maps on a compact metric space [2,16]. Any IFS system has a unique non-empty compact fixed set $\mathcal{A}$ called the attractor of the IFS. The graphical representation of this attractor is a fractal image. Conversely, each real-world image in 2D can be closely approximated by an IFS. This result, proved by Barnsley and known as the collage theorem [2], will be described in detail in Sect. 2. As a result, there has been a great interest in obtaining the parameters of the IFS representing a given image. This issue, called the *IFS inverse problem*, is the key component of the fractal image compression technique, a lossy fractal-based compression method for digital images [4,13].

Unsurprisingly, the IFS inverse problem is extremely difficult. Previous approaches to address this problem include Gröbner basis [1], wavelets transform [3], and moment matching [19]. Unfortunately, they fail to solve the general

© Springer International Publishing AG 2017
Y. Tan et al. (Eds.): ICSI 2017, Part I, LNCS 10385, pp. 543–551, 2017.
DOI: 10.1007/978-3-319-61824-1_59

problem. One of the most promising approaches is based on the application of evolutionary techniques, mostly genetic algorithms [11,18]. However, the potential of swarm intelligence for this problem has been ignored so far in the literature. To fill this gap, this work focuses on the IFS problem for the particular case of 2D binary fractal images. Our approach applies a powerful swarm intelligence technique called *cuckoo search algorithm* (see Sect. 3 for details).

The structure of this paper is as follows: Sect. 2 introduces the main concepts and definitions about the IFS and the collage theorem. Then, Sect. 3 describes the cuckoo search algorithm, the swarm intelligence approach used in this paper. The proposed method is described in detail in Sect. 4, while the experimental results are briefly discussed in Sect. 5. The paper closes with the main conclusions and some ideas about future work in the field.

2   Mathematical Concepts and Definitions

2.1   Iterated Function Systems

An *Iterated Function System* (IFS) in $\mathbb{R}^2$ is a finite set $\{w_i\}_{i=1,\ldots,N}$ of affine contractive maps $w_i : \mathbb{R}^2 \longrightarrow \mathbb{R}^2$ with the Euclidean distance d_2. We refer to the IFS as $\mathcal{W} = \{(\mathbb{R}^2, d_2); w_1, \ldots, w_N\}$. In that case, w_i are of the form:

$$\begin{bmatrix} x^* \\ y^* \end{bmatrix} = w_i \begin{bmatrix} x \\ y \end{bmatrix} = \begin{bmatrix} a_i \ b_i \\ c_i \ d_i \end{bmatrix} \cdot \begin{bmatrix} x \\ y \end{bmatrix} + \begin{bmatrix} e_i \\ f_i \end{bmatrix} \Leftrightarrow \mathbf{x}^* = \mathbf{w}_i(\mathbf{x}) = \mathbf{A}_i.\mathbf{x} + \mathbf{b}_i \quad (1)$$

where $\mathbf{b}_i$ is a vector and $\mathbf{A}_i$ is a 2×2 matrix with eigenvalues λ_1, λ_2 such that $|\lambda_i| < 1$. In fact, $s_i = |det(\mathbf{A}_i)| < 1$ meaning that w_i shrinks distances between points. The collection of 6-tuples $\{(a_i, b_i, c_i, d_i, e_i, f_i)\}_{i=1,\ldots,N}$ is called the *IFS code of the fractal*; each fractal can be uniquely described by those coefficients.

Let us now define a transformation h on a compact set $S \subset \mathbb{R}^2$ by: $h(S) = \bigcup_{i=1}^{N} w_i(S)$. If all the w_i are contractions, h is also a contraction with the induced Hausdorff metric [2,16]. Then, h has a unique fixed point, $\mathcal{A} = h(\mathcal{A})$. A procedure to generate this fixed point is to start with an initial compact set S_0 and iterate h as: $S_{n+1} = h(S_n) = \bigcup_{i=1}^{N} w_i(S_n)$. It can be proved that $\lim_{n \to \infty} S_n = \mathcal{A}$. Such a fixed point $\mathcal{A}$ is called the *attractor of the IFS*.

An important result is the *collage theorem*, which states that given a 2D image $\mathcal{I} \subset \mathbb{R}^2$, a non-negative real threshold value $\epsilon \geq 0$, and an IFS $\mathcal{W} = \{(\mathbb{R}^2, d_2); ; w_1, \ldots, w_N\}$ with contractivity factor $0 < s < 1$, if $H(\mathcal{I}, h(\mathcal{I})) = H\left(\mathcal{I}, \bigcup_{i=1}^{N} w_i(\mathcal{I})\right) \leq \epsilon$, then $H(\mathcal{I}, \mathcal{A}) \leq \dfrac{1}{1-s} H\left(\mathcal{I}, \bigcup_{i=1}^{N} w_i(\mathcal{I})\right)$ for the Hausdorff metric H. In short, any digital image $\mathcal{I}$ can be approximated by an IFS $\mathcal{W}$.

The *IFS inverse problem* can now be stated as follows: suppose that we are given a digital image $\mathcal{I}$. The goal is to obtain an IFS whose attractor has a graphical representation $\mathcal{I}'$ that approximates $\mathcal{I}$ accurately according to a given metrics Ξ. Following the collage theorem, we can solve the IFS inverse problem by means of the constrained optimization problem:

$$\min_{\{\mathbf{A}_i, \mathbf{b}_i\}_{i=1,\ldots,N}} \left[\Xi \left(\mathcal{I}, \bigcup_{i=1}^{N} w_i(\mathcal{I}) \right) \right] \quad \text{s.t.} \quad s_i = |det(\mathbf{A}_i)| < 1 \qquad (2)$$

for all $i = 1, \ldots, N$. The problem (2) is a constrained continuous optimization problem. It is also multimodal, since there can be several global or local minima of the fitness function. Therefore, we have to solve a difficult multimodal and multivariate constrained continuous optimization problem. The problem is so difficult that it still remains unsolved in the literature. In this paper, we apply the cuckoo search method to solve it for the case of 2D binary fractal images.

3   The Cuckoo Search Algorithm

Cuckoo search (CS) is a powerful metaheuristic algorithm originally proposed by Yang and Deb in 2009 [21] and applied to difficult optimization problems [7,17,20,22]. It is inspired by the behavior of some cuckoo species that lay their eggs in the nests of host birds of other species to escape from the parental investment in raising their offspring. In the CS algorithm, the eggs in the nest are seen as a pool of candidate solutions of an optimization problem while the cuckoo egg represents a new coming solution. The method uses these new (and potentially better) solutions associated with the parasitic cuckoo eggs to replace the current solution associated with the eggs in the nest. This replacement, carried out iteratively, will eventually lead to a very good solution of the problem. In addition, the CS algorithm is also based on three idealized rules [21,22]:

1. Each cuckoo lays one egg at a time, and dumps it in a randomly chosen nest;
2. The best nests with high quality of eggs (solutions) will be carried over to the next generations;
3. The number of available host nests is fixed, and a host can discover an alien egg with a probability $p_a \in [0,1]$. In this case, the host bird can either throw the egg away or abandon the nest and build a new one in a new location. This assumption can be approximated by a fraction p_a of the n nests being replaced by new nests (with new random solutions at new locations).

The basic steps of the CS algorithm are shown in Table 1. It starts with an initial population of n host nests and it is performed iteratively. The initial values of the jth component of the ith nest are given by $x_i^j(0) = rand.(up_i^j - low_i^j) + low_i^j$, where up_i^j and low_i^j are the upper and lower bounds of that jth component, respectively, and $rand$ is a standard uniform random number on the interval $(0,1)$. These boundary conditions are controlled in each iteration step.

Table 1. Cuckoo search algorithm via Lévy flights as originally proposed in [21,22].

---

begin
    Objective function $f(\mathbf{x})$, $\mathbf{x} = (x_1, \ldots, x_D)^T$
    Generate initial population of n host nests $\mathbf{x}_i$ $(i = 1, 2, \ldots, n)$
    while $(t < MaxGeneration)$ or (stop criterion)
        Get a cuckoo (say, i) randomly by Lévy flights
        Evaluate its fitness F_i
        Choose a nest among n (say, j) randomly
        if $(F_i > F_j)$
            Replace j by the new solution
        end
        A fraction (p_a) of worse nests are abandoned and new ones
            are built via Lévy flights
        Keep the best solutions (or nests with quality solutions)
        Rank the solutions and find the current best
    end while
    Postprocess results and visualization
end

For each iteration g, a cuckoo egg i is selected randomly and new solutions $\mathbf{x}_i(g+1)$ are generated through a Lévy flight as: $\mathbf{x}_i(g+1) = \mathbf{x}_i(g) + \alpha \oplus levy(\lambda)$ where $\alpha > 0$ indicates the step size, the symbol $\oplus$ indicates the entry-wise multiplication, and $levy(\lambda)$ is a transition probability modulated by the Lévy distribution as: $levy(\lambda) \sim g^{-\lambda}$, $(1 < \lambda \leq 3)$. The authors suggested to use the Mantegna's algorithm for symmetric distributions to compute the step length ς (see [22] for details). Then, the stepsize ζ is computed as $\zeta = 0.01\,\varsigma\,(\mathbf{x} - \mathbf{x}_{best})$. Finally, $\mathbf{x}$ is modified as: $\mathbf{x} \leftarrow \mathbf{x} + \zeta.\boldsymbol{\Psi}$ where $\boldsymbol{\Psi}$ is a normal random vector of the dimension of the solution $\mathbf{x}$. The CS method then evaluates the fitness of the new solution and compares it with the current one. In case the new solution brings better fitness, it replaces the current one. On the other hand, a fraction of the worse nests (according to the fitness) are abandoned and replaced by new solutions so as to increase the exploration of the search space looking for more promising solutions. The rate of replacement is given by the probability p_a, a tuning parameter of the method. At each iteration step, all solutions are ranked according to their fitness and the best solution so far is stored as $\mathbf{x}_{best}$.

4   The Proposed Method

Our approach applies the CS with Lévy flights described above to solve the IFS inverse problem for 2D binary fractal images. We assume that the images can be approximated by a fixed number M of contractive functions, but the optimal value for M is not computed; it will be part of our future work in the field.

We consider a population of size η, $\{\boldsymbol{\Phi}_i^{(k)}\}_{i=1,\ldots,\eta}$, where $\boldsymbol{\Phi}_i^{(k)}$ is a vector of M contractive functions $\boldsymbol{\Phi}_i^{(k)} = \{\mathbf{w}_{i,j}^{(k)}\}_{j=1,\ldots,M}$, and the superscript $(.)^{(k)}$ is used to indicate the generation. Each contractive function is uniquely determined by its IFS code as: $\mathbf{w}_{i,j}^{(k)} = \left(a_{i,j}^{(k)}, b_{i,j}^{(k)}, \ldots, f_{i,j}^{(k)}\right)$. This population is initialized with uniform random values in the interval $[-1,1]$ for the variables in $\mathbf{A}_{i,j}^{(0)}$ and in the interval $[-\alpha, \alpha]$ for the elements in $\mathbf{b}_{i,j}^{(0)}$, where α is determined according to the size of the bounding box of the input fractal image. Then, we compute the contractive factor $s_{i,j}^{(k)}$ and remove all functions $w_{i,j}^{(k)}$ with $s_{i,j}^{(k)} \geq 1$ to ensure that only contractive functions are included in the population for all generations.

Before applying the cuckoo search, we also need to define a suitable fitness function. The most natural choice is the Hausdorff distance. However, it is computationally expensive and inefficient for this problem. A more advisable option is to use the Hamming distance instead, so we do so in this paper. We encode any binary fractal image as a bitmap image on a grid of pixels for a given resolution driven by a parameter called the mesh size, m_s. Then, we generate its corresponding template matrix with 0s and 1s, where 1 means that the corresponding pixel is drawn and 0 otherwise. Once the reconstructed image is obtained, it is also encoded by a similar procedure. With this strategy, measuring the similarity between the initial and the reconstructed fractal images is transformed into the problem of comparing their associated binary template matrices.

The parameter tuning of swarm intelligence methods is usually troublesome and problem-dependent. Fortunately, CS is specially advantageous, as it depends on only two parameters: population size, n_p, and probability p_a. We carried out several trials for different values of these parameters, and finally set $n_p = 100$ and $p_a = 0.25$. Each run is executed for $n_{iter} = 2500$ iterations, a proper value to reach convergence in all cases without wasting too much time unnecessarily. Our method needs two more parameters: number of contractive functions M and mesh size, m_s. In this work, they are set to 3 and 40, respectively.

5   Experimental Results

The proposed method has been applied to several examples of 2D binary fractal images. We include only two here because of limitations of space: the spiral fractal and the Christmas tree, depicted in Figs. 1 and 2, respectively. Both figures show: the original fractal image in red (top-left); the best reconstructed image in blue (top-right); the combination of both pictures for better visual comparison between them (bottom-left) and the convergence diagram of the error function for the three contractive maps of each IFS. From the figures, it becomes clear that our method captures the underlying structure of the given fractal images with good visual quality. This is a very valuable (even surprising) result taking into account that our initial population is totally random, meaning that the reconstructed images at the initial generations are very far from   the

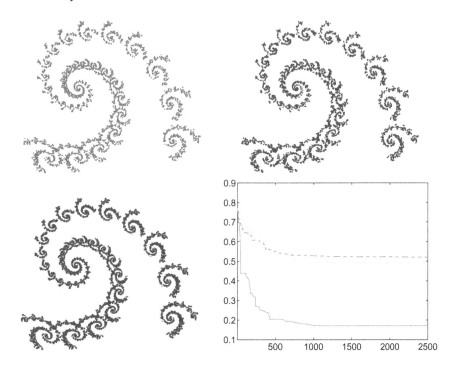

Fig. 1. Application of our method to the spiral fractal: (t-l) original image; (t-r) reconstructed image; (b-l) combination of both images for better visual comparison; (b-r) convergence diagram of the three contractive functions. (Color figure online)

target image. Even in this case, our method is able to select the best contractive functions in each iteration and improve them over the generations until reaching a final image that matches the source image pretty well. Note however, that the matching is not really optimal yet. This fact becomes more evident from our numerical results in Table 2 (expressed on a per unit basis). These results indicate that our method is good enough to replicate the general shape of the image, but there is still plenty of potential for further improvement.

Table 2. Matching error for the contractive functions of images in Figs. 1 and 2.

Example	Matching error of w_1	Matching error of w_2	Matching error of w_3
Spiral fractal	0.171484	0.490783	0.520755
Christmas tree	0.234398	0.244701	0.228202

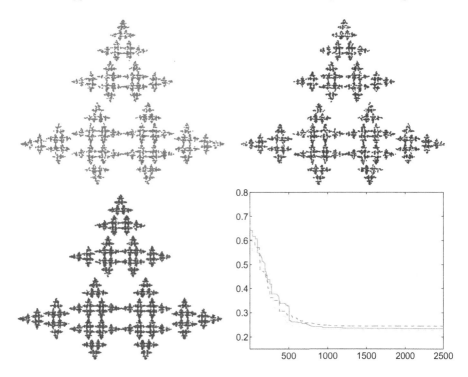

Fig. 2. Application of our method to the Christmas tree fractal: (t-l) original image; (t-r) reconstructed image; (b-l) combination of both images for better visual comparison; (b-r) convergence diagram of the three contractive functions. (Color figure online)

6   Conclusions and Future Work

This paper introduces a new method to solve the IFS inverse problem for 2D binary fractal images. Given a fractal image, the goal is to determine the IFS code of an IFS fractal whose attractor approximates the input image accurately. The proposed method is based on the cuckoo search algorithm, a powerful swarm intelligence method for continuous optimization. The method has been applied to various examples of 2D binary fractal images with satisfactory results. The method is able to recover the underlying shape of the input image with good visual quality and acceptable numerical accuracy. However, the method is not optimal yet and hence, it can be improved in several ways. We plan to hybridize the method with a local search strategy for further improvement. We also plan to analyze the different parameters of the method in order to derive optimal values for parameter tuning, and to apply other error metrics for better performance. Finally, we plan to extend this work to the case of general digital images.

Acknowledgements. This research has been kindly supported by the Computer Science National Program of the Spanish Ministry of Economy and Competitiveness, Project Ref. #TIN2012-30768, Toho University, and the University of Cantabria.

References

1. Abiko, T., Kawamata, M.: IFS coding of non-homogeneous fractal images using Gröbner basis. In: Proceedings of the IEEE International Conference on Image Processing, pp. 25–29 (1999)
2. Barnsley, M.F.: Fractals Everywhere, 2nd edn. Academic Press, San Diego (1993)
3. Berkner, K.: A wavelet-based solution to the inverse problem for fractal interpolation functions. In: Lévy Véhel, J., Lutton, E., Tricot, C. (eds.) Fractals in Engineering, pp. 81–92. Springer, London (1997). doi:10.1007/978-1-4471-0995-2_7
4. Barnsley, M.F., Hurd, L.P.: Fractal Image Compression. AK Peters, Wellesley (1993)
5. Falconer, K.: Fractal Geometry: Mathematical Foundations and Applications, 2nd edn. Wiley, Chichester (2003)
6. Gálvez, A.: IFS Matlab generator: a computer tool for displaying IFS fractals. In: Proceedings ICCSA 2009, pp. 132–142. IEEE CS Press, Los Alamitos (2009)
7. Gálvez, A., Iglesias, A.: Cuckoo search with Lévy flights for weighted Bayesian energy functional optimization in global-support curve data fitting. Sci. World J. **2014**, 11 (2014). Article ID 138760
8. Gálvez, A., Iglesias, A., Takato, S.: Matlab-based KETpic add-on for generating and rendering IFS fractals. CCIS **56**, 334–341 (2009)
9. Gálvez, A., Iglesias, A., Takato, S.: KETpic Matlab binding for efficient handling of fractal images. Int. J. Future Gener. Commun. Netw. **3**(2), 1–14 (2010)
10. Gálvez, A., Kitahara, K., Kaneko, M.: *IFSGen4*: interactive graphical user interface for generation and visualization of iterated function systems in LaTeX. In: Hong, H., Yap, C. (eds.) ICMS 2014. LNCS, vol. 8592, pp. 554–561. Springer, Heidelberg (2014). doi:10.1007/978-3-662-44199-2_84
11. Goentzel, B.: Fractal image compression with the genetic algorithm. Complex. Int. **1**, 111–126 (1994)
12. Gutiérrez, J.M., Iglesias, A.: A Mathematica package for the analysis and control of chaos in nonlinear systems. Comput. Phys. **12**(6), 608–619 (1998)
13. Gutiérrez, J.M., Iglesias, A., Rodríguez, M.A.: A multifractal analysis of IFSP invariant measures with application to fractal image generation. Fractals **4**(1), 17–27 (1996)
14. Gutiérrez, J.M., Iglesias, A., Rodríguez, M.A., Burgos, J.D., Moreno, P.A.: Analyzing the multifractal structure of DNA nucleotide sequences. Chaos Noise Biol. Med. **7**, 315–319 (1998). World Scientific, Singapore
15. Gutiérrez, J.M., Iglesias, A., Rodríguez, M.A., Rodríguez, V.J.: Generating and rendering fractal images. Math. J. **7**(1), 6–13 (1997)
16. Hutchinson, J.E.: Fractals and self similarity. Indiana Univ. Math. J. **30**(5), 713–747 (1981)
17. Iglesias, A., Gálvez, A.: Cuckoo search with Lévy flights for reconstruction of outline curves of computer fonts with rational Bézier curves. In: Proceedings of Congress on Evolutionary Computation-CEC 2016. IEEE CS Press, Los Alamitos (2016)
18. Nettleton, D.J., Garigliano, R.: Evolutionary algorithms and a fractal inverse problem. Biosystems **33**, 221–231 (1994)
19. Vyrscay, E.R.: Moment and collage methods for the inverse problem of fractal construction with iterated function systems. In: Peitgen, H.O., et al. (eds.) Fractals in the Fundamental and Applied Sciences. Elsevier, Amsterdam (1991)

20. Yang, X.-S.: Nature-Inspired Metaheuristic Algorithms, 2nd edn. Luniver Press, Frome (2010)
21. Yang, X.S., Deb, S.: Cuckoo search via Lévy flights. In: Proceedings World Congress on Nature & Biologically Inspired Computing (NaBIC), pp. 210–214. IEEE Press, New York (2009)
22. Yang, X.S., Deb, S.: Engineering optimization by cuckoo search. Int. J. Math. Model. Numer. Optim. **1**(4), 330–343 (2010)

Solving the Graph Coloring Problem Using Cuckoo Search

Claus Aranha[1(✉)], Keita Toda[2], and Hitoshi Kanoh[1]

[1] Faculty of Engineering, Information and Systems, University of Tsukuba,
Tsukuba, Japan
{caranha,kanoh}@cs.tsukuba.ac.jp
[2] Graduate School of Systems and Information, University of Tsukuba,
Tsukuba, Japan

Abstract. We adapt the Cuckoo Search (CS) algorithm for solving the three color Graph Coloring Problem (3-GCP). The difficulty of this task is adapting CS from a continuous to a discrete domain. Previous researches used sigmoid functions to discretize the Lévi Flight (LF) operator characteristic of CS, but this approach does not take into account the concept of *Solution Distance*, one of the main characteristics of LF. In this paper, we propose a new discretization of CS that maintains LF's solution distance concept. We also simplify CS's parasitism operator, reducing the number of evaluations necessary. We compare different combinations of the proposed changes, using GA as a baseline, on a set of randomly generated 3-GCP problems. The results show the importance of a good discretization of the LF operator to increase the success rate and provide auto-adaptation to the CS algorithm.

Keywords: Graph Coloring Problem · Cuckoo algorithm · Lévy flight

1 Introduction

Cuckoo Search (CS) is an swarm-based optimization meta heuristic developed by Yang and Deb [1], inspired by the breeding behaviors of cuckoo birds. One of its main characteristics is the use of the *Lévy flight distribution* as a variation operator. The CS showed excellent results solving optimization problems in the continuous domain. Recently there are efforts to adapt this approach to discrete value domains as well, such as combinatory optimization and constraint satisfaction problems. However, there is not yet an accepted general approach for the application of CS to optimization problems on the discrete domain.

In this context, we are interested in solving the Graph Coloring Problem (GCP) using CS. Former works such as Yongquan et al. [2] and Djelloul et al. [3] used a Binary representation of the Lévy Distribution that does not take into account the concept of solution distance, which we believe is a key characteristic for the performance of this algorithm. Aoki et al. introduced a formulation of PSO for the GCP which uses the Hamming distance to calculate the distance between solutions [4]. While we consider this to be a better modeling of

© Springer International Publishing AG 2017
Y. Tan et al. (Eds.): ICSI 2017, Part I, LNCS 10385, pp. 552–560, 2017.
DOI: 10.1007/978-3-319-61824-1_60

discrete distance, Aoki's model of Hamming distance is focused on the comparison between two solutions, and not appropriate to calculating distances from a single solution, making it difficult to apply it directly to CS.

In this work we propose a new discretization model of the Lévy flight distribution, allowing us to use CS for the 3-GCP problem. Additionally, we propose adjustments to the parasitism operator to obtain even better results in this problem. We compare the effectiveness of the proposed operators with standard formulations, and include GA as a baseline. Our experiment results show that the proposed methods are an effective way to discretize CS for the 3-GCP.

2 Background

2.1 Lévy Flight Distribution

Various studies have showed that the flight and feeding behavior of animals have the characteristics of the Lévy distribution [5–7]. A *Lévy Flight* (LF) is a random walk which uses the Lévy distribution. Compared with the standard random walk, the LF shows occasional long steps among the many short ones, as illustrated in Fig. 1. It has been shown that this behavior allows the Lévy flight to be more effective than the Random Walk when used in a variety of search algorithms [8,9].

Fig. 1. Lévy flight (left) versus random walk (right) on a 2-D space.

2.2 Cuckoo Search Algorithm

The Cuckoo Search (CS) algorithm is a meta-heuristic search algorithm for optimization problems in the continuous domain. In comparison with other meta-heuristic search algorithms such as GA or PSO, it has fewer control parameters which are simpler to fine-tune [10]. The CS algorithm can be described as follows. Initially, a random population is generated. Then, at each generation, up to two replacement candidates are generated for each individual x_i:

1. A modified individual u_i^1 is generated using the *Lévy Flight* operator;
2. With probability $p_a \in \{0, 1\}$, a second modified individual u_i^2 is generated;

The best among x_i, u_i^1 and u_i^2 is added to the following generation.

The *Lévy Flight* operator generates u_i^1 from x_i as follows [11,12]:

$$u_i = x_i^{(t)} + \alpha \times L(\beta), \text{ where } L(\beta) = \frac{p}{|q|^{1/\beta}}, (0.3 \leq \beta \leq 1.99). \tag{1}$$

In Eq. 1, α and β are problem-dependent constants. p and q are random variables sampled from the following normal distributions:

$$p \sim N(0, \sigma_p^2), q \sim N(0, \sigma_q^2), \text{ where } \sigma_p = \left\{ \frac{\Gamma(1+\beta)\sin(\pi\beta/2)}{\Gamma((1+\beta)/2)\beta 2^{(\beta-1)/2}} \right\}^{1/\beta}, \sigma_q = 1$$

The second replacement candidate is generated on the *Parasitism Operator* step. Random generation [1] and the Lévy flight operator [10] have both been suggested for this purpose.

2.3   Graph Coloring Problem

Given a simple, adirected graph $G = \{V, E\}$, the Graph Coloring Problem (GCP) is the problem of assigning a label (color) to every vertex $v_i \in V$ so that no two vertices which share an Edge have the same label. In the n-GCP, the set of labels is fixed with exactly n elements, while in the min-GCP, it is necessary to find the smallest label set for a given graph. In this work, we focus on the 3-GCP problem. The 3-GCP is a NP-complete problem, and often used as a benchmark to evaluate constraint satisfaction algorithms.

2.4   Related Research

Zhou et al. applied CS to the 4-GCP [2]. Instead of the Lévy flight, they used a binary expression obtained by a sigmoid function, which does not model the concept of solution distance. They apply some hybridizations. They compare this algorithm with PSO, MPSO and MTPSO, obtaining improvements against the first two, but not the third. Also, the performance of their discretization of CS without the hybridization was quite low.

Djelloul et al. applied CS to the min-GCP [3]. They also used a binary expression obtained by a sigmoid function, as in the above work. However, their solution encoding is based on the encoding for the knapsack problem, so it is not applicable to the n-GCP.

Aoki et al. used PSO for solving the 3-GCP problem [4]. They proposed the use of Hamming Distance to calculate the distance vectors between the PSO candidate solutions. This approach showed better performance than the Binary discretizations of previous works, however, their modeling of the hamming distance can only be applied between two candidate solutions, which makes it hard to use in the Lévy Flight operator.

3   Proposed Method

Our implementation of the CS algorithm for solving the 3-GCP obeys the following structure:

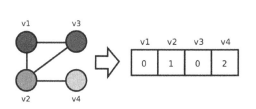

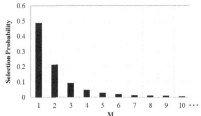

Fig. 2. Encoding of a GCP, each element in the vector represents the color assignment of one vertex in the graph. (Color figure online)

Fig. 3. Distribution of M possible values based on the modified Lévy distribution proposed

1. A candidate solution is represented using the encoding described in Fig. 2. One solution is represented as an array where each element corresponds to one vertex in the graph, and contains the color label for that vertex.
2. The Fitness of an individual is defined as the number of edges that end in two vertices of the same color.
3. At every generation, the candidate solutions are modified by the Lévy Flight operator followed by the Parasitism operator, as described below.

The resulting method is summarized in Algorithm 1.

3.1   Discrete Lévy Flight Operator

The Discrete Lévy Flight operator is used to modify each candidate x_i in the population. First, it determines a value M as follows:

$$M = \lfloor \alpha \times L(\beta) \rfloor + 1. \tag{2}$$

Then, M vertices from x_i are selected, and their values (colors) are randomly changed, with 0.5 probability for each color. The size of M as selected by the Lévi distribution models the solution distance behavior. In most cases, only a few vertices will be modified. But occasionally, a larger number of vertices will me changed at once. Figure 3 illustrates this behavior.

3.2   Modified Parasitism Operator

Originally, the parasitism operator replaces one individual from the population with a modified one with probability p_a, if the modified one has equal or better fitness. In past works, the modified individual can be fully random [1] or a modification of the original using the Uniform distribution [10]. In the 3-GCP problem, there are multiple local optima, and a smaller range of fitness values, when compared with continuous optimization problems. As a result, we observed that the Parasitism operator rarely replaced existing candidates.

To improve the exploration power of the algorithm and reduce wasted evaluations, we change the Parasitism operator so that the new solution is always adopted without comparison with the current one. To compensate for the randomness of introducing new solutions to the population, we also drastically reduce the value of parameter p_a.

Additionally, we perform the modification in three different ways: 1-Modifying a random number of vertices drawn from a uniform distribution; 2-Modifying a random number of vertices drawn from the Lévy distribution; and 3-Modifying a fixed number of vertices, decided by a fine-tuning experiment.

Algorithm 1. Proposed Cuckoo Search Algorithm for 3-GCP

Generate random initial solution set S^0
while Current Evaluations $<$ Max Evaluations **do**
    for Each candidate solution $x_i^{(t)} \in S^{(t)}$ **do** $\triangleright$ Lévy Flight operator
        Select M using the Discrete Lévi distribution
        Generate u_i^1 by randomly changing the color of M vertices in $x_i^{(t)}$
        if Fitness$(u_i^1) \geq$ Fitness$(x_i^{(t)})$ **then** replace $x_i^{(t)}$ with u_i^1
    end for
    for Each candidate solution $x_i^{(t+1)} \in S^{(t+1)}$ **do** $\triangleright$ Parasitism operator
        if random uniform number $k_i \in (0, 1) \leq p_a$ **then**
            Select M using one of {Uniform Distribution, Discrete Lévi, Fixed Value}
            Generate u_i^2 by randomly changing the color of M vertices in $x_i^{(t)}$
            if (*parasitism_comparison* is "No") or (Fitness$(u_i^2) \geq$ Fitness$(x_i^{(t)})$)) **then**
                Replace $x_i^{(t)}$ with u_i^2
            end if
        end if
    end for
end while

---

4 Experiments

We perform a computational experiment to evaluate the contributions of the various proposed changes. In this experiment we compare four algorithms composed of different combinations of the modifications proposed in the previous section, along with a GA to be used as a baseline. The four algorithms and their descriptions are detailed in Table 1.

Each algorithm is tested on a set of random graphs generated based on the formulation by Minton et al. [13]. In our formulation, the vertices in the graphs are divided in three groups, and edges between the groups are randomly added to the graphs until the necessary number is reached. Since the 3-GCP strongly depends on the local structure of the graph, using random instances allows us to avoid local optima. Note that this formulation guarantees that a solution exists for the graph.

Each experiment is generated with a fixed number of vertices, n, and a fixed edge density $d = |E|/|V|$. The edge density can be considered a difficulty index for the 3-GCP problem. Hogg et al. showed that the 3-GCP is most difficult when $2.0 \leq d \leq 2.5$ [14].

Table 1. Algorithm variations compared in the experiment.

Algorithm	CS-R-Yes	CS-R-NO	CS-L-NO	CS-CONST-NO
Parasitism variation operator	Uniform dist.	Uniform dist.	Lévy flight	Constant E
Parasitism comparison	Yes	No	No	No

4.1 Parameter Setting

To determine the parameter values for each algorithm we perform a preliminary experiment to find the optimal value for each parameter independently. This experiment was done on a problem set with $n = 120, d = 2.5$ and 100 random graphs. The parameter value with highest proportion of successes was adopted. The selected values for each parameter is listed on Table 2.

Table 2. Parameter values used for the comparison experiment

Algorithm	CS-R-Yes	CS-R-NO	CS-L-NO	CS-CONST-NO	GA
Population size	200	10	10	10	40
p_a	0.2	0.0001	0.0001	0.001	N/A
β			1.5		N/A
α			1		N/A
E	N/A	N/A	N/A	3	N/A
Mutation probability			N/A		0.011
Tournament size			N/A		2

4.2 Evaluation Experiment

To compare the performance of the proposed methods, we execute them on a series of 3-GCP data sets. The number of nodes in the graphs used are 90, 120, 150 and 180, and the edge density d goes from 1.5 to 9.0, at 0.5 intervals. For each combination of vertex number and edge density, we test each algorithm on a set of 100 random graphs, and report the proportion of solved graphs, and the average number of evaluation functions until a solution was found.

The results of the experiment can be seen on Figs. 4, 5, 6, 7, 8, 9, 10 and 11. From these results we can see that the CS variants normally outperform GA, specially when the number of vertices in the graph is higher.

Among the proposed variants, CS-L-NO and CS-CONST-NO outperform the CS-R variants, specially in terms of number of evaluations until a solution is found. This shows that the use of Levy Flight when compared to the Random mutation does indeed improve the performance of the CS.

At first glance, the similar results between the CS-L-NO and the CS-CONST-NO may seem to imply that there is no benefit in using the Lévy Flight as opposed to a constant mutation number. However, note that the preliminary

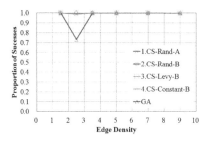

Fig. 4. Proportion of problems solved by each method (n = 90)

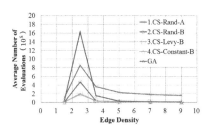

Fig. 5. Average number of evaluations to solve a problem (n = 90)

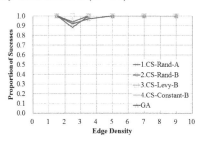

Fig. 6. Proportion of problems solved by each method (n = 120)

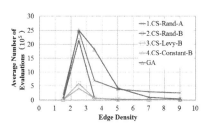

Fig. 7. Average number of evaluations to solve a problem (n = 120)

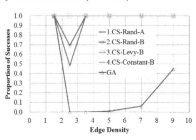

Fig. 8. Proportion of problems solved by each method (n = 150)

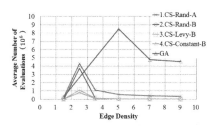

Fig. 9. Average number of evaluations to solve a problem (n = 150)

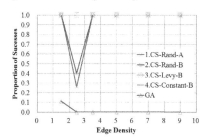

Fig. 10. Proportion of problems solved by each method (n = 180)

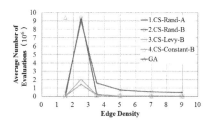

Fig. 11. Average number of evaluations to solve a problem (n = 180)

experiment necessary to find the optimal constant E is a cost that is not necessary for the Lévy Flight (the Lévy Flight showed very little sensitivity to the choices of α and β. This means that the use of the Lévy Flight gives a degree of self-adaptation to the algorithm.

5 Conclusion

In this paper, we proposed an adaptation of the Cuckoo Search (CS) algorithm for the Graph Coloring Problem with Three colors (3-GCP). The CS is characterized by the Lévy Flight mutation, so we paid special attention to the discretization of the Lévy distribution.

We compared the different proposed changes in a computational experiment with a large variety of graph sizes and difficulties, using the GA as a baseline. The results indicate that the proposed discretization of the Lévy flight allows us to use CS for the 3-GCP without having to worry with fine-tuning parameter values or hybridization with local search operators.

Acknowledgments. This work was supported by JSPS KAKENHI grand number 15K00296.

References

1. Yang, X.-S., Deb, S.: Cuckoo search via lévy flights. Nature Biol. Inspired Comput. **37**, 210–214 (2009)
2. Zhou, Y., Zheng, H., Luo, Q., Jinzhao, W.: An improved cuckoo search algorithm for solving planar graph coloring problem. Appl. Math. Inf. Sci. **7**(2), 785–792 (2013)
3. Djelloul, H., Layeb, A., Chikhi, S.: A binary cuckoo search algorithm for graph coloring problem. Int. J. Appl. Evol. Comput. **5**(3), 42–56 (2014)
4. Aoki, T., Aranha, C., Kanoh, H.: PSO algorithm with transition probability based on hamming distance for graph coloring problem. In: IEEE International Conference On Systems, Man, and Cybernetics, pp. 1956–1961 (2015)
5. Brown, C., Liebovitch, L.S., Glendon, R.: Lévy flights in Dobe Ju'hoansi foraging patterns. Hum. Ecol. **35**, 129–138 (2007)
6. Pavlyukevich, I.: Lévy flights, non-local search and simulated annealing. J. Comput. Phys. **226**, 1830–1844 (2007)
7. Reynolds, A.M., Frye, M.A.: Free-flight odor tracking in Drosophila is consistent with an optimal intermittent scale-free search. PLoS ONE **2**, e354 (2007)
8. Viswanathan, G.M., Raposo, E.P., da Luz, M.G.E.: Lévy flights and super diffusion in the context of biological encounters and random searches. Phys. Life Rev. **5**(3), 133–150 (2008)
9. Ali, A.F.: A hybrid gravitational search with lévy flight for global numerical optimization. Inf. Sci. Lett. **4**(2), 71–83 (2015)
10. Yang, X.S., Deb, S.: Engineering optimization by cuckoo search. Int. J. Math. Model. Numer. Optim. **1**, 330–343 (2010)
11. Yang, X.-S.: Nature-Inspired Metaheuristic Algorithms, 2nd edn. Luniver Press, Bristol (2010)

12. Mantegna, R.N.: Fast, accurate algorithm for numerical simulation of Lévy stable stochastic processes. Phys. Rev. E **49**(5), 4677–4683 (1994)
13. Minton, S., Johnston, M.D., Philips, A.B., Laird, P.: Minimizing conflicts: a heuristic repair method for constraint-satisfaction and scheduling problems. Artif. Intell. **58**, 161–205 (1992)
14. Hogg, T., Williams, C.: The hardest constraint problems: a double phase transition. Artif. Intell. **69**, 359–377 (1994)

A Deep Learning-Cuckoo Search Method for Missing Data Estimation in High-Dimensional Datasets

Collins Leke[✉], Alain Richard Ndjiongue, Bhekisipho Twala, and Tshilidzi Marwala

School of Electrical and Electronic Engineering, University of Johannesburg, Johannesburg, South Africa
{collinsl,arrichard,btwala,tmarwala}@uj.ac.za

Abstract. This study brings together two related areas: deep learning and swarm intelligence for missing data estimation in high-dimensional datasets. The growing number of studies in the deep learning area warrants a closer look at its possible application in the aforementioned domain. Missing data being an unavoidable scenario in present day datasets results in different challenges which are nontrivial for existing techniques which constitute narrow artificial intelligence architectures and computational intelligence methods. This can be attributed to the large number of samples and high number of features. In this paper, we propose a new framework for the imputation procedure that uses a deep learning method with a swarm intelligence algorithm, called Deep Learning-Cuckoo Search (DL-CS). This technique is compared to similar approaches and other existing methods. The time required to obtain accurate estimates for the missing data entries surpasses that of existing methods, but this is considered a worthy bargain when the accuracy of the said estimates in a high dimensional setting are taken into consideration.

Keywords: Missing data · Deep learning · Swarm intelligence · High-dimensional data · Supervised learning · Unsupervised learning

1 Introduction

Datasets nowadays such as those that record production, manufacturing and medical data may suffer from the problem of missing data at different phases of the data collection and storage processes. Faults in measuring instruments or data transmission lines are predominant causes of missing data. The occurrence of missing data results in difficulties in decision making and analysis tasks which rely on access to complete and accurate data, resulting in data estimation techniques which are not only accurate, but also efficient. Several methods exist as a means to alleviate the problems presented by missing data ranging from deleting records with missing attributes (list-wise and pair-wise data deletion)

© Springer International Publishing AG 2017
Y. Tan et al. (Eds.): ICSI 2017, Part I, LNCS 10385, pp. 561–572, 2017.
DOI: 10.1007/978-3-319-61824-1_61

to approaches that employ statistical and artificial intelligence methods such as those presented in the next paragraph. The problem though is, some of the statistical and naive approaches more often than not produce biased approximations, or they make false assumptions about the data and correlations within the data. These have an adverse effect on the decision making processes which are data dependent.

The applications of missing data estimation techniques are vast and with that said, the existing methods depend on the nature of the data, the pattern of missingness and are predominantly implemented on low dimensional datasets. The authors in [1] implemented a joint neural network-genetic algorithm framework to impute missing values in one dimension while the authors in [2] used a combination of particle swarm optimization, simulated annealing and genetic algorithm with a multi-layer perceptron (MLP) autoencoder network yielding good results in one dimension. Authors in [3] used MLP autoencoder networks, principal component analysis and support vector machines in combination with the genetic algorithm to impute missing variables. In papers such as [4–8], new techniques to impute missing data and comparisons between these and existing methods are observed. Recently in [9], a technique involving a deep network framework and a swarm intelligence algorithm was proposed to handle missing data in high dimensional datasets, a first of its nature, with promising outcomes. It is as a result of this work that we conducted the research to investigate other high dimensional missing data estimation approaches to observe whether the accuracy obtained can be improved upon.

Investigating new techniques to address the previously mentioned drawbacks led to the implementation of a deep autoencoder, which is by definition an unsupervised learning technique that tries to recall the input space by learning an approximation function which diminishes the disparity between the original data, x, and the reconstructed data, $\tilde{x}$, the same as with Principal Component Analysis (PCA). A deep autoencoder comprises of two symmetrical deep belief networks with a couple of shallow layers representing the encoding section of the network, while the second set of shallow layers represent the decoding section, with each of the individual layers being a restricted Boltzmann machine (RBM). The RBMs are stacked together to form the overall deep autoencoder network which is subsequently fine-tuned using the back-propagation algorithm. These networks are applicable in a variety of sectors, for example, deep autoencoders were used for visuomotor learning in [10] while in [11], they were used with Support Vector Machines (SVMs) to learn sparse features and perform classification tasks. The authors in [12] used an autoencoder for HIV classification, and in [13], these networks were used to map small color images to short binary codes. Autoencoders have several advantages such as being more flexible by introducing nonlinearity in the encoding part of the network contrary to the likes of PCA, which is a key property of the pretraining procedure, they require no prior knowledge of the data properties and correlations, and also they are intuitive. The main drawback of this network is its need for a significant amount of samples for training and learning, which are not always readily available.

In combination with a deep autoencoder network, the Cuckoo Search (CS) algorithm is used. It is a population based stochastic optimization technique inspired by the brood parasitism of cuckoo birds [14]. The algorithm is further enhanced by the use of Lévy flights to introduce more randomness in the motion of the birds. It has been used to minimize the generation and emission costs of a microgrid while satisfying system hourly demands and constraints [15], while in [16], it was used to establish the parameters of chaotic systems via an improved cuckoo search algorithm. The authors in [17] presented a new hybrid algorithm comprised of the cuckoo search algorithm and the nelder-mead method with the aim being to solve the integer and minimax optimization problems. This algorithm is selected due to its merits as observed in the literature, as well as its wide range of applications as mentioned above.

In this paper, a deep autoencoder network is trained and in combination with the CS algorithm (DL-CS), its performance is compared to that of an ordinary Multilayer Perceptron (MLP) autoencoder and other schemes. A description of the methodology is described followed by a presentation of the model performance and analyses. Subsequently, results and the findings from the experiments are presented, followed by concluding remarks.

2   Proposed Approach

In this section, information on the proposed approach is presented which uses a combination of a deep learning regression model, a deep autoencoder network, and an optimization technique, the CS algorithm, to approximate the missing data. Figure 1 illustrates how the regression model and optimization method will be used.

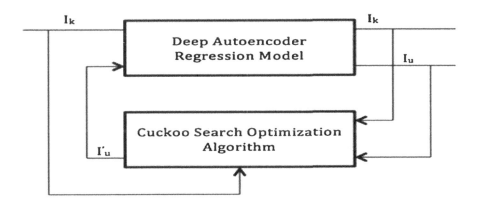

Fig. 1. Data imputation configuration.

Two predominant features of an autoencoder being; (i) its autoassociative nature, and (ii) the butterfly-like structure of the network resulting from a bottleneck trait in the hidden layers, were the reasons behind the network being used.

Autoencoders are also ideal courtesy of their ability to replicate the input data by learning certain linear and non-linear correlations and covariances present in the input space, by projecting the input data into lower dimensions. The only condition required is that the hidden layer(s) have fewer nodes than the input layer, though it is dependent on the application. Prior to optimizing the regression model parameters, it is necessary to identify the network structure where the structure depends on the number of layers, the number of hidden units per hidden layer, the activation functions used, and the number of input and output variables. After this, the parameters can then be approximated using the training set of data. The parameter approximation procedure was done for a given number of training cycles, with the optimal number of this being obtained by analysing the validation error. The aim of this was to avoid overfitting the network and also to use the fastest training approach without compromising on accuracy. The optimal number of training cycles was found to be 500. The training procedure estimated weight parameters such that the network output was as close as possible to the target output.

The CS algorithm was used to estimate the missing values by optimizing an objective function which has as part of it the trained network. It used values from the population as part of the input to the network, and the network recalled these values which subsequently form part of the output. The complete data matrix containing the estimated values and observed values was fed into the autoencoder as input. Some inputs were considered as being known, with others unknown and to be estimated using the regression method and the CS algorithm as described at the beginning of the paragraph. The symbols I_k and I_u as used in Figs. 1 and 2 represent the known and unknown/missing values, respectively.

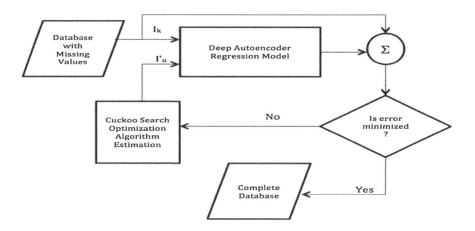

Fig. 2. Missing data estimator structure.

Considering that the approach made use of a deep autoencoder, it was imperative that the autoencoder architecture match the output to the input. This trait is expected when a dataset with familiar correlations recorded in the network is used. The error used is the disparity between the target output and the network output, expressed as:

$$\delta = \overrightarrow{I} - f(\overrightarrow{W}, \overrightarrow{I}), \tag{1}$$

where $\overrightarrow{I}$ and $\overrightarrow{W}$ represent the inputs and the weights, respectively.

The square of Eq. (1) was used to always guarantee that the error is positive. This results in the following equation:

$$\delta = \left(\overrightarrow{I} - f(\overrightarrow{W}, \overrightarrow{I})\right)^2. \tag{2}$$

Courtesy of the fact that the input and output vectors contain both I_k and I_u, the error function is rewritten as:

$$\delta = \left(\begin{bmatrix} I_k \\ I_u \end{bmatrix} - f\left(\left\{\begin{matrix} I_k \\ I_u \end{matrix}\right\}, w\right)\right)^2. \tag{3}$$

Equation (3) is the objective function used and minimized by the CS algorithm in order to estimate I_u, with f being the regression model function. The stopping criteria of the CS algorithm were either a maximum of 40,000 function evaluations being attained, or no change observed in the objective/error value during the estimation procedure. From the above descriptions of how the deep autoencoder and CS algorithm were used, the equation below summarizes the function of the proposed approach, with f_{CS} being the CS algorithm estimation operation and f_{DAE} being the function of the deep autoencoder.

$$y = f_{DAE}(W, f_{CS}(\overrightarrow{I})), \tag{4}$$

where $\overrightarrow{I} = \begin{bmatrix} \overrightarrow{I_k} \\ \overrightarrow{I_u} \end{bmatrix}$ represent the input space of known and unknown features.

3   Performance Analysis and Results

To investigate and validate the proposed model against other existing approaches, the Mixed National Institute of Standards and Technology (MNIST) dataset of handwritten digits was used. It consists of 60,000 training samples and 10,000 test samples, each with 784 features representing the pixel values of the image. The black pixels are represented by 0, while the white pixels have values of 255, as such, all the variables could be considered as being continuous. The data in its current form has been cleaned of outliers, therefore, each image used in the dataset should contribute in generalizing the system. The data are normalized to being in the range $[0, 1]$, and randomized to improve the network performance. The training set of data was further split into training and validation sets (50,000 and 10,000, respectively).

The neural network architecture was optimized using the validation set of data with the correctness in performance of the proposed system being tested using the test data. It is worth mentioning that the objective of this research is not to propose a state-of-the-art classification model which rivals existing methods, but rather to propose an approach to reconstruct an image in the event of missing data (missing pixels). The individual network layers were pre-trained using RBMs and the complete training set of data with no missing values, to initialize the weights and biases in a good solution space. These are stacked together and trained in a supervised learning approach using the stochastic gradient descent (SGD) algorithm. The optimized deep network architecture used is 784-1000-500-250-30-250-500-1000-784 as suggested initially by [18] and further verified by performing cross-validations using the validation set of data with complete records, with 784 input and output nodes, and seven hidden layers with 1000, 500, 250, 30, 250, 500 and 1000 nodes, respectively. The objective function value (mean squared error) obtained after training was 1.96% which was close to the lowest value obtained during cross-validation (2.12%). The Multilayer Perceptron (MLP) autoencoder used for comparison purposes has the same number of nodes in the input and output layer as the deep autoencoder, with a single hidden layer consisting of 400 nodes which was determined by varying the number of nodes in the hidden layer, and determining which architecture produces the lowest network error. It was experimentally observed that the 784-400-784 mlp autoencoder network architecture yielded a network error value of 2.5088%. The sigmoid and linear activation functions were used for the hidden and output layers of the mlp network, respectively. The sigmoid activation function was used in each layer of the deep autoencoder except for the bottle-neck layer in which the linear activation function was used. The training was done using the scaled conjugate gradient (SCG) algorithm for the mlp and the stochastic gradient descent (SGD) algorithm for the deep autoencoder network, for 500 epochs. Next, values were removed from the entire test set respecting the missing at random (MAR) and missing completely at random (MCAR) mechanisms, as well as the arbitrary missing data pattern. This was done by creating a binomial matrix of the same size as the test set $(10,000 \times 784)$ with zeros and ones adhering to the stated mechanisms and pattern, and replacing every occurrence of one with NaN (implying missingness). 100 samples were selected from the modified test set obtained from this procedure, with missing values and were used to test the performance of the missing data estimation scheme, which was then compared against existing methods.

The effectiveness of the proposed approach was determined using the mean squared error (MSE), the root mean squared logarithmic error (RMSLE), the correlation coefficient (r) and the relative prediction accuracy (RPA). Also used were the signal-noise ratio (SNR) and global deviation (GD). The mean squared and root mean squared logarithmic errors as well as the global deviation yield measures of the difference between the actual and predicted values, and provide an indication of the capability of the estimation.

$$MSE = \frac{\Sigma_{i=1}^{n}(I_i - \hat{I_i})^2}{n}, \tag{5}$$

$$RMSLE = \sqrt{\frac{\Sigma_{i=1}^{n}(log(I_i + 1) - log(\hat{I_i} + 1))^2}{n}}, \tag{6}$$

and

$$GD = \left(\frac{\sum_{i=1}^{n} \left(\hat{I}_i - I_i \right)}{n} \right)^2. \tag{7}$$

The correlation coefficient provides a measure of the similarity between the predicted and actual data. The output value of this measure lies in the range $[-1, 1]$ where the absolute value indicates the strength of the link, while the sign indicates the direction of said link. Therefore, a value close to $1(100\%)$ signifies a strong predictive capability while a value close to $-1(-100\%)$ signifies otherwise. In the equation below, '$-$' represents the mean of the data.

$$r = \frac{\sum_{i=1}^{n}(I_i - \bar{I}_i)(\hat{I}_i - \bar{\hat{I}}_i)}{\left[\sum_{i=1}^{n} \left(I_i - \bar{I}_i \right)^2 \sum_{i=1}^{n} \left(\hat{I}_i - \bar{\hat{I}}_i \right)^2 \right]^{1/2}}. \tag{8}$$

The relative prediction accuracy on the otherhand measures the number of estimates made within a specific tolerance, with the tolerance dependent on the sensitivity required by the application. The tolerance was set to 10% as it seemed favorable for the application domain. This measure is given by:

$$A = \frac{n_\tau}{n} * 100. \tag{9}$$

The signal-noise ratio used in this paper is obtained by:

$$SNR = \frac{var(I - \hat{I})}{var(\hat{I})}. \tag{10}$$

In Eqs. (5)–(8), n represents the number of samples, while I and $\hat{I}$ represent the real test set values and estimated missing output values from the modified test set, respectively. In Eq. (9), n_τ represents the number of correctly predicted outputs. Taking into consideration the above mentioned metrics, the performance of the estimation method was evaluated and compared against existing methods (refer to [1] (MLP-GA), [2] (MLP-GA, MLP-SA and MLP-PSO), and [9] where a deep learning-swarm intelligence method was proposed, herein denoted by DL-FA) by estimating the missing attributes concurrently, wherever missing data may be ascertained. This led to scenarios whereby any sample/record could have at least 62, and at most 97 missing attributes (dimensions) to be approximated. It will be seen that the approach proposed in this research outperforms that implemented in [9] which was part of the aim of this work.

Figures 3, 4, 5, 6 and 7 show the performance and comparison of DL-CS with DL-FA, MLP-PSO, MLP-SA and MLP-GA. Before analysing the results, MSE is a deviation that measures the average of the squares of the errors. The optimal performance of an estimator corresponds to a minimum MSE value. Another measure is the RMSLE which measures the difference between samples predicted by the models. Similar to MSE, an estimator gives better performance when the lowest RMSLE value is obtained. In evaluating data mining approaches, we also look at the correlation coefficient. It quantifies the type of dependence between samples. An estimator performs better when the correlation coefficient is higher. As with the correlation coefficient, the RPA is also given in percentage and represents the mean absolute deviation with the optimum accuracy corresponding to a 100%. Figures 3 and 4 are bar charts that show the MSE and RMSLE values for DL-CS when compared to DL-FA, MLP-PSO, MLP-SA and MLP-GA.

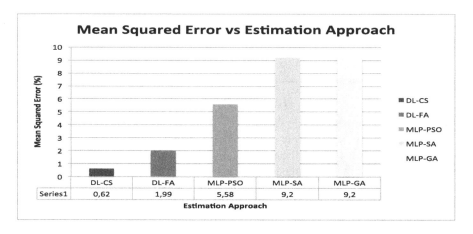

Fig. 3. Mean squared error vs estimation approach.

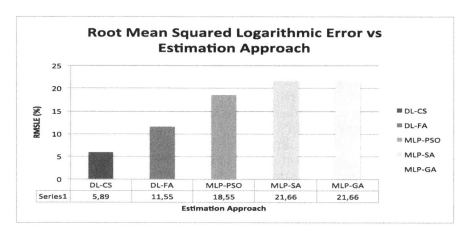

Fig. 4. Root mean squared logarithmic error vs estimation approach.

We recorded 0.62%, 1.99%, 5.58%, 9.2% and 9.2% of MSE and 5.89%, 11.55%, 18.55%, 21.66% and 21.66% of RMSLE for DL-CS, DL-FA, MLP-PSO, MLP-SA and MLP-GA, respectively. Both DL-CS and DL-FA yielded low MSE when compared to the others due to the fact that their algorithms are based on the same principle. The DL-CS method showed better performance over DL-FA. These results are validated by the correlation coefficient whose bar chart is given in Fig. 5.

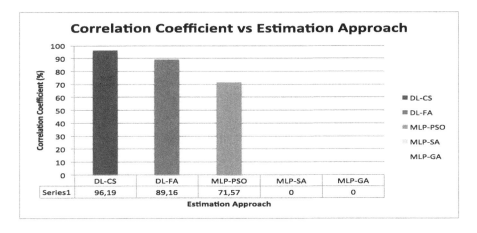

Fig. 5. Correlation coefficient vs estimation approach.

DL-CS, DL-FA and MLP-PSO yielded 96.19%, 89.16% and 71.57% correlation values, respectively, while MLP-SA and MLP-GA showed no correlation. MLP-SA and MLP-GA yielded 79% of RPA while DL-FA and MLP-PSO respectively yielded 60.83% and 56.33%, as shown in Fig. 6. However, DL-CS remained the approach that outperformed all the others as it exhibited an 87% of RPA.

Lastly, we consider the execution time. MLP-PSO, MLP-SA and MLP-GA have always been outperformed by DL-CS and DL-FA. This was justified by

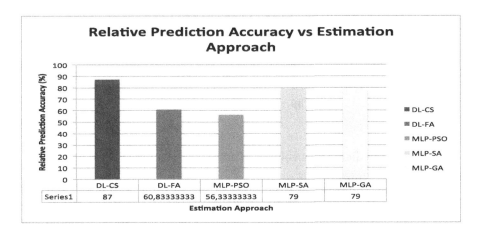

Fig. 6. Relative prediction accuracy vs estimation approach.

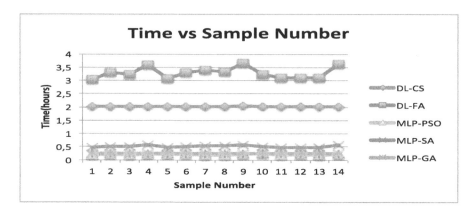

Fig. 7. Time vs sample number.

their execution times which were relatively short when compared to DL-CS and DL-FA as shown in Fig. 7. Nevertheless, this cannot be set as a rule owing to the fact that the DL-CS approach, which presents better characteristics when compared to DL-FA, recovers the same amount of data as DL-FA with less time, better accuracy, better MSE and RMSLE, and higher correlation coefficient.

Table 1. Mean squared error objective value per sample.

Sample	Dimensions	DL-CS	DL-FA	MLP-PSO	MLP-SA	MLP-GA
1	73	**0.71**	2.05	6.02	7.62	7.62
2	75	**0.44**	1.78	5.77	2.97	2.97
3	78	**1.59**	2.86	7.01	12.94	12.94
4	78	**1.03**	2.22	4.75	8.82	8.82
5	84	**1.21**	2.21	4.42	10.16	10.16
6	97	**0.93**	1.75	5.44	6.35	6.35
7	83	**0.69**	2.42	4.69	5.32	5.32
8	75	**2.02**	3.12	8.49	13.63	13.63
9	85	**0.84**	2.62	6.15	9.07	9.07
10	74	**2.39**	3.42	9.91	11.25	11.25

In Table 1, the dimensions column refers to the number of missing values in a sample/record.

Tables 1 and 2 further back the findings from Figs. 3, 4, 5 and 6 showing that the proposed approach yielded the lowest objective function value in the estimation of missing values in each sample, as well as the lowest SNR and GD values.

Table 2. Signal-noise ratio and global deviation values.

	DL-CS	DL-FA	MLP-PSO	MLP-SA	MLP-GA
SNR	**0.0836**	0.2815	0.6035	∞	∞
GD	**1.48E-04**	0.0033	0.0123	0.0118	0.0118

4   Conclusion

This paper investigated the estimation of missing data via a novel approach. The estimation method comprised of a deep autoencoder network to replicate the input data, in combination with the cuckoo search algorithm to estimate the missing data. The performance of the model was investigated and compared against existing methods including an MLP autoencoder with Genetic Algorithm, Simulated Annealing and Particle Swarm Optimization. The proposed method was also compared to a recent technique suggested to estimate missing data in high dimensional datasets which makes use of a deep autoencoder network in combination with the firefly algorithm. The results obtained revealed that the proposed system yielded more accurate estimates not only when compared against the MLP hybrid systems, but also when compared to the recent deep autoencoder-firefly algorithm system. This was made evident when the mean squared error, root mean squared logarithmic error, correlation coefficient, relative prediction accuracy, signal-noise ratio and global deviation were taken into account, with the proposed approach yielding the best values of these. Also, when the objective function value was considered during the estimation process, it was observed that the proposed approach resulted in the lowest values for this for each sample. An obstacle faced was the computational time required to estimate the missing data which can be addressed by parallelizing the estimation procedure to observe whether this approach does speed up the process while maintaining efficiency and accuracy. Generalization of the model could also be achieved by implementing it on other datasets. It is also worth experimenting with different proportions of missing data.

References

1. Abdella, M., Marwala, T.: The use of genetic algorithms and neural networks to approximate missing data in database. In: 3rd International Conference on Computational Cybernetics. (ICCC), pp. 207–212. IEEE (2005)
2. Leke, C., Twala, B., Marwala, T.: Modeling of missing data prediction: computational intelligence and optimization algorithms. In: International Conference on Systems, Man and Cybernetics (SMC), pp. 1400–1404. IEEE (2014)
3. Vukosi, M.N., Nelwamondo, F.V., Marwala, T.: Autoencoder, principal component analysis and support vector regression for data imputation. arXiv preprint arXiv:0709.2506 (2007)
4. Jerez, J.M., Molina, I., García-Laencina, P.J., Alba, E., Ribelles, N., Martín, M., Franco, L.: Missing data imputation using statistical and machine learning methods in a real breast cancer problem. Artif. intell. Med. **50**(2), 105–115 (2010). Elsevier

5. Liew, A.W.-C., Law, N.-F., Yan, H.: Missing value imputation for gene expression data: computational techniques to recover missing data from available information. Brief. Bioinform. **12**(5), 498–513 (2011). Oxford University Press

6. Myers, T.A.: Goodbye, listwise deletion: presenting hot deck imputation as an easy and effective tool for handling missing data. Commun. Methods Meas. **5**(4), 297–310 (2011). Taylor & Francis

7. Schafer, J.L., Graham, J.W.: Missing data: our view of the state of the art. Psychol. Methods **7**(2), 147 (2002). American Psychological Association

8. Van Buuren, S.: Flexible Imputation of Missing Data. CRC Press, Boca Raton (2012)

9. Leke, C., Marwala, T.: Missing data estimation in high-dimensional datasets: a swarm intelligence-deep neural network approach. In: Tan, Y., Shi, Y., Niu, B. (eds.) ICSI 2016. LNCS, vol. 9712, pp. 259–270. Springer, Cham (2016). doi:10. 1007/978-3-319-41000-5_26

10. Finn C., Tan, X., Duan, Y., Darrell, T., Levine, S., Abbeel, P.: Deep spatial autoencoders for visuomotor learning. In: International Conference on Robotics and Automation (ICRA), pp. 512–519 (2016)

11. Ju, Y., Guo, J., Liu, S.: A deep learning method combined sparse autoencoder with SVM. In: 2015 International Conference on Cyber-Enabled Distributed Computing and Knowledge Discovery (CyberC), pp. 257–260, September 2015

12. Brain, L.B., Marwala, T., Tettet, T.: Autoencoder networks for HIV classification. Curr. Sci. **91**(11), 1467–1473 (2006)

13. Krizhevsky, A., Hinton, G.E.: Using very deep autoencoders for content-based image retrieval. In: 19th European Symposium on Artificial Neural Networks (ESANN), Bruges, Belgium, 27–29 April 2011

14. Yang, X.S., Debb, S.: Cuckoo search: recent advances and applications. Neural Comput. Appl. **24**(1), 169–174 (2014)

15. Vasanthakumar, S., Kumarappan, N., Arulraj, R., Vigneysh, T.: Cuckoo search algorithm based environmental economic dispatch of microgrid system with distributed generation. In: International Conference on Smart Technologies and Management for Computing, Communication, Controls, Energy and Materials (ICSTM), pp. 575–580. IEEE (2015)

16. Wang, J., Zhou, B., Zhou, S.: An improved cuckoo search optimization algorithm for the problem of chaotic systems parameter estimation. Comput. Intell. Neurosci. **2016**, 8 (2016)

17. Ali, F.A., Mohamed, A.T.: A hybrid cuckoo search algorithm with Nelder Mead method for solving global optimization problems. SpringerPlus **5**(1), 473 (2016). Springer International Publishing

18. Hinton, G.E., Osindero, S., Teh, Y.W.: A fast learning algorithm for deep belief nets. Neural Comput. **18**(7), 1527–1554 (2006)

Strategies to Improve Cuckoo Search Toward Adapting Randomly Changing Environment

Yuta Umenai[✉], Fumito Uwano, Hiroyuki Sato, and Keiki Takadama

The University of Electro-Communications, Tokyo, Japan
{umenai,uwano}@cas.hc.uec.ac.jp, sato@hc.uec.ac.jp,
keiki@inf.uec.ac.jp

Abstract. Cuckoo Search (CS) is the powerful optimization algorithm and has been researched recently. Cuckoo Search for Dynamic Environment (D-CS) has proposed and tested in dynamic environment with multi-modality and cyclically before. It was clear that has the hold capability and can find the optimal solutions in this environment. Although these experiments only provide the valuable results in this environment, D-CS not fully explored in dynamic environment with other dynamism. We investigate and discuss the find and hold capabilities of D-CS on dynamic environment with randomness. We employed the multi-modal dynamic function with randomness and applied D-CS into this environment. We compared D-CS with CS in terms of getting the better fitness. The experimental result shows the D-CS has the good hold capability on dynamic environment with randomness. Introducing the *Local Solution Comparison* strategy and *Concurrent Solution Generating* strategy help to get the hold and find capabilities on dynamic environment with randomness.

Keywords: Dynamic environment · Cuckoo Search · Swarm intelligence

1 Introduction

Cuckoo Search (CS) [10] is a meta-heuristic algorithm inspired from a breeding behavior of cuckoo. [11] shows that CS is one of the effective optimization algorithm with a strong search capability as well as other modern meta-heuristic algorithms such as Particle Swarm Optimization (PSO) [6], Artificial Bee Colony Algorithm (ABC) [2,4]. CS employs *Lévy flight* as a search strategy, which usually enables CS to search solutions locally and globally with the same mechanism. It is known that *Lévy flight* is more effective search strategy than a random walk or the Gaussian distribution [11]. Lately, CS have been applied on many kind of optimization problems, e.g., the scheduling problem [5,8], the steel frame design problem [3], and the clustering problem [13]. Cuckoo Search for Dynamic Environment (D-CS) proposed by Umenai et al. [9] in 2016 is introduced three modification: (1) *Short-range Searching strategy*; (2) *Local Solution Comparison strategy*; and (3) *Concurrent Solution Generating strategy*, in order to adapt to

© Springer International Publishing AG 2017
Y. Tan et al. (Eds.): ICSI 2017, Part I, LNCS 10385, pp. 573–582, 2017.
DOI: 10.1007/978-3-319-61824-1_62

the dynamic environment which the global optimum can be changeable with passage of time. However the dynamic environment changes too simple to adapt the real world problem, the environment change with the certain rules without randomness. To tackle this issue, we test and discuss the search capability of the D-CS in the dynamic environment with randomness. Concretely, we employ the dynamic function, which is modified problem based on the multimodal function employed in [9] has 3 features: cyclically, multi-modality and randomly (added in this paper). The position of optimum solution moves randomly in this problem. In other words, we cannot predict the next position of optimum solution correctly. In this paper, we study how D-CS can be working well for dynamic environment with randomness. We compare D-CS with CS in terms of getting better fitness in order to verify these modification help to search on this environment. If we can know the effective searching strategy on this environment, we may apply to the other randomly environment and algorithms. Then we analyze the capability of D-CS on dynamic environment with randomness in terms of the distribution of candidate solutions and the combination of employed strategies.

This paper is organized as follows. We briefly explain the mechanism of CS in Sect. 2. We describe the mechanisms of D-CS in Sect. 3. We explain the dynamic optimization problem and the experimental settings in Sect. 4. Section 5 show the experimental results on dynamic environments with randomness and discuss how D-CS adapt dynamic environment and what mechanisms improve to find and hold capabilities. Finally, we conclude this paper with future work in Sect. 6.

2   Cuckoo Search

CS is inspired from a breeding behavior of cuckoo consisting of the following procedures; a new solution is stochastically generated from an current solution with a distribution controlled by *Lévy flight*; the best solution with the highest fitness remains in the population to the next generation; to keep a number of solutions (i.e., the population size) constant, worse solutions are deleted with the deletion probability P_a.

In brief, CS generates a new solution x_{new} based on an current solution $x_{current}$ selected randomly;

$$x_{new} = x_{current} \oplus \alpha \cdot cos(\theta) \cdot Levy(\lambda), \tag{1}$$

where α represents a step size which should be set according to the scale of the problem; θ ($0 < \theta < \pi$) decides the moving direction to the new solution from the current one; $\oplus$ represents the *Hadamard product* which multiples the elements in the same entry in the same size matrixes (note that the $\alpha \cdot cos(\theta) \cdot Levy(\lambda)$ is represented by the same size matrix of the solution $x = (x_1, x_2, x_3, \ldots, x_D)^T$); and $Levy(\lambda)$ is a Lévy distribution. The value of α is recommended to set to $L/100$ for enhancing the search capability [12]. The parameter L is the characteristic scale of the problem of interest. In this paper, L is set to the size of problem as $L = Domain_{max} - Domain_{min}$, where $Domain_{max}$ and $Domain_{min}$

indicate the maximum and minimum value of variables x defined by a problem (see Sect. 4). $Levy(\lambda)$ is represented by Eq. (2);

$$Levy(\lambda) \sim step = t^{-\lambda} \tag{2}$$

where, t is a uniform random value. The Lévy distribution has an infinite variance and an infinite mean. As described in [10], the steps form a random walk process with a power-law step-length distribution with a heavy tail, which triggers an intensive local search.

3   Cuckoo Search for Dynamic Environment

D-CS is modified in terms of the following three mechanisms: (i) *Short-range Searching strategy*; (ii) *Local Solution Comparison strategy*; (iii) *Concurrent Solution Generating strategy*. We describe the detail of these modification and flow of D-CS bellow.

3.1   Short-Range Searching strategy

It is known that adjusting stepsize promote precision of searching [1]. Here, we suppose that the effect as same by shorting step length. The solutions with high fitness value employ this strategy, we expect that good candidate solutions find better solution by searching near the good solution. In this paper, we regard the top 20% solutions in the population as good solutions. The new solutions are generated from the selected solutions by Eq. (3);

$$x_{new} = x_{current} \oplus \alpha \cdot cos(\theta) \cdot rnd/100, (0 \leq \theta \leq \pi) \tag{3}$$

where $rnd \in [0, 1]$ is a random value. Equation (3) is designed so that the new solution is slightly different (moved) from the existing solutions. We use the value "100" in order to promote local search. Hence, the new solutions with Eq. (3) do not employed the Lévy distribution and only search near the location of them. Note that for the not-selected solutions, new solutions are generated with the original mechanism (Eq. (1)).

3.2   Local Solution Comparison Strategy

In CS, the solution are replaced the current solution with new solution when the new solution are generated. CS may delete the better fitness solutions because CS must delete the current solution. CS mechanism decreases a diversify of the solutions because the new solutions generated in the near of the optimum solution In other words, CS attempts to find the better solutions by exploring around current search space but not the whole search space. In contrast, our modification attempts to preserve where the solutions are located by comparing each generated solution with the current ones. This modification attempts not only to preserve the good solutions at the current time since it does not delete a solution with comparison of whole solutions (the inserted new solution is inherited from a specific location of the existing one).

3.3 Concurrent Solution Generating Strategy

In order to more effectively explore the search space near the preserved good solutions, we generate n solutions from all current solution in 1 generation, where n is population size. That is, each current solution is used to generate its own new solution for each generation. Note that CS mechanism generates one new solution based on the randomly-selected current solution for each generation. This modification increases an opportunity of exploring the search space.

3.4 Flow of D-CS

We described the flow of D-CS in detail as follows. n, P_a and G indicate the population size, the deletion probability and iteration respectively.

Step1: Initialize
   The n number of solutions are generated in the search space and their fitness are calculated.

Step2: Solution Generation
   The top of $X\%$ solutions with high fitness value generate their new solutions by Eq. (3), while the other solutions by Eq. (1). After all new solutions are generated, their fitness are calculated.

Step3: Solution Comparison
   Then we compared new solutions with current solutions, we preserve the better solutions. In other words, the current solutions remain their place when new solutions are worse than current solutions.

Step4: Solution Deletion
   All solutions are deleted with P_a. If high fitness value solutions are deleted, generate new solutions with *Short-range Searching strategy*. The others are deleted as same and generate new solutions by using Lévy flight. Then we compare the solutions same as Step3.

Step5: Ranking Solutions
   All solutions are ranked to find the current best solution, and the current best one is carried over to the next generation.

Step6: Judge the Completeness
   G is increased by 1 and return to Step2 when G is less than max iteration.

4   Experiment

Here, we apply D-CS to dynamic environment with randomness. We introduce the modification into the function employed in [7] in order to introduce randomness into this function.

4.1 Problem

We employ the function employed in [7]. In this problem, the optimal solution locates on the circumference of circle decided with parameters.

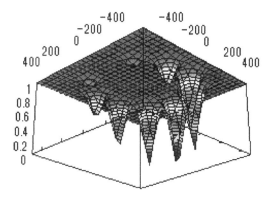

Fig. 1. Function landscape of $f_2(x_1, x_2, k)$

$$f_1(x_1, x_2, k) =$$

$$1 - \sum_{n=0}^{N} \left[\frac{\cos(\beta k + \frac{2n\pi}{N})}{2} \exp\left\{ -\frac{1}{2} \right. \right.$$

$$\left. \left. \left(\frac{(x_1 - r\cos(\frac{2n\pi}{N}))^2}{40^2} + \frac{(x_2 - r\sin(\frac{2n\pi}{N}))^2}{40^2} \right) \right\} \right] \tag{4}$$

Equation (4) shows the definition of this function, where x_1 and x_2 are variables, r decide the radius of the circle, β controls speed of function change, and N indicates the number of negative peaks. When the β is set to large value, the value of function $f_1(x_1, x_2, k)$ changes largely. If N is set 10, the number of local and global optimal solutions is 10. Figure 1 shows the outline of this function when $r = 350$, $N = 10$. This function has the N negative peaks. In this function, the position of optimum solution can be changeable depending on the parameter k. The parameter k set randomly in every generation in order to introduce randomness into this function. In other words, optimum solution locate on the predetermined peaks randomly. If the parameter k increase like as $k = k+1$, the location of negative peaks are moving counterclockwise on circumference of this circle. When β is set 0.01, the position of the global optimum changes cyclically with the period of about 300 generations.

4.2 Experimental Settings

- *Comparison.*
  We compare CS (described in Sect. 2) with D-CS. In order to fairly compare all methods in terms of the generating solution procedure, we slightly modify CS. Specifically, as explained in Sect. 2, CS generates one solution for each generation; while D-CS generates n solutions based on all current n solutions for each generation. Hence, in order to have the same opportunity of the generating solution procedure, we modify CS so that it generates n solutions

for each generation. The difference between CS and D-CS is the selecting solution. In D-CS, all solution are selected and moving in one generation. On the other hands, in CS, the solution may not be selected and moving in one generation (the one solution may be selected twice or more).

• *Evaluation criterion.*
  In this experiment, we evaluate utility of CS and D-CS by Eq. (5). In Eq. (5), $F_{optimum}$ indicates the fitness of the global optimum and F_{best} indicates the fitness of the best solution in the population. When difference between $F_{optimum}$ and F_{best} is lower, the solution is evaluated as better solution.

$$\Delta Fitness = |F_{optimum} - F_{best}| \qquad (5)$$

We experienced both method on this environment (indicates Eq. (4)) in 30 trials each. Here, comparing CS with D-CS in terms of transition of $\Delta Fitness$ through 1seed and average of 30 trials.

• *Parameter settings.*
  For the parameter settings of CS and D-CS, we use the same parameter settings [1] as follows; $\lambda = 1.5$ and $P_a = 0.25$. For the population size n, we use values $n = 15, 50, 100$ in order to investigate the effect of population size on searching performance. In Eq. (4), we use the parameter setting: $Domain_{min} = -800, Domain_{max} = 800, N = 10$. We set the parameter $r = 350$ and $\beta = 0.01\ Generation_{max} = 5000$. Here, $Generation_{max}$ represents the maximum iteration.

5   Results and Discussion

Figures 2 and 3 show the transition of $\Delta Fitness$ from 1 seed and average of 30 trials. The vertical axis indicates the $\Delta Fitness$. The horizontal axis indicates the iteration, where iteration means cycle of algorithm: the parameter G. The gray straight lines indicate CS and the black dotted lines indicate D-CS. From Fig. 2, we can say that D-CS keeps lower $\Delta Fitness$ than CS through iterations in all population size n. CS cannot find and hold the negative peaks since CS cannot get the lower value of $\Delta Fitness$. The lager population size is, the performance of D-CS better. Especially, when population size n is 100, D-CS keeps lower $\Delta Fitness$ than 0.1. Figure 3 shows that D-CS keeps low $\Delta Fitness$ when population size n is 15. Although CS also decline the $\Delta Fitness$ as population size increases, the $\Delta Fitness$ of D-CS better than CS. From these result, D-CS has the hold and find capabilities.

Next, we analyze and discuss the find and hold capabilities of D-CS. First, we investigate why D-CS keeps lower $\Delta Fitness$. D-CS has the better performance to get the optimum solution than CS from Figs. 2 and 3. We analyze this results in terms of location of candidate solutions. Figure 4 shows the location of candidate solutions and negative peaks of CS and D-CS when $k = 5000$ and $n = 15$. In these figure, vertical axis and horizontal axis indicate the variables (x_1, x_2) in (4). The circles indicate the candidate solutions and the diamonds indicate the location of the negative peaks. From Fig. 4a, we can see that candidate

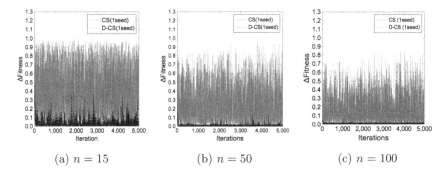

(a) $n = 15$ (b) $n = 50$ (c) $n = 100$

Fig. 2. The transition of $\Delta Fitness$ (1 seed)

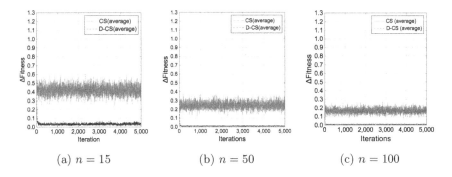

(a) $n = 15$ (b) $n = 50$ (c) $n = 100$

Fig. 3. The transition of $\Delta Fitness$ (Average of 30 trials)

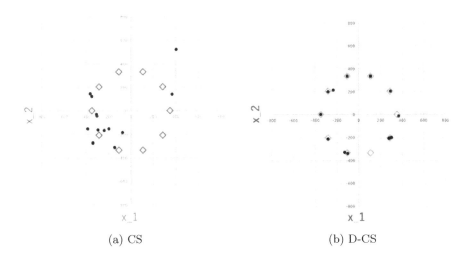

(a) CS   (b) D-CS

Fig. 4. The location of candidate solutions and negative peaks

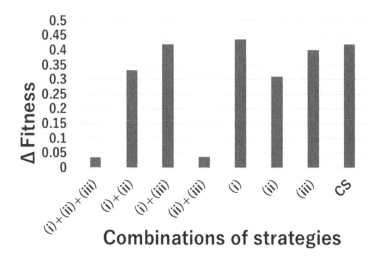

Fig. 5. The average of $\Delta Fitness$ of combinations of strategies

solutions do not locate near the location of optimum. We can see that most of the candidate solutions locate lower left from Fig. 4a. This distribution of the candidate solutions prevent from finding the optimum on dynamic environment with randomness. On the other hand, not only candidate solutions locate the near the optimum solution, D-CS also hold the most of the locations of negative peaks. If the candidate solutions hold the negative peaks, the candidate solutions find the global optima quickly when the global optima move randomly. Thus hold capability is effective for random change (not sequential change).

Then, we investigate what strategy or combination of strategies are effective for this environment. Figure 5 shows the average $\Delta Fitness$ through iterations on each combinations of strategies. The horizontal axis indicates the combinations of strategies, where (i), (ii) and (iii) indicate the *Short-range Searching strategy*, *Local Solution Comparison strategy* and *Concurrent Solution Generating strategy* respectively. The vertical axis indicates the average of $\Delta Fitness$ (average of 30 trials) through iterations. The lower value of average of $\Delta Fitness$ indicates that the combination of strategy contribute to keep the lower $\Delta Fitness$. From Fig. 5, we can see the combination of (ii) and (iii) contribute to keep lower $\Delta Fitness$. Although candidate solutions tend to hold the better solutions, this mechanism prevent from candidate solutions searching globally when *Local Solution Comparison strategy* is introduced, While, *Concurrent Solution Generating strategy* helps to search globally, however, do not have the mechanism to hold the good solutions. Thus, the combination of both strategies mainly contribute to improve the performance of D-CS.

Here, we focus on the difference of combination of (ii) + (iii) and (i) + (ii) + (iii) in Fig. 5. The average of $\Delta Fitness$ of both combinations different slightly: the combination of (i) + (ii) + (iii) is better than the combination of (ii) + (iii). We verified the significant difference of both combinations. Figure 6 shows the aver-

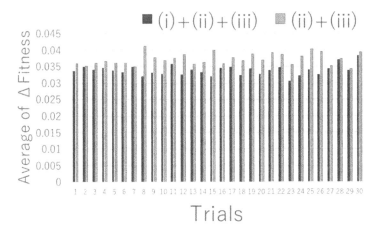

Fig. 6. The average of $\Delta Fitness$ of both combinations on 30 trials

age of $\Delta Fitness$ of both combination on 30 trials. The vertical axis indicate the average of $\Delta Fitness$ and the horizontal axis indicate the each trial. From Fig. 6, shows the performance of the combination of (i) + (ii) + (iii) is better than combination of (ii) + (iii). In addition, we conduct a t-test (one side test), which guaranteed the significant difference between both strategies ($t(29) = 8.139, p < .01$). The candidate solutions are searching with short step when *Short-range Searching strategy* with elite is introduced. This mechanism promote searching near the location of optimum solutions. Hence, the combination of (i) + (ii) + (iii) is the best combination of strategies from these combinations on this environment.

6   Conclusion

In this paper, we investigated the capability of D-CS on dynamic environments with randomness. D-CS is introduced three strategies into CS for dynamic environment. We analyzed the effective strategies toward the randomness as dynamism. We employed the dynamic environment, the optimal solution moves randomly on predetermined locations on the circle in order to verify the find and hold capabilities of D-CS. This environment includes the two dynamism: cyclically and randomness since this environment requires to find and hold the locations of negative peaks. We compared the D-CS with CS on this environment with various population size: $n = 15, 50, 100$ in order to investigate the effect of population size on searching performance. From the experimental results, it is clear that D-CS has find and hold capabilities on dynamic environment with randomness. Moreover, D-CS has these capabilities in case of a small population size ($n = 15$). Needless to say, the performance of D-CS is better with large population size. We carried out t-test to combinations of three strategies. This result suggests the most effective strategy is combinations of *Local Solution Comparison strategy* and *Concurrent Solution Generating strategy*. Introducing these strategies into CS (possibly any other algorithm) can improve the performance on dynamic environment with randomness. Furthermore, adding more *Short-range Searching strategy* improve the performance more.

For future work, we should apply D-CS to different types of dynamic environment where is not cyclically and high dimensional problems in order to explore the capability of this. Furthermore, we should investigate how the strategies work effectively on the dynamisms, and proposed more effective other approach for dynamic environment with previous results.

References

1. Jamil, M., Zepernick, H.J.: 3 levy flights and global optimization. In: Swarm Intelligence and Bio-Inspired Computation: Theory and Applications, p. 49 (2013)
2. Karaboga, D., Basturk, B.: A powerful and efficient algorithm for numerical function optimization: artificial bee colony (ABC) algorithm. J. Global Optim. **39**(3), 459–471 (2007)
3. Kaveh, A., Bakhshpoori, T.: Optimum design of steel frames using cuckoo search algorithm with lévy flights. Struct. Design Tall Spec. Build. **22**(13), 1023–1036 (2013)
4. Ong, P.: Adaptive cuckoo search algorithm for unconstrained optimization. Sci. World J. **2014**, 8 (2014)
5. Ouaarab, A., Ahiod, B., Yang, X.S.: Discrete cuckoo search algorithm for the travelling salesman problem. Neural Comput. Appl. **24**(7–8), 1659–1669 (2014)
6. Poli, R., Kennedy, J., Blackwell, T.: Particle swarm optimization. Swarm Intell. **1**(1), 33–57 (2007)
7. Takano, R., Harada, T., Sato, H., Takadama, K.: Artificial bee colony algorithm based on local information sharing in dynamic environment. In: Handa, H., Ishibuchi, H., Ong, Y.-S., Tan, K.C. (eds.) Proceedings of the 18th Asia Pacific Symposium on Intelligent and Evolutionary Systems. PALO, vol. 1, pp. 627–641. Springer, Cham (2015). doi:10.1007/978-3-319-13359-1_48
8. Tein, L.H., Ramli, R.: Recent advancements of nurse scheduling models and a potential path. In: Proceedings of the 6th IMT-GT Conference on Mathematics, Statistics and its Applications (ICMSA 2010), pp. 395–409 (2010)
9. Umenai, Y., Uwano, F., Tajima, Y., Nakata, M., Sato, H., Takadama, K.: A modified cuckoo search algorithm for dynamic optimization problems. In: 2016 IEEE Congress on Evolutionary Computation (CEC), pp. 1757–1764. IEEE (2016)
10. Xin-She, Y., Suash, D.: Cuckoo search via levy flight. In: World Congress on Nature and Biologically Inspired Computing, NaBIC 2009, pp. 210–214 (2009)
11. Yang, X.S., Deb, S.: Cuckoo search: recent advances and applications. Neural Comput. Appl. **24**(1), 169–174 (2014)
12. Yang, X.S., Karamanoglu, M.: Swarm intelligence and bio-inspired computation: an overview. In: Swarm Intelligence and Bio-inspired Computation-Tehory and Applications, pp. 3–23. Elsevier (2013)
13. Zaw, M.M., Mon, E.E.: Web document clustering using cuckoo search clustering algorithm based on levy flight. Int. J. Innov. Appl. Stud. **4**(1), 182–188 (2013)

Firefly Algorithm

Firefly Algorithm Optimized Particle Filter for Relative Navigation of Non-cooperative Target

Dali Zhang[1($\boxtimes$)], Chao Zhong[2], Changhong Wang[1], Haowei Guan[3], and Hongwei Xia[1($\boxtimes$)]

[1] Harbin Institute of Technology, Harbin 150001, People's Republic of China
15b904016@hit.edu.cn, simonxhw@163.com
[2] Shanghai Institute of Spaceflight Control Technology,
Shanghai 201109, People's Republic of China
zhongchao0327@163.com
[3] Shanghai Institute of Satellite Engineering,
Shanghai 201109, People's Republic of China
davyfeng@sina.com

Abstract. Particle filter (PF) has been proved to be an effective tool in solving relative navigation problems. However, the sample impoverishment problem caused by resampling is the main disadvantage of PF, which strongly affect the accuracy of navigation. To solve this problem, an improved PF based on firefly algorithm (FA) is proposed. Combine with the operation mechanism of PF, the optimization mode of FA is revised, and a new update formula of attractiveness is designed. By means of firefly group's mechanism of survival of the fittest and individual firefly's attraction and movement behaviors, this algorithm enables the particles to move toward the high likelihood region. Thus, the number of meaningful particles can be increased, and the particles can approximate the true state of the target more accurately. Simulation results show that the improved algorithm improves the navigation accuracy and reduces the quantity of the particles required by the prediction of state value.

Keywords: Firefly algorithm · Particle filter · Relative navigation

1 Introduction

Relative navigation of spacecraft is the foundation of autonomous rendezvous, on-orbit servicing and formation flying missions, which affect the accuracy of control and guidance directly [1]. At present, research on relative navigation of cooperative target is continuing to mature, while tracking non-cooperative target like inactive satellites and space debris is the subject of active research [2].

Classical method is based on Extended Kalman Filter (EKF), which is the best solution under the assumption of a linear system and Gaussian noise [3]. Using the feedback information from space based sensor, the true relative state can be estimated precisely. However, the movement of the non-cooperation target

© Springer International Publishing AG 2017
Y. Tan et al. (Eds.): ICSI 2017, Part I, LNCS 10385, pp. 585–592, 2017.
DOI: 10.1007/978-3-319-61824-1_63

cannot be described by a linear model in the actual situation, and the measurement noise usually doesn't satisfy the Gaussian distribution. Given this, the Unscented Kalman Filter (UKF) was present to deal with nonlinear filtering problem, but still lack of adaptability to non-Gaussian noise [4].

Particle filter is considered to be one of the most powerful tools to estimate Bayesain models with non-linear and non-Gaussian noise [5]. Based on the Monte Carlo method, it approximates the posterior density of hidden state with particles, and no assumption on the functional form of the posterior density [6]. However, the standard particle filter using resampling strategy, which lead to the sample impoverishment problem [7].

What is noteworthy is that, more and more researchers have started to use intelligent optimization algorithm to prevent the sample impoverishment problem [8]. Oshman proposed a Genetic Algorithm (GA) embedded quaternion particle filter in attitude estimation from vector observations [9]. Saini improved the particle filter with particle swarm optimization (PSO) for articulated $3D$ human motion tracking [10]. Recent years, many new modern intelligence optimization algorithms are being developed, and the Firefly Algorithm (FA) which is inspired by the idealized behavior of the flashing characteristics of fireflies is one of the best [11]. In this paper, FA is implemented to improve the resampling process of PF by ensuring the effectiveness of the particle sets. Meanwhile, the navigation performance of the proposed algorithm is studied comparatively.

The remainder of this paper is organized as follows: The process and measurement model of relative navigation is presented in Sect. 2. In Sect. 3, the shortcoming of PF is being analysis. Section 4 describes the firefly algorithm and proposes an improved method. Section 5 presents simulation results and subsequently concluded in Sect. 6.

2   Relative Navigation

2.1   Process Model

With reference to Fig. 1(a), suppose there is a chaser vehicle C in a circular orbit of angular velocity n, and T is the non-cooperation target vehicle. Attached to the vehicle center of mass o is a right-handed curvilinear coordinate frame $o - xyz$ with the x axis along the geocentric radius vector direction pointing to the satellite, the y axis pointing at chaser velocity vector, while z axis completing the right-handed frame.

In this paper, the distance between C and T is so small that the difference between local vertical curvilinear (LVC) frame and the local vertical/local horizontal ($LVLH$) can be ignored, less than the sensor noise [12]. Therefore, this paper will not distinguish the two frames. In this frame, the motion of T located at a station (x, y, z), where x, y, and z are much smaller than the target orbit radius, is governed by the Clohessy-Wiltshire equations [13]. The discrete state form can be written as Eq. 1:

$$X(k) = \Phi X(k-1) + \Gamma Q(k) \tag{1}$$

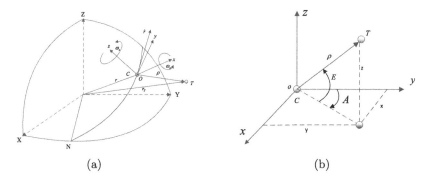

(a)   (b)

Fig. 1. Process and measurement model

where n is the angular velocity of C, Φ is the state transition matrix, $Q(k)$ is the process noise which is assumed to be Gaussian distributed. The concrete form of Φ is:

$$\Phi = \begin{bmatrix} 4 - 3\cos nt & 0 & 0 & \sin nt/n & 2(1 - \cos nt)/n & 0 \\ 6(\sin nt - nt) & 1 & 0 & 2(\cos nt - 1) & 4\sin nt/n - 3t & 0 \\ 0 & 0 & \cos nt & 0 & 0 & \sin nt/nt \\ 3n\sin nt & 0 & 0 & \cos nt & 2\sin nt & 0 \\ 6n(\cos nt - 1) & 0 & 0 & -2\sin nt & 4\cos nt - 3 & 0 \\ 0 & 0 & -n\sin nt & 0 & 0 & \cos nt \end{bmatrix} \quad (2)$$

The corresponding 6×6 discrete process noise matrix Γ can be shown to be

$$\Gamma = \begin{bmatrix} \sigma_x^2(T^3/3) & 0 & 0 & \sigma_x^2(T^2/2) & 0 & 0 \\ 0 & \sigma_y^2(T^3/3) & 0 & 0 & \sigma_y^2(T^2/2) & 0 \\ 0 & 0 & \sigma_z^2(T^3/3) & 0 & 0 & \sigma_z^2(T^2/2) \\ 0 & 0 & 0 & \sigma_x^2 T & 0 & 0 \\ 0 & 0 & 0 & 0 & \sigma_y^2 T & 0 \\ 0 & 0 & 0 & 0 & 0 & \sigma_z^2 T \end{bmatrix} \quad (3)$$

2.2 Measurement Model

As is shown in Fig. 1(b), the Azimuth (A), Elevating (E) angle and relative distance (ρ) between C and T are measured by optical angle measuring camera and laser range finder. The measurement model can be written as Eq. 4:

$$H(k) = \begin{bmatrix} A \\ E \\ \rho \end{bmatrix} = \begin{bmatrix} \arctan(x/y) \\ \arcsin(z/\sqrt{x^2 + y^2 + z^2}) \\ \sqrt{x^2 + y^2 + z^2} \end{bmatrix} \quad (4)$$

where $H(k)$ is the observation matrix at $t = k$. In general, the standard observation equation can be written as:

$$Z(k) = H(k)X(k) + V(k) \quad (5)$$

where $V(k)$ is glint noise. It arises due to the interference between reflections from the target surface and induces angular errors in line-of-sight measurement, particularly for radar seekers [14]. The mixture distribution is represented as:

$$p(x) = (1 - \epsilon)p_G + \epsilon p_L \qquad (6)$$

where ϵ is the glint probability. $p_G \sim N(0, \sigma^2)$ is zero-mean Gaussian distribution, and $p_L \sim Laplace(0, b), b > 0$ is the scaling parameter.

3   Particle Filter Analysis

Particle filter (PF) is based on the Monte-Carlo simulation technique. The key idea of PF is using a cloud of particles to represent the system uncertainty distributions [15]. The steps of standard PF contain importance sampling, weight updating and resampling. $\{X_0^i\}_{i=1}^N$ represents the set of N particles which are sampling based on the prior probability distribution $p(X_0)$ at the beginning. According to the importance sampling density function $q(X_k^i | X_{k-1}^i)$, new particle sets $\{X_{0:k}^i\}_{i=1}^N$ are being established at time k. The particle weight updating formula can be calculated as Eq. 7 [15]:

$$\omega_k^i = \omega_{k-1}^i \frac{p(Y_k | X_k^i) p(X_k^i | X_{k-1}^i)}{q(X_k^i | X_{k-1}^i)} \qquad (7)$$

After resampling, the quantity of meaningful particles was reduced owing to the high weighted particles' over-replication, which leads to the information capacity of new particle set seriously reduced. Although resampling can eliminate the smaller weighted particles' impact, it introduces a new negative issue, which is known as sample impoverishment.

4   Firefly Algorithm Optimized Particle Filter

4.1   Firefly Algorithm

The Firefly Algorithm (FA) mimics the idealized behaviour of the flashing characteristics of fireflies: points in the searching space are abstract as fireflies individual. For simplicity, the attract and moving process is simulated by the optimization process, and the optimization object is abstract to the merits of the individual position. These flashing characteristics can be idealized by following three rules:

(1) All fireflies are unisex so that one firefly will be attracted to other fireflies regardless of their sex. (2) Attractiveness is proportional to their brightness. The attractiveness is proportional to the brightness and they both decrease as their distance increases. If there is no brighter one than a particular firefly, it will move randomly. (3) The brightness or light intensity of a firefly is affected or determined by the land-scape of the objective function.

There are two important issues in FA: the variation of light intensity I and the attractiveness β. In general, the brightness I at a particular location x can be chosen as $I(x) \propto f(x)$, and β is vary with the distance r_{ij} between firefly i and j. In the implementation, the actual form of attractiveness function can be any monotonically decreasing functions. As it is often faster to calculate $1/(1 + r^2)$ than an exponential function, the attractiveness can be defined as

$$\beta = \beta_0/(1 + \gamma r_{ij}^2) \tag{8}$$

where β_0 is the original attractiveness, γ is the light absorption coefficient. For many cases, the parameter $\beta_0 = 1$, and $\gamma = 1$.

The movement of firefly i attracted to another more attractive firefly j is determined by

$$X_i = X_i + \beta(X_j - X_i) + \alpha\epsilon_i \tag{9}$$

where the third term is randomization with α being the randomization parameter, and ϵ_i is a vector of random numbers drawn form a uniform distribution [16].

4.2   Improved Particle Filter Based on FA

This paper adopts the FA algorithm to mitigate the particle impoverishment. It is worth noting that FA algorithm have a lot in common with PF. The former seeking the optimal value by updating the light intensity and position of firefly, while the latter approximating the true posterior probability density of the sample by updating the weights and location of particles.

The key idea of the improvement is using the attractiveness and position updating formula of FA to simulate the movement process of particles with different weights. Schematically, improved Particle Filter based on Firefly Algorithm (FA-PF) can be summarised as the pseudo code.

5   Experiments and Discussions

In this section, the simulations based on the relative navigation scenario are presented to illustrate the effectiveness of the proposed improving estimation method. The chaser is assumed to be flying at a 780 km altitude. Tracking filter starts at $r = [-510, 0, 1], v = [0, 0, 0]$, while the real initial value is $r = [-520, 10, 20], v = [-0.0324, -0.02, 0.02]$. According to a practical system, the ranging accuracy is set to 0.6 m (3σ) and the standard deviation of angle measurement is 0.3° (3σ). The angle measurements are corrupted by glint noise. With the glint probability $\epsilon = 0.05$, the standard deviation of ranging and angle measurement is 15 m and 0.2°, respectively. To demonstrate the ability of FA-based particle filter algorithm in relative navigation, the results are compared with EKF, UKF and PF of 1000 s with 200 particles.

Table 1 shows the Root-Mean-Square Error (RMSE) and simulation time of four kinds of algorithms. Under the effect of glint noise, PF and FA-PF show superior performance compared with EKF and UKF. The FA-PF has the highest precision and it reduced by approximately 39% compared with PF. In addition, the improved algorithm needs less computing time.

Initialization $x_k^i \sim q(x_k^i | x_{k-1}^i, y_k) = p(x_k^i | x_{k-1}^i)$
for $i = 1$ to N **do**   ▷ calculate the state and observe value
    $X(k) = \Phi X(k-1) + Q(k)$
    $Z = H(x) + V(k)$
    $I_i = \omega_k^i = p(Z_k | X_k^i)$
end for
for $i = 1$ all $n$fireflies **do**   ▷ Firefly Algorithm module
    for $j = 1 : i$ all $n$fireflies **do**
        Light intensity I_i at x_i is determined by $f(x_i)$
        if $I_j > I_i$ **then**
            Move firefly i towards j in all d dimensions
        end if
        Evaluate new solutions and update light intensity
    end for
end for
if $N_{eff} < N_{thr}$ **then**   ▷ $N_{eff} = 1/\sum_{i=1}^{N}(\omega_k^i)^2$
    Conducting resampling process
end if
Obtain the optimized particle weights $\omega = I$
Normalization $\omega_k^i = \omega_k^i / \sum_{i=1}^{N} \omega_k^i$
State output $\hat{x}_k = \sum_{i=1}^{N} \omega_k^i x_k^i$

Table 1. RMSE and required time of different algorithms

Algorithm	RMSE (m)	Time (sec)
EKF	11.6100	0.1006
UKF	6.4285	0.7806
PF	2.6889	6.6527
FA-PF	1.6470	2.6947

Figures 2 and 3 show the results of different algorithms within 100 steps. It demonstrate that FA-PF has faster convergence speed and better estimation performance than other algorithms. Figure 4 presents the distribution of particles

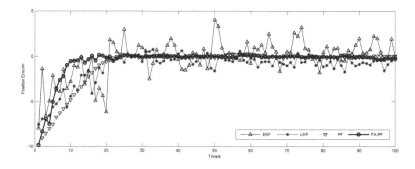

Fig. 2. Position estimation error comparisons of different algorithms

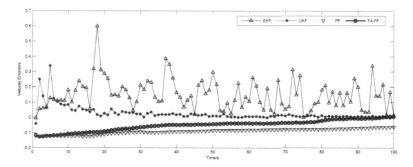

Fig. 3. Velocity estimation error comparisons of different algorithms

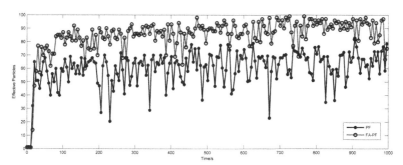

Fig. 4. Comparison of the number of meaningful particles

before state output. After simulation, the number of meaningful particles of FA-PF remains at a higher level than PF, and does not have large fluctuations. It is obvious that the firefly algorithm can restrain the problem of particle impoverishment effectively.

6   Conclusions

This paper proposes an improved particle filter based on firefly algorithm to improve the accuracy of relative navigation. In this research, the process and measurements model of relative navigation problem is given. Disadvantages of PF are being analysed, and pointing out that the resampling strategy caused the particle impoverishment. The firefly algorithm is introduced to solve this problem, and the update mechanism of fluorescence intensity and attractiveness is being improved. In the improved algorithm, particles are considered as fireflies and the fireflies exchange information frequently based on the current observational information during the movement. Therefore, meaningful particles are increased and the particles approximate the true state more accurately. Experimental results demonstrate that the FA-PF outperforms the standard PF and other classical methods, and restrain the problem of particle impoverishment phenomenon effectively. In general, the FA-PF improved the precision of relative navigation significantly, and there are of great value of engineering application.

References

1. Wu, D., Wang, Z.: Strapdown inertial navigation system algorithms based on geometric algebra. Adv. Appl. Clifford Algebras **22**(4), 1151–1167 (2012)
2. Gaias, G., D'Amico, S., Ardaens, J.S.: Angles-only navigation to a noncooperative satellite using relative orbital elements. J. Guid. Control Dyn. **37**(2), 439–451 (2014)
3. Grewal, M.S., Henderson, V.D., Miyasako, R.S.: Application of Kalman filtering to the calibration and alignment of inertial navigation systems. IEEE Trans. Autom. Control **1**(1), 65–72 (1991)
4. Tang, X., Yan, J., Zhong, D.: Square-root sigma-point Kalman filtering for spacecraft relative navigation. Acta Astronaut. **66**(56), 704–713 (2010)
5. Arulampalam, M.S., Maskell, S., Gordon, N., Clapp, T.: A tutorial on particle filters for online nonlinear/non-Gaussian Bayesian tracking. IEEE Trans. Signal Process. **50**(2), 174–188 (2002)
6. Okuma, K., Taleghani, A., Freitas, N., Little, J.J., Lowe, D.G.: A boosted particle filter: multitarget detection and tracking. In: Pajdla, T., Matas, J. (eds.) ECCV 2004. LNCS, vol. 3021, pp. 28–39. Springer, Heidelberg (2004). doi:10.1007/978-3-540-24670-1_3
7. Breitenstein, M.D., Koller-Meier, E., Reichlin, F., Leibe, B.: Robust tracking-by-detection using a detector confidence particle filter. In: IEEE International Conference on Computer Vision, pp. 1515–1522 (2009)
8. Park, S., Hwang, J.P., Kim, E., Kang, H.J.: A new evolutionary particle filter for the prevention of sample impoverishment. IEEE Trans. Evol. Comput. **13**(4), 801–809 (2009)
9. Oshman, A., Carmi, Y.: Attitude estimation from vector observations using a genetic-algorithm-embedded quaternion particle filter. J. Guid. Control Dyn. **29**(4), 879–891 (2006)
10. Saini, S., Rambli, D.R., Sulaiman, S.B., Zakaria, M.N.B.: Hierarchical approach for articulated 3D human motion tracking using PF-based PSO. WIT Trans. Inf. Commun. Technol. **59**, 789–795 (2014)
11. Yang, X.S.: Firefly algorithm, stochastic test functions, design optimisation. Int. J. Bio-Inspired Comput. **2**(2), 78–84(7) (2010)
12. Li, J., Baoyin, H., Vadali, S.R.: Autonomous rendezvous architecture design for lunar lander. J. Spacecr. Rockets **52**(3), 1–10 (2015)
13. Jezewski, D.J., Donaldson, J.D.: An analytic approach to optimal rendezvous using Clohessy-Wiltshire equations. J. Astronaut. Sci. **27**(3), 293–310 (1979)
14. Siouris, G.M.: Missile Guidance and Control Systems. Springer, Berlin (2004). **57**(6), B32
15. Li, H.W., Wang, J.: Particle filter for manoeuvring target tracking via passive radar measurements with glint noise. IET Radar Sonar Navig. **6**(3), 180–189 (2012)
16. Gao, M.L., He, X.H., Luo, D.S., Jiang, J.: Object tracking using firefly algorithm. IET Comput. Vis. **7**(4), 227–237 (2013)

An Improved Discrete Firefly Algorithm Used for Traveling Salesman Problem

Liu Jie, Lin Teng$^{(\boxtimes)}$, and Shoulin Yin

Software College, Shenyang Normal University, No. 253, HuangHe Bei Street,
HuangGu District, Shenyang 110034, China
nan127@sohu.com, 1532554069@qq.com, 352720214@qq.com

Abstract. In this paper, we propose a novel method based on discrete firefly algorithm for traveling salesman problem. We redefine the distance of firefly algorithm by introducing swap operator and swap sequence to avoid algorithm easily falling into local solution and accelerate convergence speed. In addition, we adopt dynamic mechanism based on neighborhood search algorithm. Finally, the comparison experiment results show that the novel algorithm can search perfect solution within a short time, and greatly improve the effectiveness of solving the traveling salesman problem.

Keywords: Firefly algorithm · Traveling salesman problem · Swap operator · Swap sequence · Neighborhood search algorithm

1 Introduction

Traveling salesman problem (TSP) [1] is a typical optimization problem in the area of operational research. It belongs to Non-deterministic Polynomial (NP) problem and is used in many areas such as PCB drilling, goods delivery routes and workshop scheduling. Although there are some precise algorithms which can be used to solve the problem, the principle of precise algorithms is complex. For small size TSP, Branch and bound method, greedy method and cutting plane method are often used to solve these problems. But for big size TSP, it is difficult to use above methods. Therefore, the domestic and foreign scholars have been trying to seek a highly efficient and stable algorithm for solving this complex problem. Firefly algorithm (FA) [2] is one of heuristic search algorithms based on swarm intelligence, which was proposed by YANG in 2010. FA [3] is based on the self-organization of the swarm simulation model with the advantages of less setting parameters, strong robustness, and has been applied in many fields. Li et al. [4] showed that firefly algorithm was introduced to solve this problem, and a series of discrete operations were conducted to adapt to it. Meng et al. [5] proposed discrete artificial bee colony algorithm for TSP. Zhang [6] presented a novel firefly algorithm for solving the nonlinear equations problem.

The above methods almost are used for solving continuous domain optimization problems. However, FA is relatively few used in the aspect of discrete domain application. To improve the performance of TSP solution, we propose improved discrete FA through redefining distance of firefly algorithm by introducing swap operator and swap sequence which better coordinates and balances the search process of FA algorithm.

© Springer International Publishing AG 2017
Y. Tan et al. (Eds.): ICSI 2017, Part I, LNCS 10385, pp. 593–600, 2017.
DOI: 10.1007/978-3-319-61824-1_64

To facilitate the description, this paper also gives some definitions. We present a new discrete FA for traveling salesman problem. This new scheme takes neighborhood search algorithm into consideration. Finally, we conduct some experiments to verify its performance. The results show that the new algorithm has a good effect on solving TSP.

2 The Firefly Algorithm

Fireflies emit fluorescence signal in the nature. From the point of biology view, its role is to attract the opposite sex and get copulation, reproduction chance, sometimes for predation. Firefly uses fluorescence signal within a certain range to attract companion in FA, eventually it makes most fireflies gather in one or more locations, and realize the position optimization.

Fireflies attract accompanies mainly depending on fluorescence intensity and attraction degree. Fluorescence intensity depends on the target value of its own location. The better target value is, the high fluorescence intensity is. Attraction degree has closely relationship with fluorescence intensity. Brighter fireflies have higher attraction intensity, which can attract weaker fireflies moving towards to it. In addition, it should take physical properties into consideration in FA. Fluorescence intensity and attraction degree are in inverse proportion of the distance among fireflies. Namely, fluorescence intensity and attraction degree will decrease with the increase of distance.

Fireflies mainly obey the following rules:

- firefly moving direction is determined by the fluorescence intensity, it always move towards to the brighter fireflies.
- the firefly moving distance is determined by the attraction degree.

This paper defines firefly fluorescence intensity and attraction degree as follows.

Definition 1. I: Fluorescent brightness of fireflies.

$$I = I_0 e^{-\gamma r_{ij}}. \tag{1}$$

Where r_{ij} is the distance between firefly i and j. γ is light intensity absorption coefficient, it denotes that fluorescence gradually decreases with the increase of distance, it can be set as a constant. I_0 is fluorescent brightness of firefly when $r = 0$. I_0 not only is its own fluorescence intensity, but is the maximum fluorescence intensity, which has relationship with function value.

Definition 2. β_{ij}: Attraction degree of fireflies.

$$\beta_{ij} = \beta_0 e^{-\gamma r_{ij}^2}. \tag{2}$$

Where β_0 (maximum attraction degree) is attraction degree of firefly when $r = 0$. According to Definition 1, every firefly uses roulette method to move towards to the individual with higher fluorescence intensity. On the basis of Definition 2, each firefly can determine the movement distance. By constantly updating its location, it achieves the purpose of optimization.

Definition 3. Location updating formula of firefly i attracted by firefly j.

$$x_i = x_i + \beta(x_j - x_i) + \alpha(rand - 0.5). \tag{3}$$

This formula is composed of three parts. First part, x_i denotes current position of firefly i. Second part, $\beta(x_j - x_i)$ denotes the distance firefly i moving to j. Third part, there is disturbance mechanism when location updating to avoid fireflies prematurely falling into local optimal solution. $\alpha \in [0, 1]$ is step length factor. $rand \in [0, 1]$ is random factor and obeys uniform distribution.

3 The Improved Firefly Algorithm for TSP

When we solve TSP by FA, we should know distance and light intensity absorption coefficient. Then it updates location through formula (1, 2, 3) to realize optimizing. We apply serial number coding method into FA to solve TSP in this paper. Because TSP is a combinatorial optimization problem, we should first definite discrete variable distance, then we present discrete variable location updating formula and its disturbance mechanism. Finally, we give the solution step based on discrete firefly algorithm for solving TSP.

In this section, we introduce the following new definitions.

Definition 4. Assuming that solution sequence of n nodes TSP is $S = (a_i)$, swap operator $S_0(i_1, i_2)$ is the point a_{i_1} and a_{i_2} in swap solution S. So $S' = S + S_0(i_1, i_2)$ is the new solution after operation with swap operator.

Definition 5. Ordered queue of one or multiple swap operators is called a swap sequence. $S' = (S_{O_1}, S_{O_2}, \ldots, S_{O_n})$, $S_{O_1}, S_{O_2}, \ldots, S_{O_n}$ is swap operator.

Definition 6. Different swap sequences act on the sane solution that may generate the common new solution. Therefore, all the same effect swap sequences set is called equivalent set of swap sequence.

Definition 7. In the equivalent set of swap sequence, the swap sequence with minimum swap operators is the basic swap sequence in equivalent set.

Space distance between fireflies can be calculated by Euclidean distance when solving optimization problems of continuous variables. However, when solving TSP, solution is discrete variable, so it cannot use Euclidean distance. In this paper, we definite space distance:

$$r = A/N. \tag{4}$$

Where A is swap operator number of two fireflies in solution vector basic swap sequence. N is the cities number. After getting r, fluorescent brightness of firefly can be calculated by formula (1). Therefore, each firefly can move towards to other individuals with higher fluorescent brightness based on roulette wheel method.

Firefly movement distance can be calculated according to attraction degree when solving optimization problems of continuous variables. However, when solving TSP, movement is discrete variable, so we definite the distance between firefly i and j:

$$\beta_{ij} = rand\,\text{int}(0, A_{ij}). \tag{5}$$

Where $rand\,\text{int}(0, A_{ij})$ is random integer. A_{ij} is swap operator number of two fireflies i and j in solution vector basic swap sequence. β_{ij} is random swap operator number from 0 to A_{ij}. Location updating by formula (6):

$$x_i = x_i + \beta_{ij}. \tag{6}$$

When solving the TSP, in order to avoid the firefly algorithm falling into local optimum, this paper defines the disturbance mechanism of discrete variables based on variable neighborhood search algorithm. It adopts several neighborhood structure forms to search optimization solution, systematically changes its neighborhood so as to expand the search scope, enhance the ability to jump out of local optimal.

In this paper, we adopt three neighborhood structures to solve TSP.

a. *Insert* neighborhood. We randomly select different position x and y in solution sequence, and insert the city in x into y, as *Insert*(x, y).
b. *Swap* neighborhood. We randomly select different position x and y in solution sequence, and exchange the cities in x and y, as *Swap*(x, y).
c. *2-opt* neighborhood. We use side (i, j) and $(i+1, j+1)$ to replace $(i, i+1)$ and $(j, j+1)$. This change will affect the direction of mid-edge $(i+1, j)$.

The detailed process of the proposed discrete firefly algorithm is as follows:

- Step1. Initializing the swarm. Set parameter firefly number m, light intensity absorption coefficient γ, maximum iteration number $T_{\max}$.
- Step2. Randomly initializing solution sequence of each firefly. Calculating objective function P_i and maximum fluorescent brightness I_0.

$$I_0 = P_g / P_i. \tag{7}$$

Where P_g is the current searched optimal solution.

- Step3. According to formula (1), (4), (5), (7), calculating relative brightness I of firefly and attraction degree β_{ij}. So determine the movement direction and distance.
- Step4. According to formula (6), updating position of firefly.
- Step5. Choosing the three neighborhood in variable neighborhood search method based on a certain probability. Repeat executing n iteration. Selecting position with optimal objective function as the current position of firefly.
- Step6. According to updated position of firefly, re-compute objective function P_i and make a comparison with current optimal position P_g. If function is better, then it updates P_g.
- Step7. Whether the process satisfies termination conditions or not. If YES, then finish and output results. If NO, return back step3.

4   Experiments and Results

4.1   The Effect of Light Intensity Absorption Coefficient γ on Firefly Algorithm

The light intensity absorption coefficient γ is the main parameter in firefly algorithm (FA). We make simulation experiments to analyze the effect of γ on performances of algorithm solution under the MATLAB platform. We adopt different γ to conduct experiments under the other same parameters conditions. Setting maximum iterations $T_{max} = 500$, number of firefly is $m = 50$, the three neighborhood in variable neighborhood search method are same: $1/3$, algorithm runs 30 times independently. Table 1 records the results with different γ.

Table 1. Effect of different γ on FA.

r	Best value $\times 10^3$	Worst value $\times 10^3$	Average value $\times 10^3$	Standard deviation $\times 10^2$
0.01	7.54443	8.48821	8.01094	2.34474
0.03	7.54443	8.48449	8.00774	2.07446
0.05	7.54466	8.58599	8.10785	2.48271
0.07	7.67768	8.68004	8.07167	2.38243
0.09	7.54436	8.59042	8.10332	2.17915
0.11	7.54436	8.69427	8.22529	3.19748
0.13	7.74988	8.78536	8.17088	2.49888
0.15	7.79758	8.86203	8.35733	2.68301

From Table 1 we can know that when γ gradually increases, the solution performance is little-changed. When $\gamma = 0.01, 0.02, 0.03, 0.04, 0.06, 0.09$ and 0.11, FA gets the best solution, 7.54436×10^3. And $\gamma = 0.03$, average value is 8.00774×10^3 and standard value 2.07446×10^2 which is the minimum value. But as a whole, when $\gamma = 0.03$, FA has a better solving performance, which is the best value for FA. The effect of different γ on FA is not obvious. So this is one of advantages for FA.

4.2   The Comparison Experiment

In order to further verify the feasibility and effectiveness of this paper's method, we choose multiple instances in international universal test library for testing, and use testing results to make a comparison to spanning tree algorithm (STA) [7], Improved Ant Colony Optimization [8] (IACO).

In the experiment, digit indicates number of cities. "O" shows calculation result of the index is not given in the comparison literature. Experimental environment is the same as above. Parameters set as follows: maximum number of iterations $T_{max} = 2000$, $\gamma = 0.03$, $k_1 : k_2 : k_3 = 2 : 1 : 2$, number of firefly $m = 20$ when city number is less than 48, otherwise $m = 50$. The two algorithms run 20 times. Calculation results are as shown in Table 2.

Table 2. Calculation results.

Instance	Calculation index	IABC	STA	FA
Bays29	Best value	30.8788	30.8788	30.8788
	Average value	30.8788	30.8788	30.8788
	Iteration number	2000	200	200
	Running time	5.11	O	O
Oliver30	Best value	73.987	73.9875	73.9875
	Average value	74.0079	74.0077	74.0075
	Iteration number	2000	200	200
	Running time	5.36	O	O
Dantzig42	Best value	76.1236	76.1149	75.9098
	Average value	75.3069	75.3069	75.3069
	Iteration number	2000	200	200
	Running time	8.95	O	O
Berlin52	Best value	7680.77	7554.38	7542.13
	Average value	7941.26	7562.81	7573.49
	Iteration number	2000	200	200
	Running time	10.76	O	O

From Table 2, STA algorithm gets the known optimal solutions in Berlin52, Oliver30 and Eil51, which is close to FA. But iteration number of FA has reduced by 98%. What's more, steps of FA are not complicated. The average value has increased by 0.6% (Not exceed 0.6%). Compared with IABC algorithm, the best value and average value is relatively small, and for Bays29, Oliver30 and Dantzig42, the standard deviation is smaller. For the four instances, the running time of FA has saved by 97.1%, 98.2%, 98.8% and 98.99% respectively.

FA algorithm also can obtain the results closely to known optimal solutions for all instances. For Oliver30 and Bays29, FA algorithm gets the same results as known optimal solutions. For Dantzig42, calculation result of FA algorithm has decreased by 19.3% compared with IABC algorithm. Figures 1, 2 and 3 show the experimental simulation diagram of Oliver30, Dantzig42 and Berlin52.

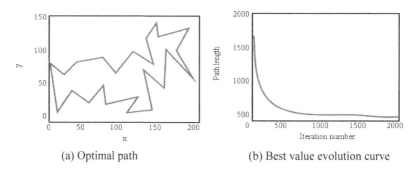

(a) Optimal path   (b) Best value evolution curve

Fig. 1. Experimental simulation diagram of Oliver30

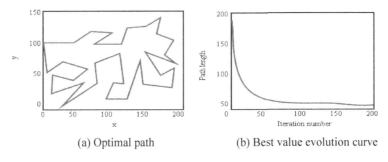

(a) Optimal path (b) Best value evolution curve

Fig. 2. Experimental simulation diagram of Dantzig42

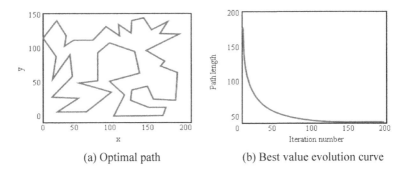

(a) Optimal path (b) Best value evolution curve

Fig. 3. Experimental simulation diagram of Berlin52

The average value of FA algorithm is relatively small compared with IABC algorithm for Oliver30, Dantzig42 and Bays29. IABC algorithm is mainly based on the STA algorithm and has combined with other operations. When the number of population in IABC is equal to that in STA, the running time in IABC is closely to that of STA at least. Nevertheless, the running time in IACO is relatively long. Therefor, FA algorithm has the better solution performance in terms of solution time and solution quality compared with STA and IABC algorithm.

5 Conclusions

This paper proposes a new discrete firefly algorithm for solving TSP problem. It extends continuous artificial swarm (especially FA) optimization algorithm to discrete domain. The new algorithm makes some definitions based on swap operator and swap sequence. It also adopts disturbance strategy based on variable neighborhood search method to increase search areas and improve the local refinement ability of the algorithm and the optimal speed. The experimental results show that the algorithm can find relatively satisfactory solution in a short time and improve the efficiency of solving the TSP. In the future, we would find more effective firefly algorithms to improve optimum searching method and solve other combinatorial optimization problems.

References

1. Carrabs, F., Cordeau, J.F., Laporte, G.: Variable neighborhood search for the pickup and delivery traveling salesman problem with LIFO loading. Informs J. Comput. **19**(4), 2007 (2015)
2. Yang, X.S.: Cuckoo search and firefly algorithm. Eng. Optim. **516**, 1–26 (2014)
3. Massan, S.U.R., Wagan, A.I., Shaikh, M.M., et al.: Wind turbine micrositing by using the firefly algorithm. Appl. Soft Comput. **27**(2), 450–456 (2014)
4. Li, M., Ma, J., Zhang, Y., et al.: Firefly algorithm solving multiple traveling salesman problem. J. Comput. Theor. Nanosci. **12**(7), 1277–1281 (2015)
5. Meng, M., Shoulin, Y., Xinyuan, H.: A new method used for traveling salesman problem based on discrete artificial bee colony algorithm. TELKOMNIKA **14**(1), 342–348 (2016)
6. Zhang, Y.F., Li, Y.G.: An improve firefly algorithm and its application in nonlinear equation groups. Adv. Mater. Res. **1049**(5), 1670–1674 (2014)
7. Iwata, S., Newman, A., Ravi, R.: Graph-TSP from Steiner cycles. In: Kratsch, D., Todinca, I. (eds.) WG 2014. LNCS, vol. 8747, pp. 312–323. Springer, Cham (2014). doi:10.1007/978-3-319-12340-0_26
8. Yu, L., Li, M., Yang, Y., et al.: An improved ant colony optimization for vehicle routing problem. In: Logistics: The Emerging Frontiers of Transportation and Development in China, ASCE, pp. 3360–3366 (2015)

Firefly Clustering Method for Mining Protein Complexes

Yuchen Zhang[1], Xiujuan Lei[1(✉)], and Ying Tan[2]

[1] School of Computer Science, Shaanxi Normal University,
Xi'an 710119, China
xjlei@snnu.edu.cn
[2] School of Electronics Engineering and Computer Science,
Peking University, Beijing 100871, China

Abstract. It is a hot research to explore protein complexes which are closely related to biological processes from the biological network. As a novel swarm intelligence optimization algorithm, the firefly algorithm (FA) has been verified to solve many optimization problems. In this study, we transform the protein clustering problem into an optimization problem in protein-protein interaction (PPI) network. A new method for mining protein complexes based on the firefly algorithm was proposed, called FC. A new objective function was proposed to find the high cohesion and low coupling clusters. A thorough comparison completed for different protein clustering methods has been carried out. The clustering results show that FC method outperforms the other state-of-the-art methods in accuracy of detecting complexes from PPI network.

Keywords: Firefly clustering · Protein complexes · Dynamic PPI network · Clustering objective function

1 Introduction

Proteins play an important role in biological processes. Studying proteins helps us understand genes, disease mechanisms, and so on. However, proteins usually work by interacting with each other. The study of a single protein does not reflect its significance in Biology. The proteins and interactions of proteins form a biological network, protein-protein interaction network (PPI) [1]. At the same time, a group of proteins that works in same space can be used as a protein complex. Protein complexes often have specific functions that can reflect some of the protein properties. So, it is important to explore and research protein complexes. In recent years, a large amount of protein-protein interactions were generated by high-throughput experimental techniques such as yeast two-hybrid and mass spectrometry [2, 3]. These techniques provide a basis for the identification of protein complexes.

Many scholars have proposed a lot of methods to identify the protein complexes. Most of the methods are based on graph theory and dense region discovery. Bader and Hogue proposed the molecular complex detection (MCODE) [4]. MCL [5] was also used to identify protein complexes. There are two main operations, called expansion and inflation. Wang *et al.* [6] used the gene expression data to establish the sequential

© Springer International Publishing AG 2017
Y. Tan et al. (Eds.): ICSI 2017, Part I, LNCS 10385, pp. 601–610, 2017.
DOI: 10.1007/978-3-319-61824-1_65

dynamic PPI network. The clustering result of MCL were optimized. Because of the core-attachment characteristics of protein complexes, CORE [7] and COACH [8] were proposed to predict protein complexes. In identifying overlapping protein complexes, Nepusz et al. introduced ClusterONE [9]. In order to make clustering method take into account the biological characteristics of protein complexes, some scholars considered other biological information. Such as CSO [10], it used the gene ontology (GO) annotation data to find complex cliques.

With the development of swarm intelligence algorithm, more and more scholars also begin to apply swarm intelligence algorithm to graph mining. Emad et al. try to detect protein complexes by using genetic algorithm [11]. And FA algorithm [12] is also applied to network clustering with significant performance. The algorithm simulates the behaviors of the fireflies that the darker fireflies move to the bright fireflies to solve the optimal solution. A community detecting algorithm [13] was proposed by Amiri et al. based on a multi-objective enhanced firefly algorithm. In our previous research, we also used FA to improve the parameters of the MCL algorithm [14].

In the paper, we used the FA algorithm to detect protein complexes from PPI network. To find a corresponding relationship between the behaviors of fireflies and the clustering process. And a new objective function is proposed to transform the PPI clustering into an optimization problem. Finally, in order to verify the performance of the proposed method, we compared it with other clustering methods on different PPI datasets.

2   PPI Network Preprocessing

In most studies, the PPI network is used as an undirected graph $G = (V, E)$, where V is a set of proteins and E represents all interactions. Protein complexes are a set of dense subgraphs with high cohesion and low coupling. Construction of protein network is very important for identifying protein complexes. Wang et al. have proved that the dynamic PPI network based on biological characteristics is better than the static PPI network. Therefore, we also used protein gene expression data to construct a dynamic PPI network. Using the 3-sigma method [6] to identify the activity of protein at the different time points.

However, there are a lot of false positives and false negatives in high throughput protein interaction data. So not all interaction relationships are reliable. In order to optimize the network, we weighted the edges of the dynamic network to distinguish their contribution for the task of detecting protein complexes. First, the topology score of edge e_{ij} [15] is defined

$$topology\_score_{ij} = \frac{|N_i \cap N_j| + 1}{max\{avg(G), |N_i|\} + max\{avg(G), |N_j|\}} \tag{1}$$

where N_i and N_j denote the neighbors of v_i and v_j. $|N_i \cap N_j|$ denotes the number of common neighbors of v_i and v_j. And $avg(G)$ is the average degree of the network G.

On the other hand, Gene Ontology annotations are also considered [10]. If there are some common GO annotations between interacting proteins, the interaction is believed to be reliable. They are expressed as follows:

$$GSM_{ij} = \frac{|GSM_i \cap GSM_j|}{|GSM_i| \times |GSM_j|} \qquad (2)$$

where the $|GSM_i|$, $|GSM_j|$ represent the number of GO annotations for v_i and v_j, respectively. $|GSM_i \cap GSM_j|$ denotes the number of common GO annotations for both v_i and v_j. Based on the topology score and GO annotations, the weight of an edge e_{ij} is given as:

$$W_{ij} = topology\_score_{ij} \cdot GSM_{ij} \qquad (3)$$

The value range of W_{ij} is [0, 1]. If weight of an edge is 0, it will be considered to be false data and deleted from the dynamic network.

3   Firefly Clustering Method

3.1   Firefly Representation

In the process of clustering protein complexes, we first defined the representation of a firefly. A firefly corresponds to a clustering result. In other words, a firefly contains a set of clusters. Obviously, if a firefly is directly represented by a set of clusters, the subsequent steps will be not east to operate. Therefore, the locus-based adjacency representation [16] was used. In the graphic representation, a set of clusters (a firefly) are considered to be a N-dimensional vector. N is the number of nodes in a timestamp network. For a firefly $X = \{x_1, x_2, \ldots, x_N\}$, there is a set of possible range of values based on the adjacency matrix of PPI network. For example in Fig. 1. the node v_3 connect the nodes v_2, v_4, v_5, so the possible range of values $\bar{r}\ v_3$ for v_3 is {2, 4, 5}.

A firefly represents a group of protein complexes. For a firefly, if the value of ith element is j, node v_i and node v_j will be contained in same protein complex. In addition, the existence of independent nodes as clusters in the process of clustering. We added 0 into the value range of each element. So the finally range $\bar{r}\ v_3$ is {0, 2, 4, 5}. If the value of an element is 0 and no other values of elements are equal to its corresponding node, the corresponding node of the element is an independent cluster. After the clustering results are expressed by fireflies, a decoding operation is used to identify all the components in the timestamp network. We can build an adjacency matrix that contained at most N edges according to a firefly. We used the breadth first traversal method to find out connected subgraphs. This representation method does not need to be given a number of clusters in advance. In Fig. 1, a firefly is decoded into two clusters $\{v_1, v_2, v_3, v_4\}$ and $\{v_5, v_6, v_7, v_8, v_9\}$, v_{10} is an independent cluster and is excluded. In the initial stage, we randomly generate m fireflies as initial population according to the range of each element.

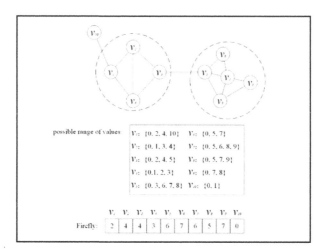

Fig. 1. Correspondence between a firefly and a set of clusters

3.2 Clustering Objective Function

In order to translate the protein clustering problem into an optimization problem, the method needs a reasonable objective function. In FA algorithm, the brightness of firefly is the value of objection function. The objective function need to reflect the properties of protein complexes. Therefore, the objective function should be able to distinguish between high cohesion—low occasional clusters and generic clusters. For objective clusters, there are many edges inside, and the edges between them are less. In this paper, we given the following objective function by combing definition of the density of the cluster and considering the appropriate number of clusters. We also used weight sum of edges to replace the number of edges in clusters.

$$F(\{C^1, C^2, \dots, C^k\}) = \frac{\sum_{i=1}^{k} \frac{C_{in}^i}{C_{in}^i + C_{out}^i + W_{ave} \times |C^i| \times (|C^i|-1)/2} \times |C^i|}{\sum_{i=1}^{k} |C^i|} \tag{4}$$

$$C_{in}^i = \sum_{p,q \in C^i} W_{pq} \tag{5}$$

$$C_{out}^i = \sum_{p \in C^i, q \notin C^i} W_{pq} \tag{6}$$

where $\{C^1, C^2, \dots, C^k\}$ represents a set of clusters determined by a firefly. $|C^i|$ represents the number of proteins in a cluster. $\sum_{i=1}^{k} |C^i|$ is the total number of proteins found in the protein complexes detected. W_{ave} is the average weight in the network G. And $|C^i| \times (|C^i|-1)/2$ is the maximum possible number of edges in the cluster C^i. C_{in}^i is the weight sum of all edges in the cluster C^i. C^i_{out} is the weight sum of edges whose one endpoint is in C^i and another endpoint is not. The goal of proposed method is to find the maximum value of F.

3.3 Firefly Movement Strategy

After generating the initial firefly population, fireflies will find adaptively the optimal solution and generate a set of clusters. However, due to the complexity of the PPI network and dimension of the problem, FC method is easy to fall into the local optimal solution. In order to avoid such problems, we introduced some random search fireflies to improve the method. The FC randomly selected r fireflies from the population, and mutated randomly their some element value with mutation probability mp in each iteration. After mutating, the fitness values of all new fireflies are calculated by clustering objective function F. If the fitness values of new fireflies are greater the fitness values of original fireflies, the new fireflies will replace the original fireflies. The mutation probability mp_i of firefly i is defined as follows:

$$mp_i = \frac{F_{max} - F_i + \alpha}{F_{max}} \qquad (7)$$

where F_i is fitness value of firefly i. F_{max} is the fitness of the brightest firefly. α is a constant to avoid that the probability is 0.

In each iteration, the fireflies will automatically move to the better solution through the exchange of information between them. If the brightness of a firefly is greater, it will attract the lesser brightness fireflies in the surrounding. In function optimization problem, a firefly moves to all the higher brightness fireflies. However, in the processes of mining protein complexes, the value range of each element is not continuous. So the firefly can only move into one direction. Therefore, we can estimate the probabilities that the firefly will move to the next positions to make the firefly close to the brightest firefly (the optimum solution). Where Fireflyi_kth is kth element of firefly i. The corresponding probability of movement position is $\bar{p}_{ik}$ in the next generation. We used roulette to determine the direction of movement of fireflies in next generation. For example, in Fig. 2, the brightness of firefly 2, 3, 4 are greater than the brightness of firefly 1. Firefly 1 will move to one of firefly 2, 3, 4. It can be found that is better clustering results, when node v_3 and v_5 is not connected. The number of values in $\bar{r}\ v_3 = \{0, 2, 4, 5\}$ that appear in Firefly2_3th, Firefly3_3th, Firefly4_3th are 0, 1, 2, 0, respectively. FC used that the occurrences number of each value divided by number of the brighter fireflies as occurrences probability of each value in next generation. It can be found that 2 appeared one times, and 4 appeared two times. So the probability of movement position $\bar{p}_{13}$ is $(0, \frac{1}{3}, \frac{2}{3}, 0)$. The firefly will move in the brighter firefly with the probability in each iteration.

For the brightest firefly, it will be randomly perturbed to jump out of the local optimum. The process stops until the algorithm convergence or the maximum number of iterations is reached. Finally, the set of clusters decided by the brightest firefly are the predicted protein complexes. Since the method is run on the dynamic network,

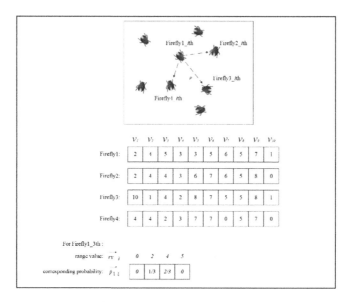

Fig. 2. Firefly movement strategy

there are overlapping protein complexes in results. We removed the protein complexes that are contained by other complexes. Table 1 shows the corresponding relation between firefly biological characteristics and FC method.

Table 1. The corresponding relation of firefly biological characteristics in FC

Firefly characteristics and behavior	Firefly clustering (FC) method
Firefly	A group of protein complexes
Position of firefly (element)	Two proteins in a same complex
Firefly brightness	Clustering objective function value
Movement	Detecting protein complexes
The brightest firefly	Protein complexes (optimal solution)

4 Experiment Results

In order to test the performance of the algorithm, we used three Saccharomyces cerevisiae PPI datasets, DIP [17], Krogan [18], MIPS [19]. And the gene expression data are provided by Gene Expression Omnibus (GEO) [20], accession number of the data is GSE3431. Gene ontology data is the most commonly used data can represent the functions of proteins. In this paper, we used GO-slims data. This data is cut-down version of the GO ontology data [10], which can be acquired at (http://www. yeastgenome.org/download-data/curation). And we used CYC2008 [21] as a known protein complexes set. There are 408 protein complexes.

In the evaluations of predicted protein complexes, the algorithm used several commonly evaluation methods. The Overlapping Score (OS), Sensitivity (Sn), Specificity (Sp), *f-measure* [4] and *p-value* [22] both are used commonly. According to Literature [4], when OS is greater than 0.2, we considered that the predicted protein complexes is matched. If OS is equal to 1, the predicted complexes is perfectly matched. The *p-value* denotes the probability that a predicted protein complex is enriched by a given functional group with random chance. It is generally believed that the smaller the *p-value* (less than 0.01) is, the more significant the predicted protein complex is.

In order to verify the superiority of FC method, we compared with other method such as MCODE [4], MCL [5], CORE [7], CSO [10], ClusterONE [9] and COACH [8] in a same dynamic PPI network. The comparison results are showed in the Table 2. *PC* denotes the total number of protein complexes by predicted. *MPC* represents the

Table 2. Performance comparsion with other methods

Dataset	Method	Sn	Sp	f-measure	PC	MPC	MKC	Perfect	AS
DIP	MCODE	0.2318	0.6182	0.3372	165	102	70	6	6.7212
	MCL	0.7031	0.2505	0.3694	1541	386	245	14	4.4361
	CORE	0.7381	0.2769	0.4027	1517	420	259	39	2.443
	CSO	0.4403	0.6257	0.5169	342	214	136	11	4.652
	ClusterONE	0.6093	0.3385	0.4352	972	329	197	15	3.5422
	COACH	0.5009	0.5591	0.5284	474	265	144	13	4.9789
	FC-*best*	**0.655**	**0.4612**	**0.5413**	**774**	**357**	**220**	**39**	**3.4406**
	FC-*worst*	**0.6345**	**0.4329**	**0.5147**	**790**	**342**	**211**	**31**	**3.4278**
	FC-*ave*	**0.6359**	**0.4422**	**0.5217**	**778**	**344**	**211**	**36**	**3.4511**
Krogan	MCODE	0.2749	0.7937	0.4084	160	127	73	10	5.125
	MCL	0.566	0.4559	0.5051	658	300	178	40	3.9544
	CORE	0.5417	0.4121	0.4681	677	279	172	39	2.6041
	CSO	0.3284	0.8254	0.4699	189	156	89	10	5.2646
	ClusterONE	0.5232	0.4632	0.4914	585	271	161	28	3.935
	COACH	0.3566	0.81	0.4952	221	179	85	11	5.3575
	FC-*best*	**0.4271**	**0.7537**	**0.5452**	**272**	**205**	**133**	**36**	**3.6765**
	FC-*worst*	**0.4008**	**0.7559**	**0.5239**	**254**	**192**	**121**	**34**	**3.8307**
	FC-*ave*	**0.4131**	**0.7493**	**0.5325**	**265**	**199**	**126**	**39**	**3.7542**
MIPS	MCODE	0.1714	0.5333	0.2595	135	72	60	4	5.437
	MCL	0.5451	0.2017	0.2945	1259	254	196	17	4.7434
	CORE	0.6235	0.249	0.3558	1217	303	225	29	2.5859
	CSO	0.2835	0.5163	0.366	246	127	87	6	4.5528
	ClusterONE	0.4483	0.2796	0.3444	744	208	152	17	3.1317
	COACH	0.3145	0.3662	0.3384	396	145	92	5	6.5253
	FC-*best*	**0.51**	**0.4205**	**0.461**	**604**	**254**	**164**	**32**	**3.2897**
	FC-*worst*	**0.4896**	**0.3865**	**0.432**	**608**	**235**	**163**	**27**	**3.2599**
	FC-*ave*	**0.4989**	**0.4147**	**0.453**	**590**	**245**	**162**	**30**	**3.2989**

number of matched predicted complexes. *MKC* is the number of matched known protein complexes. *AS* denotes average size of predicted protein complexes. Since the FC method has random characteristics, we run the FC method 10 times to analyze the results. FC-*best*, FC-*worst*, and FC-*ave* represent the best result, the worst result, the average result according to *f-measure*, respectively. On DIP data, the *f-measure* of FC-*best* is the highest. And the *f-measure* of FC-*ave* is slightly lower than the value of CSO. And the *f-measure* of FC is the highest on Krogan and MIPS data. In addition, the number of perfect matched complexes is the largest.

Similarly, we compared the proposed method with other methods on function enrichment analysis. We calculated the *p-values* of the protein complexes mined by the algorithms in Biological Process (BP). The result are showed in Table 3. On the Krogan data, percentage of protein complexes whose *p-value* are greater than 0.01 in all complexes identified by FC is the smallest. And on DIP, MIPS data, the *p-value* of FC-*best* are the smallest. And the *p-value* of FC-*worst* are slightly high than the value of CSO and MCODE, respectively. Therefore, in the biological significance terms of predicted proteins complexes, the performance of the proposed method also can be accepted.

Table 3. Function enrichment analysis of predicted protein complexes from different methods

Dataset	Algorithms	PC (size ≥ 3)	<E-15	[E-15, E-10)	[E-10, E-5)	[E-5, 0.01)	≥0.01
DIP	MCODE	165	12 (7.27%)	17 (10.30%)	80 (48.48%)	38 (23.03%)	18 (10.91%)
	MCL	1053	19 (1.80%)	47 (4.46%)	183 (17.38%)	362 (34.38%)	442 (41.98%)
	CORE	344	1 (0.29%)	3 (0.87%)	78 (22.67%)	114 (33.14%)	148 (43.02%)
	CSO	342	26 (7.6%)	42 (12.28%)	148 (43.27%)	90 (26.32%)	36 (10.53%)
	ClusterONE	574	21 (3.66%)	52 (9.06%)	177 (30.84%)	184 (32.06%)	140 (24.39%)
	COACH	474	33 (6.96%)	44 (9.28%)	205 (43.25%)	126 (26.58%)	66 (13.92%)
	FC-*best*	**393**	**23 (5.85%)**	**51 (12.98%)**	**179 (45.55%)**	**106 (26.97%)**	**34 (8.65%)**
	FC-*worst*	**404**	**21 (5.20%)**	**49 (12.13%)**	**172 (42.57%)**	**119 (29.46%)**	**43 (10.64%)**
Krogan	MCODE	160	8 (5.00%)	28 (17.50%)	68 (42.50%)	46 (28.75%)	10 (6.25%)
	MCL	403	16 (3.97%)	43 (10.67%)	103 (25.56%)	119 (29.53%)	122 (30.27%)
	CORE	255	3 (1.18%)	10 (3.92%)	60 (23.53%)	102 (40.00%)	80 (31.37%)
	CSO	189	20 (10.58%)	36 (19.05%)	79 (41.80%)	42 (22.22%)	12 (6.35%)
	ClusterONE	399	13 (3.26%)	43 (10.78%)	98 (24.56%)	120 (30.08%)	125 (31.33%)
	COACH	221	23 (10.41%)	37 (16.74%)	91 (41.18%)	54 (24.43%)	16 (7.24%)
	FC-*best*	**157**	**14 (8.92%)**	**27 (17.20%)**	**73 (46.50%)**	**36 (22.93%)**	**7 (4.46%)**
	FC-*worst*	**155**	**14 (9.03%)**	**25 (16.13%)**	**75 (48.39%)**	**34 (21.94%)**	**7 (4.52%)**
MIPS	MCODE	135	5 (3.70%)	10 (7.41%)	70 (51.58%)	39 (28.89%)	11 (8.15%)
	MCL	606	5 (0.83%)	13 (2.15%)	94 (15.51%)	220 (36.30%)	274 (45.21%)
	CORE	340	0 (0.00%)	4 (1.18%)	65 (19.12%)	107 (31.47%)	164 (48.24%)
	CSO	246	7 (2.85%)	27 (10.98)	110 (44.72%)	73 (29.67%)	29 (11.79%)
	ClusterONE	372	7 (1.88%)	16 (4.30%)	117 (31.45%)	126 (33.87%)	106 (28.49%)
	COACH	396	16 (4.04%)	46 (11.62%)	145 (36.62%)	149 (37.63%)	40 (10.10%)
	FC-*best*	**285**	**7 (2.46%)**	**25 (8.77%)**	**127 (44.56%)**	**106 (37.19%)**	**20 (7.02%)**
	FC-*worst*	**290**	**8 (2.76%)**	**25 (8.62%)**	**133 (45.86%)**	**97 (33.45%)**	**27 (9.31%)**

5 Conclusion

In this paper, a novel protein complexes clustering method based on firefly algorithm was proposed. The proposed method has high accuracy in predicting protein complexes. Combined with the FA algorithm, the clustering problem is abstracted as an optimization problem. Because of the adaptability of the algorithm, it is not necessary to set the number of clusters in advance. In the clustering process, through the firefly searching, the algorithm does not need to consider other aspects and its implementation is simple. And the experiment results show that FC method outperforms the other method for mining protein complexes.

Acknowledgments. This paper is supported by the National Natural Science Foundation of China (61672334, 61502290, 61401263), Industrial Research Project of Science and Technology in Shaanxi Province (2015GY016).

References

1. Zhang, A.: Protein Interaction Networks. Cambridge University Press, New York (2009)
2. Uetz, P., Giot, L., Cagney, G., Mansfield, T.A., Judson, R.S., Knight, J.R., Lockshon, D., Narayan, V., Srinivasan, M., Pochart, P.A.: Comprehensive analysis of protein–protein interactions in Saccharomyces cerevisiae. Nature **403**(6770), 623–627 (2000)
3. Gavin, A.C., Bösche, M., Krause, R., Grandi, P., Marzioch, M., Bauer, A., Schultz, J., Rick, J.M., Michon, A.M., Cruciat, C.M.: Functional organization of the yeast proteome by systematic analysis of protein complexes. Nature **415**(6868), 141–147 (2002)
4. Bader, G.D., Hogue, C.W.: An automated method for finding molecular complexes in large protein interaction networks. BMC Bioinform. **4**(1), 2 (2003)
5. Dongen, S.M.V.: Graph clustering by flow simulation. Ph.D. thesis. University of Utrecht, The Netherlands (2000)
6. Wang, J., Peng, X., Li, M., Luo, Y., Pan, Y.: Active protein interaction network and its application on protein complex detection. In: IEEE International Conference on Bioinformatics and Biomedicine, Atlanta, pp. 37–42, November 2011
7. Leung, H.C., Xiang, Q., Yiu, S.M., Chin, F.Y.: Predicting protein complexes from PPI data: a core-attachment approach. J. Comput. Biol.: J. Comput. Mol. Cell Biol. **16**(2), 133–144 (2009)
8. Wu, M., Li, M., Kwoh, C.K., Ng, C.K.: A core-attachment based method to detect protein complexes in PPI networks, BMC Bioinform., **10**(1, article 169), 1–16 (2009)
9. Nepusz, T., Yu, H., Paccanaro, A.: Detecting overlapping protein complexes in protein-protein interaction networks. Nat. Methods **9**(5), 471–472 (2012)
10. Zhang, Y., Lin, H., Yang, Z., Wang, J, Li, Y., Xu, B.: Protein complex prediction in large ontology attributed protein-protein interaction networks. IEEE/ACM Trans. Comput. Biol. Bioinform. (TCBB) **10**(3), 729–741 (2013)
11. Emad, R., Ahmed, N., Moataz, A.: Protein complexes predictions within protein interaction networks using genetic algorithms. BMC Bioinform. **17**(7), 269 (2016)
12. Yang, X.S.: Firefly Algorithm. Nature-Inspired Metaheuristic Algorithms, vol. 2. Luniver Press, Bristol (2010)
13. Amiri, B., Hossain, L., Crawford, J.W., Wigand, R.T.: Community detection in complex networks: multi–objective enhanced firefly algorithm. Knowl.-Based Syst. **46**, 1–11 (2013)

14. Lei, X., Wang, F., Wu, F.X., Zhang, A., Pedrycz, W.: Protein complex identification through Markov clustering with firefly algorithm on dynamic protein–protein interaction networks. Inf. Sci. **329**(6), 303–316 (2016)
15. Liu, G., Wong, L., Chua, H.N.: Complex discovery from weighted PPI networks. Bioinformatics **25**(15), 1891–1897 (2009)
16. Handl, J., Knowles, J.: An evolutionary approach to multiobjective clustering. IEEE Trans. Evol. Comput. **11**, 56–76 (2007)
17. Xenarios, I., Salwinski, L., Duan, X.J., Higney, P., Kim, S.M., Eisenberg, D.: DIP, the database of interacting proteins: a research tool for studying cellular networks of protein interactions. Nucleic Acids Res. **30**(1), 303–305 (2002)
18. Krogan, N.J., Cagney, G., Yu, H., Zhong, G., Guo, X., Igatchenko, A., Li, J., Pu, S., Datta, N., Tikuisis, A.P., et al.: Global landscape of protein complexes in the yeast Saccharomyces cerevisiae. Nature **440**(7084), 637–643 (2006)
19. Guldener, U., Munsterkotter, M., Oesterheld, M., Pagel, P., Ruepp, M., Mewes, H.W., Stumpflen, V.: MPact: the MIPS protein interaction resource on yeast. Nucleic Acids Res. **34**(Suppl. 1), 436–441 (2006)
20. Tu, B.P., Kudlicki, A., Rowicka, M., McKnight, S.L.: Logic of the yeast metabolic cycle: temporal compart mentalization of cellular processes. Science **310**, 1152–1158 (2005)
21. Pu, S., Wong, J., Turner, B., Cho, E., Wodak, S.J.: Up-to-date catalogues of yeast protein complexes. Nucleic Acids Res. **37**(3), 825–831 (2009)
22. Altaf-Ul-Amin, M., Shinbo, Y., Mihara, K., Kurokawa, K., Kanaya, S.: Development and implementation of an algorithm for detection of protein complexes in large interaction networks. BMC Bioinform. **7**, 207–219 (2006)

Improved Two-Dimensional Otsu Based on Firefly Optimization for Low Signal-to-Noise Ratio Images

Li Li[1], Jianwei Liu[2], Mingxiang Ling[3], Yuanyuan Wang[1(✉)], and Hongwei Xia[3]

[1] Harbin Institute of Technology (Weihai), Weihai 264209,
People's Republic of China
wangyuanyuan@hitwh.edu.cn
[2] Jiangsu Automation Research Institute, Lianyungang 222061,
People's Republic of China
[3] Harbin Institute of Technology, Harbin 150001, People's Republic of China

Abstract. To improve two-dimensional (2D) Otsu thresholding's performance in both computation speed and segmentation quality, an improved 2D Otsu algorithm is proposed for low Signal-to-noise Ratio (SNR) images. A new 2D histogram is defined based on median gray-scale and Gaussian average gray-scale. By meeting better to the assumption of that the object's probability and the background's probability sum up to 1, the new 2D histogram enhances the thresholding algorithm's robustness to severe noise. Then a scheme of calculating the fitness function based on firefly optimization algorithm is employed to search for optimal thresholds. The proposed algorithm is applied to typical low SNR images–microscopic images of ocean plankton, and to Lenna test image. Experiment results show that with better thresholding quality, the running time of the proposed algorithm is reduced to 2.5% of the conventional 2D Otsu.

Keywords: Two dimensional Otsu thresholding · Firefly optimization · Plankton microscopic image analysis

1 Introduction

Two-dimensional Otsu [1] is a robust thresholding technique combining both 2D histogram of images and Otsu thresholding. In the last decades, many works have been devoted to improving 2D Otsu's speed and segmentation quality. Recursive techniques [2–6] remove redundant computations by recursively evaluating the fitness function. However, the exhaustive searching schemes still introduces many computation cost. And dimension reduction [7, 8] causes a decline in quality for information lost. Swarm intelligence methods [9–11] perform competitively to recursive techniques and dimension reduction with quick convergence. Curve thresholding segmentation [12] and local grid box filter [13] improves the segmentation quality by re-dividing the 2D histogram. Besides that, image denoising [14] implies a valid approach to prompt 2D Otsu's robustness.

In this paper, to improve the segmentation quality of 2D Otsu thresholding, a 2D histogram is established based on median filter and Gauss filter. To reduce the

© Springer International Publishing AG 2017
Y. Tan et al. (Eds.): ICSI 2017, Part I, LNCS 10385, pp. 611–617, 2017.
DOI: 10.1007/978-3-319-61824-1_66

computation cost, a scheme of evaluating the fitness function based on firefly optimization algorithm [15] is employed. The proposed algorithm is tested on both microscopic images of ocean planktons and Lenna test image. The algorithm is presented in Sect. 2. Section 3 shows the experiments and analysis.

2 The Proposed Algorithm

In an image f, with the size of $H \times W$, suppose the median gray scale of pixel $f(x, y)$ is denoted as the following:

$$h(x, y) = med \left\{ \sum_{i=-(n+1)/2}^{(n+1)/2} \sum_{j=-(n+1)/2}^{(n+1)/2} f(x+i, y+j) \right\}. \tag{1}$$

The Gaussian average gray scale of $f(x, y)$ is defined by the Gaussian filter as the following:

$$gs(x, y) = \sum_{i=-(n+1)/2}^{(n+1)/2} \sum_{j=-(n+1)/2}^{(n+1)/2} f(x+i, y+i) \frac{1}{2\pi\sigma^2} e^{-\left(\frac{i^2+j^2}{2\sigma^2}\right)}. \tag{2}$$

Then a threshold vector (s,t) is obtained to denote the pair of $h(x, y)$ and $gs(x, y)$, and the number of its occurrences in the image f, is denoted as $M_{s,t}$. The joint probability can be defined by:

$$P(s, t) = \frac{M_{s,t}}{H \times W}. \tag{3}$$

Two dimensional Otsu thresholding classifies pixels of an image into two groups: the object and the background, by a threshold vector (s,t). The probabilities of the object and the background are given by:

$$\omega_0 = \sum_{i=0}^{s} \sum_{j=0}^{t} p_{i,j}, \omega_1 = \sum_{i=s+1}^{L-1} \sum_{j=t+1}^{L-1} p_{i,j}. \tag{4}$$

Due to the introduction of median filter and Gaussian average filter, the sum of the probability of both the object and the background can be assumed to be 1. The average gray value vectors of two classes are:

$$\mu_0 = (\mu_{0i}, \mu_{0j})^T = \left(\sum_{i=0}^{s} \sum_{j=0}^{t} i \frac{p_{i,j}}{\omega_0}, \sum_{i=0}^{s} \sum_{j=0}^{t} j \frac{p_{i,j}}{\omega_0} \right)^T$$

$$\mu_1 = (\mu_{1i}, \mu_{1j})^T = \left(\sum_{i=s+1}^{L-1} \sum_{j=t+1}^{L-1} i \frac{p_{i,j}}{\omega_1}, \sum_{i=s+1}^{L-1} \sum_{j=t+1}^{L-1} j \frac{p_{i,j}}{\omega_1} \right)^T. \tag{5}$$

The total average gray value vector of the image is:

$$\mu_T = (\mu_{Ti}, \mu_{Tj})^T = \left(\sum_{i=0}^{L-1} \sum_{j=0}^{L-1} i p_{i,j}, \sum_{i=0}^{L-1} \sum_{j=0}^{L-1} j p_{i,j} \right)^T. \tag{6}$$

Then we have the 2D Otsu thresholding criterion as:

$$S_B(s,t) = \sum_{r=0}^{1} \omega_r \left[(\mu_{ri} - \mu_{Ti})^2 + (\mu_{rj} - \mu_{Tj})^2 \right]. \tag{7}$$

The best threshold vector (s^*, t^*) is determined by maximizing the fitness function $S_B(s,t)$.

To accelerate the searching procedure, firefly algorithm [15] is applied to determine the best thresholds. As a heuristic searching strategy, firefly algorithm simulates that brighter firefly has more attractiveness to draw other fireflies near. The attractiveness is defined by:

$$\beta = \beta_0 \times e^{-\gamma r_{ij}^2}. \tag{8}$$

Where γ is the absorption coefficient of the optical transmission medium, r is the spatial distance between firefly i and j, is the attractiveness when $r = 0$. The movement that firefly i was attracted to firefly j is defined by the formula:

$$x_i = x_i + \beta \times (x_j - x_i) + \alpha \varepsilon_i. \tag{9}$$

Where ε_i is a random variable, α is the weight coefficient, x_i and x_j are firefly i and j's location. The searching procedure is specified by the following steps:

1. Initialize parameters of the firefly algorithm including: the number of fireflies $m = 20$, initial location for every firefly $x_i = (s_i, t_i), i = 1, 2, \cdots, m$, attractiveness $\beta_0 = 1$, the absorption coefficient $\gamma = 1$ and the weight coefficient $\alpha = 0.2$.
2. Calculate the fitness function S_B for each firefly.
3. Sort all fireflies decently by results in step (2), and move the less bright firefly to the brighter one by formula (9), while the brightest one has a random walk.
4. If the end condition is satisfied, the searching procedure is finished, and the optimal threshold vector is the brightest firefly's location (s, t); if not, return to step 2.

3   Experiment Results and Analysis

To verify the performances of the proposed algorithm, experiments are implemented on Windows 7 with 1.6 GHz CPU and 3G RAM. The algorithm is coded in C ++ of Visual Studio 2010. Test images are a group of microscopic images of Prorocentrum donghaiense and Lenna image with 512×512 pixels. Binarization is implied by setting the pixels below the threshold to zero and the pixels above the threshold to 255.

The proposed algorithm is applied to threshold microscopic images of Prorocentrum donghaiense, which is one of the dominant plankton in red tide bloom occurs in China. And the results were compared with conventional 2D Otsu. The original microscopic images of Prorocentrum donghaiense in Fig. 1 is collected by an optical microscope connected to a CMOS camera with 576 × 768 resolution, and thresholded by both the proposed algorithm and the conventional 2D Otsu.

Plankton images thresholded by the conventional 2D Otsu are hardly distinguished from surroundings as shown in Fig. 2. But the results obtained by the proposed algorithm show that plankton have distinct contours in contrast backgrounds as shown in Fig. 3. Comparing results of two approaches, our method is more effective than the conventional one, for plankton microscopic images.

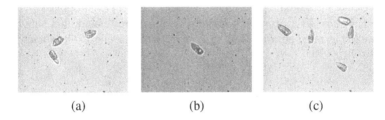

(a)　　　　　　　　(b)　　　　　　　　(c)

Fig. 1. (a)–(c) Original microscopic images of planktons.

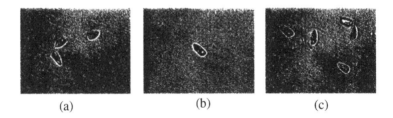

(a)　　　　　　　　(b)　　　　　　　　(c)

Fig. 2. (a)–(c) Thresholding results of plankton's microscopic images by the 2D Otsu [1].

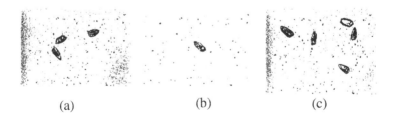

(a)　　　　　　　　(b)　　　　　　　　(c)

Fig. 3. (a)–(c) Results of plankton's microscopic images by the proposed method.

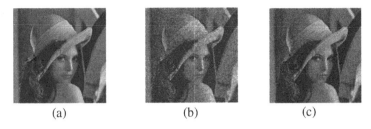

Fig. 4. (a) Original Lenna. (b) Gaussian noise. (c) Salt&Peppernoise.

Lenna images in Fig. 4 are all thresholded by both the conventional 2D Otsu (see Fig. 5) and the proposed algorithm (see Fig. 6). Obviously, the proposed algorithm presents more stable thresholding quality for both Gaussian noise and Salt&Pepper noise, while the conventional 2D Otsu's segmentation is badly influenced by the two kinds noises in Fig. 5.

Fig. 5. Thresholding results by the traditional 2D Otsu. (a) Original Lenna. (b) Gaussian noise. (c) Salt&Pepper noise.

Fig. 6. Thresholding results by the proposed algorithm. (a) Original Lenna. (b) Gaussian noise. (c) Salt&Pepper noise.

Table 1 gives the threshold results of two methods. The thresholds evidently show that the proposed algorithm has more robust performance in searching the optimal threshold under noise disturbance, no matter Gaussian noise or Salt&Pepper noise, resulting from the 2D histogram based on the median gray-scale and the Gaussian average gray-scale.

Table 1. Thresholds obtained by the proposed algorithm and the conventional 2D Otsu.

	Original	Gaussian	Salt&Pepper
The conventional 2D Otsu	(125, 71)	(72, 128)	(72, 126)
The proposed algorithm	(130, 88)	(130, 88)	(130,88)

Time consuming in Table 2 indicates that the proposed algorithm cuts the time cost of thresholding to nearly 2.5% of the conventional 2D Otsu method, which is significant.

Table 2. Time consuming of the proposed algorithm and the conventional 2D Otsu.

	Original	Gaussian	Salt&Pepper
The conventional 2D Otsu	28.398 s	28.409 s	28.461 s
The proposed algorithm	0.662 s	0.657 s	0.687 s

4 Conclusion

An improved two-dimensional Otsu thresholding method is presented, through defining a 2D histogram based on median gray-scale and Gaussian average gray-scale, and employing firefly optimization in optimal thresholds searching. The algorithm is applied to both microscopic images of ocean plankton and Lenna test image. Experiment results show that the proposed algorithm has better and more stable thresholding quality under noise disturbance compared with conventional 2D Otsu, and it cut the time consuming to 2.5% of that of conventional 2D Otsu. Practical image thresholding results for ocean plankton microscopic images give more accurate segmentation effect, indicating that the approach has applicability in ocean microscopic image thresholding.

Acknowledgments. This work was supported by the National Scientific Foundation of China (No. 61304108) and the Discipline Guidance Foundation of Harbin Institute of Technology (Weihai) (No. IDOA 1000290131).

References

1. Liu, J.Z., Li, W.Q., Tian, Y.P.: Automatic thresholding of gray-level pictures using two-dimension Otsu method. Acta Automatica Sinica **19**(1), 101–105 (1993)
2. Gong, J., Li, L.Y., Chen, W.N.: Fast recursive algorithms for two-dimensional thresholding. Pattern Recogn. **32**(3), 295–300 (1998)
3. Wang, H.Y., Pan, D.L., Xia, D.S.: A fast algorithm for two-dimensional Otsu adaptive threshold algorithm. Acta Automatica Sinica **33**(9), 968–971 (2007)
4. Chen, Q., Zhao, L., Lu, L., Kuang, G., Wang, N., Jiang, Y.: Modified two-dimensional otsu image segmentation algorithm and fast realization. IET Image Proc. **6**(4), 426–433 (2012)

5. Zhang, X.M., Sun, Y.J., Zheng, T.B.: Precise two-dimensional Otsu's image segmentation and its fast recursive realization. Dianzi Xuebao (Acta Electronica Sinica) **39**(8), 1778–1784 (2011)
6. Wu, Y.Q., Fan, J., Wu, S.H.: Fast iterative algorithm for image segmentation based on an improved two-dimensional Otsu thresholding. J. Electron. Meas. Instrum. **25**(3), 218–225 (2011)
7. Hao, Y.M., Zhu, F.: Fast algorithm for two-dimensional Otsu adaptive threshold algorithm. J. Image Graph. **10**(4), 484–488 (2005)
8. Chen, J.W., Wu, B.: An Otsu threshold segmentation method based on rebuilding and dimension reduction of the two-dimensional histogram. J. Graph. **36**(4), 570–575 (2015)
9. Suresh, S., Lal, S.: An efficient cuckoo search algorithm based multilevel thresholding for segmentation of satellite images using different objective functions. Expert Syst. Appl. **58**, 184–209 (2016)
10. Horng, M.H.: A multilevel image thresholding using the honey bee mating optimization. Appl. Math. Comput. **215**, 3302–3310 (2010)
11. Chen, K., Chen, F., Dai, M., Zhang, Z.S., Shi, J.F.: Fast image segmentation with multilevel threshold of two-dimensional entropy based on firefly algorithm. Opt. Precis. Eng. **22**(2), 517–523 (2014)
12. Fan, J.L., Zhao, F.: Two-dimensional Otsu's curve thresholding segmentation method for gray-level images. Acta Electron. Sinica **35**, 751–755 (2007)
13. Guo, W., Wang, X., Xia, X.: Two-dimensional Otsu's thresholding segmentation method based on grid box filter. Optik **125**, 5234–5240 (2014)
14. Sha, C.S., Hou, J., Cui, H.X.: A robust 2D Otsu's thresholding method in image segmentation. J. Vis. Commun. Image Represent. **41**, 339–351 (2016)
15. Yang, X.S.: Firefly Algorithms for Multimodal Optimization. In: Nature-Inspired Metaheuristic Algorithms. Luniver Press, Frome (2010)

3D-FOAdis: An Improved Fruit Fly Optimization for Function Optimization

Kejie Wang, Huiling Chen, Qiang Li, Junjie Zhu, Shubiao Wu, and Hui Huang[(⊠)]

College of Physics and Electronic Information,
Wenzhou University, Wenzhou, China
huanghui@wzu.edu.cn

Abstract. In fruit fly optimization algorithm (FOA), the search speed of each fruit fly is fast, but when it traps into the local optimum, it is difficult to re-find a better solution. In order to overcome this drawback, we propose an improved version of FOA, termed as 3D-FOAdis. In the proposed method, three-dimensional coordinates and the disturbance mechanism were both introduced. We firstly extends the original two-dimensional coordinates to three-dimensional coordinates, where fruit flies can fly more widely so that it is more likely to jump out of the local optimum. Then we introduce a disturbance mechanism force the FOA to find better solutions when the fruit flies fall into the local optimums. The effectiveness of 3D-FOAdis has been rigorously evaluated against the nine benchmark functions. The experimental results demonstrate that the proposed approach outperforms the other two counterparts.

Keywords: Fruit fly optimization · Function optimization · Disturbance mechanism

1 Introduction

Fruit fly optimization algorithm (FOA) [1] is a new method for global optimization based on the foraging behavior of fruit flies. The fruit fly is superior to other species in sensory perception, especially in the sense of smell and vision. The olfactory organs of fruit fly can well collect all kinds of smell floating in the air, and even smell the food source 40 km away. Then, after flying near the food location, the fruit fly can also use sharp vision to find food and mates to gather and fly in that direction.

As a new member of the swarm intelligence algorithms, FOA has been applied to many fields, such as model parameters optimization [2, 3], function optimization [4], solving the multidimensional knapsack problem [5], building the financial distress model [1], tuning of PID controller [6], constructing the operating performance of enterprises model [7] and so on.

The location coordinates of each fruit fly are two-dimensional in the original FOA. The FOA firstly determines the distance between the fruit fly and the origin through the two-dimensional coordinates, and then puts the reciprocal of the distance as the fruit fly's smell, lastly inserts the smell as the intermediate element into the objective function to obtain the smell concentration of the fruit fly. The dependent variable

© Springer International Publishing AG 2017
Y. Tan et al. (Eds.): ICSI 2017, Part I, LNCS 10385, pp. 618–625, 2017.
DOI: 10.1007/978-3-319-61824-1_67

(distance) is changed by the coordinates of the fruit fly, and the smell concentration of each fruit fly is determined by establishing a two-dimensional plane. In the course of searching for the optimal solution, each fruit fly shares information so that other fruit flies fly to the current best fruit fly and update their positions around it. The search speed of fruit fly is very fast, but if the FOA traps into the local optimum, it is difficult to re-find a better solution. Through many tests, it is found that the original FOA is trapped in local optimum, which makes it difficult to find the optimal solution. In order to overcome this shortcoming of the FOA and further enhance its performance, this paper firstly extends the original two-dimensional coordinates to three- dimensional coordinates and then introduces a disturbance mechanism. In the three-dimensional coordinates, fruit flies can fly more widely, which makes it more likely to jump out of the local optimum. With the guidance of the disturbance mechanism, if the fruit flies fall into the local optimums, the fruit flies will be forced to find better solutions.

The remainder of this paper is structured as follows. In Sect. 2 the detailed implementation of the three-dimensional fruit fly algorithm with disturbance (3D-FOAdis) is presented. Section 3 describes the experimental results. Finally, conclusions are summarized in Sect. 4.

2 Three-Dimensional Fruit Fly Algorithm with Disturbance (3D-FOAdis)

Though the original FOA has fast speed but it runs in a two-dimensional coordinates, which may hampered the fruit flies move freely. So we try to extend the two-dimensional coordinates to three-dimensional coordinates where fruit flies are more likely to shift freely. Additionally, in the searching process, fruit flies are inclined to falling into the local minima. In order to deal with it, we have introduced a disturbance mechanism, namely, if the current smell is no better than the previous one over a certain number of times, the current positions of fruit flies will be changed with pulsing or subtracting a given value. The 3D-FOAdis can be divided into several steps as follows.

Step 1. Parameters initialization.
Initialize the parameters of the FOA, such as the maximum iteration number, the population size, the initial fruit fly swarm location (X_axis, Y_axis, Z_axis), the random flight distance range and the time to record how long does the smell not change.

Step 2. Population initialization.
Give the random location (Xi, Yi, Zi) and distance for the food search of an individual fruit fly, where i represents the population size.

Step 3. Population evaluation.
Firstly, calculate the distance of the food location to the origin (D). Then, compute the smell concentration judgment value (S), which is the reciprocal of the distance of the food location to the origin.

Step 4. Replacement.
Replace the smell concentration judgment value (S) with the smell concentration judgment function (also called the Fitness function) so as to find the smell concentration (Smelli) of the individual location of the fruit fly.

Step 5. Find the maximal smell concentration.
Determine the fruit fly with the maximal smell concentration and the corresponding location among the fruit fly swarm.

Step 6. Keep the maximal smell concentration.
Retain the maximal smell concentration value and coordinates X, Y and Z. Then, the fruit fly swarm flies towards the location with the maximal smell concentration value. If the current smell is not better than the last one, making the timer of disturbance (time) plus one. Once time reaches to a certain value, the coordinates of fruit flies are forced to change for achieving the purpose of disturbance.

Step 7. Iterative optimization.
Enter the iterative optimization to repeat the implementation of step 2 to 6. The circulation stops when the smell concentration is no longer superior to the previous iterative smell concentration or when the iterative number reaches the maximal iterative number.

Table 1. The pseudo code of the 3D-FOAdis

Randomly set the initial population position and set the time to 0;

Initialize X_axis, Y_axis, Z_axis;

time=0;

Caculate bestSmell;

X(bestIndex,:)= Optimum position X;

Y(bestIndex,:)= Optimum position Y;

Z(bestIndex,:)= Optimum position Z;

bestSmell= Optimum smell;

while(*i <Max number of iterations*)

  for *each search agent*

    if *(smell concentration is better than that of the previous iteration)*

      Update smell concentration values and fruit fly coordinates;

    else

      time=time+1;

      if *(time>=threshold)*

        Disturb the fruit flies;

      end if

    end if

    end for

  end while

return *bestSmell;*

---

3   Experimental Results and Discussions

In the experiments, the number of iterations and fruit flies was set to 500 and 50, respectively. When the timer of disturbance reached 40, the course of iteration would be disturbed. Three models including original FOA, three-dimensional fruit fly algorithm (3D-FOA) and 3D-FOAdis were built in MATLAB R2014a. The experiments are done on Windows 10 Operation System with i5-5200U Dual-Core CPU (2.7 GHZ) and 4 GB RAM.

In order to verify the effectiveness of the proposed algorithm, three methods were employed to compared against nine commonly used multidimensional functions as shown in Table 2, of which $f1$–$f5$ are unimodal benchmark functions and $f6$–$f8$ are the multimodal benchmark functions. We tested each function for 10 times, and calculated the average (Ave) and standard deviation (Std) respectively.

Table 3 shows the average and standard deviation of the nine benchmark functions in the dimension 10, 20, 30, 50, 100 and 200, respectively. From the table we can see that the proposed 3D-FOAdis not only performs better on unimodal functions than 3D-FOA and FOA, but also obtains better experimental results when it is aimed at multimodal functions. It can be seen from Table 2 that the 3D-FOA and the 3D-FOAdis are better able to find the minimum value than the original FOA when the function dimension is relatively low, such as $f1$, $f3$ and $f7$. When the dimension of the benchmark function is increased, the 3D-FOA and the 3D-FOAdis will be more likely

Table 2. Benchmark functions

Function	Range	Minimum
$f_1(x) = \sum_{i=1}^{n} x_i^2$	$[-100, 100]$	0
$f_2(x) = \sum_{i=1}^{n} |x_i| + \prod_{i=1}^{n} |x_i|$	$[-10, 10]$	0
$f_3(x) = \sum_{i=1}^{n} \left(\sum_{j-1}^{i} x_j \right)^2$	$[-100, 100]$	0
$f_4(x) = \max_i \{|x_i|, 1 \leq i \leq n\}$	$[-100, 100]$	0
$f_5(x) = \sum_{i=1}^{n} ([x_i + 0.5])^2$	$[-100, 100]$	0
$f_6(x) = \sum_{i=1}^{n} [x_i^2 - 10\cos(2\pi x_i) + 10]$	$[-5.12, 5.12]$	0
$f_7(x) = \dfrac{\pi}{n} \left\{ 10\sin(\pi y_1) + \sum_{i=1}^{n-1} (y_i - 1)^2 [1 + 10\sin^2(\pi y_{i+1})] + (y_n - 1)^2 \right\}$ $+ \sum_{i=1}^{n-1} u(x_i, 10, 100, 4)$ $y_i = 1 + \dfrac{x_i + 1}{4} \quad u(x_i, a, k, m) = \begin{cases} k(x_i - a)^m & x_i > a \\ 0 & -a < x_i < a \\ k(-x_i - a)^m & x_i < -a \end{cases}$	$[-50, 50]$	0
$f_8(x) = [1 + (x_1 + x_2 + 1)^2 (19 - 14x_1 + 3x_1^2 - 14x_2 + 6x_1x_2 + 3x_2^2)]$ $\times [30 + (2x_1 - 3x_2)^2 \times (18 - 32x_1 + 12x_1^2 + 48x_2 - 36x_1x_2 + 27x_2^2)]$	$[-2, 2]$	0

Table 3. Results of testing benchmark functions

Function	Dim	FOA		3D-FOA		3D-FOAdis	
		Ave	Std	Ave	Std	Ave	Std
f_1	10	0.00015	7.753E−05	0.00013	5.930E−05	0.00012	4.709E−05
	20	0.00015	0.00010	0.00011	5.658E−05	0.00012	6.228E−05
	30	0.00015	9.484E−05	0.00013	6.723E−05	0.00013	5.568E−05
	50	0.00014	7.578E−05	0.00012	6.943E−05	0.00013	6.457E−05
	100	0.00015	9.251E−05	0.00012	5.813E−05	0.00013	7.795E−05
	200	0.00016	8.057E−05	0.00012	6.103E−05	0.00012	7.007E−05
f_2	10	0.04509	0.03566	0.04825	0.03676	0.01823	0.00420
	20	0.03399	0.02860	0.05403	0.03698	0.01839	0.00365
	30	0.05014	0.04056	0.04995	0.03291	0.01828	0.00517
	50	0.04288	0.03798	0.05426	0.03324	0.01872	0.00478
	100	0.04799	0.03611	0.05210	0.03521	0.01772	0.00399
	200	0.05555	0.03661	0.04675	0.03456	0.01865	0.00392
f_3	10	0.00052	0.00046	0.00039	0.00018	0.00041	0.00025
	20	0.00050	0.00033	0.00047	0.00046	0.00038	0.00024
	30	0.00056	0.00048	0.00050	0.00034	0.00039	0.00022
	50	0.00047	0.00041	0.00035	0.00023	0.00036	0.00017
	100	0.00057	0.00039	0.00051	0.00054	0.00036	0.00017
	200	0.00054	0.00039	0.00039	0.00019	0.00039	0.00017
f_4	10	0.01047	0.00229	0.00941	0.00172	0.00937	0.00163
	20	0.01030	0.00172	0.00961	0.00184	0.00940	0.00211
	30	0.01017	0.00220	0.00944	0.00191	0.00932	0.00187
	50	0.01045	0.00210	0.00982	0.00212	0.00955	0.00143
	100	0.01059	0.00221	0.00951	0.00184	0.00944	0.00194
	200	0.01033	0.00215	0.00921	0.00195	0.00961	0.00192
f_5	10	0.52963	0.02582	0.53930	0.03166	0.51810	0.00463
	20	0.52947	0.02859	0.54468	0.03265	0.51848	0.00432
	30	0.53576	0.03043	0.54700	0.03623	0.51883	0.00523
	50	0.53334	0.02953	0.54018	0.03066	0.51788	0.00406
	100	0.53372	0.02563	0.54138	0.03231	0.51840	0.00478
	200	0.52759	0.02475	0.54105	0.03050	0.51871	0.00435
f_6	10	0.07494	0.14846	0.02684	0.01367	0.02253	0.00930
	20	0.05303	0.10173	0.02610	0.01572	0.02394	0.01001
	30	0.02920	0.01401	0.02580	0.01598	0.02847	0.01391
	50	0.03710	0.02401	0.02838	0.01677	0.02406	0.01234
	100	0.06665	0.12623	0.02399	0.01081	0.02744	0.01418
	200	0.07222	0.14275	0.02590	0.01448	0.02173	0.00990
f_7	10	8.76910	0.14378	8.74606	0.08218	8.68973	0.04091
	20	8.79768	0.15100	8.74864	0.10519	8.70688	0.04612
	30	8.79177	0.12987	8.74119	0.09622	8.69905	0.04319

(*continued*)

Table 3. (*continued*)

Function	Dim	FOA		3D-FOA		3D-FOAdis	
		Ave	Std	Ave	Std	Ave	Std
	50	8.78118	0.12435	8.75185	0.09837	8.69657	0.04233
	100	8.75954	0.10604	8.75203	0.09818	8.70811	0.04878
	200	8.76134	0.12022	8.71994	0.07522	8.69495	0.04593
f_8	10	614.63085	7.39771	614.39719	9.17471	611.67830	2.86448
	20	613.09317	5.84888	615.76907	11.87001	612.63875	3.86299
	30	613.46286	13.56586	615.06322	11.07727	611.90166	3.51743
	50	613.80883	7.02099	617.52934	13.91489	611.61879	3.26522
	100	576.06481	95.71655	614.48933	22.14465	611.67610	2.67643
	200	611.00024	25.60832	615.86818	12.31299	611.99624	2.64636

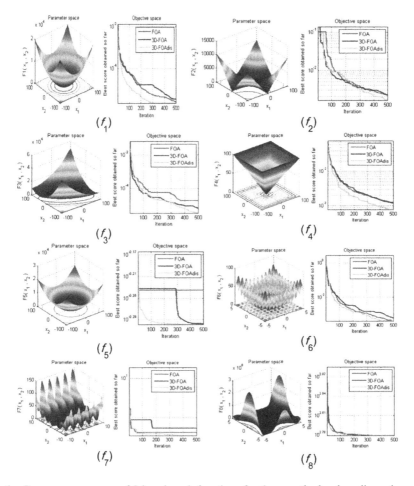

Fig. 1. Convergence curves of 8 benchmark functions for three methods when dimension = 10

to go beyond the original FOA. Moreover, Table 2 also shows that the 3D-FOAdis has better stability than the 3D-FOA, indicating that its probability of finding a better solution is bigger than the 3D-FOA. The superiority results suggest that the third dimension can make the FOA more likely to jump out of the local optimum, and the introduced disturbance mechanism can force the fruit flies to fly away from their current positions to find better smell.

Limited to the space, the example is taken only in the dimension of 10 to describe the convergence of the function. Figure 1 records the convergence curve of each function for above three methods. The 3D-FOAdis algorithm proposed in this paper has much higher convergence speed than 3D-FOA and FOA as these graphs shown. The 3D-FOA inherits the characteristic of the FOA to find the optimal value quickly, and it also avoids the situation of falling into a local optimum and never comes out.

4   Conclusions

This work has proposed an improved FOA, 3D-FOAdis for function optimization. The main novelty of this paper lies in the proposed 3D-FOAdis approach, which aims at obtaining the better solution quality and converging at a faster speed. In order to achieve this purpose, we have introduced three-dimensional coordinates and disturbance mechanism into the original FOA. The empirical experiments on a set of multidimensional functions have demonstrated the superiority of the proposed method over the other two counterparts. In future works, we plan to apply the proposed method to optimize the parameters of machine learning methods such as support vector machines and kernel extreme learning machine.

Acknowledgements. This research is funded by the Zhejiang Provincial Natural Science Foundation of China (LY17F020012), the Science and Technology Plan Project of Wenzhou, China (Y20160070).

References

1. Pan, W.-T.: A new Fruit Fly Optimization Algorithm: taking the financial distress model as an example. Knowl.-Based Syst. **26**, 69–74 (2012)
2. Shen, L., et al.: Evolving support vector machines using fruit fly optimization for medical data classification. Knowl.-Based Syst. **96**, 61–75 (2016)
3. Yu, Y., Li, Y., Li, J.: Parameter identification and sensitivity analysis of an improved LuGre friction model for magnetorheological elastomer base isolator. Meccanica **50**, 2691–2707 (2015)
4. Wu, L., Zuo, C., Zhang, H.: A cloud model based fruit fly optimization algorithm ☆. Knowl.-Based Syst. **89**, 603–617 (2015)
5. Wang, L., Zheng, X.-L., Wang, S.-Y.: A novel binary fruit fly optimization algorithm for solving the multidimensional knapsack problem. Knowl.-Based Syst. **48**, 17–23 (2013)

6. Han, J., Wang, P., Yang, X.: Tuning of PID controller based on Fruit Fly Optimization Algorithm. In: 2012 International Conference on Mechatronics and Automation (ICMA). IEEE Press, New York (2012)
7. Wen-Chao, P.: Using Fruit Fly Optimization Algorithm optimized general regression neural network to construct the operating performance of enterprises model. J. Taiyuan Univ. Technol. (Soc. Sci. Edn.) **4**, 2 (2011)

Author Index

Printed in the United States
By Bookmasters